World Energy Outlook 2010

INTERNATIONAL ENERGY AGENCY

The International Energy Agency (IEA), an autonomous agency, was established in November 1974. Its mandate is two-fold: to promote energy security amongst its member countries through collective response to physical disruptions in oil supply and to advise member countries on sound energy policy.

The IEA carries out a comprehensive programme of energy co-operation among 28 advanced economies, each of which is obliged to hold oil stocks equivalent to 90 days of its net imports. The Agency aims to:

■ Secure member countries' access to reliable and ample supplies of all forms of energy; in particular, through maintaining effective emergency response capabilities in case of oil supply disruptions.

■ Promote sustainable energy policies that spur economic growth and environmental protection in a global context – particularly in terms of reducing greenhouse-gas emissions that contribute to climate change.

■ Improve transparency of international markets through collection and analysis of energy data.

■ Support global collaboration on energy technology to secure future energy supplies and mitigate their environmental impact, including through improved energy efficiency and development and deployment of low-carbon technologies.

■ Find solutions to global energy challenges through engagement and dialogue with non-member countries, industry, international organisations and other stakeholders.

IEA member countries:

Australia
Austria
Belgium
Canada
Czech Republic
Denmark
Finland
France
Germany
Greece
Hungary
Ireland
Italy
Japan
Korea (Republic of)
Luxembourg
Netherlands
New Zealand
Norway
Poland
Portugal
Slovak Republic
Spain
Sweden
Switzerland
Turkey
United Kingdom
United States

International
Energy Agency

The European Commission
also participates in
the work of the IEA.

Three of the thousands of numbers in the *World Energy Outlook 2010*, despite their disparity, are worth putting alongside each other:

- $312 billion — the cost of consumption subsidies to fossil fuels in 2009.
- $57 billion — the cost of support given to renewable energy in 2009.
- $36 billion per year — the cost of ending global energy poverty by 2030.

Adding under two percent to electricity tariffs in the OECD would raise enough money to bring electricity to the entire global population within twenty years; while, in the past year, the prospective cost of the additional global energy investment to 2035 to curb greenhouse-gas emissions has risen by $1 trillion because of the caution of the commitments made at Copenhagen.

My chief economist, Fatih Birol, and his team have again met our high expectations. We have new projections, fuel by fuel, extending now to 2035; a special focus on renewable energy; a stock-taking on energy and climate change in the aftermath of Copenhagen; a look at the cost of achieving universal access to electricity and clean cooking fuels; detailed information on the energy demand and resources of the countries in the Caspian region; and insights into the scale of fossil-fuel subsidies and the implications of phasing them out.

The basis of our projections this year has changed. The old Reference Scenario is dead (though reborn as the Current Policies Scenario). The centrepiece of our presentation is now the New Policies Scenario. This departs from our previous practice of building our projections only on the measures governments had already taken.

Predicting what governments might do is a hazardous business. We have gone no further than to take governments at their word, interpreting the intentions they have declared into implementing measures and projecting the future on that basis. More commitments and more policies will surely follow. We have not attempted to guess what they might be; but the 450 Scenario remains as a measure of how much more must be done to realise a sustainable future and how it could be done.

One point is certain. The centre of gravity of global energy demand growth now lies in the developing world, especially in China and India. But uncertainties abound. Is our emergence from the financial crisis of 2008-2009 a solid enough basis for our assumptions about economic growth? Will China sustain and intensify the four-fold improvement in energy intensity it has achieved in the last thirty years? Would a three-fold increase in oil revenues in real terms satisfy OPEC producers in a world committed to keep the global temperature rise below 2°Celsius? What will be the upshot of the controversy about the sustainablility of biofuels production? Will carbon capture and storage become a commercially available technology within a decade?

We cannot know. But, with the invaluable financial and analytical support of our member countries and others who rely on the *WEO*, we can and do ensure, through this new edition of the *WEO*, that responsible and rigorous information is available to help decision-makers discharge their responsibilities to shape the energy future.

Nobuo Tanaka
Executive Director

This publication has been produced under the authority of the Executive Director of the International Energy Agency. The views expressed do not necessarily reflect the views or policies of individual IEA member countries.

ACKNOWLEDGEMENTS

This study was prepared by the Office of the Chief Economist (OCE) of the International Energy Agency in co-operation with other offices of the Agency. It was designed and directed by **Fatih Birol**, Chief Economist of the IEA. **Laura Cozzi** and **Marco Baroni** co-ordinated the analysis of climate policy and modelling; **Trevor Morgan** co-ordinated the analysis of oil and natural gas and the Caspian outlook; **Amos Bromhead** co-ordinated the analysis of fossil-fuel subsidies. **Maria Argiri** led the work on renewables, **John Corben** and **Paweł Olejarnik** (oil, gas and coal supply), **Christian Besson** (unconventional oil), **Alessandro Blasi** (Caspian and oil), **Raffaella Centurelli** (energy poverty and modelling), **Michael-Xiaobao Chen** (fossil-fuel subsidies and China), **Michel D'Ausilio** (power sector and renewables), **Dafydd Elis** (power sector and renewables), **Matthew Frank** (fossil-fuel subsidies and power sector), **Tim Gould** (Caspian and oil), **Timur Gül** (transport and modelling), **Kate Kumaria** (climate policy), **Qiang Liu** (China), **Bertrand Magné** (climate policy and modelling), **Teresa Malyshev** (energy poverty), **Timur Topalgoekceli** (oil), **David Wilkinson** (power sector and modelling) and **Akira Yanagisawa** (fossil-fuel subsidies and modelling). **Sandra Mooney** provided essential support. For more information on the OCE team, please see *www.worldenergyoutlook.org.*

Robert Priddle carried editorial responsibility.

The study benefited from input provided by IEA experts in different offices. Paolo Frankl, Milou Beerepoot, Hugo Chandler and several other colleagues of the Renewable Energy Division made valuable contributions to the renewable energy analysis. Ian Cronshaw provide very helpful input to the gas and power sector analysis. Other IEA colleagues who provided input to different parts of the book include, Jane Barbière, Madeleine Barry, Ulrich Benterbusch, Rick Bradley, Aad van Bohemen, Pierpaolo Cazzola, Anne-Sophie Corbeau, Bo Diczfalusy, David Elzinga, Lew Fulton, David Fyfe, Rebecca Gaghen, Jean-Yves Garnier, Grayson Heffner, Christina Hood, Didier Houssin, Brian Ricketts, Bertrand Sadin, Maria Sicilia, Sylvie Stephan and Cecilia Tam. Experts from a number of directorates of the OECD also made valuable contributions to the report, particularly Helen Mountford, Ronald Steenblik, Jean-Marc Burniaux, Jean Château and Dambudzo Muzenda. Thanks also go to Debra Justus for proofreading the text.

The work could not have been achieved without the substantial support and co-operation provided by many government bodies, international organisations and energy companies worldwide, notably:

Department of Energy, United States; Enel; Energy Research Institute, China; Foreign Affairs and International Trade, Canada; Foreign and Commonwealth Office, United Kingdom; HM Treasury, United Kingdom; IEA Coal Industry Advisory Board (CIAB); Intergovermental Panel on Climate Change (IPCC); Ministry of Economic Affairs, The Netherlands; Ministry of Economy, Trade and Industry, Japan; Ministry of Foreign Affairs, Norway; Ministry of the Economy, Poland; National Renewable Energy

Laboratory (NREL), United States; Natural Resources, Canada; Navigant Consulting; Norwegian Agency for Development Co-operation; Renewable Energy and Energy Efficiency Partnership (REEEP); Schlumberger; Statoil; The Energy and Resources Institute (TERI), India; Toyota Motor Corporation; United Nations Development Programme (UNDP), the United Nations Industrial Development Organization (UNIDO) and the World Health Organisation (WHO).

Many international experts provided input, commented on the underlying analytical work and reviewed early drafts of each chapter. Their comments and suggestions were of great value. They include:

Asset Abdualiyev	Consultant, Kazakhstan
Saleh Abdurrahman	Ministry of Energy and Mineral Resources, Indonesia
Kalle Ahlstedt	Fortum
Jun Arima	Ministry of Economy, Trade and Industry, Japan
Polina Averianova	Eni
Georg Bäuml	Volkswagen
Paul Bailey	Department of Energy and Climate Change, United Kingdom
Jim Bartis	RAND Corporation
Chris Barton	Department of Energy and Climate Change, United Kingdom
Vaclav Bartuska	Ministry of Foreign Affairs, Czech Republic
Paul Baruya	IEA Clean Coal Centre, United Kingdom
Morgan Bazilian	UNIDO
Carmen Becerril Martinez	Acciona
Rachid Bencherif	OPEC Fund for International Development, Austria
Osman Benchikh	UN Educational Scientific and Cultural Organisation, France
Kamel Bennaceur	Schlumberger
Bruno Bensasson	GDF SUEZ
Edgard Blaustein	Ministry of Foreign Affairs, France
Roberto Bocca	World Economic Forum
Jean-Paul Bouttes	Electricite de France
Julien Bowden	BP
Albert Bressand	Columbia School of International and Public Affairs, United States

Nigel Bruce	World Health Organisation, Switzerland
Peter Brun	Vestas
Kenny Bruno	Corporate Ethics International
Guy Caruso	Center for Strategic and International Studies, United States
Martin Child	British Embassy, Kazakhstan
Ed Chow	Center for Strategic and International Studies, United States
Jan Cloin	Ministry of Foreign Affairs, The Netherlands
Janusz Cofala	International Institute for Applied Systems Analysis, Austria
Michael Cohen	Department of Energy, United States
Ben Combes	Committee on Climate Change, United Kingdom
Jennifer Coolidge	CMX Caspian and Gulf Consultants
Joel Couse	Total
Kevin Covert	United States Embassy, Kazakhstan
Christian De Gromard	Agence Française de Développement
Jos Delbeke	European Commission
Carmen Difiglio	Department of Energy, United States
Andrew Dobbie	Department of Energy and Climate Change, United Kingdom
Joanne Doornewaard	Ministry of Economic Affairs, The Netherlands
Nick Douglas	Department of the Interior, United States
Jens Drillisch	KfW Bankengruppe, Germany
Stanislas Drochon	PFC Energy
Simon Dyer	The Pembina Institute, Canada
Ottmar Edenhofer	Intergovernmental Panel on Climate Change, Switzerland
Koffi Ekouevi	World Bank, United States
Mike Enskat	Deutsche Gesellschaft für Technisch Zusammenarbeit (GTZ) GmbH Germany
Hideshi Emoto	Development Bank of Japan
Mikael Eriksson	Ministry for Foreign Affairs, Sweden

Jean-Pierre Favennec	Institut Français du Pétrole
Roger Fairclough	Neo Leaf Global
Herman Franssen	International Energy Associates
Peter Fraser	Ontario Energy Board, Canada
Irene Freudenschuss-Reichl	Ministry for European and International Affairs, Austria
Dario Garofalo	Enel
Carlos Gascò-Travesedo	Iberdrola
Holger Gassner	RWE
Claude Gauvin	Natural Resources Canada
John German	International Council on Clean Transportation
Dolf Gielen	UNIDO
Guido Glania	Alliance for Rural Electrification, Belgium
José Goldemberg	Instituto de Eletrotécnica e Energia, Brazil
Rainer Görgen	Federal Ministry of Economics and Technology, Germany
Irina Goryunova	Central Asia Regional Economic Cooperation, Kazakhstan
Alex Greenstein	Department of State, United States
Sanjeev Gupta	International Monetary Fund, United States
Antoine Halff	Newedge, United States
Kirsty Hamilton	Royal Institute of International Affairs, United Kingdom
Antonio Hernandez Garcia	Ministry of Industry, Tourism and Trade, Spain
James Hewlett	Department of Energy, United States
Masazumi Hirono	Tokyo Gas
Ray Holland	EU Energy Initiative Partnership Dialogue Facility, Germany
Takashi Hongo	Japan Bank for International Cooperation
Trevor Houser	Peterson Institute for International Economics, United States
Tom Howes	European Commission
Mustaq Hussain	Delegation of the European Union to the Republic of Kazakhstan

Catherine Inglehearn	Foreign and Commonwealth Office, United Kingdom
Fumiaki Ishida	New Energy and Industrial Technology Development Organization, Japan
Peter Jackson	IHS CERA
C.P. Jain	World Energy Council
James Jensen	Jensen Associates
Jan-Hein Jesse	Heerema Marine Contractors
David Jhirad	Johns Hopkins University, United States
Robert Johnston	Eurasia Group
Leanne Jones	Department for International Development, United Kingdom
Marianne Kah	ConocoPhillips
John Sande Kanyarubona	African Development Bank
Mahama Kappiah	ECOWAS Regional Centre for Renewable Energy and Energy Efficiency, Cape Verde
Tor Kartevold	Statoil
Ryan Katofsky	Navigant Consulting
Paul Khanna	Natural Resources Canada
Hisham Khatib	Honorary Vice Chairman, World Energy Council; and former Minister of Energy, Jordan
Mohamed Hafiz Khodja	Consultant, Algeria
David Knapp	Energy Intelligence
Kenji Kobayashi	Asia Pacific Energy Research Centre, Japan
Yoshikazu Kobayashi	Institute of Energy Economics, Japan
Hans-Jorgen Koch	Ministry of Transportation and Energy, Denmark
Masami Kojima	World Bank, United States
Doug Koplow	Earth Track
Edward Kott	LCM Commodities
Ken Koyama	Institute of Energy Economics, Japan
Natalia Kulichenko-Lotz	World Bank, United States
Akihiro Kuroki	Institute of Energy Economics, Japan
Takayuki Kusajima	Toyota Motor Corporation

Sarah Ladislaw	Center for Strategic and International Studies, United States
Gordon Lambert	Suncor Energy
Michael Levi	Council on Foreign relations, United States
Steve Lennon	Eskom
Michael Liebreich	Bloomberg New Energy Finance
Vivien Life	Foreign and Commonwealth Office, United Kingdom
Qiang Liu	Energy Research Institute, China
Agata Łoskot-Strachota	Centre for Eastern Studies, Poland
Gunnar Luderer	Potsdam Institute for Climate Impact Research, Germany
Michael Lynch	Strategic Energy & Economic Research, United States
Gordon Mackenzie	UNEP Risø Centre, Denmark
Joan MacNaughton	Alstom Power Systems
Claude Mandil	Former Executive Director, International Energy Agency
David McColl	The Canadian Energy Research Institute
Hilary McMahon	World Resources Institute
Neil McMurdo	HM Treasury, United Kingdom
Albert Melo	Centro de Pesquisas de Energia Elétrica, Brazil
Emanuela Menichetti	Observatoire Méditerranéen de l'Energie, France
Angus Miller	Foreign and Commonwealth Office, United Kingdom
Tatiana Mitrova	Energy Research Institute of the Russian Academy of Sciences, Russia
A. Tristan Mocilnikar	Mission Union pour la Méditerranée, France
Arne Mogren	Vattenfall
Lucio Monari	World Bank, United States
Jacob Moss	Environmental Protection Agency, United States
Nebojsa Nakicenovic	International Institute for Applied Systems Analysis, Austria
Julia Nanay	PFC Energy
Aldo Napolitano	Eni

Fernando Naredo	Westinghouse Electrical Company
Brian Nicholson	Department of Energy, Government of Alberta
Kare Riis Nielsen	Novozymes
Petter Nore	Norwegian Agency for Development Cooperation
Patrick Nussbaumer	UNIDO
Martha Olcott	Carnegie Endowment for International Peace, United States
Patrick Oliva	Michelin
Simon-Erik Ollus	Fortum
A. Yasemin Örücü	Ministry of Energy and Natural Resources, Turkey
Shonali Patchauri	International Institut for Applied System Analysis, Autralia
Binu Parthan	Renewable Energy & Energy Efficiency Partnership, Austria
Brian Pearce	International Air Transport Association, Switzerland
Serge Perineau	World CTL Association
Christian Pichat	AREVA
Roberto Potì	Edison
Ireneusz Pyc	Siemens
Ibrahim Hafeezur Rehman	The Energy and Resources Institute, India
David Renné	National Renewable Energy Laboratory, United States
Gustav Resch	Vienna University of Technology, Austria
Teresa Ribera	Secretary of State for Climate Change, Spain
Christoph Richter	SolarPACES, Spain
Kamal Rijal	UNDP
Wishart Robson	Nexen Inc
Hans-Holger Rogner	International Atomic Energy Agency
David Rolfe	Department of Energy and Climate Change, United Kingdom
Simon Rolland	Alliance for Rural Electrification, Belgium
Ralph D. Samuelson	Asia Pacific Energy Research Centre, Japan
Catharina Saponar	Nomura

Steve Sawyer	Global Wind Energy Council, Belgium
Hans-Wilhelm Schiffer	RWE
Glen Schmidt	Laricina Energy
Philippe Schulz	Renault
Adnan Shihab-Eldin	Former Acting Secretary General of the Organization of Petroleum Exporting Countries (OPEC)
P.R. Shukla	Indian Institute of Management
Adam Sieminski	Deutsche Bank
Ron Sills	Consultant
Ottar Skagen	Statoil
Bob Skinner	Statoil
Robert Socolow	Princeton University, United States
Virginia Sonntag-O'Brien	REN21, France
Leena Srivastava	Tata Energy Research Institute, India
Robert Stavins	Harvard University, United States
Till Stenzel	Nur Energie
Stephanie Sterling	Shell Canada Services
Jonathan Stern	Oxford Institute for Energy Studies, United Kingdom
Pau Stevens	Royal Institute of International Affairs, United Kingdom
Ulrik Stridbaek	Dong Energy
Goran Strbac	Imperial College, United Kingdom
Greg Stringham	Canadian Association of Petroleum Producers
Minoru Takada	UNDP
Bernard Terlinden	GDF Suez
Anil Terway	Asian Development Bank
Sven Teske	Greenpeace International
Wim Thomas	Shell
Douglas Townsend	Consultant
Simon Trace	Practical Action, United Kingdom
Oras Tynkkynen	Prime Minister's Office, Finland
Fridtjof Unander	The Research Council of Norway

Bernd Utz	Siemens
Maria Vagliasindi	World Bank, United States
Christof Van Agt	Clingendael Institute, The Netherlands
Noé Van Hulst	International Energy Forum, Saudi Arabia
Frank Verrastro	Center for Strategic and International Studies, United States
Roland Vially	Institut Français du Pétrole, France
Peter Wells	Cardiff Business School, United Kingdom
Roger Wicks	Anglo American
Francisco Romário Wojcicki	Ministry of Mines and Energy, Brazil
Peter Wooders	International Institute for Sustainable Development, Switzerland
Henning Wuester	UN Framework Convention on Climate Change, Germany
Annabel Yadoo	Centre for Sustainable Development at the University of Cambridge
Shigehiro Yoshino	Nippon Export and Investment Insurance, Japan
Dimitrios Zevgolis	Global Environment Facility, United States

The individuals and organisations that contributed to this study are not responsible for any opinions or judgements contained in this study. All errors and omissions are solely the responsibility of the IEA.

Comments and questions are welcome and should be addressed to:

Dr. Fatih Birol
Chief Economist
Director, Office of the Chief Economist
International Energy Agency
9, rue de la Fédération
75739 Paris Cedex 15
France

Telephone: (33-1) 4057 6670
Fax: (33-1) 4057 6509
Email: weo@iea.org

More information about the World Energy Outlook is available at
www.worldenergyoutlook.org.

ANNEXES

List of figures

Part A: GLOBAL ENERGY TRENDS

Chapter 3: Oil market outlook

Chapter 4: The outlook for unconventional oil

Chapter 15: Implications for oil markets

Part D: OUTLOOK FOR CASPIAN ENERGY

Chapter 16: Caspian domestic energy prospects

Part E: FOCUS ON ENERGY SUBSIDIES

Chapter 19: Analysing fossil-fuel subsidies

Chapter 20: Country subsidy profiles

List of tables

Part A: GLOBAL ENERGY TRENDS

Part D: OUTLOOK FOR CASPIAN ENERGY

List of boxes

Part B: OUTLOOK FOR RENEWABLE ENERGY

Part C: ACHIEVING THE 450 SCENARIO AFTER COPENHAGEN

Part D: OUTLOOK FOR CASPIAN ENERGY

Chapter 17: Hydrocarbon resources and supply potential

Chapter 18: Regional and global implications

Part E: FOCUS ON ENERGY SUBSIDIES

Chapter 19: Analysing fossil-fuel subsidies

List of spotlights

Part A: GLOBAL ENERGY TRENDS

Part B: OUTLOOK FOR RENEWABLE ENERGY

Part C: ACHIEVING THE 450 SCENARIO AFTER COPENHAGEN

Part D: OUTLOOK FOR CASPIAN ENERGY

Part E: FOCUS ON ENERGY SUBSIDIES

The energy world faces unprecedented uncertainty. The global economic crisis of 2008-2009 threw energy markets around the world into turmoil and the pace at which the global economy recovers holds the key to energy prospects for the next several years. But it will be governments, and how they respond to the twin challenges of climate change and energy security, that will shape the future of energy in the longer term. The economic situation has improved considerably over the past 12 months, more than many dared to hope for. Yet the economic outlook for the coming years remains hugely uncertain, amid fears of a double-dip recession and burgeoning government budget deficits, making the medium-term outlook for energy unusually hard to predict with confidence. The past year has also seen notable steps forward in policy making, with the negotiation of important international agreements on climate change and on the reform of inefficient fossil-fuel subsidies. And the development and deployment of low-carbon technologies received a significant boost from stepped-up funding and incentives that governments around the world introduced as part of their fiscal stimulus packages. Together, these moves promise to drive forward the urgently needed transformation of the global energy system. But doubts remain about the implementation of recent policy commitments. Even if they are acted upon, much more needs to be done to ensure that this transformation happens quickly enough.

The outcome of the landmark UN conference on climate change held in December 2009 in Copenhagen was a step forward, but still fell a very long way short of what is required to set us on the path to a sustainable energy system. The Copenhagen Accord — with which all major emitting countries and many others subsequently associated themselves — sets a non-binding objective of limiting the increase in global temperature to two degrees Celsius (2°C) above pre-industrial levels. It also establishes a goal for the industrialised countries of mobilising funding for climate mitigation and adaptation in developing countries of $100 billion per year by 2020, and requires the industrialised countries to set emissions targets for the same year. This followed a call from G8 leaders at their July 2009 summit to share with all countries the goal of cutting global emissions by at least 50% by 2050. But the commitments that were subsequently announced, even if they were to be fully implemented, would take us only part of the way towards an emissions trajectory that would allow us to achieve the 2°C goal. That does not mean that the goal is completely out of reach. But it does mean that much stronger efforts, costing considerably more, will be needed after 2020. Indeed, the speed of the energy transformation that would need to occur after 2020 is such as to raise serious misgivings about the practical achievability of cutting emissions sufficiently to meet the 2°C goal.

The commitment made by G-20 leaders meeting in the US city of Pittsburgh in September 2009 to "rationalize and phase out over the medium term inefficient fossil-fuel subsidies that encourage wasteful consumption" has the potential to, at least partly, balance the disappointment at Copenhagen. This commitment was made in recognition that subsidies distort markets, can impede investment in clean energy sources and can thereby undermine efforts to deal with climate change. The analysis we have carried out in collaboration with other international organisations at the request of G-20 leaders, and which is set out in this *Outlook*, shows that removing fossil-fuel consumption subsidies, which totalled $312 billion in 2009, could make a big contribution to meeting energy-security and environmental goals, including mitigating carbon-dioxide (CO_2) and other emissions.

Recently announced policies, if implemented, would make a difference

The world energy outlook to 2035 hinges critically on government policy action, and how that action affects technology, the price of energy services and end-user behaviour. In recognition of the important policy advances that have been made recently, the central scenario in this year's *Outlook* — the *New Policies Scenario* — takes account of the broad policy commitments and plans that have been announced by countries around the world, including the national pledges to reduce greenhouse-gas emissions and plans to phase out fossil-energy subsidies even where the measures to implement these commitments have yet to be identified or announced. These commitments are assumed to be implemented in a relatively cautious manner, reflecting their non-binding character and, in many cases, the uncertainty shrouding how they are to be put into effect. This scenario allows us to quantify the potential impact on energy markets of implementation of those policy commitments, by comparing it with a *Current Policies Scenario* (previously called the *Reference Scenario*), in which no change in policies as of mid-2010 is assumed, *i.e.* that recent commitments are not acted upon. We also present the results of the *450 Scenario*, which was first presented in detail in *WEO-2008*, which sets out an energy pathway consistent with the 2°C goal through limitation of the concentration of greenhouse gases in the atmosphere to around 450 parts per million of CO_2 equivalent (ppm CO_2-eq).

The policy commitments and plans that governments have recently announced would, if implemented, have a real impact on energy demand and related CO_2 emissions. In the New Policies Scenario, world primary energy demand increases by 36% between 2008 and 2035, from around 12 300 million tonnes of oil equivalent (Mtoe) to over 16 700 Mtoe, or 1.2% per year on average. This compares with 2% per year over the previous 27-year period. The projected rate of growth in demand is lower than in the Current Policies Scenario, where demand grows by 1.4% per year over 2008-2035. In the 450 Scenario, demand still increases between 2008 and 2035, but by only 0.7% per year. Energy prices ensure that projected supply and demand are in balance throughout the *Outlook* period in each scenario, rising fastest in the Current Policies Scenario and slowest in the 450 Scenario. Fossil fuels — oil, coal and natural gas — remain the dominant energy sources in 2035 in all three scenarios, though their share of the overall primary fuel mix varies markedly. The shares of renewables and nuclear power are correspondingly highest in the 450 Scenario and lowest in the Current Policies Scenario. The range of outcomes — and therefore the uncertainty with respect to future energy use — is largest for coal, nuclear power and non-hydro renewable energy sources.

Emerging economies, led by China and India, will drive global demand higher

In the New Policies Scenario, global demand for each fuel source increases, with fossil fuels accounting for over one-half of the increase in total primary energy demand. Rising fossil-fuel prices to end users, resulting from upward price pressures on international markets and increasingly onerous carbon penalties, together with policies to encourage energy savings and switching to low-carbon energy sources, help to restrain demand growth for all three fossil fuels. Oil remains the dominant fuel in the primary energy mix during the *Outlook* period, though its share of the primary fuel mix, which stood at 33% in 2008, drops to 28% as high prices and government measures to promote fuel efficiency lead to further switching away from oil in the industrial and power-generation sectors, and new opportunities emerge to substitute other fuels for oil products in transport. Demand for coal rises through to around 2020 and starts to decline towards the end of the *Outlook* period. Growth in demand for natural gas far surpasses that for the other fossil fuels due to its more favourable environmental and practical attributes, and constraints on how quickly low-carbon energy technologies can be deployed. The share of nuclear power increases from 6% in 2008 to 8% in 2035. The use of modern renewable energy — including hydro, wind, solar, geothermal, modern biomass and marine energy — triples over the course of the *Outlook* period, its share in total primary energy demand increasing from 7% to 14%. Consumption of traditional biomass rises slightly to 2020 and then falls back to just below current levels by 2035, with increased use of modern fuels by households in the developing world.

Non-OECD countries account for 93% of the projected increase in world primary energy demand in the New Policies Scenario, reflecting faster rates of growth of economic activity, industrial production, population and urbanisation. China, where demand has surged over the past decade, contributes 36% to the projected growth in global energy use, its demand rising by 75% between 2008 and 2035. By 2035, China accounts for 22% of world demand, up from 17% today. India is the second-largest contributor to the increase in global demand to 2035, accounting for 18% of the rise, its energy consumption more than doubling over the *Outlook* period. Outside Asia, the Middle East experiences the fastest rate of increase, at 2% per year. Aggregate energy demand in OECD countries rises very slowly over the projection period. Nonetheless, by 2035, the United States is still the world's second-largest energy consumer behind China, well ahead of India (in a distant third place).

It is hard to overstate the growing importance of China in global energy markets. Our preliminary data suggest that China overtook the United States in 2009 to become the world's largest energy user. Strikingly, Chinese energy use was only half that of the United States in 2000. The increase in China's energy consumption between 2000 and 2008 was more than four times greater than in the previous decade. Prospects for further growth remain strong, given that China's per-capita consumption level remains low, at only one-third of the OECD average, and that it is the most populous nation on the planet, with more than 1.3 billion people. Consequently, the global energy projections in this *Outlook* remain highly sensitive to the underlying assumptions for the key variables that drive energy demand in China, including prospects for economic growth, changes in economic structure, developments in energy and environmental policies, and the rate of urbanisation. The country's growing need to import fossil fuels to meet its rising domestic demand will have an increasingly large impact on international markets. Given the sheer scale of China's domestic market, its push to increase the share of new low-carbon energy technologies could play an important role in driving down their costs through faster rates of technology learning and economies of scale.

Will peak oil be a guest or the spectre at the feast?

The oil price needed to balance oil markets is set to rise, reflecting the growing insensitivity of both demand and supply to price. The growing concentration of oil use in transport and a shift of demand towards subsidised markets are limiting the scope for higher prices to choke off demand through switching to alternative fuels. And constraints on investment mean that higher prices lead to only modest increases in production. In the New Policies Scenario, the average IEA crude oil price reaches $113 per barrel (in year-2009 dollars) in 2035 — up from just over $60 in 2009. In practice, short-term price volatility is likely to remain high. Oil demand (excluding biofuels) continues to grow steadily, reaching about 99 million barrels per day (mb/d) by 2035 — 15 mb/d higher than in 2009. All of the net growth comes from non-OECD countries, almost half from China alone, mainly driven by rising use of transport fuels; demand in the OECD falls by over 6 mb/d. Global oil production reaches 96 mb/d, the balance of 3 mb/d coming from processing gains. Crude oil output reaches an undulating plateau of around 68-69 mb/d by 2020, but never regains its all-time peak of 70 mb/d reached in 2006, while production of natural gas liquids (NGLs) and unconventional oil grows strongly.

Total OPEC production rises continually through to 2035 in the New Policies Scenario, boosting its share of global output to over one-half. Iraq accounts for a large share of the increase in OPEC output, commensurate with its large resource base, its crude oil output catching up with Iran's by around 2015 and its total output reaching 7 mb/d by 2035. Saudi Arabia regains from Russia its place as the world's biggest oil producer, its output rising from 9.6 mb/d in 2009 to 14.6 mb/d in 2035. The increasing share of OPEC contributes to the growing dominance of national oil companies: as a group, they account for all of the increase in global production between 2009 and 2035. Total non-OPEC oil production is broadly constant to around 2025, as rising production of NGLs and unconventional oil offsets a fall in that of crude oil; thereafter, total non-OPEC output starts to drop. The size of ultimately recoverable resources of both conventional and unconventional oil is a major source of uncertainty for the long-term outlook for world oil production.

Clearly, global oil production will peak one day, but that peak will be determined by factors affecting both demand and supply. In the New Policies Scenario, production in total does *not* peak before 2035, though it comes close to doing so. By contrast, production does peak, at 86 mb/d, just before 2020 in the 450 Scenario, as a result of weaker demand, falling briskly thereafter. Oil prices are much lower as a result. The message is clear: if governments act more vigorously than currently planned to encourage more efficient use of oil and the development of alternatives, then demand for oil might begin to ease soon and, as a result, we might see a fairly early peak in oil production. That peak would not be caused by resource constraints. But if governments do nothing or little more than at present, then demand will continue to increase, supply costs will rise, the economic burden of oil use will grow, vulnerability to supply disruptions will increase and the global environment will suffer serious damage.

Unconventional oil is abundant but more costly

Unconventional oil is set to play an increasingly important role in world oil supply through to 2035, regardless of what governments do to curb demand. In the New Policies Scenario, output rises from 2.3 mb/d in 2009 to 9.5 mb/d in 2035. Canadian oil sands and Venezuelan extra-heavy oil dominate the mix, but coal-to-liquids, gas-to-liquids and, to a lesser extent, oil shales also make a growing contribution in the second half of the *Outlook* period. Unconventional oil resources are thought to be huge — several times larger than conventional oil resources. The rate at which they will be exploited will be determined by economic and environmental considerations, including the costs of mitigating their environmental impact. Unconventional sources of oil are among the more expensive available: they require large upfront capital investment, which is typically paid back over long periods. Consequently, they play a key role in setting future oil prices.

The production of unconventional oil generally emits more greenhouse gases per barrel than that of most types of conventional oil, but, on a well-to-wheels basis, the difference is much less, as most emissions occur at the point of use. In the case of Canadian oil sands, well-to-wheels CO_2 emissions are typically between 5% and 15% higher than for conventional crude oils. Mitigation measures will be needed to reduce emissions from unconventional oil production, including more efficient extraction technologies, carbon capture and storage and, with coal-to-liquids plants, the addition of biomass to the coal feedstock. Improved water and land management, though not unique to unconventional sources, will also be required to make the development of these resources and technologies more acceptable.

China could lead us into a golden age for gas

Natural gas is certainly set to play a central role in meeting the world's energy needs for at least the next two-and-a-half decades. Global natural gas demand, which fell in 2009 with the economic downturn, is set to resume its long-term upward trajectory from 2010. It is the only fossil fuel for which demand is higher in 2035 than in 2008 in all scenarios, though it grows at markedly different rates. In the New Policies Scenario, demand reaches 4.5 trillion cubic metres (tcm) in 2035 — an increase of 1.4 tcm, or 44%, over 2008 and an average rate of increase of 1.4% per year. China's demand grows fastest, at an average rate of almost 6% per year, and the most in volume terms, accounting for more than one-fifth of the increase in global demand to 2035. There is potential for Chinese gas demand to grow even faster than this, especially if coal use is restrained for environmental reasons. Demand in the

Middle East increases almost as much as projected in China. The Middle East, which is well-endowed with relatively low-cost resources, leads the expansion of gas production over the *Outlook* period, its output doubling to 800 billion cubic metres (bcm) by 2035. Around 35% of the global increase in gas production in the New Policies Scenario comes from unconventional sources — shale gas, coalbed methane and tight gas — in the United States and, increasingly, from other regions, notably Asia-Pacific.

The glut of global gas-supply capacity that has emerged as a result of the economic crisis (which depressed gas demand), the boom in US unconventional gas production and a surge in liquefied natural gas (LNG) capacity, could persist for longer than many expect. Based on projected demand in the New Policies Scenario, we estimate that the glut, measured by the difference between the volumes actually traded and total capacity of inter-regional pipelines and LNG export plants, amounted to about 130 bcm in 2009; it is set to reach over 200 bcm in 2011, before starting a hesitant decline. This glut will keep the pressure on gas exporters to move away from oil-price indexation, notably in Europe, which could lead to lower prices and to stronger demand for gas than projected, especially in the power sector. In the longer term, the increasing need for imports — especially in China — will most likely drive up capacity utilisation. In the New Policies Scenario, gas trade between all *WEO* regions expands by around 80%, from 670 bcm in 2008 to 1 190 bcm in 2035. Well over half of the growth in gas trade takes the form of LNG.

A profound change in the way we generate electricity is at hand

World electricity demand is expected to continue to grow more strongly than any other final form of energy. In the New Policies Scenario, it is projected to grow by 2.2% per year between 2008 and 2035, with more than 80% of the increase occurring in non-OECD countries. In China, electricity demand triples between 2008 and 2035. Over the next 15 years, China is projected to add generating capacity equivalent to the current total installed capacity of the United States. Globally, gross capacity additions, to replace obsolete capacity and to meet demand growth, amount to around 5 900 gigawatts (GW) over the period 2009-2035 — 25% more than current installed capacity; more than 40% of this incremental capacity is added by 2020.

Electricity generation is entering a period of transformation as investment shifts to low-carbon technologies — the result of higher fossil-fuel prices and government policies to enhance energy security and to curb emissions of CO_2. In the New Policies Scenario, fossil fuels — mainly coal and natural gas — remain dominant, but their share of total generation drops from 68% in 2008 to 55% in 2035, as nuclear and renewable sources expand. The shift to low-carbon technologies is particularly marked in the OECD. Globally, coal remains the leading source of electricity generation in 2035, although its share of electricity generation declines from 41% now to 32%. A big increase in non-OECD coal-fired generation is partially offset by a fall in OECD countries. Gas-fired generation grows in absolute terms, mainly in the non-OECD, but maintains a stable share of world electricity generation at around 21% over the *Outlook* period. The share of nuclear power in generation increases only marginally, with more than 360 GW of new additions over the period and extended lifetime for several plants.

Globally, the shift to nuclear power, renewables and other low-carbon technologies is projected to reduce the amount of CO_2 emitted per unit of electricity generated by one-third between 2008 and 2035.

The future of renewables hinges critically on strong government support

Renewable energy sources will have to play a central role in moving the world onto a more secure, reliable and sustainable energy path. The potential is unquestionably large, but how quickly their contribution to meeting the world's energy needs grows hinges critically on the strength of government support to make renewables cost-competitive with other energy sources and technologies, and to stimulate technological advances. The need for government support would increase were gas prices to be lower than assumed in our analysis.

The greatest scope for increasing the use of renewables in absolute terms lies in the power sector. In the New Policies Scenario, renewables-based generation triples between 2008 and 2035 and the share of renewables in global electricity generation increases from 19% in 2008 to almost one-third (catching up with coal). The increase comes primarily from wind and hydropower, though hydropower remains dominant over the *Outlook* period. Electricity produced from solar photovoltaics increases very rapidly, though its share of global generation reaches only around 2% in 2035. The share of modern renewables in heat production in industry and buildings increases from 10% to 16%. The use of biofuels grows more than four-fold between 2008 and 2035, meeting 8% of road transport fuel demand by the end of the *Outlook* period (up from 3% now). Renewables are generally more capital-intensive than fossil fuels, so the investment needed to provide the extra renewables capacity is very large: cumulative investment in renewables to produce electricity is estimated at $5.7 trillion (in year-2009 dollars) over the period 2010-2035. Investment needs are greatest in China, which has now emerged as a leader in wind power and photovoltaic production, as well as a major supplier of the equipment. The Middle East and North Africa region holds enormous potential for large-scale development of solar power, but there are many market, technical and political challenges that need to be overcome.

Although renewables are expected to become increasingly competitive as fossil-fuel prices rise and renewable technologies mature, the scale of government support is set to expand as their contribution to the global energy mix increases. We estimate that government support worldwide for both electricity from renewables and for biofuels totalled $57 billion in 2009, of which $37 billion was for the former. In the New Policies Scenario, total support grows to $205 billion (in year-2009 dollars), or 0.17% of global GDP, by 2035. Between 2010 and 2035, 63% of the support goes to renewables-based electricity. Support per unit of generation on average worldwide drops over time, from $55 per megawatt-hour (MWh) in 2009 to $23/MWh by 2035, as wholesale electricity prices increase and their production costs fall due to technological learning. This does not take account of the additional costs of integrating them into the network, which can be significant because the variability of some types of renewables, such as wind and solar energy. Government support for renewables can, in principle, be justified by the long-term economic, energy-security and environmental benefits they can bring, though attention needs to be given to the cost-effectiveness of support mechanisms.

The use of biofuels — transport fuels derived from biomass feedstock — is expected to continue to increase rapidly over the projection period, thanks to rising oil prices and government support. In the New Policies Scenario, global biofuels use increases from about 1 mb/d today to 4.4 mb/d in 2035. The United States, Brazil and the European Union are expected to remain the world's largest producers and consumers of biofuels. Advanced biofuels, including those from ligno-cellulosic feedstocks, are assumed to enter the market by around 2020, mostly in OECD countries. The cost of producing biofuels today is often higher than the current cost of imported oil, so strong government incentives are usually needed to make them competitive with oil-based fuels. Global government support in 2009 was $20 billion, the bulk of it in the United States and the European Union. Support is projected to rise to about $45 billion per year between 2010 and 2020, and about $65 billion per year between 2021 and 2035. Government support typically raises costs to the economy as a whole. But the benefits can be significant too, including reduced imports of oil and reduced CO_2 emissions — if sustainable biomass is used and the fossil energy used in processing the biomass is not excessive.

Unlocking the Caspian's energy riches would enhance the world's energy security

The Caspian region has the potential to make a significant contribution to ensuring energy security in the rest of the world, by increasing the diversity of oil and gas supplies. The Caspian region contains substantial resources of both oil and natural gas, which could underpin a sizeable increase in production and exports over the next two decades. But potential barriers to the development of these resources, notably the complexities of financing and constructing transportation infrastructure passing through several countries, the investment climate and uncertainty over export demand, are expected to constrain this expansion to some degree. In the New Policies Scenario, Caspian oil production grows strongly — especially over the first 15 years of the projection period; it jumps from 2.9 mb/d in 2009 to a peak of around 5.4 mb/d between 2025 and 2030, before falling back to 5.2 mb/d by 2035. Kazakhstan contributes all of this increase, ranking fourth in the world for output growth in volume terms to 2035 after Saudi Arabia, Iraq and Brazil. Most of the incremental oil output goes to exports, which double to a peak of 4.6 mb/d soon after 2025. Caspian gas production is also projected to expand substantially, from an estimated 159 bcm in 2009 to nearly 260 bcm by 2020 and over 310 bcm in 2035. Turkmenistan and, to a lesser extent, Azerbaijan and Kazakhstan drive this expansion. As with oil, gas exports are projected to grow rapidly, reaching nearly 100 bcm in 2020 and 130 bcm in 2035, up from less than 30 bcm in 2009. The Caspian has the potential to supply a significant part of the gas needs of Europe and China, which emerges as a major new customer, enhancing their energy diversity and security.

Domestic energy policies and market trends, beyond being critical to the Caspian's social and economic development, have an influence on world prospects by determining the volumes available for export. Despite some improvement in recent years, the region remains highly energy-intensive, reflecting continuing gross inefficiencies in the way energy is used (a legacy of the Soviet era), as well as climatic and structural economic factors. If the region were to use energy as efficiently as OECD countries, consumption of primary energy in the Caspian as a whole would be cut by one-half. How quickly this energy-efficiency potential might be exploited hinges largely on government policies, especially on energy pricing (all the main Caspian countries subsidise at least one form of fossil energy), market reform and financing. In the New Policies Scenario, total Caspian primary energy demand expands progressively through the *Outlook* period, at an average rate of 1.4% per year, with gas remaining the predominant fuel. Kazakhstan and Turkmenistan see the fastest rates of growth in energy use, mainly reflecting more rapid economic growth.

Copenhagen pledges are collectively far less ambitious than the overall goal

The commitments that countries have announced under the Copenhagen Accord to reduce their greenhouse-gas emissions collectively fall short of what would be required to put the world onto a path to achieving the Accord's goal of limiting the global temperature increase to 2°C. If countries act upon these commitments in a cautious manner, as we assume in the New Policies Scenario, rising demand for fossil fuels would continue to drive up energy-related CO_2 emissions through the projection period. *Such a trend would make it all but impossible to achieve the 2°C goal*, as the required reductions in emissions after 2020 would be too steep. In that scenario, global emissions continue to rise through the projection period, though the rate of growth falls progressively. Emissions jump to just under 34 gigatonnes (Gt) in 2020 and over 35 Gt in 2035 — a 21% increase over the 2008 level of 29 Gt. Non-OECD countries account for all of the projected growth in world emissions; OECD emissions peak before 2015 and then begin to fall. These trends are in line with stabilising the concentration of greenhouse gases at over 650 ppm CO_2-eq, resulting in a likely temperature rise of more than 3.5°C in the long term.

The 2°C goal can only be achieved with vigorous implementation of commitments in the period to 2020 and much stronger action thereafter. According to climate experts, in order to have a reasonable chance of achieving the goal, the concentration of greenhouse gases would need to be stabilised at a level no higher than 450 ppm CO_2-eq. The 450 Scenario describes how the energy sector could evolve were this objective to be achieved. It assumes implementation of measures to realise the more ambitious end of target ranges announced under the Copenhagen Accord and more rapid implementation of the removal of fossil-fuel subsidies agreed by the G-20 than assumed in the New Policies Scenario. This action results in a significantly faster slowdown in global energy-related CO_2 emissions. In the 450 Scenario, emissions reach a peak of 32 Gt just before 2020 and then slide to 22 Gt by 2035. Just ten emissions-abatement measures in five regions — the United States, the European Union, Japan, China and India — account for around half of the emission reductions throughout the *Outlook* period needed in this scenario compared with the Current Policies Scenario. While pricing carbon in the power and industry sectors is at the heart of emissions reductions in OECD countries and, in the longer term, other major economies (CO_2 prices reach $90-120 per tonne in 2035), fossil-fuel subsidies phase-out is a crucial pillar of mitigation in the Middle East, Russia and parts of Asia. The power-generation sector's share of global emissions drops from 41% today to 24% by 2035, spearheading the decarbonisation of the global economy. By contrast, the transport sector's share jumps from 23% to 32%, as it is more costly to cut emissions rapidly than in most other sectors.

Cutting emissions sufficiently to meet the 2°C goal would require a far-reaching transformation of the global energy system. In the 450 Scenario, oil demand peaks just before 2020 at 88 mb/d, only 4 mb/d above current levels, and declines to 81 mb/d in 2035. There is still a need to build almost 50 mb/d of new capacity to compensate for falling production from existing fields, but the volume of oil which has to be found and developed from new sources by 2035 is only two-thirds that in the New Policies Scenario, allowing the oil industry to shelve some of the more costly and more environmentally sensitive prospective projects. Coal demand peaks before 2020, returning to 2003 levels by 2035. Among the fossil fuels, demand for natural gas is least affected, though it too reaches a peak before the end of the 2020s. Renewables and nuclear make significant inroads in the energy mix, doubling their current share to 38% in 2035. The share of nuclear power in total generation increases by about 50% over current levels. Renewable-based generation increases the most, reaching more than 45% of global generation — two-and-a-half times higher than today. Wind power jumps to almost 13%, while the combined share of solar PV and CSP reaches more than 6%. Carbon capture and storage plays an important role in reducing power-sector emissions: by 2035, generation from coal plants fitted with CCS exceeds that from coal plants not equipped with this technology, accounting for about three-quarters of the total generation from all CCS fitted plants. Biofuels and advanced vehicles also play a much bigger role than in the New Policies Scenario. By 2035, about 70% of global passenger-car sales are advanced vehicles (hybrids, plug-in hybrids and electric cars). Global energy security is enhanced by the greater diversity of the energy mix.

Failure at Copenhagen has cost us at least $1 trillion...

Even if the commitments under the Copenhagen Accord were fully implemented, the emissions reductions that would be needed after 2020 would cost more than if more ambitious earlier targets had been pledged. The emissions reductions that those commitments would yield by 2020 are such that much bigger reductions would be needed thereafter to get on track to meet the 2°C goal. In the 450 Scenario in this year's *Outlook*, the additional spending on low-carbon energy technologies (business investment and consumer spending) amounts to $18 trillion (in year-2009 dollars) more than in the Current Policies Scenario in the period 2010-2035, and around $13.5 trillion more than in the New Policies Scenario. The additional spending compared with the Current Policies Scenario to 2030 is $11.6 trillion — about $1 trillion more than we estimated last year. In addition, global GDP would be reduced in 2030 by 1.9%, compared with last year's estimate of 0.9%. These differences are explained by the deeper, faster cuts in emissions needed after 2020, caused by the slower pace of change in energy supply and use in the earlier period.

...though reaching the Copenhagen goal is still (just about) achievable

The modest nature of the pledges to cut greenhouse-gas emissions under the Copenhagen Accord has undoubtedly made it less likely that the 2°C goal will actually be achieved. Reaching that goal would require a phenomenal policy push by governments around the world. An indicator of just how big an effort is needed is the rate of decline in carbon intensity — the amount of CO_2 emitted per dollar of GDP — required in the 450 Scenario. Intensity would have to fall in 2008-2020 at twice the rate of 1990-2008; between 2020 and 2035, the rate would have to be almost four times faster. The technology that exists today could enable such a change, but such a rate of technological transformation would be unprecedented. And there are major doubts about the implementation of the commitments for 2020, as many of them are ambiguous and may well be interpreted in a far less ambitious manner than assumed in the 450 Scenario. A number of countries, for instance, have proposed ranges for emissions reductions, or have set targets based on carbon or energy intensity and/or a baseline of GDP that differs from that assumed in our projections. Overall, we estimate that the uncertainty related to these factors equates to 3.9 Gt of energy-related CO_2 emissions in 2020, or about 12% of projected emissions in the 450 Scenario. It is vitally important that these commitments are interpreted in the strongest way possible and that much stronger commitments are adopted and acted upon after 2020, if not before. Otherwise, the 2°C goal would probably be out of reach for good.

Getting rid of fossil-fuel subsidies is a triple-win solution

Eradicating subsidies to fossil fuels would enhance energy security, reduce emissions of greenhouse gases and air pollution, and bring economic benefits. Fossil-fuel subsidies remain commonplace in many countries. They result in an economically inefficient allocation of resources and market distortions, while often failing to meet their intended objectives. Subsidies that artificially lower energy prices encourage wasteful consumption, exacerbate energy-price volatility by blurring market signals, incentivise fuel adulteration and smuggling, and undermine the competitiveness of renewables and more efficient energy technologies. For importing countries, subsidies often impose a significant fiscal burden on state budgets, while for producers they quicken the depletion of resources and can thereby reduce export earnings over the long term. Fossil-fuel consumption subsidies worldwide amounted to $312 billion in 2009, the vast majority of them in non-OECD countries. The annual level fluctuates widely with changes in international energy prices, domestic pricing policy and demand: subsidies were $558 billion in 2008. Only a small proportion of these subsidies go to the poor. Considerable momentum is now building globally to cut fossil-fuel subsidies. In September 2009, G-20 leaders committed to phase out and rationalise inefficient fossil-fuel subsidies, a move that was closely mirrored in November 2009 by APEC leaders. Many countries are now pursuing reforms, but steep economic, political and social hurdles will need to be overcome to realise lasting gains.

Reforming inefficient energy subsidies would have a dramatic effect on supply and demand in global energy markets. We estimate that a universal phase-out of all fossil-fuel consumption subsidies by 2020 — ambitious though it may be as an objective — would cut global primary energy demand by 5%, compared with a baseline in which subsidies remain unchanged. This amounts to the current consumption of Japan, Korea and New Zealand combined. Oil demand alone would be cut by 4.7 mb/d by 2020, equal to around one-quarter of current US demand. Phasing out fossil-fuel consumption subsidies could represent an integral building block for tackling climate change: their complete removal would reduce CO_2 emissions by 5.8%, or 2 Gt, in 2020.

Energy poverty in the developing world calls for urgent action

Despite rising energy use across the world, many poor households in developing countries still have no access to modern energy services. The numbers are striking: we estimate that 1.4 billion people — over 20% of the global population — lack access to electricity and that 2.7 billion people — some 40% of the global population — rely on the traditional use of biomass for cooking. Worse, our projections suggest that the problem will persist in the longer term: in the New Policies Scenario, 1.2 billion people still lack access to electricity in 2030 (the date of the proposed goal of universal access to modern energy services), 87% of them living in rural areas. Most of these people will be living in sub-Saharan Africa, India and other developing Asian countries (excluding China). In the same scenario, the number of people relying on the traditional use of biomass for cooking *rises* to 2.8 billion in 2030, 82% of them in rural areas.

Prioritising access to modern energy services can help accelerate social and economic development. The UN Millennium Development Goal of eradicating extreme poverty and hunger by 2015 will not be achieved unless substantial progress is made on improving energy access. To meet the goal, an additional 395 million people need to be provided with electricity and an additional one billion provided with access to clean cooking facilities. To meet the much more ambitious goal of achieving universal access to modern energy services by 2030, additional spending of $36 billion per year would be required. This is equal to less than 3% of the global investment in energy-supply infrastructure projected in the New Policies Scenario to 2030. The resulting increase in energy demand and CO_2 emissions would be modest: in 2030, global oil demand would be less than 1% higher and CO_2 emissions a mere 0.8% higher compared with the New Policies Scenario. To get close to meeting either of these goals, the international community needs to recognise that the projected situation is intolerable, commit itself to effect the necessary change and set targets and indicators to monitor progress. The Energy Development Index, presented in this *Outlook*, could provide a basis for target-setting and monitoring. A new financial, institutional and technological framework is required, as is capacity building at the local and regional levels. Words are not enough — real action is needed now. We can and must get there in the end.

PREFACE

Part A of this *WEO* presents a comprehensive summary of our energy projections for three scenarios to 2035. Our central scenario this year is called the *New Policies Scenario*. It takes account of the broad policy commitments and plans that have been announced by countries around the world, to tackle either environmental or energy-security concerns, even where the measures to implement these commitments have yet to be identified or announced. This scenario allows us to quantify the potential impact on energy markets of implementation of those policy commitments, by comparing it with a *Current Policies Scenario* (previously called the Reference Scenario), in which no change in policies as of mid-2010 is assumed. We also present the results of the *450 Scenario*, (first presented in detail in *WEO-2008*), which sets out an energy pathway consistent with the goal agreed at the UN climate meeting in Copenhagen in December 2009 to limit the increase in global temperature to 2°C.

Chapter 1 describes the methodological framework and the assumptions that underpin the projections in each of the scenarios. Chapter 2 summarises the global trends in energy demand and supply, as well as the implications for investment and emissions of carbon dioxide. It also puts the spotlight on the increasing importance of China. The detailed projections for oil, gas, coal and electricity are then set out in Chapters 3-7, with a special focus on unconventional oil in Chapter 4.

Chapter 8 investigates the key strategic challenge of energy poverty. It quantifies the number of people without access to modern energy services in developing countries and the scale of the investments required in order to achieve the proposed goal of universal access. It also presents an Energy Development Index and a discussion of the path to improving access to modern energy services, as well as financing mechanisms and the implications for government policy.

CHAPTER 1

CONTEXT AND ANALYTICAL FRAMEWORK

What will shape the energy future?

H I G H L I G H T S

- Three scenarios are presented in this year's *Outlook*, differentiated by the underlying assumptions about government policies. The New Policies Scenario, presented here for the first time, takes account of the broad policy commitments that have already been announced and assumes cautious implementation of national pledges to reduce greenhouse-gas emissions by 2020 and to reform fossil-fuel subsidies.

- The Current Policies Scenario (equivalent to the Reference Scenario of past *Outlooks*) takes into consideration only those policies that had been formally adopted by mid-2010. The third scenario, the 450 Scenario, assumes implementation of the high-end of national pledges and stronger policies after 2020, including the near-universal removal of fossil-fuel consumption subsidies, to achieve the objective of limiting the concentration of greenhouse gases in the atmosphere to 450 parts per million of CO_2-equivalent and global temperature increase to 2° Celsius.

- Assumptions about population and economic growth are the same in each scenario. World population is assumed to expand from an estimated 6.7 billion in 2008 to 8.5 billion in 2035, an annual average rate of increase of about 1%. Population growth slows progressively, in line with past trends. The population of non-OECD countries continues to grow most rapidly. Most of the growth occurs in cities.

- GDP — a key driver of energy demand in all regions — is assumed to grow worldwide by 3.2% per year on average over the period 2008-2035. In general, the non-OECD countries continue to grow fastest. The world economy contracted by 0.6% in 2009, but is expected to rebound by 4.6% in 2010. India, China and the Middle East remain the fastest growing economies.

- In the New Policies Scenario, the IEA crude oil import price, a proxy for international prices, is assumed to rise steadily to $99/barrel (in year-2009 dollars) in 2020 and $113 in 2035, reflecting rising production costs. The price rises more rapidly in the Current Policies Scenario, as demand grows more quickly, and more slowly in the 450 Scenario, on lower demand. Natural gas prices are assumed to remain low relative to oil prices in all scenarios, notably in North America, under pressure from abundant supplies of unconventional gas. North American prices nonetheless converge to some degree with prices in Europe and Asia-Pacific over the projection period, as the cost of production climbs. Coal prices rise much less than oil and gas prices, and fall in the 450 Scenario. CO_2 trading becomes more widespread and CO_2 prices rise progressively in the New Policies and 450 Scenarios.

Scope and methodology

This year's edition of the *World Energy Outlook* (WEO) sets out long-term projections of energy demand and supply, related carbon-dioxide (CO_2) emissions and investment requirements. The IEA's World Energy Model (WEM) — a large-scale mathematical construct designed to replicate how energy markets function — is the principal tool used to generate the projections, sector-by-sector and region-by-region.[1] The model has been updated, drawing on the most recent data, and parts of it enhanced, notably the transport and power-generation modules, including more detailed coverage of renewables. New models for selected countries and regions have also been developed, including separate models for the main Caspian countries. The projections have been extended from 2030 to 2035. The last year for which comprehensive historical data is available is 2008; however, preliminary data are available in some cases for 2009 and have been incorporated into the projections.

Future energy trends will be the interplay of a number of different factors, most of which are hard to predict accurately. For this reason, this *World Energy Outlook* adopts its customary scenario approach to analysing the long-term evolution of energy markets. In the near to medium term, economic factors are the main source of uncertainty surrounding energy prospects. There is also enormous uncertainty about the outlook for energy prices, the size of energy resources and their cost, and the prospects for new energy-related technology, especially in the longer term. But government policies are arguably the biggest source of uncertainty to 2035. Governments around the world have expressed a will to take decisive action to steer energy use onto a more environmentally and economically sustainable course, although the measures needed to bring this about, the way in which they are to be implemented and their timing are often unclear. We know that most governments will act, but how, when and how vigorously are far from clear. For these reasons, the scenarios set out in this year's *Outlook*, as in past editions, derive from different underlying assumptions about policy. In this way, the *Outlook* provides insights into what policy can achieve and what the absence of policy action or delay in implementing policies would mean for energy markets, energy security and the environment.

The past twelve months have seen some important developments in international climate policy, preparing the ground for the adoption of new measures in the coming years. The UN negotiations on climate change held in December 2009 in Copenhagen did not result in a legally-binding agreement on limiting emissions of greenhouse gases. However, the Copenhagen Accord — the agreement that was reached at the meeting and with which all major emitting countries and many others subsequently associated themselves — does set a non-binding objective of limiting the increase in global temperature to two degrees Celsius (2°C) above pre-industrial levels. It also establishes a goal for the industrialised countries to mobilise funding for climate mitigation and adaptation in developing countries of $100 billion per year by 2020, and requires the industrialised countries (Annex I countries) to set emissions targets for 2020.

---

1. A detailed description of the WEM can be found at www.worldenergyoutlook.org/model.asp.

By the middle of 2010, nearly 140 countries, including many non-Annex I countries, had associated themselves with the Accord, either setting caps on their emissions for 2020 or announcing actions to mitigate emissions. However, the actual measures that would need to be taken to achieve these pledges had, in many cases, not yet been decided. Some targets are conditional on funding by Annex I countries or comparable emissions reductions across a set of countries, while other commitments involve a range. In addition, how much of the financing set out in the Accord is to be used for emissions mitigation is not specified. Some pledges relate to energy or carbon intensity, rather than emissions. As a result, it is far from certain what these commitments would mean for emissions, even if they were met fully. Since the Accord is not legally binding, the extent to which those commitments will be fulfilled remains highly uncertain. Similarly, it is uncertain what new action governments may decide to take in the coming years to deal with other concerns, such as threats to energy security, and what implications these might have for greenhouse-gas emissions.

Another important development has been the commitment made by G-20 leaders meeting in the US city of Pittsburgh in September 2009 to "rationalize and phase out over the medium term inefficient fossil fuel subsidies that encourage wasteful consumption". This commitment was made in recognition that subsidies distort markets, can impede investment in clean energy sources and can thereby undermine efforts to deal with climate change. G-20 leaders called upon the International Energy Agency, together with the Organisation for Economic Co-operation and Development (OECD), the Organization of Petroleum Exporting Countries (OPEC) and the World Bank to provide an analysis of the extent of energy subsidies and suggestions for the action necessary to implement this commitment. The results were presented in a joint report to the subsequent G-20 summit in June 2010.[2] At that summit, the leaders encouraged continued and full implementation of country-specific strategies.

In this year's *Outlook*, our central scenario, taking account of these political developments, takes a new form. It is called the **New Policies Scenario**. This scenario takes account of the broad policy commitments and plans that have been announced by countries around the world, to tackle either environmental or energy-security concerns, even where the measures to implement these commitments have yet to be identified or announced. These policies and plans include the national pledges to reduce greenhouse-gas emissions (communicated formally under the Copenhagen Accord) as well as plans to phase out fossil-energy subsidies. This scenario allows us to quantify the potential impact on energy markets of implementation of those policy commitments. But this scenario does not assume that they are all fully implemented. How governments strive to meet their policy commitments and the strength of their policy action to achieve them remains uncertain, for the reasons described above. For the purposes of this scenario, therefore, whereas we take into account action extending beyond existing policies alone (the basis of our former Reference Scenario) where there is a high degree of uncertainty, we have adopted a relatively narrow set

---

2. The report is available at www.worldenergyoutlook.org/subsidies.asp.

of policy assumptions corresponding to a cautious interpretation and implementation of the climate pledges and planned subsidy reforms. Countries that have set a range for a particular target are assumed to adopt policies consistent with reaching the less ambitious end of the range. In countries where uncertainty over climate policy is very high, it is assumed that the policies adopted are insufficient to reach their target. Financing for mitigation actions is also assumed to be limited and carbon markets are assumed to grow only moderately. These assumptions may be regarded as contentious. Their adoption is not a judgment on the countries concerned, but rather a means of illustrating the implications for world energy and emissions should these assumptions prove accurate.

Most of the formal national climate commitments that have been made relate to the period to 2020. For the period 2020-2035, we have assumed that additional measures are introduced that maintain the pace of the global decline in carbon intensity — measured as emissions per dollar of gross domestic product, in purchasing power parity terms — established in the period 2008-2020. The assumption of additional, but not necessarily ambitious further measures, reflects the absence of a binding international agreement to reduce global emissions. It is nonetheless assumed that each OECD country introduces an emission-reduction target across all sectors of the economy and establishes a harmonised emissions cap-and-trade scheme covering the power and industry sectors, which results in an acceleration of the decline in carbon intensity. Non-OECD countries are assumed to continue to implement national policies and measures, maintaining the pace of decline in domestic carbon intensity of 2008-2020. International sectoral agreements are assumed to be implemented across several industries, including cement and light-duty vehicles. In addition, we assume that fossil-fuel consumption subsidies are fully removed in all importing regions and are removed in exporting regions where specific policies have already been announced (Box 1.1).

We continue to present, as in previous *WEO*s, projections for a scenario, which we now call the **Current Policies Scenario**, in which no change in policies is assumed. This scenario, previously called the Reference Scenario, is intended to serve as a baseline against which the impact of new policies can be assessed. It takes into account those measures that governments had formally adopted by the middle of 2010 in response to and in pursuit of energy and environmental policies, but takes no account of any *future* changes in government policies and does not include measures to meet any energy or climate policy targets or commitments that have not yet been adopted or fully implemented. The Current Policies Scenario should in no sense be considered a forecast: it is certain that energy and climate policies in many — if not most — countries will change, possibly in the way we assume in the New Policies Scenario.

We also present updated projections for the **450 Scenario**, which was first presented in detail in *WEO-2008*. According to climate experts, there is a reasonable chance of limiting the global temperature increase to 2°C if the concentration of greenhouse gases in the atmosphere is limited to around 450 parts per million of carbon-dioxide equivalent (ppm CO_2-eq). The 450 Scenario sets out an energy pathway consistent with that objective, albeit involving initial overshooting of the target (see Chapter 13).

For the period to 2020, the emissions path reflects an assumption of vigorous policy action to implement fully the Copenhagen Accord, including achieving the maximum emissions reductions pledged, relatively limited use of emissions-reduction credits and no use of banked allowances from earlier periods. Thus, the policies assumed are collectively consistent with the high-end of the range of commitments, resulting in a lower emissions path than in the New Policies Scenario. A summary of the policy targets and measures for 2020 taken into account in the 450 and New Policies Scenarios is set out in Table 1.1; more detailed assumptions can be found in Annex B.

Box 1.1 ● Summary of fossil-fuel consumption subsidy assumptions by scenario

- In the *New Policies Scenario*, we assume that fossil-fuel subsidies are completely phased out in all net-importing regions by 2020 (at the latest) and in net-exporting regions where specific policies have already been announced.

- In the *Current Policies Scenario*, we assume that fossil-fuel subsidies are completely phased out in countries that already have policies in place to do so.

- In the *450 Scenario*, we assume fossil-fuel subsidies are completely phased out in all net-importing regions by 2020 (at the latest) and in all net-exporting regions by 2035 (at the latest), except the Middle East where it is assumed that the average subsidisation rate declines to 20% by 2035.

After 2020, OECD countries and Other Major Economies (defined here as Brazil, China, Russia, South Africa and the countries of the Middle East) are assumed to set economy-wide emissions targets for 2035 and beyond that collectively ensure an emissions trajectory consistent with stabilisation of the greenhouse-gas concentration at 450 ppm. OECD countries and Other Major Economies are assumed to establish separate carbon markets, and buy offsets in other countries. Fossil-fuel consumption subsidies are assumed to be completely phased out in all regions, except the Middle East, by 2035. The emissions and energy trajectories in the period to 2020 are higher than those shown in *WEO-2009* (IEA, 2009), which assumed stronger policy action in the near term, but the decline in emissions after 2020 is correspondingly faster.[3]

In this *Outlook*, we deliberately focus more attention on the results of the New Policies Scenario to provide a clear picture of where currently planned policies, if implemented in a relatively cautious way, would take us. Yet this scenario should not be interpreted as a forecast: even though it is likely that many governments around the world will take firm policy action to tackle climate and other energy-related problems, the policies that are actually put in place in the coming years may deviate markedly from those assumed in this scenario. On the one hand, governments may decide to take stronger action to implement their current commitments than assumed in this scenario and/or may adopt more stringent targets, possibly as a result of negotiations in the coming months and years on a more robust global

---

3. Details of the projections for the 450 Scenario are set out in Chapter 13.

climate agreement. In particular, a firmer deal may emerge on financing of emissions reductions in developing countries by the industrialised countries. On the other hand, it is possible that governments will fail to implement the policies required to meet even their current pledges, especially as the Copenhagen Accord is not legally binding and contains no provision for penalising countries that fail to meet their commitments. Policy action after 2020 may also falter, putting the world on a course that takes us closer to the Current Policies Scenario.

Table 1.1 ● **Principal policy assumptions by scenario and major region, 2020**

	New Policies Scenario	450 Scenario
OECD		
United States	15% share of renewables in electricity generation; push for domestic supplies, including gas and biofuels.	17% reduction in greenhouse-gas emissions compared with 2005 (with access to international offset credits).
Japan	Implementation of the Basic Energy Plan.	25% reduction in greenhouse-gas emissions compared with 1990 (with access to international offset credits).
European Union	25% reduction in greenhouse-gas emissions compared with 1990 (including Emissions Trading Scheme).	30% reduction in greenhouse-gas emissions compared with 1990 (with access to international offset credits).
Non-OECD		
Russia	15% reduction in greenhouse-gas emissions compared with 1990.	25% reduction in greenhouse-gas emissions compared with 1990.
China	40% reduction in CO_2 intensity compared with 2005 (low-end of targeted range).	45% reduction in CO_2 intensity compared with 2005 (high-end of targeted range); 15% share of renewables and nuclear power in primary demand.
India	20% reduction in CO_2 intensity compared with 2005.	25% reduction in CO_2 intensity compared with 2005.
Brazil	36% reduction in greenhouse-gas emissions compared with business-as-usual.	39% reduction in greenhouse-gas emissions compared with business-as-usual.

Main non-policy assumptions

Population

Population growth is an important driver of the amount and type of energy use. The rates of population growth assumed in this *Outlook* for each region and in all three scenarios are based on the most recent projections by the United Nations (UNPD, 2009). World population is projected to grow by 0.9% per year on average, from an estimated 6.7 billion in 2008 to 8.5 billion in 2035. Population growth slows progressively over the projection period, in line with the long-term historical trend, from 1.1% per year in 2008-2020 to 0.7% in 2020-2035 (Table 1.2). Population expanded by 1.5% per year from 1980 to 2008 and 1.3% per year from 1990.

Table 1.2 ● **Population growth by region** (compound average annual growth rates)

	1980-1990	1990-2008	2008-2020	2010-2015	2020-2035	2008-2035
OECD	0.8%	0.7%	0.5%	0.5%	0.3%	0.4%
North America	1.2%	1.2%	0.9%	0.9%	0.6%	0.7%
United States	*0.9%*	*1.1%*	*0.9%*	*0.9%*	*0.6%*	*0.7%*
Europe	0.5%	0.5%	0.3%	0.4%	0.1%	0.2%
Pacific	0.8%	0.4%	0.0%	0.1%	-0.3%	-0.1%
Japan	*0.5%*	*0.2%*	*-0.2%*	*-0.2%*	*-0.6%*	*-0.4%*
Non-OECD	**2.0%**	**1.5%**	**1.2%**	**1.2%**	**0.8%**	**1.0%**
E. Europe/Eurasia	0.8%	-0.2%	-0.1%	0.0%	-0.2%	-0.2%
Caspian	*n.a.*	*0.8%*	*1.0%*	*1.0%*	*0.6%*	*0.7%*
Russia	*n.a.*	*-0.2%*	*-0.4%*	*-0.3%*	*-0.5%*	*-0.4%*
Asia	1.8%	1.4%	1.0%	1.1%	0.6%	0.8%
China	*1.5%*	*0.9%*	*0.6%*	*0.6%*	*0.1%*	*0.3%*
India	*2.1%*	*1.6%*	*1.2%*	*1.3%*	*0.7%*	*1.0%*
Middle East	3.6%	2.3%	1.8%	1.8%	1.3%	1.5%
Africa	2.9%	2.5%	2.2%	2.2%	1.7%	1.9%
Latin America	2.0%	1.5%	1.0%	1.0%	0.6%	0.8%
Brazil	*2.1%*	*1.4%*	*0.7%*	*0.8%*	*0.3%*	*0.5%*
World	**1.7%**	**1.3%**	**1.1%**	**1.1%**	**0.7%**	**0.9%**
European Union	*n.a.*	*0.3%*	*0.2%*	*0.2%*	*0.0%*	*0.1%*

Note: The assumed rates of population growth are the same for all three scenarios presented in this *Outlook*.
Sources: UNPD and World Bank databases; IEA analysis.

The increase in global population is expected to occur overwhelmingly in non-OECD countries, mainly in Asia and Africa (Figure 1.1). Non-OECD population expands from 5.5 billion in 2008 to 7.2 billion in 2035, an average rate of increase of 1% per year, their share of the world's population rising from 82% to 85%. The only major non-OECD country that experiences a decline in its population is Russia, where the population falls from 142 million in 2008 to 126 million in 2035. Africa sees the fastest rate of growth, averaging 1.9% per year between 2008 and 2035. The population of non-OECD Asia rises from 3.5 billion to 4.3 billion. India overtakes China towards the end of the projection period to become the world's most heavily populated country, with 1.47 billion people in 2035. The population of the OECD increases by only 0.4% per year on average over 2008-2035. Most of the increase in the OECD occurs in North America; Europe's population increases slightly, while the population in the OECD Pacific region falls marginally.

All of the overall increase in world population will occur in urban areas; the rural population will decline in most regions, with the notable exception of Africa (UNPD,

2010). In 2009, for the first time in history, the world's urban population was larger than the rural population. The population living in urban areas is projected to grow by 1.9 billion, passing from 3.3 billion in 2008 to 5.2 billion 2035, with most of this increase occurring in non-OECD countries. Continuing rapid urbanisation will push up demand for modern energy services, as they are more readily available in towns and cities. Providing access to modern energy for poor urban and rural households will remain an increasingly pressing challenge (see Chapter 8).

Figure 1.1 ● **Population by major region**

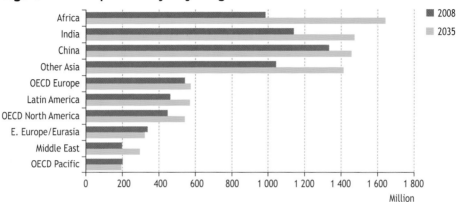

Sources: UNPD and World Bank databases; IEA analysis.

Economic growth

Economic activity is the principal driver of demand for each type of energy service. Thus, the projections in all three scenarios described in this *Outlook* are highly sensitive to the underlying assumptions about the rate of growth of gross domestic product (GDP). Energy demand tends to grow in line with GDP, though typically at a lower rate. For example, between 1980 and 2008, world primary energy demand increased by 0.59% each year on average for every percentage point of GDP growth (expressed in real purchasing power parity, or PPP, terms[4]). This (gross) income elasticity of demand, as it is known, has fluctuated over time, falling from 0.64 in the 1980s to 0.46 in the 1990s and then rebounding to 0.67 in 2000-2008, mainly because of a rapid expansion of energy-intensive manufacturing in China. In general,

---

4. Purchasing power parities (PPPs) measure the amount of a given currency needed to buy the same basket of goods and services, traded and non-traded, as one unit of the reference currency — in this report, the US dollar. By adjusting for differences in price levels, PPPs, in principle, can provide a more reliable indicator than market exchange rates of the true level of economic activity globally or regionally and, thus, help in analysing the main drivers of energy demand and comparing energy intensities across countries and regions. However, GDP and GDP-related indicators based on market exchange rates are used to compare trends over time, as no projections of PPPs are available.

the income elasticity of demand tends to be higher for countries at an early stage of economic development than for the more mature economies, where saturation effects curb income-driven increases in demand.

The global economy is now thought to be on the road to recovery, having endured the worst recession since the Second World War, though the threat of a double-dip recession persists. The International Monetary Fund (IMF) estimates that world GDP in PPP terms contracted by 0.6% in 2009, having expanded by 3.0% in 2008. But these figures disguise some very big differences in economic performance across the world. The recession was generally worse among the OECD economies, with most non-OECD economies experiencing a slowdown in growth rather than an outright contraction. Overall, the recession turned out to be less severe than originally expected, in part because of the strength of the policy response. Most of the world's largest economies introduced fiscal stimulus packages between late 2008 and mid-2009, in many cases involving tax reductions or spending increases worth several percentage points of GDP. While these packages helped to counter the effects of the global financial and economic crisis, they led to a ballooning of budget deficits and a sharp rise in national debt in many countries, especially in the OECD. Many countries are now faced with a need to tackle these problems, but most want to ensure that the recovery is well-established before undertaking fiscal tightening: over-zealous action to cut deficits could, it is feared, stall the recovery and tip the economy into a downward recessionary and debt spiral.

In many parts of the developing world, economies are growing rapidly once again, allowing the countries concerned to begin to rein in their expansionary macroeconomic policies as they experience growing capital inflows and a rebound in asset prices, notably property. With growth prospects in the OECD countries likely to remain relatively weak for several years as they grapple with rising national debt, the emerging economies will remain the main drivers of the global economic recovery. However, sustained rapid growth in the non-OECD countries will hinge on their ability to absorb rising inflows of capital and to nurture domestic demand without triggering a new boom-bust cycle (IMF, 2010a).

The IMF now projects global GDP growth to reach 4.6% in 2010 and 4.3% in 2011 (IMF, 2010a). The advanced economies (essentially the OECD) are projected to expand by 2.6% in 2010 and by 2.4% in 2011, following a decline in output of more than 3% in 2009. Growth in the rest of the world is projected to top 6% during 2010-11, following a modest expansion of 2.5% in 2009. Nonetheless, the IMF acknowledges that the outlook for economic activity remains unusually uncertain, and risks are generally to the downside. The risks to growth associated with the surge in public debt in the advanced economies are the most obvious, especially with respect to market concerns about sovereign liquidity and solvency in, for example, Greece and other European countries, and the danger that these concerns could evolve into a full-blown and contagious sovereign debt crisis (IMF, 2010b). Bank exposure to toxic assets, including mortgages and household debt, also threatens further turmoil in financial markets, particularly in the United States and Europe. There could be knock-on effects for growth prospects for the non-OECD countries.

Table 1.3 ● **Real GDP growth by region** (compound average annual growth rates)

	1980-1990	1990-2008	2008-2020	2010-2015	2020-2035	2008-2035
OECD	3.0%	2.5%	1.8%	2.4%	1.9%	1.8%
North America	3.1%	2.8%	2.1%	2.7%	2.2%	2.2%
United States	*3.2%*	*2.8%*	*2.0%*	*2.4%*	*2.1%*	*2.1%*
Europe	2.4%	2.2%	1.5%	2.1%	1.8%	1.6%
Pacific	4.3%	2.1%	1.7%	2.6%	1.2%	1.5%
Japan	*3.9%*	*1.2%*	*1.0%*	*1.9%*	*1.0%*	*1.0%*
Non-OECD	3.3%	4.7%	5.6%	6.7%	3.8%	4.6%
E. Europe/Eurasia	4.0%	0.8%	3.0%	4.4%	3.1%	3.1%
Caspian	*n.a.*	*2.0%*	*4.6%*	*5.4%*	*3.2%*	*3.8%*
Russia	*n.a.*	*0.6%*	*2.9%*	*4.1%*	*3.1%*	*3.0%*
Asia	6.6%	7.4%	7.0%	8.3%	4.2%	5.4%
China	*9.0%*	*10.0%*	*7.9%*	*9.5%*	*3.9%*	*5.7%*
India	*5.6%*	*6.4%*	*7.4%*	*8.1%*	*5.6%*	*6.4%*
Middle East	-1.3%	3.9%	4.0%	4.3%	3.8%	3.9%
Africa	2.3%	3.8%	4.5%	5.5%	2.8%	3.5%
Latin America	1.2%	3.5%	3.3%	4.0%	2.7%	3.0%
Brazil	*1.5%*	*3.0%*	*3.6%*	*4.1%*	*3.1%*	*3.3%*
World	3.1%	3.3%	3.6%	4.4%	2.9%	3.2%
European Union	*n.a.*	*2.1%*	*1.4%*	*2.1%*	*1.7%*	*1.6%*

Note: Calculated based on GDP expressed in year-2009 dollars at constant purchasing power parity (PPP) terms.
Sources: IMF and World Bank databases; IEA databases and analysis.

This *Outlook* assumes that the world economy grows on average by 4.4% over the five years to 2015.[5] In the longer term, the rate of growth is assumed to temper, as the emerging economies mature and their growth rates converge with those of the OECD economies. World GDP is assumed to grow by an average of 3.2% per year over the period 2008-2035, the same rate as in 1980-2008 (Table 1.3). Growth slows over the projection period, averaging 3.1% per year in the period 2015-2035. The non-OECD countries as a group are assumed to continue to grow much more rapidly than the OECD countries, driving up their share of world GDP. In several leading non-OECD countries,

---

5. The GDP growth assumptions to 2015 are based primarily on the latest IMF projections from the July 2010 update of its *World Economic Outlook* (IMF, 2010a), with some adjustments according to more recent information available for the OECD (OECD, 2010) and other countries from national and other sources. The assumptions are the same for eagch scenario, because of the uncertainty surrounding the relationships between policy-driven changes in energy-related investment, the resulting impact on climate change and the pace of economic growth.

a combination of important macro- and micro-economic reforms, including trade liberalisation, more credible economic management, and regulatory and structural reforms have improved the investment climate and the prospects for strong long-term growth. India overtakes China in the 2020s to become the fastest-growing WEO region, the result of demographic factors and its earlier stage of economic development. India's growth nonetheless slows from 7.9% in 2008-2015 to 5.9% in 2015-2035. China's growth rate slows to 4.4% in 2015-2035, less than half the rate at which it has been growing in recent years (and in 2009, when it still grew by 9.1% despite the global recession). Among the OECD regions, North America continues to grow fastest, at 2.2% per year on average over the projection period, buoyed by more rapid growth in its population and labour force, and lower debt than in Europe and the Pacific region.

Energy prices

As with any good, the demand for a given energy service depends on the price, which in turn reflects the price of the fuel as well as the technology used to provide it. The price elasticity of demand, *i.e.* the sensitivity of demand to changes in price, varies across fuels and sectors, and over time, depending on a host of factors, including the scope for substituting the fuel with another or adopting more efficient energy-using equipment, the need for the energy service and the pace of technological change. In each scenario, projections are based on the average retail prices of each fuel used in end uses, power generation and other transformation sectors. These prices are derived from assumptions about the international prices of fossil fuels (Table 1.4), and take account of any taxes, excise duties and carbon-dioxide emissions penalties (see below), as well as any subsidies. Final electricity prices are derived from marginal power-generation costs (which reflect the price of primary fossil-fuel inputs to generation, and the cost of hydropower, nuclear energy and renewables-based generation) and the non-generation costs of supply. The fossil-fuel-price assumptions reflect our judgment of the prices that will be needed to stimulate sufficient investment in supply to meet projected demand over the projection period.[6] Although the price paths follow smooth trends, prices are likely, in reality, to fluctuate.

Having rebounded through much of 2009, international crude oil prices settled into a range of around $70-85 per barrel in the first half of 2010. Prices are assumed to rise steadily over the entire projection period in all but the 450 Scenario, as rising global demand requires the development of increasingly more expensive sources of oil (see Chapter 3). The level of prices needed to match oil supply and demand varies with the degree of policy effort to curb demand growth and differs markedly across the three scenarios. In the New Policies Scenario, the average IEA crude oil import price reaches $105/barrel (in real 2009 dollars) in 2025 and $113/barrel in 2035 (Figure 1.2).[7] In nominal terms, prices more than double to $204/barrel in 2035.[8] In the Current Policies Scenario, substantially higher prices

---

6. This methodology differs from that used in the IEA's *Medium Term Oil and Gas Market Report*, which assumes the prices prevailing on futures markets (IEA, 2010a).

7. In 2009, the average IEA crude oil import price was $1.52/barrel lower than the first-month forward spot price of West Texas Intermediate (WTI) and $1.27/barrel lower than spot dated Brent.

8. The dollar exchange rates used were those prevailing in 2009 (€0.720 and ¥93.6), which were assumed to remain unchanged over the projection period.

Does rising prosperity inevitably push up energy needs?

That energy use typically rises with incomes is incontrovertible and widely understood. As economies grow, they require more energy to fuel factories and trucks, to heat and cool buildings and to meet growing personal demand for mobility, equipment and electrical appliances. Over the last several decades, energy use has tended to rise proportionately with GDP at the global level and, in most cases, at the national level too, though the relationship is usually less than one to one: in other words, energy needs usually grow somewhat less rapidly in percentage terms than the size of the economy, because of changes in economic structure towards less energy-intensive activities and because of technological change that gradually improves the efficiency of providing energy-related services.

But will this relationship persist far into the future and do rising incomes, therefore, make increased energy use inevitable? This *Outlook* and previous editions predict that the relationship will indeed remain strong — at least for the next quarter of a century — *unless governments intervene to change it,* through measures that lead to a shift in behaviour and/or in the way in which energy needs are met. For as long as the global economy continues to expand — and no-one doubts that it will, in the longer term, in the absence of a catastrophic event — and population expands, then the world's overall energy needs will undoubtedly rise. But just how quickly, and in what way those needs are met, is far from certain. The energy projections in this *Outlook* — and experience in many countries over the past three decades — show very clearly that the link between GDP and energy use can be loosened, if not entirely broken, through a combination of government action and technological advances.

What matters to users of energy, whether they be businesses or individuals, is the ultimate energy-related services that they receive: mobility, heating, cooling or a mechanical process. Today, these services are often provided in ways that involve unnecessarily large amounts of energy, much of it derived from fossil fuels. The technology exists today to increase greatly the efficiency with which those services are provided and that technology will surely continue to improve in the future. The commercial incentives for manufacturers to make available more efficient equipment, appliances and vehicles, and for consumers to buy them, are set to increase with rising energy costs. But commercial factors alone will be not sufficient. Governments need to act to reinforce those incentives so as to encourage even faster improvements in energy efficiency and to discourage energy waste, confident in the environmental, energy-security and broader economic benefits that would follow. Experience has shown what governments can achieve through determined action; our projections show what more can be achieved in the future.

Table 1.4 • Fossil-fuel import price assumptions by scenario (dollars per unit)

	Unit	New Policies Scenario						Current Policies Scenario					450 Scenario				
		2009	2015	2020	2025	2030	2035	2015	2020	2025	2030	2035	2015	2020	2025	2030	2035
Real terms (2009 prices)																	
IEA crude oil imports	barrel	60.4	90.4	99.0	105.0	110.0	113.0	94.0	110.0	120.0	130.0	135.0	87.9	90.0	90.0	90.0	90.0
Natural gas imports																	
United States	MBtu	4.1	7.0	8.1	9.1	9.9	10.4	7.0	8.2	9.3	10.4	11.2	7.0	8.0	8.9	9.4	9.7
Europe	MBtu	7.4	10.6	11.6	12.3	12.9	13.3	10.7	12.1	12.9	13.9	14.4	10.4	10.6	10.7	10.9	11.0
Japan	MBtu	9.4	12.2	13.4	14.2	14.9	15.3	12.4	13.9	14.9	15.9	16.5	11.9	12.2	12.3	12.5	12.6
OECD steam coal imports	tonne	97.3	97.7	101.7	104.1	105.6	106.5	97.8	105.8	109.5	112.5	115.0	92.5	85.8	75.8	66.3	62.1
Nominal terms																	
IEA crude oil imports	barrel	60.4	103.6	127.1	151.1	177.3	204.1	107.7	141.3	172.7	209.6	243.8	100.7	115.6	129.5	145.1	162.6
Natural gas imports																	
United States	MBtu	4.1	8.0	10.4	13.1	15.9	18.9	8.0	10.5	13.3	16.7	20.3	8.0	10.3	12.8	15.1	17.5
Europe	MBtu	7.4	12.2	14.9	17.8	20.9	24.1	12.3	15.5	18.6	22.4	26.0	11.9	13.6	15.4	17.5	19.8
Japan	MBtu	9.4	14.0	17.2	20.4	24.0	27.6	14.2	17.8	21.4	25.7	29.8	13.6	15.6	17.7	20.1	22.7
OECD steam coal imports	tonne	97.3	112.0	130.6	149.8	170.2	192.4	112.1	135.9	157.6	181.4	207.8	106.0	110.2	109.0	106.8	112.1

Note: Natural gas prices are weighted averages, expressed on a gross calorific-value basis. All prices are for bulk supplies exclusive of tax. The US gas import price is used as a proxy for prices prevailing on the domestic market. Nominal prices assume inflation of 2.3% per year from 2009.

are needed to balance supply with the faster growth in demand. The average crude oil price rises more briskly, especially after 2020, reaching $120/barrel in 2025 and $135/barrel ten years later. In the 450 Scenario, by contrast, prices increase more slowly, levelling off at about $90/barrel by 2020, as demand peaks and then begins to decline by around 2015 (see Chapter 15 for details of the drivers of oil demand in this scenario). Falling demand is assumed to outweigh almost entirely the rising cost of production (see Chapter 3). Higher CO_2 prices contribute to lower demand and, therefore, lower international prices (see below). In reality, whatever the policy landscape, oil prices are likely to remain volatile.

Figure 1.2 ● **Average IEA crude oil import price by scenario** (annual data)

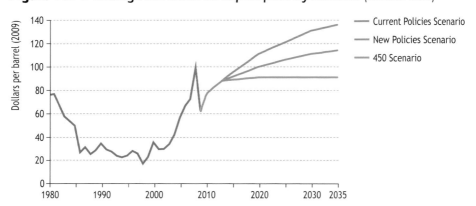

Traditionally, natural gas prices have moved in fairly close tandem with oil prices, either because of indexation clauses in long-term supply contracts or indirectly through competition between gas and oil products in power generation and end-use markets. In recent years, gas prices have tended to decouple from oil prices, as a result of relatively abundant supplies of unconventional gas in North America, which have driven gas prices there down relative to oil, increased availability of spot supplies of cheaper liquefied natural gas in Europe and Asia-Pacific, and some provisional changes to contractual terms in Europe, which have lessened the role of oil prices and increased the importance of gas-price indexation in long-term contracts. There is considerable uncertainty about whether this tentative move away from oil indexation will prove permanent and, even if it does, whether this will herald an era of lower gas prices relative to oil (see Spotlight in Chapter 5). One uncertainty is the length of time that long-term contracts in bulk gas supply will remain dominant in Europe and Asia-Pacific. Yet, even if direct gas-to-gas competition becomes more widespread and allowing for the fact that the underlying cost drivers for oil and gas differ, the potential for substitution between oil products and gas will ensure that changes in the price of one will continue to affect the price of the other.[9] In all three scenarios, the ratio of gas prices to oil prices in North America is assumed to rise modestly through to 2035 as the cost of unconventional gas production rises, but the ratio remains well below the

---

9. See IEA (2009) for a detailed discussion of the prospects for gas pricing.

historical average. In Europe and Japan (a proxy for Asia-Pacific), we assume that the ratio of gas prices remains broadly unchanged to 2035 (Figure 1.3). The ratio of gas to oil prices throughout the projection period remains well below the average for the period 1980-2009 in all regions.

International steam-coal prices have fallen from record levels attained in mid-2008, with the slowdown in demand and weaker prices for gas, the main competitor to coal (especially in the power sector). The price of coal imported by OECD countries averaged slightly over $95 per tonne in 2009. In the New Policies Scenario, coal prices are assumed to remain at about this level in real terms to 2015 and then, with rising demand to 2020 and higher prices of gas to rise to $107/tonne by 2035. Coal prices rise less in percentage terms than oil or gas prices, partly because coal production costs are expected to remain low and because coal demand flattens out by 2020. Coal prices rise more quickly in the Current Policies Scenario on stronger demand growth, but fall in the 450 Scenario, reflecting the impact of policy action to cut demand.

Figure 1.3 ● *Ratio of average natural gas and coal import prices to crude oil in the New Policies Scenario*

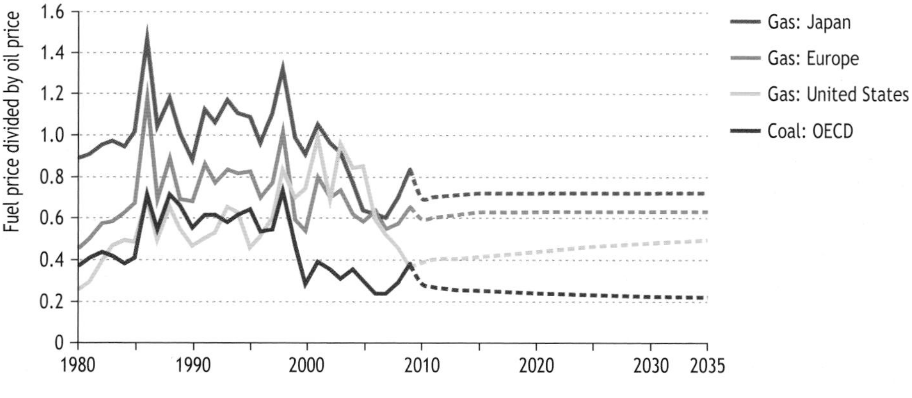

Note: Calculated on an energy-equivalent basis.

CO$_2$ prices

The pricing of carbon emissions could play an increasingly important role in driving energy markets in the long term. For now, only the European Union and New Zealand have adopted formal cap-and-trade schemes, which set caps on carbon-dioxide emissions by the power generation and industry sectors and provide for trading of CO$_2$ certificates, yielding prices of CO$_2$ for specific time periods. Thus, in the Current Policies Scenario, carbon pricing is assumed to be limited to EU countries and to New Zealand. The price of CO$_2$ under the EU Emission Trading System is projected to reach $30/tonne in 2020 and $42/tonne in 2035 (Table 1.5).

Table 1.5 • CO$_2$ prices by main region and scenario ($2009 per tonne)

	Region	2009	2020	2030	2035
New Policies	European Union	22	38	46	50
	Japan	n.a.	20	40	50
	Other OECD	n.a.	-	40	50
Current Policies	European Union	22	30	37	42
450	OECD+	n.a.	45	105	120
	Other Major Economies	n.a.	-	63	90

Note: OECD+ includes all the OECD countries plus non-OECD EU countries. The CO$_2$ price in the European Union is assumed to converge with that in OECD+ by 2020 in the 450 Scenario. Other Major Economies comprise Brazil, China, the Middle East, Russia and South Africa.

Carbon pricing is assumed to be adopted in other regions in the New Policies and 450 Scenarios. In the New Policies Scenario, cap-and-trade systems covering the power and industry sectors are assumed to be established in Australia, Japan and Korea as of 2013, and in OECD countries (see note to Table 1.5) after 2020, where it reaches $50/tonne in 2035. In the 450 Scenario, cap-and-trade covering power generation and industry is assumed to start in 2013 in OECD+ and after 2020 in the Other Major Economies category (see note to Table 1.5). In this scenario, we assume that CO$_2$ is traded in these two groups separately. To contain emissions at the levels required in the 450 Scenario, we estimate that the price of CO$_2$ in OECD+ would need to reach $45/tonne in 2020 and $120/tonne in 2035. The price rises to $63/tonne in 2030 and to $90/tonne in 2035 in the Other Major Economies. The prices are set by the most expensive abatement option, for example, carbon capture and storage in industry in the OECD+ in 2035. It is assumed that OECD+ countries have access to international offsets, up to a limit of one-third of total abatement in 2020. Further details of carbon pricing and how it is modelled in the 450 Scenario can be found in Chapter 13.

Technology

Technology has an important impact on both the supply and use of energy. Our projections are, therefore, very sensitive to assumptions about developments in technology and how quickly new technologies are deployed. Those assumptions vary for each fuel, each sector and each scenario, according to our assessment of the current stage of technological development and commercialisation and the potential for further improvements and deployment, taking account of economic factors and market conditions.[10] Government policies and energy prices have an important impact on the pace of development and deployment of new technologies. As a consequence, more rapid technological advances are seen in the 450 Scenario.

In all three scenarios, the performance of currently available categories of technology is assumed to improve on various operational criteria, including energy efficiency,

---

10. See *Energy Technology Perspectives 2010* (IEA, 2010b) for a detailed assessment of the long-term prospects for energy-related technologies.

practicality, environmental impact and flexibility. But the pace of improvement varies: it is fastest in the 450 Scenario, thanks to the effect of various types of government support, including economic instruments (such as carbon pricing, taxes and subsidies), regulatory measures (such as standards and mandates) and direct public-sector investment. These policies stimulate increased spending on research, development and deployment. Technological change, in general, is slowest in the Current Policies Scenario, because no new public policy actions are assumed. Yet, even in this scenario, significant technological improvements occur, aided by higher energy prices. In the New Policies Scenario, the pace of technological change lies between that in the two other scenarios. Crucially, no completely new technologies on the demand or supply side, beyond those known today, are assumed to be deployed before the end of the projection period, as it cannot be known whether or when such breakthroughs might occur and how quickly they may be commercialised.

The critical factor with respect to energy use concerns how the introduction of more advanced technologies affects the average energy efficiency of equipment, appliances and vehicles in use, and, therefore, the overall intensity of energy consumption (the amount of energy needed to provide one dollar of gross domestic product). Practical and financial constraints on how quickly energy-related capital stock[11] can be replaced affect the rate at which new technologies can be introduced and, consequently, the rate of improvement in energy efficiency. Some types of capital stock, such as power stations (which have a long design life), are so costly and difficult to install that they are replaced only after a very long time. Indeed, much of the capital stock in use today falls into this category. As a result, much of the impact of recent and future technological developments that improve energy efficiency will not be felt until towards the end of the projection period. Rates of capital-stock turnover differ greatly: most cars and trucks, heating and cooling systems, and industrial boilers in use today will be replaced by 2035. But most existing buildings, roads, railways and airports, as well as many power stations and refineries will still be in use then, unless strong government incentives and/or a change in market conditions encourage or force early retirement. The extent to which this happens (or the stock is modernised to reduce energy needs) is limited in the Current Policies Scenario; it is greater in the New Policies Scenario and especially in the 450 Scenario.

On the supply side, technological advances are assumed to improve the technical and economic efficiency of producing and supplying energy. In some cases, they result in lower unit costs, lead to cleaner ways of producing and delivering energy services, or make available resources that are not recoverable commercially or technically today. Many emerging renewable energy technologies, such as wind and photovoltaic energy, fall into this category. In other cases, where technologies are relatively mature, such as conventional oil and gas drilling, the impact of technological advances on unit costs is expected to be at least partially offset by the rising cost of raw materials and labour. Some major new supply-side technologies that are approaching the

---

11. Any type of asset that affects the amount and the way in which energy is supplied or used, such as oil wells, power stations, pipelines, buildings, boilers, machinery, appliances and vehicles.

commercialisation phase are assumed to become available and to be deployed to some degree before the end of the projection period. These include carbon capture and storage, advanced biofuels, large-scale concentrating solar power and smart grids. Details about how fast these technologies are deployed can be found in the relevant chapters.

CHAPTER 2

ENERGY PROJECTIONS TO 2035

Twilight in demand?

H I G H L I G H T S

- Global primary energy demand continues to grow in the New Policies Scenario, but at a slower rate than in recent decades. By 2035, it is 36% higher than in 2008. Non-OECD countries account for 93% of the increase. The OECD share of world demand falls from 44% today to 33% in 2035. Energy demand in the other scenarios diverges over the period: by 2035, it is 8% higher in the Current Policies Scenario and 11% lower in the 450 Scenario than in the New Policies Scenario.

- Fossil fuels maintain a central role in the primary energy mix in the New Policies Scenario, but their share declines, from 81% in 2008 to 74% in 2035. Oil demand is up by 18%, from 84 mb/d in 2009 to 99 mb/d in 2035. Coal demand is around 20% higher in 2035 than today, with almost all of the growth before 2020. The 44% increase in natural gas demand surpasses that for all other fuels due to the favourable environmental and practical attributes of gas. Electricity demand grows by around 80% by 2035, requiring 5 900 GW of total capacity additions.

- The importance of China in global energy markets continues to grow. In 2000, China's energy demand was half that of the United States, but preliminary data indicate it is now the world's biggest energy consumer. Growth prospects remain strong, given China's per-capita energy use is still only one-third of the OECD average and it is the most populous nation.

- Investment in energy-supply infrastructure to meet demand to 2035 in the New Policies Scenario amounts to $33 trillion (in year-2009 dollars). Power sector investment accounts for $16.6 trillion, or just over half of the total. Almost two-thirds of total investment is in non-OECD countries.

- The New Policies Scenario implies a persistently high level of spending on energy imports by many countries. Total spending on oil and gas imports more than doubles from $1.2 trillion in 2010 to $2.6 trillion in 2035. The United States is overtaken by China around 2025 as the world's biggest spender on oil imports; India overtakes Japan around 2020 as the world's third-largest spender.

- In the New Policies Scenario, energy-related CO_2 emissions rise from 29.3 Gt in 2008 to 35.4 Gt in 2035, consistent with an eventual increase in global average temperature of over 3.5°C. All of the growth in emissions comes from non-OECD countries; emissions in the OECD drop by 20%. Chinese emissions exceed those from the entire OECD by 2035.

Overview of energy trends by scenario

What governments do to tackle critical energy-related problems holds the key to the outlook for world energy markets over the next quarter of a century. Our projections of energy demand and supply accordingly vary significantly across the three scenarios presented in this *Outlook* (Box 2.1). In the New Policies Scenario, which takes account of both existing policies and declared intentions, world primary energy demand is projected to increase by 1.2% per year between 2008 and 2035, reaching 16 750 million tonnes of oil equivalent (Mtoe), an increase of 4 500 Mtoe, or 36% (Figure 2.1). Demand increases significantly faster in the Current Policies Scenario, in which no change in government policies is assumed, averaging 1.4% per year over 2008-2035. In the 450 Scenario, in which policies are assumed to be introduced to bring the world onto an energy trajectory that provides a reasonable chance of constraining the average global temperature increase to 2° Celsius, global energy demand still increases between 2008 and 2035, but by a much reduced 22%, or an average of 0.7% per year. Energy prices ensure that projected supply and demand are in balance throughout the *Outlook* period in each scenario (see Chapter 1).

Figure 2.1 ● **World primary energy demand by scenario**

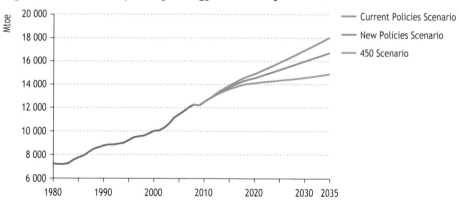

Fossil fuels remain the dominant energy sources in 2035 in all three scenarios, though their share of the overall primary fuel mix varies markedly, from 62% in the 450 Scenario to 79% in the Current Policies Scenario, compared with 74% in the New Policies Scenario and 81% in 2008 (Table 2.1 and Figure 2.2). These differences reflect the varying strength of policy action assumed to address climate-change and energy-security concerns. The shares of renewables and nuclear power are correspondingly highest in the 450 Scenario and lowest in the Current Policies Scenario. The range of outcomes — and therefore the uncertainty with respect to future energy use — is largest for coal and non-hydro renewable energy sources

Box 2.1 ● Understanding the three *WEO-2010* scenarios

WEO-2010 presents detailed projections for three scenarios: a New Policies Scenario, a Current Policies Scenario and a 450 Scenario. The scenarios differ with respect to what is assumed about future government policies related to the energy sector. There is much uncertainty about what governments will actually do over the coming quarter of a century, but it is highly likely that they will continue to intervene in energy markets. Indeed, many countries have announced formal objectives; but it is very hard to predict with any degree of certainty what policies and measures will actually be introduced or how successful they will be. The commitments and targets will undoubtedly change in the course of the years to come.

Given these uncertainties, we present projections for a *Current Policies Scenario* as a baseline in which only policies already formally adopted and implemented are taken into account. In addition, we present projections for a *New Policies Scenario*, which assumes the introduction of new measures (but on a relatively cautious basis) to implement the broad policy commitments that have already been announced, including national pledges to reduce greenhouse-gas emissions and, in certain countries, plans to phase out fossil-energy subsidies. We focus in this *Outlook* on the results of this New Policies Scenario, while also referring to the outcomes in the other scenarios, in order to provide insights into the achievements and limitations of the important developments that have taken place in international climate and energy policy over the past year.

The *450 Scenario*, which was first presented in detail in *WEO-2008* and for which updated projections are presented here, sets out an energy pathway consistent with the goal of limiting the global increase in average temperature to 2°C, which would require the concentration of greenhouse gases in the atmosphere to be limited to around 450 parts per million of carbon-dioxide equivalent (ppm CO_2-eq). Its trajectory to 2020 is somewhat higher than in *WEO-2009*, which started from a lower baseline and assumed stronger policy action before 2020. The decline in emissions is, by necessity, correspondingly faster after 2020.

Global energy intensity — the amount of energy needed to generate each unit of GDP — has fallen steadily over the last several decades due to several factors including improvements in energy efficiency, fuel switching and structural changes in the global economy away from energy-intensive industries. The implications for global energy consumption and environmental pollution have been significant: if no improvements in energy intensity had been made between 1980 and 2008, global energy consumption would be 32% higher today, roughly equivalent to the combined current consumption of the United States and the European Union.

Table 2.1 ● World primary energy demand by fuel and scenario (Mtoe)

	1980	2008	New Policies Scenario		Current Policies Scenario		450 Scenario	
			2020	2035	2020	2035	2020	2035
Coal	1 792	3 315	3 966	3 934	4 307	5 281	3 743	2 496
Oil	3 107	4 059	4 346	4 662	4 443	5 026	4 175	3 816
Gas	1 234	2 596	3 132	3 748	3 166	4 039	2 960	2 985
Nuclear	186	712	968	1 273	915	1 081	1 003	1 676
Hydro	148	276	376	476	364	439	383	519
Biomass and waste*	749	1 225	1 501	1 957	1 461	1 715	1 539	2 316
Other renewables	12	89	268	699	239	468	325	1 112
Total	7 229	12 271	14 556	16 748	14 896	18 048	14 127	14 920

* Includes traditional and modern uses.

Figure 2.2 ● Shares of energy sources in world primary demand by scenario

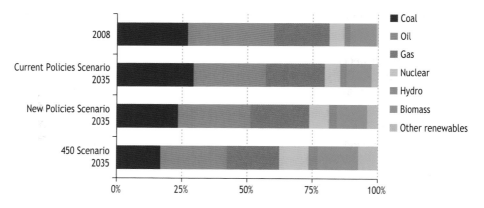

The policies that are assumed to be introduced in the New Policies and 450 Scenarios have a significant impact on the rate of decline in energy intensity. In the Current Policies Scenario, energy intensity continues to decline gradually over the projection period, but at a much slower rate than in the other scenarios. By 2035, energy intensity declines compared to 2008 are: 28% in the Current Policies Scenario, 34% in the New Policies Scenario and 41% in the 450 Scenario. By comparison, between 1981 and 2008 global energy intensity fell by 23% (Figure 2.3). Over the period 2008 to 2035, the annual average improvement in energy intensity is 1.2% in the Current Policies Scenario, 1.5% in the New Policies Scenario and 1.9% in the 450 Scenario.

Figure 2.3 ● **Change in global primary energy intensity by scenario**

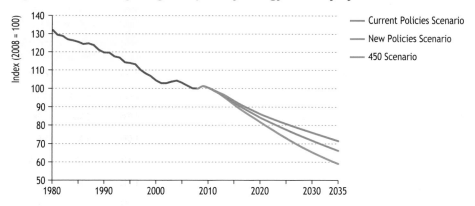

Note: Calculated based on GDP expressed in year-2009 dollars at market exchange rates (MER).

Energy trends in the New Policies Scenario

Primary energy demand

In this chapter, we deliberately focus more attention on the results of the New Policies Scenario.[1] This is done to provide a clear picture of where planned policies, assumed to be implemented in a cautious way, would take us. As indicated, the New Policies Scenario projects global energy consumption to increase by 36% from 2008 to 2035, rising from 12 300 Mtoe to 16 750 Mtoe (Table 2.2). Growth in demand slows progressively, from an average of 1.4% per year in the period 2008-2020 to 0.9% per year in 2020-2035, as measures introduced to combat climate change and meet energy-security objectives take effect.

Over the *Outlook* period, demand for each fuel source increases (Figure 2.4). Fossil fuels (oil, coal and natural gas) account for 53% of the increase in energy demand. They continue to supply the bulk of global energy consumption, though their share falls from 81% in 2008 to 74% in 2035. Rising fossil-energy prices to end-users, resulting from upward price pressures on international markets and increasing costs of carbon, together with policies to encourage energy savings and switching to low-carbon energy sources, help to restrain demand growth for all three fossil fuels.

Oil remains the dominant fuel in the primary energy mix during the *Outlook* period in the New Policies Scenario, with demand increasing from 85 million barrels per day (mb/d) in 2008 (84 mb/d in 2009) to 99 mb/d in 2035. Its share of the primary fuel mix, which stood at 33% in 2008, drops to 28% as high prices lead to further switching away from oil in the industrial and power-generation sectors and opportunities emerge

---

1. Annex A provides detailed projections of energy demand by fuel, sector and region for all three scenarios.

to substitute other fuels for oil products in transport. Demand for coal increases from 4 736 million tonnes of coal equivalent (Mtce) in 2008 to just over 5 600 Mtce in 2035, with most of the growth before 2020.[2] Growth in demand for natural gas far surpasses that of all other fossil fuels due to its more favourable environmental and practical attributes and constraints on how quickly low-carbon energy technologies can be deployed. Global natural gas consumption increases from 3 149 billion cubic metres (bcm) in 2008 to just above 4 500 bcm in 2035. By the end of the *Outlook* period, natural gas is close to overtaking coal as the second most important fuel in the primary energy mix.

Table 2.2 ● **World primary energy demand by fuel in the New Policies Scenario** (Mtoe)

	1980	2008	2015	2020	2030	2035	2008-2035*
Coal	1 792	3 315	3 892	3 966	3 984	3 934	0.6%
Oil	3 107	4 059	4 252	4 346	4 550	4 662	0.5%
Gas	1 234	2 596	2 919	3 132	3 550	3 748	1.4%
Nuclear	186	712	818	968	1 178	1 273	2.2%
Hydro	148	276	331	376	450	476	2.0%
Biomass and waste**	749	1 225	1 385	1 501	1 780	1 957	1.7%
Other renewables	12	89	178	268	521	699	7.9%
Total	7 229	12 271	13 776	14 556	16 014	16 748	1.2%

* Compound average annual growth rate. ** Includes traditional and modern uses.

The share of nuclear power increases over the projection period, from 6% in 2008 to 8% in 2035. Government policies are assumed to boost the role of nuclear power in several countries. Furthermore, it is assumed that a growing number of countries implement programmes to extend the lifetime of their currently operating nuclear plants, thereby reducing the capacity that would otherwise be lost to retirement in the period to 2035.

The use of modern renewable energy — including wind, solar, geothermal, marine, modern biomass and hydro — triples over the course of the *Outlook* period, growing from 843 Mtoe in 2008 to just over 2 400 Mtoe in 2035. Its share in total primary energy demand increases from 7% to 14%. Consumption of traditional biomass drops from 746 Mtoe in 2008 to a little over 720 Mtoe in 2035, after a period of modest increase to 2020. Demand for renewable energy increases substantially in all regions, with dramatic growth in some areas, including China and India. Power generation from renewables triples from 2008 to 2035, with its share of the generation mix increasing from 19% in 2008 to 32% in 2035.

---

2. 1 Mtce is equal to 0.7 Mtoe.

2

How do the energy demand projections in *WEO-2010* compare with *WEO-2009*?

Though this chapter concentrates on the results of the New Policies Scenario, it is also informative to compare the level of world primary energy demand in this year's Current Policies Scenario with the results projected in the Reference Scenario of *WEO-2009*, using a similar methodology. Total primary energy demand in 2015 is 3% higher compared with last year's projections, but it is less than 1% higher by 2030 (the last year of the projection period in *WEO-2009*). This small divergence masks important changes among regions: projected demand in OECD countries in 2030 is lower than projected last year, but this is more than offset by higher projected demand in the rest of the world. Projected demand for all fuels, with the exception of oil, is higher in absolute terms in 2030 in this year's report. The biggest increase is for natural gas, with demand 4.4%, or 192 bcm, higher than projected last year, while global oil demand is 2.4%, or 2.5 mb/d, lower. Compared with the projections in *WEO-2009*, projected electricity generation this year is essentially unchanged, but there are some notable shifts in the generating mix, with both natural gas and nuclear seeing sizeable increases.

These differences result from the combined effect of many changes. Numerous new policies enacted between mid-2009 and mid-2010, aimed at encouraging a transition to a cleaner, more efficient and more secure energy system, have been incorporated into the Current Policies Scenario and act to dampen growth in projected demand. However, these new policies are insufficient to offset other factors that drive projected demand higher. Most importantly, the global economy appears to be emerging from the economic and financial crisis faster than expected. Therefore, our assumed rate of growth in world GDP — the main driver of energy demand — is now higher than in WEO-2009, particularly in non-OECD countries, which are coming out of the recession more strongly than OECD countries. Compared with the *WEO-2009*, which assumed a more protracted recovery, the upward revision in GDP plays a key role in boosting demand growth in the early stages of the projection period (hence the big differences between the two scenarios to 2015).

Adjustments to the assumptions about energy prices, including changes to relative pricing that affect the energy mix, further explain some of the differences. The price assumptions vary across the different scenarios presented in *WEO-2010* in line with the degree of policy effort needed to curb demand growth. In the Current Policies Scenario, higher oil prices are needed (compared with *WEO-2009*) to choke off demand to bring it into balance with supply, while coal prices also increase slightly. In contrast, natural gas price assumptions have been scaled back, in North America by as much as 10% after 2020, as the substantial rise in unconventional gas production drives prices lower.

This year's 450 Scenario depicts a somewhat higher trajectory for CO_2 emissions to 2020 than in *WEO-2009*, due to less ambitious action in the early period to curb emissions. This is offset by a faster decline in emissions after 2020. The main reason for the change in trajectory is that the opportunity for concerted, immediate action to slow the growth in emissions was missed as the United Nations climate meeting in Copenhagen in December 2009 did not achieve a comprehensive agreement on limiting emissions of greenhouse gases.

Figure 2.4 ● World primary energy demand by fuel in the New Policies Scenario

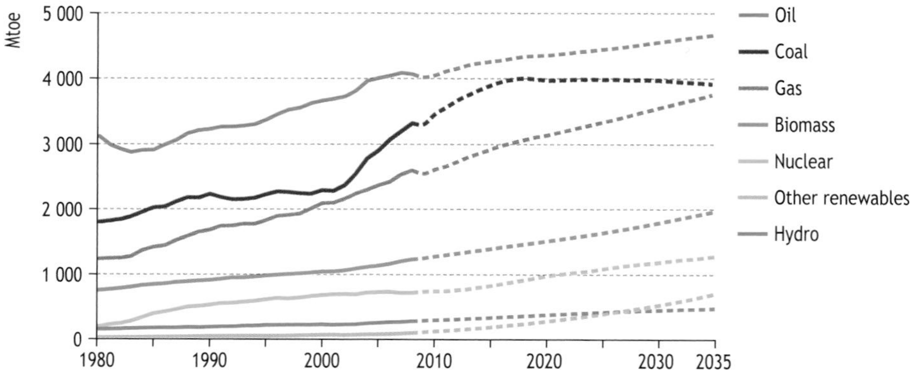

Regional trends

The faster pace of growth in primary energy demand that has occurred in non-OECD countries over the last several decades is set to continue, reflecting faster rates of growth of population, economic activity, urbanisation and industrial production. In the New Policies Scenario, total non-OECD energy consumption increases by 64% in 2008-2035, compared with a rise of just 3% in OECD countries. Nonetheless, annual average growth in non-OECD energy demand slows through the *Outlook* period, from 2.4% in 2008-2020 to 1.4% in 2020-2035. The OECD share of global primary energy demand, which declined from 61% in 1973 to 44% in 2008, falls to just 33% in 2035 (Table 2.3).

The increase in non-OECD energy consumption is led by brisk growth in China, where primary demand surges by 75% in 2008-2035, a far bigger increase than in any other country or region (Figure 2.5). China accounts for 36% of the global increase in primary energy use between 2008 and 2035, with its share of total demand jumping from 17% to 22%. India is the second-largest contributor to the increase in global demand to 2035, accounting for 18% of the rise. India's energy consumption more than doubles by that date, growing on average by 3.1% per year, a rate of growth significantly higher than in any other region. Outside Asia, the Middle East experiences the fastest rate of increase, at 2.0% per year. After a modest increase to 2020, aggregate energy demand in OECD countries stagnates. Nonetheless, by 2035 the United States is still the world's second-largest energy consumer, well ahead of India, which is a distant third.

Table 2.3 ● **Primary energy demand by region**
in the New Policies Scenario (Mtoe)

	1980	2000	2008	2015	2020	2030	2035	2008-2035*
OECD	4 050	5 233	5 421	5 468	5 516	5 578	5 594	0.1%
North America	2 092	2 670	2 731	2 759	2 789	2 836	2 846	0.2%
United States	*1 802*	*2 270*	*2 281*	*2 280*	*2 290*	*2 288*	*2 272*	*0.0%*
Europe	1 493	1 734	1 820	1 802	1 813	1 826	1 843	0.0%
Pacific	464	829	870	908	914	916	905	0.1%
Japan	*345*	*519*	*496*	*495*	*491*	*482*	*470*	*-0.2%*
Non-OECD	3 003	4 531	6 516	7 952	8 660	10 002	10 690	1.9%
E.Europe/Eurasia	1 242	1 019	1 151	1 207	1 254	1 344	1 386	0.7%
Caspian	*n.a*	*128*	*169*	*205*	*220*	*241*	*247*	*1.4%*
Russia	*n.a*	*620*	*688*	*710*	*735*	*781*	*805*	*0.6%*
Asia	1 067	2 172	3 545	4 609	5 104	6 038	6 540	2.3%
China	*603*	*1 107*	*2 131*	*2 887*	*3 159*	*3 568*	*3 737*	*2.1%*
India	*208*	*459*	*620*	*778*	*904*	*1 204*	*1 405*	*3.1%*
Middle East	128	381	596	735	798	940	1 006	2.0%
Africa	274	502	655	735	781	868	904	1.2%
Latin America	292	456	569	667	723	812	855	1.5%
Brazil	*114*	*185*	*245*	*301*	*336*	*386*	*411*	*1.9%*
World**	7 229	10 031	12 271	13 776	14 556	16 014	16 748	1.2%
European Union	*n.a*	*1 682*	*1 749*	*1 722*	*1 723*	*1 719*	*1 732*	*0.0%*

* Compound average annual growth rate.
** World includes international marine and aviation bunkers (not included in regional totals).

Figure 2.5 ● **World primary energy demand by region**
in the New Policies Scenario

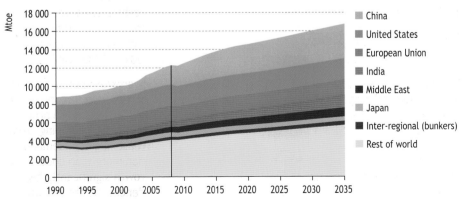

Non-OECD countries generate the bulk of the increase in global demand for all primary energy sources (Figure 2.6). OECD oil demand falls by 6 mb/d in 2009-2035, but this is offset by a 19-mb/d increase in the non-OECD (international bunker demand also rises by almost 3 mb/d). Oil demand increases the most in China (7.1 mb/d), India (4.5 mb/d) and the Middle East (2.7 mb/d) as a consequence of rapid economic growth and, in the case of the Middle East, the continuation of subsidies on oil products. By 2035, China overtakes the United States to become the largest oil consumer in the world. Having reached a peak of 46 mb/d in 2005, oil demand in the OECD continues to decline, reaching 35 mb/d in 2035, due to further efficiency gains in transport and continued switching away from oil in other sectors. Oil demand in the United States declines from 17.8 mb/d in 2009 to 14.9 mb/d in 2035.

Non-OECD regions are responsible for the entire net increase in coal demand to 2035. China alone accounts for 54% of the net increase; although coal's share of China's energy mix continues to decline, more than half of its energy needs in 2035 are still met by coal. Most of the rest of the growth in coal demand comes from India and other non-OECD Asian countries. Driven by policies to limit or reduce CO_2 emissions, coal use falls sharply in each of the OECD regions, particularly after 2020. By 2035, OECD countries consume 37% less coal than today.

Unlike demand for the other fossil fuels, demand for natural gas increases in the OECD. where it remains the leading fuel for power generation and an important fuel in the industrial, service and residential sectors. Collectively, the OECD countries account for 16% of the growth in natural gas consumption to 2035. Developing Asia, again led by China and India, accounts for 43% of the incremental demand, as gas use increases rapidly in the power sector and in industry. The Middle East, which holds a considerable share of the world's proven natural gas reserves, is responsible for one-fifth of the global increase in gas consumption.

Figure 2.6 ● **Incremental primary energy demand by fuel and region in the New Policies Scenario, 2008-2035**

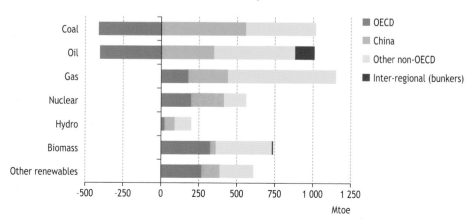

World Energy Outlook 2010 - **GLOBAL ENERGY TRENDS**

Box 2.2 ● China becomes the world's largest energy consumer

Preliminary data suggest that China overtook the United States in 2009 to become the world's largest energy user. This comes just two years after China overtook the United States as the world's largest emitter of energy-related CO_2. Preliminary IEA data, which align closely with those of most of the other main sources of international energy statistics, indicate that in 2009 China consumed about 4% more energy than the United States.

China's emergence as the world's largest energy consumer is not a surprise. Its phenomenal rate of demand growth over the last decade meant it was destined to become the top energy consumer. This has occurred slightly earlier than expected, however, because of China's continuing strong economic performance and its quick recovery from the global financial crisis compared to the United States. Since 2000, China's energy demand has doubled. Growth prospects remain robust considering the country's low per-capita consumption levels (it is still only around one-third of the average in OECD countries), and the fact that China is the most populous nation on the planet, with more than 1.3 billion people.

Today, energy demand in China would be even higher had it not made remarkable progress in reducing its energy intensity (the energy input required per dollar of output). In 2009, China consumed about one-quarter of the energy per unit of economic output than it did in 1980. China has also become a world leader in renewable energy and is pursuing a 10-year programme aimed at boosting the share of low-carbon energy to 15% of total consumption by 2020 and meeting ongoing carbon emissions reduction targets. These efforts are being backed by a development plan entailing planned investment of 5 trillion yuan (approximately $735 billion) in nuclear, wind, solar and biomass projects. Given the sheer scale of China's domestic market, its push to increase the share of new low-carbon energy technologies (both on the supply side and the demand side, such as advanced vehicle technologies) could play an important role in driving down their costs by contributing to improvements in technology learning rates.

Under the assumptions of the New Policies Scenario, nuclear power expands in both OECD and non-OECD regions between 2008 and 2035, the increase in the non-OECD being almost twice as big in absolute terms. The increase in nuclear power generation in China alone (215 Mtoe) exceeds that of the entire OECD (198 Mtoe). Within the OECD, Japan, Korea, France and the United States are responsible for almost all of the growth. In aggregate, the supply of nuclear power in OECD Europe remains flat. This is consistent with the general assumptions for the New Policies Scenario, in which countries with declared plans to discontinue their nuclear programmes are assumed to pursue them.

Non-OECD countries account for 56% of the global increase in the use of non-hydro renewable energy between 2008 and 2035. Biomass, mostly fuel wood, crop residues and charcoal for cooking and heating, represents 38% of incremental energy demand in Africa (see Chapter 8). Demand for biomass and waste, consumed mostly in modern applications in power generation and transport, also increases rapidly in the OECD. Non-OECD countries account for almost 90% of the increase in hydropower generation, as considerable potential exists, particularly in Asia and Latin America. By contrast, in the OECD the most suitable sites, especially for large hydro, have already been developed.

Sectoral trends

The power sector (which includes both heat and electricity generation) accounts for 53% of the increase in global primary energy demand in 2008-2035. Its share of the primary mix reaches 42% in 2035, compared with 38% in 2008. Total capacity additions of 5 900 GW are required in 2008-2035, or around six times current US capacity. Coal remains the leading fuel for power generation, although its share of total power output peaks at about 42% soon after 2010, and declines to 32% in 2035. This declining coal share benefits non-hydro renewables (including biomass and waste) as their share increases from 3% to 16% by 2035. The shares of total power output of natural gas (21%), nuclear (14%) and hydro (16%) remain relatively constant throughout the *Outlook* period, while the share of oil continues to decline, to less than 2% in 2035.

Total final consumption[3] is projected to grow by 1.2% per year throughout the *Outlook* period (Figure 2.7). Industry demand grows most rapidly, at 1.4% per year, having overtaken transport in 2008 to once again become the second-largest final-use sector, after the buildings sector. By 2035, the industrial sector consumes around 30% of the world's total final energy consumption. Over three-fifths of the growth in industrial energy demand comes from China and India, while the Middle East and Latin America also see strong growth in demand. OECD industrial energy demand increases through to 2020 before dropping back to levels similar to today by the end of the *Outlook* period.

In aggregate, growth in global transport energy demand averages 1.3% per year in 2008-2035. This is a sharp decline in the rate of growth observed over the last several decades, thanks largely to measures to improve fuel economy. Transport's share of total final consumption remains flat at around 27% through the *Outlook* period. All of the growth in transport demand comes from non-OECD regions and inter-regional bunkers; transport energy demand declines slightly in the OECD. Although biofuels, and, to a lesser extent, electricity for plug-in hybrid and electric vehicles take an increasing share of the market for road-transport fuels, oil-based fuels continue to dominate transport energy demand.

In the buildings sector, energy use grows at an average rate of 1.0% per year through the *Outlook* period. The sector's share of total final energy consumption remains at around one-third throughout the period to 2035.

Electricity consumption is projected to increase at an annual average rate of 2.2% in the period 2008-2035, resulting in overall growth of around 80%. Electricity's share of total final consumption grows from 17% to 23%. More than 80% of the growth in

---

3. Total final consumption includes total energy delivered to end-users to undertake activities in industry, transport, agriculture, buildings (including residential and services) and non-energy use.

electricity demand takes place in non-OECD countries as a result of increased demand for household appliances and industrial and commercial electrical equipment, in line with rising prosperity. The shares of biomass and natural gas in total final consumption remain essentially constant through to 2035, while those for oil and coal decline, principally to the benefit of electricity.

Figure 2.7 ● **Incremental energy demand by sector and region in the New Policies Scenario, 2008-2035**

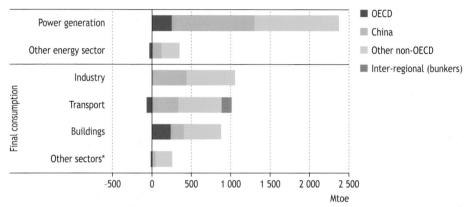

* Includes agriculture and non-energy use.

Per-capita energy consumption and energy intensity

Even though emerging economies experience markedly higher growth in energy demand during the *Outlook* period, a significant gulf still exists between rich and poor countries in the amount of energy used per capita. Today, the average per-capita energy consumption for the world as a whole is 1.8 tonnes of oil equivalent (toe) per year, but, in most cases, there is a great difference between developing and developed countries. There are also significant variations between countries at similar stages of economic development. Per-capita consumption in Japan, for example, is around half that of the United States.

Per-capita global energy consumption rises at 0.3% per year, on average, over the projection period (one-third of the rate experienced since 1995) reaching 2 toe in 2035. Large geographical discrepancies in energy consumption remain. In 2035, the average per-capita level in the OECD, despite having already peaked and now being in steady decline, is still more than twice the global average (Figure 2.8). The most rapid increase in per-capita consumption is in India, but at 1.0 toe in 2035, use per capita is still less than one-quarter that of the OECD. Although China's per-capita energy consumption is currently below the world average, in 2035 it is 40% higher than today's global average (or 30% higher than the 2035 global average), thanks to strong economic growth and relatively slow population growth. By 2035, Russia has the world's highest per-capita energy consumption, at 6.4 toe. This results from the combination of a harsh climate, continuing population decline, the importance of heavy industry in the economy and relatively inefficient energy production and consumption practices (a legacy of the Soviet era). Per-capita consumption remains lowest in sub-Saharan Africa at only 0.4 toe in 2035, *down* 23% from 2008 and only one-twelfth of the average OECD per-capita consumption. This trend results from sub-Saharan Africa's rapid population growth and the shift from traditional to modern energy, which is used more efficiently.

Figure 2.8 ● **Per-capita primary energy demand by region as a percentage of 2008 world average in the New Policies Scenario**

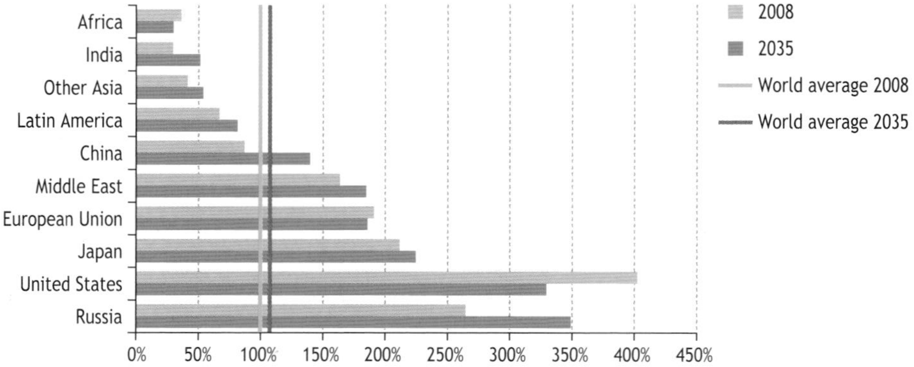

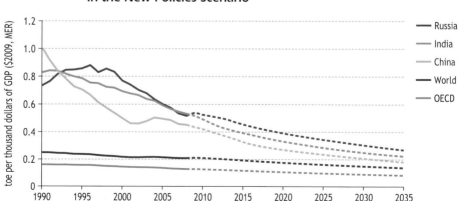

As with per-capita energy consumption, large differences in energy intensity exist among countries, primarily due to differences in energy efficiency, economic structure and climate. In most cases, non-OECD countries have much higher levels of energy intensity than those of the OECD, but they are also experiencing much faster reductions. Energy intensity in the OECD declines at 1.6% per year between 2008 and 2035, while the rate of decline in the non-OECD is 2.5% (Figure 2.9). China achieves the strongest improvement in its energy intensity at 3.3% per year on average, reaching 0.18 toe per thousand dollars of GDP at market exchange rates (MER) in 2035.

Figure 2.9 ● **Energy intensity in selected countries and regions in the New Policies Scenario**

Energy production and trade

Resources and production prospects[4]

Estimates of the world's total endowment of economically exploitable fossil fuels and hydroelectric, uranium and renewable energy resources indicate that they are more than sufficient to meet the projected increase in consumption to 2035. There is, however, some uncertainty about whether energy projects will be developed quickly enough to bring these resources to market in a timely manner, as many factors may act to defer investment spending. These include uncertainty about the economic outlook, developments in climate change and other environmental policies, depletion policies in key producing regions and changes to legal, fiscal and regulatory regimes.

Coal is the world's most abundant fossil fuel by far, with proven reserves of 1 000 billion tonnes (BGR, 2009). At present coal production levels, reserves would meet demand for almost 150 years. Remaining recoverable resources are even larger and a resource shortage is unlikely to constrain coal production. Coal is also the most widely distributed of fossil-fuel resources, with 43% of proven reserves in OECD countries, compared to natural gas (10%) and oil (16%). Proven reserves of oil amounted to 1.35 trillion barrels at the end of 2009, or 46 years production at current levels (O&GJ, 2010). Other economically recoverable resources that are expected to be found will support rising production. Today, proven gas reserves, at around 60 years of current production, far exceed the volume needed to satisfy demand to 2035 and undiscovered conventional gas resources are also sizeable. Moreover, there is huge potential to increase supply from unconventional resources of both oil and gas. Although these resources are generally more costly to exploit, rising fossil-fuel prices throughout the *Outlook* period and advances in technology and extraction methods are set to make them increasingly important sources of supply. Resources of uranium, the raw material for nuclear fuel, are sufficient to fuel the world's nuclear reactors at current consumption rates for at least a century (NEA and IAEA, 2009). Significant potential also remains for expanding energy production from hydropower, biomass and other renewable sources (see Chapters 9).

In the New Policies Scenario, non-OECD regions account for all of the net increase in aggregate fossil-fuel production between 2009 and 2035 (Figure 2.10). The world's total oil production reaches 96 mb/d by 2035. Total non-OPEC oil production peaks before 2015 at around 48 mb/d and falls to 46 mb/d by the end of the *Outlook* period. By contrast, OPEC oil production continues to grow, pushing up the group's share of world production from 41% in 2009 to 52% in 2035. Projected global gas production in 2035 in the New Policies Scenario increases by 43% compared with 2008. Non-OECD countries collectively account for almost all of the projected increase in global natural gas production in 2008-2035. The Middle East, with the largest reserves and lowest production costs, sees the biggest increase in absolute terms, though Eurasia remains the largest producing region and Russia the single biggest producer. Coal production is projected to rise by 15% between 2008 and 2035. All of the growth comes from

---

4. Resource and production prospects for each fuel are discussed in more detail in later chapters.

non-OECD countries, with production in the OECD falling by more than one-quarter. China sees the biggest increase in coal output in absolute terms, although the rate of increase in production is much higher in both India and Indonesia.

Figure 2.10 ● *World incremental fossil-fuel production in the New Policies Scenario, 2008-2035*

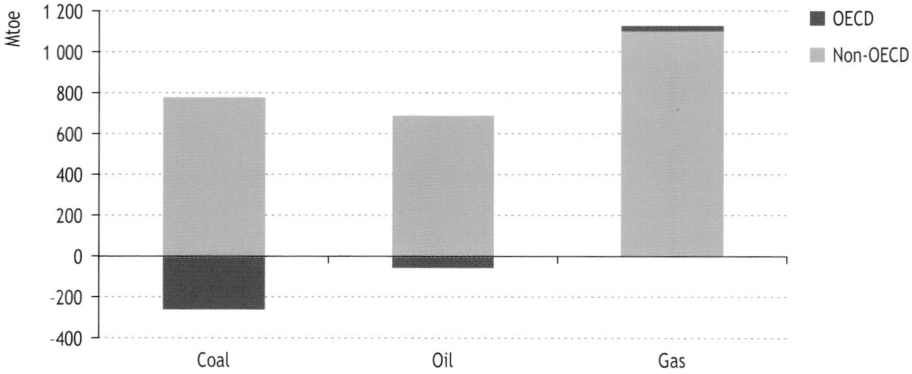

Inter-regional trade

The New Policies Scenario sees growing international trade in energy, due to the regional mismatch between the location of demand and production. The share of global oil consumption traded between *WEO* regions reaches 49% in 2035, compared with 44% today. In absolute terms, net trade rises from 37 mb/d in 2009 to 48 mb/d in 2035. Net imports into the OECD increase slightly to 2015, before gradually falling as OECD oil production declines at a slower rate than the fall in its demand, reducing the need for imports. By 2035, the OECD in aggregate is importing almost 18 mb/d, compared with 23 mb/d in 2009. Developing Asia, led by China and India, sees the biggest jump in oil imports in absolute terms. China's imports rise from 4.3 mb/d in 2009 to close to 13 mb/d by 2035; India's jump from 2.2 mb/d to 6.7 mb/d. Total oil exports from the Middle East continue to grow steadily, with the region's share of global trade increasing from 50% today to 60% in 2035.

Inter-regional natural gas trade rises from 670 bcm in 2008 to around 1 200 bcm in 2035, an increase of 77%. Developing Asia, led by China and India, is responsible for the bulk of the increase in gas imports. Of the OECD regions, Europe sees by the far the biggest increase in reliance on imports.

International trade in hard coal among *WEO* regions is projected to rise from 728 Mtce today to just under 870 Mtce before 2020, before decreasing to settle at a level around 840 Mtce as global demand for coal stabilises over the second half of the projection period. Over the course of the *Outlook* period demand for increased imports of coal into non-OECD Asia is offset by a sharp drop in demand for imports into OECD Europe, Japan and Korea. By 2035, inter-regional trade meets 15% of global hard coal demand, a level similar to today.

Spending on imports

Even with the measures that are assumed to be introduced to cut growth in energy demand, the New Policies Scenario implies a persistently high level of spending on oil and gas imports by many importing countries (Figure 2.11). India's projected spending is highest as a proportion of GDP, reaching 5.1% of GDP at market exchange rates by 2035, followed by China's at 3.1%. In aggregate, spending in the OECD as a proportion of GDP is set to decline through the *Outlook* period with the fall in the volume of its imports.

Figure 2.11 ● **Expenditure on net imports of oil and gas as a share of real GDP in the New Policies Scenario**

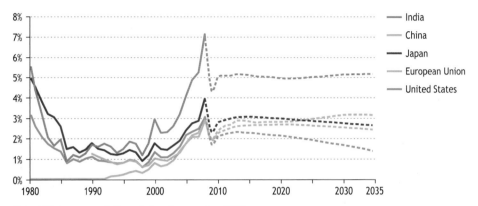

Note: GDP is measured at market exchange rates (MER).

Annual expenditure on oil and gas imports in dollar terms continues to increase throughout the *Outlook* period in most importing countries. Total expenditure at the global level on oil and gas imports more than doubles, from approximately $1.2 trillion in 2010 to $2.6 trillion in 2035, with the share of natural gas in total spending steadily increasing. On a country basis, China overtakes the United States around 2025 to become the world's biggest spender on oil imports, while India overtakes Japan around 2020 to become the world's third-largest spender. By 2025, China also surpasses Japan to become the world's biggest spender on natural gas imports.

Investment in energy-supply infrastructure

Cumulative investment of $33 trillion (year-2009 dollars) over 2010-2035 is needed in energy-supply infrastructure in the New Policies Scenario (Table 2.4). The projected investment is equal to around 1.4% of global GDP on average to 2035. This investment enables the replacement of reserves and production facilities that are retired, as well as the expansion of production and transport capacity to meet demand growth. The projected investment does not include demand-side investments, such as expenditure on purchasing cars, air conditioners, refrigerators, etc.

Although aggregate energy demand in OECD countries only increases by 3%, they require 35% of the projected investment (Figure 2.12). This disproportionally high share results from several factors, including the OECD need to retire and replace significant amounts of ageing energy infrastructure, its more capital-intensive energy mix and the higher average unit costs of its capacity additions. Almost 64% of total energy investment will take place in non-OECD countries, where production and demand are expected to increase most. China alone will need to invest $5.1 trillion, or 16% of the world total.

The energy mix in the New Policies Scenario has a higher share of energy technologies that are more capital intensive than those adopted in the WEO-2009 Reference Scenario. This factor, together with the extension of the period to 2035, more than offsets the lower rate of projected energy demand, leading to an investment requirement which is some $150 billion higher per year on average over the projection period.

Table 2.4 ● Cumulative investment in energy-supply infrastructure in the New Policies Scenario, 2010-2035 (billion $ in year-2009 dollars)

	Coal	Oil	Gas	Power	Biofuels	Total
OECD	201	1 811	2 875	6 477	211	11 574
North America	110	1 358	1 746	2 777	120	6 111
Europe	34	373	751	2 730	86	3 974
Pacific	57	80	378	970	5	1 490
Non-OECD	474	6 001	4 152	10 130	124	20 881
E. Europe/Eurasia	47	1 270	1 213	1 073	5	3 608
Russia	*20*	*676*	*792*	*570*	*1*	*2 060*
Asia	375	904	1 136	7 197	62	9 673
China	*263*	*475*	*360*	*4 000*	*32*	*5 130*
India	*56*	*207*	*216*	*1 883*	*17*	*2 380*
Middle East	1	965	586	597	0	2 149
Africa	34	1 313	764	559	3	2 674
Latin America	16	1 549	452	704	54	2 776
Inter-regional transport	46	241	74	n.a	n.a	361
World	721	8 053	7 101	16 606	335	32 816

The power sector requires $16.6 trillion or 51% of the total energy-supply investment projected to 2035 in the New Policies Scenario. If the investments in the oil, gas and coal industries that are needed to supply fuel to power stations are included, the share increases to 62%. Expenditures to develop transmission and distribution systems account for 42% of the total investment in the electricity industry, with the remainder going to power generation.

Investment to meet projected demand for oil in 2010-2035 amounts to $8.1 trillion, or one-quarter of total energy investment. The upstream oil sector accounts for 85% of the total, with the rest needed in downstream oil activities. Capital spending gradually declines over the course of the *Outlook* period, in line with the slowdown in global oil demand growth and as production shifts increasingly to lower-cost regions. On an annual average basis, investment is $310 billion per year. Investment in the OECD is high relative to its production capacity because unit costs are higher than other regions, particularly in the upstream segment of the supply chain.

Figure 2.12 ● **Cumulative investment in energy-supply infrastructure by region and fuel in the New Policies Scenario, 2010-2035**

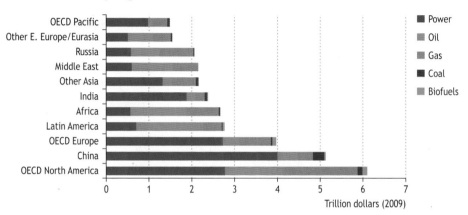

Cumulative investment in the natural gas supply chain in 2010-2035 is projected at $7.1 trillion, slightly less than for oil. Annual expenditures will increase over time with the increase in demand. Exploration and development of gas fields, including bringing new fields on stream and sustaining output at existing fields, will absorb 64% of total gas investment. In the period 2010-2035, some $720 billion needs to be invested in the coal sector, or 2% of total energy investment. Investment in production of coal is much less capital-intensive than investment in oil or natural gas.

Energy-related CO$_2$ emissions in the New Policies Scenario

Rising demand for fossil fuels continues to drive up energy-related carbon dioxide (CO$_2$) emissions through the projection period (Figure 2.13). Additional government policies that are assumed to be adopted, including action to implement pledges to reduce greenhouse-gas emissions announced under the Copenhagen Accord and moves to phase out fossil-energy subsidies in certain regions, help to slow the rate of growth in emissions, but do not stop the increase. Global energy-related CO$_2$ emissions jump by 21% between 2008 and 2035, from 29.3 gigatonnes (Gt) to 35.4 (Gt). Nonetheless, the average rate of growth of 0.7% per year represents a notable improvement on the Current Policies Scenario, in which emissions grow at 1.4% per year on average, reaching 42.6 Gt in 2035.

Figure 2.13 ● World energy-related CO_2 emissions by fuel in the New Policies Scenario

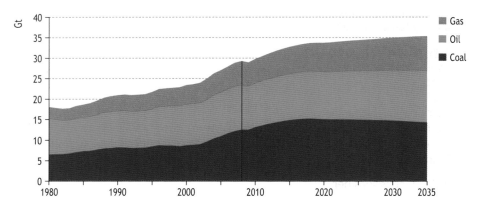

Non-OECD countries account for all of the projected growth in energy-related CO_2 emissions to 2035 in each of the three scenarios. In the New Policies Scenario, emissions from non-OECD countries continue to rise steadily and are 53% higher in 2035 than today. By 2035, non-OECD energy-related emissions of CO_2 are nearly two-and-a-half times those of the OECD. By the end of the *Outlook* period, emissions from China alone slightly exceed those from the OECD as a whole. All sectors contribute to overall growth in CO_2 emissions in 2008-2035: at 2.2 Gt, transport adds the largest amount (and has the highest growth rate), while power generation accounts for a rise of 1.8 Gt.

Energy-related CO_2 emissions in the OECD peak before 2015 and decline to 11.8 Gt in 2020, 7% above 1990 levels. OECD countries finance almost 500 million tonnes (Mt) of reductions in non-Annex I countries through purchases of offset emissions credits to comply with their own targets. Direct financing from OECD countries to non-OECD countries is also provided, in order to assist with low-carbon technology investment and to achieve additional abatement. Given the assumption that OECD countries step up domestic abatement efforts after 2020, OECD emissions steadily decline to 10 Gt in 2035.

Energy-related CO_2 emissions in non-OECD countries are projected to grow from 15.7 Gt in 2008 to 20.8 Gt by 2020 and 24 Gt by 2035. This increase occurs despite the assumed implementation of measures in China and India to significantly reduce their energy intensity, as well as policies in Indonesia, Brazil and South Africa to improve upon the business-as-usual situation (see Chapter 13 for a discussion of the uncertainty around non-Annex I targets). The low end of the intensity improvement targets set by China and India are achieved in the Current Policies Scenario through measures already enacted. This means that in the New Policies Scenario, these targets are exceeded, though much of the additional effort is assumed to be supported through an international offset mechanism or direct finance. With respect to domestically-financed actions, non-OECD countries are assumed to maintain the same level of effort to combat climate change over the projection period.

While the projection for greenhouse-gas emissions in the New Policies Scenario is a marked improvement on current trends, much more would need to be done to realise the Copenhagen Accord objective of limiting the average rise in global temperature to 2°C. The New Policies Scenario puts the world onto a trajectory consistent with stabilising the concentration of greenhouse gases at just over 650 ppm CO_2-eq, resulting in a likely temperature rise of over 3.5°C in the long term (see Chapter 13).

Energy-related CO_2 emissions by fuel exhibit a broadly similar pattern to that of fuel demand, in that the share of oil and coal falls across the period, while the share of gas increases. In 2008, coal had the largest share of total emissions, at 43%, with oil at 37% and gas at 20%. In 2035, this order remains the same in the New Policies Scenario, though the share of coal falls to 41% and that of oil to 36%, while the share of gas increases to 24%. Emissions from bunker fuels change by less than half a percentage point from 2008 to 2035, accounting for 3.5% of emissions in 2008 and 4.0% in 2035.

World CO_2 emissions per capita have been increasing sharply since 2000. In the New Policies Scenario, this upward trend continues until they reach a peak of 4.5 tonnes around 2015 and then decline to less than 4.2 tonnes by the end of the *Outlook* period. Large discrepancies remain between regions. Although average per capita emissions continue to fall in the OECD, by 2035 they are still 1.7 times the current global average (Figure 2.14). The fastest growth in per-capita emissions occurs in China; from 4.9 tonnes in 2008, they grow by 41% to 6.9 tonnes in 2035. Africa's per-capita emissions decline through the *Outlook* period, reaching less than one-sixth of the world average in 2035.

Figure 2.14 ● **Per-capita energy-related CO_2 emissions by region as a percentage of 2008 world average in the New Policies Scenario**

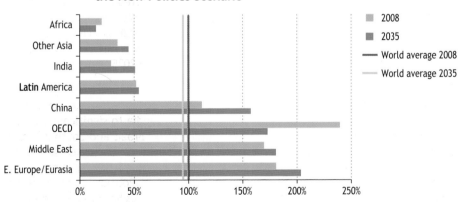

The crucial role of China in global energy markets

The increase in China's energy consumption between 2000 and 2008 was more than four times greater than in the previous decade. The prospects for further growth remain very strong: energy demand per capita in China is still only 35% of the OECD average. Future developments in China's energy system, therefore, have major implications for global supply and demand trends for oil, natural gas and coal, as well as the prospects for limiting climate change. Consequently, the global energy projections in this *Outlook* remain highly sensitive to the underlying assumptions for the key variables that drive energy demand in China. These include prospects for economic growth, changes in economic structure, developments in energy and environmental policies and the rate of urbanisation.

The rapid expansion in China's energy demand since 2000 is the result of extremely rapid GDP growth and a structural shift in its economy towards energy-intensive heavy industry and exports, especially following its accession to the World Trade Organization in 2001. China now accounts for 28% of global industrial energy demand, a sharp increase on its 16% share in 2000. The rising share of industry in China's economy led to an increase in the country's energy intensity. China's energy intensity increased on average by 2.5% per year between 2002 and 2005, reversing average gains of 6.4% per year between 1990 and 2002. Recognising the adverse implications of rising energy intensity on the economy and energy security, China's 11th Five-Year Plan set a target to reduce energy intensity by 20% between 2005 and 2010. Government reports indicate that the country's energy intensity fell by 15.6% from 2005 to 2009 but then edged up slightly in early 2010 (NBS, 2010), suggesting that it will be difficult to achieve the full 20% target. Nonetheless, gains realised over such a short period of time represent a very impressive achievement.

The momentum of economic development looks set to generate strong growth in energy demand in China throughout the *Outlook* period. In the New Policies Scenario, China's primary energy demand is projected to climb by 2.1% per year between 2008 and 2035, reaching two-thirds of the level of consumption of the entire OECD (Figure 2.15). China's total final energy consumption increases at a similar rate, expanding by 2.0% per year between 2008 and 2035. In absolute terms, industry accounts for the single biggest element in the growth in final energy demand. Industry's share declines marginally, however, as demand is increasingly driven by domestic consumption. This reflects the emergence of a sizeable middle class whose aspirations for modern lifestyles and comfort levels creates a surge in demand for motor vehicles, electrical appliances and other energy-using equipment. China's electricity demand is projected to almost triple in 2008-2035, requiring capacity additions equivalent to 1.5 times the current installed capacity of the United States.

During much of the period of its economic expansion, China was able to meet all of its energy needs from domestic production. A growing share is now being met by imports. China has extensive coal resources, but in recent years has become a net importer. It has struggled to expand its mining and rail-transport infrastructure quickly enough to move coal from its vast inland reserves to the prosperous coastal areas where demand has been growing most rapidly. In the New Policies Scenario, China's net imports of coal increase to 2015, but the country once again becomes a net exporter towards the end of the *Outlook* period. Its oil imports jump from 4.3 mb/d in 2009 to 12.8 mb/d in 2035, the share of imports in demand rising from 53% to 84%. Natural gas imports also increase substantially to reach a share of 53% of demand in 2035, requiring a major expansion of pipeline and liquefied natural gas (LNG) regasification infrastructure.

Figure 2.15 ● Total primary and per-capita energy demand in China and the OECD in the New Policies Scenario

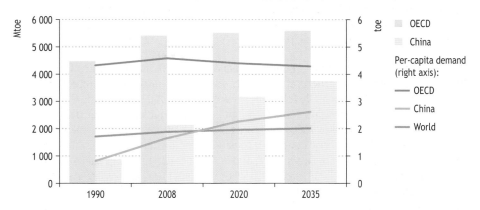

The projected rise in China's energy demand has implications for the local and global environment. In the New Policies Scenario, 58% of the global increase in CO_2 emissions to 2035 comes from China alone (Figure 2.16). China's emissions increase by 54%, to 10.1 Gt, surpassing the emissions from the entire OECD by 2035. One contribution to the strong increase in China's emissions is that as it has become the world's biggest export manufacturer, and given its significant reliance on fossil energy, a proportion of its emissions are caused by the manufacturing of goods for export to other countries. This "embedded carbon" far outweighs the carbon embedded in its imports.

Figure 2.16 ● China's share of the projected net global increase for selected indicators

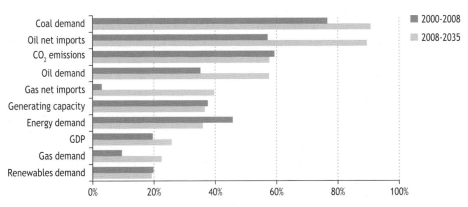

Although China's per-capita emissions are much lower than those in most industrialised countries, they are increasing rapidly. China already emits 12% more per capita than the global average and is set to overtake the per-capita level of the European Union soon after 2020 in the New Policies Scenario. China is currently one of the world's highest emitters of CO_2 per unit of GDP, but our projections indicate an improvement in emissions intensity (3.8% per year) between 2008 and 2035, which is faster than improvements achieved elsewhere.

CHAPTER 3

OIL MARKET OUTLOOK

A peak at the future?

H   I   G   H   L   I   G   H   T   S

- The global outlook for oil remains highly sensitive to policy action to curb rising demand and emissions. In the Current Policies and New Policies Scenarios, global primary oil use increases in absolute terms between 2009 and 2035, driven by population and economic growth, but demand falls in the 450 Scenario in response to radical policy action to curb fossil-fuel use.

- The prices needed to balance the oil market differ markedly across the three scenarios — reflecting the growing insensitivity of demand and supply to price. In the New Policies Scenario, the average IEA crude oil import price (in year-2009 dollars) reaches $113/barrel in 2035. In the Current Policies Scenario, much higher prices — reaching $135/barrel in 2035 — are needed to bring demand into balance with supply. Prices in the 450 Scenario are much lower, as demand peaks before 2020 and then falls. The weaker the response to the climate challenge, the greater the risk of oil scarcity and the higher the economic cost for consuming countries.

- In the New Policies Scenario, demand continues to grow steadily, reaching about 99 mb/d (excluding biofuels) by 2035 — 15 mb/d higher than in 2009. All of the growth comes from non-OECD countries, 57% from China alone, mainly driven by rising use of transport fuels; demand in the OECD falls by over 6 mb/d.

- Global oil production reaches 96 mb/d in the New Policies Scenario, the balance of 3 mb/d coming from processing gains. Crude oil output reaches a plateau of around 68-69 mb/d by 2020 — marginally below the all-time peak of about 70 mb/d reached in 2006, while production of natural gas liquids and unconventional oil grows strongly.

- Total OPEC production rises continually through to 2035 in the New Policies Scenario, its share of global output increasing from 41% to 52%. Total non-OPEC oil production is broadly constant to around 2025, as rising production of NGLs and unconventional production offsets a fall in that of crude oil; thereafter, production starts to drop. Increased dependence on a small number of producing countries would intensify concerns about their influence over prices.

- Worldwide upstream oil investment is set to bounce back in 2010, but will not recover all of the ground lost in 2009, when lower oil prices and financing difficulties led oil companies to slash spending. Upstream capital spending on both oil and gas is budgeted to rise by around 9% to about $470 billion in 2010; it fell by 15% in 2009. Projected oil supply in the New Policies Scenario calls for cumulative investment along the entire oil-supply chain of $8 trillion (in year-2009 dollars) in 2010-2035.

Demand

Primary oil demand trends

The global outlook for oil remains highly sensitive to policy action to curb rising demand and emissions, especially in the developing world. In the Current Policies and New Policies Scenarios, global primary oil use increases in absolute terms between 2009 and 2035, driven by population and economic growth, but demand falls in the 450 Scenario in response to the counter-balancing effects of radical policy action to curb fossil-energy use (Figure 3.1). The global economic recovery is expected to drive oil demand back up, following two consecutive years of decline in 2008 and 2009 that resulted from previously surging oil prices and the subsequent global financial and economic crisis.[1] Nonetheless, the effect of the recession on demand was slightly less than was expected in last year's *Outlook*: global demand bottomed out at an estimated 84 million barrels per day (mb/d) in 2009 — 1 mb/d down on 2008. The share of oil in total primary energy demand is nonetheless projected to fall progressively in each scenario, most sharply in the 450 Scenario, where it reaches 26% in 2035 — down from 33% in 2009. In the New Policies Scenario, the share falls to 28%.

Figure 3.1 ● **World primary oil demand by scenario**

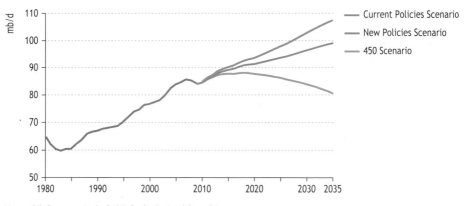

Note: Oil does not include biofuels derived from biomass.

There are big differences in the trajectory of oil demand across the three scenarios. In the New Policies Scenario, demand continues to grow steadily, reaching about 99 mb/d by 2035 — a level that is still 15 mb/d higher than in 2009. A combination of policy action to promote more efficient oil use and switching to other fuels and higher prices (resulting from price rises on international markets, reduced subsidies in some major consuming countries and increased taxes on oil products) partially offsets growing demand for mobility, especially in non-OECD countries. In the Current Policies

---

1. Preliminary data on oil demand are available for 2009. Because of methodological differences, the oil projections in this report are not directly comparable with those published in the IEA's monthly *Oil Market Report* or annual *Medium Term Oil and Gas Market Report*.

Scenario, oil demand rises more quickly through to 2035, reaching about 107 mb/d. In the 450 Scenario, demand reaches a peak of about 88 mb/d soon after 2015 and then falls steadily to about 81 mb/d by 2035 — 3 mb/d down on the 2009 level.

Table 3.1 ● **Primary oil demand* by scenario** (mb/d)

	1980	2009	New Policies Scenario		Current Policies Scenario		450 Scenario	
			2020	2035	2020	2035	2020	2035
OECD	41.3	41.7	39.8	35.3	40.5	38.7	38.2	28.0
Non-OECD	20.0	35.8	44.1	54.6	45.4	59.4	42.2	45.6
Bunkers**	3.4	6.5	7.5	9.1	7.5	9.3	7.2	7.3
World	64.8	84.0	91.3	99.0	93.5	107.4	87.7	81.0
Share of non-OECD	*33%*	*46%*	*53%*	*61%*	*53%*	*61%*	*52%*	*62%*

* Excludes biofuels demand, which is projected to rise from 1.1 mb/d (in energy-equivalent volumes of gasoline and diesel) in 2009 to 2.3 mb/d in 2020 and to 4.4 mb/d in 2035 in the New Policies Scenario.
** Includes international marine and aviation fuel.

The prices needed to balance oil demand — which varies with the degree of policy effort to curb demand growth — with supply differ markedly across the three scenarios. In the New Policies Scenario, the average IEA crude oil import price reaches $105/barrel in real terms in 2025 on average and $113/barrel in 2035. In the Current Policies Scenario, in which no change in government policies is assumed, substantially higher prices are needed to bring demand into balance with supply. Prices rise more briskly, especially after 2020. The crude oil price reaches $120 per barrel in 2025 and $135/barrel ten years later. Our analysis suggests that the rate of increase in production capacity is relatively insensitive to price, as net capacity additions are constrained by the steep decline in output from existing fields, particularly in non-OPEC countries, problems of access to undeveloped resources and logistical constraints (see the supply section below). Similarly, the increasing dominance of transport in overall oil demand will tend to lower the sensitivity of demand to price, as the alternatives to conventional oil-based fuels struggle to compete in that sector (see the section on sectoral trends below). Prices in the 450 Scenario are considerably lower, levelling off at $90 after 2020, as demand increases much less, peaking by around 2015. The oil demand and supply peak in this scenario is, thus, driven entirely by policy rather than by any geological constraint. The message from this analysis is clear: the weaker and slower the response to the climate challenge, the greater the risk to oil-importing countries of oil scarcity and higher prices.

Economic activity is expected to remain the principal driver of oil demand in all regions in every scenario, but the relationship weakens in the New Policies Scenario and, to an even greater extent, in the 450 Scenario. On average, since 1980, each 1% increase in gross domestic product (GDP) has been accompanied by a 0.3% rise in primary oil demand (Figure 3.2). This ratio — the oil intensity of GDP, or the amount of oil needed to produce one dollar of GDP — has fallen progressively since the 1970s, though in an uneven fashion.[2] Oil intensity fell more sharply after 2004, mainly as a result of higher oil prices, which have encouraged conservation, switching to other fuels and spending on more efficient equipment and vehicles. In 2009, global oil intensity (expressed in purchasing power parities, or PPP) was only about half the level of the early 1970s. This downward trend continues in the New Policies Scenario, with intensity falling to one-half of its 2009 level by 2035, boosted by policies to promote more efficient oil use in end-use sectors and switching to lower carbon fuels, including vehicle fuel-efficiency standards and the phase-out of subsidies (see Part E).

Figure 3.2 ● *Annual change* in global real GDP and primary oil demand in the New Policies Scenario*

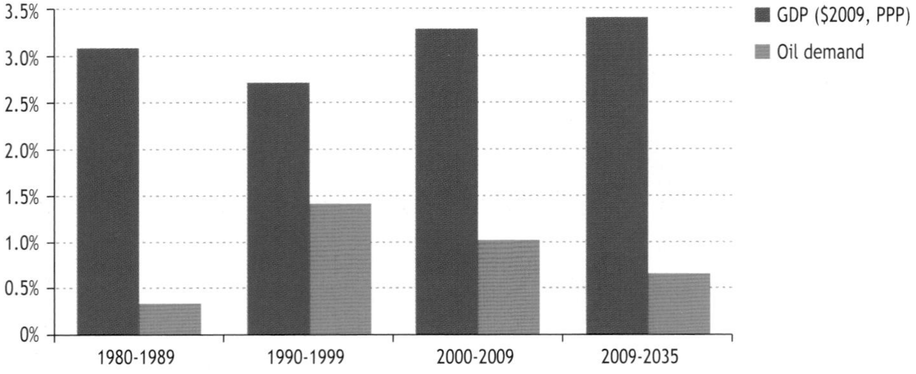

*Compound average annual growth rate.

Regional trends

The outlook for oil demand differs markedly across regions. All of the increase in world oil demand between 2009 and 2035 comes from non-OECD countries in every scenario, as OECD demand drops. In the New Policies Scenario, OECD demand falls by over 6 mb/d between 2009 and 2035, but this is offset by an almost 19-mb/d increase in the non-OECD (international bunker demand also rises by almost 3 mb/d). Demand drops in all three OECD regions: progressive improvements in vehicle fuel efficiency, spurred by higher fuel costs as international prices increase as well as government fuel-economy mandates, more than offset the effect of rising incomes (Table 3.2). By contrast, in non-OECD regions, strong economic and population growth, coupled with the enormous latent demand for mobility, more than outweighs efficiency gains in transport.

The biggest increase in demand in absolute terms occurs in China, where it jumps from just over 8 mb/d in 2009 to more than 15 mb/d in 2035 — an increase of 2.4% per year on average in the New Policies Scenario. China accounts for 57% of the global increase

---

2. Oil prices also affect GDP, by altering energy costs. The rapid run-up in oil prices in the period 2003 to mid-2008 undoubtedly played a role, albeit a secondary one, in provoking the financial and economic crisis of 2008-2009. It follows that a sharp rise in oil prices in the years to come would threaten the global economic recovery.

in demand. Demand could grow even more if the rising international prices of oil assumed in this scenario were offset by an appreciation of the yuan against the dollar. High as it is, the projected growth rate in the New Policies Scenario is still significantly lower than in the past: Chinese oil use more than quadrupled between 1980 and 2009. Other emerging Asian economies, notably India, and the Middle East also see rapid rates of growth. The latter region has emerged as a major oil-consuming as well as oil-producing region, on the back of a booming economy (helped by high oil prices) and heavily subsidised prices in domestic markets. Middle East countries account for one-fifth of the growth in oil demand over the projection period. Demand in all three OECD regions, by contrast, falls, most heavily in relative terms in the Pacific region and Europe. As a result of these trends, the non-OECD countries' share of global oil demand (excluding international marine bunkers) rises from 46% in 2009 to 61% in 2035.

Table 3.2 ● Primary oil demand* by region in the New Policies Scenario (mb/d)

	1980	2009	2015	2020	2025	2030	2035	2009-2035**
OECD	**41.3**	**41.7**	**41.1**	**39.8**	**38.2**	**36.7**	**35.3**	**-0.6%**
North America	20.8	21.9	21.9	21.4	20.8	20.1	19.4	-0.5%
United States	*17.4*	*17.8*	*17.7*	*17.2*	*16.5*	*15.8*	*14.9*	*-0.7%*
Europe	14.4	12.7	12.4	11.9	11.4	10.8	10.4	-0.8%
Pacific	6.1	7.0	6.9	6.4	6.1	5.8	5.6	-0.9%
Japan	*4.8*	*4.1*	*3.8*	*3.5*	*3.2*	*3.0*	*2.9*	*-1.3%*
Non-OECD	**20.0**	**35.8**	**41.1**	**44.1**	**47.5**	**51.1**	**54.6**	**1.6%**
E. Europe/Eurasia	9.1	4.6	4.9	5.0	5.2	5.2	5.4	0.6%
Caspian	*n.a.*	*0.6*	*0.7*	*0.8*	*0.8*	*0.9*	*0.9*	*1.6%*
Russia	*n.a.*	*2.8*	*2.8*	*2.9*	*3.0*	*3.0*	*3.0*	*0.4%*
Asia	4.4	16.3	19.7	21.8	24.4	27.3	30.0	2.4%
China	*1.9*	*8.1*	*10.6*	*11.7*	*13.0*	*14.3*	*15.3*	*2.4%*
India	*0.7*	*3.0*	*3.6*	*4.2*	*5.1*	*6.2*	*7.5*	*3.6%*
Middle East	2.0	6.5	7.5	8.0	8.5	8.9	9.2	1.3%
Africa	1.2	3.0	3.1	3.3	3.4	3.6	3.8	0.9%
Latin America	3.4	5.3	5.8	5.9	6.0	6.1	6.2	0.6%
Brazil	*1.3*	*2.1*	*2.4*	*2.5*	*2.5*	*2.5*	*2.6*	*0.8%*
Bunkers***	**3.4**	**6.5**	**7.0**	**7.5**	**7.9**	**8.5**	**9.1**	**1.3%**
World	**64.8**	**84.0**	**89.2**	**91.3**	**93.6**	**96.4**	**99.0**	**0.6%**
European Union	*n.a.*	*12.2*	*11.8*	*11.3*	*10.7*	*10.1*	*9.6*	*-0.9%*

*Excludes biofuels demand, which is projected to rise from 1.1 mb/d (in energy-equivalent volumes of gasoline and diesel) in 2009 to 2.3 mb/d in 2020 and to 4.4 mb/d in 2035. **Compound average annual growth rate. ***Includes international marine and aviation fuel.

Sectoral trends

The transport sector is expected to continue to drive the growth in global oil demand. In the New Policies Scenario, transport accounts for almost all of the increase in oil demand between 2009 and 2035, with oil use in power generation falling and consumption in other sectors in aggregate expanding only modestly (Figure 3.3). Transport's share in global primary oil consumption (including bunker fuels) rises from 53% in 2009 to 60% in 2035. China alone accounts for half of the global increase in oil use for transport. Oil remains the dominant source of energy for transportation, by road, rail, air and sea, though it comes under increasing competition from alternative fuels, notably biofuels and electricity for cars and trains, and natural gas for buses and trucks. The share of oil-based fuels (primarily gasoline and diesel) in total road transportation energy use falls from 96% in 2009 to 89% by 2035, mainly due to increased use of conventional biofuels and, increasingly, advanced biofuels (see Chapter 12).

Figure 3.3 ● Change in primary oil demand by sector and region in the New Policies Scenario, 2009-2035

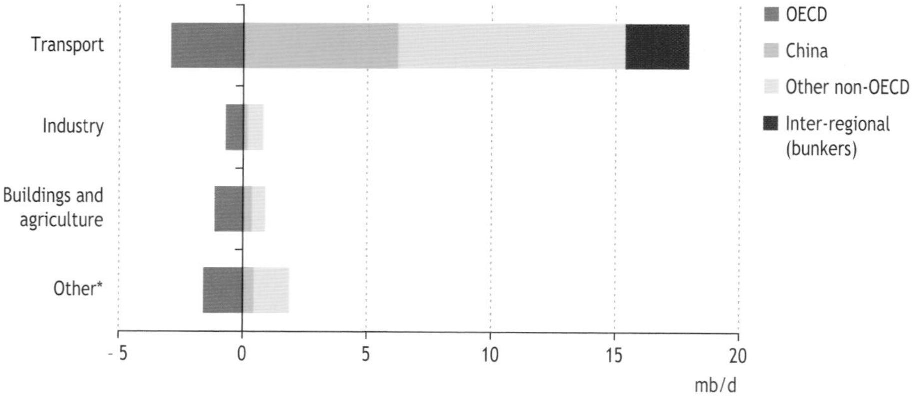

*Includes power generation, other energy sector and non-energy use.

Demand for road transport fuels is set to continue to expand rapidly in the emerging economies in line with rising incomes, which boost car ownership and usage as well as freight, and expanded road networks. In contrast to the OECD regions, these factors more than offset the effect of continuing improvements in vehicle fuel efficiency, a modest expansion of biofuels use and the deployment of full-electric vehicles in the longer-term. Trucks and passenger light-duty vehicles (PLDVs) account for most of the increase in transport-related oil use (Figure 3.4).

The passenger-car and truck fleet is growing faster in China than anywhere else: preliminary data show that new car sales topped 13.6 million in 2009, overtaking for the first time sales in the United States. The total car fleet in China is now estimated at almost 40 million — more than twice as big as just three years ago. Car and truck sales are growing rapidly in many other non-OECD countries as well, particularly in Asia.

Figure 3.4 ● Transport oil consumption by type in the New Policies Scenario

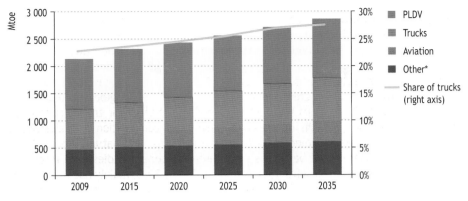

*Includes other road, rail, pipelines, navigation and non-specified.

The potential for continued brisk expansion of the vehicle fleet in those countries remains large, as vehicle ownership rates are still well below those in the OECD: there are only 30 cars for every thousand people in China, compared with around 700 in the United States and almost 500 in Europe. In the New Policies Scenario, the total stock of passenger light-duty vehicles in non-OECD countries is projected to quadruple over the projection period to about 850 million, overtaking that of OECD countries soon after 2030 (Figure 3.5). The vehicle fleet of China overtakes that of the United States by around 2030.

Figure 3.5 ● Passenger light-duty vehicle fleet and ownership rates by region in the New Policies Scenario

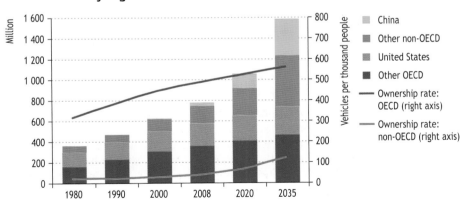

The rate of growth in car ownership in non-OECD countries in general and in China in particular is a critical uncertainty for the prospects for global oil use. Holding all other factors equal, a 1% per year faster rate of growth in car ownership in China alone (compared with the global average of 1.8% in the New Policies Scenario) would result in around 95 million more cars on the road in 2035 and 0.8 mb/d of additional oil demand — an increase of 0.8% in world demand. Were this faster growth rate applied to all non-OECD countries, demand would, in theory, be about 3.6 mb/d, or 4%, higher. To avoid such an increase, oil prices would have to rise much faster than assumed in this scenario, unless there were faster improvements in vehicle efficiency, fewer kilometres driven per vehicle and/or faster penetration of biofuels and alternative fuel and vehicle technologies.

Fuel economy — the amount of fuel consumed in driving one kilometre — is another key uncertainty. Rising incomes will tend to encourage people to opt for larger, more energy-intensive vehicles, though this phenomenon is expected to be more than offset by continuing fuel economy improvements in each vehicle category. Conventional internal combustion engine vehicles are expected to continue to become more efficient, the result of higher oil prices as well as policy initiatives to encourage vehicle manufacturers to develop and market more efficient vehicles and motorists to buy them. A number of countries, including the United States and EU members, have adopted regulations to increase the average vehicle fuel efficiency; others such as China or Korea are also discussing standards (these are taken into account in the New Policies Scenario). Other measures include programmes to encourage fuel-efficient driving, such as the EU-funded Ecodrive programme. In addition, hybrid cars and plug-in hybrids, with significantly better fuel efficiency than conventional cars, together with full-electric vehicles that consume no oil at all directly, account for a growing share of light-duty vehicle sales. In the New Policies Scenario, these new vehicle technologies collectively account for 6% of new passenger vehicle sales by 2020 and 19% by 2035, the bulk of which are hybrids (Figure 3.6).

Figure 3.6 ● *Passenger light-duty vehicle sales by type in the New Policies Scenario*

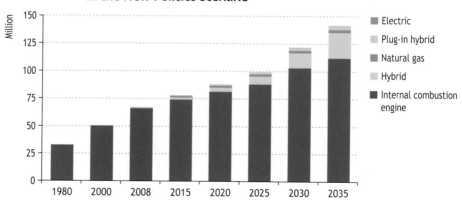

The combination of more efficient conventional vehicles and the growing contribution of new vehicle technologies results in a drop in the average fuel consumption of new light-duty vehicles sold worldwide from 9.7 litres/100 kilometres (km) of fuel in 2009 to 7.6 litres/100 km in 2020 and 6.7 litres/100 km in 2035 (Figure 3.7). The improvement in fuel economy is greatest in the period to 2015, mainly as a result of stringent new government measures that are assumed to be introduced and the relatively rapid increase in oil prices. In the period to 2020, the improved efficiency of conventional cars is the main driver. Thereafter, hybrid and, to a lesser extent, plug-in hybrid cars play an increasingly important role. A significant part of the potential efficiency gains from conventional cars is exploited within the first half of the projection period. It is possible to reduce the fuel consumption of a conventional internal combustion engine vehicle of medium size on average worldwide by about 40% within the next two decades, compared with the year 2000 (IEA, 2009). Beyond this, the only way that average vehicle fuel efficiency can be further reduced significantly without reducing the size of the vehicle is through the deployment of alternative technologies.

Figure 3.7 ● *Average fuel economy of new passenger light-duty vehicle sales by region in the New Policies Scenario*

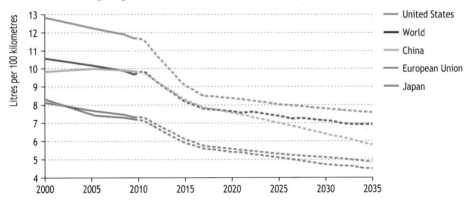

The net result of the projected trends in vehicle ownership, fuel economy and technology is a rise in per-capita oil use for road transportation in all non-OECD regions and a fall in all three OECD regions in each scenario. Yet average per-capita demand remains much lower in the non-OECD by 2035, mainly because incomes and, therefore, vehicle ownership rates remain significantly lower. In the New Policies Scenario, per-capita road-transport-related oil demand is on average four times higher in the OECD than in non-OECD regions by the end of the *Outlook* period, down from seven times in 2009 (Figure 3.8).

Given the limitations on further improving the efficiency of conventional vehicles, how quickly new vehicle technologies penetrate the car market will have a major impact on oil demand for road transport. The pump price of oil-based fuels and advances in alternative vehicle technologies to lower their cost and improve their operational performance are the main factors. For now, alternative technologies are struggling to compete on cost, which is holding back their deployment. However, a relatively modest but sustained rise in the price of oil-based fuels and/or a drop in the cost of these new technologies could make them attractive to end users and lead to rapid growth in their uptake. In the United States, for example, low fuel taxes and, hence, low pump prices mean that conventional hybrids pay back their much higher purchase cost to motorists only after 120 000 km at 2009 fuel prices (Figure 3.9). At an average of 20 000 km per year, the payback period is therefore around six years – far too high to persuade most motorists to opt for this type of vehicle. However, a 30% fall in the difference in the cost of buying a hybrid would cut the payback period to four years, increasing significantly the attractiveness of such a car to motorists.

Figure 3.8 ● Road transportation per-capita oil consumption by region in the New Policies Scenario

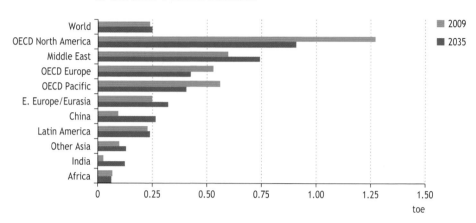

Figure 3.9 ● Comparative running cost of conventional and hybrid light-duty vehicles in the United States

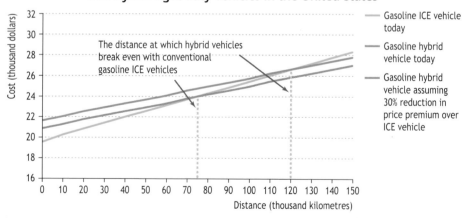

Note: Assumes vehicle life of 15 years and average 2009 gasoline price of $0.65 per litre ($2.46 per US gallon). ICE is internal combustion engine.

Pump prices of gasoline and diesel vary enormously across countries, because of differences in tax rates and – in some countries – subsidies (see Part E). There are also differences in the relative prices of hybrids and conventional cars. These factors result in a big variation in the attractiveness to motorists of buying hybrids today. The payback period is currently shortest in Germany and France, where fuel taxes are highest (Figure 3.10). In China, the payback period is relatively long, at close to eight years (assuming average mileage there of 9 000 km a year). Yet even the quickest paybacks are too long to appeal to most motorists. In practice, there are many other factors that come into play in determining a motorist's decisions about which car to buy, so that the payback period on a more efficient car typically has to be very short to swing the decision. But higher fuel prices and lower purchase costs would reduce the payback period and greatly increase the appeal of hybrids. For example, an increase in international oil prices of one-third would reduce the payback period of a hybrid in China from about eight to seven years; a 30% drop in the premium for a hybrid car over a conventional car would cut the payback period to slightly less than six years.

Achieving cost-competitiveness for other alternative vehicle options, such as plug-in hybrids and electric cars, is likely to require more than just higher oil prices. Despite the current strong momentum towards deployment of these vehicles, a number of issues that raise doubts about their long-term viability remain open. Technical aspects would need to be addressed for global mass manufacturing of electric cars, such as standardisation of batteries and differences in voltage by country, and, even then, it is unclear whether consumers would be prepared for the prospective limitations on driving range and the length of the necessary recharging time. It is not likely that high oil prices alone will suffice to create a global market for electric cars; policy intervention will probably be required too. In light of all these factors, we conservatively project that electric cars and plug-in hybrids account for only 2.6% of car sales by 2035 in the New Policies Scenario.

Figure 3.10 ● **Payback period for hybrid light-duty vehicles in selected countries at current costs**

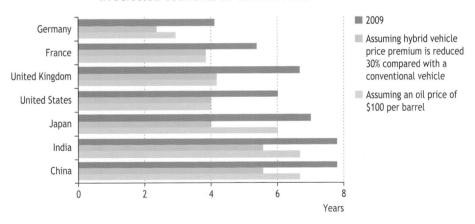

There is also considerable scope for reducing the amount of oil-based fuels used in road freight — a major contributor to the growth of road-transport oil demand in non-OECD countries — through more efficient vehicles and the use of alternative fuels. Medium and heavy freight traffic, is responsible for 30% of all transport oil demand worldwide today and this share is projected to increase to 35% by 2035 (Figure 3.4, above). One uncertainty for road-freight oil use is the outlook for compressed natural gas as a fuel, which could displace diesel. The recent fall in gas prices relative to oil prices, especially in North America, has led to greater interest in promoting compressed natural gas (CNG) as a road fuel for fleet vehicles, including lorries, trucks and buses, as a way of reducing costs, improving energy security and reducing emissions of local pollutants and, to a limited degree, greenhouse gases. CNG already makes a significant contribution to meeting road-transport fuel needs in several countries, notably in Pakistan and Argentina, but in most major economies CNG use is marginal. This could change, especially if gas prices remain low relative to oil prices. However, there are major barriers to the expansion of natural gas use, including the cost and practicalities of on-board fuel storage, the cost of installing the infrastructure for delivering and distributing the fuel at existing refuelling stations and the risk that prices might move against gas in the future.[3] Nonetheless, the prospects — especially as a fuel for fleet vehicles (as the infrastructure costs are lower) — have certainly improved in recent years.

In the New Policies Scenario, CNG use worldwide more than triples between 2009 and 2035, from almost 20 billion cubic metres (bcm) to over 60 bcm. The amount of oil saved as a result increases from about 300 thousand barrels per day (kb/d) to over 1 mb/d. Most of the increase in oil savings comes from non-OECD countries, but North America, where wholesale gas prices are lowest, makes a significant contribution (Figure 3.11). By 2035, around 4% of the heavy-duty vehicle fleet in North America runs on CNG — up from almost nil today. Oil savings could be much greater; if CNG took a 5% share of the global freight vehicle fleet by 2035, compared with 1.5% in the New Policies Scenario, oil consumption would be reduced by a further 0.6 mb/d.

Figure 3.11 • **Oil savings from use of natural gas in road transport by region in the New Policies Scenario**

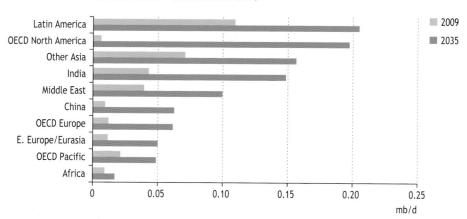

Another important factor in the future oil demand increase is the rate of growth of fuel use in the aviation sector. Combined, jet fuel and aviation gasoline demand grew at

---

3. See, for example, IEA (2010a) and Box 10.1 in IEA (2009).

a similar pace as total oil demand in transport between 1980 and 2009, a steady 2.1% per year, making up 12% of all transport oil demand in 2009. This share is projected to increase over the projection period to 14% by 2035 in the New Policies Scenario, mainly driven by non-OECD countries. The largest single contributor to growth in aviation oil demand is China, where demand is projected to expand by 2.6% per year (Figure 3.12). In the OECD, the aviation sector is the only major sector that sees any significant growth in oil demand. Government measures aimed at curbing aviation-fuel demand have been limited to date, in sharp contrast to the action taken in the road-transport sector. The inclusion of aviation to the EU Emission Trading Scheme from 2012 is one of the few policy actions undertaken. However, the industry itself has made significant efforts to reduce fuel use, through operational changes and investments in more efficient aircraft.

Figure 3.12 • Aviation oil consumption by region in the New Policies Scenario

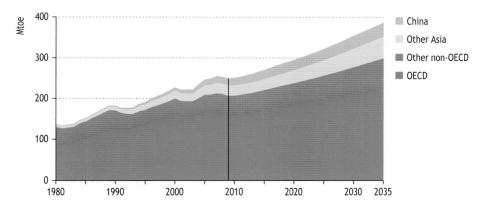

There is little prospect of any significant long-term increase in oil demand in non-transport uses, as oil is expected to lose market share to coal, gas and other fuels. Globally, the use of oil in other sectors in aggregate remains flat over the projection period in the New Policies Scenario, at around 39 mb/d; an increase in non-OECD countries (mainly in the industry, residential and services sectors, and as a feedstock in the petrochemical industry) is more than outweighed by a drop in OECD demand (reflecting energy efficiency gains and some switching to gas in buildings). Oil use in power generation falls in every region bar the Middle East.

Production

Resources and reserves

According to the *Oil and Gas Journal* (O&GJ, 2009), proven reserves of oil worldwide at the end of 2009 amounted to 1 354 billion barrels — a marginally higher volume than estimated a year earlier and the highest level ever attained (see Box 3.1 for definitions). Reserves have more than doubled since 1980 and have increased by one-third over the last decade. Half of the increase since 2000 is due to Canadian oil sands reserves; most of the remainder is due to revisions in OPEC countries, particularly in

Iran, Venezuela and Qatar. There are continuing question-marks over the estimates for some OPEC countries and their comparability with the figures for other countries.[4] Notwithstanding these uncertainties, OPEC countries account for about 70% of the world total reserves, with Saudi Arabia holding the largest volume (Figure 3.13).

Figure 3.13 ● **Proven oil reserves in the top 15 countries, end-2009**

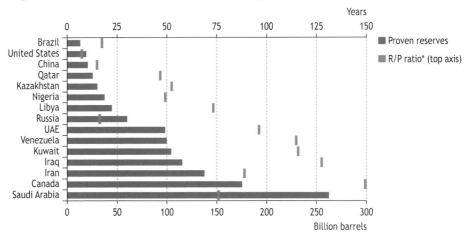

*See footnote 5 on reserves to production (R/P) ratios.
Sources: Proven reserves — O&GJ (2009); production — IEA databases.

Box 3.1 ● **Defining and measuring oil and gas reserves and resources**

In the *WEO*, we use the following definitions, drawing on the Petroleum Resources Management System (SPE, 2007) and US Geological Survey (USGS, 2000):

● **A proven reserve** (or 1P reserve) is the volume of oil or gas that has been discovered and for which there is a 90% probability that it can be extracted profitably on the basis of prevailing assumptions about cost, geology, technology, marketability and future prices.

● **A proven and probable reserve** (or 2P reserve) includes additional volumes that are thought to exist in accumulations that have been discovered and have a 50% probability that they can be produced profitably.

● **Reserves growth** refers to the typical increases in 2P reserves that occur as oil or gas fields that have already been discovered are developed and produced.

● **Ultimately recoverable resources** are latest estimates of the total volume of hydrocarbons that are judged likely to be ultimately producible commercially, including initial 1P reserves, reserves growth and as yet undiscovered resources.

---

4. Our modelling of oil supply is based on recoverable resources rather than proven reserves (see Box 3.3).

- **Remaining recoverable** resources are ultimately recoverable resources less cumulative production to date.

- **Oil originally in place** refers to the total amount of oil or gas contained in a reservoir before production begins.

- **The recovery factor** is the share of the oil or gas originally in place that is ultimately recoverable (*i.e.* ultimately recoverable resources/original hydrocarbons in place).

Definitions of reserves and resources, and the methodologies for estimating them, vary considerably around the world, leading to confusion and inconsistencies. In addition, there is often a lack of transparency in the way reserves are reported: many national oil companies in both OPEC and non-OPEC countries do not use external auditors of reserves and do not publish detailed results. OPEC figures of proven reserves may be more comparable to figures of proven and probable reserves in other parts of the world. The IEA continues to work with the UN Economic Commission for Europe, the Society of Petroleum Engineers and other organisations on harmonising the way reserves and resources are defined and estimated in order to provide a clearer picture of how much oil and gas remains to be produced.

In 2009, the US Securities and Exchange Commission (SEC) introduced updated guidelines for evaluating oil and gas reserves to take account of recent technological and market developments. US-quoted companies are now able to use seismic and numerical modelling techniques and data from down-hole tools in estimating reserves. They can now use an average 12-month price to value reserves, rather than the year-end price, and can provide sensitivity analyses of reserves estimates, using different price outlooks. The SEC also now permits companies to report probable and possible reserves, as well as proven reserves. Producers can now also report reserves of unconventional oil. The aim of these changes is to provide a better insight into the reporting companies' long-term production potential.

The bulk of proven reserves, which include all types of oil (Box 3.2), are conventional: the only significant volumes of unconventional oil included in the figure from O&GJ for end-2009 are an official estimate of 170 billion barrels for Canadian oil sands reserves, of which some 16% are currently "under active development". Globally, conventional and unconventional reserves combined are equal to about 46 years of current production. The reserves to production ratio[5] has increased in the last two years as a result of the recession-induced drop in demand for oil and continuing modest increases in reserves.

---

5. R/P ratios are commonly used in the oil and gas industry as indicators of production potential, but do not imply continuous output for a certain number of years, nor that oil production will stop at the end of the period. They can fluctuate over time as new discoveries are made, reserves at existing fields are reappraised, and technology and production rates change.

Box 3.2 ● Definitions of different types of oil in the WEO

For the purposes of this chapter (and Chapter 4), the following definitions are used:

● Oil comprises crude, natural gas liquids, condensates and unconventional oil, but does not include biofuels (for the sake of completeness and to facilitate comparisons, relevant biofuels quantities are separately mentioned in some sections and tables).

● Crude makes up the bulk of oil produced today; it is a mixture of hydrocarbons that exist in liquid phase under normal surface conditions. It includes condensates that are mixed-in with commercial crude oil streams.

● Natural gas liquids (NGLs) are light hydrocarbons that are contained in associated or non-associated natural gas in a hydrocarbon reservoir and are produced within a gas stream. They comprise ethane, propane, butane, isobutene, pentane-plus and condensates.[6]

● Condensates are light liquid hydrocarbons recovered from associated or non-associated gas reservoirs. They are composed mainly of pentane (C_5) and higher carbon number hydrocarbons. They normally have an API gravity of between 50° and 85°.

● Conventional oil includes crude and NGLs.

● Unconventional oil includes extra-heavy oil, natural bitumen (oil sands), oil shale, gas-to-liquids (GTL), coal-to-liquids (CTL) and additives (see Chapter 4).

● Biofuels are liquid fuels derived from biomass, including ethanol and biodiesel (see Chapter 12).

Almost half of the increase in proven reserves in recent years has come from revisions to estimates of reserves in fields already in production, rather than new discoveries. Although discoveries have picked up in recent years with increased exploration activity (prompted by higher oil prices), they continue to lag production by a considerable margin: in 2000-2009, discoveries replaced only one out of every two barrels produced — slightly less than in the 1990s (even though the amount of oil found increased marginally) — the reverse of what happened in the 1960s and 1970s, when discoveries far exceeded production (Figure 3.14). The contribution of offshore discoveries, including deepwater, has increased significantly since the early 1990s. Since 2000, more than half of all the oil that has been discovered is in deep water. Although some giant fields have been found, the average size of fields being discovered has continued to fall. The New Policies Scenario requires average annual development of 9 billion barrels of new discoveries from 2015 onwards (see the section on oil production prospects below).

---

6. See IEA (2010c) for a detailed analysis of the medium-term prospects for NGLs.

Figure 3.14 ● Conventional oil discoveries and production worldwide

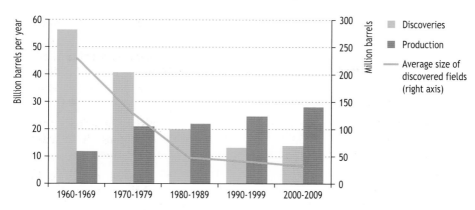

The volume of ultimately recoverable resources, comprising proven and probable reserves, plus oil that is yet to be discovered and additional volumes of oil in existing fields that could be "proven up" in the future, is estimated to be much bigger than proven reserves. Yet there is uncertainty about this figure and, therefore, about just how much oil remains to be produced. The main uncertainties lie in estimating how much oil was originally in place in the world and in evaluating how much of this resource can be recovered profitably (the recovery factor). The latter is heavily influenced by future trends in oil prices and oilfield development costs, which will hinge on assumptions about technology and the underlying cost of various inputs to oil production, as well as geological considerations.

The leading source of estimates of ultimately recoverable resources of conventional crude oil and NGLs is the US Geological Survey (USGS). It last carried out a major assessment of global resources in 2000, but has carried out partial updates covering specific basins since then, including a major reassessment of the Arctic region in 2008 (USGS, 2008). Based on those assessments, we estimate that around 2.5 trillion barrels of conventional oil remain to be produced worldwide as of the beginning of 2010, taking account of cumulative production to date and mean estimates of ultimately recoverable resources. Of this total, 900 billion barrels are in deposits that are yet to be found. At the start of 2010, the proportion of remaining recoverable resources classified as proven reserves varied widely across regions: proven reserves accounted for 68% of remaining recoverable resources in the Middle East, but only 17% in North America. As with reserves, the bulk of the remaining resources are in the Middle East and the former Soviet Union countries (Figure 3.15). In the New Policies Scenario, around half of the conventional resources are produced by 2035, but the share reaches 61% for non-OPEC countries as a group compared with only 47% for OPEC. By end-2009, only 32% of global ultimately recoverable resources had been produced. However, these estimates do not include unconventional resources — oil sands, extra-heavy oil and oil shales. The size of these resources is uncertain, as they have been studied much less than conventional resources, but they are certainly very large; potentially around 2 to 3 trillion barrels of unconventional oil may be economically recoverable.

Figure 3.15 ● Proven reserves, recoverable resources and production of conventional oil by region in the New Policies Scenario

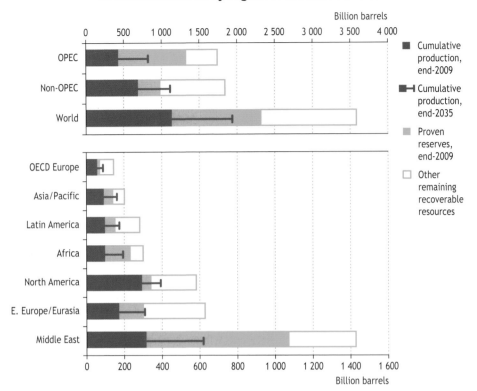

Sources: BGR (2009); O&GJ (2009); USGS (2000 and 2008) and information provided by the USGS directly to the IEA; IEA estimates and analysis.

Oil production prospects

Oil supply follows the same trajectory as demand in each of the three scenarios, though production of oil (crude, NGLs and unconventional oil) rises marginally less than overall supply, due to increasing processing gains.[7] In the New Policies Scenario, total oil production reaches 96 mb/d by 2035 (Table 3.3). In the Current Policies Scenario, production continues to expand through to 2035, though the pace slows over the second half of the projection period. In the 450 Scenario, production peaks before 2020 and then declines steadily to 2035. The breakdown of production between OPEC and non-OPEC, and between conventional and unconventional oil differs across the three scenarios. The share of OPEC in overall production by the end of the projection period is highest in the 450 Scenario, at more than 53%, as lower oil prices inhibit investment

---

7. Oil refining involves the upgrading of heavy oil into lighter products, which reduces their density and gives rise to an increase in volume for a given amount of energy content. Processing gains as a share of overall supply increase slightly in all three scenarios as a result of more upgrading of oil feedstocks in response to the shift in demand towards lighter products such as diesel and gasoline.

in high-cost resources, mainly in non-OPEC countries. The share of unconventional oil is highest in the Current Policies Scenario, as higher oil prices stimulate more investment in developing those higher-cost resources.

Table 3.3 ● Oil production and supply by source and scenario (mb/d)

	1980	2009	New Policies Scenario		Current Policies Scenario		450 Scenario	
			2020	2035	2020	2035	2020	2035
OPEC	**25.5**	**33.4**	**40.5**	**49.9**	**41.9**	**54.2**	**40.1**	**41.7**
Crude oil	24.7	28.3	30.9	35.8	32.0	38.6	31.4	31.8
Natural gas liquids	0.9	4.6	8.0	11.1	8.2	12.3	7.1	7.6
Unconventional	0.0	0.5	1.6	3.0	1.7	3.2	1.6	2.3
Non-OPEC	**37.1**	**47.7**	**48.2**	**46.1**	**48.9**	**49.9**	**45.1**	**36.7**
Crude oil	34.1	39.6	37.6	32.8	38.2	35.0	35.1	25.9
Natural gas liquids	2.8	6.2	6.8	6.8	6.9	7.1	6.5	5.7
Unconventional	0.2	1.8	3.7	6.5	3.9	7.8	3.4	5.1
World production	**62.6**	**81.0**	**88.7**	**96.0**	**90.8**	**104.1**	**85.2**	**78.5**
Crude oil	58.8	67.9	68.5	68.5	70.1	73.6	66.5	57.7
Natural gas liquids	3.7	10.8	14.8	17.9	15.1	19.5	13.6	13.3
Unconventional	0.2	2.3	5.4	9.5	5.5	11.0	5.0	7.4
Processing gains	1.2	2.3	2.6	3.0	2.7	3.3	2.5	2.5
World supply	**63.8**	**83.3**	**91.3**	**99.0**	**93.5**	**107.4**	**87.7**	**81.0**
*World liquids supply**	*63.9*	*84.4*	*93.6*	*103.4*	*95.7*	*110.9*	*90.3*	*89.1*

*Includes biofuels (see Chapter 12 for details of biofuels projections).

There is also a marked difference in the profile of crude oil production across the three scenarios, with global output rising in the Current Policies Scenario to 74 mb/d by 2035, but reaching a plateau by 2020 in the New Policies Scenario (Figure 3.16). The increase in production in the former scenario comes with the higher prices that are needed to bring forth more investment in productive capacity. Slower global demand growth and lower prices in the New Policies Scenario mean that crude oil resources can be developed in a steadier fashion, keeping crude oil production in that scenario at a plateau of around 68-69 mb/d from 2015 (marginally below the all-time peak of about 70 mb/d reached in 2006). In the 450 Scenario, the strong greenhouse-gas emissions-reduction policies assumed quickly send oil demand growth into reverse, causing prices to level off, resulting in less investment in conventional oilfields, a marginal drop in oil output to 2020 and accelerating decline thereafter (see Chapter 15).

Overall, worldwide production of both NGLs and unconventional oil increases much more than crude oil between 2009 and 2035 (Figure 3.17). The increase in output of all three types of oil is highest, unsurprisingly, in the Current Policies Scenario and lowest in the 450 Scenario. Conversely, the increase in production of biofuels (not included in our definition of oil — see Box 3.3) is highest in the 450 Scenario, adding more to liquids supply than any of the other sources.

Figure 3.16 ● World crude oil production by scenario

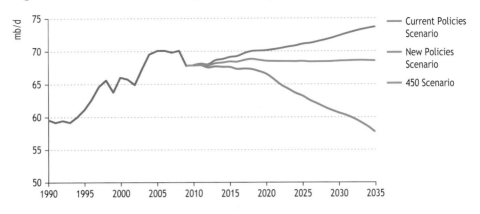

Figure 3.17 ● Change in world oil and biofuels production by scenario, 2009-2035

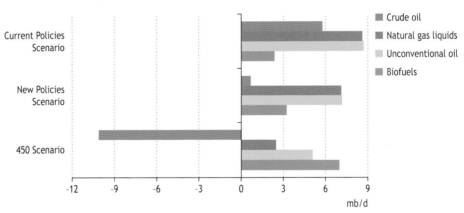

In the New Policies Scenario, non-OPEC production in total peaks before 2015 at around 48 mb/d and then begins to decline, falling to 46 mb/d by the end of the projection period (Figure 3.18). Conventional oil production goes into decline before 2015 but, until around 2025, this decline is offset by rising unconventional production – chiefly oil sands in Canada, supplemented by about 500 kb/d of oil from coal-to-liquids (in China, South Africa and the United States), gas-to-liquids and oil shales. OPEC oil production, by contrast, continues to grow throughout the projection period, on the assumption that the requisite investment is forthcoming. OPEC share of world production rises from 41% in 2009 to 52% in 2035. The shares of NGLs and unconventional oil in world production also grow markedly over the projection period.

Box 3.3 ● Enhancements to the oil-supply model for *WEO-2010*

The IEA oil supply model has been improved for this year's *Outlook*, to allow for more complex modelling of global supply scenarios, with more detailed assumptions per country and resource category. This modelling includes simulating the impact of different assumptions about resource endowment and accessibility, oil prices, costs (finding and development and lifting), fiscal terms and investment risks, logistical constraints on the pace of resource exploration and development, production profiles and decline rates, carbon emission regulations and CO_2 prices, and technological developments. The model projects supply, investment in exploration and production, and company and government revenues by country/region and by resource category. The projections are underpinned by current field production profiles and decline rates, drawing on the detailed results of the field-by-field analysis of *WEO-2008* (IEA, 2008), and take into account specific near-term project development plans (IEA, 2010b). OPEC production projections take into account stated policies on resource depletion and investment.

Figure 3.18 ● World oil production by source in the New Policies Scenario

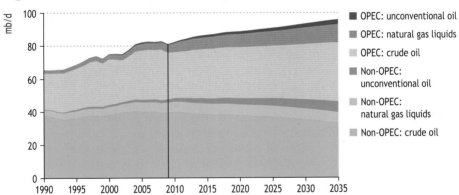

Although global oil production in the New Policies Scenario increases by only 15 mb/d between 2009 and 2035, the need for new capacity is much larger because of the need to compensate for the decline in production at existing fields as they pass their peak and flow-rates begin to drop. Crude oil output from those fields that were in production in 2009 drops from 68 mb/d in 2009 to 16 mb/d by 2035, a fall of three-quarters (Figure 3.19). This projection takes account of the build-up and decline rates of different types of fields in each region, drawing on the detailed field-by-field analysis carried out in 2008 (IEA, 2008). On average, the production-weighted rate of decline in production year-on-year accelerates through the projection period, as more and more fields pass their peak and enter their decline phase and as the share of smaller and offshore fields, with higher decline rates, grows. By 2035, aggregate output from fields already in production in 2009 is declining at a rate of 8.3% per year.[8] We calculate that, over the *Outlook* period, there is a need to add a total of 67 mb/d of gross capacity in order to compensate for the decline at existing conventional oilfields and to meet the growth in demand. The gross new capacity required by 2020 is 28 mb/d. Just under 60% of the crude oil produced from new fields in 2035 is from fields that have already been found, most of which are in OPEC countries. The bulk of the oil that is produced in 2035 from new fields that are yet to be found is in non-OPEC countries, largely in deep water.

Figure 3.19 ● **World oil production by type in the New Policies Scenario**

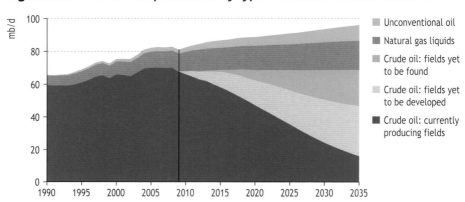

As noted above, slightly more than half of the world's ultimately recoverable resources of conventional oil are produced by the end of the projection period in the New Policies Scenario (see Figure 3.15, above). Cumulative production reaches 1.9 trillion barrels by the end of 2035, up from 1.1 trillion barrels at end-2009. The share of unconventional oil resources that are produced by 2035 is much lower, at less than 3% (based on a conservative estimate of 1.9 trillion barrels). The size of ultimately recoverable resources of both conventional and unconventional oil is obviously crucial in determining how soon global oil production peaks and at what level. However, the estimate of their size inevitably changes over time, as advances in technology open up new sources or areas of production and lower their cost of development, shifting more of the oil originally in place worldwide into the category of recoverable resources (see the Spotlight). Higher prices — as we assume in all three scenarios in this *Outlook* — would also effectively increase the recovery factor. Non-OPEC production is particularly sensitive to the estimated size of conventional resources, as there are fewer constraints on the development of those resources.

In order to test the sensitivity of the level of production in non-OPEC countries to the level of ultimately recoverable resources, we have modelled the impact of both higher and lower levels of conventional oil resources, based broadly on the upper and lower bounds estimated by the USGS (corresponding to 5% and 95% probability) and restrictions on resource access, particularly for volumes in environmentally sensitive areas, deep water and the Arctic (Figure 3.20). In the New Policies Scenario, the lower

---

8. This takes account of enhanced oil recovery projects that are implemented at currently producing fields.

resource case would lead to a much faster decline in non-OPEC production compared with the mean case, with production falling a further 6 mb/d by 2035. Assuming unchanged supplies of NGLs and unconventional oil, this would increase the call on OPEC oil by the same amount. In reality, it is far from certain that OPEC would be willing or able to produce this much oil within this timeframe. Were OPEC producers unwilling or unable to make up the difference, oil prices would rise, stimulating more investment in unconventional non-OPEC supplies and choking off demand.

Figure 3.20 ● *Sensitivity of non-OPEC crude oil production to ultimately recoverable resources*

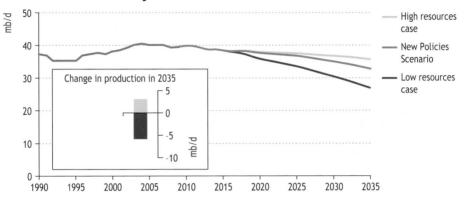

Offshore fields are expected to account for a slightly growing share of crude oil production, especially during the first half of the projection period, when a number of new deepwater projects are brought online in non-OPEC countries (Figure 3.21). In the long term, the offshore share levels off, as large new increments to onshore production in the Middle East play an increasingly important role. In aggregate, worldwide crude oil production from offshore fields rises marginally, from 21.6 mb/d in 2009 to a peak of 23 mb/d by 2025, falling back slightly by 2035 in the New Policies Scenario. Their share in world crude oil production rises from 32% in 2009 to 34% in 2025 and then drops back to 33% in 2035. The contribution from deepwater fields (at depths of more than 400 metres) rises from around 5 mb/d in 2009 to nearly 9 mb/d in 2035. In non-OPEC countries, the share of offshore fields in total crude oil production rises from just over one-third to almost half.

NGLs account for almost half of the increase in overall global oil production between 2009 and 2035 in the New Policies Scenario, their output rising from 10.8 mb/d to nearly 18 mb/d (Table 3.4). Production increases particularly sharply in the near term, jumping by more than one-quarter already by 2015, as a result of a number of major gas projects coming on stream. The strong rise in natural gas production, particularly in the Middle East, where gas generally has higher liquids content than in most other regions, is the main driver, but other factors, including reduced flaring, which will make available more associated gas (which tends to be relatively wet), and the increasing wetness of gas reservoirs now being developed in other areas helps boost NGLs supplies. These factors more than offset the projected increase in the share of non-associated gas in total production (Figure 3.22).

Figure 3.21 ● World crude oil production by physiographical location in the New Policies Scenario

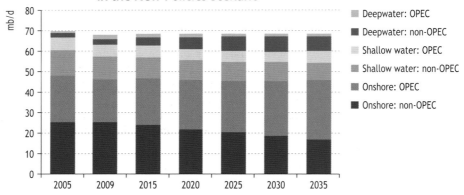

Table 3.4 ● Natural gas liquids production by region in the New Policies Scenario (mb/d)

	1980	2009	2015	2020	2025	2030	2035	2009-2035*
OPEC	0.9	4.6	7.1	8.0	9.0	10.1	11.1	3.5%
Middle East	0.5	3.3	5.4	5.8	6.1	6.8	7.3	3.1%
Other	0.3	1.3	1.6	2.3	2.8	3.3	3.8	4.3 %
Non-OPEC	2.8	6.2	6.6	6.8	6.9	6.9	6.8	0.3%
North America	2.2	2.9	2.7	2.7	2.6	2.5	2.4	-0.7%
Europe	0.1	0.7	0.7	0.7	0.7	0.7	0.7	0.0%
Pacific	0.1	0.1	0.1	0.1	0.1	0.1	0.2	2.0%
E. Europe/Eurasia	0.2	1.1	1.4	1.5	1.6	1.7	1.8	1.9%
Asia	0.1	0.7	0.8	0.9	0.8	0.8	0.8	0.7%
Middle East	0.0	0.2	0.2	0.2	0.2	0.2	0.2	1.1%
Africa	0.0	0.3	0.3	0.3	0.3	0.3	0.2	-1.0%
Latin America	0.0	0.3	0.4	0.5	0.5	0.5	0.5	1.8%
World	3.7	10.8	13.7	14.8	15.9	17.0	17.9	2.0%

*Compound average annual rate of growth.

Figure 3.22 ● Drivers of natural gas liquids production

Declining share of associated gas (which tends to be wetter) in world gas production

Increasing share of unconventional gas, which tends to have lower liquids content

Growth in natural gas supply with large developments ongoing

Increasing share of associated gas is being marketed (through reduced flaring)

Increasing wetness of non-associated gas

Peak oil revisited: is the beginning of the end of the oil era in sight?

Public debate about the future of oil tends to focus on when conventional crude oil production is likely to peak and how quickly it will decline as resource depletion passes a certain point. Those who argue that an oil peak is imminent base their arguments largely on the indisputable fact that the resource base is finite. It is held that once we have depleted half of all the oil that can ever be recovered, technically and economically, production will enter a period of long-term decline.

What is often missing from the debate is the other side of the story – demand – and the key variable in the middle – price. How much capacity is available to produce oil at any given moment depends on past investment. Decisions by oil companies on how much and where to invest are influenced by a host of factors, but one of the most important is price (at least relative to cost). And price is ultimately the result of the balance between demand and supply (setting aside short-term fluctuations that may have as much to do with financial markets than with oil-market fundamentals). In short, if demand rises relative to supply capacity, prices typically rise, bringing forth more investment and an expansion of capacity, albeit usually with a lag of several years.

Another misconception is that the amount of recoverable oil is fixed. The amount of oil that was ever in the ground – oil originally in place, to use the industry term – certainly is a fixed quantity, but we have only a fairly vague notion of just how big that number is. But, critically, how much of that volume will eventually prove to be recoverable is also uncertain, as it depends on technology, which will certainly improve, and price, which is likely to rise: the higher the price, the more oil can be recovered profitably. An increase of just 1% in the average recovery factor at existing fields would add more than 80 billion barrels to recoverable resources (IEA, 2008). So, the chances are that the volume of resources that prove to be recoverable will be bigger than the mean estimate we use to project production, especially since that estimate does not include all areas of the world. Even if conventional crude oil production does peak in the near future, resources of NGLs and unconventional oil are, in principle, large enough to keep total oil production rising for several decades.

Clearly, global oil production will peak one day. But that peak will be determined by factors on both the demand and supply sides. We project a peak before 2020 in the 450 Scenario. In the New Policies Scenario, production in total does not peak before 2035, though it comes close to doing so, conventional crude oil production in that scenario holding steady at 68-69 mb/d over the entire projection period and never attaining its all-time peak of 70 mb/d in 2006. In other words, if governments put in place the energy and climate policies to which they have committed themselves, as we assume in this scenario, then our analysis suggests that crude oil production has probably already peaked.

If governments act vigorously now to encourage more efficient use of oil and the development of alternatives, then demand for oil might begin to ease quite soon and we might see a fairly early peak in oil production. That peak would not be caused by any resource constraint. But if governments do nothing or little more than at present, then demand will continue to increase, the economic burden of oil use will grow, vulnerability to supply disruptions will increase and the global environment will suffer serious damage. The peak in oil production will come then not as an invited guest, but as the spectre at the feast.

The strong growth in NGLs supply will lighten the overall product mix, although this effect is expected to be at least partially offset by a rise in the share of extra-heavy oil and natural bitumen in overall oil production (Figure 3.23). This changing production mix will require more investment in upgraders for the heavier crudes and bitumen, and condensate and NGL processing facilities for the lighter fluids. Much of the increase in the supply of NGLs is likely to be used a petrochemical feedstock, notably in the Middle East.

Figure 3.23 ● **World oil production by quality in the New Policies Scenario**

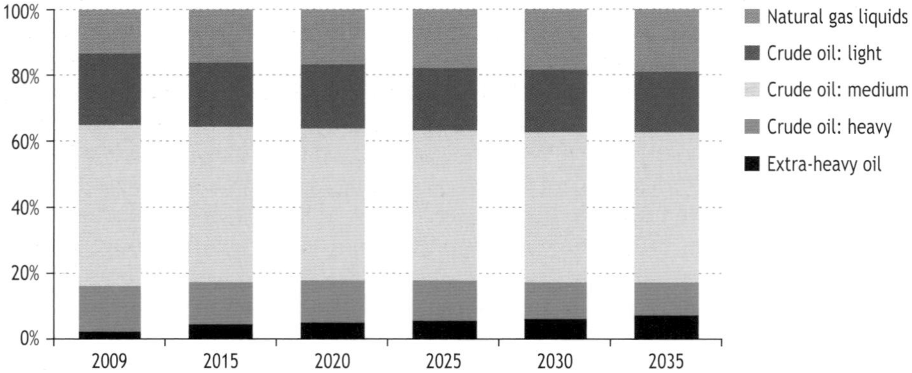

Note: Light crude oil has an API gravity of at least 35°; medium between 26° and 35°; heavy between 10° and 26°; and extra-heavy less than 10°.

Sources: Data provided to the IEA by the Italian oil company, Eni; IEA estimates and analysis.

The structure of the global oil industry is set to change strikingly in the coming decades, as production shifts to countries dominated by national oil companies, which control most of the world's remaining oil resources. In the New Policies Scenario, national companies as a group are projected to contribute all of the growth in global oil production over the projection period, their share rising from 58% in 2009 to about 66% in 2035, based on their current resource ownership (Figure 3.24). These projections assume sufficient investment

World Energy Outlook 2010 - **GLOBAL ENERGY TRENDS**

is made in exploration, development and production to meet demand at the assumed price. The major resource-rich countries may favour slower depletion of their hydrocarbon resources. In some cases, there are also doubts about the financial and technical ability of national companies to bring new capacity on stream in a timely manner.

Figure 3.24 ● **World oil production by type of company in the New Policies Scenario**

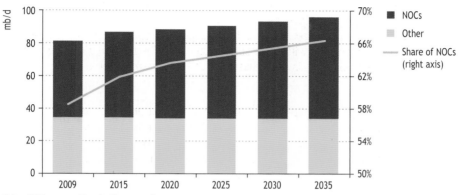

Note: NOCs are national oil companies.

Non-OPEC production outlook in the New Policies Scenario

North America will remain an important non-OPEC producing region, with output projected to rise over the next quarter of a century in the New Policies Scenario (Table 3.5). In *Canada*, conventional oil production declines steadily, but this is more than offset by rapid growth in output from oil sands (see Chapter 4). As new policies to mitigate climate change take hold, the increasing amount of carbon dioxide (CO_2) captured during oil-sands production is accompanied by growth in CO_2 enhanced oil recovery projects in the ageing conventional fields of Alberta, slowing their production declines. In the eastern seaboard and Arctic regions, production holds steady, with slow declines in established projects such as Hibernia, Terra Nova and White Rose being offset by new projects. Arctic developments are expected to be slow and provide only small volumes, due to the relatively modest resource endowment, high costs and tighter environmental regulations in the aftermath of the Macondo disaster offshore of the US Gulf Coast. With the short drilling season and strict requirements for same-season relief-well drilling in case of an accident, costs may well increase in the first half of the projection period, outstripping the impact of technological advances.

Oil production in the *United States* is projected to continue to fall slowly in the medium term, but then recovers towards the end of the projection period as higher oil prices spur growth in enhanced recovery and unconventional oil. In recent years, increased production offshore in the Gulf of Mexico has helped offset the continuing decline in older producing areas. But with the rapid decline rates characteristic of deep offshore projects with large upfront capital expenditures, new offshore regions will need to be opened to drilling to limit the overall decline in production. In the aftermath of the Macondo disaster, such opening of new areas to drilling, which was part of proposed legislation, is likely to proceed only slowly, if at all (Box 3.4). Production of NGLs in the United States is projected to remain high, as indigenous production of gas increases gradually, driven by the shale-gas revolution. Additional volumes of unconventional oil, mainly from coal-to-liquids plants, supplement supply, especially towards the end of the projection period.

Mexico continues to struggle to bring new fields on-line to offset the rapid decline of the Cantarell super-giant field. Production from Cantarell dropped from its peak of 2.2 mb/d in 2003 to an estimated 0.5 mb/d by the middle of 2010. This precipitous decline is linked to the way production has been augmented using nitrogen injection and the highly fractured geology of the field, where most of the producible oil was contained in natural fractures and so was produced quickly. Pemex, the national oil company, has implemented various tertiary recovery technologies and now expects the rate of decline to moderate. Production from new fields has not been able to keep pace with Cantarell's decline, with production from new projects such as Chicontepec rising much more slowly than expected. Nonetheless, significant resources are thought to be present offshore in the Mexican waters of the Gulf of Mexico, so after a continued decline in the first part of the *Outlook* period, overall Mexican oil production is expected to inch back up as new projects come on stream. With rising domestic demand, Mexico's role as an exporter to the United States is set to continue to diminish.

Table 3.5 ● **Non-OPEC oil production in the New Policies Scenario** (mb/d)

	1980	2009	2015	2020	2025	2030	2035	2009-2035**
OECD	17.3	18.7	17.4	17.0	16.9	17.2	17.5	-0.3%
North America	14.1	13.6	13.1	13.3	13.7	14.3	15.0	0.4%
Canada	*1.7*	*3.2*	*3.8*	*4.0*	*4.5*	*4.9*	*5.3*	*2.0%*
Mexico	*2.1*	*3.0*	*2.5*	*2.4*	*2.4*	*2.5*	*2.5*	*-0.7%*
United States	*10.3*	*7.4*	*6.9*	*6.9*	*6.8*	*6.9*	*7.1*	*-0.1%*
Europe	2.6	4.5	3.5	3.1	2.7	2.4	2.1	-2.9%
Pacific	0.5	0.7	0.7	0.6	0.5	0.5	0.5	-1.4%
Non-OECD	19.9	28.9	30.8	31.2	31.4	30.3	28.6	-0.0%
E. Europe/Eurasia	12.5	13.4	14.1	14.2	14.7	14.7	14.5	0.3%
Caspian	*0.9*	*2.9*	*3.7*	*4.4*	*5.3*	*5.4*	*5.2*	*2.2%*
Russia	*11.1*	*10.2*	*10.2*	*9.5*	*9.2*	*9.2*	*9.1*	*-0.4%*
Asia	4.5	7.4	7.4	7.0	6.7	5.9	5.0	-1.5%
China	*2.1*	*3.8*	*3.8*	*3.7*	*3.6*	*3.1*	*2.4*	*-1.7%*
India	*0.2*	*0.8*	*0.9*	*0.8*	*0.8*	*0.8*	*0.8*	*-0.2%*
Middle East	0.5	1.7	1.5	1.3	1.2	1.1	1.0	-1.9%
Africa	1.0	2.5	2.5	2.3	2.1	2.0	1.8	-1.2%
Latin America	1.3	3.9	5.3	6.4	6.7	6.5	6.2	1.8%
Brazil	*0.2*	*2.0*	*3.1*	*4.4*	*5.0*	*5.2*	*5.2*	*3.7%*
Total non-OPEC	37.1	47.7	48.2	48.2	48.2	47.4	46.1	-0.1%
Non-OPEC market share	*59%*	*59%*	*56%*	*54%*	*53%*	*51%*	*48%*	*-*
Conventional	37.0	45.8	45.1	44.4	43.6	41.9	39.6	-0.6%
Crude oil	34.1	39.6	38.4	37.6	36.7	35.0	32.8	-0.7%
Natural gas liquids	2.8	6.2	6.6	6.8	6.9	6.9	6.8	0.3%
Unconventional	0.2	1.8	3.1	3.7	4.6	5.6	6.5	5.0%
Share of total non-OPEC	*0%*	*4%*	*6%*	*8%*	*10%*	*12%*	*14%*	*-*
Canada oil sands	0.1	1.3	2.4	2.8	3.3	3.7	4.2	4.5%
Gas-to-liquids	-	0.0	0.0	0.0	0.1	0.2	0.3	7.7%
Coal-to-liquids	0.0	0.2	0.2	0.3	0.6	0.8	1.1	7.6%

* Compound average annual rate of growth.

Production in *Europe*, mainly in the North Sea, continues its steady decline from 4.5 mb/d in 2009 to 2.1 mb/d in 2035. Recovery rates are likely to continue to rise as tertiary recovery technologies are deployed, partially offsetting the impact of dwindling new discoveries. Elsewhere in the OECD, production in the *Pacific*, already only 0.7 mb/d, continues to decline, the fall in crude oil production more than offsetting rising output of NGLs and CTL in Australia (see Chapter 4).

Box 3.4 ● Impact of the Gulf of Mexico oil spill

The tragic accident that occurred at the end of April at the Macondo well in the Gulf of Mexico will have both short-term and long-lasting consequences for the oil industry. Although not all the facts are known at the time of writing, it appears that a series of human errors and equipment failures led to an uncontrolled blow-out while the well was being completed. The resulting explosion killed 11 people and sank the drilling rig, provoking a major oil spill. Over 4 million barrels of oil are reported to have been released into the Gulf of Mexico during the four months that it took to cap the well.

The accident has led to a *de facto* moratorium on drilling in the Gulf of Mexico with floating rigs; the US Administration announced a six-month moratorium in May, but this decision was initially over-ruled and is now being reviewed in court. In any event, deepwater drilling activity there has more or less come to a halt. Drilling is expected to resume only after an extensive review of regulations and contingency procedures. One plausible scenario is for drilling in moderate water depths to resume gradually over the next few months, while deeper water operations may not resume until new technologies to mitigate the consequences of such an accident are put in place. The medium-term effect on production will obviously depend on the duration of the moratorium: we estimate that the drop in production (in the Gulf of Mexico) would be of the order of 100 to 200 kb/d per year of stopped activity. In the longer term, tighter regulations on deepwater drilling are likely to curb the growth of production in other parts of the United States — particularly those areas that have not yet been opened to drilling.

A full moratorium is unlikely to be declared in other regions with deepwater production, notably Brazil, West Africa, the North Sea and Canada. However, they have already started reviewing their regulations and will continue to do so when all the facts from the Macondo accident are known. Corporate policies on deepwater operations are also undergoing changes, reflecting potentially increased liabilities in the event of an accident; it is likely that some smaller companies will withdraw from deepwater activities. Overall, new regulations are likely to result in some delays to deepwater projects all over the world. This is taken into account in our modelling of oil production in this *Outlook*. But the capital planned to be spent by oil companies for deepwater projects would probably be at least partly re-allocated to other locations, bringing production from other projects forward, so the net impact on global oil supply is expected to be small.

In principle, tighter regulatory requirements would lead to higher costs for developing deepwater resources. However, the main cost driver will remain drilling rig day-rates, themselves driven by the utilisation rates of available rigs. A moderate slowdown in deepwater developments could constrain any cost increases. Coupled with improvements in technology prompted by the lessons learned from the accident, deepwater developments are likely to continue to play a key role in the world supply/demand balance at the oil price trajectories projected in the three scenarios.

Russia has consolidated its position as the world's leading oil producing country with increases in production in 2009 and 2010, driven by a more favourable tax regime, particularly for new fields in eastern Siberia. Although resources are thought to be plentiful in the vast, remote regions of eastern Siberia, high development costs will probably mean that the region is developed only slowly. Allowing for a possible tightening of the fiscal regime, at least in the early part of the projection period, as the Russian government needs to replenish its coffers after the economic downturn of the last two years, Russian oil production is projected to remain relatively flat to 2015, with new projects slowly coming online to offset decline in the mainstay producing region of western Siberia. However, in the longer term, oil production falls steadily, to slightly over 9 mb/d by 2035, despite a projected increase in NGLs production as natural gas output expands (from around 580 bcm in 2009 to over 800 bcm by 2035).

Oil production in the leading Caspian oil-producing country, *Kazakhstan,* is projected to increase throughout the projection period, before decline sets in at the major new offshore fields and production stabilises at nearly 4 mb/d (see Chapter 17). Oil production in *Azerbaijan,* the only other significant producer in the region, levels out at 1.3 mb/d in the next few years and then starts to decline as 2020 approaches, reaching 0.9 mb/d by 2035. Exports from both countries will depend on policies to improve energy efficiency, in order to rein-in the growth of demand with growing prosperity.

China is projected to maintain production close to the current level of 3.8 mb/d to 2015, followed by a steady decline as resource depletion sets in. A similar situation holds for other non-OPEC Asian countries, with production in the region as a whole dropping from 7.4 mb/d in 2009 to 5 mb/d by 2035.

Africa still has substantial scope to increase oil production, but with the slow pace of development in recent years and political instability in some countries, a steady decline in non-OPEC production is projected over the *Outlook* period. The deepwater offshore West Africa region is in the early phases of its development, and production there is expected to steadily increase in spite of the rapid decline rates characteristic of projects in such areas. New producing countries, such as Ghana or Uganda, are projected to make a growing but modest contribution to the oil production of the region. Oil development in Sudan has been halted by political risks, but the country has the potential to increase production in the longer term.

Latin America sees the second-fastest rate of increase in oil production of any non-OPEC region in the New Policies Scenario. Output growth is led by *Brazil*, where, thanks to several major deep water offshore discoveries in the last few years in pre-salt layers (so called because the hydrocarbon reservoirs are located underneath thick salt deposits and were therefore difficult to spot on 3D seismic data before recent advances in that technology), including the Tupi and Jupiter fields, production increases to 5 mb/d by 2025 and then levels off through to the end of the projection period. The Tupi field, a probable super-giant found in 2006, with recoverable resources estimated to be as much as 8 billion barrels, is due to enter production in 2011. Total production from the pre-salt projects (including Tupi) is projected to reach about 1.4 mb/d by 2020. Discoveries of other big fields in the pre-salt layer would allow for higher peak production and extend the plateau for a longer period. The pre-salt area is thought to contain as much as 30 billion barrels of recoverable resources — twice the current proven reserves of Brazil. The deposits are also gas rich, so NGLs production is also set to increase.

OPEC production outlook in the New Policies Scenario

OPEC accounts for *all* of the projected growth in global oil production between 2009 and 2035 in the New Policies Scenario (see Table 3.3 above).[9] Roughly 16% of the increase in OPEC output goes to meet the growth in local consumption. The growth in OPEC output is expected to come from four main sources (Table 3.7).

- Further expansion of Saudi crude oil production and increased NGLs supply as the country's gas production expands substantially.

- The re-emergence of Iraq as one of the world's leading oil-producing countries (Box 3.5), commensurate with its large resource base.

- A large increase in NGLs production, linked to increased gas production, especially in OPEC Middle East countries (where most of the increased gas supply goes to meeting booming domestic demand), and increasing exports from Qatar and Algeria.

- The emergence of unconventional oil production from the Orinoco belt in Venezuela and from gas-to-liquids plants, notably in Qatar and Nigeria (see Chapter 4).

Saudi Arabia is projected to regain from Russia its place as the world's biggest oil producer, its combined output of crude oil and NGLs rising from 9.6 mb/d in 2009 to 11.5 mb/d in 2020 and 14.6 mb/d in 2035 (including its share of output from the Neutral Zone). Sustainable crude oil production capacity has been raised to a little over 12 mb/d with the recent completion of the 1.2-mb/d Khurais field development. The next major development, the 900-kb/d Manifa field, will be completed by around 2016, but this will probably not increase overall capacity, due to declines in output at other fields (IEA, 2010b). The Kingdom has stated for several years that it is capable and willing, if there is sufficient market demand, to increase crude oil production capacity to 15 mb/d and to sustain that level for 50 years, though it has no plans to exceed that capacity. NGLs production is projected to rise from 1.3 mb/d in 2009 to 2.2 mb/d

---

9. Our projections of OPEC production are based on assumptions that adequate investment is forthcoming. See IEA (2008) for a detailed discussion of the uncertainties surrounding future OPEC investment and production policies.

by 2035 in line with the expansion of gas production. The projected level of overall production, even in 2035, would still leave Saudi Arabia with a modest amount of spare capacity. The stated policy goal in this respect is to maintain around 1.5 to 2.0 mb/d of spare capacity on average, which would enable Saudi Arabia to continue to play a vital role in balancing the global oil market.

Oil production in *Qatar* will continue to be driven by gas exports, thanks to its super-giant North gas/condensate field. We expect more LNG export capacity to be added and to see a resurgence of interest in GTL, beyond the current Oryx and Pearl plants, as a hedge against decoupling of gas and oil prices. As a result of increased gas production, NGLs production will exceed crude oil production in Qatar from 2010 onwards.

Box 3.5 ● The renaissance of Iraqi oil production

Over the last two years, the gradual normalisation of the political situation and improved security in Iraq have enabled the country to stabilise oil production at around 2.5 mb/d and to hold two bidding rounds for licenses, which provide for the participation of foreign oil companies in the development of the country's abundant oil resources (IEA, 2010b). Eleven different field development projects have been agreed so far, including the rehabilitation of some existing fields, notably the Rumaila field in the south of the country, and the more intensive development of fields that have as yet barely been exploited, including the super-giant Majnoon field – the 25th largest field in the world (Table 3.6).

Were all these projects to proceed on schedule, Iraqi oil production capacity would reach more than 12 mb/d by 2017. This would involve more than $160 billion of investment. The sheer scale of this, coupled with political and security-related uncertainties, suggests that the expansion of capacity will, in practice, be much slower. In the New Policies Scenario, we expect that it will take until the 2030s for Iraqi oil production to exceed even 6 mb/d. Although ambitious work has started on several of the projects, much basic infrastructure, including roads, bridges, airports, power and water supply is in need of repair and expansion. Existing export routes are fully utilised and a major expansion of the shipping ports will be needed even to reach the projected level of production. Iraq's crude oil production nonetheless overtakes that of Iran soon after 2015 and total oil production (including NGLs) by around 2020.

Iran has significant upside production potential, both for crude oil and NGLs. However, the current political isolation of the country makes it unlikely that this potential will be realised quickly. We project a slow increase in overall oil output during the projection period, in large part driven by NGLs.

Kuwait has been making plans for boosting production capacity to 4 mb/d for the last 20 years. These plans, originally known as "Project Kuwait", called for the involvement of international companies in developing the country's large heavy oil resources under service contracts, but this approach was halted in the face of political opposition. Officially, the country aims to reach the targeted production level by 2020 – 1 mb/d above current capacity – but achieving this will be contingent on securing the technical assistance of foreign firms. Emphasis has now shifted away from heavy oil to developing the country's lighter oil reserves. We project gradually increasing production for most of the period, reaching 3.6 mb/d only by 2035. The *United Arab Emirates* is also projected to increase production steadily throughout the projection period, remaining an important contributor to the global supply/demand balance.

Table 3.6 ● Oil production technical services contracts issued in Iraq in 2010

Field	Companies	Target capacity (mb/d)	Time period (years)
Rumaila	BP/CNPC	2.85	7
West Qurma 1	Exxon/Shell	2.32	7
West Qurma 2	Lukoil/Statoil	1.80	13
Majnoon	Shell/Petronas/Missan	1.80	10
Zubair	ENI/Oxy/Kogas	1.20	7
Halfaya	CNPC/Total/Petronas	0.53	13
Garraf	Petronas/Japex	0.23	13
Badra	Gazprom/Kogas/Petronas/TPAO	0.17	7
Qayara	Sonangol	0.12	9
Najmah	Sonangol	0.11	9
Missan	CNOOC/Turkish Petroleum	0.45	7
Total		11.59	

Table 3.7 ● OPEC oil production in the New Policies Scenario (mb/d)

	1980	2009	2015	2020	2025	2030	2035	2009-2035*
Middle East	**18.0**	**23.1**	**28.1**	**30.0**	**31.6**	**34.1**	**37.1**	**1.8%**
Iran	*1.5*	*4.3*	*4.7*	*4.8*	*5.0*	*5.1*	*5.3*	*0.8%*
Iraq	*2.6*	*2.5*	*3.6*	*4.8*	*5.3*	*6.1*	*7.0*	*4.1%*
Kuwait	*1.4*	*2.5*	*2.9*	*3.0*	*3.1*	*3.3*	*3.6*	*1.5%*
Qatar	*0.5*	*1.5*	*2.2*	*2.3*	*2.3*	*2.5*	*2.5*	*1.9%*
Saudi Arabia	*10.0*	*9.6*	*11.2*	*11.5*	*12.2*	*13.2*	*14.6*	*1.6%*
United Arab Emirates	*2.0*	*2.8*	*3.5*	*3.5*	*3.6*	*3.9*	*4.2*	*1.6%*
Non-Middle East	**7.6**	**10.3**	**10.4**	**10.6**	**11.1**	**11.9**	**12.8**	**0.8%**
Algeria	*1.1*	*1.9*	*2.0*	*2.1*	*2.1*	*2.2*	*2.2*	*0.6%*
Angola	*0.2*	*1.8*	*1.5*	*1.6*	*1.7*	*1.5*	*1.4*	*-1.1%*
Ecuador	*0.2*	*0.5*	*0.4*	*0.3*	*0.3*	*0.3*	*0.2*	*-2.5%*
Libya	*1.9*	*1.7*	*1.7*	*1.7*	*1.8*	*1.9*	*2.1*	*1.0%*
Nigeria	*2.1*	*2.1*	*2.1*	*2.1*	*2.3*	*2.5*	*2.8*	*1.1%*
Venezuela	*2.2*	*2.4*	*2.8*	*2.7*	*2.9*	*3.4*	*4.0*	*2.0%*
Total OPEC	**25.5**	**33.4**	**38.5**	**40.5**	**42.7**	**46.0**	**49.9**	**1.6%**
OPEC market share	*41%*	*41%*	*44%*	*46%*	*47%*	*49%*	*52%*	*-*
Conventional oil	**25.5**	**32.9**	**37.1**	**38.9**	**40.7**	**43.6**	**46.9**	**1.4%**
Crude oil	24.7	28.3	30.0	30.9	31.7	33.5	35.8	0.9%
Natural gas liquids	0.9	4.6	7.1	8.0	9.0	10.1	11.1	3.5%
Unconventional oil	**0.0**	**0.5**	**1.4**	**1.6**	**2.0**	**2.4**	**3.0**	**7.1%**
Venezuela extra-heavy oil	*0.0*	*0.4*	*1.2*	*1.3*	*1.5*	*1.8*	*2.3*	*6.9%*
Gas-to-liquids	*-*	*0.0*	*0.2*	*0.2*	*0.3*	*0.4*	*0.5*	*14.5%*

* Compound average annual growth rate.

Nigeria, where the complex political situation and sporadic civil conflicts over oil resources have hampered investment for several years, also has significant potential for higher production. We project a drop in production in the early part of the *Outlook* period, but, in the longer term, a rebound in output on the assumption that the investment climate improves. An increase in NGLs production contributes to higher production, as efforts to reduce gas flaring slowly bear fruit.

Venezuela sees a modest decline in conventional oil production over the projection period, as its relatively limited resources are depleted and a lack of investment and modern technology take their toll. However, this decline is more than offset by rapid growth in unconventional, extra-heavy oil from the Orinoco belt (see Chapter 4).

Other OPEC countries are expected to maintain more or less steady levels of production for a large part of the projection period, variations reflecting their individual resource endowments. *Angola's* output, in particular, is limited by its currently estimated ultimately recoverable resources, though new discoveries could alter this picture.

Inter-regional trade and supply security

Inter-regional trade in oil (crude oil, NGLs, unconventional oil and refined products) is set to grow markedly over the next quarter of a century in the New Policies Scenario. Rising demand outstrips indigenous production in the main non-OECD importing regions, more than offsetting the drop in demand and imports in the OECD. The volume of trade between the main regions modelled in this *Outlook* expands from 37 mb/d in 2009 to 42 mb/d in 2020 and 48 mb/d in 2035 (Table 3.8). Over the projection period, the share of inter-regional trade in world oil production rises from 44% to 49%. China and India see the biggest jump in imports in absolute terms: China's net imports reach almost 13 mb/d in 2035 — up from 4.3 mb/d in 2009. Oil imports in the United States drop from 10.4 mb/d to 7.8 mb/d over the same period; moreover, a growing share of these imports come from Canada (much as synthetic crude, or diluted bitumen, derived from oil sands), so the country's dependence on suppliers outside the region diminishes even more. The Middle East sees the biggest jump in exports, with much of the increase going to non-OECD Asia.

The rise in inter-regional trade does not *necessarily* make oil supplies less secure. But the growing reliance on supplies from a small number of producers, using vulnerable supply routes, could increase the risk of a supply disruption. Moreover, the growing concentration of the sources of exports would increase the exporters' market power, and could lead to lower investment and higher prices. Policies to tackle climate change would make a big difference: policy-driven reductions in oil demand in the 450 Scenario cut substantially import needs, though the share of OPEC oil in total supply to importing countries increases slightly (see Chapter 15).

Table 3.8 ● Inter-regional oil net trade in the New Policies Scenario

	2009		2020		2035	
	mb/d	Share of primary demand*	mb/d	Share of primary demand*	mb/d	Share of primary demand*
OECD	-23.0	55%	-22.8	57%	-17.8	50%
North America	-8.4	38%	-8.1	38%	-4.4	23%
United States	*-10.4*	*59%*	*-10.3*	*60%*	*-7.8*	*52%*
Europe	-8.2	64%	-8.9	74%	-8.3	80%
Pacific	-6.4	91%	-5.8	91%	-5.1	92%
Japan	*-4.0*	*100%*	*-3.4*	*99%*	*-2.8*	*99%*
Non-OECD	26.5	43%	27.6	39%	23.9	30%
E. Europe/Eurasia	8.8	66%	9.1	65%	9.2	63%
Caspian	*2.3*	*80%*	*3.7*	*83%*	*4.3*	*83%*
Russia	*7.5*	*73%*	*6.6*	*70%*	*6.1*	*67%*
Asia	-9.0	55%	-14.8	68%	-25.0	83%
China	*-4.3*	*53%*	*-8.0*	*68%*	*-12.8*	*84%*
India	*-2.2*	*73%*	*-3.4*	*81%*	*-6.7*	*90%*
Middle East	18.3	74%	23.3	74%	28.9	76%
Africa	7.0	70%	6.5	66%	6.5	63%
Latin America	1.4	21%	3.5	37%	4.3	41%
Brazil	*-0.1*	*2%*	*1.9*	*43%*	*2.7*	*51%*
World**	36.7	44%	42.1	46%	48.1	49%
European Union	*-10.0*	*82%*	*-10.1*	*89%*	*-9.0*	*94%*

Note: Positive numbers denote exports; negative numbers imports.

*Per cent of production for exporting regions/countries. **Total net exports for all *WEO* regions/countries (some of which are not shown in this table), not including trade within *WEO* regions.

Oil investment

Current trends

Worldwide upstream oil investment is set to bounce back in 2010, but will not recover all of the ground lost in 2009, when sharply lower oil prices and financing difficulties led oil companies to slash spending. Worldwide, total upstream capital spending on both oil and gas[10] is budgeted to rise in 2010 by around 9% to $470 billion, compared with a fall of 15% in 2009. These investment trends are based on the announced plans of 70 oil and gas companies. Total upstream investment is calculated by adjusting upwards the spending of the 70 companies, according to their share of world oil and gas production for each year. Our survey points to a faster increase in upstream spending in 2010 than in downstream spending (Table 3.9).

---

10. Upstream investment is not reported separately for oil and gas.

Table 3.9 ● Oil and gas industry investment (nominal dollars)

Company	Upstream			Total		
	2009 ($ billion)	2010 ($ billion)	Change 2009/2010	2009 ($ billion)	2010 ($ billion)	Change 2009/2010
Petrobras	18.4	23.8	29%	35.1	44.8	28%
Petrochina	18.9	23.1	22%	39.1	42.9	10%
ExxonMobil	20.7	27.5	33%	27.1	28.0	3%
Royal Dutch Shell	20.3	19.4	-5%	26.5	26.0	-2%
Gazprom	11.5	12.9	13%	15.2	23.7	55%
Chevron	17.5	17.3	-1%	19.8	21.6	9%
Pemex	16.8	16.0	-4%	18.6	19.5	5%
BP	14.7	13.0	-12%	20.7	18.0	-13%
Total	13.7	14.0	2%	18.5	18.0	-3%
Sinopec	7.5	8.2	9%	15.9	16.4	3%
Eni	13.2	13.8	5%	19.0	14.6	-23%
Statoil	11.8	11.1	-6%	12.4	13.0	5%
ConocoPhillips	8.9	9.7	9%	10.9	12.0	10%
Rosneft	5.9	6.5	11%	7.3	9.5	31%
Lukoil	4.7	5.5	17%	6.5	8.0	22%
CNOOC	6.4	7.8	22%	6.4	7.9	24%
Repsol YPF	2.5	3.4	36%	12.1	7.9	-35%
BG Group	4.4	6.2	41%	6.5	7.0	8%
Chesapeake	4.8	4.5	-7%	6.1	6.8	12%
Apache	3.1	4.7	49%	3.8	6.0	58%
Anadarko	4.0	4.5	12%	4.6	5.5	20%
Suncor Energy	4.2	4.5	8%	4.9	5.3	8%
Devon Energy	4.2	4.7	12%	4.9	4.7	-4%
EnCana	3.7	4.4	19%	4.6	4.5	-3%
Occidental	3.0	3.6	21%	3.6	4.5	26%
Sub-total 25	244.7	270.0	10%	350.1	376.0	7%
Total 70 companies	345.9	378.4	9%	n.a	n.a.	n.a.
World	428.0	468.1	9%	n.a.	n.a.	n.a.

Note: The world total for upstream investment was derived by prorating upwards the spending of the 70 leading companies, according to their share of oil and gas production in each year.

Sources: Company reports and announcements; IEA analysis.

Private companies will continue to dominate upstream spending, though national oil companies are set to increase their spending more quickly in 2010 (Figure 3.25). The five super-majors (ExxonMobil, Shell, BP, Chevron and Total) alone account for almost one-fifth of total spending, rising 5% in 2010, with other private companies' capital expenditures rising 11%. Spending by the national oil companies is set to rise by 10%, taking their share of world upstream investment to 39%. The trends in investment for 2010 should be treated as indicative only, as they are based on announced plans, which could change were oil prices and costs to differ markedly from our assumptions. Global upstream investment in 2009 is now estimated to have totalled $40 billion *more* than was budgeted in the middle of the year. The upward revision reflects a surge in spending in the second half of the year, prompted by rising oil prices and a sharp drop in the value of the dollar against most currencies (which automatically increased investment outside North America, expressed in dollars).

Figure 3.25 ● **Worldwide upstream oil and gas capital spending by type of company**

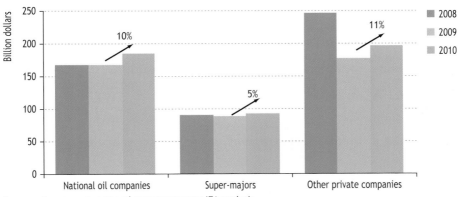

Sources: Company reports and announcements; IEA analysis.

Annual upstream investment more than quadrupled between 2000 and 2008, before falling back in 2009. But most of this increase was needed to meet the higher unit costs of exploration and development, as the prices of cement, steel and other materials used in building production facilities, the cost of hiring skilled personnel and drilling rigs, and the prices of oil-field equipment and services soared. According to our Upstream Investment Cost Index, costs doubled on average over the eight years to 2008 (Figure 3.26). They fell back by about 9% in 2009, but are poised to rebound in 2010 by about 5%.

Adjusted for changes in costs, annual global upstream investment only doubled between 2000 and 2008. With nominal investment falling more heavily than costs in 2009, real investment was 90% higher than in 2000 (Figure 3.27). On current plans and cost trends, capital spending in real terms is set to increase by more than 4% in 2010.

Recent trends in upstream investment and knowledge of projects now under way — if completed to schedule — point to continuing growth in total oil production capacity (including unconventional sources). Between 2009 and 2015, capacity is set to expand in net terms by around 5 mb/d (IEA, 2010b). In the New Policies Scenario, demand rises by 5.7 mb/d, implying a modest reduction in the amount of effective spare capacity, all of which is in OPEC countries, from above 5 mb/d in 2009 to less than 4 mb/d in 2015.

Figure 3.26 ● IEA Upstream Investment Cost Index and annual inflation rate

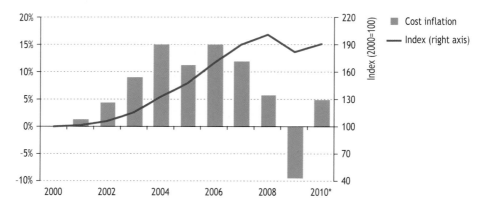

* Preliminary estimate based on trends in the first half of the year.

Note: The Upstream Investment Cost Index, set at 100 in 2000, measures the change in underlying capital costs for exploration and production. It uses weighted averages to remove effects of changes in spending on different types and locations of upstream projects.

Sources: Company reports and announcements; IEA analysis.

Figure 3.27 ● Worldwide upstream oil and gas capital spending

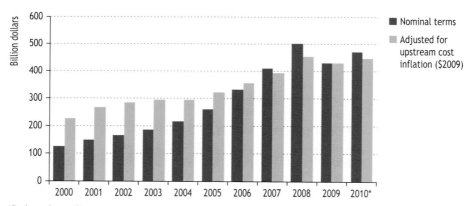

*Budgeted spending.

Sources: Company reports and announcements; IEA analysis.

Upstream investment and operating costs vary with the physiographical location of resources, the geological characteristics of the deposits and multiple regional factors. Finding and development costs and lifting (or operating) costs per barrel of reserves developed and produced are generally lowest for crude oil in the Middle East (Figure 3.28). The future trajectory of these costs will be affected by opposing factors: the development and use of new technologies will facilitate access to more resources and will help reduce unit costs in certain cases, while the depletion of basins in production increases the effort and expense needed to extract more oil. Cyclical cost variations will also occur as short-term fluctuations in activity and the oil price affect the availability of services and other resources.

Figure 3.28 ● **Upstream oil and gas investment and operating costs by region**

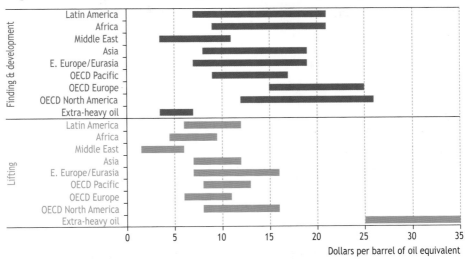

Note: Finding and development (F&D) costs are initial capital investments; lifting costs are ongoing operating costs. The profitable price of oil is determined not just by F&D and lifting costs, but also by the cost and rate of capital repayment, taxes, royalties and profit margin. Cost ranges represent average regional values over the three-year period to 2009 per barrel of oil equivalent developed and produced. Some projects fall outside these ranges. Extra-heavy oil includes Canadian oil sands and deposits in the Venezuelan Orinoco belt.

Source: IEA databases and analysis.

Investment needs to 2035

The projected trends in oil supply in the New Policies Scenario call for cumulative infrastructure investment along the oil-supply chain of around $8 trillion over 2010-2035, or $310 billion per year. About 85% of this investment is needed in the upstream. Including upstream investment needs for gas (see Chapter 5) yields a total annual upstream oil and gas capital spending requirement of about $440 billion – slightly less than the $470 billion the industry is planning to spend in 2010. This fall in the overall level of upstream investment, mainly in the latter part of the projection period, is caused by the shift in investment towards the Middle East and other regions, where finding and development costs are generally lower. This, together with lower unit costs as technology progresses, more than offsets cost increases due to resource depletion. Around three-quarters of global cumulative oil investment to 2035 is needed in non-OECD countries in the New Policies Scenario (Table 3.10). Investments in OECD countries are large, especially in the upstream, despite the small and declining share of these countries in world production. In contrast, investment in Middle East countries – the biggest contributor to production growth – accounts for only 12% of total investment, because costs are lowest in this region.

Table 3.10 ● Cumulative investment in oil-supply infrastructure by region and activity in the New Policies Scenario, 2010-2035

($ billion in year-2009 dollars)

	Conventional production	Unconventional production	Refining	Total*	Annual average
OECD	1 284	283	244	1 811	70
North America	973	263	121	1 358	52
United States	*721*	*51*	*95*	*868*	*33*
Europe	286	2	85	373	14
Pacific	25	17	38	80	3
Non-OECD	5 004	262	735	6 001	231
E. Europe/Eurasia	1 173	15	81	1 270	49
Caspian	*539*	*4*	*13*	*555*	*21*
Russia	*624*	*9*	*44*	*676*	*26*
Asia	396	58	450	904	35
China	*222*	*34*	*220*	*475*	*18*
India	*57*	*11*	*139*	*207*	*8*
Middle East	821	39	105	965	37
Africa	1 254	20	39	1 313	51
Latin America	1 361	129	60	1 549	60
Brazil	*984*	*5*	*30*	*1 019*	*39*
World*	6 288	545	979	8 053	310
European Union	*117*	*0*	*81*	*198*	*8*

*World total includes an additional $241 billion investment in inter-regional transport infrastructure.

There is considerable uncertainty about the prospects for upstream investment, costs and, therefore, the rate of capacity additions, especially after 2015. Few investment decisions that will determine new capacity additions after that time have yet been taken. Government policies in both consuming and producing countries are a particular source of uncertainty. Periodic underinvestment in bringing new capacity on stream, together with time lags in the way demand and investment respond to price signals, tends to result in cyclical swings in price and investment (Figure 3.29). Under-investment in producing countries, where national companies control all or a large share of reserves, could initially lead to shortfalls in capacity, driving prices higher and increasing price volatility. But this effect is likely to be countered by consuming government policies, aimed at curbing oil-demand growth for reasons of energy security and/or climate change (see Chapter 15). In our judgment, the policies, regulatory frameworks and prices assumed in the New Policies Scenario together provide an investment environment that is consistent with the level of investment projected over 2010-2035, but there will undoubtedly be short periods when investment falls short of that required to balance supply with projected demand.

Figure 3.29 ● How government policy action affects the oil investment cycle

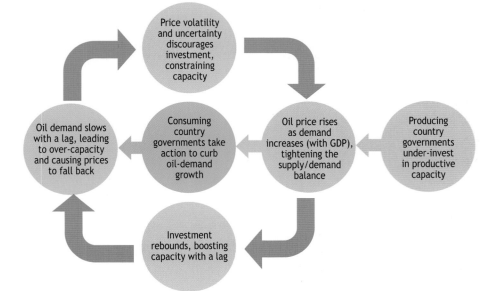

3

Sources: Deutsche Bank (2009); IEA analysis.

CHAPTER 4

THE OUTLOOK FOR UNCONVENTIONAL OIL

Are alternatives to crude coming of age?

H I G H L I G H T S

- The role of unconventional oil is expected to expand rapidly, enabling it to meet about 10% of world oil demand in all three scenarios by 2035. Canadian oil sands and Venezuelan extra-heavy oil dominate the mix, but coal-to-liquids (CTL), gas-to-liquids (GTL) and, to lesser extent, oil shales also make a growing contribution in the second half of the *Outlook* period. In the New Policies and 450 Scenarios, this growth is predicated on the introduction of new technologies that mitigate the environmental impact of these sources of oil, notably their relatively high CO_2 emissions.

- Unconventional oil resources are huge — several times larger than conventional oil resources — and will not be a constraint on production rates over the projection period, nor for many decades beyond that. Most of these resources are concentrated in Canada, Venezuela and a few other countries. Production will be determined by economic and environmental factors, including the costs of mitigating emissions.

- The cost of production puts unconventional oil among the more expensive sources of oil available over the *Outlook* period; unconventional oil projects require large upfront capital investment, typically paid back over long periods. Nonetheless, its exploitation is economic at the oil prices in all three scenarios and unconventional oil, together with deepwater and other high-cost sources of non-OPEC conventional oil, is set to play a key role in setting future oil prices.

- The production of unconventional oil generally emits more greenhouse gases per barrel than that of most types of conventional oil. However, on a well-to-wheels basis, the difference is much less, since most emissions occur at the point of use. In the case of Canadian oil sands, CO_2 emissions are between 5% and 15% higher. Mitigation measures will be needed to reduce emissions from unconventional oil production, including more efficient extraction technologies, carbon capture and storage (CCS) and, in the case of CTL, the addition of biomass to the coal feedstock. Improved water and land management will also be required to make the development of these resources and technologies socially acceptable.

- CTL, if coupled with CCS, has the potential to make a sizeable contribution in all three scenarios; many of the large coal-producing countries are investigating new projects, but clarification of the legal framework for CCS will most likely be required before they can proceed. Renewed interest in new GTL plants is expected, with major gas producers seeing GTL as a way to hedge the risks of gas prices remaining weak relative to oil prices.

Introduction

Unconventional oil is set to play a key role in the oil supply and demand balance and so in determining future oil prices (Chapter 3). However there are many challenges surrounding the development of unconventional oil supplies:

- Total development costs are often higher than those for conventional oil resources.

- Developments are capital-intensive with payback over long time periods, so the timely availability of enough capital has been questioned.

- Resources are relatively localised, casting doubts on the availability of labour and a supporting social infrastructure.

- CO_2 emissions for extracting and upgrading oil from unconventional sources are currently larger than those from most conventional sources, so production will be affected by climate policies.

- A large fraction of the world's unconventional resources is located in environmentally sensitive areas, where water and land use could constrain new developments.

The uncertainties surrounding the response to these challenges are reflected in large differences in the share of unconventional oil in world oil supply in the three scenarios (Table 4.1). In particular, the attractiveness of investing in unconventional oil is highly sensitive to the outlook for oil prices, the extent of the introduction of penalties on CO_2 emissions and the level of development costs relative to conventional oil. In the New Policies Scenario, unconventional sources play an increasingly important role in supplying the world's oil needs. The main sources of unconventional oil today — Canadian oil sands and Venezuelan extra-heavy oil — continue to dominate over the projection period, with other sources just beginning to play a role near the end of the projection period. Unconventional oil supply grows more rapidly in the Current Policies Scenario, in line with higher oil prices (which boost the economic attractiveness of the high-cost unconventional sources). In the 450 Scenario, oil demand is relatively weak and the large CO_2 penalty further depresses demand for unconventional oil, though production from Canadian oil sands and of Venezuelan extra-heavy oil, nonetheless increases beyond current levels. Coal prices, being depressed even more than oil prices, make coal-to-liquids production (with carbon capture and storage) relatively attractive.

Table 4.1 • World unconventional oil supply by type and scenario (mb/d)

	1980	2008	New Policies Scenario		Current Policies Scenario		450 Scenario	
			2020	2035	2020	2035	2020	2035
Canadian oil sands	0.1	1.3	2.8	4.2	2.8	4.6	2.5	3.3
Venezuelan extra-heavy	0.0	0.4	1.3	2.3	1.3	2.3	1.3	1.9
Oil shales	0.0	0.0	0.1	0.3	0.1	0.5	0.1	0.2
Coal-to-liquids	0.0	0.2	0.3	1.1	0.4	1.6	0.3	1.0
Gas-to-liquids	-	0.1	0.2	0.7	0.3	1.0	0.2	0.5
Other*	0.0	0.4	0.6	0.9	0.7	1.0	0.6	0.6
Total	0.2	2.3	5.3	9.5	5.5	11.0	5.0	7.4

* Refinery additives and blending components (see the discussion at the end of this chapter).

What is unconventional oil?

There is no universally agreed definition of unconventional oil, as opposed to conventional oil. Roughly speaking, any source of oil is described as unconventional if it requires production technologies significantly different from those used in the mainstream reservoirs exploited today. However, this is clearly an imprecise and time-dependent definition. In the long-term future, in fact, "unconventional" heavy oils may well become the norm rather than the exception.

Some experts use a definition based on oil density, or American Petroleum Institute (API) gravity. For example, all oils with API gravity below 20 (*i.e.* a density greater than $0.934\,g/cm^3$) are considered to be unconventional. This definition includes "heavy oil", "extra-heavy oil" (with API gravity less than 10) and bitumen deposits. While this classification has the merit of precision, it does not always reflect the technology used for production. For example, some oils with 20 API gravity located in deep offshore reservoirs in Brazil are extracted using entirely conventional techniques. Other classifications focus on the viscosity of the oil, treating as conventional any oil which can flow at reservoir temperature and pressure without recourse to viscosity-reduction technology. But such oils may still need special processing at the surface if they are too viscous to flow at surface conditions.

Oil shales are generally regarded as unconventional, although they do not fit into the above definitions (more details on oil shales can be found later in this chapter). Also classified as unconventional are both oil derived from processing coal with coal-to-liquids (CTL) technologies and oil derived from gas through gas-to-liquids (GTL) technologies. The raw materials in both cases are perfectly conventional fossil fuels. These oil sources are discussed briefly later in this chapter. Oil derived from biomass, such as biofuels, or biomass-to-liquids (BTL, whereby oil is obtained from biomass through processes similar to CTL and GTL) are sometimes included in unconventional oil, but not always.

Another approach, used notably by the United States Geological Survey (USGS), is to define unconventional oil (or gas) on the basis of the geological setting of the reservoir. The hydrocarbon is considered conventional if the reservoir sits above water-bearing sediments and if it is relatively localised. If neither is the case, for example if the hydrocarbon is present continuously over a large area, the hydrocarbon is defined as unconventional. This type of definition has a sound geological basis, but does not always reflect the technology required for production, nor the economics of exploitation.

For the purpose of this *Outlook*, we define as unconventional the following categories of oil:[1]

- Bitumen and extra-heavy oil from Canadian oil sands.
- Extra-heavy oil from the Venezuelan Orinoco belt.

---

1. This definition differs from that used in the IEA *Oil Market Report (OMR)*, which includes some but not all of the Canadian oil sands and Venezuelan Orinoco production (it includes upgraded "synthetic" oil, but not raw bitumen or extra-heavy oil). The *OMR* also includes biofuels, but these are included in biomass in the *WEO*. The *OMR* definition is driven primarily by the way the production data is reported by various countries and the short time available for making adjustments to monthly figures. The definitions we have adopted here are primarily to facilitate the discussion of long-term issues.

- Oil obtained from kerogen contained in oil shales.

- Oil obtained from coal through coal-to-liquids technologies.

- Oil obtained from natural gas through gas-to-liquids technologies, as well as refinery additives and gasoline blending additives originating primarily from gas or coal, such as methyl tertiary butyl ether (MTBE), or methanol for blending.

There are bitumen and extra-heavy oil deposits in countries other than Canada and Venezuela (Table 4.2), but only Canada and Venezuela are likely to play a significant role in the exploitation of these resources in the timescale of these projections. This is because of the size of their resources and the facts that they are already in production, plans exist for their further development, significant reserves are considered as proven and they are geographically concentrated; their decline is not an issue over the 25-year horizon of these projections. Their development is much more like a manufacturing operation than a traditional upstream oil industry project. Whether or not they will be exploited is mainly a matter of economics and capital spending dynamics, not one of geology. By contrast, the resources in Russia and Kazakhstan, which are also sizeable, are more geographically dispersed and, with large conventional oil resources still available, there is little incentive to develop these heavy oils quickly. Their production potential in the next 25 years is not large enough to affect world supply significantly. They are briefly discussed in this chapter, but do not feature as part of our unconventional oil production estimates up to 2035.

Table 4.2 ● **Natural bitumen and extra-heavy oil resources by country**
(billion barrels)

	Proven reserves	Ultimately recoverable resources	Original oil in place
Canada	170	≥ 800	≥ 2 000
Venezuela	60*	500	≥ 1 300
Russia	-	350	850**
Kazakhstan	-	200	500
United States	-	15	40
United Kingdom	-	3	15
China	-	3	10
Azerbaijan	-	2	10
Madagascar	-	2	10
Other	-	14	30
World	230	≥ 1 900	≥ 5 000

* As reported by the *Oil & Gas Journal* (O&GJ, 2009); the national oil company, PDVSA, currently reports 130 billion barrels as proven (as discussed later in this chapter).
** From BGR (2009); Russian authors report significantly smaller resources, of the order of 250 billion barrels; the same applies for Kazakhstan. Bitumen resources in particular are poorly known, as a high percentage is located in the vast and poorly explored region of eastern Siberia. BGR reports 345 billion barrels recoverable, which is more in line with Russian publications.

Sources: BGR (2009); USGS (2009a); IEA analysis.

Box 4.1 ● How oil is formed

A basic understanding of the formation of oil reservoirs is helpful in understanding the differences between the types of unconventional oil presented in this chapter. Oil deposits result from the burial and transformation of biomass over geological periods during the last 200 million years or so. The biomass is typically contained in a type of sediment called shale (though its mineral composition can vary), deposited at the bottom of the ocean or lake basins. As those sediments get buried, the biomass is transformed into complex solid organic compounds called kerogen. When the sediments are deeply buried, the temperature may be sufficient for the kerogen to be transformed into oil and gas.

Under pressure, the oil (or gas) can be expelled from the shale sediments where they were created (known as source rocks) and begin to migrate upwards (due to their low density) into other sedimentary rocks, such as sandstone or carbonates. This upward migration stops when the oil encounters a low permeability rock that acts as a barrier to its movement (cap rock). In this way, a conventional oil reservoir is formed. When the oil does not encounter any significant barrier until it gets near the surface, it can become more and more viscous, as the temperature decreases and some of the lighter components of the oil seep to the surface, where they are degraded by bacteria and escape to the atmosphere. The remaining very viscous oil can become almost solid and stop migrating, even in the absence of a strong cap rock, forming relatively shallow deposits of very viscous, extra-heavy oil or natural bitumen. Occasionally, it can even seep out to the surface, as seen in tar pits, for example.

Canadian oil sands

Production from Canadian oil sands is set to continue to grow over the projection period, making an important contribution to the world's energy security. Just how rapidly will depend on a number of factors, including whether the environmental impact can be mitigated through the use of new technology without rendering the oil uneconomic. Extraction involving the injection of steam via wells into the oil-sands deposit to reduce the viscosity of the oil and allow it to flow to the surface (in-situ projects, see below) is economically and environmentally preferable, but mining is an alternative and significant mining capacity is under construction which will ensure mining remains a substantial contributor to production growth. In the New Policies Scenario, oil-sands production climbs from about 1.3 million barrels per day (mb/d) in 2009 to 4.2 mb/d in 2035,[2] with around two-thirds of the increase coming from in-situ projects (Figure 4.1). The 450 Scenario projects only modest additions to current capacity: projects currently under construction or being planned would suffice to match supply to demand. The Current Policies Scenario calls for rapid growth in

2. This is marketed production, actually part raw bitumen, part upgraded synthetic crude oil. Raw bitumen production is higher, due to volume loss during upgrading; for example in 2009, raw bitumen production was 1.49 mb/d.

oil sands production, although still below what could be achieved with the projects already proposed. The critical drivers and uncertainties surrounding the prospects for oil-sands production are discussed in detail below.

Figure 4.1 ● Canadian oil-sands production by type in the New Policies Scenario

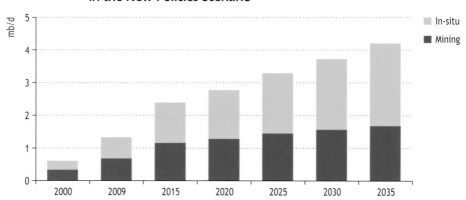

Resources and production technology

Very large deposits of very viscous oil and bitumen — oil sands — exist in Canada at relatively shallow depth. They cover a vast region of Alberta and, to a lesser extent, Saskatchewan. The term "oil sands" is a slight misnomer, as the oil or bitumen is found not only in sand formations, but also in carbonates. The main centres of activity are the Athabasca, Cold Lake and Peace River districts (Figure 4.2), though there are also significant resources in neighbouring regions of Saskatchewan. The total oil in place is estimated to be in excess of 2 trillion barrels, as much as the remaining technically recoverable conventional oil in the entire world. However, because of its very high viscosity, this oil is difficult to produce and, with current technology and oil prices, only part of this volume is thought to be recoverable. The Alberta provincial government currently recognises 170 billion barrels as established reserves, *i.e.* currently economically and technically recoverable.

Because they outcrop over a large area, the presence of bitumen in the Canadian oil sands has been known for centuries. Various early attempts at industrial exploitation took place during the 20th century, leading to the refinement of the techniques for mining and bitumen/sand separation. The modern era for the oil sands started in 1967 with the opening of the Great Canadian Oil Sands base mine, the first large-scale mining operation. It has since been expanded to what is now the Suncor Corporation Steepbank/Millenium mine. In-situ primary production, began in the 1970s and the first steam-stimulation projects in the 1980s. Quantification of reserves in the 1990s, as well as the new oil sands royalty regime introduced in Alberta in 1997, paved the way for the boom of the 2000-2008 period, when many new projects were launched and extensive exploration/appraisal land leases were granted.

Figure 4.2 ● Main Canadian oil-sands districts

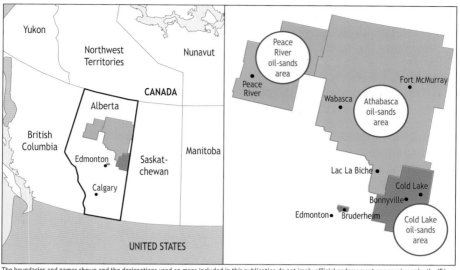

The boundaries and names shown and the designations used on maps included in this publication do not imply official endorsement or acceptance by the IEA.

There are two main methods used to produce oil sands:

- **Mining:** Part of the Canadian oil sands outcrop to the surface and therefore can be mined by essentially conventional strip-mining techniques. Some 7% of the total oil originally in place is estimated to be mineable, *i.e.* some 130 billion barrels. Of the 170 billion barrels of the total Canadian oil-sands established reserves, about 20%, or 35 billion barrels, is recoverable by mining. The "ore", a mixture of bitumen and sand, is treated with hot water to separate out the bitumen. The remaining sludge of slightly oily sand/clay/water mixture is left to settle in large tailing ponds. Some of the solids may eventually be used as part of land reclamation programmes, while some of the water is recycled.

- **In-Situ:** Deeper deposits (75 metres and below) cannot be mined from the surface. A small part can be produced by conventional oil-production techniques. For the very viscous oil found in the Canadian oil sands, these techniques can be applied only to the deepest deposits of slightly lower viscosities, and even there recovery is proportionately small, typically less than 5%. However, production costs can be very low. In some fields, polymer flooding is also applied, with a polymer solution being injected through wells to help push the viscous oil towards the producing wells. A variant on primary recovery is called Cold Heavy Oil Production with Sand (CHOPS), in which the production rate is large enough to entrain sand with the oil, with the oil and sand then being separated at the surface using technologies similar to those used in mining. These "cold" recovery techniques currently produce close to 250 thousand barrels per day (kb/d).

Most of the oil in the oil sands is too viscous to be produced naturally by such primary, or even polymer-flooding, approaches. The temperature of the oil needs to be increased, so that its viscosity decreases, before it begins to flow out of the reservoirs. The method of choice to heat-up the reservoir is to inject hot steam (at a temperature of 250-350°C). There are numerous variants on steam-injection technologies. Cycling Steam Stimulation (CSS) injects steam in a well for a while then, when the reservoir temperature around the well has risen sufficiently, it turns the well into a producer, produces the heated oil, and then starts again — an approach sometimes dubbed "huff-and-puff". Steam Assisted Gravity Drainage (SAGD), which has become the most popular technology for new in-situ projects, uses a pair of horizontal wells, one above the other in the reservoir. Steam is injected in the top well and oil accumulates by gravity in the bottom well. Other approaches to providing heat are at an early stage of experimentation, for example, driving an electrical current through the reservoirs or injecting air to burn some of the oil in-situ (toe-to-heel air injection, [THAI] using horizontal wells; combustion overhead gravity drainage [COGD] using a combination of vertical and horizontal wells; or the older fire-flood technique, using vertical wells). Other experimental approaches use solvents (the so-called VAPEX process), or a combination of steam and solvents, to reduce the viscosity of the bitumen.

At the beginning of 2010, there were more than 80 oil-sands projects in operation, with total raw bitumen capacity of 1.9 mb/d (Table 4.4). Total production in 2009 averaged 1.5 mb/d of raw bitumen. Projects under construction will add a further 0.9 mb/d capacity by 2015. If all proposed and announced projects were to be completed, another 4.5 mb/d capacity would be added. Production will continue to be dominated by a few large projects, operated by large companies. Mining and in-situ current capacities are about equal, but more incremental capacity will derive from in-situ projects, which are regarded as providing better financial returns and facing fewer environmental problems. Very few new projects are planned using primary production only: although financially attractive, they provide only short-term returns, as the recovery rate is low and production declines rapidly.

Production costs depend on the production method, the quality of the reservoir, the size of the project and the location (Table 4.3). Generally, expansions of existing projects cost less than new green-field developments. The profitability of oil-sands projects depends on many variables, including the bitumen/conventional oil price spread, gas prices, construction costs and the prices of steel and oilfield services and labour. At mid-2010 values for these variables, most new oil-sands projects are thought to be profitable at oil prices above $65 to $75 per barrel.

Table 4.3 ● **Typical costs of new Canadian oil sands projects**

	Capital cost ($ per b/d capacity)	Operating cost ($/barrel)	Economic WTI price ($/barrel)
Mining (without upgrader)	50 000-70 000	25-35	50-80
In-situ primary	10 000	5-10	25-50
In-situ SAGD	30 000-40 000	20-30	45-80

The current narrow price spread between conventional light oil (such as West Texas Intermediate [WTI]) and Canadian bitumen blends is likely to persist, as refineries in the United States are geared to process relatively heavy crude and will continue to need Canadian bitumen to balance their crude input slate. The construction of a pipeline from Alberta to the Pacific coast in British Columbia, currently under consideration, would give support to the price of bitumen by opening the Asian market for Canadian bitumen. However, both the proposed pipeline to the Pacific coast and another proposed pipeline to the United States face strong opposition on environmental grounds. Delays or outright cancellation of these projects could affect the marketability of Canadian bitumen. As oil prices increase, as assumed in each of the three scenarios presented in this *Outlook*, some of the costs, notably of gas and services, will also rise, so the price threshold for profitability will also increase; but analysis suggests internal rates of return could continue to increase over the next 25 years (Biglarbigi *et al.*, 2009, where a similar analysis is done for oil shales). Technological progress and learning would further boost profitability. Most projects are economic while oil (West Texas Intermediate) is priced at more than $80/barrel, but many become uneconomic when the price drops below $50/barrel. This is why many new projects were delayed at the end of 2008 and the beginning of 2009. By mid-2010, when the oil price had rebounded to around $70/barrel, many projects were being reactivated. Overall, the breakeven oil price for Canadian oil-sands projects is comparable to that of deepwater offshore conventional oil projects, but production, and therefore investment payback periods, is spread over a much longer time period.

Upgrading

As the oil produced, whether by mining or by in-situ techniques, is extremely viscous (several 100 000 cP,[3] or 100 000 times the viscosity of water, is typical), it cannot be transported economically to refineries without pre-treatment. Two solutions are used in the Canadian oil sands: dilution and upgrading.

In the dilution approach, the viscous bitumen is mixed with light hydrocarbons, for example, the NGLs associated with gas production or synthetic crude oil (SCO) from the upgraders. This yields a mixture, sometimes called Dilbit (for "diluted bitumen"), or SynDilBit if diluted with SCO, that can be transported by pipeline to a refinery in the same way as conventional oil. Not all refineries are equipped to process Dilbit, as the bitumen contains a high concentration of sulphur and asphaltenes, beyond the specifications of some refineries. When the Dilbit is delivered to a nearby refinery, the diluting fluid can often be recycled, transported back to the diluting plant and reused. When the diluted bitumen goes to refineries farther away, reuse of the diluting fluid may not be economic. Availability of enough diluting fluid to cater for a significant rise in production of bitumen is likely to require new long-distance pipelines and increased imports, as NGLs production in western Canada is set to decline (IEA, 2010).

---

3. A centipoise (cP) is a unit of measurement for dynamic viscosity (equal to one-hundredth of a poise). Water at 20°C has a viscosity of 1 centipoise.

Table 4.4 ● Current and planned Canadian oil sands projects (as of mid-2010)

	Project name	Operator	Raw bitumen capacity (kb/d)	Start year	Area	Technology
Producing mining	Steepbank/Millenium	Suncor	320	1967	Athabasca	
	Syncrude 21 (Mildred Lake)	Syncrude	135	1978	Athabasca	
	Aurora North	Syncrude	215	2001	Athabasca	
	Muskeg River	Shell	155	2002	Athabasca	
	Horizon	CNRL	110	2009	Athabasca	
	Total mining producing		**935**			
Producing in-situ	Primrose	CNRL	120	1985	Cold Lake	CSS
	Cold Lake	Imperial	147	1985	Cold Lake	CSS
	Others (< 100 kb/d)	*Various*	721	*Various*	*Various*	*Various*
	Total in-situ producing		**988**			
Total producing			**1 923**			
In construction mining	Muskeg river expansion	Shell	100	2011	Athabasca	
	Jackpine 1	Shell	100	2010	Athabasca	
	Kearl 1	Imperial	100	2012	Athabasca	
	Jackpine 2	Shell	100	2013	Athabasca	
	Total mining construction		**400**			
In construction in-situ	Firebag 3	Suncor	63	2011	Athabasca	SAGD
	Others (< 40 kb/d)	*Various*	85	2011	*Athabasca*	*Various*
	Firebag 4	Suncor	63	2012	Athabasca	SAGD
	Terre de Grace	Value Creation/BP	50	2012	Athabasca	SAGD
	Christina Lake C	Cenovus	40	2012	Athabasca	SAGD
	Sunrise	Husky	200	2014-2018	Athabasca	SAGD
	Total in-situ construction		**501**			
Total in construction			**901**			

World Energy Outlook 2010 - **GLOBAL ENERGY TRENDS**

Table 4.4 ● Current and planned Canadian oil sands projects (as of mid-2010) (continued)

	Project name	Operator	Raw bitumen capacity (kb/d)	Start year	Area	Technology
Proposed Mining	Voyageur South	Suncor	120	2013	Athabasca	
	Joslyn North mine	Total	100	2014	Athabasca	
	Horizon phase 2 and 3	CNRL	130	2015	Athabasca	
	Kearl 2	Imperial	100	2015	Athabasca	
	Jackpine expansion	Shell	100	2015	Athabasca	
	Aurora South	Syncrude	215	2016	Athabasca	
	Fort-Hills	Suncor	190	2017	Athabasca	
	Equinox	UTS energy/Teck	50	2017	Athabasca	
	Pierre River	Shell	200	2018	Athabasca	
	Frontier	UTS energy/Teck	160	2018	Athabasca	
	Joslyn South mine	Total	100	2018	Athabasca	
	Horizon phase 4 and 5	CNRL	260	2020	Athabasca	
	Kearl 3	Imperial	100	2021	Athabasca	
	North Steepbank expansion	Suncor	180	no date	Athabasca	
	Northern Lights	Total/Sinopec	100	no date	Athabasca	
	Total mining proposed		**2 105**			
Proposed in-situ	Foster Creek expansion	Cenovus	50	2011	Athabasca	SAGD
	Others (< 40 kb/d)	Various	45	2012	Athabasca	SAGD
	Kirby	CNRL	45	2013	Athabasca	SAGD
	Christina Lake D	Cenovus	40	2013	Athabasca	SAGD
	Others (<40 kb/d)	Various	67	2013	Various	Various
	Carmon Creek	Shell	80	2014	Peace River	SAGD
	Christina lake 3a	MEG energy	50	2014	Athabasca	SAGD
	KKDL commercial	Statoil	40	2014	Athabasca	SAGD
	Others (< 40 kb/d)	Various	145	2014	Various	Various
	Surmont 2	Conoco Philips	83	2015	Athabasca	SAGD
	Long Lake 2	Nexen	70	2015	Athabasca	SAGD
	Others (< 40 kb/d)	Various	95	2015	Various	Various

4

Table 4.4 ● Current and planned Canadian oil sands projects (as of mid-2010) (continued)

Project name	Operator	Raw bitumen capacity (kb/d)	Start year	Area	Technology
Birch Mountain east	CNRL	60	2016	Athabasca	CSS/SAGD
Grouse	CNRL	60	2016	Athabasca	CSS/SAGD
Corner	Statoil	60	2016	Athabasca	SAGD
Others (< 40kbd)	Various	95	2016	Various	Various
Narrows lake	Cenovus	120	2017	Athabasca	Vapex
Christina lake 3b	MEG energy	50	2017	Athabasca	SAGD
Others (< 40 kb/d)	Various	40	2017	Athabasca	SAGD
Christina Lake E/F/G	Cenovus	120	2018	Athabasca	SAGD
Poplar Creek	E-T Energy	100	2018	Athabasca	Electrothermal
Foster creek addition	Cenovus	90	2018	Athabasca	SAGD
Gregoire	CNRL	60	2018	Athabasca	CSS/SAGD
Primrose expansion	CNRL	30	2018	Cold Lake	CSS
Long Lake 3	Nexen	70	2020	Athabasca	SAGD
Leismer	CNRL	30	2020	Athabasca	CSS/SAGD
Firebag 5-6	Suncor	130	no date	Athabasca	SAGD
Meadow Creek	Suncor	80	no date	Athabasca	SAGD
Lewis	Suncor	66	no date	Athabasca	SAGD
Hangingstone	Excelsior	50	no date	Athabasca	COGD
Surmont	MEG energy	50	no date	Athabasca	SAGD
Christina lake 3 c	MEG energy	50	no date	Athabasca	SAGD
Others (≤ 45 kb/d)	Various	244	no date	Various	Various
Total in-situ proposed		**2 465**			
Total proposed		**4 570**			
Total producing + construction + proposed		**7 394**			

Sources: IEA analysis of public documents. Project names and operators may change as properties are traded. Capacity and start dates are indicative only; at the end of 2008 and beginning of 2009, many projects were put on-hold, as companies needed to reduce investments rapidly as the financial crisis deepened. Since the end of 2009, many of those projects have been revived, but exact new schedules are still being elaborated. Actual production is typically 20% below capacity due to maintenance, downtime and incidents.

In the upgrading approach, the bitumen is processed locally in an upgrader to produce synthetic crude oil (SCO), with a composition similar to that of conventional crude oil. This can be used by most refineries. An upgrader is basically a refinery with limited functionality; its role is to reduce the carbon content of the bitumen, either by removing carbon (coking), or by adding hydrogen (hydrocracking). In the former process the excess carbon is recovered as solid coke that can be sold as such or burnt to provide energy locally (with corresponding CO_2 generation). In hydrocracking, hydrogen (originally coming from steam or from natural gas) is added to the hydrocarbon chains, increasing the energy content of the oil. This requires energy inputs, emitting CO_2 in the process (if the energy comes from fossil fuels).

Upgraders require very large capital investment, typically in excess of $60 000 per barrel per day (b/d) of capacity. Most mining operations and a few in-situ projects have an associated upgrader. Smaller in-situ operations cannot justify this level of capital investment and use the dilution approach or send bitumen to off-site upgraders. New experimental technologies for small-scale upgraders, such as the Ivanhoe HTL (Heavy-to-Light) system, are being tested on a pilot scale and may allow more of the smaller in-situ projects to produce SCO, a higher value product. Integrating upgraders into the in-situ operation promises to reduce the need for natural gas to produce the steam required by CSS or SAGD processes. Indeed the Nexen/Opti Long Lake project has developed a process in which the heavy residues from the upgrader are gasified to provide energy for the steam generators. Availability of natural gas for the steam generators, otherwise, is one factor that could limit the growth of production from the oil sands, although the current gas glut, linked to the shale gas revolution in North America, has reduced these concerns. Certainly, producing 2.5 mb/d from in-situ SAGD technology, as projected for 2035 in the New Policies Scenario, with a steam-oil-ratio (the volumetric ratio of injected steam to produced oil) of 3 (typical of most projects today) would consume more than 28 bcm/year of gas, if all the energy required came from gas. This compares with total gas production in Canada of 161 bcm in 2009. Alternatives to the Nexen/Opti approach are to use nuclear, wind, or geothermal energy. Various projects along these lines have been discussed, but none are nearing the point of decision. New in-situ production technologies, such as the use of solvents, have the potential to reduce significantly the need for steam. THAI and other experimental in-situ combustion approaches can even do away with any steam usage, in a sense obtaining the equivalent energy from the bitumen itself.

Availability of capital and labour

Before the financial crisis of 2008-2009, many oil-sands projects were planned. Several new small companies were created, borrowing capital to develop those projects. This led to an overheating of the economy in the Alberta Fort McMurray region, with rapid cost inflation, labour shortages and saturated infrastructure. The financial crisis has put many projects on hold, with smaller companies now considering alternative business approaches. Some consolidation has taken place already. However, as the economic recovery takes hold, many projects are being revived. This raises the possibility that overheating, cost inflation, labour shortage and competition for capital may return to the oil sands region, leading to another down-cycle. By mid-2010, the number of workers living in camps in the Fort McMurray region had already passed the previous peak in 2008 and housing costs were at an all-time high. It is estimated that each 1 mb/d of new capacity requires an additional 20 000 direct employees. Attracting enough skilled labour for new projects is likely to be a challenge in the coming years, with a risk of cost overruns and project delays.

Assuming a weighted average of mining, in-situ and upgrader investment costs of $70 000 per b/d, the capital required to bring production from the oil sands to the level of 3.3 mb/d in 2025 projected in the New Policies Scenario is estimated at around $11 billion per year. This is in-line with what has been spent in the last few years, but remains large compared with total current investment in Canada of about $230 billion/year [4] (though it is relatively small as a percentage of capital investment in the global upstream oil and gas industry of around $470 billion in 2010). Growing investment in the oil sands by Japanese, Korean and, to a larger extent, Chinese oil companies (for example, in 2010, Sinopec acquired the 9% of Syncrude previously owned by ConocoPhillips) is likely to alleviate capital availability constraints.

CO_2 emissions

CO_2 emissions from oil-sands production are higher than those associated with conventional oil production for two reasons:

- Large amounts of energy are used to produce the steam for in-situ production or the hot water for bitumen/sand separation in mining operations. Most of this energy is currently supplied by burning natural gas.[5]

- Added CO_2 emissions per energy unit supplied result from the fact that the process starts from a carbon-rich fluid. These additional emissions come from energy used during upgrading (if supplied by natural gas), the use of coke (produced in upgraders), or higher energy use in refineries during processing of bitumen to produce the same amount of standard gasoline, diesel or naphtha.

Life-cycle emissions, taking account of all stages of the supply of oil, comprise:

- Emissions during the oil-production processes (upstream emissions).

- Emissions during upgrading and transport to the refinery gate.

The sum of these first two components makes up the "well-to-refinery" emissions.

- Emissions incurred in refineries and in transporting finished products to market.

---

4. Yearly investment in non-residential construction, machinery and equipment.

5. Several of the steam plants actually co-generate heat and electricity. This provides electricity with a lower carbon footprint than the average electricity mix of Alberta. We do not account for the corresponding reduction of CO_2 emissions in the discussion in this chapter, as the electricity mix of the region could change in the future, independently of the production from oil sands.

The sum of these first three components makes up the "well-to-tank" emissions.

- Emissions during the use of the products (typically combustion in an engine).

The sum of all four components makes up the "well-to-wheels" emissions.

Our analysis of independent estimates is presented in terms of well-to-wheels emissions of carbon dioxide equivalent (CO_2-eq) per barrel of crude (Figure 4.3). Box 4.2 discusses various ways to present life-cycle emissions. Emissions from oil-sands production vary with the maturity of the project: for example, in the early phases of SAGD projects, the steam-oil ratio can be very high and, therefore, the CO_2 emissions per barrel also high, but they tend to fall as the project matures. CO_2 emissions are, of course, different for mining projects and for in-situ projects. Our analysis shows that the well-to-wheels emissions of oils sands are slightly higher than for most other oils, the relatively-low difference being explained by the fact that emissions are dominated by the end-use (combustion) of the fuel. The difference ranges from zero to about 15%.

Figure 4.3 ● Well-to-wheels greenhouse-gas emissions of various oils

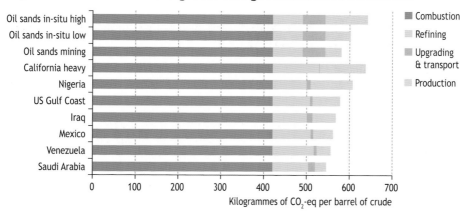

Note: Transport emissions are based on delivery to the United States. The bottom seven bars are examples of specific conventional crudes; they do not imply an average value for the countries of origin. The range of values for in-situ production of oil sands is indicated by the high and low cases.

Sources: Jacobs Consultancy (2009); TIAX (2009); US DOE (2009a); CERA (2009); Charpentier (2009).

In principle, emissions from production of oil sands ought to be compared with those from alternative sources of oil that oil sands might be displacing, such as conventional oil from Arctic locations or deepwater. These are likely to be at the high end of the current range of emissions for conventional oils. The production-related (upstream) emissions from those conventional sources can vary greatly, ranging from 10 kilogramme (kg) of CO_2-equivalent/barrel of crude for Arabian Light from Saudi Arabia to 100 kg CO_2-eq/barrel of crude for Nigerian Bonny Light. These figures compare to typical production-related emissions of about 80 kg CO_2-eq/barrel for crude from in-situ oil sands and 40 kg for oil-sands mining. It is arguably more meaningful to compare the well-to-tank emissions, *i.e.* to include emissions from upgraders and refineries (for oil sands, only the sum of the two is meaningful, as the degree of upgrading before transport to refineries can vary). Well-to-tank emissions range from 100 to 190 kg CO_2-eq/barrel for conventional oil, compared with typical values of 160 kg for oil-sands mining and 200 kg for oil sands in-situ production (Californian heavy crude generates even higher emissions than oil sands). It is clear that some regions produce conventional oil with CO_2 emissions similar to those of unconventional oil from Canadian oil sands. The large CO_2 emitters are primarily regions, such as Nigeria, where large amounts of associated gas are flared, due to the lack of markets for such gas. Reduced gas flaring would lead to a convergence of CO_2 emissions from conventional oil production towards the value typical of the more mature production areas, though increased production of heavy conventional oil resources would offset this factor to a small degree.

Box 4.2 ● Life-cycle emissions

Life-cycle emissions analysis can be carried out for specific oil products, such as gasoline or diesel, or for the barrel of oil as a whole. Analysis of a product looks at emissions incurred during its production and use, including emissions from the production of the crude needed as feedstock. They can be reported in kilogrammes of CO_2-eq per barrel of diesel or gasoline or per mile driven, or kg of CO_2-eq per megajoule (MJ) of product. Actual emissions depend on the final product in question; for example, they differ for gasoline and diesel. This is useful when looking at fuel standards. One can compare diesel coming from different crude feedstocks: those coming from oil sands feedstock typically have 10% higher well-to-wheels emissions than those coming from average conventional oil.

Analysis of crude oil looks at emissions incurred during production and subsequent transformation and use of a barrel of crude. It is also reported in kg of CO_2-eq per barrel or per MJ of crude. Crudes of different origins differ according to the emissions incurred during production and refining, but they also differ in emissions coming from end use, because different crudes give different product slates at the refinery exit door. For example, bitumen from Canadian oil sands could have low life-cycle emissions because it produces a lot of coke that is used for landfill (as is sometimes practised for coke produced in upgraders) rather than burnt. Similarly a light oil could have high emissions because it produces mostly gasoline and middle distillates and little tar or petrochemicals. So life-cycle emissions comparisons between different crudes can be difficult to interpret. Instead of a full life-cycle analysis, we present emissions per barrel of crude, assuming the emissions from end-use are the same for each crude and equal to those of the combustion of an average crude (Figure 4.3). A similar approach is used in *CERA (2009)*.

Of course, technological improvements are likely to reduce CO_2 emissions per barrel of oil produced from oil sands over the projection period, for example through more efficient use of steam in SAGD or CSS, solvent-based technologies, replacement by nuclear or renewable energy of the natural gas used to supply the energy for steam generation, in-situ combustion techniques and carbon capture and storage (CCS) of the concentrated CO_2 emissions from upgraders. Some of these technologies (for example, nuclear power or CCS) could bring the CO_2 emissions close to zero. However, they require large investments and long construction times and are therefore likely to have a significant impact on emissions only towards the end of the projection period. Other approaches, mainly involving efficiency improvements, will undoubtedly be implemented progressively and achieve reductions in production-and-upgrading related emissions.

Taking account of the evolution of the mix between mining and in-situ, we project the average differential in well-to-tank emissions between oil sands and conventional oil to fall from about 50 to about 40 kg CO_2-eq/barrel over the period 2009-2035. This represents "extra" CO_2 emissions of 60 Mt CO_2 annually for the 4 mb/d of oil-sands production projected in the New Policies Scenario for 2035 (vis-à-vis conventional oil production).[6] Although this is not large compared with current worldwide CO_2 emissions of 30 Gt/year, it is significant on the scale of Canada's emissions of 550 Mt/year and clearly creates a significant national challenge. The extra 60 Mt would be equivalent to 4% of the projected US transport-related emissions of more than 1.5 Gt CO_2 in 2035.

Our estimated emissions differential of 40 kg CO_2-eq/barrel can be translated into an extra "cost" for bitumen from the oil sands. At $50/tonne of CO_2 (the projected price of CO_2 in 2035 in the New Policies Scenario) the higher emissions represent an extra $2/barrel, which does not significantly affect the economics of oil sands at the oil prices assumed in this scenario. At $120/tonne of CO_2 (the projected price in the 450 Scenario in 2035) the additional production cost would be $5/barrel, which, coupled with the reduced oil price of the 450 Scenario, would make the economics of new oil-sands projects marginal and cast doubt over the most expensive projects. However, it is likely that the cost of CCS with CO_2 captured from some concentrated sources, such as the upgraders or the hydrogen plants, would be significantly less than $120/tonne (although it is early to attempt estimates, figures of around $50/tonne have been suggested). Pilot projects are planned, with support from the Alberta public authorities and the Canadian government; in particular the Quest project will capture 1 Mt of CO_2 per year from the Shell Scotford upgrader. Part of the CO_2 will be stored in a nearby deep aquifer and part may be made available for CO_2 enhanced-oil-recovery (EOR) projects elsewhere in Alberta. In another pilot, CNRL plans to capture CO_2 at its hydrogen plant and use it in management of tailings ponds at its Horizon mining facilities, effectively storing it as carbonate mineral. Success of these pilot projects in the next three to four years could lead other upgraders to follow the same approach, with CCS removing up to 20 Mt/year from oil-sands emissions by 2025.

Water usage

Another potential constraint on future production from Canadian oil sands is the availability of water. Mining operations use hot water to separate the bitumen from

---

6. For the purpose of projecting CO_2 emissions from conventional oil production, we assume an average of 30 kg CO_2eq/barrel over the projection period.

the sand. Current operations use 2 to 3 barrels of water per barrel of bitumen produced (bw/bo), net of recycling of tailings ponds water. The water is typically taken from local rivers. Water can also be used during the upgrading process, bringing the mining plus upgrading use to 3 to 5 barrels of water per barrel of SCO.

In-situ operations use water to produce steam for CSS or SAGD processes. Typical usage is 8 bw/bo produced, but a large part of the water is recycled, so that for ongoing production the average net water requirement is closer to 1 bw/bo. Currently, about 50% of that water comes from freshwater, but more and more projects take water from underground saline aquifers.

Water extraction from local rivers is regulated and limited to 3% of river flow (and less at times of low water flow); but even that amount is considered by some to be potentially damaging to the river ecosystems. Clearly, large increases in production from the oil sands will depend upon significant reductions in river-water usage. Reductions in water needs could come from:

■ An improved steam-oil ratio in SAGD/CSS production.

■ Increased production from steam-less processes, such as primary, solvent-based, or in-situ combustion.

■ Increased reliance on underground saline aquifers. The impact of large-scale pumping out of shallow saline aquifers has not yet been fully assessed and more studies are underway to ensure this can be done without harmful ecological effects.

■ Increased recycling of water. For example, the possibility of recycling water from mining operations into the in-situ operations is being considered. Improvements in tailings management, such as more rapid separation of solids and water, would ease recycling.

In addition to water usage, pollution of rivers and water tables has been attributed to production of oil sands. Rigorous monitoring is required by regulation and performed. However, recently, abnormal concentrations of (unregulated) polycyclic aromatic compounds have been detected downstream of mining operations and even near some in-situ operations (Kelly *et al.*, 2009). These compounds, possibly toxic to water wildlife, are naturally present in the outcropping oil sands but may be released during extraction operations and land disturbance. Proper monitoring and prevention of seepage from tailings ponds, or bird deterrence near tailing ponds, are required components of proper protection of ecosystems.

Land usage

Most of the Canadian oil-sands deposits are located in the environmentally sensitive Canadian boreal forest. The total oil-sands area occupies about 140 000 square kilometres (km^2) of northern and eastern Alberta. The Alberta boreal forest occupies about 380 000 km^2 (part of the 3 million km^2 total Canadian boreal forest).

Mining operations have a large impact on the landscape. A typical mine clears about 80 km² of land per billion barrels of production. Sustained mining production of 1.5 mb/d over 20 years, as projected in the New Policies Scenario, would require about 900 km² of land to be cleared. Mining companies are required to reclaim the land after 20 years, though there is controversy over the impact on ecosystems even after reclamation. Mining activities in the oil sands have so far disturbed 602 km², of which 65 km² have been reclaimed (and only 1 km² has so far been certified as reclaimed by the regulatory authorities).

In-situ projects have a smaller footprint, but still require some clearing for basic infrastructure, including roads, landing strips, steam plants, steam lines and well pads. Estimates range from 10 to 15 km² per billion barrels. Sustained production of 2 mb/d for 20 years, again as projected in the New Policies Scenario, would, therefore, disturb about 200 km². As they tend to be more geographically dispersed, a large number of small projects could give rise to significant concerns for ecosystems, through forest fragmentation and wildlife disturbance.

Prospects for reducing the amount of land disturbed are limited, as this is more linked to the density of the resources per km² than to the technology used for production. Efforts are likely to focus on accelerated reclamation and improvements in reclamation technologies in order to better reconstitute the original ecosystems. Some aspects of land disturbance, such as the tailing ponds created by mining operations, could be alleviated by novel technologies to accelerate the separation of solids and water, a number of which are being tested.

Venezuelan Orinoco Belt

With the assumption of no interference from political events, the production of extra-heavy oil from the Orinoco Belt in Venezuela is projected to grow to over 2.3 mb/d in the New Policies Scenario (Figure 4.4).[7] The growth in output to 2020 could be derived from current capacity and additions that have already been announced. In the Current Policies Scenario, with its larger demand for oil, the Orinoco could compensate for slower growth in Canadian oil-sands production, if Canadian projects were delayed by environmental concerns, provided Venezuela was more ready to accept international capital. Total Venezuelan production does not increase as strongly, as the rise in extra-heavy oil production is offset by the decline in ageing conventional oil fields.

The Venezuela Orinoco oil belt is the second-largest deposit of extra-heavy oil (with an API gravity of less than 10) in the world, after the Canadian oil sands (Table 4.2). The amount of oil in place is estimated to be 1.3 trillion barrels, over an area of about 50 000 km². Although the deposits are deeper than in Canada, typically 500 to 1000 metres, and therefore the oil is somewhat less viscous at

---

7. There is uncertainty on the status of Orinoco production with respect to future OPEC production quotas. We have assumed that it would be included in future Venezuelan quotas as per current agreements, but a different approach might allow larger production growth.

reservoir temperatures (typically of about 55°C), it is still not generally amenable to conventional production techniques. The primary recovery rate with vertical wells is less than 5%; multilateral horizontal wells allow a recovery rate of 10 to 15%; higher recovery rates require thermal methods, such as Cyclic Steam Stimulation or SAGD.

Figure 4.4 ● *Venezuelan oil* production by type in the New Policies Scenario*

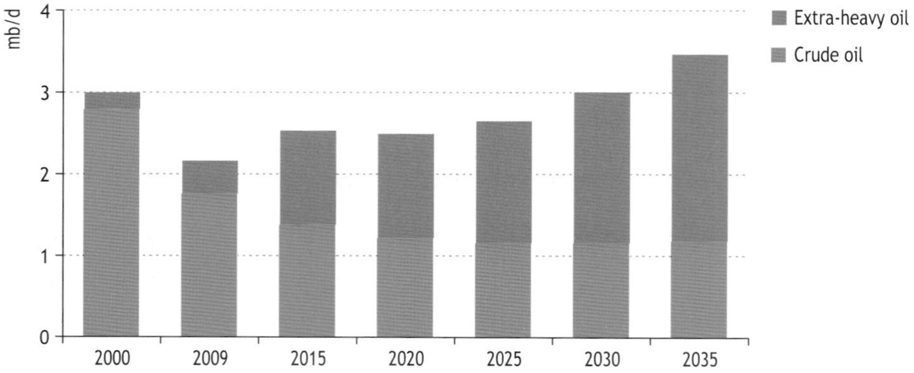

*NGLs are not included.

A recent evaluation by the USGS estimated the technically recoverable oil from the Orinoco province to be about 500 billion barrels. Although the USGS has not given any estimate of *economically* recoverable resources, it is likely that a large fraction of that volume is economically recoverable at current prices. Petroleos de Venezuela (PDVSA), the national oil company, launched in 2006 the Magna Reserva project to certify reserves in the Orinoco. By early 2010, 133 billion barrels had been certified, though the *Oil & Gas Journal* currently reports only 60 billion barrels. PDVSA expects around 230 billion barrels to be proven by the end of the project.

Orinoco production started in earnest at the beginning of the 2000s, with several projects contributing to total production of about 700 kb/d in 2005, about two-thirds from primary production from vertical or multilateral horizontal wells and the rest produced with steam stimulation. Capacity remains near that level (Table 4.5), but production fell to around 400 kb/d in 2009 (see Table 4.1, above). Early projects emulsified the extra-heavy oil with water to create a mix, dubbed Orimulsion, which could be transported by pipeline and used as fuel oil in power generation; but all of the production is now upgraded into synthetic crude oil (SCO). Several new projects have been announced which, collectively, would add about 2.3 mb/d capacity by around 2017. Taking into account project lead times and delays, total capacity is unlikely to exceed 2.0 mb/d by 2020. Most of the announced projects involve the construction of upgraders, although they are not always large enough to treat the full production. Deliveries will be a mix of SCO and extra-heavy oil diluted with light hydrocarbons.

Table 4.5 ● Venezuelan Orinoco Belt extra-heavy oil projects

Project name	Foreign partners	Status	Capacity (kb/d)	Planned start
PetroAnzoategui (PetroZuata)	None (100% PDVSA)	Producing	120	n.a.
Petrocedeno (Zuata)	Total (30%)/Statoil (10%)	Producing	200	n.a.
Petropiar (Hamaca)	Chevron (30%)	Producing	190	n.a.
Petromonagas (Cierro Negro)	BP (17%)	Producing	110	n.a.
Sinovensa	CNPC	Producing	80	n.a.
Total producing			**700**	
Junin 2	Petrovietnam	Announced	200	2012
Junin 5	ENI	Announced	240	2013
Carabobo 1	Repsol/India/Petronas	Announced	480	2015
Carabobo 3	Chevron/Inpex/Mitsubishi/ Suelopetrol	Announced	400	2015
Junin 4	CNPC	Announced	400	2017
Junin 6 (Petromiranda)	Russian companies	Announced	450	2017
Junin 10	Total/Statoil	Under negotiation	200	
Total proposed			**2 370**	
Total producing + proposed			**3 070**	

Note: Dates and production capacity are somewhat uncertain, as PDVSA, which owns a majority interest in all projects, does not publish detailed plans.

In principle, production from the Orinoco will face similar challenges to those of in-situ Canadian oil-sands projects, notably the availability of energy for steam generation, the availability of water and CO_2 emissions. But there is very little information available on current performance and future plans for reducing the environmental impact. This is an area in which open, joint work between PDVSA and environmental non-governmental organisations would be beneficial.

Little recent information is available on the costs of new developments in the Orinoco belt. For steam stimulation projects, technologies are similar to those used at Canadian oil sands in-situ projects, so it can be assumed that the capital and operating costs are similar (Canadian capital costs are around $30 000 to $40 000 per b/d of capacity) (Table 4.3). These costs are roughly in line with the capacity and investment costs quoted at the signing of recent new joint ventures, such as the Junin 6, or Carabobo 1 and 3 agreements. Primary production with multilateral horizontal wells, which gives higher recovery rates than in Canada, due to lower oil viscosity, is significantly cheaper. So, overall, assuming a mix of primary and steam stimulation, new projects would be expected to cost on average about one-third less than Canadian oil-sands projects on a per-barrel basis.

Other extra-heavy oil provinces

Heavy oil has been produced in other parts of the world for many years, using either primary or thermal techniques (steam stimulation). For example, the Kern River heavy oil area in California has used steam stimulation since 1965, producing more than 1 billion barrels from this technology, and the area still produces around 250 kb/d. The recovery rate in this heavy oil field, typically around 5% with primary production alone, can reach 50% to 70% with steam stimulation. A similar situation applies in the Duri field in Indonesia, the largest steam-stimulation project in the world, which has produced close to 2 billion barrels since 1975 and still produces around 200 kb/d.

Heavy oil projects are active or planned in Brazil, in the North Sea, in the Neutral Zone between Saudi Arabia and Kuwait (where Chevron plans production of up to 300 kb/d from steam enhanced oil recovery in the Wafra field) and several other places in the world. China and East Venezuela also have some active steam injection projects. The Pungarayacu heavy oil field in Ecuador may have close to 20 billion barrels of oil originally in place, according to operator, Ivanhoe Energy, which plans to apply its small scale upgrading technology to development of this remote field.

In the United States, there are deposits similar to, though much smaller than, the Canadian oil sands, in Utah (with 16 billion barrels of oil originally in place). Congo, Madagascar and a few other countries have small projects in "oil-sands-like" deposits. However, none of these are large enough to have a significant impact on world oil supply. For example the Bemolanga oil sands in Madagascar could produce 200 kb/d, with mining technology, at an oil price above $80 per barrel, according to the operator, Total.

Russia is thought to have several hundred billion barrels of technically recoverable extra-heavy oil and bitumen. The large bitumen resources thought to be present in Eastern Siberia are poorly known and difficult to exploit, due to their remoteness from infrastructure. Some of the reported heavy oil is, in fact, medium-viscosity and is exploited by conventional methods. In the more viscous reservoirs, and some of the bitumen deposits in Tatarstan, there have been pilot projects with steam stimulation, more recently with SAGD technology, but no clear plan exists for large scale development. Current economics favour the exploitation of large conventional oil resources. A similar situation exists in Kazakhstan. The Tatarstan Republic region of Russia, which is thought to have more than 20 billion barrels of extra-heavy oil and bitumen ultimately recoverable resources and an economy highly dependent on very depleted conventional fields, is the most likely location for the start of larger scale development. China has some heavy and extra-heavy oil reservoirs which are yet to be tapped, with probably a total of a few billion barrels of recoverable oil.

The projections for these other countries are included in the conventional oil projections in this *Outlook*, as there is a continuum and no clear boundary between the categories (Figure 4.5). Only Canadian oil sands and Venezuela Orinoco extra-heavy oil have been separated out as unconventional oil on the basis of the very large resources involved.

Figure 4.5 ● Continuum from conventional to unconventional oil resources

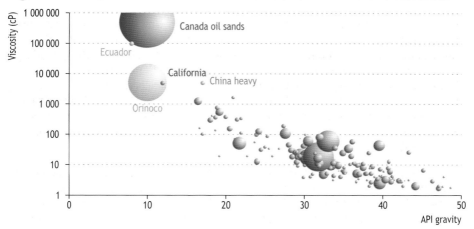

Note: The size of the bubbles indicates recoverable resources. Reservoirs with similar properties in each geographical area have been grouped; the smallest bubbles each represent approximately 1 billion barrels of recoverable resources.

Oil shales

Oil shales are fine sediments containing kerogen (Box 4.3). Because they are the source rocks for most conventional oil reservoirs, they are found in every oil province in the world. However, most of them are too deep to be exploited economically, as exploitation involves heating up the kerogen to temperatures between 350°C and 450°C in order to transform it into oil. So oil shales are generally considered possible sources of (unconventional) oil only when they are at shallow depth, though there can be exceptions, such as the Bazhenov shale in Russia (Box 4.4). The term shale oil is used to designate oil that has been produced through "retorting", *i.e.* industrially heating up oil shales, whether done in-situ or after mining the shale rock.

There may be the equivalent of more than 5 trillion barrels of oil in place in oil shales around the world (including deeper shales) of which more than 1 trillion barrels may be technically recoverable (Table 4.6 includes only oil shales at shallow depth). How much may be economically recoverable is not known. The Green River area in the United States where Colorado, Utah and Wyoming meet is thought to contain more than half of all the recoverable oil shale resources in the world, around 800 billion barrels, and therefore has received the most attention.

Oil shales have been exploited for centuries, mostly as a low-quality fuel for heating. Estonia has long mined oil shales for power generation. Worldwide, only a small amount (15 kb/d) is processed into liquid oil, in Estonia (4 kb/d), Brazil (4 kb/d) and China's Fushun shale oil plant (7 kb/d). Extensive studies were made of the US Green River area and some pilot projects launched in the 1970s and 1980s, when this resource was seen as a potentially important source of domestic oil supply. However, during the period of low oil prices from the early 1980s to the early 2000s, all projects were shelved; only in the last few years have some feasibility studies and pilot projects been resumed. Australia had a significant project planned in the Stuart shale near Gladstone in Queensland (up to 200 kb/d in the third phase) in the early 2000s, following a pilot plant in the 1990s, but it was shelved due to concerns about damage to the environment and rising costs.

Box 4.3 ● When oil from shales is not shale oil: the case of the Bakken

The term, oil shales, is used to designate very fine grained sediments with a high content of kerogen, be they clays, marls or carbonates. However, such rock formations sometimes also contain oil. This can happen when at least part of the oil produced by the natural maturation of kerogen under deep burial has not been expelled to higher permeability sedimentary rocks, or when the shale, normally very impermeable, is fractured and can itself serve as an oil reservoir. When this is the case, oil shales can produce oil in exactly the same way as conventional, low permeability, fractured reservoirs. This is the case, for example, in the Bakken Shale in Montana and North Dakota in the United States and Saskatchewan in Canada. For the purpose of this report, such reservoirs are classified as conventional. They tend to be relatively localised and have steep decline rates, but they can contain significant resources: the Bakken, for example, contains 4 billion barrels of technically recoverable oil.

Gas shales are analogues of such reservoirs, containing gas rather than oil. The recent "shale-gas revolution" in the United States has shown that such gas shales are quite common and can be economically exploited. This has triggered renewed interest in exploring oil shales for oil (rather than for their kerogen). Occidental Petroleum, for example, recently announced the acquisition of very large oil-shale acreage in California for the purpose of looking for oil-bearing shales similar to the Bakken. The Eagle Ford shale in Texas is also experiencing a boom in exploration for oil. The term "light tight oil" is emerging to describe these types of resources.

Table 4.6 ● Oil shale resources by country (billion barrels)

	Oil originally in place	Technically recoverable
United States	≥ 3 000	≥ 1 000
Russia	290	n.a.
Dem. Rep. of Congo	100	n.a.
Brazil	85	3
Italy	75	n.a.
Morocco	55	n.a.
Jordan	35	30
Australia	30	12
China	20*	4
Canada	15	n.a.
Estonia	15	4
Other (30 countries)	60	20
World	≥ 3 500	n.a.

* A recent Chinese study from Jilin University, performed as part of the Chinese National Petroleum Assessment, reports 350 billion barrels in place of which 80 billion is recoverable.
Sources: BGR (2009); Dyni (2005); USGS (2009b); USGS (2010).

Production methods

Oil-shale deposits near the surface can be mined, in a very similar way to mining in the oil sands: the "ore" (kerogen-rich shale) is then heated in industrial retorts, where the kerogen is transformed into oil and gas. The left-over shale is disposed of or used for land reclamation. Like all strip-mining techniques, land use is controversial, but the yield in barrels per acre can be about 10 times bigger than in Canadian oil sands mining, so the area of land disturbed will be less for a given level of production. This is primarily because deposits are thicker (which of course also results in deeper land disturbance, with possibly more impact on ground water).

Somewhat deeper deposits, typically at depths from about 100 to 700 metres in the Green River area in the United States, require in-situ retorting or underground mining. Various technologies are being investigated for in-situ retorting, using very dense well networks (typically one well every few metres), with some wells used for heating with steam or electrical power and others for producing the oil and gas. Ten pilot projects are under investigation in this area (Table 4.7).

Table 4.7 ● Proposed pilot shale-oil projects in the Green River area in the United States

Companies/projects	Basin	Partners
Shell/Mahogany (4 projects)	Piceance	none
Chevron	Piceance	none
EGL	Piceance	none
AMSO	Piceance	Total (50%)
OSEC/White river mine	Uintah	Mitsui/Petrobras
Enshale	Uintah	Bullion Monarch Mining (parent)
Red Leaf/Ecoshale	Uintah	none

In addition to the United States, there are pilot projects planned in Canada and Jordan. The Jordanian project, led by the Estonian company Eesti Energia, aims for a capacity of 38 kb/d in 2017. Plans in China include expansion of the existing Fushun plant to 15 kb/d and several small (3 to 5 kb/d) pilot mining projects in other provinces. A joint venture with Shell has been announced, an in-situ pilot using the technology developed by Shell in its Green River property in the United States, though no date nor capacity have been reported.

Environment

There have been fewer studies about the environmental issues associated with oil shales than those about Canadian oil sands. Yet the challenges are likely to be very similar. Retorting, whether done at the surface or done in-situ, requires large amounts of energy to heat the oil shales to the required temperature of 350°C to 450°C. The energy required typically represents about 20% to 25% of the heating value of the produced oil in mining and surface retorting production methods and for in-situ projects it could reach 50%, though there are very few published analyses (Brandt, 2008). Most of this energy, however, can be provided by burning the oil shale itself in a surface retort, or producing gas in in-situ retorting that is then re-used to provide the energy.

As a result of the large energy needs of shale oil production, CO_2 emissions are also very large, unless the energy can be provided by renewable sources or the CO_2 can be captured and stored. Estimates run from 180 to 250 kg CO_2-eq/barrel of produced crude (Brandt, 2008). Development of oil shales is still in its infancy from a technological point of view, so some reductions can be expected in the future. However, the very nature of the process is likely to leave a differential with conventional oil of the order of 150 kg CO_2-eq/barrel. At a price of CO_2 of $50/ tonne, as in the New Policies Scenario in 2035, this represents $7.50/barrel, which significantly increases the required break-even oil price for these resources. CCS is probably the best option for mitigating these large emissions. The CO_2 sources would be localised, so capture should be possible; and CO_2 enhanced oil recovery in the Rocky Mountains area could provide a natural market for the CO_2 from Green River shale projects.

The rate of water use during retorting is estimated at two barrels of water per barrel of oil produced. Some recycling is probably possible, though the technology has not yet been deployed. The availability of water to sustain large scale production is likely to be a constraint in the Green River area, a relatively dry environment. Concern over pollution of surface and underground water is even greater than for Canadian oil sands, as the Green River deposits are much thicker than the oil-sands deposits. Shell has worked on a "freeze-wall" technique, in which the water table is fully isolated from the shale submitted to in-situ retorting by a frozen wall surrounding the entire volume of shale. But this type of technology is still in its infancy and it remains to be seen whether it can achieve the objective of full isolation.

Land use for shale mining should be less than that involved in exploiting the Canadian oil sands, because of the higher hydrocarbon content per acre due to the thick layer of kerogen-rich shales in the Green River area. But the need for proper land reclamation will be just as strong. In-situ production may have similar land disturbance effects to in-situ projects in Canadian oil sands, the large number of wells required for heating the shale formation offsetting any benefit from the greater concentration of resources. Large-scale development of the Green River deposits in the United States is likely to face strong opposition on environmental grounds.

Costs and production prospects

Cost estimates based on the various pilot projects in the pipeline in the United States indicate that oil shales investment and operating costs should be similar to, and possibly even slightly lower than, those of Canadian oil sands, with commercial exploitation possible at oil prices of the order of $60 per barrel at current costs. Adding a CO_2 penalty corresponding to 150 kg CO_2/barrel (compared to conventional oil) and taking into account the likely link between costs and the oil price (Biglarbigi, 2009) makes oil shale exploitation economic in both the Current Policies and New Policies Scenarios, though this is the most costly of our unconventional fuel sources, together with CTL. In the 450 Scenario, the lower oil prices and higher prices of CO_2 make oil shales marginal from an economic point of view. Costs in China have been reported to be much lower — less than $25 per barrel (Qian, 2008) — but there is no recent confirmation of this figure.

There is long way to go from pilot projects producing a few thousand b/d to an industrial scale activity able to produce quantities that are significant in terms of world oil supply. For example Shell has indicated it will not take a decision on a commercial scale project in the Green River area before 2015 and such a project would then probably take 10 years to reach large scale operation, say in excess of 100 kb/d. These long time scales, together with the small number of projects being piloted, explain why we foresee only slow growth of oil shale exploitation even in the New Policies Scenario; oil shales begin to play a small role only at the end of the projection period (Figure 4.6).

Figure 4.6 ● *Shale-oil production by country in the New Policies Scenario*

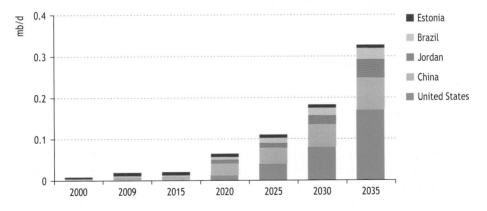

Faster exploitation of oil shales in the United States could result from strong government policies, motivated by energy security. However, even a massive government programme is unlikely to lift production above 1 mb/d by 2035, still amounting to only a fraction of projected US imports of close to 8 mb/d. China could see faster development of its oil-shale industry, but currently planned projects are all on a small scale, suggesting that slow growth is likely there as well.

Box 4.4 ● Exploiting deep shales: the case of the Bazhenov formation in Russia

Oil shales are generally considered of interest only when they are at shallow depths. Most analyses of resources contained in oil shales include only such shallow deposits. Deeper source rocks, even if they contain a very large amount of kerogen, are generally not considered exploitable economically. However there are still places where they could play a significant role. An example is the Bazhenov shale in western Siberia. The Bazhenov is the source rock for all the oil fields of western Siberia. It underlies the entire western Siberia basin, an area of about 1 million km². It is estimated to contain kerogen corresponding to 1 trillion barrels of oil. However it lies at depths from 2 500 to 3 000 metres, too deep for mining, but also too deep to be economically recoverable with the in-situ recovery techniques being developed in the US Green River area.

But Russia has a unique geography: most of its oil and gas resources lie in remote regions, scarcely populated and with a harsh climate. Development of such resources requires large investments in infrastructure, such as housing, roads, air strips, water supplies and energy supplies. In western Siberia, such infrastructure was developed in the 1970s and 1980s, at the time when the Soviet Union began to develop the western Siberian oil fields. The conventional exploitation of the basin is now mature and decline will soon set-in. To maintain its oil production, Russia has started to explore and develop the huge area of eastern Siberia. However this is an even more remote province, which will require very large investment in infrastructure to build up significant production. As a result there is considerable interest in developing technology in western Siberia that would allow exploitation of the Bazhenov oil shale formation, which would make use of the existing infrastructure and extend the life of the basin as a producing area.

So how could it be done? Probably the most promising approach involves in-situ combustion, similar to the THAI or COGD technologies being piloted in the Canadian oil sands (see the oil sand section earlier in this chapter). How well such technology could work in oil shales is unknown at this time, but pilot projects are likely to be undertaken in the next few years. However, even if they are successful, large-scale implementation is probably a couple of decades away, allowing for the time necessary to build-up experience from small-scale pilots and then scaling-up the process.

Some parts of the Bazhenov formation are fractured and contain oil in addition to kerogen, like the Bakken shale in the United States. These localised reservoirs are likely to be exploited earlier, with Bakken-shale-like horizontal wells, prolonging the life of some of the oil towns of western Siberia.

Coal-to-liquids

Although economical at assumed oil prices in each of the three *WEO* scenarios, oil derived from coal-to-liquids processes (CTL) and oil shales is the most expensive of the unconventional oil sources. Provided carbon capture and storage (CCS) is accepted (both by regulators and by public opinion), CTL is likely to develop faster than oil shales because the technology is more mature and less risky and the environmental impact is less controversial: the plants will mostly be located near active coal mines that are already being exploited, so land use is likely to be more acceptable to the local communities. Coal-and-biomass-to-liquids (CBTL) with CCS, with its smaller carbon footprint, is particularly attractive. Taking into account the current slow build-up of announced projects, the time it takes to approve large investments and the time required to build large scale plants, most of the growth in CTL in the New Policies Scenario will take place in the second half of the projection period (Figure 4.7). The Current Policies Scenario, which assumes higher oil prices, sees faster growth (Table 4.1). The 450 Scenario follows a trajectory very similar to that of the New Policies Scenario: although oil demand is weaker, demand for coal is even more reduced, making the price differential between oil and coal larger and therefore making it more economically attractive to build CTL plants; in addition, acceptance of CCS is assumed to be faster.

Figure 4.7 ● Coal-to-liquids production by country in the New Policies Scenario

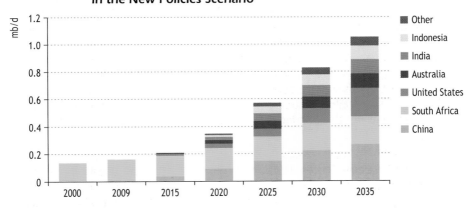

Note: Production is assumed to average 80% of installed capacity.

CTL technology

CTL, a process involving synthesising liquid hydrocarbons from coal, has a long history. First used industrially in Germany during the Second World War, it was then extensively applied in South Africa. Sasol started its famous CTL plant there in 1955 and has since produced more than 1.5 billion barrels of synthetic liquid fuel.

There are several routes to turn coal into liquid hydrocarbons. The most popular starts with gasification of the coal to turn it into "syngas", a mixture of hydrogen and carbon monoxide. This is similar to the old "town gas" that was used before natural gas became widely available. The same process of gasification is used in integrated gasification combined-cycle (IGCC) power plants. In a second step, the syngas is turned into a liquid hydrocarbon, typically high quality diesel, using the Fischer-Tropsch catalysis technique with an iron or cobalt catalyser. This was the technology used in Germany during the Second World War and it is still used by Sasol in its 160 kb/d capacity plant in South Africa. An alternative to the second step is to first turn the syngas into methanol and then the methanol into gasoline. This process was piloted by ExxonMobil; a plant operated in New Zealand for ten years before closing. Methanol can also be converted to DME (dimethyl ether), which is being commercialised in Asia as a liquified petroleum gas (LPG) blend stock and being developed as a diesel alternative, or used as petrochemical feedstock as in the Baotou plant of the Shenhua coal company in China.

Finally there is the "direct" route, in which the coal is directly reacted with hydrogen, in the presence of suitable catalysers, to produce liquid oil that can be used in a standard refinery to produce commercial hydrocarbon products. This is the technology used by the Shenhua coal company in China in its plant in Inner Mongolia. The plant has a nameplate capacity of 24 kb/d, but is still in the start-up phase. Similar technology was also used in Germany during the Second World War.

Most projects under study plan to use one of the two indirect routes, since the technology is more mature. Even though no new plant has been built recently, there is considerable experience with the key components (gasification and Fischer-Tropsch) in other applications (power generation, GTL, chemical plants). It also provides more flexibility: syngas can be used for power generation, as chemical feedstock and to produce methane, in addition to being used as an input to the second stage of liquid hydrocarbon synthesis. As the gasification unit represents the largest capital investment, this offers a useful diversification of the investment risks. There is also some flexibility in the feedstock to the gasification process: biomass can be mixed with the coal in CBTL (coal and biomass-to-liquids), or even used by itself (BTL, biomass-to-liquids), without major changes to the equipment.

Projects and economics

A number of projects have been announced in the past five years, some ten in the United States, half a dozen in China, a few in Indonesia, India and Australia, one in Canada and a second plant in South Africa. However many of them are in a very early pre-feasibility phase and little information is available about plant capacity and timing. Several have also been put on hold, due to uncertainty about oil prices and CO_2 costs. Several projects announced the intended use of CBTL. The most advanced seem to be:

- The Clinton project in Australia, with a capacity of 13 kb/d scheduled for 2015.

- The Felton/Ambre project in Australia, with a capacity of 18 kb/d, scheduled for 2014, based on the ExxonMobil methanol-to-gasoline process.

- The DKRW Medicine Bow project in the United States, with a capacity of 20 kb/d expected in 2015, also based on the Exxon-Mobil process.

- The Rentech Natchez project in the United States, with a capacity of 30 kb/d.

- Three 4-kb/d projects in China, in Lu'An and Yitai with the Fischer-Tropsch route and the ExxonMobil-Jincheng Anthracite Mining Co (JAMG) project using the methanol-to-gasoline route. These are all in the start-up phase.

The largest projects are those being investigated by Sasol, one with a possible site in China, one in India, one in Indonesia and a second site in South Africa. Each would have 80 kb/d capacity. No dates for construction or operation have been announced. Assuming four to five years for the feasibility study and design, followed by five years for construction and start-up, these plants could come on stream around 2020. Monash Energy (a Shell/Anglo-American joint venture) has announced a 60 kb/d capacity project in Australia, with start of construction possible by 2015. Russia is considering a large project in collaboration with the Chinese coal company Shenhua.

Essentially, all of the announced projects assume capture and storage of CO_2 emissions (more on this below). Uncertainty surrounding the regulatory framework for CCS is probably one of the key reasons for the slow pace of development of new projects.

The Linc Energy Chinchilla project in Australia is also worth mentioning. It combines Underground Coal Gasification (UCG) to produce the syngas, with a Fischer-Tropsch plant to transform the syngas into liquid hydrocarbon. The project aims at a capacity of 20 kb/d of liquid hydrocarbons. In principle, UCG provides the syngas at much lower capital costs and allows deeper, un-mineable, coal beds to be exploited. UCG has been piloted in various places in the world, with mixed success; although in principle very attractive, it is considered an immature technology (see Box 6.1 in Chapter 6).

Because no large plant has been built recently, there is a range of estimates for the capital costs associated with CTL technology: capital costs range from $80 000 to $120 000 per b/d of capacity. Syngas/FT plants offer significant economies of scale and are in this range of capital costs only for capacities above 50 kb/d. The capital costs of plants using the methanol and direct routes are less dependent on size.

The equivalent oil price required to make CTL economical is in the range $60 to $100/barrel, depending on the location of the projects (China being in the lower part of the range) and the cost and quality of the feedstock. These prices include CCS, which typically represents only a small addition to the cost, as explained below. CTL is economical at the assumed oil price trajectories in all three scenarios even though, together with oil shales, it constitutes the most expensive source of unconventional oil in our models.

Environment

CO_2 emissions are the main disadvantage of CTL. These emissions are different from those of a coal-based power plant. Basically, to turn coal into diesel or gasoline means adding hydrogen and making it react with the coal to form hydrocarbon chains. The CO_2 emissions arise primarily from generating the hydrogen. In the direct CTL approach, it is in principle possible to generate the hydrogen using renewable energies, although this may be expensive.

In the indirect routes, it is intrinsic to the syngas generation process that the energy comes from the coal itself. CO_2 is produced with the syngas. However it must be separated from the syngas prior to the Fischer-Tropsch (or methanol) process. So the bulk of the CO_2 is, in any case, captured. This is why CCS is a relatively inexpensive addition: only transport and storage need to be added, and these are normally much less expensive than capture. Estimates for the cost of adding CO_2 purification to a CTL plant, as required for sequestration, range from $3 to $5/barrel of oil produced.

Various studies have shown that, without CCS (or with conventional hydrogen production for the direct CTL route), the well-to-wheels emissions of CTL are 80% to 100% higher than those of conventional oil. This is why most proposed projects include CCS from the start. With CCS, a CTL plant can produce diesel with well-to-wheels emissions 5 to 10% lower than conventional oil (as production and refinery emissions are not captured for conventional oil). Adding from 10% to 30% of biomass to the coal feedstock (CBTL) can make well-to-wheels emissions 20% lower than those of conventional oil (US DOE, 2009b), with only moderate impact on the economics. This is why several of the announced projects plan to use CBTL.

Water usage in existing plants is reported to be quite significant: more than 10 barrels of water per barrel of oil produced. At this level, water availability could be a constraint on the location of CTL plants. The quality of used water released back to the environment also needs to be carefully monitored. However, most of the water can in principle be recycled and it should be possible to restrict actual use of water to less than two barrels per barrel of oil produced. Coal mining itself uses water, with one barrel of water per tonne of coal being typical. A typical CTL plant would produce 2 to 3 barrels of liquid hydrocarbon per tonne of coal. It is expected that the mines feeding the CTL plants would also be used to provide coal for power generation, so the actual increase in water usage would depend on what fraction of the mined coal is used in the CTL plants.

Gas-to-liquids

Gas-to-liquids (GTL) is a relatively mature technology, but experienced an upsurge in interest in the early to mid-2000s as a result of technological advances and higher oil prices. However, some technical problems with the commissioning of a new plant in Qatar and a sharp rise in construction costs, together with increased interest in LNG, which competes with GTL for gas feedstock, have led to many planned GTL projects being shelved in the last few years. Some projects are, nonetheless, under construction and we assume that several others, now at the planning stage, will also be commissioned. The current low price of gas and the persistent large price differential between gas and oil prices that we assume in our projections could lead to a resurgence of interest in GTL, with producers diversifying their portfolios with more ways of monetising gas in order to mitigate the risks of price fluctuations. However, the lengthy time scales involved in design, approval, construction and start-up of new large plants are likely to lead to slow growth in production. In the New Policies Scenario, GTL production rises from about 50 kb/d in 2009 to almost 200 kb/d in 2015 and to nearly 750 kb/d in 2035 (Figure 4.8).

Figure 4.8 ● Gas-to-liquids production by source in the New Policies Scenario

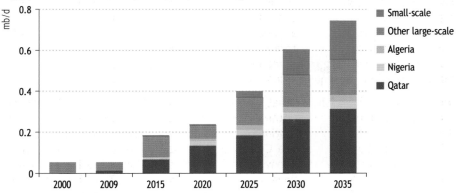

Note: Production is assumed to average 80% of installed capacity.

Gas-to-liquids (GTL) technology is similar to the CTL indirect route: natural gas (primarily methane) is reacted with steam and oxygen to form syngas (a mixture of carbon monoxide and hydrogen) and, in a second step, the syngas is turned into liquid hydrocarbon using the Fischer-Tropsch synthesis, typically yielding high-quality diesel and naphtha. The technology has a long history, dating back to the Second World War. Just as with CTL, it is possible to turn the syngas into methanol and then the methanol into gasoline, using the ExxonMobil process. Methanol can also be converted to DME (dimethyl ether) which is being developed as a diesel alternative, due to its high quality and clean burning characteristics.

Currently, GTL plants are economical only on a large scale, due to economies of scale. With capital costs of $60 000 to $100 000 per b/d of capacity and low operating costs, large-scale GTL projects (30 kb/d and above) are estimated to be economical at crude oil prices as low as $50 to $70/barrel.[8] However many efforts are being made around the world to design GTL processes that would be economical at smaller scales. The prize is enormous as small scale GTL (or, for that matter, economical small-scale LNG) would make it possible to produce the enormous amount of "stranded gas" (known gas fields that have no economical way to bring the gas to market) or to avoid flaring the associated gas produced with oil in places where there is no way to transport the gas economically. It is estimated that about 140 billion cubic meters of gas are flared every year, about one third of the gas consumption of Europe and 5% of world-wide gas production. Turning just the flared gas into liquids would produce as much liquid fuel as 1.4 mb/d of crude. Several pilot facilities with new micro-channel technologies are being built, for example by CompactGTL or by Velocys for Petrobras. Such technologies are expected to be deployed in significant numbers in the 2020s and applications to grow rapidly in the 2030s, driven in part by efforts to eliminate flaring completely.

---

8. The lower part of the range may apply to wet gas (gas rich in NGLs) for which the NGLs provide additional revenue.

Two projects have been producing for many years: the Sasol 25 kb/d Mossgas facility in South Africa and the Shell 15 kb/d Bintulu facility in Malaysia. In the early 2000s, Qatar proposed a large number of GTL projects (totalling as much as 700 kb/d) to commercialise gas from its giant North field. However, many of these proposed projects were shelved after Qatar declared a moratorium on GTL in 2006. Only one new project has been built, the Oryx 30 kb/d plant, which started operation in 2007 and has now reached its nameplate capacity. The Shell 140 kb/d Pearl project is under construction, with first production expected in 2011.

Other large projects are the Escravos 33 kb/d plant in Nigeria, currently under construction and expected to start production in 2013, and the recently announced Sasol project in Uzbekistan, with a 35 kb/d capacity and no target completion date yet announced (it is assumed to be commissioned before 2020 in the New Policies Scenario). Several other proposed projects, such as the Sonatrach Tinrhert project in Algeria and the Ivanhoe project in Egypt have been shelved, though the growing disconnect between gas and oil prices could lead to their revival in the future. Interest in GTL has been expressed in Russia, as a hedge against low gas prices, and in Turkmenistan, to help the country diversify its market outlets.

Although it tends to benefit from the positive image of gas as a greener hydrocarbon, the CO_2 footprint of GTL is not small. In modern plants, about a quarter of the carbon content of the natural gas is turned into CO_2 during the synthesis process. As a result, the well-to-wheels CO_2 emissions of GTL diesel are about 10% higher than those of diesel refined from conventional crude (just as for oil sands, this has to be qualified: some conventional crudes also have higher emissions than average and their emissions can be higher than GTL). A number of technical solutions exist, either involving storage (completing the CCS process) of the fairly concentrated CO_2 stream coming out of the process, or improved reforming processes that can recycle a large part of the CO_2. Future plants are likely to apply some of these technologies and achieve a CO_2 footprint similar to or better than that of conventional oil. Water usage is not a serious issue for GTL, with the newer plants (*e.g.* the Pearl project in Qatar) planning to recycle close to 100% of the water required in the process. Similarly, the physical size of the plant is similar to that of a refinery of equivalent capacity and does not give rise to specific land usage issues.

Additives

A variety of chemicals are added to crude oil as it enters refineries, or are blended into finished products. For example, anti-knocking agents, such as methyl tertiary butyl ether (MTBE), ethyl tertiary butyl ether (ETBE), or tertiary amyl methyl ether (TAME), are added to gasoline and methanol or ethanol can be blended with gasoline. Such chemicals are produced by the petrochemical industry from varying original feedstocks: oil, natural gas, coal and biomass. Since they contribute to both the volume and energy content of oil products, these additives must be accounted for in the balance between demand and supply. The part that originates from natural gas or coal is, quite reasonably, usually reported as unconventional oil, as they can be classified as a variation on GTL or CTL. However, it is not easy to separate the contributions of gas and coal from the other feedstocks. As an example, MTBE is obtained from reacting methanol with iso-butene. The methanol is generally obtained from gas or coal (although bio-methanol is now coming onto the market), but the iso-butene can be made by a variety of different routes from varying feedstocks, such as gas, NGLs and oil-refinery products. Similarly, ETBE can be made from petrochemicals or from bio-ethanol.

In the United States, MTBE usage has essentially been eliminated, being replaced by bio-ethanol. In Europe, a mix of MTBE and ETBE (coming from bio-ethanol) is used; MTBE is expected to continue to make up between 30% and 50% of these fuel ethers, as a compromise between cost and biofuel content. MTBE consumption is growing in the rest of the world. Blending of methanol in gasoline is rapidly growing, particularly in China, where a 15% methanol mix (M-15) is common and M-85 (85% methanol) is being introduced. This requires engine modifications that have been agreed between car manufacturers and the Chinese government. DME (dimethyl ether, a compound obtained from methanol) usage as an LPG blendstock is growing rapidly in a number of countries. Methanol is also used as a trans-esterification agent in the manufacture of biodiesel; one tonne of biodiesel incorporates about 0.1 tonne of methanol. With at least part of this methanol coming from gas or coal feedstock, the growing use of biofuels will create an increase in this "unconventional oil" supply.

With the expected decrease in oil demand in OECD countries and growth in demand in emerging economies, our projections (Table 4.1) show an increase in additives as a percentage of total oil supply in both the New Policies and the Current Policies Scenarios. In the 450 Scenario, the large reduction in overall demand for gasoline offsets the percentage growth in content of additives to result in a stable supply of additives. The supply of additives is reported as "oil equivalent" barrels, as the additives have lower energy content per barrel than oil (for example, less than half for methanol, about 60% for DME and about 75% for MTBE).

CHAPTER 5

NATURAL GAS MARKET OUTLOOK
Are we entering the golden age of gas?

H I G H L I G H T S

- Global natural gas demand is set to resume its long-term upward trajectory from 2010, following an estimated 2% drop in demand in 2009 — the biggest since the 1970s. It is the only fossil fuel for which demand is higher in 2035 than in 2008 in all scenarios, though it grows at markedly different rates. In the New Policies Scenario, demand reaches 4.5 tcm in 2035, an increase of 1.4 tcm, or 44%, over 2008 at an average rate of increase of 1.4% per year. Demand grows more quickly, by 1.6% per year, in the Current Policies Scenario; in the 450 Scenario, demand rises by a more modest 0.5% per year, peaking in the late 2020s.

- In the New Policies Scenario, non-OECD countries account for 84% of the increase in demand between 2008 and 2035. China's demand grows fastest, at an average rate of almost 6% per year, and the most in volume terms, accounting for almost a quarter of the rise in global demand to 2035. Demand in the Middle East, which is well-endowed with relatively low-cost resources, increases almost as much.

- In that scenario, the Middle East also leads the expansion of gas production over the *Outlook* period, its output almost doubling to 800 bcm by 2035. Two-thirds of this increase is consumed locally. China sees a sizeable expansion of capacity too, with most of the increase in the longer term coming from tight gas deposits, coalbed methane and shale gas. Around 35% of the *global* increase in gas production in this scenario comes from such unconventional sources.

- International trade in natural gas is set to grow. In the New Policies Scenario, gas trade between all *WEO* regions expands by around 80%, from 670 bcm in 2008 to 1 190 bcm in 2035. China's imports grow the most, from just 5 bcm in 2008 to 200 bcm in 2035. In fact, China accounts for a stunning 40% of the growth in inter-regional trade over the *Outlook* period. Most of the growth in gas trade takes the form of LNG; LNG trade doubles between 2008 and 2035. LNG supply will expand rapidly in the next few years as a wave of projects are completed.

- A sizeable glut of global gas-supply capacity has developed, a result of the economic crisis, which depressed gas demand, together with unexpectedly strong growth in unconventional gas production in the United States in the last few years and a surge in LNG capacity. Based on projected demand in the New Policies Scenario, we estimate that the glut, measured by the difference between the volumes actually traded and total capacity of inter-regional pipelines and LNG export plants, is set to reach over 200 bcm in 2011, before starting a hesitant decline. This glut will keep the pressure on gas exporters to move away from oil-price indexation.

Demand

Primary gas demand trends

To say that natural gas is entering a golden age may be an exaggeration, but it is certainly set to play a central role in meeting the world's energy needs for at least the next two-and-a-half decades. Global natural gas demand grows across the three scenarios, especially after 2015, though the rates of growth are markedly different, reflecting the differing impact of government energy and environmental policies. Nonetheless, demand is significantly higher in 2035 than in 2008 in each scenario (Figure 5.1). In the New Policies Scenario, demand growth slows progressively over the *Outlook* period, total demand reaching 4.5 trillion cubic metres (tcm) in 2035 (Table 5.1) — an increase of 1.4 tcm, or 44%, over 2008 and an average rate of increase of 1.4% per year. Demand grows more quickly — by 1.6% per year — in the Current Policies Scenario, attaining 4.9 tcm by 2035, with only a modest slowdown in the rate of demand growth towards the end of the projection period. In the 450 Scenario, gas demand peaks towards the end of the 2020s and then begins to decline, reaching 3.6 tcm in 2035 — a 15% increase over 2008 but about 5% down on its peak. In fact, gas is the only fossil fuel for which demand is higher in 2035 than in 2008 in this scenario. The share of gas in overall primary energy demand worldwide rises marginally over the projection period in the Current and New Policies Scenarios, but falls slightly after 2025 in the 450 Scenario, as the market penetration of renewables and nuclear power increases.

Figure 5.1 ● **World primary natural gas demand by scenario**

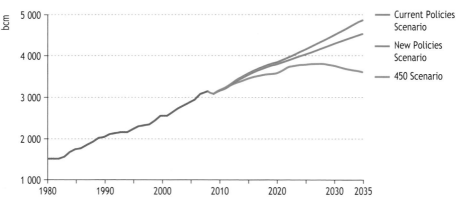

There is only a modest difference in gas demand growth rates across the three scenarios in the period to 2015, with global demand in every case recovering steadily following a drop in demand in 2009 — the biggest since the 1970s. According to preliminary data, demand in 2009 plunged by around 2% as a result of the global economic crisis, the decline occurring mainly in the OECD (averaging more than 3%). Trends diverged more in non-OECD countries, with demand plummeting in Russia, but continuing to grow strongly in China, India and the Middle East. In the OECD and Russia, demand was affected most by falling industrial output, which reduced gas needs for heat and process energy, and falling demand for electricity, which reduced gas needs for power generation. However, power sector gas demand did not fall, or at least not much, in *all* OECD countries: gas managed to increase its share of the power generation mix in some cases, notably the United States, usually because of competitive pricing.

Table 5.1 ● **Primary natural gas demand by region and scenario** (bcm)

	1980	2008	New Policies Scenario		Current Policies Scenario		450 Scenario	
			2020	2035	2020	2035	2020	2035
OECD	958	1 541	1 625	1 758	1 637	1 840	1 528	1 330
Non-OECD	559	1 608	2 169	2 777	2 198	3 047	2 055	2 279
World	1 517	3 149	3 794	4 535	3 835	4 888	3 584	3 609
Share of non-OECD	37%	51%	57%	61%	57%	62%	57%	63%

There are signs that gas demand is already starting to rebound, with OECD demand in the first quarter of 2010 up by an estimated 7% on the same quarter a year earlier (though demand was boosted by exceptionally cold weather). Demand rose by an estimated 5% in the second quarter. Over the whole of 2010, demand worldwide is expected to climb by more than 2%, though this will depend on near-term economic prospects as well as gas pricing, which can have a major impact on demand for gas in the power sector. For example, gas use for power generation actually increased by 4% in 2009 in the United States, because gas was more competitive than coal in some locations (IEA, 2010). On the assumption that the global economic recovery continues (see Chapter 1), demand is projected to resume its long-term upward path. It grows by 12% between 2008 and 2015 in the New Policies Scenario (compared with 13% in the Current Policies Scenario and 10% in the 450 Scenario).

Regional trends

Non-OECD countries will continue to drive gas demand growth over the next quarter of a century. In the New Policies Scenario, they account for 84% of the increase in demand between 2008 and 2035 (Table 5.2). China's demand grows faster than in any other region, at an average of almost 6% per year in 2008-2035, and the most in volume terms, reaching nearly 400 billion cubic metres (bcm) per year by the end of the *Outlook* period. China accounts for 22% of the increase in global demand over the projection period. Projected growth in the medium term is spectacular, with demand jumping from around 85 bcm in 2008 (and an estimated 98 bcm in 2009, based on preliminary data) to almost 170 bcm in 2015 and 215 bcm in 2020, the result mainly of booming demand in the power, residential and industrial sectors. In the longer term, gas demand is driven increasingly by the power sector, which accounts for almost half of total gas use in China in 2035. Yet gas still accounts for only 8% of all inputs to power generation by 2035 and the share of gas in China's overall primary energy mix reaches only 6% in 2020 and 9% in 2035, compared with 3% in 2008. By 2035, China's gas market is still 20% smaller than that of Russia and 40% smaller than that of the United States — the world's largest.

Table 5.2 ● **Primary natural gas demand by region in the New Policies Scenario** (bcm)

	1980	2008	2015	2020	2025	2030	2035	2008-2035*
OECD	958	1 541	1 568	1 625	1 666	1 713	1 758	0.5%
North America	659	815	817	844	864	886	913	0.4%
United States	*581*	*662*	*641*	*645*	*646*	*655*	*664*	*0.0%*
Europe	264	555	562	582	601	620	628	0.5%
Pacific	35	170	189	199	200	206	216	0.9%
Japan	*25*	*100*	*107*	*112*	*112*	*112*	*117*	*0.6%*
Non-OECD	559	1 608	1 969	2 169	2 367	2 584	2 777	2.0%
E. Europe/Eurasia	438	701	744	771	802	826	838	0.7%
Caspian	*n.a.*	*124*	*150*	*162*	*175*	*182*	*185*	*1.5%*
Russia	*n.a.*	*453*	*468*	*479*	*491*	*503*	*503*	*0.4%*
Asia	36	341	497	585	676	800	934	3.8%
China	*14*	*85*	*169*	*216*	*266*	*331*	*395*	*5.9%*
India	*2*	*42*	*80*	*97*	*117*	*143*	*177*	*5.4%*
Middle East	35	335	424	466	523	573	608	2.2%
Africa	13	100	136	149	155	161	164	1.9%
Latin America	36	131	168	197	212	223	232	2.1%
Brazil	*1*	*25*	*44*	*60*	*67*	*71*	*77*	*4.2%*
World	1 517	3 149	3 536	3 794	4 033	4 297	4 535	1.4%
European Union	*n.a.*	*536*	*540*	*558*	*574*	*591*	*598*	*0.4%*

*Compound average annual growth rate.

The Middle East, which is well-endowed with large and relatively low-cost resources, sees an increase in gas demand almost as big as that of China in absolute terms. This is driven by rising needs for power generation (the result of rapid growth in electricity demand and policies to replace oil with gas to free up more oil for export) and by use in heavy industry and as a feedstock for petrochemicals. Demand in non-OECD Asia and Latin America also grows rapidly.

India's demand grows almost as fast as China's, at 5.4% per year, but reaches only about 180 bcm by the end of the *Outlook* period, as it starts from a lower level (demand barely exceeded 40 bcm in 2008). Nonetheless, India's gas market would still be bigger than that of any OECD country except the United States. Increased availability of gas from the Krishna Godavari field, which came into production in 2009, is set to fuel an expansion of demand to more than 60 bcm in the very near future. Other developing Asian countries also see rapid growth. Among the non-OECD regions, demand in Russia grows least rapidly, by only 11% between 2008 and 2035, mainly because of continuing improvements in energy efficiency (as out-of-date technologies are replaced) and less waste — in part the consequence of higher prices as subsidies are phased out. Demand in Caspian countries grows more quickly, by 50% between 2008 and 2035, mainly for power generation (see Chapter 16). Brazil's demand grows strongly, tripling by 2035, drawing on the rapid development of the large offshore resources that have been discovered in the last few years.

The prospects for demand in the mature OECD markets are generally much weaker, largely because economic growth — the main determinant of gas demand — is assumed to be lower than in the rest of the world. In addition, there is much less scope for increased residential demand in OECD countries, because of saturation effects (most homes that can economically be heated with gas already are, and the number and size of households will barely grow). Industrial demand actually falls marginally between 2008 and 2035 in the New Policies Scenario, as slow growth in industrial production is outweighed by improved end-use efficiency. Power-sector demand will also be constrained by the growth in renewables-based generating capacity, which is always given priority in dispatching power ahead of gas-fired plants (as renewables often have low or zero operating costs). In that scenario, total OECD gas demand grows by only 0.5% per year on average to 2035, with growth slowing progressively over the projection period as higher prices and policies to curb gas and electricity demand take effect. In the United States, gas use in total declined by an estimated 1.7% in 2009, but is projected to recover slowly to 2035, due to rising demand for power generation (which averages 0.4% per year); the share of gas in power output remains flat at about 20%.

Sectoral trends

The power sector is set to remain the leading contributor to gas-demand growth in most regions. Yet, just how fast gas-fired generation will grow in the coming decades is very uncertain for several reasons, including relative fuel prices, the capital costs of building different types of generating plant, the ease of financing new power plants, government policies on renewables and nuclear power, and environmental policies and measures to deal with emissions of pollutants and greenhouse gases, including plans for CO_2-emissions trading. In the New Policies Scenario, power and heat generation account for more than 45% of the global increase in gas use between 2008 and 2035 (Figure 5.2). Gas-burning in power stations and heat plants (including co-generation plant) increases by more than half over that period — an average annual rate of growth of 1.6%. As a result, the power sector's share of the world gas market increases marginally, from 39% in 2008 to 41% in 2035.

Despite rising prices, natural gas used mainly in combined-cycle gas turbines (CCGTs) is expected to remain the preferred option for new power stations in many parts of the world, because of its inherent environmental advantages over coal (notably its much lower carbon content and smaller contribution to local air pollution), the higher thermal efficiency and lower capital costs and construction lead-times of CCGTs, and their operational flexibility (see Chapter 7). The expansion of carbon trading and rising CO_2 prices enhance the competitiveness of gas against coal in power generation, though renewables and nuclear power are favoured even more. For this reason, gas is often the lowest-cost generating option at CO_2 prices that are neither very low nor very high: low prices typically favour coal, while high prices (for example, in excess of \$100/tonne as assumed after 2030 in the 450 Scenario) favour renewables and nuclear power.

Figure 5.2 ● *World primary natural gas demand by sector in the New Policies Scenario*

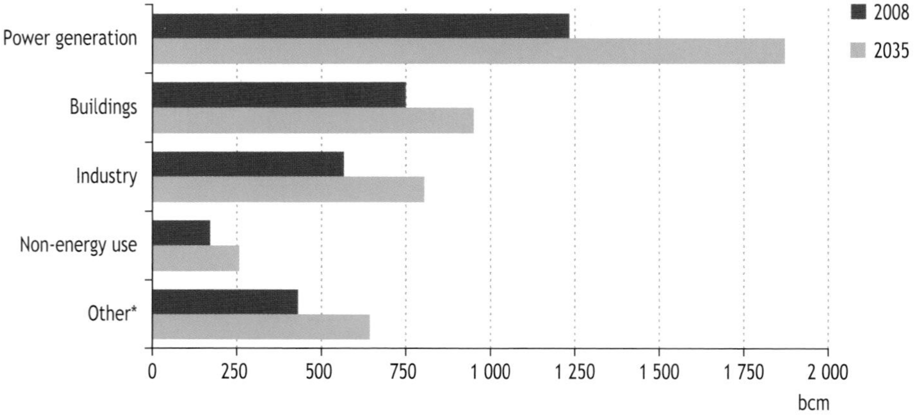

* Includes other energy sector, transport and agriculture.

Demand for gas in industry is set to grow faster than in any other end-use sector other than transport (where gas use remains small, globally). In the New Policies Scenario, industrial demand rises by 1.3% per year on average over the projection period, with most of the increase coming from non-OECD countries (mainly in Asia and the Middle East). Direct use of gas by industry in OECD countries barely grows, as industrial output expands only slowly, electricity accounts for much of the increase in industrial energy needs and efficiency gains limit the need to burn more gas. Worldwide, gas demand in other end-use sectors — mainly residential and services — grows by 1.1% per year. Growth in the use of gas in buildings — which remains the largest end-use sector — for space and water heating is limited by saturation effects in many OECD countries. In much of the rest of the world, the potential for using gas for space heating and hot water is generally lower, because of climatic factors and the high cost of building local distribution networks. Nonetheless, some countries see rapid growth in gas use in buildings. China is in the midst of one of the largest residential construction booms in history, with thousands of new housing estates being connected to local gas distribution grids every month, increasing demand massively and accounts for almost one-third of the global increase in gas use in buildings between 2008 and 2035.

Oil and gas prices: a temporary separation or a divorce?

Spot gas prices weakened significantly in 2009 and the first half of 2010, relative to oil prices, reflecting two revolutions on the supply side: the surge in LNG capacity, which will see liquefaction capacity growing by 47% between end-2008 and end-2013, and the unexpected boom in unconventional gas production in North America. With demand for gas dropping heavily in the face of recession, a sizable glut of gas has emerged. Gas demand is expected to recover in 2010, but less rapidly than oil demand, which is being driven mainly by China and other large non-OECD economies that tend to be much less dependent on gas. The result of this gas-market imbalance is that a large and unprecedented gap has opened up between the prices prevailing in the competitive markets of North America and Great Britain, on the one hand, and those in continental Europe and Asia-Pacific, where gas prices remain largely indexed to oil prices under long-term contracts, on the other. In 2009, the spot price averaged $4 per million British thermal units (MBtu) at Henry Hub in the United States and $5/MBtu at the National Balancing Point in Britain, compared with around $9/MBtu in Japan and continental Europe.

This regional gas price decoupling is already putting pressure on buyers of gas under oil-linked contracts in Europe to seek changes from their suppliers to their pricing terms — a development that we predicted in last year's *Outlook*. Gas buyers are caught between their long-term contractual obligations and the pressure from their customers, in particular industrial, to supply gas at more competitive prices. Russia's Gazprom has already granted some important concessions on pricing, partially moving from oil to spot gas price indexation over a three-year period, with prices falling as a result in key markets like Germany. This has led to a narrowing of the gap between spot and contract prices in Europe. Take-or-pay clauses have also been eased, giving more flexibility to buyers as to when they are required to lift contracted volumes.

The 64-million-dollar question now is: what will happen to the traditional oil-gas price linkage on European continental and Asian markets? The suppliers claim that recent pricing concessions are merely temporary. Whether the use of spot gas price indexation remains beyond the three years, and is extended to other contracts, or traditional oil indexation fully returns will depend on the global supply/demand balance and on the evolution of the gap between the different spot and oil-linked prices. For as long as the gas glut persists — and our analysis suggests it will for several years (see below) — the pressure to move further away from oil indexation will remain, especially for new long-term contracts. Ultimately, full contractual decoupling between gas and oil prices could occur, were sufficient momentum to build, though the dynamics of interfuel competition are likely to ensure a continuing degree of correlation between fuel prices. Contractual price decoupling would not necessarily mean weaker gas prices in the longer term: as the gas glut gradually dissipates, gas prices are likely to come under renewed upward pressure relative to oil prices, with the rising cost of supplying gas from remote and difficult locations.

Gas demand is set to expand rapidly in two emerging sectors: as feedstock for gas-to-liquids (GTL) plants and as a road-transport fuel. At present, there are only three large GTL plants in operation worldwide, the biggest of which — the 34 thousand barrels per day (kb/d) Oryx plant in Qatar — was commissioned in 2006, though production only recently approached its full capacity (see Chapter 4). The other two are Shell's 15-kb/d Bintulu plant in Malaysia and PetroSA's 25-kb/d plant in South Africa. Two more plants are under construction: Shell's 140-kb/d Pearl plant in Qatar, which is due to start operation in 2011, and the 34-kb/d Escravos plant in Nigeria being built by Chevron and the Nigerian National Oil Company, which is planned to start-up in 2012. By 2015, assuming there are no technical problems, all these plants together will consume around 20 bcm — up from 8 bcm in 2008 (when Oryx was still being commissioned) — and produce around 190 kb/d of liquids (mostly high-quality diesel and other light oil products). In the longer term, the prospects for GTL projects hinge particularly on relative oil and gas prices, and on the operational performance of the new plants. We assume that a project under development in Uzbekistan, together with some other projects in the Middle East and Africa, are completed, pushing up the volume of gas consumed in GTL production to 40 bcm (with oil production reaching 400 kb/d) by 2025 and 72 bcm (750 kb/d) by 2035.

The recent fall in the price of gas relative to oil (see Spotlight below), especially in North America, has stimulated interest in using natural gas as a road-transport fuel. Today, natural gas vehicles are common in only a few countries and the global use of compressed natural gas (CNG) as a road fuel is tiny (see Chapter 3). The biggest potential lies with heavy-duty vehicles (trucks and buses), as the costs of installing refuelling infrastructure for light-duty vehicles and adapting cars to run on gas are likely to limit the growth of CNG use in light vehicles. There is scope for increased CNG consumption in countries with an established market, notably in non-OECD Asia and Latin America. But the potential may be greatest in North America, where abundant supplies of unconventional gas are expected to hold gas prices down in the coming years, making CNG an attractive alternative to diesel for heavy-duty vehicles. Nonetheless, the take-off of CNG use even there is likely to be slow, in view of the need to develop distribution facilities. In the New Policies Scenario, we project North American gas use for road transport to grow from 0.9 bcm in 2008 to 12 bcm by 2035, with global use rising from 18 bcm to 61 bcm over the same period.

Production

Resources and reserves[1]

Remaining resources of natural gas are abundant, relative to those of oil, and are easily large enough to meet the projected increase in global demand — even in the Current Policies Scenario. The biggest uncertainty for supply over the next quarter

---

1. See Box 3.2 in Chapter 3 for our definitions of reserves and resources and *WEO-2009* for a more detailed discussion of gas resources (IEA, 2009).

of a century is whether sufficient and timely investment will be made in developing those resources and how much their exploitation will cost. Proven reserves of gas have increased steadily since the 1970s, as reserve additions have outpaced production by a wide margin. Proven reserves stood at 184 tcm at the end of 2008 — close to twice as high as 20 years ago and equivalent to 58 years of production at current rates and 42 years at our projected average annual growth rate of 1.3% in the New Policies Scenario.[2] Most of these reserves are conventional gas; unconventional gas forms a significant proportion of the total only in the United States — the leading unconventional gas producer — and Canada. The overwhelming bulk of the world's proven reserves are in the Middle East and former Soviet Union countries; just three countries — Russia, Iran and Qatar — hold 54% of the world total. Gas reserves (mostly conventional) in OECD countries amount to only 18 tcm, equal to about 10% of the world total, or 16 years of current OECD production.

Proven reserves represent only a small proportion of the total amount of gas resources that are thought to remain and that could be produced profitably at today's prices and with current technology (recoverable resources). The scale of overall gas resources is not known with certainty, as many parts of the world have been poorly explored. This is especially true for unconventional gas, including shale gas, coalbed methane, tight gas (from low permeability reservoirs) and gas (or methane) hydrates. Based on data from the US Geological Survey (USGS) and from the German Federal Institute for Geosciences and Natural Resources (BGR), we estimate that remaining recoverable resources of conventional gas alone amount to 404 tcm.[3] At end-2009, cumulative production (including flaring and venting) since gas production first began amounted to about 90 tcm, i.e. a little under one-fifth of ultimately recoverable conventional resources (the resources that existed before production began). As with proven reserves, the majority of remaining resources are in former Soviet Union countries and the Middle East (Figure 5.3). But unconventional gas resources could turn out to be even larger; excluding gas hydrates (for which commercial production technology has not yet been demonstrated), unconventional gas in place is estimated at over 900 tcm (IEA, 2009). Assuming around 380 tcm of this gas is recoverable, total recoverable gas resources would amount to close to 800 tcm — equivalent to about 250 years of current production. Unconventional gas resources are thought to be more widely dispersed geographically than conventional resources.

---

2. Preliminary data points to a 4.4% increase in proven reserves in 2009.

3. We have compiled data on resources for different basins around the world, drawing on the results of the last major resource assessment by the USGS in 2000, more recent updates of specific basins, new USGS assessments of basins not covered in the 2000 report, including a recent assessment of Arctic resources (USGS, 2008), and a 2009 study by BGR.

Figure 5.3 ● **Proven reserves, recoverable resources and production of conventional natural gas by region in the New Policies Scenario**

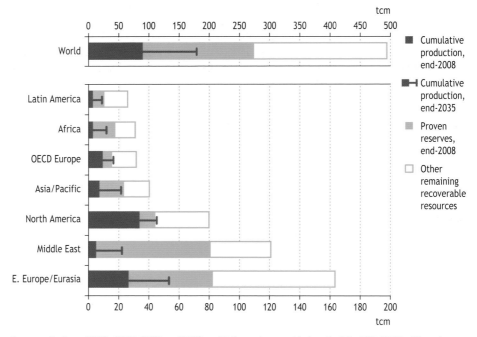

Sources: Cedigaz (2009); USGS (2000 and 2008) and information provided to the IEA; BGR (2009); IEA estimates and analysis.

Gas production prospects

Projected global gas production in 2035 ranges from some 3 600 bcm to 4 900 bcm across the three scenarios, corresponding to demand in each case (Table 5.3). In the New Policies Scenario, demand reaches over 4 500 tcm, the rate of increase being tempered by policies to curb fossil-energy use and emissions. The lower prices in the 450 Scenario, resulting from weaker demand brought about by more far-reaching policy action, result in less investment and, therefore, lower production, to balance lower demand in that scenario. Production in the 450 Scenario actually peaks by the late 2020s, before going into steady decline. In the Current Policies Scenario, production grows quickest, and in a fairly constant fashion in absolute terms, as prices rise most rapidly (see Chapter 1). In all three scenarios, most of the increase in output occurs in non-OECD countries.

Around 35% of the increase in global gas production in the New Policies Scenario comes from unconventional sources — mainly coal beds (coalbed methane), low-permeability reservoirs (tight gas) and shale formations (shale gas). Their combined share of production rises from around 12% in 2008 to about 19% in 2035 (Figure 5.4). The United States and Canada contribute more than one-quarter of the increase in absolute terms, with the bulk of the additional North American output coming from shale gas. US shale gas production has soared in recent years, from only 12 bcm in 2000 to an estimated 45 bcm in 2009, reversing the downward trend in the country's overall gas output; indeed, overall, US gas production jumped 16% in the four years to 2009. This has largely eliminated the need for the country to import liquefied natural gas (LNG) to make good a previously expected shortfall in domestic gas supplies. This evolution has contributed to existence of surplus supply capacity in the rest of the world, brought about primarily by the global recession, and has been instrumental in driving down spot prices (see Spotlight and the section on trade below).

Table 5.3 ● Natural gas production by region and scenario (bcm)

	1980	2008	New Policies Scenario		Current Policies Scenario		450 Scenario	
			2020	2035	2020	2035	2020	2035
OECD	889	1 157	1 158	1 188	1 173	1 203	1 103	1 033
Non-OECD	640	2 010	2 636	3 347	2 661	3 685	2 480	2 577
World	1 529	3 167	3 794	4 535	3 835	4 888	3 584	3 609
Share of non-OECD	42%	63%	69%	74%	69%	75%	69%	71%

Figure 5.4 ● World natural gas production by type in the New Policies Scenario

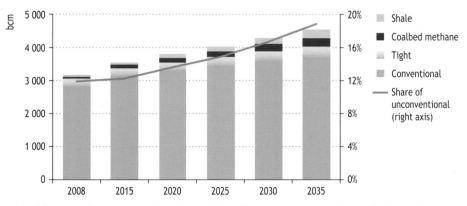

Note: Tight gas production is defined and reported in different ways across regions, so the data and projections shown here are subject to considerable uncertainty, indicated by the shading.

The prospects for unconventional gas production in the rest of the world, tiny for now, remain very uncertain, though they have improved over the past year with growing interest in several parts of the world. Output is projected to grow most in China, India and Australia (where coalbed methane production has grown rapidly in recent years). Exploration drilling for shale gas and coalbed methane has begun in Europe, notably in Poland, and some tight gas prospects have also been identified in Poland, Hungary and Germany (IEA, 2010). But unconventional production there is likely to remain relatively modest in the medium term, mainly because of the logistical and administrative difficulties in gaining access to land, and environmental concerns related to the need for large volumes of water for hydraulic fracturing and the risk of groundwater contamination (IEA, 2009). The uncertainty surrounding unconventional gas supplies outside North America is nonetheless very large. There is a risk that industry expectations of rapid expansion in unconventional supplies could inhibit investment in conventional resources, leading to a shortfall in overall gas supply and temporary upward pressure on prices. Conversely, more rapid development of unconventional gas supplies than projected here could lead to lower gas prices relative to oil, and more rapid penetration of gas in the power sector and in final uses.

In the New Policies Scenario, the Middle East makes the largest contribution to the expansion of gas production over the *Outlook* period, its output more than doubling to close to 800 bcm by 2035 (Table 5.4 and Figure 5.5). The region holds the largest reserves and has relatively low production costs, both for gas produced in association with oil and for dry gas. Four countries — Qatar, Saudi Arabia, Iran and Iraq — account for almost all of the 410-bcm increase. Around two-thirds of the increase in output, or 275 bcm, is consumed locally, mainly in power stations; the remaining 130 bcm is exported (see section on inter-regional trade). Although there is little doubt that these countries have the resources to increase production substantially, there is considerable uncertainty about when and how quickly this will happen, especially in Iraq and Iran. Qatar has declared a moratorium on new gas-export projects, pending the outcome of a study of the effects of current projects on the reservoirs of the country's North Field — the world's largest gas field. Most Middle East countries, with the exception of Qatar, have encountered shortages of gas in recent years, as exploration and development has failed to keep pace with demand.

Eastern Europe/Eurasia sees the second-biggest volume increase in output over the projection period (see Chapter 17 for a detailed discussion of Caspian gas production prospects). It remains the largest single producing region in 2035, well ahead of North America, with Russia and Turkmenistan pushing up the region's production. Asia and Africa account for most of the remaining increase in world output between 2008 and 2035. China is projected to see a sizeable expansion of its capacity, with the bulk of the increase in the longer term coming from tight gas deposits, coalbed methane and shale gas. Total gas production there reaches almost 140 bcm in 2020 and 180 bcm in 2035, up from only 80 bcm in 2008. The China National Petroleum Corporation has entered into joint ventures with a number of international companies to develop technically challenging resources. China signed an agreement with the United States in November 2009 to co-operate on shale gas development, Chinese resources of which are thought to be very large. Despite this projected increase in production, China's import dependence still rises over the projection period, especially after 2020. India is also set to increase gas output, though the pace of development is expected to slow in the medium term. Production surged in 2009, to an estimated 46 bcm, with the completion in late 2008 of Reliance's D6 block in the Krishna Godavari basin. Output is projected to grow to 60 bcm in 2015, with additional output from D6 more than offsetting declines at other, mature fields, and to just over 100 bcm by 2035, with a growing share coming from unconventional sources (notably coalbed methane) as conventional resources are depleted and development costs rise with declining field size. Most of the increase in African gas production occurs in Algeria and Nigeria.

Table 5.4 ● Natural gas production by region in the New Policies Scenario (bcm)

	1980	2008	2015	2020	2025	2030	2035	2008-2035*
OECD	**889**	**1 157**	**1 126**	**1 158**	**1 165**	**1 176**	**1 188**	**0.1%**
North America	657	797	783	809	821	834	846	0.2%
Canada	78	175	164	178	183	180	174	0.0%
Mexico	25	48	49	53	57	63	66	1.2%
United States	554	575	570	578	580	591	606	0.2%
Europe	219	307	270	259	240	222	206	-1.5%
Norway	26	102	104	110	115	119	122	0.7%
United Kingdom	37	74	42	32	23	17	13	-6.2%
Pacific	12	53	73	90	104	120	136	3.6%
Australia	9	45	67	86	101	118	134	4.2%
Non-OECD	**640**	**2 010**	**2 411**	**2 636**	**2 868**	**3 121**	**3 347**	**1.9%**
E. Europe/Eurasia	485	886	961	1 004	1 062	1 115	1 177	1.1%
Caspian	n.a.	188	224	259	278	298	314	1.9%
Russia	n.a.	662	697	704	742	772	814	0.8%
Asia	59	376	474	529	564	605	653	2.1%
China	14	80	117	137	152	167	185	3.1%
India	2	32	60	75	83	92	101	4.4%
Indonesia	17	74	85	91	95	102	110	1.5%
Malaysia	3	69	77	80	80	82	84	0.7%
Middle East	38	393	546	592	644	731	801	2.7%
Iran	4	130	144	156	179	210	235	2.2%
Iraq	1	2	13	24	34	52	65	14.0%
Qatar	3	78	162	179	192	213	225	4.0%
Saudi Arabia	11	74	95	100	105	113	124	1.9%
UAE	8	50	57	58	60	65	70	1.2%
Africa	22	207	259	307	361	409	435	2.8%
Algeria	13	82	108	120	138	152	162	2.6%
Egypt	2	60	72	79	83	82	65	0.3%
Libya	5	16	20	25	33	45	59	4.9%
Nigeria	2	32	39	54	74	95	113	4.8%
Latin America	36	148	172	204	237	260	280	2.4%
Argentina	10	47	45	46	54	53	43	-0.3%
Brazil	1	14	30	54	63	74	85	6.9%
Venezuela	15	23	25	28	34	43	64	3.8%
World	**1 529**	**3 167**	**3 536**	**3 794**	**4 033**	**4 297**	**4 535**	**1.3%**
European Union	n.a.	216	176	158	134	112	93	-3.1%
GECF market share**	n.a.	36%	38%	37%	38%	39%	40%	

* Compound average annual rate of growth. ** GECF = Gas Exporting Countries Forum.

Figure 5.5 ● *Change in natural gas production by region in the New Policies Scenario*

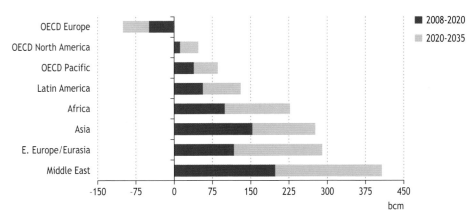

Among the OECD regions, production in North America rises by around 50 bcm between 2008 and 2035, mainly due to unconventional supplies (notably shale gas), but in Europe falls by 100 bcm to 210 bcm, as declines in North Sea production in the United Kingdom and Netherlands more than outweigh continued growth in Norway. Australian production grows strongly, more than tripling over the projection period, driven mainly by LNG export projects. Australia overtakes Norway towards the end of the projection period to become the third-biggest OECD gas producer, behind the United States and Canada. Coalbed methane accounts for a growing share of Australian supply, with the first LNG projects based on such gas likely to proceed in Queensland in the next few years.

Inter-regional trade

International trade in natural gas is set to grow rapidly in the coming quarter of a century. In the New Policies Scenario, inter-regional gas trade (between all *WEO* regions) expands by more than three-quarters from 670 bcm in 2008 to nearly 1 200 bcm in 2035 (Table 5.5), outpacing the projected 43% increase in global production. Imports into OECD North America, OECD Europe and both OECD and developing Asia grow in volume terms. China's imports grow the most, from a mere 5 bcm in 2008 to close to 80 bcm in 2020 and over 200 bcm in 2035. In fact, China accounts for a stunning 40% of the growth in inter-regional trade over the *Outlook* period. Within North America, the United States remains a net importer of gas, mainly from Canada, though its imports fall over the projection period. Net EU imports grow by 58%, from 320 bcm in 2008 (and an estimated 310 bcm in 2009) to just over 500 bcm in 2035. Africa, the Middle East, Russia, Australia and the Caspian account for the bulk of the increase in exports.

More than half of the growth in gas trade will be in the form of LNG. Trade in LNG more than doubles between 2008 and 2035, reaching 500 bcm, or 11% of world demand in the New Policies Scenario; most of the incremental LNG supply goes to Asia (Figure 5.6).

Table 5.5 ● Inter-regional natural gas net trade in the New Policies Scenario

	2008 bcm	2008 Share of primary demand*	2020 bcm	2020 Share of primary demand*	2035 bcm	2035 Share of primary demand*
OECD	-384	25%	-467	29%	-570	32%
North America	-18	2%	-35	4%	-67	7%
United States	*-87*	*13%*	*-67*	*10%*	*-58*	*9%*
Europe	-248	45%	-323	56%	-422	67%
Pacific	-118	69%	-109	55%	-80	37%
Asia	*-132*	*97%*	*-153*	*100%*	*-167*	*100%*
Oceania	*15*	*30%*	*44*	*49%*	*87*	*64%*
Non-OECD	402	20%	467	18%	570	17%
E. Europe/Eurasia	185	21%	233	23%	339	29%
Caspian	*63*	*34%*	*97*	*38%*	*129*	*41%*
Russia	*209*	*32%*	*225*	*32%*	*311*	*38%*
Asia	34	9%	-56	10%	-281	30%
China	*-5*	*5%*	*-79*	*36%*	*-210*	*53%*
India	*-10*	*25%*	*-23*	*23%*	*-75*	*43%*
Middle East	58	15%	126	21%	193	24%
Africa	108	52%	158	51%	271	62%
Latin America	16	11%	7	3%	48	17%
Brazil	*-11*	*45%*	*-7*	*11%*	*8*	*9%*
World**	670	21%	864	23%	1 187	26%
European Union	*-320*	*60%*	*-400*	*72%*	*-504*	*84%*

* Production for exporting regions/countries. ** Total net exports for all *WEO* regions/countries (some of which are not shown in this table), not including trade within *WEO* regions.

Note: Positive numbers denote exports; negative numbers imports.

The share of LNG in total gas trade rises from 31% in 2008 to 35% in 2020 and 42% in 2035 (Figure 5.7). Eight LNG liquefaction projects are under construction, all of which are due to be commissioned by 2015, adding 77 bcm to current capacity of around 360 bcm (at end-June 2010).[4] Close to 30% of this increase will come from Qatar, where two more large trains will be commissioned before the end of 2011 to supplement the four that started up between 2009 and early 2010. The rest of the capacity additions will come from Algeria (Gassi Touil and Skikda), Angola, Australia (Pluto and Gorgon) and Papua New Guinea. A number of other projects are also planned, notably in Australia.

---

4. Capacity at end-2009 was 338 bcm; one plant in Qatar and another in Peru, together with a second train in Yemen, were commissioned during the first half of 2010.

Figure 5.6 ● Inter-regional natural gas net trade flows between major regions in the New Policies Scenario (bcm)*

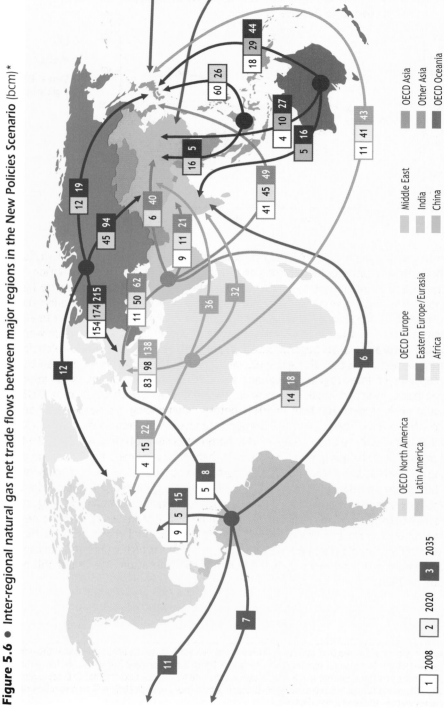

1	2008	2	2020	3	2035

OECD North America

Latin America

OECD Europe

Eastern Europe/Eurasia

Africa

Middle East

India

China

OECD Asia

Other Asia

OECD Oceania

The boundaries and names shown and the designations used on maps included in this publication do not imply official endorsement or acceptance by the IEA.

* Only flows above 4 bcm are shown.

Figure 5.7 ● World inter-regional natural gas trade by type in the New Policies Scenario

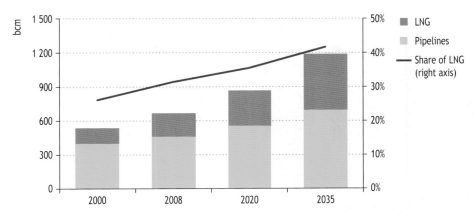

As a result of the economic crisis, which depressed gas demand around the world, together with the unexpectedly strong growth in unconventional gas production in the United States in the last few years, a sizeable glut of gas-supply capacity has developed. This has led to a sharp fall in the utilisation rate of existing pipeline and LNG capacity, which has been expanding rapidly in recent years (the investment decisions on most new projects recently completed or still under construction were taken well before the crisis began). Based on projected demand in the New Policies Scenario, we estimate that this gas glut, measured by the difference between the total capacity of inter-regional pipelines and LNG export plants and total inter-regional trade, reached about 130 bcm in 2009 (compared with 80 bcm in 2007) and could peak at over 200 bcm in 2011, before commencing a slow and hesitant decline (Figure 5.8). The capacity utilisation rate would fall from an estimated 75% in 2009 (83% in 2007) to under 70% in 2011, before recovering to about 75% in 2014. This suggests that the gas glut will last longer than many exporters believe or hope, keeping pressure on them from their major customers to modify pricing arrangements (see the earlier Spotlight). This pressure is likely to be greatest in Europe, where demand is expected to recover less quickly than in Asia-Pacific. Our analysis suggests that it may take several years for the gas glut to be fully eliminated. Even if no new pipeline or LNG project is commissioned before 2020 beyond those projects that have already obtained a final investment decision – which is highly unlikely – unused capacity would still total more than 150 bcm and the utilisation rate would still be only 80% by 2020.[5]

---

5. In part, it is to be expected that utilisation rates will not recover fully to the levels reached in the mid-2000s, as part of the incremental pipeline capacity that is being built is designed to substitute for, rather than supplement, existing capacity: this is especially the case with new Russian export lines to Europe. Also, the availability of gas to supply some existing pipelines, to which they are dedicated, will tend to fall as the source fields mature and production declines.

Figure 5.8 ● Natural gas transportation capacity between major regions in the New Policies Scenario

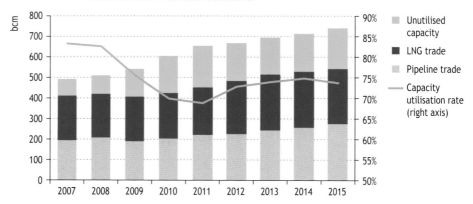

Box 5.1 ● The GECF seeks oil price parity and ponders how to achieve it

The Gas Exporting Countries Forum (GECF), which became a full-fledged international organisation in 2008, agreed at a meeting in Algeria in April 2010 to strive for gas price parity with oil and for the removal of "unjustified barriers", such as carbon taxes, to the increased use of gas. Prices of gas traded on a spot, or short-term, basis have fallen heavily relative to oil since 2008, as a result of a slump in demand and increased supplies of unconventional gas in the United States and of LNG, though the price of most internationally traded gas remains tied to oil under long-term contracts. Although no specific measures to achieve price parity were formally proposed at the meeting, Algeria had previously indicated that one option would be to agree on co-ordinated cutbacks in production, raising concerns among gas-importing countries about the prospective cartelisation of the gas market, with the GECF becoming a "Gas OPEC". GECF countries collectively control around two-thirds of the world's proven gas reserves, though several members currently make little or no contribution to international gas trade. However, such co-ordinated cutbacks would be difficult to achieve, particularly in the near term, not least because of volume commitments in long-term contracts and because of the relative ease with which other fuels could substitute for gas in power generation and end uses. The GECF will continue to emphasise information-sharing and dialogue for now, but may seek a more proactive role in market-related issues in the longer term. Bilateral co-operation between individual GECF members may prove as important as what happens under the GECF umbrella.

Investment

The projected trends in gas demand in the New Policies Scenario would require a cumulative investment along the gas-supply chain of about $7.1 trillion dollars (in year 2009 dollars), or around $270 billion per year (Table 5.6). Roughly two-thirds of that capital spending, or $175 billion per year, is needed upstream, for new greenfield projects and to combat decline at existing fields.[6] LNG facilities account for about 9% of the total, and transmission and distribution networks for the rest. Unsurprisingly, the majority of the investment is needed in non-OECD countries, where local demand and production grows the most.

Table 5.6 ● Cumulative investment in gas-supply infrastructure by region and activity in the New Policies Scenario, 2010-2035
($ billion in year-2009 prices)

	Exploration and development	Transmission and distribution	LNG*	Total	Annual average
OECD	1 863	862	150	2 875	111
North America	1 263	459	24	1 746	67
Europe	419	320	11	751	29
Pacific	180	83	114	378	15
Non-OECD	2 680	1 074	397	4 152	160
E. Europe/Eurasia	797	383	33	1 213	47
Caspian	*227*	*84*	*-*	*311*	*12*
Russia	*525*	*234*	*33*	*792*	*30*
Asia	721	321	94	1 136	44
China	*180*	*132*	*48*	*360*	*14*
India	*129*	*58*	*29*	*216*	*8*
Middle East	261	221	104	586	23
Africa	583	60	122	764	29
Latin America	319	89	44	452	17
World*	4 543	1 936	622	7 101	273
European Union	*179*	*305*	*11*	*496*	*19*

* World total includes an additional $74 billion of investment in LNG carriers.

6. Together with investment in oil, this level of gas investment yields a total upstream investment requirement of around $450 billion per year on average over 2010-2035. This compares with planned total upstream oil and gas investment worldwide in 2010 of $470 billion (see further discussion of upstream investment trends in Chapter 3). A shift in investment towards relatively low-cost regions, notably the Middle East, outweighs the effect of rising overall production over the projection period.

CHAPTER 6

COAL MARKET OUTLOOK
How fast the rise of Asia?

H I G H L I G H T S

- In the New Policies Scenario, demand for coal increases by around 20% between 2008 and 2035, with almost all of the growth before 2020. Demand is significantly higher in the Current Policies Scenario and much lower in the 450 Scenario, reflecting the varying strength of policy action assumed to address climate change and underscoring the need to significantly reduce emissions from coal use if it is to remain a mainstay for base-load power supply.

- Non-OECD countries as a group account for all of the growth in global coal demand in the three scenarios of this *Outlook*. In the New Policies Scenario their share of total demand increases from 66% today to 82% by 2035. China, India and Indonesia account for nearly 90% of the total incremental growth, highlighting their crucial influence on the future of the coal market. China remains the world's largest consumer of coal, while India becomes the second-largest around 2030; Indonesia takes fourth position (behind the United States) by 2035. Over the projection period, China installs around 600 GW of new coal-fired power generation capacity, comparable with the current combined coal-fired generation capacity of the United States, the European Union and Japan.

- Global coal production in the New Policies Scenario grows from just under 4 900 Mtce in 2008 to just above 5 600 Mtce in 2035. China accounts for half of global coal production by 2035, while Indonesia's output overtakes that of Australia. Global hard coal trade rises in the medium term, before declining to around 840 Mtce in 2035, although this is still 15% higher than today.

- Cumulative investment to meet projected coal demand through to 2035 amounts to some $720 billion (in year-2009 dollars) in the New Policies Scenario. Two-thirds takes place in non-OECD regions, with China alone needing over $260 billion. Global investment by 25 leading coal companies rose by 4.5% in 2009 to about $12 billion; this compares with a surge of 18% in 2008.

- China will continue to have a crucial influence on global coal trade. The country has been turning increasingly to imports in recent years, as domestic supply has struggled to keep up with rapidly rising demand. It is now working to overcome transportation bottlenecks and to speed-up the development of its vast coal resources in the northern and western parts of the country. Given the sheer size of China's market, the uncertainty around its future supply-demand balance will have major implications for trade patterns and prices of internationally traded coal.

Demand

Primary coal demand trends

Demand for coal remained fairly solid in 2009, despite the global economy going through an upheaval (oil and gas demand, by contrast, fell substantially). The three scenarios in this year's *Outlook* clearly demonstrate the critical influence of government policies, especially those related to climate change, on the outlook for coal demand (Figure 6.1). In the Current Policies Scenario, which assumes no change in government policies, strong global economic growth and near tripling of electricity demand in non-OECD countries lifts global coal demand to over 7 500 million tonnes of coal equivalent (Mtce) by 2035, or nearly 60% higher than in 2008. In contrast, world coal demand in the New Policies Scenario, which takes into account planned reforms of fossil-fuel subsidies, implementation of measures to meet climate targets and other planned energy-related policies, is around 1 925 Mtce, or a quarter, lower in 2035. This difference is equal to about China's current total coal demand, or 40% of global coal demand in 2008. In the 450 Scenario, which assumes more decisive implementation of policy plans and a further strengthening of policies after 2020, with the objective of limiting to 2°C the long-term rise in the global average temperature, world coal demand at about 3 565 Mtce in 2035 is a quarter lower than the level in 2008 and close to the levels of the 1990s and early 2000s.

Figure 6.1 ● World primary coal* demand by scenario

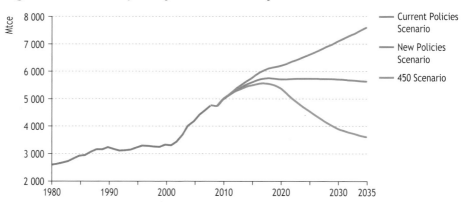

*Includes hard coal (steam and coking coal), brown coal (lignite) and peat.

Coal use in the OECD falls in all three scenarios between 2008 and 2035, as countries further decarbonise their electricity generation mix, not returning to the peak consumption levels seen before the global financial crisis that began in 2008 (Table 6.1). OECD coal demand is estimated to have contracted by 10% in 2009, with more than 50% of this decline occurring in the United States. By 2035, in the New Policies Scenario, the OECD accounts for less than one-fifth of global coal demand, compared with one-third today, its coal demand declining on average by 1.7% per year

World Energy Outlook 2010 - **GLOBAL ENERGY TRENDS**

over the projection period. Thus, non-OECD countries account for all of the growth in global coal demand, raising their share in the worldwide market from 66% today to 82% by 2035 in the New Policies Scenario. While non-OECD coal demand grows by 1.4% per year over the *Outlook* period in this scenario, on a per-capita basis it grows by only 0.4% per year. Today, annual coal consumption per head in the non-OECD is on average 0.57 tonnes of coal equivalent (tce), around 40% the level in the OECD. By 2035, in the New Policies Scenario, OECD per-capita annual coal consumption has fallen to 0.78 tce, but is still around one-fifth higher than in the non-OECD countries. By 2035 non-OECD coal intensity in the New Policies Scenario, measured as coal use per unit of GDP at market exchange rates, has more than halved relative to today's levels but is still more than double that of the OECD in 2008, leaving room for further intensity gains and lower coal demand, as demonstrated by the 450 Scenario of this *Outlook* (see Chapter 14).

Table 6.1 ● World primary coal demand by region and scenario (Mtce)

	1980	2008	New Policies Scenario		Current Policies Scenario		450 Scenario	
			2020	2035	2020	2035	2020	2035
OECD	1 379	1 612	1 452	1 021	1 596	1 507	1 348	709
Non-OECD	1 181	3 124	4 213	4 600	4 557	6 037	3 998	2 856
Total	2 560	4 736	5 665	5 621	6 153	7 544	5 347	3 566
Share of non-OECD	46%	66%	74%	82%	74%	80%	75%	80%

Regional trends

In 2008, China, the United States, the European Union, India, Russia and Japan accounted for 83% of global coal demand (Figure 6.2). These six demand centres accounted for almost 70% of global GDP and energy-related CO_2 emissions and just over half of the world's population in 2008. Within this group, the relative importance of the countries has changed significantly since 1990. Two decades ago, the United States, the European Union, Russia and Japan accounted for just over half of global coal demand: in the past decade, China, alone has become the dominant consumer. China's coal consumption, which grew by 1 120 Mtce over the last eight years, accounted for more than three-quarters of global coal demand growth in the period 2000-2008. As a result, China today accounts for 43% of global coal demand and by 2035, in the New Policies Scenario, China's share reaches 50%. China and other Asian economies with large populations and strong economic growth, such as India and Indonesia, will accordingly have a crucial influence on the future of the coal market, not only in terms of demand but also of production and trade. Among the regions where coal demand increases over the projection period in the New Policies Scenario, China, India and Indonesia together are responsible for nearly 90% of the total growth.

Figure 6.2 ● *Share of key regions in global primary coal demand in the New Policies Scenario*

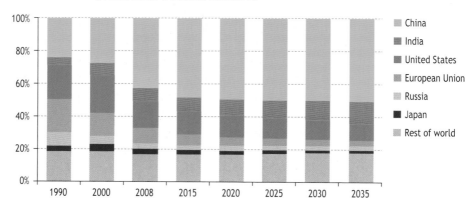

The United States, the European Union, Russia and Japan all see their coal demand decline over the projection period in the New Policies Scenario, their combined market share plunging from 33% today to 18% by 2035. The European Union's coal demand declines fastest, at 3% per annum (Table 6.2). In all four of these demand centres, the share of coal in total primary energy demand declines, as coal is displaced by gas, renewables and nuclear in electricity generation or by electricity and gas in industrial processes. By 2035, one-third of electricity in the United States is generated from coal, compared with nearly half today, as the share of renewables and, especially, that of wind grows from 1.3% today to 10% by the end of the *Outlook* period. In the European Union, the share of coal in electricity generation declines by 2035 by almost 20% compared to 2008, as the share of renewables grows from 17% to 41%. In Russia and Japan, nuclear makes strong inroads at the expense of coal in electricity generation, especially in Japan where the share of nuclear power goes from 24% to 42% by 2035 (see Chapter 7).

Coal demand in *China,* the world's largest consumer of coal, grows by 2.7% per year to 2020 in the New Policies Scenario, but then remains fairly stable through the rest of the projection period at a level of around 2 800 Mtce. The share of coal in China's total primary energy demand declines from 66% today to 53% by 2035. Continued growth in demand from the power generation sector in China, albeit at a slower pace than historically, is offset by a fall in coal demand for industry, which peaks before 2020 and soon after begins to decline. Over the projection period, China brings on-line around 600 gigawatts (GW) of new coal-fired power generation capacity, comparable with the current combined coal capacity of the United States, European Union and Japan. In China's industrial sector, about 60% of energy demand currently comes from coal, while electricity accounts for a further quarter. In the New Policies Scenario, coal's share declines to 42% by 2035. Almost two-thirds of the growth in energy use in industry is met through electricity, while gas doubles its market share in China.

Table 6.2 ● Primary coal demand by region in the New Policies Scenario (Mtce)

	1980	2008	2015	2020	2025	2030	2035	2008-2035*
OECD	1 379	1 612	1 562	1 452	1 337	1 208	1 021	-1.7%
North America	571	828	827	789	740	681	596	-1.2%
United States	*537*	*780*	*777*	*747*	*705*	*649*	*576*	*-1.1%*
Europe	663	447	392	346	312	278	226	-2.5%
Pacific	145	337	342	318	285	249	199	-1.9%
Japan	*85*	*162*	*161*	*146*	*125*	*106*	*82*	*-2.5%*
Non-OECD	1 181	3 124	3 999	4 213	4 357	4 484	4 600	1.4%
E. Europe/Eurasia	517	325	324	305	304	296	290	-0.4%
Caspian	*n.a.*	*47*	*57*	*59*	*60*	*57*	*56*	*0.7%*
Russia	*n.a.*	*167*	*170*	*163*	*163*	*159*	*158*	*-0.2%*
Asia	572	2 601	3 458	3 687	3 830	3 958	4 081	1.7%
China	*446*	*2 019*	*2 685*	*2 788*	*2 831*	*2 842*	*2 822*	*1.2%*
India	*75*	*373*	*467*	*551*	*609*	*682*	*781*	*2.8%*
Indonesia	*0*	*53*	*95*	*111*	*131*	*151*	*168*	*4.4%*
Middle East	2	14	17	16	18	23	29	2.9%
Africa	74	149	151	159	161	164	160	0.3%
Latin America	16	35	49	46	43	43	40	0.6%
Brazil	*8*	*20*	*28*	*24*	*21*	*21*	*20*	*0.2%*
World	2 560	4 736	5 561	5 665	5 694	5 692	5 621	0.6%
European Union	*n.a.*	*434*	*374*	*314*	*277*	*240*	*193*	*-3.0%*

* Compound average annual growth rate.

Over the projection period, *India* becomes the world's second-largest consumer of coal around 2030, with demand doubling from around 370 Mtce today to 780 Mtce by 2035 in the New Policies Scenario. More than half of the incremental coal demand in India comes from the power sector, as the nation strives to improve the welfare of the nearly 405 million citizens — one-third of the total population — who at present lack access to electricity and the 855 million citizens who rely on traditional biomass for cooking (see Chapter 8). Another 38% of the projected increase in India's coal demand comes from the industrial sector, raising the share of coal in that sector from around one-third today to above 40% by 2035. Despite the strong projected growth in coal demand, the share of coal in India's total primary energy demand declines from 42% today to 39% by 2035, as coal loses market share to renewables, gas and nuclear in the power generation sector.

Indonesia, traditionally considered mainly as a steam-coal exporter, sees its domestic demand tripling to nearly 170 Mtce by 2035, a rate of growth of 4.4% per year, by far the highest among all the major regions. Today, Indonesia is only the 13th-largest coal consumer; in the New Policies Scenario, it overtakes Japan, today the 5th-largest consumer, by 2025 and Russia by 2035 to become the world's 4th-largest coal-consuming country. Indonesia today is the world's fourth most populous country and by far the largest economy in the Association of Southeast Asian Nations (ASEAN). Indonesia experiences frequent electricity blackouts and only 65% of the population has access to electricity, which places a severe constraint on development (IEA, 2009). The power sector accounts for nearly 60% of the growth in projected domestic coal demand in the New Policies Scenario, as coal-fired capacity more than quadruples to 46 GW by 2035.

Coal demand grows in most other non-OECD regions, apart from Eastern Europe/ Eurasia, where it declines by 0.4% per year over the projection period in the New Policies Scenario. Within that group, the Caspian region bucks the trend by increasing its demand for coal. Kazakhstan — the world's 15th largest consumer and 10th largest producer of coal today — remains the main coal-consuming country in the Caspian region (see Chapter 16). In 2009, a sharp fall in exports to Russia, coupled with growth in domestic demand, saw coal production in Kazakhstan drop around 10%, highlighting the close link between the country's export potential and Russian demand.

Sectoral trends

In 2008, nearly two-thirds of global coal demand was consumed in the power sector and another one-fifth in the industry sector. The share of coal in industrial energy use has declined only slightly since 1990, while the share in the power sector has grown by 10 percentage points, mainly at the expense of the buildings and agriculture sector, which in 1990 consumed just over 10% of global coal demand. Over the *Outlook* period, as global coal demand grows by 0.6% per year in the New Policies Scenario, each sector's share of demand remains roughly similar. Demand in power generation accounts for almost 60% of the increase of 885 Mtce in global coal demand, while another 30% of the demand growth comes from the industry sector (Figure 6.3). Coal-to-liquids (CTL), a means of reducing oil-import dependency, emerges as an important growth sector, with demand increasing by around 125 Mtce (equivalent to 45% of the growth in global industrial coal use) as just over 1 million barrels per day (mb/d) or 1% of global oil demand by 2035 in the New Policies Scenario is obtained through CTL (see Chapter 4). Coal transformation is not limited to CTL. Coal gasification is already successfully undertaken in China, South Africa and the United States to produce syngas, and underground coal gasification holds the potential of providing a means of exploiting coal deposits which are not mineable using conventional techniques (Box 6.1).

Almost 90% of the *decline* of 590 Mtce in OECD coal demand over the projection period in the New Policies Scenario is expected to result from new policies to decarbonise the power sector in order to reach the targets proposed under the Copenhagen Accord. Over the *Outlook* period, around 390 GW of coal-fired generation capacity is expected to cease operating in the OECD, an amount greater than today's combined installed coal-fired generation capacity of OECD North America. Offsetting this to an extent, over the same time frame 255 GW of new coal-fired capacity is expected to be built in the OECD, of which 92 GW would employ highly efficient ultra-supercritical or integrated gasification combined-cycle (IGCC) technologies and an additional 33 GW would incorporate means to capture and store CO_2. By contrast, about 70% of the *growth* in non-OECD coal demand of 1 475 Mtce is projected to come from power generation, with China, India and Indonesia being responsible for 61%, 21% and 6% of the growth respectively. Just over 1 100 GW of new coal-fired generation capacity, close to double today's coal-fired generation capacity in China, is installed in the non-OECD over the projection period. A further 20% of the increase in non-OECD coal demand comes from the industrial sector, with India alone accounting for nearly half of this.

Figure 6.3 ● **Change in primary coal demand by sector and region in the New Policies Scenario, 2008-2035**

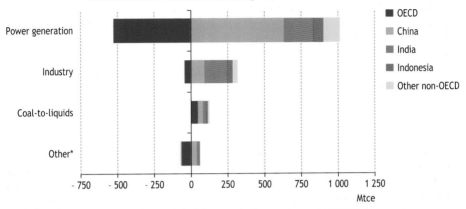

* Includes other energy sector, transport, buildings, agriculture and non-energy use.

Box 6.1 ● Coal gasification

The surge in the production of tight and shale gas in the United States over the past decade has had a profound impact on the global gas market outlook (IEA, 2009). Technologies similar to those used for shale gas production can be applied to extracting energy from coal seams. In last year's *Outlook*, the prospects for *coalbed methane* (CBM) production, extracting methane from coalbeds that are not mined due their depth or poor quality, were examined in detail. In the New Policies Scenario, CBM production is expected to contribute nearly 200 billion cubic metres (bcm), or 15%, towards global incremental production of gas (see Chapter 5). However, only a small fraction, around 1%, of the total energy stored in a coal seam is recovered during CBM production.

Underground coal gasification (UCG) has the potential to recover much more energy and is of particular interest at coal deposits which are un-mineable using conventional techniques. If successful, UCG would substantially increase the proportion of the world's coal resources that could be classified as recoverable. UCG involves an injection borehole, through which air or oxygen (and possibly steam) are injected, and a production well from which product gas (mainly hydrogen and carbon monoxide) is drawn to the surface for treatment and use. The boreholes are linked by a zone through the coal seam where coal combustion and gasification takes place in a continuously changing combustion zone that must be monitored and controlled. Commercial-scale operation would involve multiple wells. The technique has a long history in the Former Soviet Union, where it was carried out on an industrial scale, and trials have taken place in the United States, Europe and China. Recent pilot-scale tests in Australia, Canada, China and South Africa have built on developments in directional drilling and computer modelling. Successful results could be expected to spur activity in other countries rich in coal resources, including India, Poland, Russia, the United Kingdom and the United States.

Results from current pilot projects are sketchy because some knowledge is proprietary. For example, little is yet publicly known about what happens to the surrounding geology and hydrogeology when a combustion zone at 1 000°C moves through a deep coal seam. This is unfortunate because there are formidable obstacles to be overcome to integrate the knowledge from different disciplines to the point where a project can be designed with confidence that it will perform as intended. Only some 15 to 20 million tonnes (Mt) of coal have been gasified underground to date, which illustrates the limited experience with UCG.

In view of the scale of the prospective rewards, UCG project developers need to consider how to move quickly from pilot projects at carefully chosen and favourable sites to more ambitious demonstration projects that can provide the design basis for large commercial projects in a wide range of coal types and situations. Co-operation between developers and government-supported research and development could speed progress and increase confidence in UCG technology.

Coal gasification (CTG) was once the main source of town gas for use in cities. The processes used were reliable, but polluting. Technological advances mean that coal gasification is carried out today using continuous processes that produce clean synthesis gas for chemicals and liquid fuels production, or for other uses, at many plants around the world, notably in China, South Africa and the United States. As demand for gas grows, coal gasification could become a competitive source in regions with access to low-cost coal reserves, such as Xinjiang in China. According to Platts, there are 15 coal-to-gas projects in China under construction or being planned. Huineng Group's project in Inner Mongolia and Datang Power International's two 4 bcm/year projects in Chifeng and Fuxin recently won approval from China's National Development and Reform Commission.

Sources: IEA CCC (2009); Platts.

Production

Resources and reserves

According to the German Federal Institute for Geosciences and Natural Resources, coal resources make up an estimated 82% of the world's non-renewable energy resources (BGR, 2009). Of this resource, reserves totalling nearly 1 000 billion tonnes are sufficient to meet demand for many decades: at present coal production levels, reserves would meet demand for almost 150 years. Coal reserves are widespread, but the largest reserves are in a small number of countries, notably the United States, China, Russia, India and Australia. Unlike oil and gas, coal exploitation has not been generally constrained by resource nationalism, except in Venezuela. A well established international coal market ensures that demand is met from the most economic suppliers, around 15% of hard coal production was traded between countries in 2008. Undoubtedly, as demonstrated in this *Outlook*, where global coal demand grows by 1.7%, 0.6% and -1.0% on average over 2008-2035 in the three scenarios, the limit to continued growth in the use of coal does not lie in scarcity of resources, but depends rather on how coal's carbon intensity can be reconciled with the growing global momentum to stabilise greenhouse-gas emissions at a sustainable level.

Coal production prospects

In the New Policies Scenario, global coal production is projected to increase by about 740 Mtce reaching 5 620 Mtce by 2035 (Table 6.3). Most of the growth occurs in non-OECD countries. Reflecting the underlying demand trends, nearly all the incremental growth in global coal production comes in the form of steam coal; coking coal production expands by about 5% by 2035 compared to today; brown coal production declines by 20 Mtce by the end of the projection period relative to 2008 levels.

Coal production fell in most OECD countries in 2009 in reaction to weak demand, with only Australia, the Slovak Republic and the United Kingdom showing any growth. Coal output in the United States fell by 9%, in response to weak electricity demand and competition from natural gas. In OECD Europe, coal production fell by almost 7%, notably in Poland and Germany, where the government and industry have adopted an agreed plan to phase out hard coal production by 2018. In line with a projected average decline in OECD demand of 1.7% per year in the New Policies Scenario, production in most OECD regions is expected to decline over the projection period; the main exception is Australia, where growth in export demand increases production by 0.6% per year.

To meet growing electricity and industrial demand, *China's* coal production grows on average by 1.1% per year to reach 2 825 Mtce in 2035, equal to one-half of global coal output and 35% higher than in 2008. By any measure, the story of coal in China is remarkable. The annual production capacity of new coal mines under construction is estimated to be 200-300 Mt, comparable to the European Union's annual hard coal consumption. In China's latest Five-Year Plan, which envisages a rise in coal production to 3 600 Mt by 2015, Xinjiang is identified as a province for future coal exploitation (see Spotlight). Shenhua Group and other Chinese coal companies have announced plans to invest in this region and, although it is remote from demand centres, coal output there could grow to 1 000 Mt to feed coal conversion processes, such as electricity generation, chemicals production and synthetic fuels manufacturing.

Is Xinjiang destined to become the Ghawar of coal?

Xinjiang is a sparsely populated, autonomous region on the north-western frontier of China, with borders extending north to Russia and west to Central Asia. As China's largest administrative region, accounting for 17% of the nation's surface area, Xinjiang covers an area comparable to that of Iran. The region has vast mineral wealth, which could contribute crucially to China's energy needs. In addition to large oil and gas deposits, Xinjiang is significantly rich in coal, with an estimated 2.2 trillion tonnes of resources, or around 40% of China's total. However, as the region is far from major energy-consuming centres in the coastal areas, its coal resources have so far been largely untapped. Mining has been directed to meeting local demand and in 2009 output was around 90 Mt, or less than 5% of China's total production.

There is an expectation that Xinjiang will play an increasingly important role in meeting China's coal demand in the decades ahead. As part of its long-term strategy to promote economic growth in the west of the country, in order to raise living standards and shift growth away from the more prosperous coastal areas, China is promoting the development of Xinjiang's vast coal resources. This will help offset losses in production from resources in eastern regions which are being steadily depleted and smooth the way for closure and consolidation of smaller mines throughout the country for safety and environmental reasons.

The major impediment to developing Xinjiang's coal resources has been bottlenecks in transport capacity between its mines and demand centres in the east. But for a number of years now the Chinese government has been working with Xinjiang to address this constraint. Construction of a new rail link running from Xinjiang to the inland provinces of Gansu and Qingha is set to be completed in 2013 and it will allow the existing line to be dedicated exclusively to freight. Xinjiang's regional government expects that the upgraded railway network will permit an increase in the region's coal output to 500 Mt in 2015 and 1 000 Mt in 2020. By that time, Xinjiang's contribution to global coal production could be double the contribution that Ghawar — the world's largest oil field — currently makes to global oil production.

In addition to increasing coal production, Xinjiang has initiated other projects to use coal to fuel its economic development. It is rapidly expanding its power generating capacity, much of which will be dedicated to delivering electricity to the eastern provinces. Consistent with China's push to minimise reliance on costly imports of natural gas, it is also pushing ahead with the development of coal gasification projects. If Xinjiang's plans for expanding its coal production are fully realised, there would be major repercussions for global markets. It could help China revert to being a net-exporter of coal, which could be expected to put considerable downward pressure on the prices of internationally traded coal and impact the plans of other coal exporters.

Table 6.3 ● Coal production by region in the New Policies Scenario (Mtce)

	1980	2008	2015	2020	2025	2030	2035	2008-2035*
OECD	1 384	1 478	1 461	1 382	1 306	1 219	1 106	-1.1%
North America	672	883	863	825	773	709	621	-1.3%
United States	*640*	*828*	*807*	*775*	*731*	*670*	*589*	*-1.3%*
Europe	609	258	195	161	138	118	89	-3.8%
Pacific	103	337	403	396	395	392	396	0.6%
Australia	*74*	*331*	*399*	*392*	*392*	*389*	*393*	*0.6%*
Non-OECD	1 196	3 401	4 099	4 284	4 388	4 473	4 514	1.1%
E. Europe/Eurasia	519	401	376	351	344	336	325	-0.8%
Caspian	*n.a.*	*72*	*77*	*80*	*80*	*78*	*76*	*0.2%*
Russia	*n.a.*	*239*	*224*	*208*	*203*	*197*	*193*	*-0.8%*
Asia	568	2 712	3 403	3 610	3 724	3 806	3 862	1.3%
China	*444*	*2 076*	*2 605*	*2 747*	*2 814*	*2 839*	*2 825*	*1.1%*
India	*77*	*322*	*364*	*410*	*434*	*461*	*500*	*1.7%*
Indonesia	*0*	*236*	*319*	*328*	*351*	*376*	*400*	*2.0%*
Middle East	1	2	2	2	2	2	2	1.4%
Africa	100	208	217	222	221	225	226	0.3%
South Africa	*95*	*204*	*202*	*205*	*203*	*206*	*210*	*0.1%*
Latin America	9	79	101	99	97	104	99	0.8%
Colombia	*4*	*68*	*85*	*84*	*83*	*89*	*84*	*0.8%*
World	2 579	4 880	5 561	5 665	5 694	5 692	5 621	0.5%
European Union	*n.a.*	*254*	*188*	*143*	*118*	*96*	*70*	*-4.7%*

*Compound average annual growth rate.

Indonesia's production increased by an estimated 10% in 2009 over 2008 and is expected to continue growing in the future to satisfy domestic and export demand, as mining companies move to exploit reserves further inland. In 2009, China became the largest importer of Indonesian coal, having been a relatively minor importer in previous years. While the Indonesian government plans to give domestic demand priority over exports, the mining industry appears confident it can easily satisfy both growing export demand and local demand from planned new power projects. In the New Policies Scenario, Indonesian production increases by 70%, to reach 400 Mtce by 2035, a level exceeding the projected output of Australia. Production elsewhere in Asia, including India whose production increases by around 55% from today's levels, is projected to rise in response to strong domestic demand and in certain cases, like that of Mongolia, to satisfy export demand.

Russian coal production fell in 2009, reflecting the difficult economic situation, but exports increased by 20%, including to the distant Asian market through the ports of Vostochny and Siberian Coal Energy Company's (SUEK) newly expanded Vanino port. The construction of a second terminal at Muchka Bay is in progress and a new 317 kilometre rail line is under construction, linking coal reserves in the Sakha (Yakutia) Republic to the eastern ports. A projected decline in domestic demand of 0.2% per year, coupled with significant declines in demand in traditional markets in Europe, are expected to result in Russia's production declining to just under 195 Mtce by 2035 in the New Policies Scenario.

Output in *South Africa* in 2009 is estimated to have declined by 2%. Its future level of exports will depend on the relative priority given to coal production for export, given the rising domestic demand for electrification. Coal production in the New Policies Scenario for Africa as a whole is projected to grow by 0.3% per year over the projection period, South African production remaining similar to today's levels, while in Mozambique, Botswana and elsewhere new coal production prospects emerge.

Colombian coal exports are estimated to have grown by 2% in 2009. The potential to export over 100 Mt per year exists as a result of the construction by MPX, a Brazilian company, of a new port at Dibulla. Exports to Asia are expected to grow, despite the long shipping distances. Some Colombian coal will also transit the Panama Canal, which will be able to accommodate larger vessels when expansion is completed in 2014. In *Venezuela,* strikes and bad weather hindered production in 2009, which fell by 40%. The outlook is constrained, since the government has stated that production should not exceed 10 Mt and mining concessions will not be renewed, as part of the planned nationalisation of the mining industry. In line with projected domestic demand and global net-trade in the New Policies Scenario, Latin American production is expected to increase in the medium-term, before stabilising around 100 Mtce over the second-half of the projection period.

Inter-regional trade

The patterns of coal trade shifted markedly in 2009, as the Asian market consolidated its dominance of global trade. Whereas Japan and South Korea have long been the world's largest coal importers, the non-OECD economies of China, Chinese Taipei and India are now just as significant. While the overall level of global coal trade changed little from 2008, significant growth in the Pacific market was offset by a decline in the Atlantic market.

China's imports of hard coal tripled in 2009 to reach 137 Mt, while exports fell sharply from 45 Mt to 23 Mt, resulting in China becoming a net importer for the first time — a development foreseen three years ago, though the pace of growth of imports in 2009 was unexpected (IEA, 2007). Australia, Indonesia and Vietnam have been the main sources of China's imports, but China's growth has affected the international coal market as whole, with Colombian coal being shipped to China for the first time. The future extent of China's net imports of coal remains highly uncertain, hinging principally on coal demand in coastal areas and the relative competiveness of imported coal and gas against domestic sources of coal. Securing future fuel supply for power generation is of crucial importance for China's coastal region, which requires large amounts of electricity for its economic development. But as resources are scarce in the region, large amounts of fuel must be brought in from within and outside China. Imports of natural gas are expected to rise over the coming decades. Three liquefied natural gas (LNG) terminals are already in operation and six additional are being constructed in the coastal provinces; a natural gas pipeline from Turkmenistan was also commissioned in 2009. While priority for natural gas use is at present given to the residential and industry sectors, more gas is expected to be used in power generation in the future, as import capacity increases (see Chapter 5 and Chapter 18). The price of LNG imported into China, which varies at present from 4 to 12 dollars per million British thermal units ($/MBtu), will be the key factor. The generating costs of power plants using imported LNG and coal determine the mine-mouth coal costs required for domestic coal to be competitive (Figure 6.4). For example, assuming an imported coal cost of 90 dollars per tonne ($/t), power plants using indigenous coal from a mine within 500 kilometres (km) remain competitive at mine-mouth costs lower than around $65/t. However, should the imported coal cost be on the lower level of $60/t, mine-mouth costs lower than around $40/t would remain competitive.

Figure 6.4 ● Power generation costs by fuel and distances in China, 2009

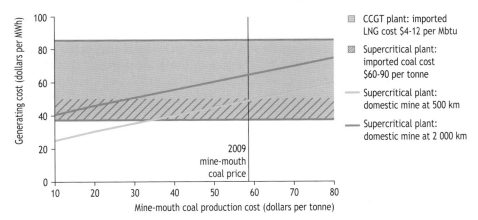

Source: IEA analysis.

In the New Policies Scenario, global trade in hard coal among *WEO* regions is projected to rise from 728 Mtce today to just under 870 Mtce before 2020, before decreasing to settle at a level around 840 Mtce as global demand for coal stabilises over the second-half of the projection period (Table 6.4). By 2035, inter-regional trade meets 16% of global hard coal demand, a level similar to today. On average the value of global hard coal trade over the period 2010-2035 is equal to $125 billion (in 2009 dollars) per year, while that for oil and gas amounts to around $1 580 billion and $410 billion, respectively. Net exports from high-cost producing countries, like the United States and Russia, decline over the projection period, while net exports from Australia and Indonesia increase, by just over 70 Mtce and 50 Mtce, respectively. India's net imports increase five-fold to reach 280 Mtce by 2035, while China swings from being a net importer in the short and medium term to a net exporter by 2035, in response to stabilisation of domestic demand at around 2 830 Mtce in the second-half of the *Outlook* period and continued strong domestic output growth.

Table 6.4 ● Inter-regional hard coal* net trade by region in the New Policies Scenario (Mtce)

	2008		2020		2035	
	Mtce	Share of primary demand**	Mtce	Share of primary demand**	Mtce	Share of primary demand**
OECD	- 154	11%	- 71	6%	86	9%
North America	51	6%	36	5%	25	4%
United States	39	5%	28	4%	13	2%
Europe	- 203	65%	- 185	77%	- 136	86%
Pacific	- 2	1%	78	21%	197	53%
Australia	247	80%	314	85%	320	86%
Non-OECD	214	7%	71	2%	- 86	2%
E. Europe/Eurasia	72	23%	46	17%	35	14%
Caspian	26	38%	21	28%	20	28%
Russia	65	33%	45	27%	35	23%
Asia	64	2%	- 77	2%	- 218	6%
China	13	1%	- 41	1%	3	0%
India	- 52	14%	- 141	26%	- 281	37%
Indonesia	181	86%	217	78%	232	78%
Middle East	- 12	88%	- 15	90%	- 27	93%
Africa	48	23%	63	28%	66	29%
South Africa	56	28%	63	31%	63	30%
Latin America	42	56%	53	55%	58	60%
Colombia	63	92%	79	94%	79	95%
World***	728	16%	844	16%	838	16%
European Union	- 194	65%	- 171	77%	- 123	88%

* Steam and coking coal (including coke). ** Production for exporting regions/countries. *** Total net imports for all *WEO* regions/countries (some of which are not shown in this table), not including trade within *WEO* regions.
Note: Positive numbers denote export; negative numbers imports.

Compared with 2007, the cost of producing steam coal for the international market rose by around $10/t across most regions in 2008, due to the higher cost of diesel, labour, steel, spare parts and other factors (IEA, 2009). But, in 2009, there appears to have been little change in the average cash cost of internationally traded coal (Figure 6.5). Based on this coal supply cash-cost curve, the weighted-average cost is around $43/t across all countries, with Indonesia, the largest exporter of steam coal, remaining one of the lowest cost producers. One key issue is the movement in costs relative to the average free-on-board (FOB) prices for 2009. The fall in FOB prices since the peak of 2008, coupled with rising costs as supply chains became stretched and infrastructure constrained, squeezed margins significantly. FOB prices in the Asian market are already rising in 2010 in response to a rise in demand from Pacific market economies, while prices in the Atlantic market have remained relatively soft due to lower demand for electricity, resulting from the economic downturn. Coal futures suggest prices will rise over the next four to five years as the world emerges from the recent economic crisis, as reflected in the underlying assumptions in this *Outlook* (see Chapter 1). During 2009, large discrepancies were observed between coal prices around the world. The highest prices could be found at Chinese ports, which translated back to high FOB prices at Australian ports. This lifted South African coal export prices, which rose above European import prices, a trend that continued during the first half of 2010, resulting in a lack of demand in Europe for South African coal. Colombia has faced the lowest export prices, because of low demand from North America and Europe. Coking coal prices have traditionally been set during annual negotiations with Japanese steel producers, with other steel producers largely accepting the outcome. This archaic system is gradually being replaced by more transparent market-based pricing.

Figure 6.5 ● **Coal supply cash-cost curve for internationally traded steam coal for 2009 and average FOB prices for 2009 and first-half 2010**

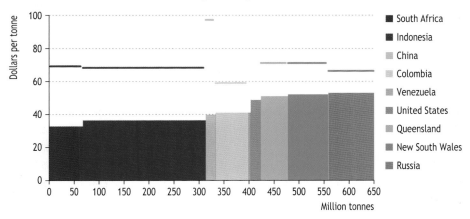

Note: Boxes represent costs and bars show FOB prices. Values adjusted to 6 000 kcal/kg.
Sources: IEA Clean Coal Centre, citing data from Marston and IHS Global Insight.

Investment

Current trends

Investments by 25 leading coal companies, which in 2009 accounted for around 35% of global coal production and 50% of global coal trade, rose by about 4.5% in 2009 (Table 6.5) to $12 billion. Nonetheless, the rise in investment was much less than in 2008, when it rose by 18%. By contrast, the top 25 oil and gas companies invested $350 billion in 2009 (see Chapter 3). Investment in the coal sector had been expected to fall significantly in 2009, with many companies announcing delayed and cancelled investments during the first half of 2009, in response to the global financial crisis. However, cash flows from higher-than-expected prices in the second-half of the year allowed many companies to maintain planned investments, even those which were forced to make production cuts and lay off workers in response to weak coal demand in OECD markets.

Table 6.5 ● Production, exports and investment of 25 leading coal companies

	Corporate base	Production (Mt)	Exports (Mt)	Investment ($ million)		
		2009	2009	2007	2008	2009
Coal India	India	431	2	426	449	630
Shenhua Group	China	254	14	2 080	2 090	1 169
Peabody Energy	United States	221	n.a.	439	264	261
Rio Tinto	United Kingdom	140	n.a.	452	653	632
Datong Coal Mining Group	China	125	3	n.a.	n.a.	n.a.
Arch Coal	United States	114	3	488	497	323
China National Coal Group	China	109	4	761	1 142	1 874
BHP Billiton	Australia	105	58	873	938	2 438
RWE Power	Germany	100	1	263	331	459
Anglo American	United Kingdom	96	45	1 052	832	496
Xstrata	Switzerland	95	79	807	1 204	1 327
SUEK	Russia	87	31	357	449	351
Shanxi Coking Coal Group	China	78	2	n.a.	n.a.	n.a.
PT Bumi Resources	Indonesia	63	49	210	567	484
Consol Energy	United States	54	3	681	446	544
Kuzbassrazrezugol	Russia	46	26	n.a.	n.a.	n.a.
Kompania W_glowa	Poland	42	6	234	371	316
PT Adaro Indonesia	Indonesia	41	32	71	151	106
Sasol	South Africa	37	3	131	121	170
Massey Energy	United States	34	5	271	737	275
Mitsubishi Development	Japan	28	28	n.a.	n.a.	n.a.
PT Kideco Jaya Agung	Indonesia	24	17	n.a.	n.a.	n.a.
Banpu	Thailand	21	21	92	120	82
Teck Cominco	Canada	19	n.a.	33	111	60
Drummond	United States	27	27	n.a.	n.a.	n.a.
Total		2 392	458	9 722	11 471	11 996

Sources: Company reports and IEA analysis.

The 25 leading coal companies saw their production drop by close to 2% in 2009. This aggregate figure hides wide differences, from a production rise of close to 20% at PT Bumi in Indonesia to a fall of 18% at coking coal producer Teck Cominco in Canada. BHP Billiton stands out in terms of its 2009 investments. Its financial year runs to 30 June, so the $2.4 billion reported includes investment made in the second-half of

2008. The figure includes a tripling of investment in Australian coking coal production, a doubling of investment in South African steam coal production and investment in a third coal terminal at Newcastle, Australia. Production at China's three largest coal companies rose 7% in 2009, in line with a rise in national production. The future investment plans of these three companies reflect China's ambition to continue the rapid expansion of its coal industry by opening large new mines. Taken together, the Shenhua and China National Coal Groups have announced 2010 investment plans that are 70% higher than in 2009.

Investment needs to 2035

Overall, the coal sector has been nimble in its response to the economic crisis that was quickly followed by a massive upturn in coal import demand from China. The investments that are being made today suggest that the industry will invest quickly enough to meet the future demand growth under the three scenarios examined in this *Outlook*. Cumulative coal-supply infrastructure investment in the period 2010-2035 amounts to around $720 billion in the New Policies Scenario, accounting for just over 2% of the cumulative investment in the world's energy-supply infrastructure (see Chapter 2). Total coal sector investment, two-thirds of which is in the non-OECD countries and nearly half within the next ten years, is mainly required for mine investments, with just under 10% required for the associated infrastructure.

6

CHAPTER 7

POWER SECTOR OUTLOOK
Evolution or revolution?

H I G H L I G H T S

- World electricity demand in the New Policies Scenario is projected to grow by 2.2% per year between 2008 and 2035, from 16 819 TWh to about 30 300 TWh, slowing toward the end of the projection period as a result of increasing economic maturity and more efficient electricity use. Demand growth is led primarily by non-OECD countries, which are responsible for more than 80% of the incremental growth that occurs between 2008 and 2035.

- Policies implemented to enhance energy security and to curb emissions underpin the transition toward low-carbon technologies in the power sector. The combined share of world electricity generation from nuclear and renewable sources is projected to increase from 32% in 2008 to 45% in 2035, with generation from renewables tripling. The shift to low-carbon technologies reduces the CO_2 intensity of the world power sector from 536 grammes of CO_2 per kWh today to less than 360 grammes of CO_2 per kWh by 2035.

- Globally, coal remains the dominant source of electricity generation in 2035, although its share declines from 41% in 2008 to 32% by 2035. In OECD countries, coal-fired generation drops by one-third between now and 2035, becoming the third-largest source of electricity generation. Growth in coal-fired generation is led by the non-OECD countries, where it doubles over the *Outlook* period. Gas-fired generation grows in absolute terms, but maintains a stable share of world electricity generation at around 21% over the *Outlook* period.

- In China, electricity demand triples between 2008 and 2035. Coal remains the cornerstone of the electricity mix, although its share of generation drops from 79% in 2008 to 55% in 2035 with expected increases in the use of renewable energy, nuclear and hydropower. In absolute terms, China sees the biggest increase in generation from both renewable sources and nuclear power over the *Outlook* period. Between 2009 and 2025, China is projected to add generating capacity equivalent to the current total installed capacity of the United States.

- Total capacity additions, to replace obsolete capacity and to meet demand growth, amount to more than 5 900 GW globally in the period 2009-2035; over 40% of this is installed by 2020. Cumulative global investment required in the power sector is $16.6 trillion (in year-2009 dollars) over 2010-2035. About $9.6 trillion of the total, or almost 60%, is needed to build new generating plants. Improvement and expansion of electricity networks accounts for the remainder, with cumulative investment in transmission and distribution totalling $2.2 trillion and $4.8 trillion, respectively.

Electricity demand

The global *Outlook* for the power sector depends heavily on the nature and extent of policy action to reduce carbon-dioxide (CO_2) emissions and enhance energy security. In all three scenarios, electricity demand increases from 2008 to 2035, driven primarily by economic and population growth. Demand growth is expected to resume, with economic recovery, in each of the scenarios, following stagnation in 2008 and 2009 as a result of the global financial crisis and subsequent recession. The Current Policies Scenario projects electricity demand to rise at an average annual growth rate of 2.5% between 2008 and 2035 (Table 7.1). Projections for electricity demand growth over the same period are lower in both the New Policies Scenario and 450 Scenario — averaging 2.2% and 1.9% per year, respectively — primarily as a result of policies aimed at improving end-use energy efficiency and curtailing CO_2 emissions.

Table 7.1 ● Final electricity consumption by region and scenario (TWh)

			New Policies Scenario		Current Policies Scenario		450 Scenario	
	1980	2008	2020	2035	2020	2035	2020	2035
OECD	4 739	9 244	10 339	11 566	10 488	12 101	10 097	10 969
Non-OECD	971	7 575	12 841	18 763	13 233	20 820	12 375	16 660
World	5 711	16 819	23 180	30 329	23 721	32 922	22 472	27 629

Note: TWh = terawatt-hours.

The rate of demand growth slows over the *Outlook* period in each of the three scenarios, reflecting increasing economic maturity and more efficient electricity use. The New Policies Scenario projects world electricity demand rising at an annual rate of 2.7% between 2008 and 2020, and 1.8% per year over the period 2020 to 2035. Increased energy efficiency causes the rate of electricity demand growth in the OECD to slow from 0.9% between 2008 and 2020 to 0.8% per year over the period 2020 to 2035. The effect of more efficient electricity use is most notable in the non-OECD, where demand growth is 4.5% per year from 2008 to 2020, but averages 2.6% annually over the remainder of the *Outlook* period. More than 80% of incremental electricity demand between 2008 and 2035 comes from non-OECD countries, led by China, where, in 2035, demand is projected to equal that of the United States and European Union combined.

Despite projections for strong demand growth outside the OCED, per-capita electricity consumption remains low in several regions in each of the scenarios. In the New Policies Scenario, electricity consumption per-capita doubles to 2 600 kilowatt-hours (kWh) in non-OECD countries from 2008 to 2035, whereas sub-Saharan African consumption only reaches 220 kWh per person by 2035, the lowest overall per-capita electricity consumption in any region. This is less than 3% of the average per-capita consumption projected for that same year in OECD countries. Some 585 million people in sub-Saharan Africa currently lack access to electricity, 79% of whom live in rural areas. The level of investment needed to achieve universal electricity access and its implications for the global energy market and CO_2 emissions are discussed in Chapter 8.

Electricity supply

Compared with today, in the New Policies Scenario the power sector undergoes a significant transition toward low-carbon technologies between 2008 and 2035, achieving a more diverse mix. This is stimulated by several major policy actions.[1] First, we assume in the New Policies Scenario that some countries (those in the OECD and in non-OECD Europe) adopt policies to curb CO_2 emissions, such as cap-and-trade systems that lead to rising prices of CO_2. Second, we assume that many countries, including large transition economies, implement policies designed to support renewable energy and nuclear power in order to diversify their fuel mix and enhance energy security.

Global electricity generation grows by 75% over the *Outlook* period, rising from 20 183 terawatt-hours (TWh) in 2008 to 27 400 TWh in 2020, and to 35 300 TWh in 2035 (Figure 7.1).[2] Coal continues to be the main source of electricity production, despite its share of the world mix declining from 41% in 2008 to 32% by 2035. In contrast, the share of generation from non-hydro renewable energy sources — wind, biomass, solar, geothermal and marine — increases more than five-fold, from 3% in 2008 to 16% by 2035. Electricity production from natural gas maintains a constant percentage of global generation at about 21%; similarly, the shares of hydro and nuclear also stay flat at 16% and 14%, respectively. Oil-fired generation, already a minor source of power generation in most countries, falls further to just 1% of total generation by 2035.

Figure 7.1 ● World electricity generation by type in the New Policies Scenario

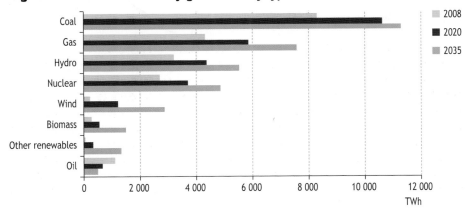

In the New Policies Scenario, coal remains the dominant fuel source in the power sector. Worldwide coal-fired generation is projected to increase from 8 273 TWh in 2008 to about 11 200 TWh by 2035, although trends differ markedly by region (Figure 7.2).[3] In OECD countries, coal-fired generation drops by one-third between

---

1. Annex B outlines key policy assumptions by region for the three different scenarios.
2. Electricity generation includes final consumption of electricity, network losses, own use of electricity at power plants and "other energy sector".
3. Annex A contains detailed projections of electricity generation by region and fuel, as well as other power sector trends.

2008 and 2035 as the price of CO_2 rises and drives up the operational costs of these plants.[4] Even in the absence of government policies to curb CO_2 emissions, many power companies have had difficulty building coal-fired plants, particularly in the United States, because of public opposition stemming from environmental concerns and uncertainties about future regulations. In the OECD, coal becomes the third-largest source of electricity generation, behind natural gas and nuclear by the end of the *Outlook* period. By contrast, coal-fired generation is projected to double in non-OECD countries between 2008 and 2035, where more favourable costs and domestic coal availability contribute to its role as a secure fuel to support economic growth.

Figure 7.2 ● Coal-fired electricity generation by region in the New Policies Scenario

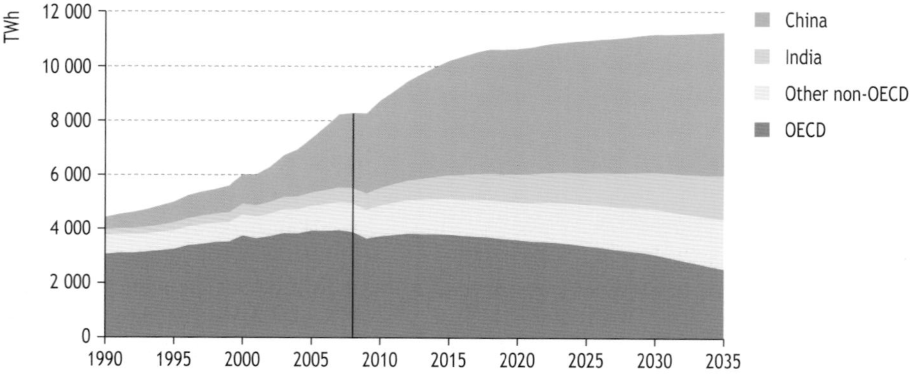

The mix of world coal-fired generation technologies evolves between 2008 and 2035. Globally, generation from less efficient subcritical plants falls off dramatically, from 73% in 2008 to 48% in 2020, and to 31% by 2035. Over the medium term, these plants are displaced, primarily by supercritical plants and a rising share of combined heat and power (CHP) plants; after 2020, more advanced technologies, such as ultra-supercritical and integrated gasification combined-cycle (IGCC) plants, are more widely deployed. These technological changes steadily improve the average efficiency of the world coal-fired fleet (excluding CHP plants), which reaches above 40% by 2035, up from 35% today. Particularly striking is the progress seen in the non-OECD countries, where the average efficiency of coal-fired generation plants rises from 33% in 2008 to 40% by 2035 (Figure 7.3).

Carbon capture and storage (CCS) technology is expected to be deployed on a limited scale in the New Policies Scenario, its share of total generation rising from zero today to 1.5% in 2035. Most of the projected generation from plants fitted with CCS equipment is in OECD countries, driven by government initiatives to build demonstration facilities. Stronger CO_2 price signals than those in the New Policies Scenario would be needed to stimulate wider adoption of CCS technology.

---

4. Cost assumptions by fuel/technology and region are available at www.worldenergyoutlook.org.

*World Energy Outlook 2010 - **GLOBAL ENERGY TRENDS***

Figure 7.3 ● **Coal-fired electricity generation by technology and region in the New Policies Scenario**

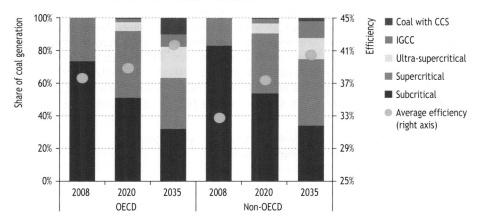

Note: Excludes coal-fired generation from CHP plants.

Gas-fired generation rises from 4 303 TWh in 2008 to almost 7 600 TWh by 2035, with about 80% of this growth occurring in non-OECD countries. Notable growth occurs in the Middle East, where gas-fired generation doubles over the *Outlook* period, rising to over 1 000 TWh by 2035. Significant gas resources are available in the region, making it an attractive fuel to meet accelerating electricity needs and also to displace oil-fired generation, thereby freeing up oil for other domestic uses or export. Gas-fired generation also rises considerably in non-OECD Asia, by 5.1% per year to 2035, driven by strong growth in both China and India.

In OECD countries, gas-fired generation continues to climb, though we project a slowing in the pace of growth (0.9% per year) from 2008 to 2035 compared to the rapid expansion (more than 6% per year) since 1990, led by the United States and Europe. Gas use in the power sector is sensitive to several factors, including the depth and duration of the shale-gas boom in North America and its impact on prices, the stringency and pace of actions to reduce CO_2 emissions and the rate of penetration by renewable energy sources. Gas plays an important role for countries making the transition to a low-carbon power sector. Emitting approximately half the CO_2 per unit of electricity produced compared with coal, gas offers a flexible source of generation that permits electricity to be quickly dispatched to meet rapid demand surges. It also provides back-up capacity to support and balance electricity markets, particularly with the increasing deployment of variable generating sources.

Projected increases in world oil prices make the economics of oil-fired generation increasingly unattractive and lead to its continued decline, with output dropping from 1 104 TWh in 2008 to around 500 TWh by the end of the *Outlook* period. By 2035, over 40% of global oil-fired generation is projected to come from the Middle East, where many countries are likely to continue to subsidise the price of oil products for electricity generation.

Low-carbon technologies increasingly penetrate the electricity mix in the New Policies Scenario. Renewable sources (including hydro) and nuclear power are projected to account for 45% of total global generation by 2035, up from 32% today (Figure 7.4). A marked shift occurs in OECD countries, where this share reaches 56% by 2035. Non-OECD countries also move towards low-carbon technologies in the power sector, albeit reaching a lower level because of a smaller base at the beginning of the *Outlook* period and less vigorous policy action to mitigate CO_2 emissions. Renewable energy and nuclear power account for 39% of generation there by 2035. In absolute terms, China sees the biggest increase in generation from both renewable sources and nuclear power between 2008 and 2035, at almost 2 000 TWh and 830 TWh.

Figure 7.4 ● *Share of nuclear and renewable energy in total electricity generation by region in the New Policies Scenario*

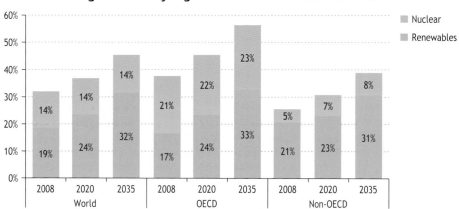

In absolute terms, global electricity generation from renewable sources triples in the New Policies Scenario, increasing from 3 772 TWh in 2008 to nearly 11 200 TWh by 2035. Rapidly expanding wind generation, rising from 219 TWh in 2008 to almost 2 900 TWh by 2035, underpins this marked growth. Electricity supply from wind grows at an average rate of 8% and 15% per year, respectively, in the OECD and the non-OECD over the period 2008-2035. Hydropower is another major source of increasing low-carbon electricity production, with generation climbing from 3 208 TWh in 2008 to about 5 500 TWh by 2035. Nearly 90% of this additional hydropower generation comes from non-OECD countries, where considerable resource potential still remains. Biomass generation increases more than five-fold over the *Outlook* period, rising to around 1 500 TWh in 2035. Other sources of renewable electricity supply — solar photovoltaics (PV), concentrating solar power (CSP) and marine energy — experience step changes in growth, but begin from a small base.

Greater deployment of renewable energy in the New Policies Scenario, while helping to achieve a lower-carbon electricity mix, has profound implications for the operation and development of the electricity system, related to security of supply, infrastructure and costs. A detailed study of renewable energy trends in the power sector and their impacts can be found in Chapter 10.

Concerns over energy security, rapidly rising demand, climate change and local pollution are driving a resurgence of interest in nuclear power in many countries. Electricity production from nuclear power is projected to climb to 4 900 TWh in 2035, up from 2 731 TWh in 2008. About 40% of this growth occurs in China alone. Rising production reflects the construction of new capacity in many other regions that are actively investing in nuclear technology or have policies in place to support nuclear power (*e.g.* policy targets, government loan guarantees), including the European Union, India, Japan, Russia, Korea and the United States. Further impetus for new nuclear construction comes from assumed rising prices of CO_2 in OECD countries.

Figure 7.5 ● Nuclear capacity under construction and additions by region in the New Policies Scenario

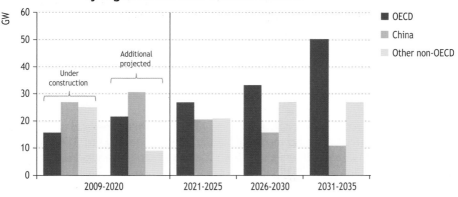

Sources: IAEA (2010); IEA analysis.

Currently, most construction of nuclear capacity is being undertaken in the non-OECD countries, where 52 gigawatts (GW) of generating capacity is being built (of which 36 GW came online in 2009-2010); about 27 GW of capacity is currently under construction in mainland China (Figure 7.5) (IAEA, 2010). Given China's plans to achieve 15% of total energy use from non-fossil-fuel sources by 2020, additional nuclear units are expected to be built between 2010 and 2020. About 16 GW of new capacity is currently under construction in OECD countries (of which 2 GW came online in 2009-2010). Of this, most is being built in Korea, Japan, France and Finland, where nuclear power development remains a core part of energy policy. Elsewhere, several projects that were previously suspended for many years have now been revived. While many OECD countries have expressed interest in and taken steps to encourage renewed development of nuclear power, new construction so far is very limited, due largely to cost uncertainties and financing limitations.

Globally, the shift to low-carbon technologies in the New Policies Scenario causes the CO_2 intensity of power generation to fall by 34%, from 536 grammes of CO_2 per kWh today to less than 360 grammes of CO_2 per kWh in 2035 (Figure 7.6). By 2035, the CO_2 intensity in the European Union and Japan declines to less than half the levels of 2008, as low-carbon power generation displaces that from retired coal plants. The use of more efficient coal technologies contributes to significant reductions in power sector CO_2 intensity in regions such as China and India, where coal-fired generation continues to grow.

Figure 7.6 ● CO_2 intensity of power generation by region in the New Policies Scenario

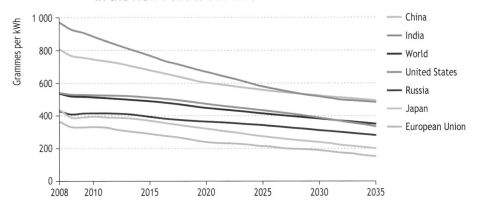

In the New Policies Scenario, worldwide CO_2 emissions from the power sector are projected to rise from 11.9 gigatonnes (Gt) in 2008 and peak at close to 14 Gt in 2030. In OECD countries, the progressive shift towards low-carbon technologies leads to declining power sector emissions from 2008 to 2035. Of total CO_2 emissions from OECD countries, the share of the power sector drops from 39% in 2008 to 33% during the *Outlook* period. Average CO_2 emissions from the power sector in non-OECD countries continue to rise through to 2035, as all forms of generation, including large amounts of coal-fired generation, increase to meet surging demand (Figure 7.7). CO_2 emissions from the world power sector increase by 1.8 Gt between 2008 and 2035, slightly less than the additional CO_2 emissions from the transport sector over the same period. In absolute terms, global CO_2 emissions from coal fall by 4.0% between 2020 and 2035, even as coal-fired generation climbs by 5.7% during that period — reflecting the growing use of more advanced technologies.

Figure 7.7 ● CO_2 emissions from the power sector by region in the New Policies Scenario

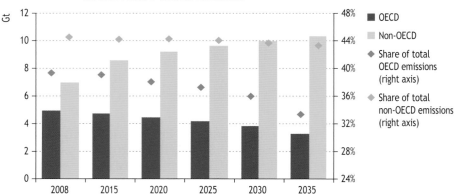

Box 7.1 ● Smart solutions to electricity system challenges

The generation of and demand for electricity is constantly evolving, with challenges in all parts of the electricity system from generation, transmission and distribution, to end use. These include managing electricity production from variable sources, meeting short-duration peak loads and accommodating the growing use of plug-in hybrid and electric vehicles. Technology change and enhancements in electricity system operation are becoming essential to ensure affordable, responsive and reliable service.

One solution to these challenges is to build what is often referred to as a "Smart Grid". A Smart Grid is an electricity network that uses digital technology to monitor and manage the generation and transport of electricity from all sources in order to meet the varying electricity demands of end users as efficiently as possible. Such a grid is able to co-ordinate the needs and capabilities of all generators, grid operators, end users and electricity market stakeholders in a way that optimises asset utilisation and operation. In the process (and with appropriate market signals in place), it can minimise both costs and environmental impacts, while maintaining system reliability, resilience and stability.

Smart Grids can enable wider deployment of variable technologies, such as wind and solar PV, by observing and responding to changing conditions throughout the entire electricity system and thereby maintaining a reliable service. Meeting peak demand for electricity requires a system to efficiently handle a load that may occur for only a very short duration. Smart Grids reduce peak demand by allowing customers, manually and/or automatically, to reduce and/or time-shift their consumption with little impact on operation or lifestyle. This permits minimisation of additional investment for peak plants and consequently lowers prices to end users (IEA, forthcoming).

New capacity additions, retirements and investment

Total global installed power generation capacity in the New Policies Scenario is projected to increase from 4 722 GW in 2008 to about 8 600 GW by 2035. Between 2009 and 2035, total gross capacity additions amount to 5 900 GW, with more than 40% installed by 2020. This equates to average capacity additions of 213 GW per year from 2009 to 2020, rising slightly to 224 GW per year over the period 2021-2035. Nuclear power and renewable energy additions respectively account for 5% and 41% of the total between 2009 and 2020, and 7% and 53% through the remainder of the *Outlook*. Investment in new plants rises more quickly from 2021 to 2035, as more capital-intensive technologies are deployed and more variable resources exploited creating a need for additional generating capacity (Figure 7.8). China is projected to install the largest amount of new capacity between 2009 and 2035, accounting for more than one-quarter of global additions.

New capacity is built to meet rising demand, projected to come mostly from non-OECD countries, and to replace retiring plants. Power plant lifetimes reflect technical limitations that arise with age and policies that influence both the economics of plants and the regulations under which they operate. Coal-fired generation has an average lifetime of 40 to 50 years before the plant becomes technically obsolete; for gas- and oil-fired generation, the average technical lifetime is about 40 years. When economically practical and technically feasible, the lifetime of some plants can be extended beyond these ranges by replacing specific parts.

Figure 7.8 ● World power-generation capacity additions and investment by type in the New Policies Scenario

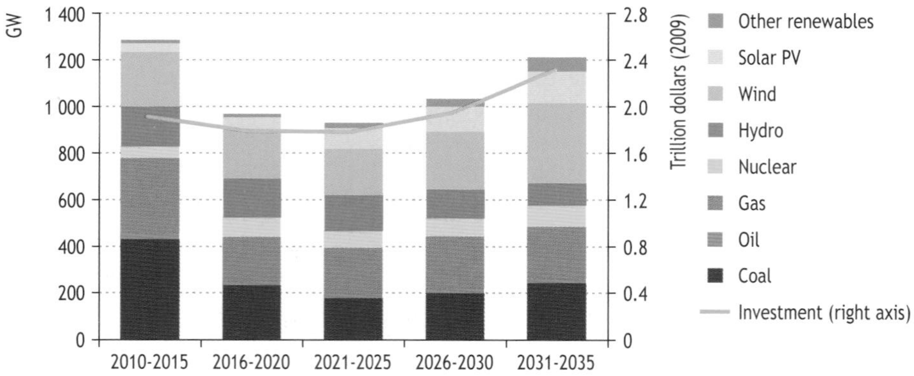

Nuclear plants, originally expected to operate for 40 years, can have their lifetimes lengthened significantly by replacing certain components. Several countries are considering extending the lifetime of nuclear plants to 60 years, with some already doing so, given adherence to safety regulations. In the United States, 20-year license renewals have already been granted for most currently operating nuclear power plants to continue operation for up to 60 years and some may have their licenses extended even further (EIA, 2010). In Germany, the average lifetime of nuclear plants is assumed to be 45 years in the New Policies Scenario.

Worldwide, over 400 GW of operational coal-, gas- and oil-fired capacity are more than 40 years old. With a further 585 GW between 30 and 40 years old, about one-third of the installed fossil-fuel capacity in 2008 will be approaching the end of its technical lifetime in the next 10 to 15 years. Further, the age distribution of power plants by region is striking. Plants in non-OECD countries are relatively young, as most have been built to respond to heightened demand growth during the past two decades. In contrast, plants in OECD countries are ageing, particularly coal plants that have long-provided base-load generating capacity (Figure 7.9). The ageing of installed thermal capacity could have implications, in both directions, on efforts to move to a less carbon-intensive electricity mix. Replacement with low-emissions technologies would work to facilitate this transition, but replacement with unmitigated thermal capacity (*i.e.* capacity that cannot be fitted for CCS) could potentially lock-in emissions for another 40 years.

In the New Policies Scenario, most power plants are retired as a result of age-related technical obsolescence, but the rising price of CO_2 in OECD countries also contribute to some early retirement of emissions-intensive plants. The impact of more aggressive CO_2 price assumptions on the early retirement of emissions-intensive capacity is discussed further in Chapter 14, together with the associated costs.

Figure 7.9 ● Age profile of installed thermal and nuclear capacity by region, 2008

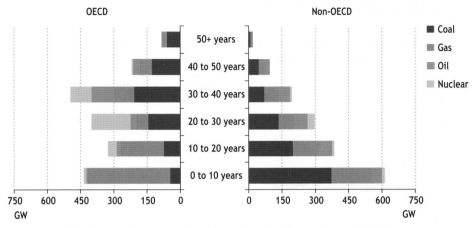

Sources: Platts World Electric Power Plants Database, December 2009 version; IAEA (2010).

Total plant retirements in the New Policies Scenario amount to some 2 000 GW over the *Outlook* period, equal to 43% of the currently installed capacity (by 2025, more than a quarter of currently installed capacity in OECD countries is retired). Fossil-fuel plants account for two-thirds of the total capacity loss from retirements between 2009 and 2035, with about 640 GW coal, 400 GW gas and 310 GW oil going offline. Additional retirements include nuclear facilities and wind installations that reach the end of their technical lifetimes, around 2030. Nearly 35% of the new capacity additions projected over the period 2009-2035 is needed to replace existing plants (Figure 7.10), while the remainder are built to meet increasing demand.

Cumulative global investment in the power sector amounts to $16.6 trillion (in year-2009 dollars) in 2010-2035 (Table 7.2). Two-thirds of the total investment comes from China, OECD Europe, the United States and India. Around $9.6 trillion of total power sector investment, or almost 60%, is invested in new generating plants (and plant refurbishments) to meet rising demand and to replace existing plants that are retired. Improvements and expansion of electricity networks account for the remainder of total power sector investment, with cumulative investment in transmission and distribution totalling $2.2 trillion and $4.8 trillion, respectively. Investment resources for transmission and distribution infrastructure can be difficult to secure given the regulatory hurdles in some countries. These are assumed to be overcome, as expanding and improving electricity networks is vital for demand management, integration of variable generation from renewable energy sources and the most efficient allocation of resources.

Table 7.2 • Capacity and investment needs in power infrastructure by region in the New Policies Scenario

	2010-2020					2021-2035				
	Capacity (GW)		Investment ($2009 billion)			Capacity (GW)		Investment ($2009 billion)		
	Additions	Retirements	New Plant	Transmission	Distribution	Additions	Retirements	New Plant	Transmission	Distribution
OECD	777	424	1 490	370	851	1 208	770	2 502	373	892
North America	322	207	585	169	363	520	324	1 039	197	424
United States	262	191	498	140	302	411	273	873	160	345
Europe	337	158	694	110	332	498	348	1 080	128	386
Pacific	118	59	211	91	156	190	98	383	48	82
Japan	74	50	120	63	105	111	61	211	28	47
Non-OECD	1 542	232	2 165	617	1 328	2 146	554	3 477	808	1 734
E. Europe/Eurasia	161	123	252	43	144	231	157	413	51	170
Caspian	29	13	35	10	33	30	18	37	8	28
Russia	91	62	143	18	60	138	94	254	22	74
Asia	1 095	74	1 526	472	975	1 494	244	2 347	613	1 265
China	773	38	1 054	306	632	760	142	1 168	274	566
India	200	22	288	102	210	428	58	679	197	407
Middle East	114	10	129	29	59	144	70	229	49	102
Africa	76	8	109	28	57	138	45	235	42	88
Latin America	95	17	149	45	93	138	38	254	53	110
Brazil	45	7	72	22	46	64	12	126	29	60
World	2 319	656	3 655	986	2 179	3 354	1 324	5 979	1 181	2 626
European Union	331	170	685	103	307	469	346	1 027	117	348

Figure 7.10 ● **World installed power-generation capacity by type in the New Policies Scenario**

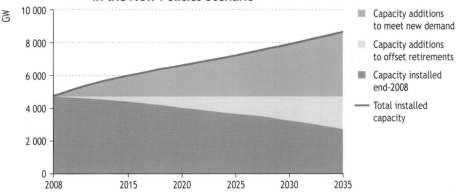

Regional trends

United States

Incentives for low-carbon energy, pricing of CO_2 and the availability of natural gas in the United States usher a major shift toward a lower-carbon electricity generation mix in the New Policies Scenario. The share of production from renewables, nuclear and plants fitted with CCS increases from 29% today to 35% by 2020, and then rises to 49% by 2035. The higher uptake of lower-carbon technologies post-2020 reflects a rising price for CO_2 in the United States, which increases to $50 per tonne by 2035. Over the *Outlook* period, coal-fired generation declines by 20% (420 TWh), as ageing capacity and escalating costs lead to the retirement of more than half of currently installed coal-fired capacity (Figure 7.11). After 2020, plants using ultra-supercritical, IGCC and CCS technology account for the majority of coal-fired capacity additions. New plants fitted with CCS are initially installed as demonstration facilities, but a climbing price for CO_2 also contribute to their reaching a 3.6% share of generation by 2035.

Figure 7.11 ● **Power-generation capacity by type in the United States in the New Policies Scenario**

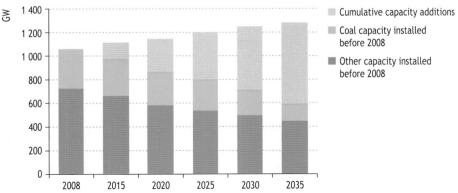

Renewable energy sources, nuclear and gas-fired generation are projected to replace generation from retired coal-fired power plants and to meet growing electricity demand in the United States. Gas-fired generation grows by about 190 TWh between 2008 and 2035 (Figure 7.14a). With the shale gas boom, and increasing gas production in North America, gas is an available resource that can enhance reliability in the power sector as more variable generation, such as wind and solar PV, is integrated. Utilisation rates for gas-fired capacity are projected to increase from 25% in 2008 to 30% in 2035.

Power generation from renewables triples from 2008 to 2035, driven partly by the assumed adoption of a federal renewable electricity standard in the New Policies Scenario. Wind power accounts for the largest additional generation, at 460 TWh. More than half of new capacity additions in the power sector over the *Outlook* period rely on renewable energy sources. After 2020, additional nuclear capacity is expected to come online, at an average rate of 1.2 GW per year, supported by loan guarantees. Few nuclear plant retirements are projected, as it is assumed most plants are granted licenses to operate for up to 80 years (most plants currently have licenses that allow operation for up to 60 years). New power plants of all types require investment of $1.4 trillion over the *Outlook* period. Nearly two-thirds of this investment comes after 2020, when a surge in deployment of low-carbon, capital-intensive technologies is anticipated. Annual CO_2 emissions from the power sector are projected to be reduced by 0.6 Gt, or more than one-quarter, compared to today.

European Union

In the European Union, the price for CO_2 in the New Policies Scenario is instituted earlier and is initially higher than in other OECD countries. It rises to $38 per tonne by 2020 and $50 per tonne by 2035. This, in combination with binding targets for renewable energy consumption, accelerates renewable electricity generation over the *Outlook* period. Coal-fired generation declines steeply, by 550 TWh between 2008 and 2035, with its share of the mix falling from 28% to 10% (Figure 7.14b). About 160 GW of coal-fired capacity (78% of currently installed capacity) is retired between 2008 and 2035, partially offset by almost 70 GW of non-subcritical and CHP coal-fired plants that come online during that period. Gas-fired power generation maintains a steady share at one-quarter of total generation mix between 2008 and 2035.

Surging generation from renewable energy sources in the European Union causes the share of electricity generation from renewables (including hydro) to climb from 17% in 2008 to 30% by 2020 and 41% by 2035 (Figure 7.12). Over the *Outlook* period, wind power accounts for more than 40% of cumulative capacity additions and supplies more incremental electricity generation than any other source. Electricity generation from nuclear power remains relatively flat in the European Union, as countries add only enough new capacity to replace plants reaching the end of their operating lifetimes (45 to 55 years). Investment in new plants of all types totals $1.7 trillion between 2010 and 2035, with more than 70% destined for renewable energy. Gradual decarbonisation of the power sector causes CO_2 emissions to fall from 1.4 Gt in 2008 to 0.8 Gt by 2035.

Figure 7.12 ● Electricity generation by fuel and region in the New Policies Scenario

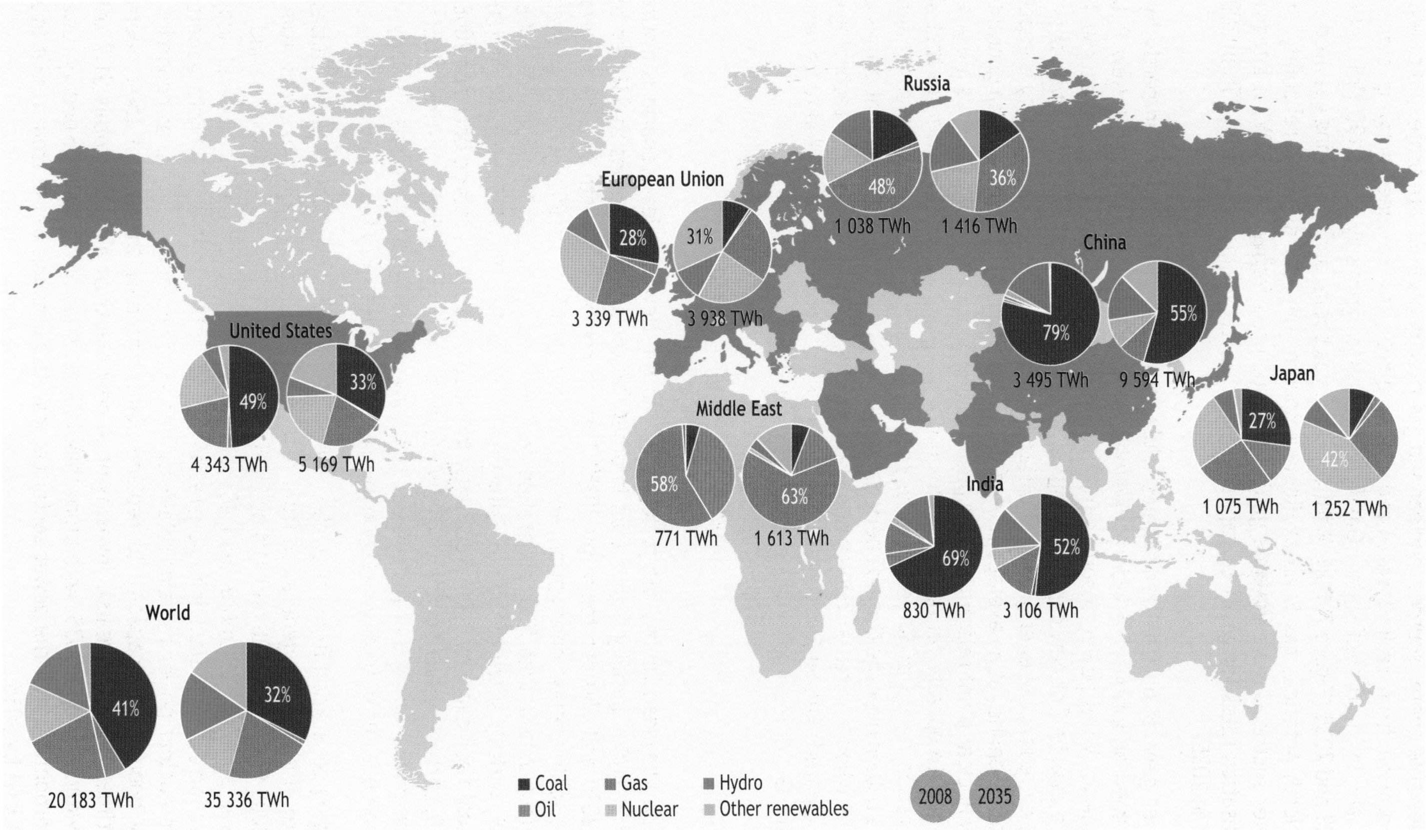

Japan
42% 1 252 TWh
27% 1 075 TWh

China
55% 9 594 TWh
79% 3 495 TWh

Russia
36% 1 416 TWh
48% 1 038 TWh

India
52% 3 106 TWh
69% 830 TWh

European Union
31% 3 938 TWh
28% 3 339 TWh

Middle East
63% 1 613 TWh
58% 771 TWh

United States
33% 5 169 TWh
49% 4 343 TWh

World
32% 35 336 TWh
41% 20 183 TWh

■ Coal   ■ Gas   ■ Hydro
■ Oil   ■ Nuclear   ■ Other renewables

2035
2008

Note: For each region, the largest source of electricity generation in 2008 and 2035 is denoted by its percentage share of the overall mix.

7

Japan

In Japan, the rising price of CO_2 in the power sector in the New Policies Scenario increases operational costs for coal-fired plants and encourages more generation from nuclear and renewables, whose combined share of total generation climbs from 34% in 2008 to 62% in 2035. Coal-fired generation in Japan drops by almost two-thirds between 2008 and 2035, its share of overall generation declining from 27% to 9% during that period (Figure 7.14c). Electricity from oil-fired plants also declines steeply over the projection period, as rising oil prices discourage their use. Lost output from coal- and oil-fired plants is partially offset by more gas-fired generation in the medium term, but this too starts to decline with a higher price for CO_2 after 2020.

Installed nuclear capacity in Japan rises from 48 GW in 2008 to around 70 GW in 2035, the share of nuclear power in electricity generation rising from 24% to 42% over the *Outlook* period. Reaching this level of nuclear capacity requires investment of about $110 billion, or one-third of the total spent by Japan on new power plants, between 2008 and 2035. The shares of wind and solar PV in the electricity mix rise to 4.5% and 2.3% by 2035. This is primarily the result of an increasing price for CO_2 and incentives in the case of solar PV. The move toward a less carbon-intensive power sector results in CO_2 emissions declining 46% by 2035, or 0.2 Gt, compared with today.

China

Electricity demand in China rises briskly in the New Policies Scenario, at an annual rate of 7.7% through 2015, and then averages 2.8% per year over the remainder of the *Outlook* period as the pace of economic growth slows and electricity use becomes more efficient. Overall, demand is projected to triple between 2008 and 2035, with China overtaking the United States in 2012 as the largest global consumer of electricity. Nonetheless, per-capita electricity consumption in China rises to only 65% of the average in OECD countries by 2035. Coal remains the cornerstone of the electricity mix during the *Outlook* period, although its share of generation drops from 79% in 2008 to 55% in 2035. Annual coal-fired electricity generation increases 2 500 TWh between 2008 and 2035, with almost 60% of the rise occurring by 2015 (Figure 7.14d). By 2035, gas-fired generation increases 20 times over current levels, supplying 9% of total electricity generation.

The share of low-carbon power generation in China — including nuclear, CCS-fitted plants, hydro and other renewables — doubles from 2008 to 2035, when it reaches 38% of total generation. This transition aims to achieve China's targets for renewables and nuclear by 2020 to diversify the energy mix away from fossil fuels and reduce local pollution. Electricity generation from hydro and wind both increase by more than 700 TWh to provide 14% and 7% of the electricity mix by the end of the *Outlook* period. With many nuclear plants already under construction, a surge of new generating capacity is expected by 2020, increasing annual generation by 800 TWh between 2008 and 2035.

The capacity additions required to meet China's electricity needs over the period 2009-2035 are staggering: between 2009 and 2025 China will have added new capacity equivalent to the current installed capacity of the United States (Figure 7.13). A total of $2.2 trillion in investment will be necessary to build new plants over the *Outlook* period, with about half required between 2010 and 2020. Of total investment in new plants from 2010 to 2035, 62% goes to renewable energy (including hydro), 20% to coal-fired facilities and 14% to nuclear power. Although the CO_2 intensity of power generation declines by 38% over the *Outlook* period, overall CO_2 emissions from the power sector increase from 3.1 Gt in 2008 to 5.1 Gt by 2035.

Figure 7.13 ● **Cumulative capacity additions in China in the New Policies Scenario from 2009 compared with the 2008 installed capacity of selected countries**

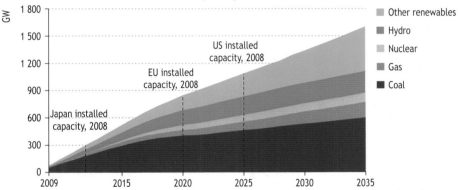

India

In the New Policies Scenario the combination of rising population and economic growth in India leads to electricity demand rising almost four-fold from 2008 to 2035, making it the third-largest consumer of electricity at the end of the *Outlook* period, behind China and the United States. Demand steadily increases through the period, with one-half of incremental growth coming from the industrial sector. Coal continues to be the main source of electricity generation, although its share declines from 69% in 2008 to 52% in 2035. In absolute terms, annual coal generation increases more than generation from any other fuel, by more than 1 000 TWh between 2008 and 2035 (Figure 7.14e). Almost all new coal-fired capacity additions between 2008 and 2020 use subcritical and supercritical technology; after this time, more advanced coal technologies begin to enter the mix. This realises a projected rise in average coal efficiency from 34% in 2020 to 40% in 2035. Gas-fired generation also increases considerably, from 82 TWh in 2008 to about 450 TWh by 2035, with its share of total generation increasing from 10% to 14%.

Total electricity generation from low-carbon energy sources in India, including nuclear, increases seven-fold from 2008 to 2035, with their share of total generation rising from 17% to 33%. New hydropower projects are expected to result in a 290 TWh increase in annual generation between 2008 and 2035. The share of nuclear generation in the electricity mix rises from 2% to 6% as 25 GW of new capacity is installed. Of non-hydro renewable energy sources, wind generation grows most in

Figure 7.14 ● Change in electricity generation relative to 2008 by type for selected countries in the New Policies Scenario

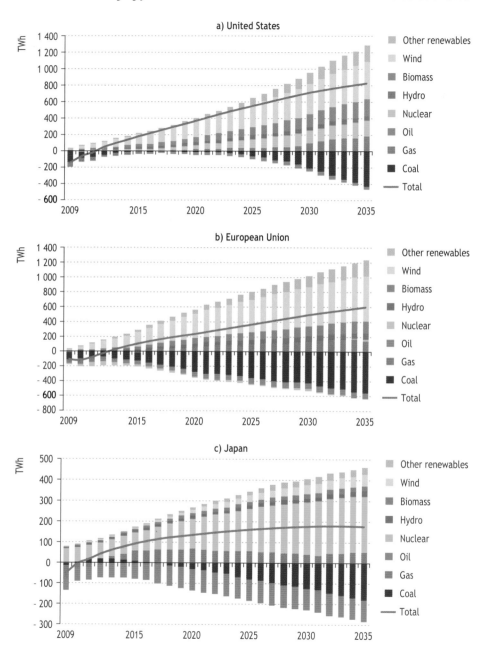

Figure 7.14 ● **Change in electricity generation relative to 2008 by type for selected countries in the New Policies Scenario (continued)**

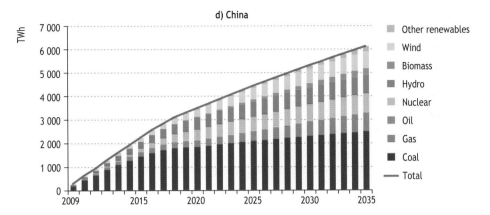

d) China

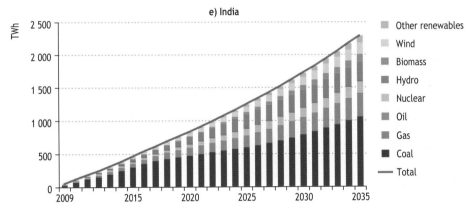

e) India

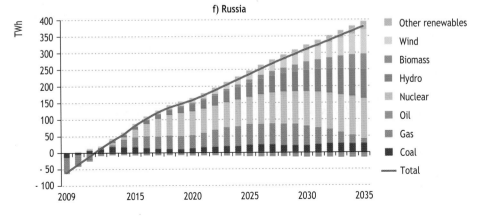

f) Russia

absolute terms, from 14 TWh in 2008 to 190 TWh in 2035. The greatest level of investment in new plants will go to coal installations, which account for almost 40% of the $1 trillion required from 2010 to 2035. The projected expansion of electricity generation from fossil-fuels causes CO_2 emissions from India's power sector to rise from 0.8 Gt in 2008 to 1.6 Gt by 2035.

Russia

In Russia, electricity generation from gas-fired plants rises through 2025 in the New Policies Scenario, declining thereafter as new nuclear and renewable capacity is brought online (Figure 7.14f). Gas-fired electricity accounted for 48% of total generation in 2008. This drops to 36% by 2035. The share of coal-fired generation in the electricity mix also falls, from 19% in 2008 to 16% by 2035.

More electricity generation from low-carbon sources — principally nuclear, hydro and wind — enters the mix, their combined share of generation reaching almost half of the total by 2035. Electricity generation from nuclear power rises more than any other source from 2008 to 2035 with an increase of 120 TWh. This is the combined result of nuclear capacity additions and projected improvements in plant operation which raise the average capacity factor of nuclear plants from 80% in 2008 to 85% by 2035. Over the *Outlook* period, electricity output from hydro climbs by 90 TWh and from wind by 70 TWh. Total investment required for new generating capacity from 2010 to 2035 amounts to $0.4 trillion, with one-third going to renewable energy (including hydro), 28% to nuclear power and 23% to gas. The introduction of low-carbon technologies slightly lowers power sector CO_2 emissions, from 0.9 Gt today to 0.8 Gt in 2035.

Middle East

In the Middle East, strong economic and population growth drive a doubling in electricity demand between 2008 and 2035 in the New Policies Scenario. As an abundant resource in the region, gas is projected to remain the dominant fuel in the power sector, with its share of total generation increasing from 58% in 2008 to 63% by 2035. In absolute terms, gas generation rises by 580 TWh over the *Outlook* period, accounting for almost 70% of growth in supply. Due to rising prices for oil, and therefore the rising value of oil exports, the share of oil-fired generation in the electricity mix is projected to decline from 36% in 2008 to 13% by 2035.

Renewable energy sees strong growth in the power sector of Middle Eastern countries, with generation from wind, CSP and solar PV rising noticeably over the *Outlook* period. As a share of electricity generation, renewable energy is projected to increase from 1% today to 16% by 2035. Nearly 280 GW of new generating capacity is added between 2008 and 2035, one-third of which is from the installation of combined water desalinisation and power plants. About 3 GW of nuclear capacity is installed in countries that have existing development plans and available capital. Total expenditure on new generating capacity in the Middle East between 2010 and 2035 amounts to $0.4 trillion, with about one-third spent on gas-fired plants. The large increase in fossil-fuel based generation leads to rising CO_2 emissions from the power sector, which increase from 0.5 Gt today to 0.7 Gt by 2035.

CHAPTER 8

ENERGY POVERTY

Can we make modern energy access universal?

H I G H L I G H T S

- We assess two indicators of energy poverty at the household level: the lack of access to electricity and the reliance on the traditional use of biomass for cooking. In sub-Saharan Africa the electrification rate is 31% and the share of people relying on biomass 80%: this is where the greatest challenge lies.

- Today, there are 1.4 billion people in the world that lack access to electricity, some 85% of them in rural areas. Without additional dedicated policies, by 2030 the number of people drops, but only to 1.2 billion. Some 15% of the world's population still lack access, the majority in sub-Saharan Africa.

- The number of people relying on biomass is projected to rise from 2.7 billion today to 2.8 billion in 2030. Using WHO estimates, linked to our projections of biomass use, it is estimated that household air pollution from the use of biomass in inefficient stoves would lead to over 1.5 million premature deaths per year (over 4 000 per day) in 2030, greater than estimates for premature deaths from malaria, tuberculosis or HIV/AIDS.

- Addressing these inequities depends upon international recognition that the projected situation is intolerable, a commitment to effect the necessary change, and setting targets and indicators to monitor progress. A new financial, institutional and technological framework is required, as is capacity building in order to dramatically scale up access to modern energy services at the local and regional levels. We provide a monitoring tool, the EDI, that ranks developing countries in their progress towards modern energy access.

- The first UN MDG of eradicating extreme poverty and hunger by 2015 will not be achieved unless substantial progress is made to improve energy access. To meet the goal, an additional 395 million people need to be provided with electricity and an additional 1 billion provided with access to clean cooking facilities. This will require annual investment in 2010-2015 of $41 billion, or only 0.06% of global GDP.

- To meet the more ambitious target of achieving universal access to modern energy services by 2030, additional investment of $756 billion in 2010-2030, or $36 billion per year, is required. This is less than 3% of the global energy investment projected in the New Policies Scenario to 2030. The resulting increase in energy demand and CO_2 emissions would be modest. In 2030, global oil demand would have risen less than 1% and CO_2 emissions would be only 0.8% higher, compared with the New Policies Scenario.

Introduction

Making energy supply secure and curbing energy's contribution to climate change are often referred to as the two over-riding challenges faced by the energy sector on the road to a sustainable future. This chapter highlights another key strategic challenge for the energy sector, one that requires immediate and focused attention by governments and the international community. It is the alarming fact that today billions of people lack access to the most basic energy services, electricity and clean cooking facilities, and, worse, this situation is set to change very little over the next 20 years, actually deteriorating in some respects. This is shameful and unacceptable.

Lack of access to modern energy services[1] is a serious hindrance to economic and social development and must be overcome if the UN Millennium Development Goals (MDGs) are to be achieved.[2] This chapter which presents the results of joint work with the United Nations Development Programme (UNDP) and the United Nations Industrial Development Organization (UNIDO) investigates the energy-access challenge. We estimate the number of people who need to gain access to modern energy services and the scale of the investments required, both in the period to 2015 and over the longer term, in order to achieve the proposed goal of universal access to modern energy services by 2030 (AGECC, 2010).[3] We also discuss the implications of universal access to modern energy services for the global energy market and for the environment and health. The chapter includes an Energy Development Index and a discussion of the path to improving access to modern energy services, as well as financing mechanisms and the implications for government policy in developing countries.

The focus of this chapter is on expanding access to modern energy services at the household level. This is but one aspect of overcoming energy poverty. Other aspects include providing access to electricity and mechanical power for income-generating activities, the reliability of the supply to households and to the wider economy and the affordability of energy expenditure at the household level. These other aspects of energy poverty are areas for future research in the *World Energy Outlook*.

The numbers related to household access to energy are striking. We estimate that 1.4 billion people — over 20% of the global population — lack access to electricity and that 2.7 billion people — some 40% of the global population — rely on the traditional

---

1. Access to modern energy services is defined here as household access to electricity and clean cooking facilities (*i.e.* clean cooking fuels and stoves, advanced biomass cookstoves and biogas systems).

2. In September 2000, at United Nations Headquarters in New York, world leaders adopted the United Nations Millennium Declaration, committing their nations to a global partnership to eradicate extreme poverty and setting out eight goals — with a deadline of 2015 — that have become known as the Millennium Development Goals (www.un.org/millenniumgoals). The MDGs do not include specific targets in relation to access to electricity or to clean cooking facilities, but universal access to both is necessary for the realisation of the Goals (see Box 8.2).

3. The Advisory Group on Energy and Climate Change (AGECC), a committee set up by UN Secretary-General Ban Ki-moon, is charged with assessing the global energy situation and incorporating this into international climate change talks. It has proposed a goal to achieve universal access to modern energy services by 2030. Because of this, the time frame for the projections in this chapter is to 2030.

use of biomass for cooking (Table 8.1).[4] Worse, our projections suggest that the problem will persist and even deepen in the longer term: in the New Policies Scenario, 1.2 billion people still lack access to electricity in 2030, 87% of them living in rural areas (Figure 8.1). Most of these people will be living in sub-Saharan Africa, India and other developing Asian countries (excluding China). In the same scenario, the number of people relying on the traditional use of biomass for cooking rises to 2.8 billion in 2030, 82% of them in rural areas.

Table 8.1 ● **Number of people without access to electricity and relying on the traditional use of biomass, 2009** (million)

	Number of people lacking access to electricity	Number of people relying on the traditional use of biomass for cooking
Africa	587	657
Sub-Saharan Africa	585	653
Developing Asia	799	1 937
China	8	423
India	404	855
Other Asia	387	659
Latin America	31	85
Developing countries*	1 438	2 679
World**	**1 441**	**2 679**

*Includes Middle East countries. **Includes OECD and transition economies.

Note: *The World Energy Outlook* maintains a database on electricity access and reliance on the traditional use of biomass, which is updated annually. Further details of the IEA's energy poverty analysis are available at www.worldenergyoutlook.org/development.asp.

Source: IEA databases and analysis.

The greatest challenge is in sub-Saharan Africa, where today only 31% of the population has access to electricity, the lowest level in the world. If South Africa is excluded, the share declines further, to 28%. Electricity consumption in sub-Saharan Africa, excluding South Africa, is roughly equivalent to consumption in New York. In other words, the 19.5 million inhabitants of New York consume in a year roughly the same quantity of electricity, 40 terawatt-hours (TWh), as the 791 million people of sub-Saharan Africa (Figure 8.2).

---

4. The traditional use of biomass refers to the basic technology used, such as a three-stone fire or an inefficient cookstove, and not the resource itself. The number of people relying on the traditional use of biomass is based on survey and national data sources, and refers to those households where biomass is the primary fuel for cooking. While the analysis in this chapter focuses on biomass, it is important to note that, in addition to the number of people relying on biomass for cooking, some 0.4 billion people, mostly in China, rely on coal for cooking. This is a highly polluting fuel when used in traditional stoves and has serious health implications.

Figure 8.1 ● Number of people without access to electricity in rural and urban areas in the New Policies Scenario (million)

China
8

India
12
281
23
381

Other developing Asia
40
212
59
328

Sub-Saharan Africa
108
544
120
465

Latin America
2
8
4
27

World population without access to electricity

2009
214
1 227
1 441

2030
161
1 052
1 213

■ Rural   ■ Urban

Note: not to scale

The boundaries and names shown and the designations used on maps included in this publication do not imply official endorsement or acceptance by the IEA.

Figure 8.2 ● Residential electricity consumption in New York and sub-Saharan Africa

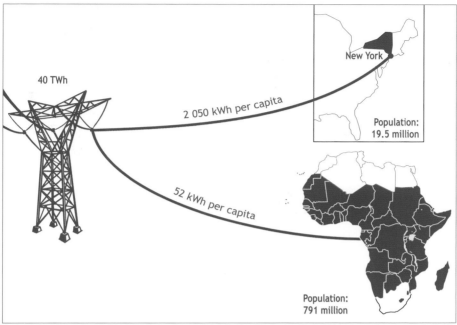

40 TWh

2 050 kWh per capita

New York

Population:
19.5 million

52 kWh per capita

Population:
791 million

The boundaries and names shown and the designations used on maps included in this publication do not imply official endorsement or acceptance by the IEA.

Energy and development

Access to modern forms of energy is essential for the provision of clean water, sanitation and healthcare, and provides great benefits to development through the provision of reliable and efficient lighting, heating, cooking, mechanical power, transport and telecommunication services.[5] The international community has long been aware of the close correlation between income levels and access to modern energy: not surprisingly, countries with a large proportion of the population living on an income of less than $2 per day tend to have low electrification rates and a high proportion of the population relying on traditional biomass (Figures 8.3 and 8.4).

As incomes increase, access to electricity rises at a faster rate than access to modern cooking fuels, largely because governments give higher priority to electrification, though access to both electricity and clean cooking facilities is essential to success in eradicating the worst effects of poverty and putting poor communities on the path to development.

---

5. Household income is the central factor linking achievement of the MDGs and access to modern energy services. Causality is mainly from income to energy access: although improved access to energy can help raise incomes. Moreover, access to electricity is not only a result of economic growth but electricity access also contributes actively to economic growth (Birol, 2007). In this regard, reliability, and not just access, is very important to sustainable economic growth.

Figure 8.3 ● Household income and electricity access
in developing countries

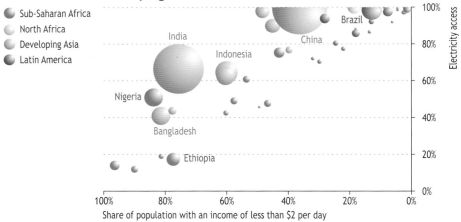

Note: The size of the bubble is proportional to population.
Sources: Electrification rate: www.worldenergyoutlook.org; and poverty rate: http://data.worldbank.org/
indicator/SI.POV.2DAY.

Figure 8.4 ● Household income and access to modern fuels*
in developing countries

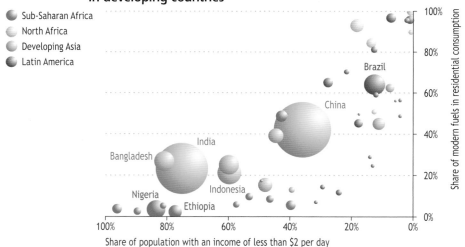

*Modern fuels exclude traditional biomass.
Note: The size of the bubble is proportional to population.
Sources: Consumption of modern fuels: IEA data and analysis; and poverty rate: http://data.worldbank.org/
indicator/SI.POV.2DAY.

The adverse consequences of the use of traditional forms of energy for health,
economic development and the environment are well illustrated by the example of
the use of traditional biomass for cooking (Hutton, Rehfuess and Tediosi, 2007; UNEP,

2003; and IEA, 2006). Currently, devices for cooking with biomass are mostly three-stone fires,[6] traditional mud stoves or metal, cement and pottery or brick stoves, with no operating chimneys or hoods (Box 8.1). As a consequence of the pollutants emitted by these devices, pollution levels inside households cooking with biomass are often many times higher than typical outdoor levels, even those in highly polluted cities. The World Health Organization (WHO) estimates that more than 1.45 million people die prematurely each year from household air pollution due to inefficient biomass combustion (thus excluding premature deaths from cooking with coal). A significant proportion of these are young children, who spend many hours each day breathing smoke pollution from the cookstove. Today, the number of premature deaths from household air pollution is greater than the number of premature deaths from malaria or tuberculosis (Figure 8.5).

Using World Health Organization projections for premature deaths to 2030,[7] the annual number of premature deaths over the projection period from the indoor use of biomass is expected to increase in the New Policies Scenario, unless there is targeted action to deal with the problem. By 2030 over 1.5 million people would die every year due to the effects of breathing smoke from poorly-combusted biomass fuels. This is more than 4 000 people per day. By contrast, the World Health Organization expects the number of premature deaths from malaria, tuberculosis or HIV/AIDS to decline over the same period.

Figure 8.5 ● Premature annual deaths from household air pollution and other diseases

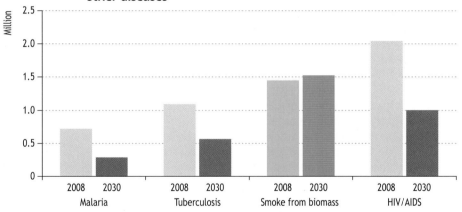

Sources: Mathers and Loncar (2006); WHO (2008); Smith *et al.*, (2004); WHO (2004); and IEA analysis.

In developing regions in which households are heavily reliant on biomass, women and children are generally responsible for fuel collection, a time-consuming and exhausting task. Women can suffer serious long-term physical damage from strenuous work without sufficient recuperation. This risk, as well as the hazards of falls, snake bites

---

6. A three-stone fire uses three stones to support the pot and firewood is placed underneath.

7. The estimations for premature deaths are based on Mathers and Loncar (2006); WHO (2008); Smith *et al.*, (2004); and WHO (2004).

or human assault, rises steeply the further from home women have to walk. Inefficient and unsustainable cooking practices also have serious implications for the environment, such as land degradation and contributing to local and regional air pollution.[8] In cities where households are primarily reliant on wood or wood-based charcoal for cooking, there is local deforestation in the surrounding areas.

Box 8.1 ● Cooking and lighting in the poorest households

The world's poorest households tend to use three-stone fires for cooking. The high moisture content of the biomass resources used and the low efficiency of the combustion process produce dangerous levels of smoke, particularly if food is cooked indoors. The efficiency of biomass can be increased through provision of improved stoves and enhanced ventilation. Adding chimneys to stoves with low combustion efficiency can be a useful improvement, as long as the chimney is kept clean and maintained. However, often there is some leakage into the room and the smoke is merely vented outside the house and will, in part, re-enter the dwelling, so this option is not as effective as a change to clean fuels or advanced biomass stoves. Experience suggests that in order for biomass gasifiers for cooking to consistently achieve emissions close to those of LPG, the stove requires assisted air flow by use of a fan. Ventilation of the home (*i.e.* eaves spaces and larger, open windows and doors) can contribute to reducing household air pollution but alone is unlikely to make a substantial difference if there is a highly polluting indoor source.

Lighting in low-income households in developing countries is generally provided by candles or kerosene/diesel lanterns. Candles and low-efficiency lanterns emit smoke. Kerosene lamps produce better light, but they are uncomfortably hot in a tropical climate and they can be difficult to light. Use of kerosene also imposes health risks, through fires and children drinking fuel stored in soft drink bottles, and there is emerging evidence of links with tuberculosis and cancer. Switching to electricity eliminates these risks and increases efficiency. A paraffin wax candle has an intensity (in lumens) of 1 and an efficiency (lumens per watt) of .01, while a 15 watt fluorescent bulb has an intensity of 600 and efficiency of 40.[9] There has been much recent success in the dissemination of compact fluorescent light bulbs (CFLs) in many developing countries. High-quality CFLs are four to five times more efficient than incandescent bulbs and last much longer. Large-scale deployment of CFLs can help reduce peak electricity needs and ameliorate infrastructure shortages.

---

8. Scientists have recently reported that soot, or black carbon, such as that emitted from the burning of biomass in inefficient stoves, plays a large role in global and regional warming. Black carbon forms during incomplete combustion, and is emitted by a wide range of sources, including diesel engines, coal-fired power plants and residential cookstoves. Warming driven by black carbon appears to be especially amplified in the high country of Asia's Tibetan Plateau, where summer melt-water provides water to more than one billion people. Glaciers on the plateau have declined by about 20% since the 1960s (Luoma, 2010).

9. Light intensity, or illuminating power of a light source, in any one direction is commonly defined as "candela", which can be thought of as "candle-power"; *i.e.* the output from a standard paraffin wax candle. The rate at which light is emitted is measured in lumens, which are defined as the rate of flow of light from a light source of one candela through a solid angle of one steradian, the Standard International unit of solid angular measure.

Effective environmental management cannot be excluded from energy and development concerns. Preventing irreversible damage to the global climate will require decarbonisation of the world's energy system (see Chapter 13). For developing countries, however, difficult choices have to be made in allocating scarce resources among pressing development needs, and climate change is often viewed as a longer-term concern that must be traded off against short-term priorities. While the poorest developing countries are not major contributors to climate change, their populations suffer acutely from its effects. For oil net importing developing countries in particular, rising and volatile prices have amplified the challenge of expanding energy access and put an extra burden on fiscal budgets. In a high-energy price and climate-conscious world, it makes sense for governments tackling the energy poverty challenge to choose a course consistent with long-term sustainable development goals, rather than choose the energy technologies and mix used by OECD countries in the 1950s and 1960s.

The World Resource Institute has defined Sustainable Development Policies and Measures (SD-PAMs) which offer an opportunity for developing countries to reduce emissions through tailored, development-focused policies, that are guided by domestic priorities.[10] Policies in the energy sector that countries would be likely to pursue as SD-PAMs include measures to promote energy efficiency, the broader use of renewable energy sources and steps to reduce energy subsidies while safeguarding the welfare of poor households.

Energy and the Millennium Development Goals

The eight Millennium Development Goals (MDGs), adopted in 2000, were designed to eradicate extreme poverty and hunger by 2015. Energy can contribute to the achievement of many of these goals (Box 8.2). But the MDGs contain no goal specifically related to energy and there are no targets or indicators associated with the MDGs that would enable governments and the international community to monitor progress towards universal access.[11] The UN Advisory Group on Energy and Climate Change has called for adoption of the goal of universal access to modern energy services by 2030.

Box 8.2 ● The importance of modern energy in achieving the MDGs

Goal 1: Eradicate extreme poverty and hunger. Access to modern energy facilitates economic development by providing more efficient and healthier means to undertake basic household tasks and means to undertake productive activities more generally, often more cheaply than by using the inefficient substitutes, such as candles and batteries. Modern energy can power water pumping, providing drinking water and increasing agricultural yields through the use of machinery and irrigation.

---

10. www.wri.org/project/sd-pams.

11. The only indicator related to energy is for CO_2 emissions: total, per capita and per \$1 GDP (PPP) under Goal 7. At the 12th International Energy Forum (IEF) Ministerial in Cancun, Mexico, in March 2010, the IEF called for the international community to set up a ninth goal, specifically related to energy, to consolidate the evident link between modern energy services and achievement of the MDGs.

Goal 2: Achieve universal primary education. In impoverished communities children commonly spend significant time gathering fuelwood, fetching water and cooking. Access to improved cooking fuels or technologies facilitates school attendance. Electricity is important for education because it facilitates communication, particularly through information technology, but also by the provision of such basic needs as lighting.

Goal 3: Promote gender equality and empower women. Improved access to electricity and modern fuels reduces the physical burden associated with carrying wood and frees up valuable time, especially for women, widening their employment opportunities. In addition, street-lighting improves the safety of women and girls at night, allowing them to attend night schools and participate in community activities.

Goals 4, 5, and 6: Reduce child mortality; Improve maternal health; and Combat HIV/AIDS, malaria and other diseases. Most staple foods require cooking: reducing household air pollution through improved cooking fuels and stoves decreases the risk of respiratory infections, chronic obstructive lung disease and lung cancer (when coal is used). Improved access to energy allows households to boil water, thus reducing the incidence of waterborne diseases. Improved access advances communication and transport services, which are critical for emergency health care. Electricity and modern energy services support the functioning of health clinics and hospitals.

Goal 7: Ensure environmental sustainability. Modern cooking fuels and more efficient cookstoves can relieve pressures on the environment caused by the unsustainable use of biomass. The promotion of low-carbon renewable energy is congruent with the protection of the environment locally and globally, whereas the unsustainable exploitation of fuelwood causes local deforestation, soil degradation and erosion. Using cleaner energy also reduces greenhouse-gas emissions and global warming.

Goal 8: Develop a global partnership for development. Electricity is necessary to power information and communications technology applications.

Source: Adapted from UN-Energy, 2005.

The Universal Modern Energy Access Case

To illustrate what would be required to achieve universal access to modern energy services, we have developed the Universal Modern Energy Access Case. This case quantifies the number of people who need to gain access to modern energy services and the scale of the investments required by 2030. It includes interim targets to 2015, related to the achievement of the Millennium Development Goals.

The energy targets adopted to 2015 are consistent with the achievement of MDG 1 — eradicating extreme poverty and hunger. We interpret this, in this context, as meaning that no more than one billion people should be without access to electricity by that date, and no more than 1.7 billion should still be using traditional biomass for cooking on open fires or primitive stoves (Table 8.2). The relationship between poverty and modern energy access has been derived from a cross-country analysis covering 100 countries and the projections are based on regression analyses, which are applied to each region.

Table 8.2 ● **Targets in the Universal Modern Energy Access Case**

	2015		2030	
	Rural	Urban	Rural	Urban
Access to electricity	Provide 257 million people with electricity access	100% access to grid	100% access, of which 30% connected to the grid and 70% either mini-grid (75%) or off-grid (25%)	100% access to grid
Access to clean cooking facilities	Provide 800 million people with access to LPG stoves (30%), biogas systems (15%) or advanced biomass cookstoves (55%)	Provide 200 million people with access to LPG stoves	100% access to LPG stoves (30%), biogas systems (15%) or advanced biomass cookstoves (55%)	100% access to LPG stoves

Note: Liquefied petroleum gas (LPG) stoves are used as a proxy for modern cooking stoves, also including kerosene, biofuels, gas and electric stoves. Advanced biomass cookstoves are biomass gasifier-operated cooking stoves which run on solid biomass, such as wood chips and briquettes. Biogas systems include biogas-fired stoves.

Our analysis shows that, compared to the projections in the New Policies Scenario, in order to achieve the stated interim goals by 2015 an additional 395 million people need to be provided with electricity and an additional 1 billion provided with access to clean cooking facilities. These are demanding targets; in the New Policies Scenario they are not achieved even in 2030 (Figure 8.6). For 2030, the Universal Modern Energy Access Case calculates what would be involved in achieving the more ambitious goal of universal access to modern energy services. Beyond the achievement of the interim 2015 target, this translates into the provision of electricity to an additional 800 million people and giving an additional 1.7 billion people access to clean cooking fuels in 2016-2030.

The investment implications are examined more closely below. But, in brief, bringing electricity to the 1.2 billion people who would otherwise not have access to it by 2030 would require additional cumulative investment, beyond that in the New Policies Scenario, of $700 billion in 2010-2030, or $33 billion per year. In addition, in order to achieve universal access to clean cooking facilities for some 2.8 billion people, additional cumulative investment of some $56 billion would be required in 2010-2030, or $2.6 billion per year. Thus $756 billion additional investment is required to achieve universal access to electricity and clean cooking facilities by 2030.

Figure 8.6 ● Access to modern energy services in the New Policies Scenario and Universal Modern Energy Access Case

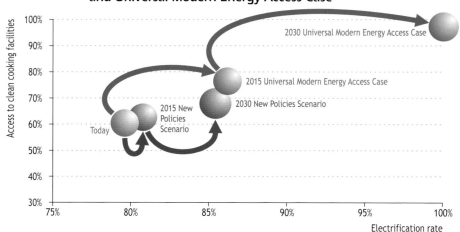

This sum is put in perspective when seen in relation to the projected global energy investment of over $26 trillion in 2010-2030 in the New Policies Scenario: it is less than 3% of global energy investment. Universal access to modern energy services would have little impact on energy demand, production or CO_2 emissions. In 2030, global electricity generation would be 2.9% higher, oil demand would have risen less than 1% and CO_2 emissions would be 0.8% higher, compared to the New Policies Scenario.

Access to electricity

Today, more than 1.4 billion people worldwide lack access to electricity: 585 million people in sub-Saharan Africa (including over 76 million in Nigeria and some 69 million in Ethiopia) and most of the rest in developing Asia (including 400 million in India and 96 million in Bangladesh). Some 85% of those without access live in rural areas.

In the New Policies Scenario, the number of people lacking access to electricity in 2015 is still around 1.4 billion, practically unchanged from today (Figure 8.7). To achieve the targets we have defined in the Universal Modern Energy Access Case to be consistent with the achievement of the first MDG of eradicating extreme poverty by 2015, the number of people without electricity in 2015 would need to be about 395 million less than this, *i.e.* about 1 billion. The global electrification rate would then be 86%, five percentage points higher than the electrification rate achieved in the New Policies Scenario in 2015.

Although electrification will progress over the period to 2030, the need will grow as the population increases.[12] In the New Policies Scenario, without additional, dedicated policies, there are still 1.2 billion people lacking access in 2030 (Table 8.3).

---

12. Electricity access occurs at a much faster rate in urban areas, as companies are often required to provide electricity service and it is more profitable. Most of the increase in the number of people with access over the projection period is in urban areas in the New Policies Scenario.

The electrification rate in developing countries increases from 73% in 2009 to 81% in 2030. China is projected to achieve universal electrification soon after 2015. In developing Asian countries apart from China and India, the electrification rate rises to 82%, but 252 million people still lack access in 2030. Electricity access in Latin America is nearly universal by 2030. In sub-Saharan Africa, the absolute number of people lacking access is projected to continue to rise, despite an increase in the electrification rate; by 2030, the region accounts for 54% of the world total, compared with 41% in 2009.

Figure 8.7 ● **Implication of eradicating extreme poverty on number of people without access to electricity by 2015**

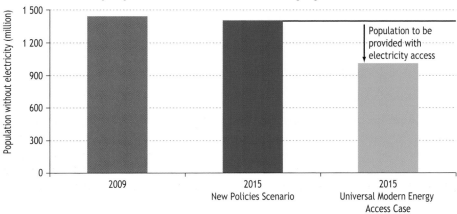

To assess the extent of the additional generating capacity required to achieve universal access, we have made assumptions about minimum levels of consumption at both the rural and urban level: rural households are assumed to consume at least 250 kilowatt-hours (kWh) per year and urban households 500 kWh per year. In rural areas, this level of consumption could provide for the use, for example, of a floor fan, two compact fluorescent light bulbs and a radio for about five hours per day. In urban areas, consumption could also include a television and another appliance, such as an efficient refrigerator or a computer. Consumption is assumed to rise every year until reaching the average national level.

This amounts to total incremental electricity output by 2030 of around 950 TWh. This additional electricity generation represents some 2.9% of the nearly 33 000 TWh generated worldwide in 2030 in the New Policies Scenario. To generate this additional electricity output would require generating capacity of 250 GW.

Various options for supplying this electricity need to be considered, including on-grid, mini-grid[13] and isolated off-grid (Table 8.4). Grid extension will contribute part of the solution, but decentralised options have an important role to play when grid extension is too expensive and will provide the bulk of the additional connections over the projection period (see also, Box 8.3, Figure 8.12 and the associated text).

---

13. Mini-grids are village- and district-level networks with loads of up to 500 kilowatts.

Table 8.3 ● Number of people without access to electricity and electrification rates by region in the New Policies Scenario (million)

	2009 Rural	2009 Urban	2009 Total	2015 Total	2030 Total	2009 %	2015 %	2030 %
Africa	466	121	587	636	654	42	45	57
Sub-Saharan Africa	465	120	585	635	652	31	35	50
Developing Asia	716	82	799	725	545	78	81	88
China	8	0	8	5	0	99	100	100
India	380	23	404	389	293	66	70	80
Other Asia	328	59	387	331	252	65	72	82
Latin America	27	4	31	25	10	93	95	98
Developing countries*	1 229	210	1 438	1 404	1 213	73	75	81
World**	1 232	210	1 441	1 406	1 213	79	81	85

*Includes Middle East countries. **Includes OECD and transition economies.

Achieving universal electricity access would have a modest impact on energy-related CO_2 emissions. Compared with the New Policies Scenario, global energy-related CO_2 emissions in the Universal Modern Energy Access Case increase by just 0.8% by 2030, or around 2% of current OECD emissions. If the generation fuel mix to supply the additional demand in the Universal Modern Energy Access Case was the same as that projected in the 450 Scenario, the increase in energy-related global CO_2 emissions would be a mere 0.6% (Figure 8.8).

Table 8.4 ● Generation requirements for universal electricity access, 2030 (TWh)

	On-grid	Mini-grid	Isolated off-grid	Total
Africa	196	187	80	463
Sub-Saharan Africa	195	187	80	462
Developing Asia	173	206	88	468
China	1	1	0	2
India	85	112	48	245
Other Asia	87	94	40	221
Latin America	6	3	1	10
Developing countries*	379	399	171	949
World**	380	400	172	952

*Includes Middle East countries. **Includes OECD and transition economies.

Figure 8.8 ● Global implications for electricity generation and CO$_2$ emissions in the Universal Modern Energy Access Case (UMEAC), 2030

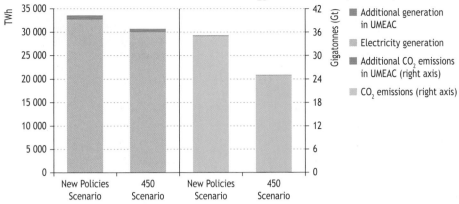

Legend:
- Additional generation in UMEAC
- Electricity generation
- Additional CO$_2$ emissions in UMEAC (right axis)
- CO$_2$ emissions (right axis)

X-axis: New Policies Scenario | 450 Scenario | New Policies Scenario | 450 Scenario

Left axis: TWh (0–35 000); Right axis: Gigatonnes (Gt) (0–42)

Access to clean cooking facilities

There are currently about 2.7 billion people in developing countries who rely for cooking primarily on biomass including wood, charcoal, tree leaves, crop residues and animal dung used in inefficient devices.[14] This number is higher than estimates in previous editions of the *World Energy Outlook,* due to population growth, rising liquid fuel costs and the global economic recession (which have driven a number of people back to using traditional biomass).[15] About 82% of those relying on traditional biomass live in rural areas, although in sub-Saharan Africa, nearly 60% of people living in urban areas also use biomass for cooking. The share of the population relying on the traditional use of biomass is highest in sub-Saharan Africa and India (Figure 8.9).

In the New Policies Scenario, the number of people relying on the traditional use of biomass for cooking increases from just under 2.7 billion in 2009 to about 2.8 billion in 2015. To achieve the Millennium Development Goals would necessitate a substantial reduction. In a similar manner to that used to define targets for universal electricity access, we have defined targets for access to clean cooking facilities, related to the MDG for poverty reduction (see Table 8.2). In the Universal Modern Energy Access Case, eradicating extreme poverty by 2015 would mean reducing the number of people still using traditional biomass to around 1.7 billion by 2015, that is, beyond the projections in the New Policies Scenario, 1 billion more people would need to gain access to clean cooking facilities, including LPG stoves, advanced biomass cookstoves and biogas systems (Figure 8.10).[16] Over 800 million of them would be living in rural areas.

---

14. In many countries, biomass is also used for space heating. The introduction of cleaner, more efficient devices for cooking does not necessarily reduce the need for traditional stoves or fires for heating.

15. For example, recent analysis by the Economic Commission for Latin America and the Caribbean (ECLAC) indicates that, while wood consumption for cooking and heating in Latin America and the Caribbean de-creased steadily in the 1990s, it has risen this decade in many countries as a result of increasing poverty (ECLAC, *et al.*, 2010).

16. For a discussion of advanced biomass stoves, see C. Venkataraman *et al.*, 2010. For a discussion of biogas digesters, see www.unapcaem.org.

Figure 8.9 ● Number and share of population relying on the traditional use of biomass as their primary cooking fuel by region, 2009

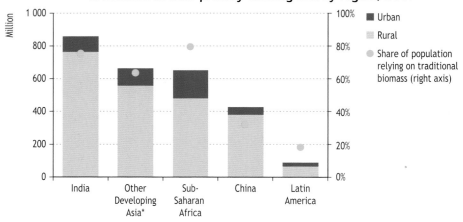

*Includes developing Asian countries except India and China.

Figure 8.10 ● Implication of reducing poverty for number of people relying on the traditional use of biomass for cooking by 2015

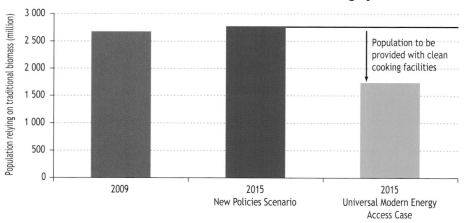

Looking further ahead to 2030 in the New Policies Scenario, the number of people relying on the traditional use of biomass remains at about 2.8 billion, one-third of whom live in sub-Saharan Africa (Table 8.5). The share of the population relying on biomass falls in all regions/countries, but the pace of decline is slowest in sub-Saharan Africa.[17] Accordingly, building on the assumed improved results in 2015, the Universal Modern Energy Access Case means that an additional 1.7 billion people must achieve access to modern cooking facilities in the period 2016-2030.

---

17. The use of traditional biomass increases only in sub-Saharan Africa over the projection period (see Chapter 11).

Table 8.5 ● Number of people relying on the traditional use of biomass and share by region in the New Policies Scenario (million)

	2009 Rural	2009 Urban	2009 Total	2015 Total	2030 Total	2009 %	2015 %	2030 %
Africa	481	176	657	745	922	67	65	61
Sub-Saharan Africa	477	176	653	741	918	80	77	70
Developing Asia	1 694	243	1 937	1 944	1 769	55	51	42
China	377	47	423	393	280	32	28	19
India	765	90	855	863	780	75	69	54
Other Asia	553	106	659	688	709	63	60	52
Latin America	60	24	85	85	79	18	17	14
Developing countries*	2 235	444	2 679	2 774	2 770	54	51	44
World**	2 235	444	2 679	2 774	2 770	40	38	34

*Includes Middle East countries. **Includes OECD and transition economies.

Expanding household access to modern fuels would inevitably increase global demand for these fuels, notably oil, but only by a small amount. In the Universal Modern Energy Access Case, 445 million people switch to LPG stoves by 2015 and another 730 million by 2030. Assuming average LPG consumption of 22 kilogrammes (kg) per person per year,[18] total world oil product demand by 2030 would be 0.9 million barrels per day (mb/d) higher than in the New Policies Scenario. This represents 0.9% of the projected 96 mb/d of global oil demand in 2030 (Figure 8.11). The additional oil demand associated with access to LPG in the Universal Modern Energy Access Case is roughly equivalent to 5% of oil demand in the United States today. In the 450 Scenario, where in 2030 global oil demand is 12.3 mb/d lower than in the New Policies Scenario, global oil demand still increases by only 1% in 2030.

The impact on greenhouse-gas emissions of switching to advanced biomass technologies or LPG is very difficult to quantify because of the diversity of factors involved, including the particular fuels, the types of stoves and whether the biomass used is replaced by new planting and that a sustainable forestry management programme is in place. But it is widely accepted that improved stoves and greater conversion efficiency would result in emissions reductions.

18. A weighted average based on WHO data for developing country households currently using LPG.

Figure 8.11 ● Global implications for oil demand in the Universal Modern Energy Access Case

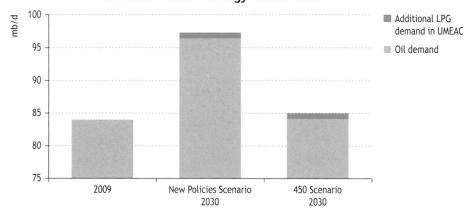

Investment needs in the Universal Modern Energy Access Case

In the Universal Modern Energy Access Case cumulative investment of $756 billion, over and above investment in the New Policies Scenario, is needed. This comprises investment to achieve universal access to electricity and clean cooking facilities by 2030. Some 30% of the investment is needed in 2010-2015 to achieve the interim target. This will require additional annual investment of $41 billion in 2010-2015, or only 0.06% of average annual global GDP over the period.

Investment needs for universal electricity access

Achievement of the targets associated with the MDG of eradicating extreme poverty and hunger by 2015 requires cumulative investment of some $223 billion in 2010-2015, and another $477 billion in 2016-2030 for access to electricity to be universal by 2030. Rural areas account for the bulk of additional household electrification in this period. The supply arrangements include grid and off-grid solutions (Figure 8.12). Consumer density is a key variable in providing electricity access: the cost per MWh delivered through an established grid is cheaper than that through mini-grids or off-grid systems, but the cost of extending the grid to sparsely populated areas can be very high and long distance transmission systems have high technical losses. Thus, decentralised solutions also have an important role to play and will, indeed, account for most of the investment over the projection period (Box 8.3).

In our calculations, all urban and peri-urban households are assumed to be connected to the grid by 2015 in the Universal Modern Energy Access Case. About a third of rural areas are assumed to be similarly connected, while other households use off-grid and mini-grid options, including solar photovoltaics, mini-hydro, biomass, wind, diesel and geothermal. In the first year of obtaining access to electricity, the minimum annual consumption per household is assumed to be 250 kWh in rural areas and 500 kWh in urban areas. Household consumption rises every year over the *Outlook* period, until reaching the national average in 2030. Average household size is assumed to be five people.

Box 8.3 ● Renewable energy for rural applications

Grid extension in rural areas is often not cost effective. Small, stand-alone renewable energy technologies can often meet the electricity needs of rural communities more cheaply and have the potential to displace costly diesel-based power generation options.

Specific technologies have their advantages and limitations. Solar photovoltaics (PV) are attractive as a source of electric power to provide basic services, such as lighting and clean drinking water. For greater load demand, mini-hydro or biomass technologies may offer a better solution, though solar PV should not be ruled out of consideration as system prices are decreasing, a trend which can be expected to continue in the years to come. Moreover, solar PV can also be easily injected in variable quantity into existing power systems. Wind energy represents a good (and available) cost-competitive resource, with mini-wind prices below those of solar PV. Wind energy systems are capable of providing a significant amount of power, including motive power. One of the main advantages of renewable energy sources, particularly for household-scale applications, is their comparatively low running costs (fuel costs are zero), but their high upfront cost demands new and innovative financial tools to encourage uptake. To combine these different sources of energy in a power system supplying a mini-grid is probably the most promising approach to rural electrification. It is important that subsidised delivery mechanisms make provision for maintenance and repair.

Improved irrigation is vital to reducing hunger and saving dwindling water resources in many developing countries. Drip irrigation is an extremely efficient mechanism for delivering water directly to the roots of plants. It increases yields and allows for introduction of new crops in regions and in seasons in which they could not be sustained by rainfall alone. Solar-powered pumps save hours of labour daily in rural off-grid areas, where water hauling is traditionally done by hand by women and children. These pumps are durable and immune to fuel shortages. In the medium term, they cost less than diesel-powered generators.[19]

The bulk of the investment for electrification by 2015 is incurred in developing Asian countries, primarily because economic growth is expected to be more rapid in these countries than in sub-Saharan Africa. The path to universal electricity access will require substantial financing in all developing regions, except Latin America, where access is already high. Cumulative investment of some $340 billion would be required to electrify all households in sub-Saharan Africa by 2030 (Table 8.6).

---

19. See, for example, www.self.org/benin.shtml.

Figure 8.12 ● Number of people gaining access to electricity and additional cumulative investment needs in the Universal Modern Energy Access Case*

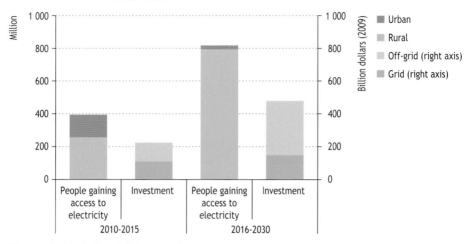

*Compared with the New Policies Scenario.

Table 8.6 ● Investment requirements for electricity in the Universal Modern Energy Access Case* ($ billion)

	2010- 2015	2016-2030	2010-2030
Africa	81	262	343
Sub-Saharan Africa	80	262	342
Developing Asia	127	214	342
China	1	0	1
India	52	130	182
Other Asia	74	84	158
Latin America	5	3	7
Developing countries**	219	478	698
World***	223	477	700

*Compared with the New Policies Scenario.
**Includes Middle East countries. ***Includes OECD and transition economies.

The additional power-sector investment, $33 billion per year on average in 2010-2030 in the Universal Modern Energy Access Case (Figure 8.13), is equivalent to just 5% of the average annual global investment in the power sector in the New Policies Scenario, or around one-fifth of the annual investment required in China's power sector in 2010-2030. Adding $0.003 per kWh, some 1.8%, to current electricity tariffs in OECD countries could fully fund the additional investment.

Figure 8.13 ● **Incremental electricity generation and investment in the Universal Modern Energy Access Case*, 2010-2030**

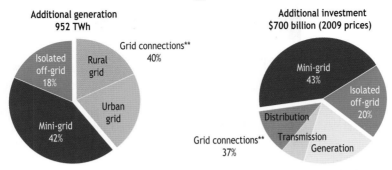

*Compared with the New Policies Scenario.
**Includes generation, transmission and distribution for both urban and rural grids.

Investment needs for universal access to clean cooking facilities

We estimate that universal access to clean cooking facilities could be achieved through additional cumulative investment of $56 billion in 2010-2030, over and above that in the New Policies Scenario. Of this investment, 38% is required in the period to 2015 (Figure 8.14). Over the entire projection period, 51% of the cumulative investment goes to biogas systems in rural areas, 23% to advanced biomass cookstoves in rural areas and 26% to LPG stoves in both rural and urban areas. The average additional annual investment over the period to 2030 is $2.6 billion. Additional cumulative investment (2010-2030) of some $16 billion is required in China, $14 billion in India and $10 billion in other developing Asian countries (Table 8.7). The necessary cumulative investment to 2030 is $14 billion in sub-Saharan Africa.

Figure 8.14 ● **Number of people gaining clean cooking facilities and additional cumulative investment needs in the Universal Modern Energy Access Case***

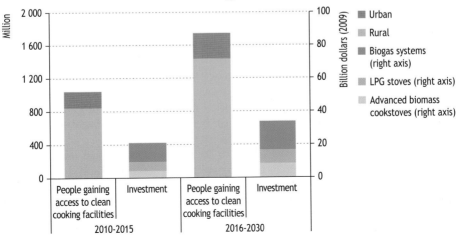

*Compared with the New Policies Scenario.

These investment allocations are derived from assumptions regarding the most likely technology solution in each region, given resource availability and government policies and measures. Advanced biomass cookstoves, with emissions and efficiencies similar to those of LPG stoves, are assumed to cost $45. The cost of a biogas digester is assumed to be $400, the middle of the range of estimated costs for household biogas systems. An LPG stove and canister is assumed to cost $60. Infrastructure, distribution and fuel costs are not included in the investment costs. We assume one stove or biogas system per household in 2010-2030, thus replacement costs are not included.

Developing Asia accounts for 80% of the total $28 billion investment needed for biogas systems, while China alone accounts for 50% of the total. In rural areas of sub-Saharan Africa, over 60% of the 645 million people that need to gain access to clean cooking facilities in 2010-2030 are provided with advanced biomass cookstoves and the remainder with LPG stoves and biogas systems. In rural areas of China, 55% of the target population are provided with biogas systems, 15% with advanced biomass cookstoves and the remainder with LPG stoves.

Table 8.7 ● **Investment requirements for clean cooking facilities in the Universal Modern Energy Access Case*** ($ billion)

	2010- 2015	2016-2030	2010-2030
Africa	4	9	14
Sub-Saharan Africa	4	9	14
Developing Asia	16	24	40
China	7	9	16
India	5	8	14
Other Asia	3	6	10
Latin America	1	1	2
Developing countries**	21	35	56
World***	21	35	56

*Compared with the New Policies Scenario.
**Includes Middle East countries. ***Includes OECD and transition economies.

Financing for universal modern energy access

Financing the $756 billion, or $36 billion per year, needed to provide universal access to modern energy services in 2010-2030, compared with the New Policies Scenario, is a major challenge. So far, investments have been far below needs, especially in sub-Saharan Africa. Investments in electrification have been greater than in clean cooking facilities.

All available sources of finance will need to be tapped: international funds, public/private partnerships, bank finance at multilateral, bilateral and local levels, microfinance, loans and targeted subsidies. The financing mechanism adopted will need to be matched to the particular characteristics of the financing need: for example, the financial mechanisms appropriate to electrification differ according to the scale of the project and also differ from those required for expanding access to clean cooking facilities.

The public sector can be expected to fund the costs of creating the necessary enabling environment, for example, establishing the appropriate policies, regulations and institutions, and will often need to finance the relatively large investments, such as additional generating capacity or transmission links. Indeed, in most developing countries, upfront public investment in developing national and local capacity is the most important ingredient in creating an environment which will encourage the private sector to assume at least part of the risk, essentially, where a commercial return can be reliably earned on the investment. Investment costs which fall to consumers are in a different category. Households will need loans (often on concessionary terms), leasing finance, grants and, even initial subsidies for both high initial investement costs as well as affordable operating costs.

Local banks, as well as bilateral and multilateral agencies, will remain important sources of finance (World Bank Group, 2010). However, those institutions are unlikely to be in a position to provide the level of financing necessary to promote universal access to modern energy services. Existing energy programmes and funds (such as the Renewable Energy and Energy Efficiency Fund (REEF), the Climate Investment Funds administered by the World Bank and implemented jointly with other development banks,[20] the Global Environment Facility and GTZ's Energising Development) can be utilised to administer and distribute finance, but will need to be scaled-up significantly.

Oil and gas-exporting countries have a source of financing that is not available to importing countries. *WEO-2008* estimated that the cost of providing electricity and LPG stoves and canisters to those households without access in the ten largest oil and gas-exporting countries in sub-Saharan Africa would be roughly equivalent to only 0.4% of the governments' cumulative take from hydrocarbon exports through to 2030 (IEA, 2008). Such resource wealth offers a significant opportunity for economic development and poverty alleviation, if managed effectively. Greater efficiency of revenue allocation and greater accountability in the use of public funds are both important.

Long-term financing for rural electrification is important. From the outset, financial provisions should extend long-term (five to ten years) support for the system, under contracts providing also for maintenance and upgrading. At least part of rural electrification should serve economic development activities as a means to generate revenue for maintenance and other operating costs with a view to the end of the support (Niez, 2010).

In contrast to investments for electrification, which are mainly funded by governments and institutional investors, cooking services involve products which are paid for by the consumer.[21] The cost of an improved cookstove ranges from a few dollars to $45 (or in some cases considerably more). Where improved combustion leads to substantial, demonstrable reductions in global warming emissions, these costs may be offset by carbon finance through the Clean Development Mechanism or other mechanisms

---

20. For example, the World Bank's Clean Technology Fund, Pilot Program for Climate Resilience and Scaling-up Renewable Energy Program.

21. The provision of cookstoves by themselves is not enough for universal access. The supply chain, including distribution and production of stoves and fuels, including biomass, also needs to be considered.

generating carbon credits.[22] To support the uptake of clean cooking facilities, governments and donors need to invest in public awareness campaigns regarding the health and other benefits of clean cooking practices.

·····················S P O T L I G H T·····················

Are fossil-fuel subsidies in developing countries crowding out investments that would expand energy access?

According to analysis for this *Outlook*, of the $312 billion of total fossil-fuel subsidies in 2009, $252 billion were incurred in developing countries. Subsidies in countries with low access to modern energy at the household level (*i.e.* electrification rates less than 90% or access to modern cooking fuels of less than 75%) amounted to some $71 billion.[23] Subsidies to kerosene, LPG and electricity in countries with low access to modern energy at the household level were less than $50 billion (see Table 19.3 in Chapter 19). Only a small share of oil-product subsidies are typically directed to cooking in the residential sector.

Subsidies impose a significant burden on national budgets, discourage efficiency of fuel use, can create shortages and result in smuggling and illicit use of subsidised petroleum products. Pressure is building in international fora for governments to phase out blanket subsidies which are not well targeted to the poorest consumers. But phase-out policies must be carefully designed to avoid depriving the poor of basic needs. Direct financial assistance to poor families is probably more efficient than a subsidy to reduce the cost of a particular energy service.

The annual average investment required to achieve universal access to modern energy services by 2030, $36 billion, is around 12% of spending in 2009 on fossil-fuel subsidies in the 37 countries analysed (Figure 8.15).

Microfinance has proved particularly valuable to poor women. They tend to obtain better credit ratings than men and value highly the improvements that can be made to the quality of family life. In Bangladesh, for example, women have shown to default on loans far less often than men. In many cases, though, the scale of microfinance is insufficient to make large inroads into energy poverty.

---

22. The Gold Standard Foundation, an international non-profit organisation based in Switzerland, operates a certification scheme for Gold Standard carbon credits.

23. 37 countries are included in the IEA subsidy database. Those countries with low access to modern energy at the household level are: Angola, Nigeria, South Africa, China, Indonesia, Philippines, Thailand, Vietnam, Bangladesh, India, Pakistan and Sri Lanka.

Figure 8.15 ● *Annual average additional investment needs in the Universal Modern Energy Access Case* compared with fossil-fuel subsidies in developing countries in 2009*

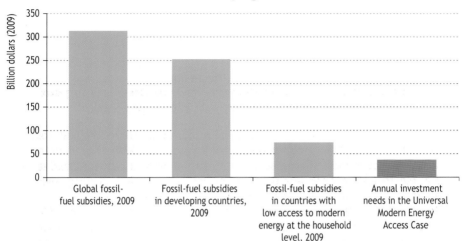

**Compared with the New Policies Scenario.

The poor often need to allocate a disproportionately high share of household budgets to energy services (Modi *et al.*, 2005) and the poorest populations accordingly need distinct forms of help, even though their per-capita consumption is low. To address this, there is a long history of using subsidies to assist affordability. But ensuring that the benefits are provided only to the people most in need is difficult and consumers ideally should have a direct stake in the investment. A contribution by the consumer is critical to successful uptake. Households that pay for even a small fraction of the cost of modern energy services, whether it is an electricity connection, advanced biomass or LPG cookstove or biogas digester, are more likely to provide for maintenance and operating costs. Upfront costs for connections to the electricity grid or for fuel canisters and clean cooking stoves, can still remain too high for the poor and, in the most extreme cases, there may be no alternative to subsidising initially even a proportion of operating costs. One example, promoted by the EU-PV working group on developing countries is a Regulatory Purchase Tariff for off-grid electrification. Under this, the user pays only part of the tariff and the rest is covered by the government. This type of subsidy is focused on people with low consumption.

Monitoring progress and the Energy Development Index

The IEA has devised an Energy Development Index (EDI) in order to better understand the role that energy plays in human development. It tracks progress in a country's or region's transition to the use of modern fuels. By publishing updates of the EDI on an annual basis the IEA hopes to raise the international community's awareness of energy poverty issues and to assist countries to monitor their progress towards modern energy access (Box 8.4). The EDI is calculated in such a way as to mirror the UNDP's Human Development Index and is composed of four indicators, each of which captures a specific aspect of potential energy poverty.[24]

■ **Per-capita commercial energy consumption:** which serves as an indicator of the overall economic development of a country.

■ **Per-capita electricity consumption in the residential sector:** which serves as an indicator of the reliability of, and consumer's ability to pay for, electricity services.

■ **Share of modern fuels in total residential sector energy use:** which serves as an indicator of the level of access to clean cooking facilities.

■ **Share of population with access to electricity.**

A separate index is created for each indicator, using the actual maximum and minimum values for the developing countries covered (Table 8.8). Performance in each indicator is expressed as a value between 0 and 1, calculated using the formula below, and the EDI is then calculated as the arithmetic mean of the four values for each country.

$$\text{Indicator} = \frac{\text{actual value} - \text{minimum value}}{\text{maximum value} - \text{minimum value}}$$

Table 8.8 ● **The minimum and maximum values used in the calculation of the 2010 Energy Development Index**

Indicator	Minimum value (country)	Maximum value (country)
Per-capita commercial energy consumption (toe)	0.03 (Eritrea)	2.88 (Libya)
Per-capita electricity consumption in the residential sector (toe)	0.001 (Haiti)	0.08 (Venezuela)
Share of modern fuels in total residential sector energy use (%)	1.4 (Ethiopia)	100 (Yemen, Lebanon, Syria, Iran)
Share of population with access to electricity (%)	11.1 (Dem. Rep. of Congo)	100 (Jordan, Lebanon)

toe = tonne of oil equivalent.

24. The choice of indicators is constrained by the type of data related to energy poverty that is currently available. For example, the per-capita commercial energy consumption figure is one indicator of overall economic development of a country, but for reasons of data deficiency it fails to take account of biomass resources, including wood, charcoal and biofuels, which are used for productive activities in developing countries. Biomass data is seldom disaggregated in a sufficient manner to capture this reality. With the introduction of low-emission, high-efficiency stoves, biomass consumption will decline in many countries. Yet the EDI cannot adequately compensate for the fact that this decline will be slower than in those countries where households switch to liquid fuels for cooking, even though the impact on energy poverty could be similar. The countries included in the EDI are those for which IEA collects energy data.

Box 8.4 • Measuring progress with energy poverty indicators

A robust set of indicators for measuring energy poverty is needed to provide a rigorous analytical basis for policy-making. Indicators:

- Improve the availability of information about the range and impacts of options for action and the actions that countries are taking to increase access to energy.

- Help countries monitor actions they take to meet their agreed target.

- Enhance the effectiveness of implementation of such policies at national and local levels.

There are numerous examples of single indictors and composite indices to measure concepts related to development and energy (Bazilian et al.,2010). The prime weakness of the various measures is related to data paucity and quality. In theory, energy development indicators should quantify not only the availability of energy — essentially a supply-side approach — but also measure to what extent the available supply is used and how much this contributes to the fulfilment of basic needs. The Earth Institute of Columbia University has pointed out that quantifying the value of some energy services, such as mechanical power or lighting, might benefit from the use of proxy indicators. Mechanical power is one of the largest energy services in terms of volume. It tends to generate a large return on investment and provides significant development leverage. Statistics on energy consumption for mechanical power, however, are not collected. An "ideal" energy development index could be based largely on the energy access recommendations set out by the UN Millennium Project.[25]

Computing a comprehensive energy development index will require the creation of new or augmented data-gathering systems and activities. A robust set of measurement indicators is crucial for informing and ensuring appropriate national policy-making, as well as effective international co-operation. Designing the right indicators and implementing a reporting system can help move energy access to the heart of a development plan. The World Energy Outlook has maintained databases on electricity access and reliance on traditional biomass in rural and urban areas since 2002 (IEA, 2002). These databases are updated annually and will be expanded with the emergence of more comprehensive data-gathering systems.

---

25. The Millennium Project was commissioned by the UN Secretary-General in 2002 to develop a concrete action plan for the world to achieve the Millennium Development Goals (see footnote 2). A common finding of the Millennium Project was the urgent need to improve access to energy services as essential inputs for meeting each MDG. The Millennium Project set out ten recommendations for priority energy interventions which national governments should take to support achieving the MDGs at the national level (Modi et al., 2005).

Figure 8.16 ● 2010 Energy Development Index*

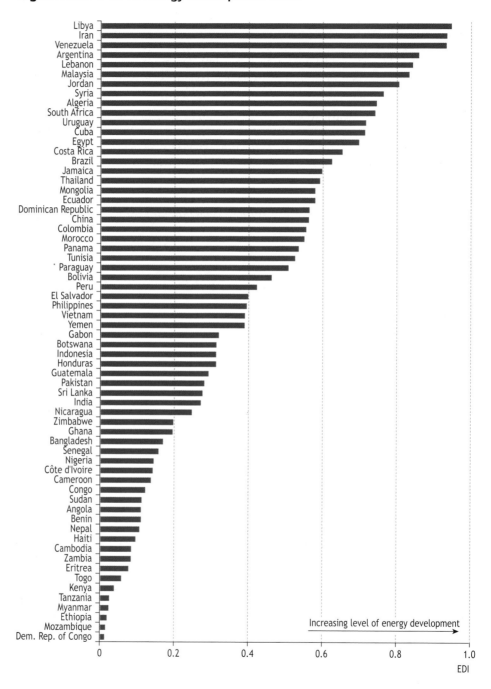

*Based on 2009 data.

Figure 8.16 ranks countries using the four energy development indicators discussed above. Except for South Africa, all sub-Saharan African countries appear in the bottom half of the EDI. Gabon ranks second in sub-Saharan Africa, behind South Africa but 23 places lower. The ranking of countries in Asia varies greatly; Myanmar and Cambodia are in the bottom ten countries, while Malaysia is in the top ten. Pakistan has the highest EDI ranking of countries in South Asia, while Venezuela has the highest ranking of Latin American countries. Oil net exporting countries, except for those in sub-Saharan Africa, are all in the top third of the EDI ranking.

Given the substantial contribution of energy services to advancing human development, it is not surprising that the EDI results are strongly correlated with those of the Human Development Index (HDI) (Figure 8.17).[26] The HDI is composed of data on life expectancy, education, per-capita GDP and other standard-of-living indicators at the national level.

Figure 8.17 ● **Comparison of the Human Development Index to the Energy Development Index**

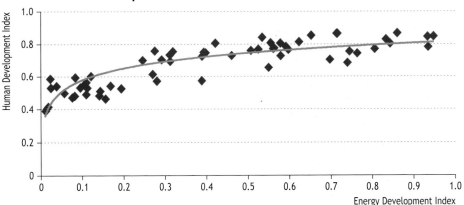

Many countries have made notable progress in improving access to electricity and clean cooking facilities since 2004, when the Energy Development Index was first created (IEA, 2004). In all countries both the absolute number with access and the share of the population with access have increased (Figure 8.18). In China, substantial progress has been made in access to modern cooking fuels. In Angola and Congo, where the share of the population with electricity and access to modern cooking fuels has expanded, most of the achievement has come from urban areas. While there has been progress on both fronts in Bangladesh, Sri Lanka and Vietnam, more progress has been made in household electrification than in the provision of access to modern cooking fuels.

26. The correlation is 0.84.

Figure 8.18 ● Evolution of household access to modern energy in selected developing countries

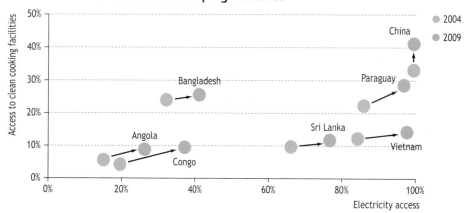

Other potential indicators

The World Energy Outlook will update the Energy Development Index on an annual basis. As more and better data become available, the EDI will also be augmented in order to enhance the monitoring of progress towards universal modern energy access. This section explores other possible indicators.

Figure 8.19 shows the relationship between fuel use and income across a range of developing countries. In low-income countries, final consumption of energy in the residential, service, industry and transport sectors is low and is comprised mainly of biomass. In high-income developing countries, the fuel mix is much more diverse and the overall amount of energy consumed is much higher.

Figure 8.19 ● The relationship between per-capita final energy consumption and income in developing countries

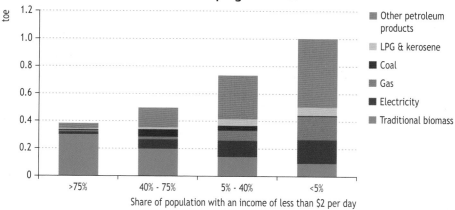

Note: Average per-capita final energy consumption is 3.1 toe in OECD countries. Other petroleum products are mostly consumed in the transport sector.

Demand for mobility, which is indicated where the share of other petroleum products in final energy consumption is high, is much greater in countries with a very low percentage of the population living on less than $2 a day.

The indicators used in the EDI capture the quantity of energy consumed as well as rates of access. Other useful indicators would capture the quality of energy consumed. Figure 8.20 provides an illustration of the quality of energy services for cooking and lighting as income rises at the household level. The figure is reflective of energy consumption in rural households, but some of the principles also apply to peri-urban and urban households. The concept of a simple "energy ladder", with households moving up from one fuel to another, does not adequately portray the transition to modern energy access, because households use a combination of fuels and technologies at all income levels. This use of multiple fuels is a result of their differing end-use efficiency, of affordability and of social preferences, such as a particular fuel for cooking. Moreover, use of multiple fuels improves energy security, since complete dependence on a single fuel or technology leaves households vulnerable to price variations and unreliable service.

Figure 8.20 ● **The quality of energy services and household income**

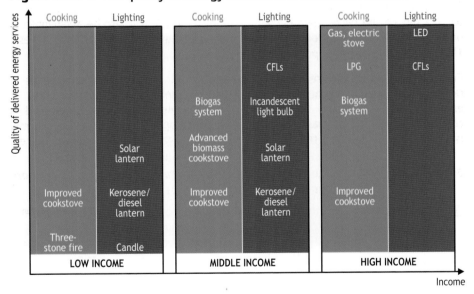

Note: CFL = compact fluorescent light bulb; LPG = liquefied petroleum gas; and LED = light-emitting diode. Improved cookstoves have higher efficiency than cooking over a three-stone fire, but emissions are not reduced considerably, while advanced biomass cookstoves have equivalent efficiency and emissions reductions as liquid-fuel, gas and electric stoves.

The indicator of the quality of delivered energy services on the vertical axis in Figure 8.20 is designed to capture a variety of dimensions, including cleanliness, efficiency and affordability. Because the amount of energy delivered from traditional technologies, such as a three-stone fire or kerosene/diesel lanterns, is much lower than that from modern services, such as electricity, poorer households pay a much higher share of their income on energy services. A study of rural energy use in Bangladesh found that, for example, the cost of each kilolumen-hour from incandescent light bulbs or fluorescent tubes is less than 2% of the cost of comparable lighting services from kerosene lamps (Asaduzzaman, Barnes and Khandker, 2009). Access to electricity accordingly can reduce total household energy costs dramatically, if upfront costs related to the connection are made affordable. In addition, successful energy efficiency initiatives reduce electricity demand, which has the secondary benefit that existing generation plants can be used to supply new households, thereby reducing the need for capacity additions.

Box 8.5 ● **Going beyond household access: indicators at the village and national level**

Village level energy services, both for electricity and mechanical power, are extremely important. In poor rural areas, providing household level electricity service is often not economically feasible. The cost of service provision is higher than in urban areas, because support infrastructures for maintenance is lacking and because low population density increases the cost per household. Where household level electrification is not feasible, providing electricity at the village level for productive activities and basic social services can be a useful stepping stone. Moreover, village level energy installations, *e.g.* mechanical power for food processing and other productive activities, irrigation, and clean water and sanitation, have a significant impact on poverty, health, education and gender equality.

While mechanical power is critical to develop industrial and productive activities necessary to local development, quantified objectives defining rates of access to mechanical power are rarely integrated into national strategies. By the end of 2009, less than 5% of developing countries had defined such targets. Those few countries that had established targets on access to mechanical power—Benin, Cameroon, Central African Republic, Mali, and Togo—are all in sub-Saharan Africa (see Table 8.10).

In addition to the impact at the household level, unreliable electricity service constrains economic activity and constitutes a severe obstacle to business operation and growth (Table 8.9). According to the World Bank, countries with underperforming energy systems may lose 1 to 2% of economic growth potential annually as a result of electric power outages, over-investment in backup electricity generators, energy subsidies and inefficient use of energy resources (World Bank, 2009).

Table 8.9 ● **Indicators of the reliability of infrastructure services**

	Sub-Saharan Africa	Developing countries
Delay in obtaining electricity connection (number of days)	79.9	27.5
Electrical outages (days per year)	90.9	28.7
Value of lost output due to electrical outages (% of turnover)	6.1	4.4
Firms maintaining own generation equipment (% of total)	47.5	31.8

Source: World Bank (2007).

Policy implications

How can countries embark on a dynamic path that will eventually lead to universal access to modern energy services? Experience shows that success can be achieved in a variety of ways. Cambodia, Mali and Madagascar have given support to private developers through rural electrification funds. Bangladesh and Nepal have developed local co-operatives, owned by consumers. Smart subsidy schemes to provide electricity to rural households, such as 'output based aid' subsidies, have been developed in some countries, *e.g.* Senegal and Mozambique, and a similar approach has been used in Colombia to connect poor households to natural gas services. In Mali, multifunctional platform[27] projects have been developed to provide mechanical power and their success has led to similar programmes being adopted in other African countries, such as Burkina Faso, Ghana, Guinea and Senegal. To meet overall universal modern energy access objectives, however, these approaches need to be scaled-up significantly and applied more widely.[28]

Increasing access to modern energy services requires, first, the integration of energy access into national development strategies, preferably with support from the UN system. Strong and sustainable financial, institutional and technology frameworks must be set up and capacity building undertaken at the local and regional levels: developing the capacity of national and local organisations, the private sector and communities themselves to provide appropriate energy technologies and services. In Nepal, for example, well over half of the total programme cost for the implementation of a programme to provide micro-hydropower and improved cooking stoves was dedicated to capacity development (UNDP and AEPC, 2010). Setting national goals and targets is important, but it is not enough, without careful monitoring of progress.

Greater regional co-operation can avoid unnecessary expansion of electricity generation capacity in the future. Coordination within a country and between regional governments can greatly enhance the efficacy of electricity projects and contribute to wider benefits: in Africa, in particular, regional power pools appear to make a valuable contribution to regional integration, which is widely perceived as one of the best engines of Africa's development.

About half of developing countries have set up electricity access targets at the national, rural and/or urban level. Objectives vary among countries. While some countries, such as Bangladesh, Bhutan, Botswana, Ghana, India, Nepal, South Africa or Swaziland aim to reach universal access within the next 5 to 17 years, others have defined intermediate goals, such as Malawi or Rwanda, that aim to achieve 30% and 35% electrification rates respectively by 2020. Both Laos and Indonesia have a target to electrify 90% of the population by 2020, in the latter case involving expanding access to some two million new subscribers each year. Cambodia has a target to increase its rural electrification rate from 12% today to 70% by 2030.

---

27. The multifunctional platform is built around a diesel engine, which can also run off jatropha oil. It can power various tools, such as a cereal mill, husker, alternator, battery charger, pump, welding and carpentry equipment. It can also generate electricity and be used to distribute water.

28. See UNDP and AEPC, 2010 and UNDP, 2006.

Worryingly, very few developing countries have set targets for access to modern cooking fuels or improved cookstoves or for reducing the share of the population relying on traditional biomass (Table 8.10).

Table 8.10 ● Number of developing countries with energy access targets

	Developing countries (total)	of which: sub-Saharan Africa
Electricity	68	35
Modern fuels	17	13
Improved cookstoves	11	7
Mechanical power	5	5

Note: Based on UNDP's classification of developing countries.
Source: UNDP and WHO (2009).

Despite the demonstrable health consequences associated with current cooking practices in many developing countries, access to clean cooking facilities has received very little high-level attention, and, not surprisingly, very little progress has been made. Adequate training and support services have been lacking, together with the market research necessary to determine the concerns of the women who would be using the stoves and their different cooking habits. Where initiatives have been taken, governments are becoming aware of the limitations of policies to encourage switching to liquid cooking fuels, such as LPG, and are putting in place strategies to increase the use of advanced biomass cookstoves and biogas systems (Box 8.6).[29]

Box 8.6 ● Initiatives to improve the efficiency of biomass for cooking

The Indian Ministry of New and Renewable Energy (MNRE) launched a "National Biomass Cookstove Initiative" in December 2009. The initiative aims to achieve for all households a quality of energy services from cookstoves comparable to that from clean energy sources, such as LPG. A large proportion of India's population, some 72% of the total population and 90% in rural areas, uses biomass for cooking. Providing a clean cooking energy option would yield enormous gains in terms of health and socio-economic welfare. Advanced biomass cookstoves also greatly reduce the products of incomplete combustion, which are greenhouse-gas pollutants, thus helping combat climate change.

The Rwandan government estimates that the value of firewood and charcoal consumed for cooking in 2007 was on the order of $122 million, or 5% of GDP (Ministry of Infrastructure, Republic of Rwanda, 2010). About 50% of this was used in rural areas. The government has devised a strategy to increase the efficiency and reduce the environmental impact of using biomass for cooking.

29. The heightened awareness of the need to improve the use of biomass for cooking is driven by different factors among countries. The most important include high oil prices, global recession, unreliable supplies of liquid fuels, and the illegal diversion of LPG and kerosene to the industry and transport sectors.

Key components are: building capacity among equipment manufacturers and importers, in order to make available modern appliances for the use of biomass; developing a quality label, promoting the use of these modern appliances; and launching a long-term publicity and awareness campaign to encourage households, institutions and businesses to adopt the new equipment.

From 2001 to mid-2010, the programme for the Development and Promotion of Biogas Utilization in Rural China (DPBURC) built some 30 million biogas systems, benefitting around 105 million people in rural areas. Measures that contributed to this achievement included: setting minimum technical and quality control standards; adapting technology to match local resources; focusing government financial support on the poorest; and providing technical support to manufacturers of biogas appliances and owners. The biogas systems are used for cooking, electricity, sanitation and the manufacture of fertiliser. On average, each household using a biogas digester saves 500 yuan ($74) every year from reduced use of fuelwood, electricity, chemical fertiliser and pesticides (Tian and Song, forthcoming). By the end of 2010, the total number of biogas systems is likely to reach 40 million, 30% of the estimated potential in China.

To summarise, providing universal access to modern energy services at the household level depends upon recognition by the international community and national governments of the urgency of the need, and long-term policy commitment as part of strategic development plans. These need to make provision for the creation of strong institutional, regulatory and legal frameworks and financing from all available sources, including the private sector. Appropriate technological choices need to be factored in. International aid will be needed to subsidise investments in the production and distribution of both electricity and clean cooking fuels, in capacity building and in creating an institutional system that integrates these different areas over the long term and addresses climate change simultaneously.[30] International development organisations can support research, design and development of appropriate technologies. Promising approaches include reliance on renewable energy in rural applications and the use of locally-produced bioenergy to generate electricity. International development organisations should take the lead in collecting, compiling and sharing knowledge and in developing tools and indicators to measure progress.

Prioritising energy access as a key driver of social and economic development is a first step towards universal modern energy access. The way forward will require:

- Commitment from the international community to the objective of achieving universal access to electricity and to clean cooking facilities by 2030.

- Establishment of national goals for access to modern energy services, supported by specific plans, targets and systematic monitoring, using appropriate indicators.

- Creation of adequate and sustainable financial, institutional and technology frameworks.

---

30. See, for example, the UNDP-UNEP Poverty-Environment Initiative (UNPEI), www.unpei.org.

PREFACE

Renewable energy has been growing rapidly in the last decade, becoming an important component of energy supply. Government intervention in support of renewables has grown, reflecting efforts to reduce carbon-dioxide emissions and to diversify energy supplies. The incentives offered, alongside rising fossil-fuel prices and the expectation that these will stay high in the future, have made renewables attractive to many investors.

This part of the report provides insights into recent and future trends in renewable energy. Chapters 10, 11 and 12 focus on their application in the electricity, heat and transport sectors respectively. Chapter 9 brings together trends across all sectors and discusses issues common to all renewables, including their costs and benefits.

Each chapter presents a brief overview of the results across the three scenarios, but with the main focus on the New Policies Scenario, which illustrates where currently planned policies, if implemented in a relatively cautious way, will take us. For ease of comparison, the main findings of the 450 Scenario are presented briefly in a box in each chapter.

The analysis of renewables for electricity in Chapter 10 includes the quantification of incentives in place to support renewables, the support needed up to 2035, and the impact on electricity prices of greater use of renewables. It also discusses how different renewables can be integrated into the network, with an estimate of the associated costs. This chapter takes a close look at two specialised topics: first, offshore wind power, with a focus on northern Europe, and second, renewables in Middle East and North Africa, a region that has some of the best solar resources in the world and could become an exporter of solar power to Europe.

Though heating is the principal energy service, as a sector it has received relatively little attention. Despite problems with data availability, we have provided an overview of the main trends in renewables for heat in Chapter 11. The chapter opens with a discussion of total demand for heat, and elaborates the large potential for renewables, including biomass, solar and geothermal heat. Chapter 12 on renewables for transport focuses on biofuels, but covers briefly renewables-based electricity and hydrogen used in transport. It also discusses biofuels-related greenhouse-gas emissions, a controversial subject in recent years. Similar to Chapter 10, it quantifies government support and looks into the costs of biofuels.

CHAPTER 9

HOW RENEWABLE ENERGY MARKETS ARE EVOLVING

How green will the future be?

H I G H L I G H T S

- The use of modern renewable energy is projected to expand rapidly to 2035 in all three scenarios presented in this *Outlook*. The rates of growth in each scenario reflect assumptions about different levels of intensity of government policies aimed at reducing greenhouse-gas emissions and diversifying the energy supply mix. The supply of modern renewable energy — including hydro, wind, solar, geothermal, modern biomass and marine energy — increases from 840 Mtoe in 2008 to between 1 900 Mtoe and nearly 3 250 Mtoe in 2035, depending on the scenario.

- In the New Policies Scenario, the share of renewables in global electricity generation increases from 19% in 2008 to almost a third in 2035. The share of modern renewables in heat production in industry and buildings increases from 10% to 16%. Demand for biofuels grows four-fold between 2008 and 2035, meeting 8% of road transport fuel demand by the end of the *Outlook* period.

- Investment needs in renewable energy to produce electricity are estimated at $5.7 trillion (in year-2009 dollars) over the period 2010-2035 in the New Policies Scenario. Biofuels need another $335 billion. Overall, renewables investment needs are greatest in China, which has now emerged as a leader in installing wind turbines and photovoltaics, as well as a major supplier of these technologies.

- We estimate that government support for electricity from renewables and for biofuels cost $57 billion in 2009, up from $44 billion in 2008 and $41 billion in 2007. This support grows to $205 billion by 2035 in the New Policies Scenario, or 0.17% of global GDP. Between 2010 and 2035, 63% of the support goes to renewable electricity and 37% to biofuels. Large-scale government support is needed to make renewables cost competitive with other energy sources and technologies and to stimulate the required technological advances.

- Several benefits may be adduced to justify government support for renewables. In the New Policies Scenario, renewables avoid 2 Gt of CO_2 emissions in 2035, relative to the Current Policies Scenario. Oil-importing countries see their bills reduced by about $130 billion in 2035. Renewables contribute to lower NO_x and SO_2 emissions.

- In the 450 Scenario, demand for modern renewables grows four-fold between 2008 and 2035. Renewables supply 45% of total electricity output by 2035 and 20% of total heat. The share of biofuels in total transport fuel supply reaches 14% in 2035.

Recent trends

Policy support for renewable energy has increased considerably over the past decade. Two drivers underpin this trend: first, the effort to constrain growth in greenhouse-gas emissions and, second, concerns to diversify the supply mix (promoted particularly by high oil prices, especially in 2005-2008). To address these concerns, more and more governments are adopting targets and taking measures to increase the share of renewables in the energy mix. Job creation through renewables has been another factor in government support, especially as a contribution to reducing unemployment following the economic and financial crisis.

Total primary renewable energy supply, including traditional biomass, grew from 1 319 million tonnes of oil equivalent (Mtoe) in 2000 to 1 590 Mtoe in 2008. Its share in total energy supply remained roughly stable during that period, at around 13%. Biomass is by far the most important source of renewable energy in this wider definition (the term "modern renewables" excludes the traditional use of biomass).[1] Biomass use amounted to 1 225 Mtoe in 2008, most of which was used in traditional ways by some 2.7 billion people in developing countries (see Table 8.1 in Chapter 8). The use of modern biomass is smaller (478 Mtoe in 2008) but is rapidly growing, particularly as it is being used more intensively to produce electricity and as feedstock for making transport fuels. Hydropower is the second-largest renewable energy source in primary energy demand (276 Mtoe) and the largest source of renewables-based electricity. Wind, solar, geothermal and marine power have been growing very quickly in recent years, but their overall contribution to primary energy supplies remains modest. The characteristics of the main forms of energy are summarised at the end of the chapter.

Renewables-based electricity output increased by nearly a third from 2000 to 2008. While most of the 900 TWh increase came from hydropower, new forms of renewables grew very rapidly, notably wind power, which expanded seven-fold. Solar photovoltaic (PV) electricity production grew 16-fold during the same period. Biomass use and geothermal power both increased too, although at a moderate pace, while marine power and concentrating solar power are just now beginning to take-off.

Growth in the use of renewables for producing heat at the point of use (and in heat from district heating systems) was much more modest, as government policies to support renewables tend to focus more on electricity and transport. The use of traditional biomass has increased since 2000, despite efforts to provide the poor with access to modern fuels.

Biofuels are supplying a growing share of transport fuels. Global consumption of biofuels, used almost exclusively in road transport, increased five-fold over the period 2000-2008, reaching 1 million barrels per day (mb/d) and meeting almost 3% of total fuel demand in road transport. While oil demand for road transport fell in 2009

---

1. Modern renewables encompass all renewable energy sources other than traditional biomass, which is in turn defined as biomass consumption in the residential sector in developing countries and refers to the use of wood, charcoal, agricultural residues and animal dung for cooking and heating. All other biomass use is defined as modern.

(for the first time since 1980), in response to higher prices and shrinking economic activity, biofuels use continued to grow, as production capacity — spurred in most countries by government support — expanded.

Outlook for renewable energy

Key parameters affecting the outlook

Despite the impressive growth in renewable energy in recent years, most of the world's energy needs are still met by fossil fuels and most of the increase in energy demand since 2000 has also been met by fossil fuels. On a global scale, 19% of electricity came from renewables in 2008, a share that has changed very little since 2000, while the shares of coal and gas have increased by 2 and 3.6 percentage points, respectively. In transportation, oil use is about fifty times greater than that of biofuels. The use of fossil fuels for heat is ten times higher than the use of modern renewables.

The renewables resource base is very large and can amply meet a large proportion of energy demand. However, most renewables are not cost competitive under present market conditions and rely on various forms of incentives. Consequently, the existence of government programmes to make renewables attractive to investors and create markets for them is the most important factor affecting the expansion of renewable energy. Such incentives already exist in many countries and are reflected in the significant rate of increase in the use of renewable energy. Often in combination with financial incentives, a number of countries have imposed a requirement on suppliers to raise the share of renewables in electricity production or in transport fuels. The use of carbon markets as a means to promote renewables is limited at present, applying, on a large-scale, only in the European Union (EU). The Clean Development Mechanism (CDM) has contributed to the expansion of renewables in developing countries. Overall, however, it is direct government support, rather than pricing of CO_2, that drives the growth in renewables at present.

Policies to facilitate the integration of variable renewables (such as wind power) into networks are important. Such policies can range from better planning for transmission projects to the development of smart grids, the creation of demand response mechanisms and the promotion of storage technologies.

Policies and strategies to support the development of large hydropower differ, but are no less important. While large hydropower is cost competitive almost everywhere in the world and does not require financial incentives, new applications demand a sensitive approach to the adverse environmental impacts, including rehabilitating populations that are displaced as a result of the construction of dams and adopting integrated water management practices.

Cost reductions are essential to large-scale development of renewable energy. Renewable energy technologies are capital-intensive, requiring significant upfront investments, and most cannot currently compete on price with conventional technologies. For many renewable energy technologies, however, costs have already come down significantly. The scope for further cost reductions for these emerging technologies is generally greater than for the more mature fossil-fuel technologies, as fossil-fuel prices are expected to increase in the future. Government support for renewables can lead to technology improvements and the widespread deployment that is necessary to make renewables cost competitive.

Large-scale development of renewable energy depends on access to finance. Because of their capital-intensive nature, these renewables projects are largely dependent on lending. Attracting finance is likely to be particularly difficult in poorer countries.

Projections by scenario

A substantial increase in modern renewable energy to 2035 is projected in all three scenarios (Figure 9.1), with government policies driving most of the growth. The largest increase in renewables occurs in the 450 Scenario, driven by policies to achieve deep cuts in CO_2 emissions (see Chapter 1 for the definitions of the scenarios). The renewable energy policies underlying the scenarios are discussed in Chapters 10 to 12. Despite the data limitations, we provide projections for heat.

Box 9.1 ● **IEA statistical conventions and renewable energy measured at primary energy level**

The choice of methodology to calculate the total primary energy demand (TPED) that corresponds to a given amount of final energy (such as electricity and heat) is important in the determination of the respective shares of each contributing energy source, but not straightforward. This is particularly true for the calculation of the shares of renewable energy sources. The IEA uses the *physical energy content* methodology to calculate TPED. For coal, oil, gas, biomass and waste, TPED is based on the calorific value of the fuels. For other sources, the IEA assumes an efficiency of 33% for nuclear and 100% for hydro, wind and solar photovoltaics (PV). For geothermal, if no country specific information is available, the primary energy equivalent is calculated using 10% for geothermal electricity and 50% for geothermal heat. As a result, for the same amount of electricity produced, the TPED calculated for biomass will be several times higher than the TPED for hydro, wind or solar PV. The IEA is in the process of determining the appropriate level of efficiency for concentrating solar power. For the purposes of this report, an average efficiency of 40% has been used.

Modern renewables grow rapidly in all scenarios, from 843 Mtoe in 2008 to between 1 900 Mtoe (in the Current Policies Scenario) and 3 250 Mtoe (in the 450 Scenario) by 2035, or up to almost four times the current level. The use of traditional biomass rises slightly to 2020 and then declines by 2035 in all three scenarios, although at different rates by region (see Chapter 8). Consequently, the share of traditional biomass in all renewables diminishes over time.

Figure 9.1 ● World primary renewable energy supply by scenario

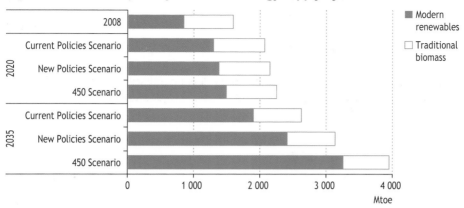

Across all scenarios, biofuels for transport grow more rapidly than renewables for heat and electricity, but from a relatively low base. They increase between three to eight times above 2008 levels by 2035 (Figure 9.2). The very large increase in the 450 Scenario is driven by higher penetration of advanced biofuels, which achieve lower overall unit costs and have lower land requirements. The biofuels share in total transport reaches between 5% and 14% in 2035, up from 2% in 2008 (Table 9.1). Most of the additional demand for biofuels comes from road transport. Renewables for heat[2] increase in absolute terms between 73% and 153%, meeting up to 21% of total heat demand. In the electricity sector, renewables output increases from about 3 800 terawatt-hours (TWh) to between 8 900 TWh and 14 500 TWh (+135% to +284%). The share of renewables in total electricity generation rises from 19% in 2008 to 23% in 2035 in the Current Policies Scenario, 32% in the New Policies Scenario and 45% in the 450 Scenario.

Table 9.1 ● Global modern renewable energy supply and
shares in total by scenario

| | 2000 | 2008 | 2035 | | |
			New Policies Scenario	Current Policies Scenario	450 Scenario
Electricity (TWh)	2 876	3 774	11 174	8 873	14 508
Share in total electricity generation	19%	19%	32%	23%	45%
Heat (Mtoe)	266	312	660	540	790
Share in total demand for heat	10%	10%	16%	12%	21%
Biofuels (Mtoe)	10	45	204	163	386
Share in total transport	1%	2%	6%	5%	14%

2. See definition of renewables for heat in Chapter 11.

Figure 9.2 ● Increase in global modern renewables by type and scenario, 2008-2035

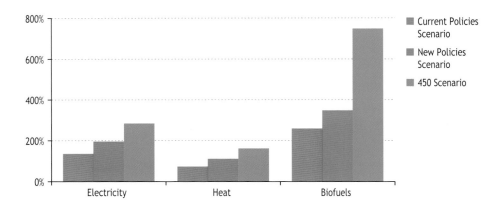

In the New Policies Scenario, the use of modern renewable energy triples over the course of the next twenty-five years, growing from 843 Mtoe in 2008 to 1 376 Mtoe in 2020 to 2 409 Mtoe in 2035. Its share in total primary energy demand increases from 7% to 9% and then 14%. Consumption of traditional biomass drops from 746 Mtoe in 2008 to 722 Mtoe in 2035, after a period of modest increase from now to 2020.

Figure 9.3 ● Modern renewables primary energy demand by region in the New Policies Scenario

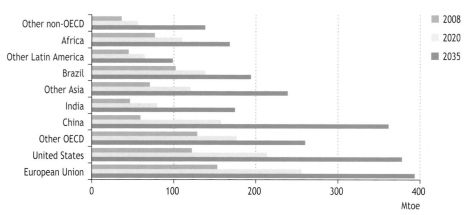

Demand for renewable energy increases substantially in all regions, with dramatic growth in some areas. Demand for renewables increases six-fold between 2008 and 2035 in China and four-fold in India (Figure 9.3). Demand remains highest in the European Union, where the increase is driven by policies to raise the share of renewables to 20% in gross final consumption in 2020, related to the commitment to cut greenhouse-gas emissions by 20% relative to 1990. The United States follows closely, as a result of large increases in renewables-based electricity generation and in biofuels use.

Global electricity generation from renewables increases from 3 800 TWh to 11 200 TWh and its share in total electricity generation grows from 19% to almost a third. The use of modern renewables for heat production in the industry and buildings sectors increases from 312 Mtoe to 660 Mtoe, with their share in total heat supply rising from 10% to 16%. Demand for biofuels grows four-fold between 2008 and 2035. Biofuels meet 8% of road transport demand in 2035, but just 1% of aviation fuel demand. Key results of the New Policies Scenario are summarised in Table 9.2.

Table 9.2 ● **Shares of renewable energy by sector and region in the New Policies Scenario**

| | Electricity | | Heat | | Biofuels | | | |
| | | | | | Road transport | | Aviation | |
	2008	2035	2008	2035	2008	2035	2008	2035
OECD	17%	33%	11%	23%	3%	12%	0%	3%
Europe	21%	44%	12%	25%	3%	12%	0%	0%
United States	9%	25%	10%	25%	4%	15%	0%	4%
Japan	10%	19%	3%	7%	0%	1%	0%	4%
Australia/ New Zealand	15%	31%	18%	41%	0%	2%	0%	0%
Non-OECD	21%	31%	9%	12%	2%	6%	0%	0%
China	17%	27%	1%	5%	1%	4%	0%	0%
India	16%	26%	24%	19%	0%	6%	n.a.	n.a.
Other Asia	16%	31%	11%	15%	1%	4%	0%	0%
Brazil	84%	75%	47%	50%	21%	41%	0%	3%
Other Latin America	52%	65%	13%	15%	0%	5%	0%	0%
Russia	16%	28%	5%	5%	0%	2%	0%	0%
Middle East	1%	16%	1%	3%	0%	0%	0%	0%
Africa	16%	39%	31%	37%	0%	2%	0%	0%
World	19%	32%	10%	16%	3%	8%	0%	1%
European Union	*17%*	*41%*	*13%*	*26%*	*3%*	*14%*	*0%*	*0%*

Note: Electricity = share of renewables in total electricity generation; heat = share of renewables for heat in total demand for heat; biofuels = share of biofuels used in road transport in total road transport and share of biofuels used in aviation in total aviation fuel.

Box 9.2 ● Renewables in the 450 Scenario

In the 450 Scenario, total primary energy demand of modern renewables grows four-fold between 2008 and 2035, from 843 Mtoe to nearly 3 250 Mtoe. Renewables supply 45% of total electricity output by 2035 and 21% of total heat. In the transport sector, 14% of transport fuel comes from biofuels in that year. Changes of this magnitude reflect the extent of government intervention assumed in this scenario, in order to limit the global temperature increase to 2°Celsius, and its dramatic implications for the renewable industry. This scenario is also accompanied by almost universal removal of fossil-fuel consumption subsidies. The main policy drivers in the electricity sector are emission trading schemes in OECD and major non-OECD economies, complemented by incentives to support those technologies that are not competitive. Growth in biofuels is underpinned by agreements to limit CO_2 emissions per car kilometre driven and in the aviation sector. The use of renewables for heat in industry increases both as a result of the emissions trading schemes that cap emissions in this sector and policies supporting renewables specifically. In buildings, renewables supply a much greater share of heat, owing to national policy plans that promote renewables alongside energy efficiency.

Total primary biomass use — both traditional and modern — in the New Policies Scenario increases from 1 225 Mtoe in 2008 to nearly 2 000 Mtoe in 2035.[3] Over 60% of total biomass used in 2008 was traditional biomass, which was consumed in developing countries (essentially in India and sub-Saharan Africa), mainly for cooking and space heating. This share drops to 37% by 2035, both because people who rely on it switch to modern fuels and technologies), and because demand for modern biomass increases substantially as a result of government policies.

Global modern primary biomass consumption nearly triples between 2008 and 2035. The pattern of use changes over time (Figure 9.4). The main application of modern biomass today is in industry, where it is mainly used in the production of process steam, while the power sector is the second-largest user. Over the period 2008-2035, most of the increase in biomass comes from the electricity sector and transportation. By 2035, power generation becomes the largest biomass-consuming sector, ahead of industry. The share of biofuels in modern biomass use grows from 10% in 2008 to 16% in 2035. Although biofuels are expected to become increasingly cost competitive with gasoline and diesel over the *Outlook* period, the allocation of biomass to the various consuming sectors is driven more by government incentives and priorities than by market economics (see Chapter 12).

---

3. Total biomass use is discussed in this section because it comprises several different uses, which may be competing for the same resource. This is not generally the case for other forms of renewable energy, so these are discussed in the subsequent chapters, which detail renewables use in particular applications.

Figure 9.4 • **World modern biomass primary demand by sector in the New Policies Scenario**

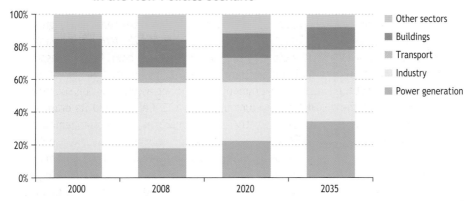

Note: Power generation includes biomass used in combined heat and power plants to produce electricity and heat, and in heat only plants.

Investment and finance

Recent trends in investment[4]

Global investment in renewable energy assets, as specifically defined in this section (Box 9.3), grew seven-fold over the period 2004-2008, from $17 billion to $126 billion. As the global financial crisis broke, credit dried up and companies postponed projects as a result of reduced liquidity and uncertainty over demand. Consequently, investment in renewables fell to $115 billion in 2009, a 9% drop from 2008. Investment in biofuels fell sharply, from $18 billion to $7 billion, a drop of over 60%. The biofuels industry was directly affected by the fall in oil prices and the lower overall demand for oil, which limited the amount of biofuels that could be absorbed by gasoline and diesel blending pools (IEA, 2009a). Regulatory changes related to the environmental benefits of conventional biofuels technology, for example, in Germany and the United States, also deterred investment. The renewables electricity sector was much less affected, mainly because of large and continued expansion in wind power projects in China. Global investment in electricity projects remained stable between 2008 and 2009, at around $108 billion.

---

4. The discussion in this section draws largely on investment data from Bloomberg New Energy Finance, which are different from the investment data used elsewhere in *WEO-2010*. The differences are outlined in Box 9.3.

When is biomass production sustainable?

Biomass is a renewable energy source so long as the growth of new crops and trees replenishes the supply. It is a carbon neutral energy source on that basis, as it releases only the CO_2 that was captured during its growth and an equivalent amount of CO_2 is recaptured in the regrowth. In that sense, biomass can greatly contribute to CO_2 emissions reductions, relative to fossil-fuel use. However, its production does give rise to several concerns.

Deforestation is a major problem in the developing world and, although it has decreased over the past decade, it continues at an alarmingly high rate in many countries (FAO, 2010). Planting crops for biofuels production — for instance, palm oil — has led to the clearance of forested land in some developing countries. Such deforestation has adverse social impacts on the local population and may lead to soil erosion and loss of biodiversity. Growing biomass crops, besides increasing water consumption, may also require intensive use of fertiliser to increase productivity, potentially resulting in water pollution.

The life-cycle greenhouse-gas emissions of biomass have also come under scrutiny. Concern has been directed particularly at biofuels, as some (e.g. corn ethanol) may provide only marginal emission savings on a life-cycle basis, or even result in an increase in emissions. The calculation of life-cycle emissions from biofuels takes into account emissions from the energy used in conversion and from land use changes (Chapter 12). The production process of the fuels gives rise, on average, to a lower level of greenhouse-gas emissions than the cultivation of the feedstock (UNEP, 2009). This is particularly true when the feedstock comes from sugar cane or ligno-cellulosic feedstocks (IEA, 2009b). The emissions attributable to feedstock cultivation are lower when no land use change is involved.

There is also some concern that diverting food crops to biofuels could increase prices and exacerbate hunger in poor countries, though some studies have indicated that there should be enough land available globally to feed the increasing world population and at the same time produce sufficient amounts of biomass feedstocks (e.g. Fischer et al., 2001; Smeets et al., 2007). However, environmental constraints relating to water and fertiliser use could reduce the amount of land that could realistically be available for biomass cultivation in the future, leading to a need to resolve the food-versus-fuel debate (Doornbosch and Steenblik, 2007).

The adverse environmental and social impacts can be minimised. Positive steps include: using marginal or under-utilised lands to avoid deforestation and competition with food production; focusing on advanced biofuels technologies that rely on ligno-cellulosic feedstocks; achieving greater productivity in growing biomass crops; making greater use of wastes, residues and surplus forestry; using high-efficiency biomass technologies for heat and power; and achieving higher standards of sustainable land use in the developing world.

Several government initiatives already address these concerns, including: the European Commission's Renewable Energy Directive; Germany's biofuels sustainability decree; the US Renewable Fuels Standard and Brazil's Agro-Ecological Zoning for Sugar Cane (IEA, 2010a). Several non-governmental initiatives promote the debate (for example, the Roundtable on Sustainable Biofuels, the mission of which is to develop standards for sustainable biofuels production).

Box 9.3 ● **Definitions of investment data**

The Bloomberg New Energy Finance (BNEF) data used throughout this section cover investment in new electricity assets (excluding hydropower projects greater than 50 MW) and biofuels. Importantly, BNEF investment data refer to finance secured for a particular new-build project or portfolio (there may be a lag from the time a contract is signed and finance is committed to when funds flow). This differs from the standard *WEO* approach, where the construction cost of projects is attributed to the year the project becomes operational. Furthermore, *WEO*-based figures include investment for all hydropower and are expressed in year-2009 dollars, while BNEF data are expressed in current dollars. Provided full account is taken of these methodological differences, the BNEF-based data presented in this section are particularly useful for the insights provided into investment in the short- to medium-term.

On a quarterly basis, investment fell to $19 billion in the first quarter of 2009, in the middle of the financial and economic crisis (Figure 9.5). Investment went up again in the following quarters and has remained broadly stable since then, at slightly above $30 billion. It has not yet regained the record level of $41 billion in the last quarter of 2007. In the first half of 2010, investment was 21% higher than over the same period in 2009.

Figure 9.5 ● **Quarterly global investment in renewable energy assets**

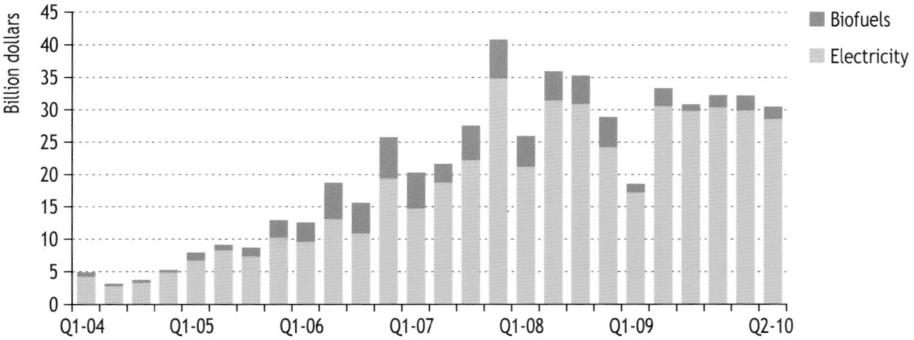

Source: Bloomberg New Energy Finance databases.

Europe leads global investment in renewable energy, while China rose to second place in 2009, overtaking the United States (Figure 9.6). Although global investment remained broadly unchanged in 2009, there were significant differences between regions. Investment went down in most regions, but the general drop was offset by a very large increase, more than 50%, in China. The most severe drop was in the United States, where investment fell to less than half the 2008 level. US financial institutions were hit hard by the crisis and credit became short. The loss of tax equity investors[5] (despite an extension of the production tax credit to 2012 and its conversion into a grant) also contributed to the collapse in investment. In addition, domestic gas prices fell from $8.35 per million British thermal units (MBtu) in 2008 to $4.12 per MBtu in 2009, which made renewable electricity projects even less attractive to investors. Investment fell less dramatically in Europe, by around 10%, owing to substantial government intervention, which facilitated lending from institutions such as the European Investment Bank.[6] Furthermore, feed-in tariffs, the main support mechanism for renewables in Europe, make renewable projects relatively more attractive to lenders, as generation leads to guaranteed revenues.

Figure 9.6 ● **Annual investment in renewable energy assets by region**

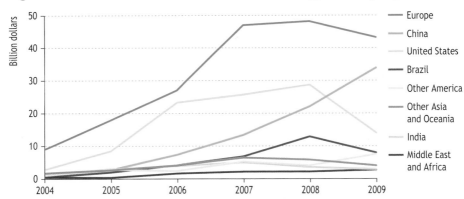

Source: Bloomberg New Energy Finance databases.

China, Europe and the United States account for most of the investment in renewables worldwide. Their combined share has been close to 80% in recent years. Outside these areas, Brazil has invested substantially in renewables in recent years, the level rising to $12.8 billion in 2008, but falling to $7.8 billion in 2009. Other countries in the American continent (outside the United States and Brazil) invested $7.4 billion in 2009. Against the general trends, investment in those countries nearly doubled between 2008 and

---

5. In the United States, the production tax credit — the main support mechanism for renewables at the federal level — can be used to finance new projects. Renewable energy developers can convert their prospective production tax credits into finance for their projects through the tax equity market. The number of *tax equity* providers fell sharply following the financial crisis (UNEP *et al.*, 2009). Lehman Brothers was one such institution.

6. In euro terms, investment went down by about 6%.

2009, driven mainly by Mexico and Canada. The Middle East and Africa also saw higher investment in 2009, although at $2.5 billion it is still rather limited. Investment in India reached $2.7 billion in 2009, 20% down on 2008.

Most renewables investment now goes into wind power, followed by solar. Global investment in wind power reached $67.3 billion in 2009, a 14% increase over 2008 and nearly 60% of the total investment in renewables. Investment in solar power fell to $24.3 billion in 2009, having climbed to $33 billion in 2008. Significantly lower PV unit costs, resulting from an oversupply of modules, contributed to this fall. Investment in biofuels boomed over 2006-2008, but collapsed in 2009, for the reasons highlighted earlier. Relative to investment in renewables for electricity, investment in biofuels is still small.

Renewable energy projects can be financed either on the balance sheet of the company or, separately, on a project finance basis. There are significant differences in practice between regions and countries (Figure 9.7). Generally, financing renewable energy projects involves a significant share of debt. In Europe, project finance has been the predominant approach. Most renewable energy projects in Europe are supported by feed-in tariffs, which guarantee revenues. This has made project finance relatively easy to obtain. In the United States, however, most projects are financed on the balance sheet of companies. This is, again, a reflection of the type of support policies used in the country. While the production tax credit and renewables portfolio standards provide an incentive to invest in renewable energy, revenues are not guaranteed unless developers can obtain the long-term contracts often necessary to secure financing. In China, on-balance-sheet deals are the most common and are done mainly by large state-owned companies securing loans from state-owned banks. Project finance is, however, becoming more common, especially as private investors enter the renewable energy market. In all three regions, project finance deals fell sharply in 2009, as they entail greater risk to financiers. Because of the capital-intensive nature of renewable energy technologies, companies that have the resources to finance renewables on their balance sheet may, nonetheless, start looking for alternative ways to finance their projects as their spending on renewables becomes a larger proportion of capital spending.

As a result of the financial crisis a shortage of credit for all purposes is expected to persist in the near term, with financing gaps in the affected areas. The International Monetary Fund (IMF) expects a credit shortfall in the Euro area of some €150 billion in 2010, with marginal improvements in 2011 (Table 9.3). In the United States, the credit shortfall is expected to be of the order of $280 billion in 2010, but the situation is expected to improve substantially in 2011, with the shortfall being reduced to $50 billion. Although there is some evidence that borrowing is now easier than in early 2009, the tight credit situation will have implications for investment in renewables in the near term. With greater competition for funds between renewables projects, the available capital is likely to be channelled towards the less risky projects in this sector. These difficulties are, hopefully, of short-term nature; additional renewables stimulus packages are expected to alleviate them. As of mid-2010, a total of $51 billion had been allocated to renewables, although most of the funding had not reached the sector at the time of writing (BNEF, 2010).

Figure 9.7 • Finance of renewables by region and type

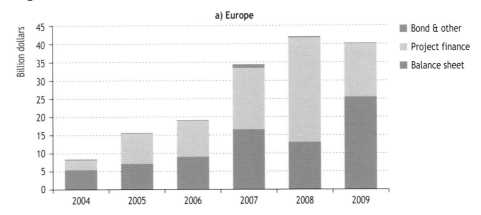

a) Europe

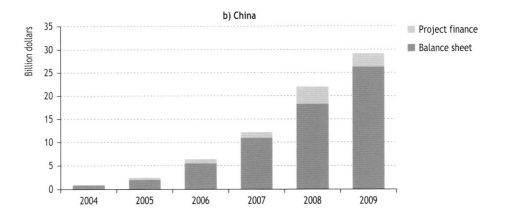

b) China

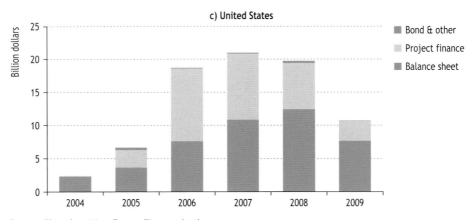

c) United States

Source: Bloomberg New Energy Finance databases.

Table 9.3 ● Credit projections for the United States and Euro area

	2010	2011
Euro area (billion euros)		
Total credit capacity	540	900
Total credit demand	690	1 040
Credit shortfall	-150	-140
United States (billion dollars)		
Total credit capacity	1 720	2 450
Total credit demand	2 000	2 500
Credit shortfall	-280	-50

Notes: Credit outside the financial sector. The Euro area comprises the countries of the European Union that use the Euro as their currency.
Source: IMF (2010).

Who invests: the structure of the renewables industry

Companies are increasingly taking note of the large growth potential in renewables and are investing more and more in renewables production (electricity or biofuels) or in the manufacturing of related equipment (notably for the production of renewables-based electricity). The largest of these companies are based mainly in Europe, the United States and China.

In the renewables electricity sector, the companies involved are generally traditional electricity generators, including some of the largest electricity producers in the world. Many are active in more than one country. For example, Iberdrola, Spain's largest energy company, is also present in the United States, France and the United Kingdom, among others, while E.ON, Germany's largest electricity producer, and Energias de Portugal are present in the United States. Three of China's largest electricity companies are among the top-ten renewable electricity producers (Table 9.4). China's largest renewable electricity generators have invested in hydropower in Southeast Asia (Box 9.4).

Box 9.4 ● China's overseas investment in renewable energy

Foreign investment by large Chinese power companies is mainly concentrated on hydro power in southeast Asian countries, in some cases, for example in the Mekong River Basin, in projects which could help enhance electricity supply in China through imports. The investment of China Huaneng Group in the Shweli I Hydropower Plant in Burma was the first of these projects. It started operation last year. Huadian Power International Corporation Limited has invested in the Asahan I Hydropower Project in Indonesia, which is about to start commercial operation, and in the Le Tour River Hydropower Project in Cambodia, which is scheduled to start up in 2012. Datang International

Power Generation Co. Ltd. has invested in Stung Atay Hydropower Project in Cambodia, which is expected to be completed in 2011. China Power Investment Corporation has been granted approval to develop the hydro resource in the upper stream of the Yi river in Burma, which could have a capacity of 20 gigawatts (GW). In addition to these power companies, others, such as Sinohydro, China Gezhouba Group Co. and China National Heavy Machinery Company, are also involved. It was estimated in 2008 that, at the time, there were 16 projects in Laos and 5 projects in Cambodia in which Chinese companies were involved as investors or developers (Heinrich Böll Stiftung Cambodia et al., 2008). While in the past Chinese companies have been involved mainly as contractors in the construction phase of projects, they now invest as the main owner.

Overseas investment by Chinese companies in other types of renewable energy projects is limited at the moment. There are only a few small projects involving investment in wind farms and wind equipment manufacturing. But the largest power companies, as well as smaller, private ones are seeking opportunities to invest in the solar and wind market abroad, especially in Africa.

Table 9.4 ● The world's ten largest owners of renewables-based electricity and biofuel producing facilities, as of June 2010

Electricity		Biofuels	
Company	Country	Company	Country
Iberdrola SA	Spain	Archer Daniels Midland Company	United States
Nextera Energy (formerly FPL Group Inc.)	United States	Valero Energy Corporation	United States
China Guodian Corporation	China	POET	United States
Enel SpA	Italy	Louis Dreyfus Group	France
Acciona SA	Spain	NTR Plc	Ireland
Energias de Portugal SA (EDP)	Portugal	Cosan Limited	Brazil
E.ON AG	Germany	Thomas H Lee Partners LP (THL Partners)	United States
China Datang Corporation	China	Sofiproteol	France
China Huaneng Group	China	Bunge Ltd	United States
Infigen Energy	Australia	Cargill Inc	United States
Share of total capacity: 24%		Share of total capacity: 18%	

Notes: Large hydro is not included. Country refers to location of the headquarters of the company (many are multinational).
Source: Bloomberg New Energy Finance databases.

Unlike renewables-based electricity, biofuels producers are not for the most part traditional energy companies. The top-ten companies in the business are mostly US companies. Many of them (for example, Archer Daniels Midland, Louis Dreyfus Group,

Cosan Limited, Sofiproteol, Bunge Ltd and Cargill) are involved in the agricultural commodities business. Most are active in more than one country. While no large oil companies appear in the top-ten, their interest in biofuels is growing: Royal Dutch Shell and Brazil's Cosan signed an agreement in August 2010 to form a joint venture in Brazil.

On the manufacturing side, the market for wind turbines and photovoltaics is becoming global and the industry is rapidly changing. While wind turbine manufacturing is still dominated by European companies, China has emerged as a major manufacturer, with three companies among the world's largest (Table 9.5). This is quite different from the market in 2000, when, outside Europe, only India's Suzlon and the United States-based GE Energy were among the top-ten manufacturers (*WEO-2009*).

Table 9.5 ● Global market shares of top-ten wind turbine manufacturers

2008			2009		
Manufacturer	Country	Market share	Manufacturer	Country	Market share
Vestas	Denmark	19.8%	Vestas	Denmark	12.5%
GE Energy	United States	18.6%	GE Energy	United States	12.4%
Gamesa	Spain	12.0%	Sinovel	China	9.2%
Enercon	Germany	10.0%	Enercon	Germany	8.5%
Suzlon	India	9.0%	Goldwind	China	7.2%
Siemens	Germany	6.9%	Gamesa	Spain	6.7%
Sinovel	China	5.0%	Dongfang	China	6.5%
Acciona	Spain	4.6%	Suzlon	India	6.4%
Goldwind	China	4.0%	Siemens	Germany	5.9%
Nordex	Germany	3.8%	Repower	Germany	3.4%

Note: Country refers to location of the headquarters of the company.
Sources: BTM Consult (2009); BTM Consult (2010).

The market for solar cells is dominated by Asian companies from China, Japan and Chinese Taipei, although the United States remains a significant producer (Table 9.6). Germany is the only European country with significant solar cell production. Many of the main players are becoming multinational, with manufacturing facilities in several countries.

Significant merger and acquisition (M&A) activity has taken place in the renewables sector in recent years, although there was a nearly 30% drop in 2009 (Table 9.7). The most important transactions now are in the solar manufacturing sector — exceeding $6 billion in 2009 — which accounted for nearly half of the total M&A activity in the production of renewables and related equipment manufacturing.[7]

---

7. The discussion of M&As in this section does not include large hydropower, as explained in Box 9.3. It should be noted, however, that substantial M&As are taking place in the hydropower sector. M&As in hydropower are estimated to have reached about $15 billion in 2009 (PWC, 2010).

Much of this activity has taken place in China, where smaller companies have suffered from an overcapacity among panel manufacturers and a plunge in global silicon prices (KPMG, 2010).

Table 9.6 ● Global market shares of top-ten solar cell manufacturers

2008			2009		
Manufacturer	Country	Market share	Manufacturer	Country	Market share
Q-Cells	Germany	7.4%	First Solar	United States	8.9%
First Solar	United States	6.4%	Suntech Power	China	5.7%
Suntech Power	China	6.3%	Sharp	Japan	4.8%
Sharp	Japan	6.0%	Q-Cells	Germany	4.8%
JA Solar	China	3.8%	Yingli	China	4.3%
Kyocera	Japan	3.7%	JA Solar	China	4.2%
Yingli	China	3.6%	Kyocera	Japan	3.2%
Motech	Chinese Taipei	3.4%	Trina Solar	China	3.2%
SunPower	United States	3.0%	SunPower	United States	3.2%
Sanyo	Japan	2.7%	Gintech	Chinese Taipei	3.0%

Note: Country refers to location of the headquarters of the company.
Sources: Hirshman (2009); Hirshman (2010).

Table 9.7 ● Mergers and acquisitions in renewable energy (billion dollars)

	2004	2005	2006	2007	2008	2009
Owners (electricity and biofuels)	1.3	4.3	5.1	9.8	11.0	6.4
Manufacturers	0.6	0.9	4.1	4.9	7.6	7.1
solar	*0.1*	*0.3*	*1.5*	*1.8*	*5.0*	*6.3*
wind	*0.4*	*0.5*	*1.9*	*2.6*	*2.4*	*0.7*
Total	1.9	5.2	9.3	14.7	18.6	13.5

Source: Bloomberg New Energy Finance databases.

Outlook for investment

In the New Policies Scenario, over 2010-2035 cumulative investment in renewables for electricity generation totals $5.7 trillion (in 2009 dollars), reverting to the normal *WEO* conventions and including large-hydro (Box 9.3). Another $335 billion goes into biofuels. China makes the largest investment in renewables electricity, followed by the European Union. The largest investment in biofuels is in the United States (Figure 9.8).

To meet the requirements of the New Policies Scenario, annual investment in 2035 needs to increase several times above current levels. There are several signs that the renewables sector will, indeed, continue to grow in the future, as discussed above: the persistent rise in investment up to 2008, the relative resilience of the sector on a global scale in 2009 despite the financial economic crisis, the involvement of a multitude of companies and increasingly of households, and intense M&A activity. But several challenges remain, both from the investors' and the lenders' perspectives.

As noted, government intervention is the main driver for the development of renewable energy. Investment will be forthcoming only if incentives are sufficient to guarantee a commercial return to power generators and biofuels producers. Further, government policies will have to address the specific risks associated with the different technologies (for example, the higher investor risk for new technologies than for mature or almost mature technologies). For industrial users, most investment in renewables is likely to be driven by the need to meet imposed emissions-reduction requirements. The potential to displace fossil fuels is large in many sectors of industry. Household investment in renewables is growing as consumers respond to environmental concerns and, in some cases, realise that they can obtain significant savings on their energy bills by switching to renewables. However, few countries incentivise renewables for heat, despite the large potential. To maximise their effect, policies to support renewables need to be clear, stable and well-publicised.

Figure 9.8 ● **Cumulative investment in renewables by type and selected country/region in the New Policies Scenario, 2010-2035**

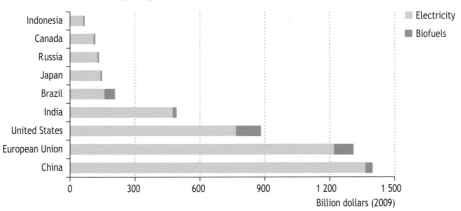

Financiers will take into account a variety of risks when considering lending to renewables, which typically include country and financial risk, policy and regulatory risk, technical and project specific risk, and market risk (UNEP et al. 2009). While such risks exist for all energy projects, some of the risks are higher for renewables. Policy and technology risks, in particular, can be significantly higher.

The challenges are likely to be the greatest in developing countries (Box 9.5). In the New Policies Scenario, these countries, excluding China, will need a total of $1.6 trillion in 2010-2035. In many developing countries, however, domestic capital markets do not have enough liquidity to cover these needs, so external financing will be necessary. Greater private sector participation is also likely to be necessary.

Box 9.5 ● Some key issues in financing renewables in developing countries

A number of financial barriers continue to constrain the development of renewable energy in developing countries (Parthan *et al.*, 2010). While their exact nature and the degree of influence varies between countries, the perception of risk, the lack of scale and higher transaction costs are all important. A number of multi-lateral, bi-lateral and national initiatives have been taken to assist in the removal of these barriers, in partnership with developing countries. Some of the conclusions to be drawn are:

- In general, significant local financing is available in most developing countries for renewable energy investments, partially from institutions such as local development banks, commercial banks and agricultural development banks and, in other cases, from specialised low-carbon energy finance institutions. Generally missing are risk mitigation instruments and retail-level institutions for channelling the finance.

- Guarantee funds can be used effectively in the early stages of market development, but they need to be replenished and sustained over the long run. Insurance products covering performance risk and the risk associated with weather fluctuations are in short supply.

- Despite prevailing misconceptions, both urban/peri-urban and rural poor people already pay significant sums for energy, both in absolute terms but, particularly, as a proportion of their total incomes. Providing renewable energy services to the poor is not just a matter for non-profit organisations but is already, in many cases, a profitable business on a commercial basis.

- Micro-finance can play a major role in the development of markets for small renewable energy systems and devices, but the achievements have so far been in market niches. The three critical factors to be addressed in order to scale-up the role of micro-finance in renewable energy are the management of transaction costs, credit risk management, and the availability of low-cost long-term financial resources at the wholesale level.

- The capacity of the finance and banking sector to evaluate and manage renewable energy projects in the commercial, development and agricultural sectors needs to be expanded.

- Supporting the establishment of dedicated finance facilities is a high-risk undertaking for development agencies. The risks can be reduced by ensuring that the projects entering the pipeline meet adequate tests of credibility, that there is evidence of serious commitment from early stage investors, and that the promoter company has a strong past track-record.

Costs of renewables

The cost of government support mechanisms

The application of renewable energy on a large scale depends on government incentives to make the unit costs competitive with conventional technologies. Incentives for renewable energy take many forms, from support to developers to support to customers. These incentives are generally described in this chapter as *government support* or *support mechanisms*, neutral terms which express no judgement on the argument that there is an economic case for intervention on the grounds that renewables are unduly disadvantaged in the energy market as it is presently constituted.

Defining government support is an uncertain undertaking. For the purposes of this analysis, government support to renewables is defined as any government measure that encourages the production or consumption of renewable energy sources. It can take a variety of forms, including mandates or portfolio standards, green certificates, feed-in-tariffs and premiums, and production, consumption and investment tax incentives. Some of these means of supporting renewables fall into the category of subsidies to consumers or producers (see Chapter 19 for a definition of subsidies). Other support mechanisms may not necessarily be a subsidy. The overall value of support to renewables is calculated here as the price paid to renewable energy producers for their output over and above the prevailing market price (or reference price), or as the incentive (price premium or tax incentives), multiplied by the quantity of energy subsidised.[8] In the case of electricity generation, the reference price is assumed to be the wholesale electricity price for all sources except solar photovoltaics in buildings, where the electricity end-user price is used. In the case of biofuels for transport, the reference price is assumed to be equal to the ex-tax price of the fuel at the pump that is substituted by ethanol and biodiesel.[9]

Measured this way, worldwide government support to renewables amounted to $57 billion in 2009 — up from $44 billion in 2008 and $41 billion in 2007 (Figure 9.9). The 29% increase in 2009 was in part due to a sharp drop in reference prices in 2009. In the New Policies Scenario, support grows throughout the period, reaching $205 billion

---

8. See chapters 10 and 12 for details of the methodology. For a discussion of subsidies to fossil-fuel consumption, see Chapter 19.

9. These calculations do not take into account spending on research and development, nor grants to households to induce them to buy renewable-energy based installations, nor spending by governments on advertising advocating the adoption of renewable energy.

by 2035.[10] It amounted to 0.08% of global GDP on average over the period 2007-2009, and grows to 0.17% of global GDP in 2035. Cumulatively, support totals $4 trillion in 2010-2035. Of this, 63% goes to renewable electricity and 37% to biofuels. While total support grows over time, it *decreases* on a per unit basis, both for electricity and biofuels, as technology costs come down.

Figure 9.9 ● Annual global support for renewables in the New Policies Scenario

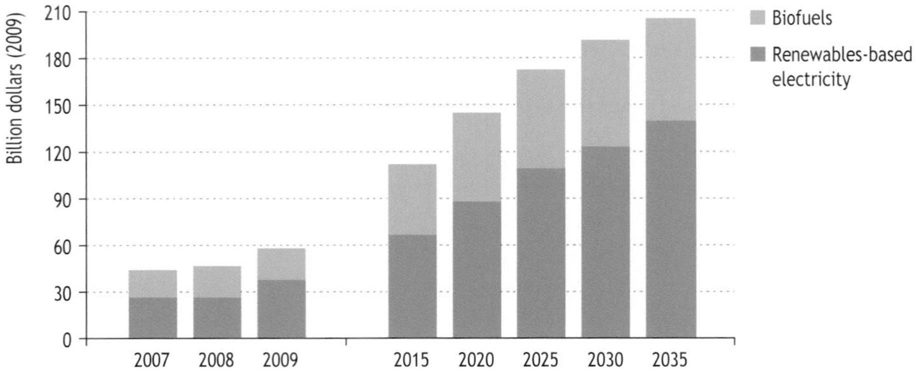

Given the array of benefits arising from greater use of renewables (not reflected in market prices) and the imperfections in the market pricing of other fuels, a degree of government support to these fuels and related technologies can be justified. Yet, governments need to ensure that the chosen mechanisms are cost effective, match the requirements of the particular technology involved and maintain competitive pressures between the different renewable technologies.

Research and development

In addition to providing support as defined above, governments are engaged in substantial continuing efforts in research and development (R&D) to bring the costs of renewable energy technologies down and to improve their performance. Some of these technologies, such as hydropower, onshore wind and biomass are mature or almost mature and do not require significant additional spending on R&D, although R&D is still needed for better wind forecasting and working variable generation into the power supply system. Photovoltaics and concentrating solar power, though commercially available, depend for their widespread diffusion on further supportive policy measures.

Total spending on R&D (using BNEF data, as explained in Box 9.3) reached $5.6 billion in 2009. Corporate R&D accounted for over 70% of this spending in recent years, but fell by 17% in 2009. Government spending rose in that year, more than compensating for the drop in corporate R&D and accounting for 45% of the total spending on R&D. More than half of current R&D spending goes into solar technologies (Figure 9.10). Spending is also significant in wind power-related research (both onshore and offshore

---

10. In the 450 Scenario, support for renewables reaches $300 billion in 2035 (see Chapter 13).

technologies) and in advanced biofuels. These three areas together absorbed 84% of total spending in 2009. Spending on R&D in the New Policies Scenario needs to rise significantly above present levels.

Figure 9.10 ● **Global spending on research and development in renewable energy by technology, 2009**

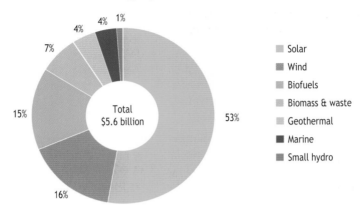

Source: Bloomberg New Energy Finance databases.

Integration costs of variable renewables

Provision also needs to be made to integrate variable renewables for electricity generation into the supply system. The additional network costs are estimated to be $13 billion in Europe and $11 billion in the United States in 2035. Cost effective ways to compensate for variable renewables supply do exist: a more integrated approach is required, planning simultaneously for the expansion of renewables production and the expansion of networks in order to keep costs low. In general, integration over large areas is more cost-effective.

Benefits of renewables

The main benefits of renewables — and the reasons for government support — are that they reduce CO_2 emissions (where used instead of fossil fuels) and reduce dependence on imported fuels, notably oil and gas. In the New Policies Scenario, renewables use cuts emissions by an extra 2 gigatonnes (Gt) CO_2 in 2035, relative to the Current Policies Scenario.[11] This is almost 30% of the total CO_2 savings in the New Policies Scenario (Figure 9.11). Most of these savings come from the power sector, where renewables displace coal and gas. Additional savings also arise from biofuels displacing oil in transport and from biomass and solar displacing fossil fuels for heat production. Renewables also reduce gas imports for power generation and oil imports for transport. Oil importing countries see their bills reduced by about $130 billion in 2035. Some reductions in gas import bills also arise, although they are much smaller.

---

11. The benefits of renewables are much larger in the 450 Scenario. See chapters 13 and 14.

Renewable energy has already created over three million jobs worldwide, of which about half are in the biofuels industry (REN21, 2010). The support for renewables included in many recent financial stimulus packages is expected to bring further employment benefits. Between 2008 and 2035, electricity generated by renewables increases three-fold, biofuels by over four times and heat from renewables by a factor of two in the New Policies Scenario, implying increases in gross employment creation (though not necessarily pro rata). Renewables are believed to create more jobs than fossil fuels per unit of output (UNEP, 2008; Fraunhofer Institute *et al.*, 2009; Greenpeace and EREC, 2010). Renewable energy has created many medium- to high-skilled jobs, particularly in the solar and wind sectors. It also helps create jobs in rural areas. However, the terms of employment there are not always favourable — currently, the bulk of biofuels jobs are found at sugar cane and palm oil plantations, where wages are low, working conditions often extremely poor and workers enjoy few rights (UNEP, 2008).

Renewables help reduce local pollution, such as sulphur dioxide (SO_2) and nitrogen oxides (NO_x) released from fossil fuels. In the New Policies Scenario, renewables reduce pollution by 4 million tonnes (Mt) SO_2 and 3 Mt NO_x in 2035. Other potential benefits of renewables include: moderating effects on rising fossil-fuel prices and reduced vulnerability to price variability; greater long-term energy supply security through supply diversification; reduced adaptation costs; trade benefits for countries that manufacture and export-related equipment; and benefits for rural development. By contrast, some renewable energy technologies may have adverse impacts that need to be addressed, such as land use, visual impacts or water consumption.[12]

Figure 9.11 ● **Contribution of renewables to the global emission and oil-import bill savings in 2035 in the New Policies Scenario vis-à-vis the Current Policies Scenario**

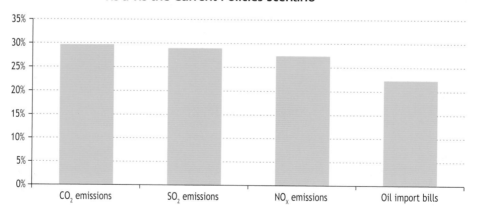

Note: Shares in total CO_2, SO_2 and NO_x emission savings and in total oil import bill reductions attributed to renewables in the New Policies Scenario, relative to the Current Policies Scenario.

---

12. See, for example, a discussion of the environmental co-impacts of emerging energy technologies in IEA, 2010b.

Characteristics of renewable energy

Hydropower

Hydropower exploits the potential energy of water by converting it into electricity, produced either in run-of-river plants or reservoirs. Hydro power can be exploited in almost all parts of the world. It is the most mature renewable energy technology. In the OECD countries the most suitable sites, especially for large hydro, have already been developed, but there is still a large potential for small-scale developments. Large potential for hydro generation still exists in Asia, Latin America and Africa.

Hydro reservoirs can be operated flexibly and therefore, especially where pumped storage is available, can meet sudden fluctuations in power demand. Depending on the volume of the reservoirs and the electrical capacity of the dam relative to the total system, some hydro plants can be operated as base load, while others serve as peaking plants. There are large differences in observed full load hours in hydro plants across the world.

Hydro developments are environmentally and socially controversial. Close attention needs to be paid to minimising the negative effects on surrounding ecosystems and to water availability and other consequences downstream. Moreover, hydro reservoirs require careful design and management in order to avoid possible emissions of methane.

9

Biomass

Biomass energy is energy produced from organic material grown, collected or harvested for energy use. At present, biomass is the only renewable energy source that can be used for electricity production, heat production and transport. The range of technologies exploiting biomass resources is very wide and the choice of technology depends not only on final use, but also on the nature of the biomass feedstock. The biomass resource can be estimated, based on the land available for dedicated crops and the available forestry and agricultural residues and waste. The main constraints on biomass exploitation are the availability of land for crops and water use (see Spotlight in this chapter and Chapter 12).

Solar

Solar energy is by far the largest energy resource available on earth. Three different technologies contribute to the capture and application of solar energy: solar photovoltaics (PV) and concentrating solar power (CSP) to provide electricity, and solar heating and cooling to provide directly usable heat (or cooling).

Solar photovoltaic systems convert direct and diffused solar radiation into electricity through a photovoltaic process using semi-conductor devices. PV systems can be developed anywhere in the world on suitable land and on buildings. PV technology is also very modular, which means that systems can be installed close to centres of demand. It represents a very suitable option for off-grid electrification. Like wind, solar PV is a variable source of power and its integration into the grid could present a challenge for system operators where it is used on a large scale. On the other hand, peak production occurs during the day, typically coinciding, in hot regions, with peak electricity demand, often driven by air conditioning loads.

Concentrating solar power (CSP) systems are designed to produce high-temperature heat for electricity generation or for co-generation of electricity and heat. CSP systems are capable only of exploiting direct normal irradiation, which is the energy received directly from the Sun (*i.e.* not scattered by the atmosphere) on a surface tracked perpendicular to the sun's rays. Areas suitable for CSP development are those with strong sunshine and clear skies, usually arid or semi arid areas. CSP is a proven technology (see Box 10.1 in Chapter 10), first commercialised in the 1980s in the United States, which has seen more widespread use in recent years. CSP technology opens up the possibility of thermal energy storage, as well as hybrid designs, for example with natural gas co-firing. CSP plants, if equipped with sufficient storage capacity, could provide base-load power.

Solar thermal collectors produce heat derived from solar radiation by heating a fluid circulated through a collector. Like PV panels, they are able to exploit both direct and diffused light and therefore can be installed anywhere in the world. The collectors produce relatively low temperature heat, suitable for space heating and hot water production in buildings and some lower temperature industrial applications. Solar thermal heat is not always available when domestic heat is needed (*e.g.* insolation is low in winter when space heat demand is the highest) and therefore solar thermal collectors have relatively limited potential to replace other sources of heat, at least until inter-seasonal storage becomes affordable. The potential for industrial heat from solar is virtually untapped for the moment.

Wind power

The kinetic energy of wind is exploited in wind turbines for electricity generation. Wind speeds suitable for electricity generation range from four metres per second to 25 metres per second. These are attainable practically all over the world, with the exception of some equatorial regions. Wind power is exploited not only onshore but also off-shore, where wind speeds are higher and the wind is typically available more regularly and for longer periods of time. The depth of water and distance from centres of demand onshore are major factors influencing the siting of off-shore developments. The availability of land enjoying suitable wind conditions is one constraint. Moreover, wind is a variable source of power: output rises and falls as wind strength fluctuates. This variability poses a challenge when integrating wind power into grids, especially once wind becomes a major component of the total system.

Geothermal energy

Geothermal energy is the energy available as heat extracted from the earth, usually in the form of hot water or steam. It can be exploited for power generation or for direct heat use. Geothermal resources of moderate or high temperature are suitable for power generation. High-temperature geothermal resources can be found typically in areas near plate boundaries or rift zones. Geothermal energy for electricity production is already exploited in a few areas of the world, while a more widespread but costlier potential exists, using moderate temperature geothermal power.

Geothermal power plants typically serve as a source of base-load power. Geothermal plants can have a long lifetime, but exploited geothermal reservoirs require constant management. Combined heat and power geothermal plants are more economical, where there is suitable heat demand. A barrier for further development exists where high-temperature geothermal sources are distant from demand centres. Where the temperature level is too low for power production, geothermal heat resources can be exploited for direct use in district heating systems and for industrial and agricultural purposes, where local markets exist. Sources of low temperature geothermal heat are found all over the world.

Marine power

Marine energy technologies exploit the kinetic energy of the tides, waves and currents of the sea, as well as temperature and salinity gradients, for the generation of electricity. The resource is, in principle, unlimited and exists in all world regions, but it is exploitable in practice only at sites that are close to demand centres and where, at the same time, damage to local ecosystems can be contained. Marine technologies are the least developed of the renewable energy technologies. Some marine technologies, namely those exploiting tides, have variable output, though this has the advantage of being predictable.

9

CHAPTER 10

RENEWABLES FOR ELECTRICITY

Ready to power the world?

H I G H L I G H T S

- The prospects for renewables-based electricity generation hinge critically on government policies to encourage their development. Worldwide, the share of renewables in electricity supply increases from 19% in 2008 to 32% in 2035 in the New Policies Scenario; it reaches only 23% in the Current Policies Scenario, but 45% in the 450 Scenario. In all three scenarios, rising fossil-fuel prices and declining costs make renewables more competitive with conventional technologies.

- In the New Policies Scenario, renewables-based electricity generation triples between 2008 and 2035, reaching almost the same level as coal-fired generation by 2035. The increase comes primarily from wind and hydropower. In 2035, renewables supply 41% of total electricity in the European Union, 27% in China and 25% in the United States. Worldwide, cumulative investment of almost $6 trillion (in year-2009 dollars) is needed over 2010-2035, close to 60% of total investment in power plants. China's investment ($1.4 trillion) exceeds that of the European Union ($1.2 trillion) and the United States ($0.8 trillion).

- The share of electricity generation from variable renewables (such as wind and solar power) is set to increase considerably, imposing additional costs on power systems. In the New Policies Scenario, integration costs amount to $16 per MWh in Europe and $17 per MWh in the United States in 2035. Generation and network planning will have to reconcile the characteristics of the new technologies with the need to maintain supply reliability.

- Government support for renewables-based electricity generation reached $37 billion in 2009 and is projected to approach $140 billion by 2035 (in year-2009 dollars) in the New Policies Scenario. Support per unit of generation falls over time, as the production costs of renewables fall, reaching a global average of $23 per MWh by 2035, down from $55 per MWh in 2009.

- The quality of its solar resource and its large uninhabited areas make the Middle East and North Africa region ideal for large-scale development of concentrating solar power, costing $100 to $135 per MWh in the New Policies Scenario in 2035. Solar power could be exported to Europe (at transmission costs of $20 to $50 per MWh) and/or to countries in sub-Saharan Africa.

- In the 450 Scenario, global renewables-based electricity generation grows from 3 800 TWh in 2008 to 14 500 TWh in 2035; its share in total output increases from 19% to 45%. Cumulative investment in renewables for electricity generation over the period 2010-2035 amounts to $7.9 trillion.

Outlook for renewables-based electricity generation

Recent trends and prospects to 2035

The prospects for electricity production from renewable energy sources in the coming decades hinge critically on government policies to encourage their development and deployment. Renewables supplied almost 3 800 terawatt-hours (TWh) of electricity worldwide in 2008, 19% of total electricity production. That share has changed only marginally since 2000. In 2008, 85% of renewables-based electricity came from hydropower. The share of other renewable energy sources combined — biomass, solar, wind, geothermal and marine power — both in total electricity and in renewables-based generation, has been rising slowly, but constantly, in recent years; their share in total electricity generation rose from 2% in 2000 to 3% in 2008, while their share in renewables-based generation rose from 9% to 15%. While hydropower has been the dominant renewable source of electricity for over a century, the strong growth recently in new technologies — particularly wind power and solar photovoltaics (PV) — has created expectations among policy makers and the industry alike that these technologies will make a major contribution to meeting growing electricity needs in the near future.

While power from renewables has been growing over the past decade, in absolute terms this growth pales beside the scale of the increase in fossil-fuel based generation. Globally, electricity from renewable energy sources increased by almost 900 TWh between 2000 and 2008, but at the same time coal-fired generation increased by about 2 300 TWh and gas-fired generation by 1 600 TWh (Figure 10.1). In the OECD region, generation based on renewables increased more than that based on coal over the same period, but much less than natural gas generation. In non-OECD countries, the increase in electricity generation from renewables was slightly lower than the corresponding increase from gas, but much lower than that from coal.

Figure 10.1 ● **World incremental electricity generation by fuel, 2000-2008**

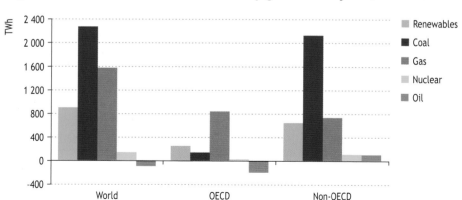

Renewables-based electricity generation is expected to continue to grow over the next 25 years, benefiting from government support, declining investment costs and rising fossil-fuel prices. But the pace of this increase varies widely across the three scenarios presented in this *Outlook*, according to the degree of government support. Worldwide, electricity based on renewable energy (including hydropower) is projected to increase from about 3 800 TWh in 2008 to about 11 200 TWh in 2035 in the New Policies Scenario; it rises less rapidly to less than 8 900 TWh in the Current Policies Scenario, but much more rapidly, to over 14 500 TWh, in the 450 Scenario (Figure 10.2). The share of renewables in total electricity generation rises from 19% in 2008 to 23%, 32% and 45% in the three scenarios respectively by 2035. In the Current Policies Scenario, renewable energy meets 28% of incremental electricity demand between 2008 and 2035. This share rises to almost 50% in the New Policies Scenario and 90% in the 450 Scenario.

Figure 10.2 ● **Electricity generation from renewables by scenario**

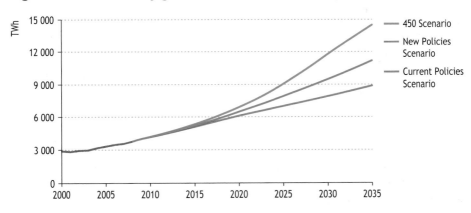

The share of hydropower in total electricity generation declines in the Current Policies Scenario, from 16% to 13%. It remains broadly unchanged in the New Policies Scenario and increases from 16% to 19% in the 450 Scenario. The shares of all other renewable energy sources increase in all three scenarios. Electricity generation from biomass, wind, solar, geothermal and marine power, grouped together, increases significantly more than hydropower.

In the New Policies Scenario, renewables-based electricity generation triples between 2008 and 2035 and in absolute terms catches up with coal-fired generation by the end of the projection period (11 200 TWh). For most renewables-based technologies and in most regions, direct government incentives are the main driver of growth rather than carbon markets in the New Policies Scenario.

While electricity generation from hydropower remains dominant over the *Outlook* period, other renewable sources collectively grow faster. By 2035, electricity generation from wind, biomass, solar, geothermal and marine energy reaches around 5 600 TWh, more than hydropower in that year. The increase in renewable electricity generation between 2008 and 2035 is derived primarily from wind and hydropower, which contribute 36% and 31% of the additional demand respectively (Figure 10.3).

Figure 10.3 ● Incremental renewables-based electricity generation by region in the New Policies Scenario, 2008-2035

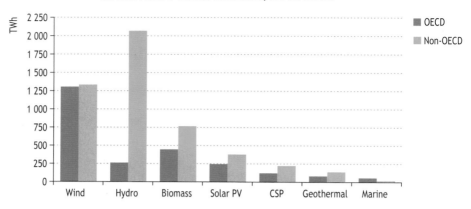

Hydropower increases from 3 200 TWh to about 5 500 TWh in 2035 and installed capacity from 945 gigawatts (GW) to 1 600 GW. The share of hydropower in total generation remains constant at around 16% throughout the *Outlook* period. Most of the increase in hydropower occurs in non-OECD countries, where the remaining potential is highest, although development continues in OECD areas, notably in Canada, the European Union (EU) and Turkey. Hydropower sees significant growth in non-OECD Asia, where it grows from 830 TWh to almost 2 200 TWh. A total of 111 GW is now under construction (out of 168 GW worldwide), of which about 80 GW are in China, 15 GW in India and 7.5 GW in Vietnam (WEC, 2010). Hydropower also grows significantly in Latin America, where it is already the most important source of electricity. Another 16 GW are under construction in this region. Africa's unexploited potential is very large, but progress in developing it is expected to be slow. Ten GW are now under construction across the continent. In the New Policies Scenario, hydropower continues to grow in Africa, but at a slower rate than in Asia and Latin America.

Wind power (both onshore and offshore) is projected to supply 8% of global electricity in 2035, up from just 1% in 2008. Electricity generation from wind farms increases by a factor of 13 between 2008 and 2035 and installed capacity increases from 120 GW to over 1 000 GW. This continues the strong trend seen in the past decade. In 2009, a total of 38 GW was added worldwide, of which about 14 GW was in China and 10 GW each in the European Union and the United States (GWEC, 2010). These three regions see the largest increases over the *Outlook* period and account for 70% of the global installed wind capacity in 2035.

While most wind power is expected to come from onshore wind farms, offshore wind installations are likely to provide a growing share. In 2008, offshore wind capacity was 1.5 GW. In the New Policies Scenario it reaches almost 180 GW in 2035 as the technology improves, costs are reduced and the current difficulties in obtaining finance dissipate.

Electricity produced from solar photovoltaics increases from 12 TWh in 2008 to 630 TWh in 2035, around 2% of global electricity. Installed PV capacity increases from 15 GW in 2008 (and an estimated 23 GW in 2009) to 410 GW in 2035. A little more than half of this is projected to be installed in buildings, meeting around 4% of their demand for electricity, while the remainder is for large-scale generation. Some PV will also be used in rural electrification projects. Over 160 GW of PV, 40% of the world total in 2035, is projected to be installed in non-OECD Asia, notably in China and India.

Box 10.1 ● Enhancements to the renewables-based power-generation module in *WEO-2010*

The renewables module, covering capacity additions and investments, electricity generation and heat production from renewable sources, has been overhauled and improved for this year's *Outlook*, allowing for more detailed and complex modelling, and tighter integration into the power generation component of the IEA World Energy Model (WEM). Government support mechanisms that encourage the development and deployment of renewable technologies are also modelled in greater detail, allowing the additional support needed for each source to become competitive to be calculated. A full review of the potential for all renewable energy sources was undertaken for this analysis, with up to 16 technologies per region incorporated into the model. The model also takes into account expected technical developments and dynamic global learning, as well as the technical and non-technical barriers that in some countries may create obstacles to the full exploitation of the potentials considered. How renewables compete with other fuels in the power-generation mix, the electricity dispatch and the electricity wholesale and end-user prices have been enhanced.

Concentrating solar power plants produce 340 TWh of electricity in 2035, from less than 1 TWh in 2008. Installed concentrating solar power (CSP) capacity increases from 1.4 GW to over 90 GW. CSP technologies have evolved rapidly over the past few years and several advanced technology systems are now being installed, mainly in the United States and Spain. CSP is a key component in India's Solar Mission. Box 10.2 discusses the main trends in CSP technology.

Geothermal power increases from 65 TWh to about 280 TWh, mainly in the United States, Indonesia and south-east Asia (notably the Philippines). These are the regions with the greatest potential as they are located around the Pacific "ring of fire". Geothermal installed capacity increases from 11 GW to over 40 GW.

Marine power, which comprises technologies that convert tidal and wave energy to electricity, increases less than other renewables technologies. This is because wave technologies are still in their infancy, requiring much further research, and because the locations in which tidal power can be used are limited. Marine power increases to some 60 TWh in 2035 and installed capacity to 17 GW.

The share of renewables in electricity generation increases in all regions except in Brazil, which has already extensively developed its hydropower resources. Nonetheless, the share of renewables in electricity generation in Brazil remains one of the highest in the world. In 2035, the share of renewables by region ranges from about one-fifth to over two-thirds of total electricity (Figure 10.4). In the European Union, renewables supply 30% of electricity in 2020 (to meet the European Union's overall target of 20% renewables in its total energy mix in 2020) and this share rises to 41% in 2035, up from 17% in 2008.

Figure 10.4 ● *Share of renewables in total electricity generation by type and region in the New Policies Scenario*

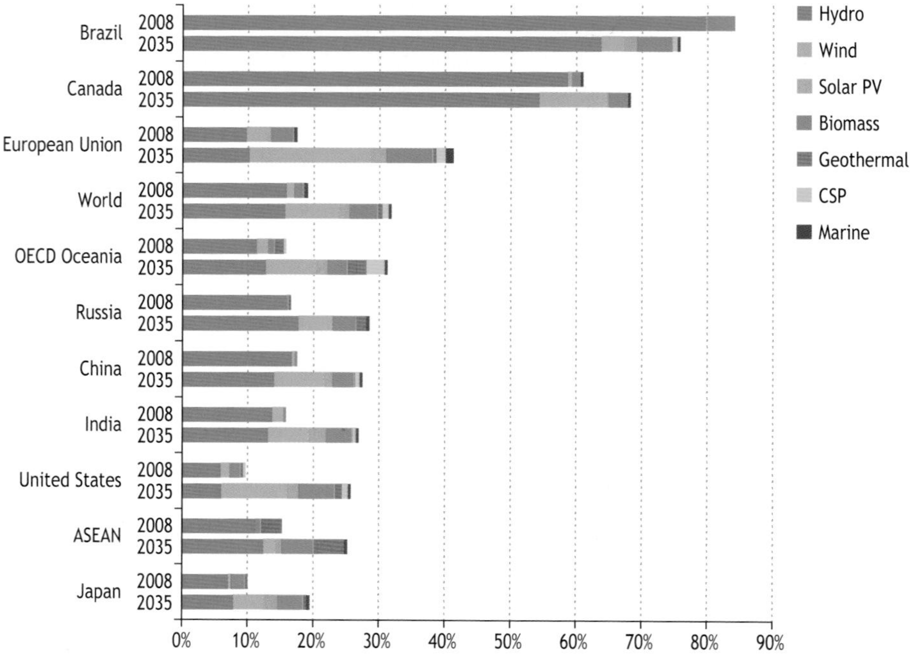

Box 10.2 ● Concentrating solar power technology

The first large CSP plants were constructed in the United States in the 1980s. Driven by technology improvements and industry initiatives in the United States, Spain and North Africa, CSP has recently gained a lot of momentum and public attention.

There are four types of CSP technology: parabolic trough systems using parabolic reflectors, which concentrate solar radiation onto a receiver pipe and heat up an absorber medium; linear Fresnel collectors, operating on the same principle but using flat mirrors; power tower systems, where several sun-tracking mirrors (heliostats) focus sunlight onto a receiver at the top of a tower for steam generation; and parabolic dish systems, which use a parabolic-shaped point focus concentrator in the form of a dish. At present, most of the projects in operation or under construction are parabolic trough systems. These are mostly located in Spain and the United States.

Further technology improvements and cost reductions are important, especially in the mirrors/reflectors, which account for around 20-40% of the overall capital costs, depending on the plant design. Power tower technologies are considered to have significant potential in this respect, with potential cost reductions for the heliostat on the order of a factor of two to three. Even more fundamental to the economics of CSP is increasing its availability, through the integration of storage (e.g. molten salt). While this significantly increases the upfront investment costs, for example due to the need for a storage tank and more reflector area, it can be more than offset by the value of the increased hours of operation per day. Provision of back-up capacity is an alternative solution.

The design of CSP stations is complex and today is still done project-by-project, given that the technology is not yet mature. Constraints to be considered include land and water availability, proximity to load centres and environmental constraints, such as safeguarding protected species in desert areas. It is widely accepted that, to achieve an adequate return, CSP is ideally located in areas with annual direct normal irradiation (DNI) in excess of 2 000 kilowatt-hours per square metre per year (kWh/m^2/year). Site selection and CSP design is a complex task which needs to consider the DNI on a daily basis and dispatchability.

In the United States, the share of renewables in total electricity generation increases from 9% in 2008 to 25% in 2035. This increase is driven by both federal and state-level incentives. Renewables increase despite strong competition from gas-fired generation, which remains very competitive in the United States owing to the abundant domestic supply of unconventional gas (see Chapter 5). Synergies also exist between gas and renewables, as gas can compensate for the irregularity of variable renewables.

In China, the share of renewables grows from 17% to 27%. China now has the largest installed hydropower capacity in the world. By 2035, China has the largest PV capacity in the world and the second-largest wind power capacity, just behind the European Union.

Renewables-based electricity generating costs

The generating costs of renewables technologies per unit of output are projected to continue to fall over the projection period (Figure 10.5). The main reason is increased deployment, which accelerates technological progress and increases the economies of scale in manufacturing the associated equipment. The costs of the more mature technologies, including geothermal and onshore wind power, are assumed to fall the least. The costs of hydropower remain broadly unchanged. The assumed technology learning rates used in this study are presented in Table 10.1.[1] They express our best judgement, based on recent research, and are assumed to be the same across the three scenarios.

---

1. Learning rates are used to represent the reductions that occur in technology costs as cumulative deployment increases. A learning rate of 5% implies that the investment cost of a technology would be expected to fall by 5% with every doubling of cumulative installed capacity.

Table 10.1 ● Generating costs of renewables-based electricity generation by technology and learning rates in the New Policies Scenario

	Generating costs						Learning rates
	2010-2020 ($2009 per MWh)			2021-2035 ($2009 per MWh)			
	Min	Max	Avg	Min	Max	Avg	(%)
Hydro - large	51	137	94	52	136	95	1%
Hydro - small	71	247	143	70	245	143	1%
Biomass	119	148	131	112	142	126	5%
Wind - onshore	63	126	85	57	88	65	7%
Wind - offshore	78	141	101	59	94	74	9%
Geothermal	31	83	52	31	85	46	5%
Solar PV - large scale	195	527	280	99	271	157	17%
Solar PV - buildings	273	681	406	132	356	217	17%
CSP	153	320	207	107	225	156	10%
Marine	235	325	281	139	254	187	14%

Note: MWh = megawatt-hour.

Figure 10.5 ● Electricity generating costs of renewable energy technologies for large-scale electricity generation in the New Policies Scenario

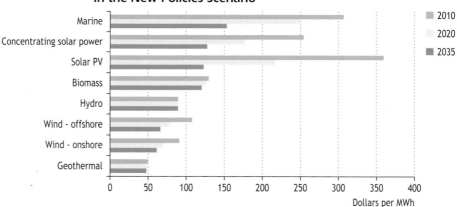

Investment needs

In the New Policies Scenario, cumulative investment in renewables-based electricity generation worldwide amounts to $5.7 trillion (in year-2009 dollars) over the period 2010-2035, close to 60% of the total investment in power plants (Table 10.2 and Figure 10.6). Totalling $1.4 trillion, China's investment exceeds that of the European Union ($1.2 trillion) or the United States ($0.8 trillion). Renewables account for a large share of total investment in power-generation plant in most regions; for example, 82% in Brazil and 71% in the European Union. Investing in renewables will pose additional financing problems, particularly in developing countries (see Chapter 9).

Table 10.2 ● **Investment in renewables-based electricity generation by technology in the New Policies Scenario** ($2009 billion)

	2010-2020	2021-2035	2010-2035
Hydro - large	689	803	1 492
Hydro - small	74	102	176
Biomass	203	484	688
Wind - onshore	598	866	1 464
Wind - offshore	99	278	376
Geothermal	24	51	75
Solar PV - large scale	99	267	366
Solar PV - buildings	212	441	653
CSP	73	274	347
Marine	4	63	67
Total	2 074	3 630	5 704

Figure 10.6 ● **Investment in renewables-based electricity generation by region in the New Policies Scenario, 2010-2035**

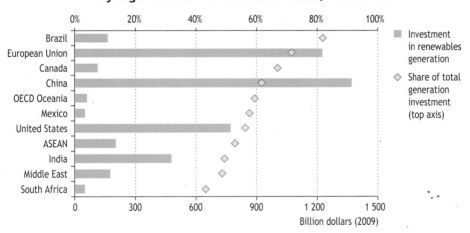

The $5.7 trillion that is invested worldwide in renewables-based generation in the New Policies Scenario would deliver almost 2 800 GW of gross renewables capacity (*i.e.* including the replacement of existing facilities). More investment goes into wind power than any other renewable source, including hydropower (Figure 10.7). A total of $1.8 trillion is spent to build over 1 200 GW of wind power (including replacement of existing facilities). Investment in hydropower totals $1.7 trillion while investment in PV is also significant, exceeding $1 trillion over the whole projection period.

Figure 10.7 • Global cumulative capacity additions and investment in renewables-based electricity generation by technology in the New Policies Scenario, 2010-2035

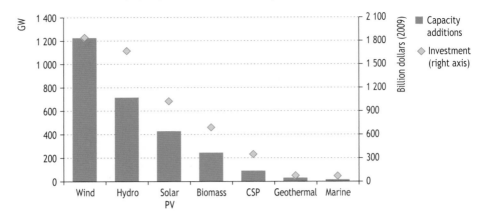

Box 10.3 • Renewables for electricity in the 450 Scenario

In the 450 Scenario, global renewables-based electricity generation is projected to grow from 3 800 TWh in 2008 to just over 14 500 TWh in 2035; its share in total output increases from 19% to 45% (see Chapter 13). By 2035, electricity generation from renewables by far exceeds generation from all fossil fuels combined. Renewables supply over 50% of the European Union's electricity in 2035, up from 17% in 2008 and one of the biggest increases in the world. While direct incentives continue to play a key role in the development of renewables in this scenario, carbon markets are increasingly a key driver.

Hydropower remains the largest source of renewables-based electricity: its share increases from 16% to 19%. The largest increase in terms of market share is in wind power, which supplies 13% of electricity worldwide in 2035, up from just 1% in 2008. Biomass supplies 6% of total electricity in 2035, solar PV 4% and CSP 3%. Cumulative investment in renewables for electricity generation over the period 2010-2035 amounts to $7.9 trillion, 65% of total investment in electricity-producing facilities and nearly 40% more than in the New Policies Scenario.

Government support for renewables

Government support[2] for renewables is becoming widespread. In early 2010, over 100 countries had some type of target, measure or programme to support renewables — almost double the number in 2005 (REN21, 2010). Policies focusing on electricity are far more common than policies for biofuels and even more so than for heat.

---

2. The term support covers all types of government policies and measures that seek to encourage the development and deployment of renewables, including, but not limited to, subsidies to production and consumption. A precise definition of subsidies is provided in Chapter 19.

Support for renewables electricity generation can be provided at the investment phase or at the operational level, or both. Investment tax credits and loan guarantees fall into the first category. The main support mechanisms at the operational level include feed-in tariffs, green certificates, premiums and production tax credits. The main categories of support measures, along with examples of countries that use them, are shown in Table 10.3. No support mechanism can be singled out as the best; each has its advantages and disadvantages. It is important to concentrate on the most cost-effective policies and, where competitive markets exist, on policies that use the strength of such markets.

Table 10.3 ● **Classification of support mechanisms for renewables-based electricity**

	Type of incentive	Countries
Price-based	Feed-in tariffs	Most EU countries; some states and few cities in the United States; China (national system from 2010) ; Japan (only for households); South Africa; Brazil; Australia (some provinces); India (certain states)
	Premiums	Denmark; Spain gives the possibility to choose between feed-in tariffs and premiums
Quantity-based	Green certificates	United States (state level); United Kingdom; Italy; Japan; India (from October 2010); Australia
	Quotas/Portfolio Standards	European Union; United States (more than half of the states + Washington D.C.); China; Japan; India; Australia; South Africa; Brazil
Tax-based	Fiscal incentives	European Union; United States; China; Japan; India; Australia; South Africa; Brazil
	Investment incentives	European Union; United States; China; Japan; India; Australia; South Africa; Brazil
Other	Loans	European Union; United States; Brazil; Canada; Korea
	Carbon offsets	China; India; Mexico

Note: Countries shown are for illustrative purposes only. In reality, many more countries than those shown in this table apply such incentives.

Recent policy developments

European Union

The 2009 EU directive on renewables set an overall binding target for 2020 to achieve a 20% share of renewables in gross final energy consumption (across electricity, heat and transport fuels). The directive set targets for each country, which then has to develop a national action plan to meet them. The directive does not specify a target for electricity generation from renewables.[3] Most countries in the European Union

3. The European Commission estimates that in order to meet the overall target, around 33% of electricity must come from renewables (CEC, 2009). Some industry sources estimate that this share could be even higher, at around 40%.

(21 out of 27) use differentiated (*i.e.* technology-specific) feed-in tariffs or premiums to support renewables (Canton and Johannesson Lindén, 2010). In most cases, these are time-limited (*i.e.* available for a fixed period of time) and are updated regularly. A few countries use green certificates and tenders.

United States

The most significant recent development in the United States is the passage of the American Recovery and Reinvestment Act (ARRA) in February 2009. ARRA provides new funding at the federal level, loan guarantees and tax credits for renewables and for energy efficiency (US DOE/EIA, 2009). The United States is now considering a federal renewable electricity (or portfolio) standard in several legislative proposals. These would require power companies to obtain an increasing share (reaching 15% to 25% in different proposals) of retail electricity to be from renewable energy sources.

The main support mechanisms at the federal level are the production tax credit (for wind, biomass, geothermal, hydro and marine power) and the investment tax credit (mainly for PV). These are complemented by federal loan programmes, such as loan guarantees or clean renewable energy bonds. Several states now have renewables portfolio standards (mandatory or not) and offer incentives.

Japan

In mid-2009, Japan enacted new legislation to support the development of renewables, nuclear power and energy efficiency (Law on the Promotion of the Use of Non-fossil Energy Sources and Effective Use of Fossil Energy Source Materials by Energy Suppliers; Amendment of the Act on the Promotion of the Development and Introduction of Alternative Energy). Based on these laws, the government started providing feed-in tariffs for PV in buildings in November 2009, along with investment grants, loans and tax reductions. In June 2010, the government revised its Basic Energy Plan, which set the target for zero-emission power (nuclear and renewables) at 50% of total generation in 2020 and 70% in 2030, compared with 34% now.

Japan has had a Renewables Portfolio Standard (RPS) in place since 2003. The current RPS runs until 2014, with a target of producing 16 TWh from solar, wind, biomass, small hydro or geothermal power. Green certificates are the main support mechanism to achieve the targets set in the RPS. The Ministry of Economy, Trade and Industry (METI) proposed in July 2010 to expand feed-in tariffs to include PV for power companies, wind power (including small-scale generation), small hydro (less than 30 megawatts [MW]) geothermal and biomass. A unique tariff of around 15 to 20 yen per kWh is proposed for all sources except PV for a period of 15 to 20 years; for PV the tariff would be higher, but for a period of ten years. This new scheme would replace the current RPS.

Australia

In June 2010, Australia passed legislation to extend and amend its mandatory renewable energy target for electricity. The original scheme ran until 2010 but was extended to 2020, with the objective of achieving 20% of electricity from renewables. The new target is expected to add a further 45 TWh of renewables-based electricity by 2020. The existing scheme will be split into two as of 2011: the small-scale renewable energy scheme and the large-scale renewable energy target. Renewable energy certificates have been in use since 2001 and are expected to remain the main mechanism for achieving the 2020 target.

China

China's most important renewable policy framework remains the Renewable Energy Law (REL), enacted in 2005. REL stipulates that grid operators must accept renewable energy power at a price higher than that of conventional generation. The Chinese government has since formulated detailed implementation rules, clarifying the levels, stages and support schemes for the development of different renewable energy technologies. A target of increasing the renewable energy share in primary energy to 15% by 2020 was set in 2009. Experts estimate that this target could increase wind, solar and biomass power generation capacity to 150 GW, 20 GW and 30 GW respectively by 2020. The government is now organising detailed surveys of renewable energy resources to provide more reliable development information, expecting that this will help the understanding of risk and encourage investors. The government is also promoting the construction of a grid to connect resource-rich areas in the west and the south to demand centres in the east and centre of the country.

The development of wind power is supported by feed-in tariffs, which recently replaced a bidding system. There are four levels of feed-in tariffs, depending on the resource. For on-grid solar power, the bidding system is still in place. The government covers part of the investment cost of building integrated PV projects. Off-grid renewable power projects are funded through the Township Electrification Programme.

India

In January 2010, the Indian government launched the Jawaharlal Nehru National Solar Mission, which aims to install 20 GW of solar power (including PV, CSP and solar lanterns) by 2022. The Solar Mission targets both large- and small-scale generation, including for rural electrification (about 400 million people in India still lack access to electricity, see Chapter 8). A three-phase roadmap has been laid out, with interim targets for the development of solar power. India launched a feed-in tariff system in 2009, to support various renewable energy technologies, and is considering introducing renewable energy certificates. In the absence of a national renewable energy incentive, 18 out of 29 Indian states have implemented renewable energy quotas and introduced preferential tariffs.

Brazil

In Brazil, capacity tenders have now replaced the PROINFA programme, which had been in place since 2004. Large hydropower is supported by a separate programme. The National Climate Change Plan, approved in 2008, provides for an increase of electricity from renewables, including greater use of hydropower (34 GW of hydropower to be added over the period 2007-2016, the current Ten Year Plan period), of wind and sugar cane bagasse and greater use of PV (on- and off-grid).

South Africa

The Renewable Energy Framework sets a target to produce 10 TWh from renewables by 2013, 60% of which would come from electricity generation and the remainder from solar water heaters. A feed-in tariff scheme was set up in 2009 to help meet the 2013 target. The scheme obliges ESKOM (the national power company) to purchase renewable energy from qualifying generators.

Quantifying government support for renewables

Most renewable technologies used to produce electricity are more expensive per kWh today than conventional power technologies. As a result, intervention to increase the use of renewables-based generation raises the cost of power generation, except in the few cases where renewables-based systems are already fully competitive (and, so, in principle, do not require any type of support). In most cases, the additional costs of renewables are passed on to the final consumer.

Methodology

In this section, we quantify the total monetary value of government support for renewables-based electricity generation worldwide. The analysis covers all support programmes and measures that we have been able to identify, and all major countries and regions (which, taken together, now account for over 99% of world renewables-based electricity generation from wind, PV, geothermal and biomass). Projected additions of small hydropower capacity are included, but existing capacity is not. Large hydropower is not included, as it is assumed that it does not, in most cases, need or receive support.

Table 10.4 ● **Government support schemes for renewables-based electricity generation and quantification method**

Support scheme	Description	How support is quantified
Feed-in tariffs (FITs)	FITs are granted to operators for the renewable electricity they feed into the grid. They take the form of a fixed price per MWh, which reflects the cost of the technology.	(FIT − wholesale electricity price) x renewable energy generated
Production tax credit (PTC)	Direct reduction in tax liability.	PTC x renewable energy generated
Investment tax credit (ITC)	Direct reduction in tax liability.	ITC x capital investment in renewables over the year
Green certificates (GC)	A green certificate is a tradable commodity proving the production and the use of a certain amount of renewable energy.	Annual average price of GC x amount of GC issued
Premiums	Premiums are a sort of bonus and are paid to the producers on top of the electricity price (market-driven or regulated).	Premium x renewable energy generated

For the purposes of this study, support for renewables electricity generation has been defined as any incentive provided by governments in order to promote the deployment and application of renewable energy (see Chapter 9). These are generally offered as part of policies to address climate change and to improve security of supply. Examples of such incentives are feed-in tariffs, green certificates, premiums and tax credits. Some are direct cash subsidies to producers or consumers, but others have a cost or value which is more complex to pin down (see Chapter 19). Recognising the limitations of the exercise, the main objective of the analysis is to seek to measure the total monetary value of the premium paid for the output of renewables-based electricity, compared with the price paid for electricity generated in other ways. For instance, with feed-in tariff mechanisms, a fixed price is paid to renewable generators for each MWh produced and supplied to the grid. The feed-in price, generally set by the government, reflects the cost of the technology and is set at a level higher than the spot price of electricity, so as to reward renewables-based electricity generators. The support given to renewable generators is, therefore, the difference between the feed-in tariff and the market price for electricity at the point of delivery. Only the additional payment above the market price is considered as support in the analysis presented here. The analysis is not fully comprehensive or definitive. The value of some forms of support, such as direct and indirect funding for research and development into innovative projects/technologies, grants and loan guarantees, has not been captured.

On this basis, global government support for wind-, geothermal-, PV- and biomass-based electricity generation is calculated to have reached $26.6 billion in 2007 (in year-2009 dollars) (Figure 10.8). Support fell slightly to $26 billion in 2008, although generation increased by 13%. The drop in support resulted from the sharp increase in wholesale electricity prices in most countries (following the fossil-fuel price hikes), which diminished the premium per unit of output paid to the renewable electricity generators. Support grew to $37 billion in 2009, almost 43% more than in 2008. The volume of electricity produced from PV, biomass, geothermal and wind combined grew by 13%, a much lower rate than the cost of support (although there were significant differences by technology). Conversely to 2008, the main reason for the significantly higher support in 2009 was the drop in wholesale electricity prices in that year. Other factors explaining the increase include changes in policies, higher quota obligations,

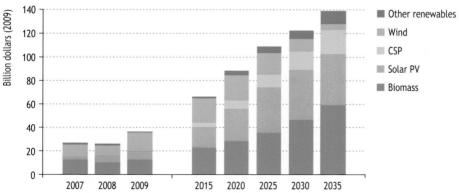

Figure 10.8 ● **Global government support for renewables-based electricity generation by technology**

Note: Other renewables include small hydro, geothermal and marine power.

greater generation output and a significant increase in electricity generation from PV, which has higher support relative to wind power or biomass. Our analysis shows that global support for PV exceeded $7 billion in 2009, representing 20% of the total spending in that year against a 3% share in the electricity produced from renewables receiving government support.

Total future support for renewables rises to nearly $140 billion by 2035 in the New Policies Scenario. Cumulative support over 2010-2035 reaches $2.5 trillion. The pattern of support differs considerably by technology. For onshore wind power, which is relatively close to being competitive with non-renewable sources in several countries and where learning will usefully reduce costs over the *Outlook* period, the total cost of support diminishes over time, from $16 billion in 2009 to $4 billion in 2035, even though electricity output from onshore wind farms increases by a factor of ten over the same period. As a result, support costs per unit of onshore wind power generation fall to a global average of $2 per MWh by 2035 in the New Policies Scenario, down from $52 per MWh in 2009 (Figure 10.9). For other technologies, including PV and biomass, technological improvements also serve to drive down unit costs, but this cost reduction does not compensate for the growth in their deployment. As a result, global support rises from $7 billion in 2009 to $43 billion in 2035 for PV and from $13 billion to $60 billion for biomass, although in both cases the cost of support per unit of renewable electricity generated falls over the *Outlook* period. Across all renewables receiving support, the cost of support falls from around $55 per MWh in 2009 to $23 per MWh in 2035.

Figure 10.9 ● **Global government support for and generation from solar PV and onshore wind in the New Policies Scenario**

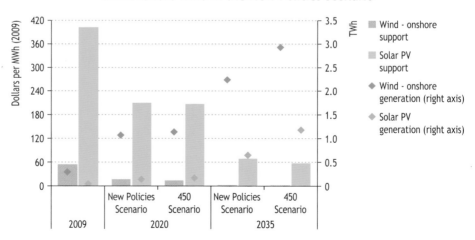

In several countries, onshore wind becomes fully competitive with conventional generation by the end of the period in the New Policies Scenario. In the United States, for example, as a result of a rising electricity prices and falling technology costs, onshore wind power becomes competitive by the late 2020s. In the European

Union onshore wind power becomes competitive earlier, around 2020. Similarly, PV in buildings becomes competitive in some regions, such as Japan and the European Union, by the mid-2020s, despite the overall growth in the costs of global support for PV over the period.

10

SPOTLIGHT
Will recent cuts in incentives for photovoltaics really harm the industry?

Strong government support has led to a boom in solar PV in recent years. Global PV capacity rose to 23 GW in 2009, from about 7 GW three years earlier. Growth was particularly strong in the European Union, where PV capacity reached almost 17 GW, nearly three-quarters of the global total, owing to generous feed-in tariffs. As a result, the total of PV government support increased rapidly in Europe in the past couple of years. At the same time, the price of PV installations decreased in the aftermath of the financial crisis. Some EU governments have now embarked on tariff cuts, causing some consternation in the industry that the cuts will be severe and affect the growth in PV.

In July, Germany — the largest PV market in the world — decided to cut tariffs by between 11% and 16%, starting in October 2010, with somewhat lower cuts in the period July-September 2010. In July 2010, Italy also passed legislation to cut tariffs by 20% on average. In both countries, the cuts were lower than originally planned. In Spain, a Royal Decree currently under discussion proposes an adjustment through a limitation on the number of hours that qualify to receive the premium. Belgium, France and Greece are also cutting tariffs.

Although these PV tariff cuts may appear at first sight to represent a weakening of government support for renewables, they are consistent with the declared intentions of most countries regularly to review and adjust feed-in tariffs, taking into account technology costs and market conditions, so as to avoid windfall profits and encourage the industry to become competitive and self-reliant. Our analysis of government support shows that the total support cost for PV in Europe grew much faster in the past few years than support for less expensive technologies, such as wind, and is set to continue to increase over the next two decades in the New Policies Scenario. The annual support cost for PV in that region begins to fall only towards the end of the *Outlook* period.

There are marked differences in the pattern of support for renewables between regions. The European Union is currently the region with the highest level of support for renewables, having spent $23 billion in 2009. A combination of a rising wholesale electricity price, falling technology costs and the particular features of Europe's renewable technology mix means that the European Union's annual support for renewables grows slowly over the decade to 2020, peaking around 2020 at almost $25 billion. It then declines gradually to a little over $21 billion by 2035. Japan shows a similar pattern, with support peaking in the early 2020s. Annual support levels in the European Union and Japan in the past have been volatile, due to the nature of the feed-in tariffs, which guarantee a steady income to producers of electricity from renewables regardless of changes in market electricity prices. In the United States, government support grew steadily over the period 2007-2009, hitting $9.6 billion in 2009. This will double to over $20 billion by the mid-2020s, and then begin to fall gradually. China's level of support over the period 2007-2009 was low compared with the European Union and the United States, but grows significantly, from around $1 billion in 2009 to almost $16 billion in 2020 and $38 billion by 2035 (Figure 10.10).

Figure 10.10 ● Global government support for renewables-based electricity generation by region in the New Policies Scenario

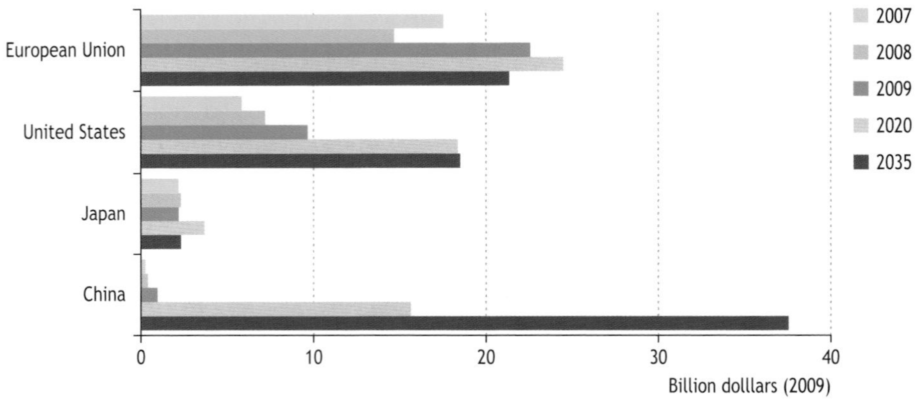

Impact of government support on electricity prices

The degree to which the additional cost of renewables that results from government support is passed through to end-users in each country depends on the details of the support mechanisms in each country. When the additional cost for renewable sources is in the form of premiums or green certificates, then the cost is passed on directly to the end-user, resulting in higher electricity tariffs. Feed-in tariffs are also usually paid for by electricity consumers. Tax-credits as a form of support result in unchanged or lower prices for the end-user, with the additional cost carried by governments.

Greater support for renewables, resulting in their increased deployment, leads to lower investment costs for renewables in the long term and ultimately to a reduction in the government support needed per unit of electricity produced. In the New Policies Scenario, the total support needed for the deployment of renewables is $1.3 trillion in the OECD countries over the *Outlook* period. The pattern of support is different in each region. In the United States, support grows as a proportion of the wholesale price until the mid-2020s, when it begins to decline, due to the falling cost of renewables and a growing wholesale price. A similar pattern applies to the European Union and Japan, but with support per MWh of electricity generation peaking earlier (around 2020) in both regions. In China support grows throughout the period, mainly due to a wholesale price that grows far more slowly than in OECD countries as there is no carbon pricing in the Chinese power sector in the New Policies Scenario.

Over the period, this support corresponds to an addition of 5% on average to the wholesale electricity price in the OECD countries. This figure is 7% in the European Union, 5% in the United States and 3% in Japan, reflecting the level of penetration of renewables in the different countries and the level of the wholesale prices (Figure 10.11).

Figure 10.11 ● *Average wholesale electricity prices and impact of renewable support in selected OECD regions in the New Policies Scenario, 2010-2035*

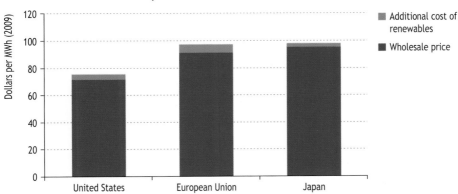

Network integration of variable renewables

Overview

The output of certain renewable electricity generation technologies, such as wind, marine, solar PV or run-of-river hydropower, is variable in nature, *i.e.* it fluctuates depending on the availability of their primary energy source, such as wind, sun, or water, which cannot be controlled, although generation based on these resources can be curtailed when necessary.[4] Growing shares of variable renewables will require modifications to the operation of the system and market, and eventually additional flexible reserves, in order to ensure system security is not impaired. There will also be a need for rules to address who carries these extra costs and how they are distributed among the different power and grid companies involved. All these need to be in place from the outset, when planning for a large-scale increase in the share of renewables.

In the New Policies Scenario, the share of electricity generation from variable renewables increases considerably in most regions over the projection period (Figure 10.12). Across the world, that share rises from just 1% in 2008 to 10% by 2035,

---

4. All generation sources are variable to an extent. However, the variability of certain renewable energy sources is significantly higher, more frequent and less predictable in nature, and generally increases with the level of their penetration in the system.

but with significant variation among regions. The share is highest now in the European Union, at almost 4%, and the EU share remains the highest in the world throughout the projection period, reaching 22% in 2035. It reaches over 10% in the United States (12%) and Canada (11%). In Australia and New Zealand, the share is 10%, while China and India reach 9%. It is 7% in South Africa and Japan.

Figure 10.12 ● Shares of variable renewables in total electricity generation by region in the New Policies Scenario

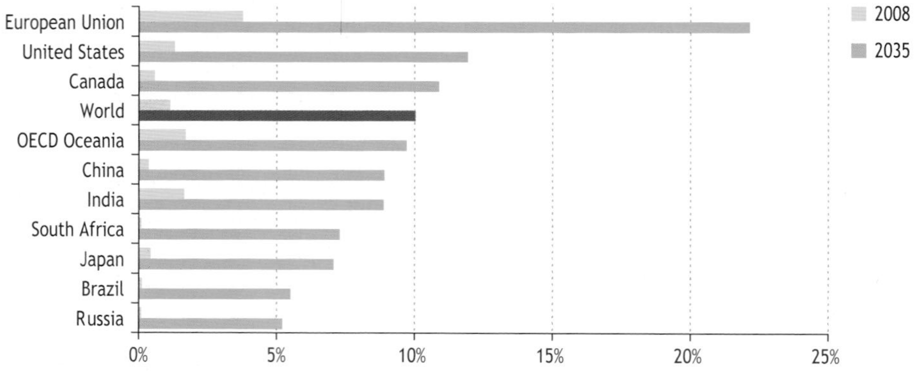

Variable generation has implications for total capacity, the design of the network and the balancing of the power system.[5] All can be addressed through greater generation flexibility and strengthening of transmission networks. Managed demand response and storage offer additional mitigation options. In order to maintain supply reliability, traditional methods of planning and operating generation and networks have to evolve to take into account the characteristics of these new technologies.

Integration costs

Among the various cost components of renewable energy generation, integration costs are perhaps the most uncertain because there is no universally accepted methodology for estimating these costs. Experts do not always agree on what constitutes an additional cost and whether it should be attributed to renewables. For example, all studies include balancing costs within integration costs, while only some also account for interconnection costs and fewer still consider adequacy costs (for definitions, see below).[6] However, an estimate of integration costs, along with information on the capital costs of generation and operating expenses (relatively

---

5. For a more comprehensive analysis of flexibility in grid systems and the major enablers of and obstacles to integrating renewables, please refer to the forthcoming results from the IEA on the Grid Integration of Variable Renewables (GIVAR) project (IEA, forthcoming).

6. See, for example: CAISO (2007); DCENR and DETI (2008); DENA (2005); EnerNex Corporation (2006); EnerNex Corporation (2010); EWEA (2005); GE Energy (2008 and 2010); Holttinen et al. (2009); Mills et al. (2009); NERC (2009); Transpower Stromübertragungs-Gmbh (2010); VTT Technical Research Centre of Finland (2009); and UK ERC (2006).

easier to obtain), is necessary to give policy makers an estimate of the total costs resulting from the adoption of renewable technologies. A better understanding of integration issues can help guide efforts to reduce these costs in the future, especially important as they become more significant with increasing penetration levels.

The various costs associated with integrating increased levels of variable generation into the system can be grouped into three major categories:

- **Network (interconnection costs):** Renewable resources may be located far from load centres and the existing transmission network. The construction of high-voltage transmission lines may be necessary to link such resources to the existing grid. Interconnection costs are incurred primarily as large upfront capital investments.

- **Balancing costs:** Matching electric power supply with demand is critical to power systems. The addition of variable renewables to the generation system increases the need for ancillary services, a term often used to refer collectively to the resources required to meet system balancing needs. These costs are mainly incurred as operational costs, on a short-term basis (seconds to days).

- **Capacity adequacy costs:** These arise from the need to maintain sufficient capacity in the grid to handle peak loads. In order to maintain system security, an adequate amount of backup generation capacity is required, which varies, depending upon the capacity value of the variable source (Box 10.4). This results in the attribution of additional capacity costs to variable generation.

10

The technical challenges and the associated integration costs vary considerably among various regions, mainly due to the different characteristics of variable renewable generation in different geographical locations, differences in the demand and generation mix of the incumbent systems, dissimilar technical (security) standards and commercial frameworks, and different ways of quantifying impacts and costs. Therefore integration costs are generally calculated on a case-by-case basis.

Despite the difficulties in assessing integration costs, we attempt in the following section to arrive at broad cost estimates for the United States and the European Union, the only two regions for which detailed cost studies have been conducted. Most of the studies focus on onshore wind power, while studies on solar are just beginning to emerge. Our estimates cover onshore and offshore wind power, CSP and PV for large-scale generation. If technology-specific costs are not available, we have used costs based on onshore wind power, because we can infer from current studies that there are similarities between onshore wind and these other technologies. We have not included distributed PV in our estimates, since we estimate that the cost impact of small dispersed systems in buildings is likely to be very small. Marine technologies are not included either, as their integration costs have not been studied and, even in 2035, they account for a very small percentage of total generation.

Figure 10.13 ● Power generation system flexibility by region in the New Policies Scenario, 2035

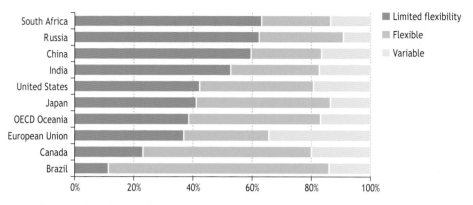

Notes: Shares are based on installed capacity. Variable capacity includes wind, solar PV, small hydro and marine power. Flexible capacity includes large hydro, combined-cycle gas turbines (without carbon capture and storage) and open-cycle gas turbines. Limited flexibility capacity includes nuclear power and coal (with and without carbon capture and storage) and combined-cycle gas turbines with carbon capture and storage. In the European Union, the level of system flexibility may vary between countries. In France, nuclear power plants are capable of load-following.

Box 10.4 ● Capacity value of variable renewables

The contribution of variable renewables to the adequacy of a system is often significantly lower (per MW of installed capacity) than that attributable to other energy options. Because only a fraction of total capacity has a high probability of running consistently, variable renewables have limited *capacity value*.

The capacity value depends on the renewable energy source and varies across different systems. It generally declines with higher penetration, eventually approaching saturation. Major factors affecting the capacity value of variable generation include the correlation between the timing of demand and variable generation output (for example, PV generation has a higher capacity value in countries where peak demand occurs during daytime, as in Japan or Spain, and wind has a higher capacity value in Denmark, because it is more generally available at the time of peak demand in the evening) and the locational diversity of the variable resource (*i.e.* a wind resource with larger distances between wind farms will generally have a higher capacity value than the same magnitude of resource concentrated in a small area; or PV in buildings has a higher capacity value than large-scale PV generation, which is more concentrated). A high frequency of zero or very low generation availability during peak demand periods of the year can also severely impact the capacity value of variable renewables.

The capacity value of a variable source, *e.g.* wind generation, has been found to differ significantly according to whether the system is dominated by thermal plants or thermal and flexible hydro power plants. Systems having a significant share of flexible hydro plant, as in Norway and New Zealand, can offer capacity support to variable generation by time shifting the available energy to meet peak demand. More generally, a system with high shares of flexible capacity and interconnections can enhance the capacity value of wind. Figure 10.13 illustrates the degree of power generation flexibility for major regions in 2035.

Interconnection costs

The location of renewable energy plants is largely determined by the geographical location of rich natural resources, which are often away from load centers and the existing transmission network. For example, in the United States, there is signifcant wind potential in sparsely populated states, such as North Dakota, Wyoming and Montana. In Europe, there is significant wind potential in the North Sea. Utilising these natural resources requires the construction of transmission lines to transport energy from the generation sites to load centres.

In the New Policies Scenario, we estimate interconnection costs to be of the order of $9 per MWh in 2035 in the European Union and $12 per MWh in the United States. These estimates are based on the Eastern Wind Integration and Transmission Study (EWITS), which focuses on a large area in the United States (EnerNex Corporation, 2010). The cost estimates given apply to integration in both the United States and Europe. The study shows transmission costs decreasing from $15 per MWh at 6% wind penetration levels to $9 per MWh at 20% wind penetration and $7 per MWh at 30% wind penetration. The decrease in unit cost with increasing penetration can be attributed to the increasing use of higher capacity transmission technology with lower costs per kW-mile, such as High Voltage Direct Current (HVDC) lines.

Balancing costs

Balancing costs can differ widely depending on factors ranging from the mix of existing generation plants in a region to the diversity of the renewable resources achieved through geographic spread or technological mix. Norway has low integration costs, due to the significant hydropower resources in their grid that greatly mitigate the balancing costs for wind. Similarly, regional studies conducted for the Eastern United States and Europe (the European Wind Integration Study, EWIS) (Transpower Stromübertragungs-Gmbh, 2010) show lower costs than those estimated for the UK (Energy Research Center, ERC study) (UK ERC, 2006), probably because of differences in the geographical spread of resources. We estimate balancing costs for onshore wind power to be of the order of $3.5 per MWh in Europe and $2.5 per MWh in the United States.

Studies conducted by the Colorado Public Service Company (CPSCo) show concentrated solar power (CSP) balancing costs to be approximately half of those for onshore wind, all other thinbgs being equal. Also, using insights from EWITS, one may estimate that the balancing costs of offshore wind could be 75% of those for onshore wind.

Adequacy costs

Adequacy costs for variable renewables arise from the lower contribution made by new renewable generation capacity to the maintenance of reliable supply in a system, compared to that provided by conventional energy sources. Actual adequacy costs incurred in a given grid system can vary widely and usually need to be evaluated on a case-by-case basis. In general, adequacy costs can become manifest as investment costs for building new generation capacity or as lost revenue for existing capacity becuase of the reduced load factor for conventional plants. A primary determinant of adequacy costs is capacity value.

Most recent studies show that capacity values for wind energy range from 10% to 25% at up to 30% wind penetration. Fewer studies have analysed the capacity value attributed to solar technologies. According to the Western Wind and Solar Integration Study (GE Energy, 2010), at low penetration levels, capacity values are around 30% for PV and 90% for CSP. In many systems, PV energy tends to be much better aligned with peak load than wind energy, leading to higher capacity values. Comparatively, CSP commands much higher capacity values for mainly two reasons. First, CSP is usually better aligned with peak load, because it is built only in areas with high direct normal irradiance (DNI), unlike PV. Second, CSP plants can include storage that contributes to avoidance of disruptions in supply and allows output to peak later in the day, when peak loads are more likely to occur. We assume adequacy costs to be of the order of $4 per MWh for onshore and offshore wind in both the United States and Europe. CSP adequacy costs are assumed to be zero, because most CSP is assumed to be equipped with storage.

Summary of integration costs

Based on the estimates above, total integration costs in 2035 in the New Policies Scenario would add, on average $16 per MWh in Europe and $17 per MWh in the United States. The total cost of integration in that year is put at $13 billion in the European Union and $11 billion in the United States. The assumed costs per MWh and total costs are summarised in Table 10.5.

Table 10.5 ● Integration costs of variable renewables in the European Union and the United States in the New Policies Scenario, 2035

	Interconnection	Balancing	Adequacy	Total
	Unit costs ($2009 per MWh)			
European Union	9	1.8 - 3.5	0 - 4.5	16.3
United States	12	1.3 - 2.5	0 - 4	17.3
	Total costs ($2009 billion)			
European Union	7.5	2.5	3.3	13.3
United States	7.8	1.4	2.1	11.3

Source: IEA analysis.

Notes: Costs have been calculated for onshore and offshore wind, CSP and large scale PV. Distributed PV costs are assumed to be zero. Adequacy costs for large PV in Europe have not been calculated as there are no relative studies and costs cannot be inferred from studies analysing costs in the United States. Balancing costs are assumed to be 50% of the costs of onshore wind for CSP and large PV and 75% for offshore wind.

Dealing with the variability of renewables

Forecasting

Improved forecasting of the output of variable generation in the coming few minutes or hours results in better utilisation of these sources and reduces the need for an operating reserve to mitigate their unpredictability. Lower operating reserve requirements enhance the capacity of the system to integrate variable generation and reduce efficiency losses and the use of high marginal cost plant. The arrival of large weather fronts, *e.g.* storms, can lead to the loss of wind generation over the entire area covered by the weather front for the duration of the storm. More accurate longer term forecasting of such phenomena contributes to bringing alternative plants online in a timely manner, but, due to the infrequent nature of these weather phenomena, the impact on balancing costs is not substantial.

Demand response

The importance of managing demand response could rise in the future.[7] The implications go well beyond the issues related to the variability of renewables, but they are also important in that area. Demand response in the form of redistribution of load (*e.g.* when load is moved from peak to off-peak periods) can help mitigate the capacity problem associated with variable generation, firming up the capacity value of variable generation and so reducing the need for peaking plant.

Demand response can reduce balancing costs because it increases the efficiency of the system operation by reducing the required operating reserve and the associated costs. Transmission related integration costs can also be reduced if demand is able to follow variable supply. Maximising the use of renewable generation locally reduces the need for interconnections to export surplus variable generation. The value of demand response in this context will depend upon the volume of surplus generation and the level of energy storage capability available.

Smart grids

A smart grid facilitates increased integration of variable renewables into the power system to increase flexibility. The smart grid makes use of enhanced system information and control to allow operational changes, such as intra-hour renewable dispatch (see also the discussion of smart grids in Box 7.1 in Chapter 7), which contribute to better management of the system, reducing system bottlenecks and congestion (IEA, forthcoming, c).

Storage

Energy storage facilities permit energy availability to be shifted across time (typically over periods of hours) by charging up during periods of low demand and/or surplus low cost generation and discharging during high demand periods, associated with high marginal cost generation. Common storage technologies include pumped hydro, compressed air energy storage and large battery energy storage systems.

---

7. A new IEA report will examine the role of demand response in OECD electricity markets (IEA, forthcoming, a).

Storage may make modest amounts of peak conventional generating capacity redundant in systems without and with variable generation. Storage facilities can also mitigate the lack of correlation between high demand and the output of variable generation, so enhancing the capacity value of the variable source. Energy storage facilities enhance system flexibility by, at least partly, decoupling fluctuating energy supply from demand. Where the building of new transmission lines is constrained, storage may offer an alternative outlet for the renewable generation produced. Currently, storage technologies have relatively high investment costs. Reducing the costs of these technologies is key to expanding the use of energy storage in the future (Inage, 2009).

Special focus: Offshore wind power

Offshore wind power is still at an early stage of commercialisation. At the end of 2008, there were 1.4 GW of installed capacity, all in European countries around the North Sea, the Baltic Sea and the Irish Sea (Table 10.6). Capacity rose to 2.1 GW in 2009. In that year, Germany, Norway and — the first country outside Europe — China installed their first offshore wind farms.

Table 10.6 ● Installed offshore wind power capacity by country (MW)

	2008	2009
Belgium	30	30
China	0	63
Denmark	398	626
Germany	0	60
Ireland	25	25
Netherlands	247	247
Norway	0	2
Sweden	133	163
United Kingdom	588	894
World	1 421	2 110

Source: BTM Consult (2010).

Compared with onshore wind power, offshore wind is still small because of its higher cost and because many technical challenges remain. The potential for offshore wind power is, however, very large. Over the *Outlook* period, offshore wind capacity is projected to increase to 115 GW in 2035 in the Current Policies Scenario, 180 GW in the New Policies Scenario and nearly 340 GW in the 450 Scenario, supplying 1%, 2% and 4% of global electricity (Figure 10.14). The largest increases are in OECD Europe, OECD North America (mostly in the United States) and in China.

OECD Europe remains the most important region for offshore wind power development in all scenarios. Installed capacity there rises to 48 GW in 2035 in the Current Policies Scenario, 64 GW in the New Policies Scenario and almost 100 GW in the 450 Scenario. Most of the development is expected to continue to be in Northern Europe, where the potential is very large. Offshore wind power is expected to be distributed across the region, requiring the construction of a major offshore grid to connect offshore wind farms to the mainland. In December 2009, the United Kingdom, Germany, France, Belgium, Netherlands, Luxembourg, Denmark, Sweden and Ireland launched the North Seas Countries' Offshore Grid initiative, providing for co-operation in the development of the grid infrastructure in the North Sea. Norway endorsed the initiative in February 2010.

Investment

Total investment in offshore wind power over 2010-2035 amounts to $260 billion (in 2009 dollars) in the Current Policies Scenario, $400 billion in the New Policies Scenario, and $640 billion in the 450 Scenario. In OECD Europe, investment ranges between $120 billion and $200 billion. Financing offshore wind farms is at present problematic, because financial institutions perceive the technology as risky and require a higher share of equity, compared with other renewables, notably onshore wind projects and PV. As the technology improves and bankers become more comfortable with it, lending should become easier. Until then, governments may have to play a role to facilitate investment in offshore wind power by, for example, increasing the role of multilateral lending institutions.

Figure 10.14 ● Offshore wind power generation capacity by region and scenario

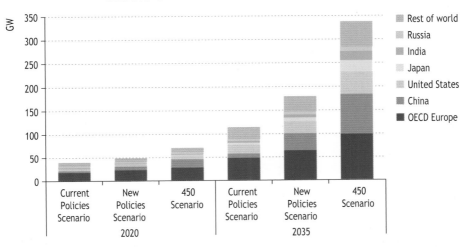

Technology

Offshore wind turbine technology needs further development. At present, most offshore turbines are based on onshore turbine technology, modified to reflect practices and experiences in other offshore industries (IEA, 2009). The reliability of offshore turbines, which is currently lower than that of onshore wind turbines, needs to improve. More robust turbines, designed from the outset to operate in offshore conditions, need to be developed for the technology to take off. This would require — among other things — a focus on the combined effects of different loads on all parts of the wind turbine and its foundations, as the marine environment interacts with waves and currents.

To date, the foundations of most offshore projects consist of a single pile driven into the seabed, called a monopile. Current monopile designs account for about a quarter of the total investment cost of an offshore wind farm. Improved foundation designs can help bring costs down. Although offshore wind turbines are currently located in shallow water areas, significant potential exists in deep waters and new designs are being developed to allow capture this potential. Floating turbines are one such design (Box 10.5).

Box 10.5 ● Floating wind turbines in Norway

The world's first large-scale prototype floating wind turbine — the 2.3 MW Hywind prototype — started operation in 2009 in Norway. The turbine is located 10 kilometres off the coast of Karmøy, near Stavanger, where the water depth reaches 220 metres. The project was developed by Statoil, the Norwegian oil and gas company, which has plans to invest a total of 400 million Norwegian kroner (about $65 million) in its construction and operation. An additional 59 million kroner ($10 million) is being funded by the Norwegian government.

The Hywind project consists of a 65 metre tall wind tower with an 82 metre rotor diameter. It weighs 138 tonnes. The turbine is anchored to the sea bed through a 100 metre long steel cylinder, weighing 3 000 tonnes, which contains a ballast of water and rocks. This allows the structure to move with the sea — a concept building on offshore oil and gas experience. A submarine cable connects the installation to the nearest power station onshore.

The facility is now in a two year test period, until the fall of 2011. A larger project, the 10 MW Sway prototype, is planned. If the design proves to be successful, floating turbines could be used in locations further offshore, in water depths of 120 to 700 metres, where wind speeds are higher and more constant than nearer the shore. Furthermore, floating wind turbines can help overcome some of the challenges that face conventional offshore turbines located near the coast, including the visual impact and the conflict with fishing and other coastal activities.

Special focus: Renewables in the Middle East and North Africa

The countries of Middle East and North Africa (MENA) are endowed with rich oil and gas resources (in particular the Gulf Cooperation Council countries, Algeria, Egypt and Libya). They also have some of the highest solar resources in the world (Table 10.7). To date the solar resources are almost totally unexploited. While solar is the most abundant resource in all countries in the region, some also have hydropower resources (*e.g.* Egypt on the Nile river, Iran on its northwestern plains, Iraq and Syria in the Tigris-Euphrates basin) and wind resources (*e.g.* along the Red Sea and on Morocco's Atlantic coast).

The main use of renewable energy is for electricity generation, mainly from hydropower. In 2008, less than 3% of the region's electricity came from renewables, but it was as high as 12% in Egypt, 7% in Syria, 6% in Morocco and 3.5% in Lebanon. In all other countries, the share of electricity from renewables was less than 2% or zero. The use of renewables for heat is very limited, except in Israel, where solar water heaters are used extensively in buildings. Biomass use for heat is limited, amounting to just 4.5 Mtoe in 2008. About half of this is traditional biomass and the other half is used in industry and commercial establishments. Liquid biofuels are not yet used in the region.

Table 10.7 ● **Technical solar potential at different levels of insolation and total electricity generation in selected MENA countries, 2008**

	> 7.5 kWh/m²/day (TWh)	> 5 kWh/m²/day (TWh)	Total electricity generation in 2008 (TWh)
Algeria	162	2 962	40
Egypt	108	1 437	131
Libya	32	2 173	29
Morocco	61	516	21
Saudi Arabia	29	2 194	204
Tunisia	17	222	15

Note: Technical potentials based on direct normal irradiation. Resources of above 5 kWh per m² per day (or 1 825 kWh per m² annually) are considered as very good. Few countries in the world have resources above 7.5 kWh per m² per day.

Source: IEA analysis using data provided by the United States National Renewable Energy Laboratory.

Domestic policies and initiatives

Support for renewables has grown in recent years and policies to promote renewables in the region are spreading. A growing number of countries have set targets for renewables, which are summarised in Table 10.8, along with the main programmes, measures and incentives involved. Most of the countries involved are in North Africa.

Table 10.8 ● Renewable energy policies and targets in selected MENA countries

	Renewable energy targets	Programmes, measures and incentives (examples)
Algeria	2015: 6% electricity from renewables; 100 MW wind; 170 MW CSP; 5.1 MW solar PV; 450 MW co-generation	Feed-in premium for all renewable electricity and co-generation; investment tax credits for solar water heaters
Egypt	2010 (non-binding): 3% electricity from renewables 2020 (binding); 4% renewables in energy consumption (of which 20% wind); 20% non-hydro renewables (12% wind, approx. 7 200 MW capacity)	Planned New Electricity Law: priority dispatch for renewables; competitive tenders and feed-in tariff (small & medium-sized projects); investment tax credits for solar water heaters
Jordan	2015: 7% in primary energy; 600 MW wind 2020: 10% in primary energy share; 1 200 MW wind; 300-600 MW solar PV and CSP; solar water heaters in 50% of households	Planned tax exemptions and cost subsidies
Libya	2020: 10% in primary energy share; 1 500 MW wind; 800 MW CSP; 150 MW solar PV; 300 MW solar water heaters	Medium-term plan 2008-2012: 610 MW wind; 5-10 MW grid-connected PV; 2 MW off-grid PV; 500 roof-top PV systems; 100 MW CSP; PV and solar water heater manufacturing
Morocco	2012: 10% in primary energy and 20% in electricity incl. 200 MW wind 2015: 400 000 m² solar water heaters 2020: 2 000 MW solar capacity installed	VAT reduction on equipment for electricity production; negotiated purchase tariff for electricity; investment tax credits and VAT reduction on equipment for solar water heaters
Tunisia	2011: 10% in primary energy; 180 MW wind; 10 MW CSP; 10 MW biogas 500 000 m² solar water heaters	Demonstration plant incentives, tax exemptions and reductions for electricity production; investment tax credits; building codes mandating use of solar water heaters
United Arab Emirates	7% in electricity (in Abu Dhabi)	No measures nor incentives introduced as yet

Note: VAT = value-added tax.
Source: IEA databases and analysis.

The region has seen a number of other renewables-related initiatives in recent years:

■ The Masdar initiative, headed by the Abu Dhabi Future Energy Company (Masdar), is the most prominent. Its focus is on clean energy, including renewables and cleaner fossil fuels (including energy efficiency and carbon capture and storage), with an investment target of $22 billion. Masdars' activities span all stages of renewable energy development from research to commercialisation. The company is currently building a zero-carbon city (Masdar City) which will make extensive use of solar power.

- A regional centre was created in Cairo in 2008 with the aim of promoting renewables and energy efficiency (Regional Center for Renewable Energy and Efficiency, [RCREE]). Its members are Algeria, Egypt, Jordan, Lebanon, Libya, Morocco, Palestine, Syria, Tunisia and Yemen.

- Saudi Arabia recently passed a decree establishing the King Abdullah bin Abdulaziz City for Atomic and Renewable Energy in Riyadh.

There are also several intra-regional or inter-regional initiatives between MENA and Europe which relate either to energy in general (including renewables) or to renewable energy specifically. These include MEDENER (Mediterranean Energy, the Mediterranean association of national agencies for energy conservation), MEDREG (Mediterranean Regulators, the association of the Mediterranean regulators for electricity and gas), MEDELEC (Mediterranean Electricity, a group of regional electricity associations), MENAREC (the Middle East and North Africa Renewable Energy Conference, with a focus on renewables for energy and water) and MEDREP (the Mediterranean Renewable Energy Programme, which aims at providing sustainable energy to rural areas and at increasing the share of renewables in the region's energy mix).

Outlook

The use of renewable energy in total grows significantly in all three scenarios. Most of the increase comes from the electricity sector. Total electricity generation from renewables increases from 26 TWh in 2008 to 222 TWh (9% of electricity generation) in the Current Policies Scenario, to about 380 TWh (18% of electricity generation) in the New Policies Scenario and 610 TWh (33% of electricity generation) in the 450 Scenario in 2035. The share of renewables in electricity generation in 2035 increases to 26% in the Middle East and up to 58% in North Africa (Table 10.9). These projections assume only domestic use of renewables.

Investment in renewables electricity generation in MENA amounts to $155 billion (in 2009 dollars) over the period 2010-2035 in the Current Policies Scenario, increasing to $260 billion in the New Policies Scenario and just over $400 billion in the 450 Scenario. Current electricity tariff systems in several countries in the region do not pass full costs on to consumers. Governments are now assuming the extra costs of renewables. Some projects could benefit from the Clean Development Mechanism. Greater involvement of the private sector, to which countries in the region and more particularly in North Africa are becoming more and more open, is likely in the future.

Policies to support greater use of solar water heaters yield useful results: the share of solar energy in heat demand in buildings stays at around 1% in the Current Policies Scenario and grows to 2% in the New Policies Scenario and 3% in the 450 Scenario in 2035. The absence of policies relating to industrial energy use keeps demand for modern biomass in industry low in all scenarios. For the same reason, demand for biofuels stays close to nil in the Current Policies and New Policies Scenarios, though biofuels supply 6% of road transport demand in 2035 in the 450 Scenario, mainly through biofuels imports to the Middle East, where governments are assumed to participate in a global agreement to improve the efficiency of road transport.

10

Table 10.9 • Renewables-based electricity generation in MENA by scenario

	2008	2035		
		New Policies Scenario	Current Policies Scenario	450 Scenario
Middle East				
Renewables electricity generation (TWh)	9	256	137	383
Share in total electricity generation	1%	16%	7%	26%
Installed renewables capacity (GW)	12	94	55	137
Hydro (GW)	*12*	*26*	*25*	*26*
Wind (GW)	*0*	*27*	*9*	*44*
Solar PV (GW)	*0*	*20*	*10*	*30*
CSP (GW)	*0*	*17*	*9*	*32*
North Africa				
Renewables electricity generation (TWh)	17	120	85	226
Share in total electricity generation	7%	26%	17%	58%
Installed renewables capacity (GW)	5	39	28	76
Hydro (GW)	*5*	*11*	*10*	*12*
Wind (GW)	*1*	*9*	*7*	*21*
Solar PV (GW)	*0*	*8*	*7*	*15*
CSP (GW)	*0*	*8*	*3*	*23*

Large-scale development of renewables in MENA

The strong interest in European countries in renewable energy has revived European interest in MENA's vast solar resources and has given rise to two major initiatives: the government-led Mediterranean Solar Plan (MSP) and the private sector-led Desertec industrial initiative Dii.

The objective of the MSP, launched in 2008, is to promote a sustainable energy future in the Mediterranean region.[8] The plan proposes to increase the use of solar and other forms of renewable energy, to improve energy efficiency, to develop electricity grid interconnections[9] and to stimulate technology transfer to developing countries in the region. MSP targets the development of 20 GW of renewables by 2020, of which 5 GW could be exported to Europe. Total investment would be of the order of 60 billion euros. More than 150 projects have been proposed (mostly from European developers) and about 70 have been selected. Developing interconnections between North Africa and Europe would cost another 4 to 5 billion euros. Within this framework, an industrial

---

8. See Guarrera *et al.* (2010) for a detailed description of these initiatives.

9. This objective is supported by the European Commission. Interconnecting the northern and southern shores of the Mediterranean is one of the European Union's four major projects for developing electricity networks. The other three involve strengthening the south-east interconnections; the interconnection of the Baltic grid to other grids; and the construction of undersea cables to link North Sea and Baltic Sea wind installations.

initiative has been set up — the Transgreen project — with the aim of co-ordinating efforts to develop such network links.

Desertec was initiated by the German Association of the Club of Rome, with the vision of developing a CSP grid in MENA, connected to Europe. The *Dii Desertec industrial initiative* was launched in 2009 by a group of large private companies, with the aim of accelerating and implementing the Desertec concept. The focus of Dii is on solar and wind power generation from the deserts of MENA countries, both to meet local demand and for export to Europe. The ultimate objective is to produce enough power by 2050 to meet 15% of Europe's electricity demand and a substantial proportion of the needs of producing countries. To realise this objective, Dii envisages the construction of a supergrid that would connect renewable energy resources with demand centres.

The economics of concentrating solar power

How best to utilise the vast potential of solar energy is a current policy focus in many MENA countries. Export to Europe is the dominant objective of the above initiatives. CSP is currently not competitive with conventional electricity generation, but significant potential for technology improvements exists (see Box 10.2) and the pace of development will very much depend on the degree to which the adoption of CSP is supported by policy measures.

To illustrate the prospects for the export of CSP-generated electricity from MENA regions to Europe, the individual cost components of CSP technologies (parabolic trough and power tower technologies) are examined in-depth below, together with the costs of transmitting electricity to Europe using high-voltage direct current (HVDC) transmission lines — the most efficient option for transmitting electricity over long distances. The cost assumptions used have additionally been reviewed by industry experts outside the IEA.

For the analysis, the maximum annual average direct normal irradiation (DNI) per day and country, as provided by the US National Renewable Energy Laboratory, have been used to identify the maximum average DNI for Northern African countries (about 7.8 kWh per m^2 per day) and Middle East countries (about 6.9 kWh per m^2 per day). This is an approximation, as each CSP plant will be optimised individually according to local solar resource conditions at different times of the day. However, it provides sufficient insight into the potential of the region as a whole to generate CSP electricity cost-competitively.

In the New Policies Scenario by 2035, CSP electricity can be produced at costs of around $100 to $120 per MWh at good sites in Northern Africa and $110 to $135 per MWh in the Middle East (Figure 10.15). Efficient storage (assumed at a level sufficient to provide electricity for eight hours in our analysis) is important to achieving sufficiently low generating costs, as it increases the capacity value of CSP plants. Lower generating costs are feasible by further increasing the capacity value through the use of larger storage tanks or additional gas backup. However, the inclusion of storage increases investment costs significantly by 50% to 90% on a per kW basis.

Figure 10.15 ● CSP electricity generating costs in MENA in the New Policies Scenario, 2035

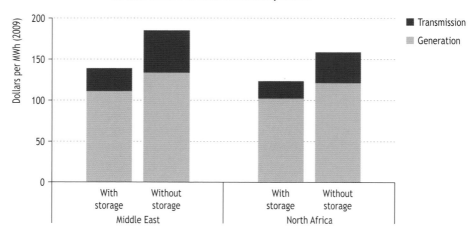

Note: These costs, which reflect the solar resource at the best sites, are lower than the average global costs shown in Table 10.1.

Potential transmission to Europe involves additional costs for HVDC lines as well as converter stations. For the purposes of the analysis, the closest geographical connection point in the European Union relative to the exporting regions was considered, distinguishing overhead and submarine cables and their respective costs. Transmission lines are assumed to be used solely for the export of electricity from CSP and so capacity factors are comparatively modest (up to around 60%). This results in transmission costs of $20 to $40 per MWh for Northern Africa, and $30 to $50 per MWh for the Middle East. Transmitting electricity further, to central European countries, entails significant additional costs. Additional cost reductions could be achieved if the use of the cables could be increased. If capacity factors were 90%, transmission costs to the borders of the European Union could be as low as $10 to $12 per MWh. Capacity factors could be increased through the construction of additional storage and/or backup capacity using, for example, natural gas combined-cycle plants.

In the New Policies Scenario, large-scale electricity from CSP in MENA countries does not become competitive with European wholesale electricity prices, but remains about 20% more expensive even in 2035 (Figure 10.16). Nevertheless, these prices are annual averages, and CSP import could be profitable at individual times of the day and year, in particular where it would be competing with other more expensive renewable electricity. The prospect of cost reductions for CSP achieved through global learning-by-doing, together with increasing wholesale electricity prices in Europe in this scenario, show that the potential is there. In Northern Africa, every country has significant solar potential in excess of 7.5 kWh per m² per day, over an area of 220 000 square kilometres. The largest areas with such solar potential considered here are located in Algeria, followed by Egypt and Morocco. In the Middle East, only Saudi Arabia and Yemen have a solar potential similar to that of Northern African

World Energy Outlook 2010 - **OUTLOOK FOR RENEWABLE ENERGY** countries. The total land area available at above 7 kWh per m² per day in the Middle East is roughly 60 000 square kilometres, of which more than one-third is located in Saudi Arabia.

Figure 10.16 ● **CSP generating costs in North Africa and European wholesale electricity price in the New Policies Scenario**

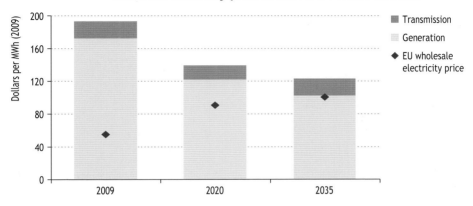

Additional policy support could further increase the competitiveness of CSP from MENA. In the 450 Scenario, where action to achieve climate stabilisation targets results in increased CSP use globally and leads to further cost reductions, CSP costs can fall to below $100 per MWh in 2035 (excluding transmission costs). At the same time, rapidly increasing CO_2 prices in the European Union — in particular after 2020 — drive up wholesale electricity prices, which reach $106 per MWh in 2035. In this case, CSP from MENA would be competitive in Europe, depending on transmission costs and how much these can be lowered through increased utilisation of the cables.

In summary, the quality of its solar resource and its large uninhabited areas make MENA ideal for large-scale development of solar power. But there are many challenges at the political, technical and market level that must first be overcome. For European countries, the main benefit would be cost-effective reductions in greenhouse gas emissions, using dispatchable (and thereby more reliable) renewable energy from MENA, and greater diversity of electricity supply. For MENA countries, such a large-scale development of solar power would both help meet their rapid growing electricity demand and expand their own transmission networks to provide reliable electricity access to all. Many of the poorer countries in the region are struggling to attract foreign capital for developing their own power sector and cross-border co-operation with Europe in a mutually beneficial manner would certainly help. Large-scale CSP development could also create jobs in the region in the power plants and, potentially, in manufacturing solar plants or components. It could also generate export revenues from selling electricity.

The success of large-scale CSP development in MENA, entailing exports to Europe, will largely depend on public acceptance in the exporting countries. A situation where CSP electricity is committed solely to export would be unacceptable in MENA countries. Since poorer countries of the sub-Saharan region are among those with the least access to modern energy services (see Chapter 8), public acceptance (also in Europe) of large-scale CSP electricity export from MENA might be greater if the benefits of such development could be seen to be shared with neighbouring countries. One way to achieve this would be to extend grids so as to provide not only for export to Europe but also to sub-Saharan Africa, where additional distribution grid capacity is required to make use of it.

CHAPTER 11

RENEWABLES FOR HEAT

The sleeping giant?

H I G H L I G H T S

- Heat — defined as the consumption of non-electrical energy for producing heat for use in stationary applications — accounted for 47% of global final energy consumption in 2008 (transport and electricity accounted for the rest). In the buildings sector, heat is needed for cooking, and water and space heating. In the industry sector, the heat produced in boilers and co-generation facilities is used for process applications.

- Worldwide, traditional and modern renewables together supplied 27% of total demand for heat, or 1 059 Mtoe, in 2008. This increases to nearly 1 400 Mtoe in 2035 in the New Policies Scenario, meeting 29% of total demand for heat. The share of modern renewables in total renewables for heat grows from 29% to 48%.

- Demand for traditional biomass falls in non-OECD Asian and Latin American countries, but increases in sub-Saharan Africa, due to rising population and the region's slower economic growth. Globally, the use of traditional biomass falls from 746 Mtoe in 2008 to just over 720 Mtoe in 2035 in the New Policies Scenario.

- Heat from modern renewables more than doubles in the New Policies Scenario, from 312 Mtoe in 2008 to over 650 Mtoe in 2035. Modern renewables account for 16% of global heat demand in 2035, up from 10% in 2008. In the OECD, most of the growth is in the European Union, the United States, Australia and New Zealand. Outside of the OECD, growth is largest in China and Brazil.

- Biomass remains the main source of renewables-based heat, both in industry (where the pulp and paper industry is the largest user) and in buildings. Its share in industrial energy demand increases from 11% in 2008 to 15% in 2035 in the New Policies Scenario. In the buildings sector, heat produced from modern biomass doubles over the projection period.

- The use of solar heat is expected to remain concentrated in buildings. In the New Policies Scenario, solar heat demand in buildings increases from 9 Mtoe in 2008 to 65 Mtoe in 2035. Most of the growth takes place in China, followed by the United States and the European Union.

- China is projected to remain the world's largest user of solar water heaters. In 2008, about 80% of the world's installed solar collector area was in China. The use of solar heat there is projected to increase from 4 Mtoe in 2008 to 18 Mtoe in 2035 in the New Policies Scenario.

- In the 450 Scenario, the share of modern renewables in total heat increases sharply, from 10% in 2008 to 21% in 2035. The most significant increase is in buildings, where renewables supply over one-quarter of the need for heat in 2035, up from 8% now. In industry, the share of renewables in total heat consumption grows from 11% to 18%.

Recent trends

This chapter discusses key trends in heat produced from renewable energy sources, whether it is produced on-site or delivered as a commercial service. It starts with an overview of total needs for heat, defined here as the consumption of energy sources (excluding electricity) to produce heat used in stationary applications. It then focuses on the fraction of this that comes from renewables. It sets out scenario projections of the consumption of renewable fuels for producing heat and presents in detail the results of the New Policies Scenario. This is followed by a brief discussion of the key technologies and the characteristics of government policies to promote renewables for heat. The last section takes a qualitative look at renewables for cooling (without quantitative analysis in our scenarios because data are not available).

Heat is the main energy service, accounting for close to half of global final energy demand. In the buildings sector, the heat produced from gas, oil, coal or renewable energy sources provides cooking, and water and space heating services. In the industrial sector, the heat produced in boilers and co-generation facilities (along with electricity) is used for process applications. Heat is also used in agriculture, for example to heat greenhouses. Heat can be produced on-site in buildings and industrial facilities or it can be purchased on a network. The latter is termed here "commercial heat", reflecting the delivery of heat as a commercial service; it does not refer to heat *used* in commercial undertakings.[1] Renewables as the energy source for heat include biomass, solar and geothermal energy used to produce heat on-site in industry (including through co-generation facilities) and buildings, as well as the renewables fraction of commercial heat. Unlike renewables for the transportation and electricity sectors, in which a large number of policies exists to promote the use of biofuels and renewables-based electricity, renewables for heat receive little policy attention today.

Demand for heat dominates final energy consumption, even when traditional biomass[2] is not included (Figure 11.1). The share of heat in global final energy consumption (excluding traditional biomass) was 47% in 2008, a far higher share than that of transport (27%), electricity (17%) or non-energy use (9%). Because of the large share of heat in final energy demand, expanding the use of modern biomass, geothermal and solar energy to produce heat could make a substantial contribution to meeting climate change and energy security objectives.

---

1. The term commercial heat as used throughout this chapter refers solely to heat produced in a heat plant or a co-generation plant (also referred to as combined heat and power) and sold through a network to industrial facilities, households or commercial establishments (district heat). In this chapter and in Chapter 9, the definition of heat is broader than the one in Annex C, which applies to the rest of book.

2. Traditional biomass is defined as biomass consumption in the residential sector in developing countries and refers to the often unsustainable use of wood, charcoal, agricultural residues and animal dung for cooking and heating. All other biomass use is defined as modern.

Figure 11.1 ● Final energy consumption by energy service, 2008

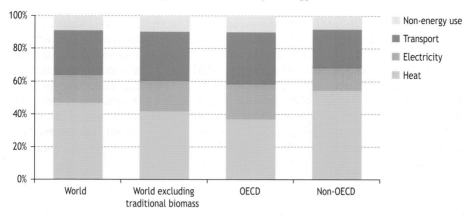

Demand for heat is, unsurprisingly, higher in colder climates; Russia and Canada, for example, have very high per-capita heat consumption (Figure 11.2).[3] Demand for heat is not, however, only climate-dependent. Some warm-climate countries also have a large share of heat in total final energy consumption. In such cases, this often stems from using significant amounts of process heat in industry or heavy reliance on traditional biomass (for example, in developing countries like Indonesia). The share of heat in final demand is particularly high in China, owing mainly to its large industrial sector.

Figure 11.2 ● Share of heat in total final energy consumption in selected countries, 2008

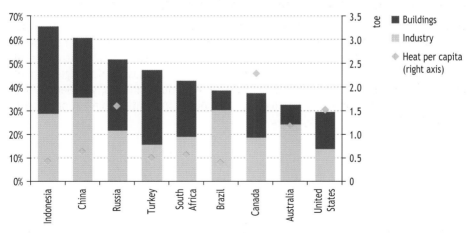

Renewable energy sources play an important role in heat supply. Worldwide, traditional biomass and modern renewables together fuelled 27% of the total demand for heat, or 1 059 million tonnes of oil equivalent (Mtoe), in 2008. Traditional biomass, including

---

3. Personal income is another important determinant of per-capita heat demand.

wood, charcoal, crop residues and animal dung, accounts for the bulk of total heat supply. It is mostly used for cooking and water heating in developing countries but, in colder climates, biomass stoves also provide space heating. The use of these biomass resources is considered traditional because they are most often burned at very low efficiencies and release many pollutants that have a serious health impact. In 2008, 746 Mtoe of traditional biomass was consumed in the residential sector in developing countries, with consumption in sub-Saharan African countries accounting for 32%.[4] Due to the large population of China and India and their heavy reliance on traditional biomass, these countries also account for a significant share of the global population relying on traditional biomass. Demand for traditional biomass worldwide increased by 12% between 2000 and 2008.

The global use of modern renewables for producing heat reached 312 Mtoe in 2008, 10% of total demand for heat. Although the use of modern renewables for heat increased by 18% between 2000 and 2008, its share in total heat demand did not increase. At 278 Mtoe in 2008, the main modern renewable energy source for producing heat is biomass, (including wood products, such as pellets and briquettes that have been made to burn efficiently, industrial biogas and bioliquids). Solar and geothermal contributed 10 Mtoe and 5 Mtoe to heat supply in 2008; commercial heat produced from modern renewables accounted for 19 Mtoe.

The share of renewable energy in total demand for heat varies widely in OECD countries (data for non-OECD countries is of low quality) (Figure 11.3). In Sweden, 63% of total heat demand in 2008 was supplied by renewables, whereas in the United Kingdom renewables contributed only 1%. Commercial heat is important in some countries, notably in Sweden, Iceland and Austria. Use of geothermal energy is considerable in Iceland and New Zealand. Greece and Austria make extensive use of solar water heaters, relative to other countries.

Figure 11.3 ● **Share of renewables in total heat demand by type in selected OECD countries, 2008**

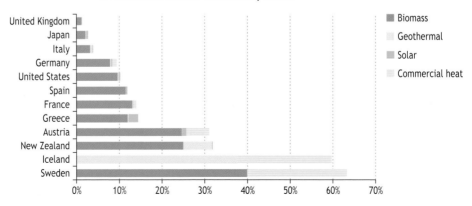

---

4. Chapter 8 provides more information on the traditional use of biomass in developing countries.

A variety of technologies are used to produce heat from renewables: solar collectors, biomass stoves using pellets or wood, anaerobic gas digesters and co-generation plants. These technologies are discussed in more detail later in the chapter.

Outlook for renewables for heat production

In the New Policies Scenario, global renewable heat demand rises from 1 059 Mtoe in 2008 to nearly 1 400 Mtoe in 2035. Traditional biomass meets the largest share of this demand, although it falls from 71% to 52% over the *Outlook* period. In the Current Policies Scenario, demand for renewables for heat increases to over 1 250 Mtoe in 2035, with the share of traditional biomass in total renewables for heat declining to 57% in 2035. In the 450 Scenario, demand for modern renewables for heat ratchets up to almost 1 500 Mtoe in 2035. In this scenario, the traditional use of biomass accounts for only 47% in 2035. The rest of this section presents more detailed results for renewable heat demand in the New Policies Scenario.

Traditional biomass

In the New Policies Scenario, traditional biomass continues to be the main source of heat in the residential sector in many developing countries, particularly in sub-Saharan Africa.[5] Nonetheless, a significant decline in the use of traditional biomass in China results in a fall in global demand from 746 Mtoe in 2008 to a little over 720 Mtoe in 2035. Reliance on traditional biomass for heat declines as incomes rise. Low-income households use a three-stone fire[6] or can usually only afford a basic cookstove (which is marginally more efficient). At higher incomes, households can afford more efficient biomass cooking and heating devices or conventional stoves and the use of traditional biomass declines. While demand for traditional biomass falls in developing Asian and Latin American countries, it *increases* in sub-Saharan Africa on the assumption of slower economic growth.

Demand for traditional biomass climbs to almost 300 Mtoe in 2035 in Africa, mainly in sub-Saharan countries (Figure 11.4). In China, traditional biomass demand drops from some 200 Mtoe in 2008 to 120 Mtoe in 2035, as a large number of households switch to conventional stoves or modern biomass, such as biogas, for cooking. Traditional use of biomass also falls in India, from 128 Mtoe to about 120 Mtoe over the *Outlook* period; a steeper decline in traditional biomass demand is tempered by the "National Biomass Cookstove Initiative", a programme that aims to improve the efficiency of cooking and heating with biomass.[7]

11

---

5. See Chapter 8 for an analysis of the number of people relying on the traditional use of biomass over the projection period and the health implications.

6. A three-stone fire uses three stones to support the pot and firewood is placed underneath.

7. See Box 8.6 in Chapter 8.

Figure 11.4 ● Traditional biomass demand by region in the New Policies Scenario

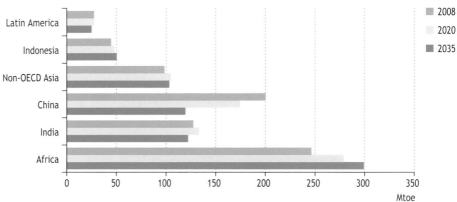

Modern renewables

In the New Policies Scenario, global demand for modern renewables for heat more than doubles over the *Outlook* period, growing from 312 Mtoe in 2008 to over 650 Mtoe by 2035. By 2035, 16% of total demand for heat comes from renewables, compared with 10% in 2008. Demand for renewables increases at an annual average growth rate of 2.8% over the projection period in the New Policies Scenario, higher than the 2% annual growth rate over 2000-2008. While solar energy grows seven-fold over the projection period, from 10 Mtoe to 70 Mtoe, modern biomass continues to dominate modern renewables for heat. Geothermal production of heat on-site increases from 5 Mtoe to 26 Mtoe.

In the OECD, much of the current building stock is likely to remain in use for many decades. Most of the potential for increased penetration of modern renewables into the supply of heat for buildings, therefore, lies in retrofitting existing buildings. In developing countries, where new building growth will be very rapid, opportunities exist to install modern renewable technologies from the outset. Similarly, the industrial and service sectors experience rapid growth in developing countries, creating large opportunities for renewables. Even though industrial demand for heat declines in OECD countries over the projection period, significant opportunities still remain to replace ageing fossil-fuel based technologies with renewables.

The share of modern renewables in total heat demand rises more substantially in OECD countries than non-OECD countries over the projection period, from 11% to 23%, in the New Policies Scenario (Table 11.1 and Figure 11.5). Nearly all of the increase occurs in the United States, European Union, and Australia and New Zealand, where policies to promote heat from renewable energy are expected to bear fruit. Modern renewables, mostly biomass, accounted for about one-fifth of total industrial sector heat demand in Australia and New Zealand in 2008, the highest share among OECD countries, which increases to 41% in 2035 in the New Policies Scenario. Most of the additional demand comes from industry, which accounts for 60% of the increase in renewables for heat between 2008 and 2035 in these two countries.

Table 11.1 ● **Share of modern renewables for heat in total heat demand by region in the New Policies Scenario**

	2008	2020	2035
OECD	**11%**	**15%**	23%
United States	*10%*	*16%*	*25%*
Australia and New Zealand	*18%*	*26%*	*41%*
Non-OECD	**9%**	**10%**	12%
China	*1%*	*2%*	*5%*
Brazil	*47%*	*49%*	*50%*
World	**10%**	**12%**	16%
European Union	*13%*	*17%*	*26%*

Outside of the OECD, the share of renewables for heat in total heat demand increases from 9% in 2008 to 12% in 2035. Demand for renewables increases more in China and Brazil than elsewhere in the non-OECD group. In China, demand for modern renewables for heat increases from 6 Mtoe in 2008 to nearly 50 Mtoe in 2035, resulting mainly from growth in biomass-based industrial co-generation and even greater use of solar water heaters in buildings. Use of modern renewables for heat in Brazil rises from 36 Mtoe to more than 65 Mtoe over the *Outlook* period, mainly in the form of bagasse (a by-product of the sugar industry) co-generation in various industries, charcoal use in steel-making and solar heat in buildings.

11

Figure 11.5 ● **Modern renewables for heat in the industry and buildings sectors in the New Policies Scenario**

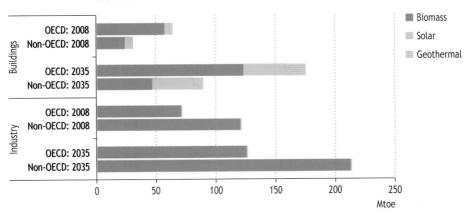

Box 11.1 ● Expanding the production of heat from biomass in the industry sector

Roughly three-quarters of industrial energy demand arises from the production of energy-intensive commodities, such as metals, chemicals and petrochemicals, non-metallic mineral materials, and pulp and paper. Because of the high share of energy in total production costs, industrial energy efficiency levels are much higher than in the buildings and transport sectors, and the potential for further improvements is lower (Taibi *et al.*, forthcoming). Thus, if significant carbon dioxide (CO_2) reductions are to be achieved in the industrial sector, switching to renewables has to be a favoured course.

Heat demand in industry falls into different temperature ranges, and so needs to be matched with the appropriate renewables-based heat technology. Solid biomass and biogas have the advantage that they can provide heat across all temperature ranges, although high temperatures cannot be achieved economically with current technologies. The industries that have significant biomass potential include chemicals and petrochemicals and cement. For chemicals and petrochemicals, successful deployment of biomass depends primarily on building biorefineries that produce a range of products. Once the logistics are in place, low-grade biomass can be procured specifically for the production of process heat. In the cement sector, waste and low-grade biomass can be used to produce heat. Overall, there is significant potential to increase the use of renewables in industry, but its development depends on government support and, in the long run, a price for greenhouse-gas emissions.

Modern biomass is used to produce process heat in the industry sector, and for space and water heating in the buildings sector. In the New Policies Scenario, global biomass use for heat increases from 278 Mtoe in 2008 to over 520 Mtoe in 2035. Industry remains the main user of modern biomass over the *Outlook* period; in absolute terms, its use for heat production increases from 191 Mtoe in 2008 to nearly 340 Mtoe in 2035. The pulp and paper sector is, by far, the largest industrial consumer of biomass for heat (Figure 11.6). In 2035, nearly 80% of the biomass-fed heat demand in the chemicals sector and around 80% in the paper industry arises in OECD countries. Due to its reliance on charcoal, Brazil accounts for 94% of global demand for biomass for heat in the iron and steel industry in 2035.[8] Modern biomass use in buildings doubles over the projection period, from 81 Mtoe to 169 Mtoe, meeting a growing share of their energy needs.

---

8. For example, ArcelorMittal Bioenergetica produces charcoal from eucalyptus forestry operations. This charcoal is used to fuel iron furnaces in Juiz de Fora or to be exchanged for pig iron with local producers (Taibi *et al.*, 2010).

Figure 11.6 ● *Global modern biomass for heat in selected industries in the New Policies Scenario*

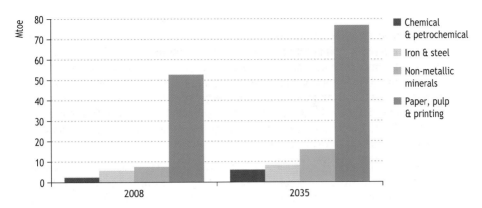

Solar energy to produce heat is used mainly in buildings. The use of solar energy for the production of heat is very small in the industry sector today (though not all of it is captured in statistics). Some uptake of solar energy for heat is projected in OECD countries, yet it still accounts for less than 1% of total global heat in the industry sector in 2035. Global solar heat demand in buildings increases from 9 Mtoe in 2008 to 65 Mtoe in 2035, growing at 7.4% per year on average, in the New Policies Scenario. Most of the growth takes place in China (alone representing 56% of non-OECD demand in 2035), followed by the United States and European Union (Figure 11.7). The United States and the European Union combined represent nearly 80% of solar heat demand in OECD countries in 2035.

Figure 11.7 ● *Solar heat consumption in the buildings sector by region in the New Policies Scenario*

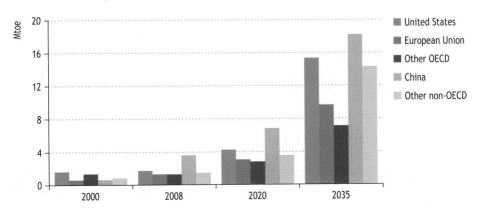

Box 11.2 ● **The impact of technology development on the uptake of solar for heat**

Higher penetration of modern renewables in heat demand will depend on technology developments related to temperature and storage. Heat demand differs by temperature levels according to the application (IEA, 2007). Water temperatures for space heating vary from 45°C (for under-floor heating) to 90°C (for heating by conventional radiators). Domestic hot water requires a temperature of 60°C, whereas industrial process heat can demand a temperature ranging from 60°C to more than 400°C. The temperature levels provided by renewable heating technologies vary from 45°C (from ground source heat pumps) to 80°C (from conventional solar thermal flat panels) and up to 400°C (from concentrating solar technologies). Biomass heat can provide all required temperature levels, whereas geothermal heat levels differ by location: they can exceed 150°C in specific areas.

Conventional solar thermal panels, providing low-temperature heat up to 80°C, have considerable potential in providing industrial process heat. This is the case for the European Union, as 30% of European industrial heat demand is estimated to consist of heat below 100°C (EcoHeatCool, 2006). Several industrial processes, such as pasteurisation, sterilisation, cooking, bleaching, dyeing, pre-heating water and washing, require temperatures of 60°C to 90°C. Solar thermal collectors currently are used for low-temperature processes in the brewing industry. Concentrating solar power (CSP) plants, which produce electricity and heat, offer another potential avenue to expand the use of solar heat. Investments in several CSP projects, including in China, India, Morocco, Spain and the United States, are expected to stimulate development of the technology. These will also amplify its potential for applications in industrial process heat, and as heat sold on the network. In the New Policies Scenario, however, CSP is projected to be used only for the production of electricity.

Geothermal energy for heat production is used mainly in buildings. Global geothermal heat use is projected to grow from 5 Mtoe in 2008 to slightly more than 25 Mtoe in 2035. Most of the increase is in OECD countries, notably in the United States and certain countries in OECD Europe (*e.g.* Turkey, Iceland and Switzerland). Almost all the growth is in the buildings sector.

Commercial heat is increasingly supplied by modern renewables, their share in total commercial heat rising from 7% in 2008 to 14% in 2035. Biomass continues to account for the lion's share of renewable energy used for commercial heat. Growth in demand for modern renewables for commercial heat is strongest in the European Union and China.

Many countries in the European Union have a high share of commercial heat in overall heat demand. While northern European countries, such as Iceland, Finland, Sweden and Denmark, supply large amounts of renewable heat for their district heating systems, countries like Poland, Czech Republic, Hungary and Slovakia rely mainly on fossil-fuel based combined heat and power plants (CHP) plants, and, in some cases, considerable amounts of coal. These countries have wide scope for replacing fossil-fuelled heat by renewables-based heat. Biomass has the greatest potential and is most efficiently used in CHP plants, supplying a district heating network. Biomass combustion to produce electricity and heat in CHP plants is a mature technology and in many cases is already competitive with fossil fuels.

In the New Policies Scenario, the share of modern renewables in commercial heat demand doubles in the European Union, from 17% to 34% over the *Outlook* period. In China, there is a switch from coal and oil to biomass for commercial heat. Modern renewables supplied just 0.5% of commercial heat in China in 2008, but this share is projected to climb to 13% by 2035.

·················S P O T L I G H T·················

How big is the potential for solar water heating in China?

Given China's abundant solar resources, we project solar technologies to make an important contribution to reducing the country's greenhouse-gas emissions, particularly in the buildings sector. In urban areas, the market share of solar water heaters in China increased from about 15% in 2001 to over 50% in 2008. Although the upfront capital cost of solar water heaters is higher than electric or gas water heaters, the average annual investment over the lifetime of the heater is considerably lower (Table 11.2). The use of solar thermal collectors in China has grown rapidly, from 15 million square kilometres (km^2) of total collector area in 1998 to 135 million km^2 in 2008, accounting for about 80% of the world total in that year (Weiss, 2010). China is also a major exporter of solar water heaters, with the value of exports increasing nearly six-fold from 2001 to 2007. In terms of industry development, production of solar water heaters in China increased nearly eight-fold from 1998 to 2008. Sales were 43 billion yuan ($6.3 billion) in 2008. In 2007, there were more than 3 000 manufacturers of solar water heaters in China. In the New Policies Scenario, solar energy use in buildings grows five-fold between 2008 and 2035.

11

Table 11.2 ● Cost comparison of water heaters in China

	Electric water heater	Gas water heater	Solar water heater
Hot water supply (litres per day)	100	100	100
Equipment investment ($)	176	146	264
Annual operating cost ($)	73	51	0.73
Lifetime (years)	8	8	10
Average annual investment over lifetime ($)	95	82	27

Note: Cost figures have been converted to dollars from yuan, using the 2009 average annual exchange rate of $1 = 6.83 yuan.
Source: REN21 (2009).

Box 11.3 ● Renewables for heat in the 450 Scenario

In the 450 Scenario, demand for traditional biomass falls from 746 Mtoe in 2008 to just under 700 Mtoe in 2035. By contrast, demand for modern renewables increases sharply, from 312 Mtoe to nearly 800 Mtoe; its share in total heat demand increases from 10% to 21%. The most significant increase is in buildings, where demand almost quadruples over the projection period. Renewables supply over one-quarter of the heat needs in buildings in 2035, up from 8% now. This increase is underpinned by concerted government action to promote energy efficiency and renewables in buildings. In industry, the share of renewables in total demand for heat grows from 11% to 18%, with growth encouraged by cap-and-trade schemes.[9]

In the 450 Scenario, biomass use more than doubles, from 278 Mtoe to almost 600 Mtoe; demand for solar increases from 10 Mtoe to nearly 120 Mtoe; and geothermal use rises from 5 Mtoe to more than 40 Mtoe. Renewables supply 20% of commercial heat in 2035, a share three times higher than in 2008. Over the period 2010-2035, the incremental investment in renewables relative to the Current Policies Scenario is $680 billion.

Renewable energy technologies for heat

Biomass

Modern biomass combustion to produce heat is a mature technology and in many cases is competitive with fossil fuels (IEA, 2007).[10] Modern on-site biomass technologies include efficient wood burning stoves, municipal solid waste (MSW) incineration, pellet boilers and biogas. Biomass is also used in CHP production, which is more efficient than production of electricity or heat alone; where the heat can be usefully employed, overall conversion efficiencies of around 70% to 90% are possible. Common feedstocks in biomass-fired CHP plants are forestry and agricultural residues, and the biogenic component of municipal residues and wastes. Sweden is the largest consumer of wood and wood waste for district heating, followed by Finland and the United States. Denmark, Germany and Sweden are the largest users of MSW for district heating.

Solar

Solar thermal collectors produce heat derived from solar radiation by heating a fluid circulated through a collector. Solar thermal panels producing low-temperature heat (less than 80°C) are a commercial technology. Rooftop solar thermal panels producing medium-temperature heat (up to 150°C) are still in the early stages of development, although some are available on the market. By the end of 2008, worldwide installed solar thermal (low- and medium-temperature) capacity totalled 152 GW$_{th}$ (Figure 11.8). Almost 90% of this capacity was in China (88 GW$_{th}$), Europe (29 GW$_{th}$) and OECD North America (16 GW$_{th}$).

---

9. A detailed overview of the 450 Scenario across all energy sectors and technologies is presented in Chapter 13. An analysis of the costs and benefits of the scenario is presented in Chapter 14.

10. Traditional cookstoves are discussed in Chapter 8.

Figure 11.8 • Total solar heat capacity by region, 2008

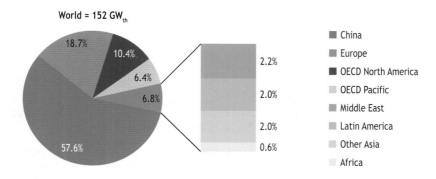

Source: Weiss, 2010.

Geothermal

Direct-use geothermal applications include mature technologies to provide heat for industrial processes, space conditioning, district networks, swimming pools, greenhouses and aquaculture ponds. In Iceland, where there are favourable geologic conditions and efficient hot water distribution networks, 88% of all households use geothermal (produced mostly in CHP plants). Other OECD countries using geothermal for district heating are Germany, Austria, Denmark, Hungary, Slovakia and Belgium.

Box 11.4 • Heat pumps

Heat pumps provide a highly efficient means of cooling, and space and water heating (IEA, 2010a). They upgrade low-temperature heat, available in ambient energy sources (air, water or ground), to useful higher temperature heat that can be used for low-temperature heating systems (*e.g.* water temperatures of up to 45°C for under-floor heating). In specific applications, heat pumps can also be used to provide domestic hot water, usually in combination with a relatively high-temperature heat source, such as exhaust air.

Heat pumps are most commonly powered by electricity. As a result, the energy output of heat pumps has a renewable energy component (the ambient energy source) and a fossil-fuel component (from the electricity requirements). Heat pumps are not included in IEA renewable energy statistics. In this report, they are modelled as energy-efficiency improvements rather than renewables. Globally, there were some three million ground-source heat pumps installed at the beginning of 2010, using around 5 Mtoe of geothermal energy (IEA 2010b). About one-quarter of these are in the European Union, mainly in Sweden, Germany and France (EurObserv'ER, 2009).

Policies to support renewables for heat

Policy support for renewables for heat is low compared with renewables-based electricity or biofuels for transport. Most renewables for heat policies have focused on solar technologies. Moreover, the policies in place have not been very effective (IEA, 2008). The stop-and-go nature of support to renewable heat in some countries has inhibited growth. In the Netherlands, for example, a capital-cost subsidy for solar thermal systems was introduced in 1988 (and subsequently adjusted in 1992, 1995, 1997 and 2000), but ended in 2003. A subsidy scheme was reintroduced in 2009. In some cases, production of heat from renewable energy increased faster in countries without incentives.

In the United States, the Energy Policy Act of 2005 established a 30% federal tax credit (up to $2 000) for the purchase and installation of residential solar water heating. Initially scheduled to expire at the end of 2007, the tax credit was extended in 2008 until December 31, 2016. Under the National Climate Change Plan (2008), Brazil plans to increase the sustainable use of charcoal in the iron and steel industry, primarily by the support of forestation in degraded areas. Brazil's plan also includes an incentive to encourage the use of solar water heating, aimed at reducing electricity consumption by 2 200 gigawatt-hours (GWh) per year by 2015. South Africa has targets for the use of solar water heaters in its Renewable Energy Framework. In 2007, the government of Shandong Province in China created a fund to support solar hot water supply systems in hotels, schools and other establishments. Other examples of policies to support renewables-based heat are found mainly in the European Union and Australia (Table 11.3).

Table 11.3 ● **Examples of policies for renewable heat in OECD countries**

Financial mechanisms		Regulatory mechanisms	
Levies (e.g. CO_2 tax)	Feed-in tariff	Use obligations	White certificate scheme
Sweden, Denmark, Norway, Finland (early 1990s): CO_2 tax	United Kingdom: Renewable Heat Incentive (expected implementation April 2011)	50% of United Kingdom's local authorities (Merton Rule), to be superseded in 2011 by UK's Renewable Heat Incentive	Italy (2005) White certificate scheme + solar thermal and biomass benefit 5-year tax deduction
California (2008): CO_2 tax		Germany (2009): new buildings must cover part of heat demand with renewables (grants and stimulus for district heating)	Australia (2001): Renewable Energy Certificates with tradable solar thermal output

Notes: Table does not include capital-cost subsidies, which are nearly ubiquitous in European Union countries. In September 2010, the United Kingdom's Committee on Climate Change, which advises the government, suggested that the proposed target in the Renewable Heat Incentive (from around 1.6% in 2009 to 12% in 2020) may be too costly to achieve, and that a slightly lower level of ambition for heat may be appropriate.

Box 11.5 ● Renewable heat obligations and feed-in tariffs in the European Union

Recently, more renewable heat policies have had a regulatory component, while still drawing on the experience of successful support mechanisms for renewable electricity policies. The Spanish government developed a national solar obligation policy in 2006. Since a solar obligation incentivises one specific technology, such a policy should be introduced only where there is no competition with other renewable technologies for the same market. The procedure for checking compliance and the absence of an incentive to exceed the required level of the obligation are weaknesses of the solar obligation. Another regulatory approach consists of requiring a defined share of a building's heat to be supplied by renewable energy, such as in the London "Merton Rule" (Table 11.3) and the German 2009 building regulations. This type of obligation allows for competition between renewable (heating) technologies, but still lacks any incentive to exceed the required renewable share in heating demand which, in the case of the Merton Rule, is a modest 10% share. When applied to new buildings only, the effect, in many cases, will be limited, as annual construction rates in OECD countries are, on average, about 1% of the total building stock. In both of these examples, the regulation applies at the individual building level, discouraging more ambitious approaches.

The United Kingdom aims to introduce a Renewable Heat Incentive by April 2011, a first initiative in designing a feed-in tariff policy for the heat market. In Germany, the introduction of a renewable heat feed-in tariff policy has been explored, but the approach has been dropped in favour of an obligation policy (Bürger et al., 2008). Introducing a feed-in tariff scheme, as used for renewable electricity, to the renewable heat market gives rise to complications, due to key differences between the delivery of heat and electricity (Connor et al., 2009). The more heterogeneous nature of the fuels used for heat production and the relatively small scale of operation means that there is a far more diverse group of companies supplying the market. The mechanism must be designed to treat all supply companies equitably. A key problem in a renewable heating feed-in tariff scheme is assessing the generated heat output. Heat metering is costly relative to any available subsidy, suggesting that an alternative is needed. Moreover, as there generally is no "grid" to which excess domestic heat can be delivered, provisions must be included to avoid rewarding the production of unused heat.

11

In the European Union, a direct capital-cost subsidy to support for the purchase of renewable heating systems is the most widely adopted financial mechanism to support renewable heat technologies. In general, capital-cost subsidies are the most successful way to encourage higher penetration of renewable technologies when they are in the prototype and demonstration phase (IEA, 2008). Solar thermal technology continues to benefit from capital-cost subsidies in many countries, even though it is a relatively mature technology.

Capital-cost subsidies incur low transaction costs, especially if an administrative entity accustomed to handling subsidy schemes is already operational. They also appeal to consumers, who are used to paying a one-time upfront investment for heating or hot water installations. In the case of renewable heat, a considerable share of the market is expected to consist of consumers buying individual heating systems. The capital subsidy in many countries is provided upfront and there is no monitoring of compliance with installation guidelines.

Renewable energy for cooling

Cooling is a service that meets demand for individual comfort and refrigeration in the buildings sector and process cooling in the industrial sector. Unlike heating, cooling demand is highly correlated to income. Energy-use data for cooling, however, are not collected. Electricity use in cooling systems, for example, is included in aggregated electricity use in the buildings and industry sectors.

Renewable cooling technologies range widely, consisting of passive cooling, storing heat in the ground for extraction during winter, using renewable heat for cooling and using renewable power for cooling. During the warm season, passive cooling uses relatively constant low temperatures of deep seawater, deep lake water or the ground (ideally between 0°C and 10°C on average), to circulate a working fluid through floor heating pipes or to cool the air in large-scale air-conditioning systems. Cooling can be provided in combination with a ground source heat pump, where the (renewable) heat of the building is transported to the ground, perhaps to be stored in aquifers for extraction during winter. This technology has already proven to be commercially competitive with conventional cooling systems in large office buildings, commercial buildings, hospitals, housing, industry and agriculture.[11] Cooling can also be provided in a district system where cold water is distributed through the network.

Solar-assisted cooling technologies match peak cooling demands with maximum solar radiation, and, hence with peak electricity loads for conventional air conditioners. The thermally-driven process in solar-assisted cooling is complex, being based on a thermo-chemical sorption process or a thermally-driven open cooling cycle. The technology has not been widely applied and needs more research and development to achieve competitive levels of reliability and cost with conventional cooling technologies. Another route is to generate electricity, for example using solar photovoltaics, to power a conventional refrigeration device.

---

11. IEA Implementing Agreement on Energy Storage (www.energy-storage.org).

CHAPTER 12

RENEWABLES FOR TRANSPORT

How much will biofuels contribute?

H   I   G   H   L   I   G   H   T   S

- Biofuels demand is expected to increase rapidly over the projection period, thanks to rising oil prices and government support, prompted by energy-security and environmental concerns. In the New Policies Scenario, global biofuels consumption increases from 1.1 mb/d today to 4.4 mb/d in 2035. Biofuels meet 8% of world road-transport fuel consumption by 2035, up from 3% in 2009. Over 2009-2035, biofuels meet about 20% of global incremental demand for total road-transport fuels. In the 450 Scenario, biofuels account for 4% of the CO_2 emissions reductions, compared with the New Policies Scenario.

- The United States and Brazil are expected to remain the world's largest producers and consumers of biofuels. The United States accounts for 38% of total biofuels use by 2035 in the New Policies Scenario (down from 45% today), followed by Brazil with 20% (28% today). The share of non-OECD Asian countries, mainly China and India, increases most, from 6% in 2009 to 19% in 2035. Biofuels use in non-OECD Asia outstrips that in EU countries by the end of the projection period.

- Today, almost all commercial biofuels production uses conventional technology. Advanced biofuels, including those from ligno-cellulosic feedstocks, are assumed to enter the market by around 2020 in the New Policies Scenario, mostly in OECD countries. In that scenario, advanced biofuels account for 36% of biofuels use in OECD countries in 2035, but only 5% in non-OECD countries.

- The projected expansion of biofuels supply in the New Policies Scenario requires cumulative investment in production capacity of $335 billion over 2010-2035. More than half of this, some $180 billion, is for conventional production of ethanol, 10% for conventional biodiesel and the remainder for advanced biofuels. Around 60% of total investment is in OECD countries.

- Biofuels receive more government support than any other renewable energy source or carrier. Total support in 2009 was $20 billion, with the highest levels in the United States and the European Union. The production of ethanol receives most of this, at $13 billion in 2009. Support is projected to average $45 billion per year between 2010 and 2020, further increasing to about $65 billion per year between 2021 and 2035, with some 60% of it directed at ethanol and 40% at biodiesel. Government support typically raises costs to motorists and to the economy as a whole. But the benefits can be significant too, including reduced imports of oil and reduced CO_2 emissions — if biomass is used sustainably and the fossil fuels used to process the biomass is not excessive.

Overview

Biofuels, electricity and hydrogen are widely regarded as renewable forms of energy, competing for application in the transport sector. All are, strictly speaking, energy carriers rather than sources of energy; but, more important, the extent to which they are genuinely renewable is open to question (Box 12.1). This chapter concentrates on biofuels for transport, biomass – the feedstock for making biofuels – being unquestionably renewable when produced in a sustainable way.

Global production of biofuels was 52 million tonnes of oil equivalent (Mtoe), or 1 112 thousand barrels per day (kb/d), in 2009 (Table 12.1). The United States and Brazil, the world's largest producers, accounted for almost three-quarters of global production on an energy-adjusted basis.[1] Ethanol accounted for about 75% of global production of biofuels for transport. Investment in biofuels was severely affected by the economic and financial crisis in 2008-2009, falling by over 60% compared with 2008 as a result of lower oil prices and a drop in demand for transport fuels, but is likely to recover over the next few years (see Chapter 9).

Table 12.1 ● World biofuels production, 2009

	Ethanol		Biodiesel		Total	
	Mtoe	kb/d	Mtoe	kb/d	Mtoe	kb/d
United States	21.5	470	1.6	33	23.1	503
Brazil	12.8	287	1.2	25	14.1	312
European Union	1.7	38	7.0	140	8.7	178
China	1.1	24	0.3	6	1.4	30
Canada	0.6	13	-	-	0.6	13
India	0.1	3	0.1	2	0.2	5
Other	0.9	20	2.7	51	3.6	72
World	38.7	855	12.9	257	51.6	1 112

Despite rapid growth in their use over the past decade in some countries, biofuels accounted for only 3% of global road-transport fuel demand in 2009. Production in the United States has grown strongly over the past few years, almost 30% per year on average in 2002-2009, and the country overtook Brazil as the largest producer in 2005 (Figure 12.1). Production in the United States reached 503 kb/d in 2009, but the share of biofuels in road-transport fuel use was still only 3%. Brazil has the highest share of biofuels in its road-transport fuel mix, 20% in 2009. Currently, biofuels are used almost exclusively for road transport, but interest in the use of biofuels for aviation is growing (see Spotlight).

---

1. All biofuels-related volumetric data is presented on a gasoline- and diesel-equivalent basis in the entire chapter for better comparability with oil, unless specified otherwise.

Box 12.1 ● Renewable transport fuels

The question as to which transport fuels can be deemed renewable is not straightforward. Transportation fuels are energy carriers, not energy sources, and so the question as to which fuels can be classified "renewable" depends on how they are produced. Despite questions about their sustainability, the fact that biofuels are produced from biomass, which is a renewable energy source, clearly means that they can be considered as at least partially renewable.

Electricity and hydrogen, the two other transport fuels that might be considered renewable, can be produced through different processes and from many different feedstocks, including fossil fuels, nuclear power and renewables. Hydrogen can be produced from a variety of renewable energy sources, including solar thermal applications, electrolysis powered by renewable energy, or the gasification of biomass. But, over the *Outlook* period, hydrogen is expected to be produced mostly from fossil fuels (natural gas, for the most part).

Electricity used in electric vehicles or in plug-in hybrids plays an important role in meeting transport energy demand in all three scenarios in this *Outlook*, especially in the 450 Scenario, in which the use of low-carbon electricity is essential to reaching climate goals (see Chapter 14). In turn, electric cars can help mitigate problems over the variable nature of renewable energy. At times of excess supply, they can act as a storage medium: with vehicle-to-grid systems, electric cars could feed electricity back to the grid when renewable electricity production is low. Total battery capacity of electric cars and plug-in hybrids in the 450 Scenario is about 20 terawatt-hours (TWh) by 2035. But electricity, like hydrogen, cannot be simply designated a renewable fuel. Even in the 450 Scenario, renewables account for only 45% of world electricity generation in 2035, *i.e.* the majority of electricity generation is fossil and nuclear. Consequently, this chapter focuses on biofuels that are derived from renewable energy sources and, in aggregate, have a much larger renewable energy component than either electricity or hydrogen in the New Policies Scenario.

12

Figure 12.1 ● Biofuels production in key regions

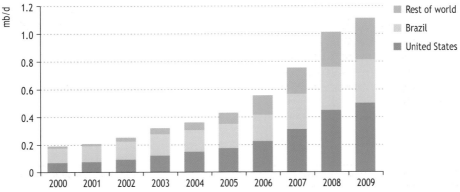

Biofuels consumption trends

Many countries have strengthened policies and measures, or introduced new ones, in recent years to encourage biofuels production and use, despite concerns about the environmental sustainability of biofuels and their associated greenhouse-gas emissions. The surge in oil prices up to 2008 also increased the attractiveness of investing in biofuels production. Oil prices are assumed to rise steadily in the Current and New Policies Scenarios, further boosting the role of biofuels over the projection period, particularly in the United States, the European Union and non-OECD Asia. New government measures also increase biofuels production and use in the New Policies and 450 Scenarios, partly driven by energy security and environmental concerns. In the New Policies Scenario, the global use of biofuels in 2035 is almost four times higher than in 2009. Biofuels expand slightly less rapidly in the Current Policies Scenario (as policies are assumed not to change). Their use grows fastest in the 450 Scenario — more than seven-fold between 2009 and 2035 — thanks to much stronger government measures. In that scenario, global biofuels consumption grows on average by about 8% per year over the *Outlook* period (Box 12.3), more than two percentage points faster than in the New Policies Scenario and mostly a result of increased use of advanced biofuels and sugar cane ethanol. Biofuels account for 4% of the carbon-dioxide (CO_2) emissions reductions in the 450 Scenario, compared with the New Policies Scenario.

Table 12.2 ● **World biofuels consumption by scenario** (mb/d)

	2000	2009	2035		
			New Policies Scenario	Current Policies Scenario	450 Scenario
OECD	0.08	0.70	2.46	1.84	4.24
North America	0.07	0.52	1.73	1.28	3.03
United States	*0.07*	*0.50*	*1.64*	*1.29*	*2.62*
Canada	*0.00*	*0.01*	*0.05*	*0.05*	*0.19*
Europe	0.01	0.17	0.69	0.52	0.97
Pacific	-	0.01	0.03	0.03	0.24
Japan	-	*0.00*	*0.01*	*0.01*	*0.05*
Non-OECD	-	0.00	1.90	1.65	3.19
E. Europe/Eurasia	-	0.00	0.06	0.05	0.11
Russia	-	-	*0.02*	*0.02*	*0.03*
Asia	0.00	0.06	0.81	0.78	1.54
China	-	*0.03*	*0.38*	*0.36*	*0.82*
India	*0.00*	*0.00*	*0.25*	*0.24*	*0.27*
Middle East	-	-	-	-	0.23
Africa	-	0.00	0.04	0.04	0.11
Latin America	0.11	0.35	0.99	0.78	1.20
Brazil	*0.11*	*0.31*	*0.90*	*0.69*	*0.99*
World*	0.19	1.11	4.38	3.50	8.11

*World includes international aviation bunkers (not included in regional totals).

Box 12.2 ● Biofuels definitions

There is a lot of discussion on the terminology and definitions used to classify biofuels. They are commonly referred to as "first-" or "second-generation biofuels", but the distinction is unclear. The reason is that the same fuel might be classified as first- or as second-generation, depending on whether the determining criterion is the maturity of the technology, the greenhouse-gas emissions balance or the applied feedstock. This year's *Outlook* classifies biofuels as "conventional" and "advanced" according to the technologies used to produce them and their respective maturity.

Conventional biofuels include well-established technologies that are producing biofuels on a commercial scale today. These biofuels are commonly referred to as first-generation and include sugar cane ethanol, starch-based ethanol, biodiesel, Fatty Acid Methyl Esther (FAME) and Straight Vegetable Oil (SVO).

Typical feedstocks used in these mature processes include sugar cane and sugar beet, starch-bearing grains, like corn and wheat, and oil crops, like canola and palm, and in some cases animal fats.

Advanced biofuels, sometimes referred to as second- or third-generation biofuels comprise different conversion technologies that are currently in the research and development, pilot or demonstration phase. More specifically, this category includes emerging biofuel technologies, such as hydrogenated biodiesel, which is based on vegetable oil, as well as all those based on ligno-cellulosic biomass, such as cellulosic-ethanol, biomass-to-liquids (BTL) diesel and bio-derived synthetic natural gas (bio-SNG), among others. The category also includes novel biofuel technologies that are mostly in the research and development and pilot stage, such as algae-based biodiesel or butanol, as well as the conversion of sugar into diesel-type biofuels using micro-organisms (such as yeast). This definition differs from the one used for "Advanced Biofuels" in the US legislation, which is based on a minimum 50% life-cycle greenhouse-gas reduction and which, therefore, includes sugar cane ethanol.

12

In the New Policies Scenario, biofuels consumption rises from 1.1 million barrels per day (mb/d) in 2009 to 2.3 mb/d in 2020 and 4.4 mb/d in 2035. The United States continues to dominate global biofuels use over the projection period (Table 12.2). This projection is, nonetheless, subject to important uncertainties, notably with respect to the pace of development and deployment of advanced biofuels, which are assumed to become more commercially viable, and the controversial question of the sustainability of conventional biofuels.[2]

---

2. There are numerous international forums such as the IEA's Implementing Agreement on Bioenergy, the Global Bioenergy Partnership (GBEP) or the Roundtable on Sustainable Biofuels (RSB) looking into developing criteria and indicators regarding the sustainability of biofuels.

Box 12.3 ● Renewables in transport in the 450 Scenario

In the 450 Scenario, use of biofuels increases from 1.1 million barrels per day (mb/d) in 2009 to 8.1 mb/d in 2035, equivalent to 15% of all transport fuels on an energy-equivalent basis in that year. This compares with an increase in the use of electricity in transportation from about 270 terawatt-hours (TWh) in 2008 to some 1 500 TWh in 2035, or 4% of all transport fuels, resulting from a significant increase in the fleet of electric cars. Hydrogen use in transport remains marginal.

The brisk expansion of biofuels in this scenario results mainly from the rapid market penetration of advanced biofuels and sugar cane ethanol, both of which can emit substantially lower levels of greenhouse gases than fossil fuels on a well-to-wheels basis (see the section on biofuels emissions), assuming that biomass is grown sustainably. Advanced biofuels account for around two-thirds of biofuels consumption by 2035 in the 450 Scenario.

The United States remains the dominant market throughout the projection period, accounting for 38% of global biofuels consumption in 2035. European consumption also grows strongly. The use of biofuels nearly triples in Latin America, from 0.35 mb/d in 2009 to 1 mb/d in 2035. Over the projection period, legal restrictions related to sustainability, such as those already introduced in the United States, are assumed to be introduced in the European Union, allowing blending targets to be met only with biofuels that substantially reduce greenhouse gases relative to fossil fuels.

Consumption of biofuels in non-OECD Asia grows to about 800 kb/d in 2035, from only 62 kb/d in 2009, resulting principally from measures aimed at addressing concerns about oil-supply security and from the assumed phase-out of fossil-fuel subsidies (Figure 12.2). Growth in China accounts for nearly half of the increase in demand in this

Figure 12.2 ● Biofuels consumption by region in the New Policies Scenario

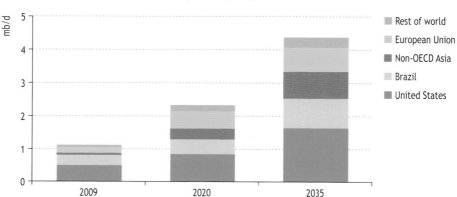

region, and India for one-third. While average annual growth hovers around 5% in more mature markets, such as the United States, Brazil and the EU countries, biofuels use grows by about 10% per year in China, albeit from a much lower base. Consequently, the share of non-OECD Asia in global demand increases from 6% in 2009 to 19% in 2035. Biofuels use in non-OECD Asia outstrips use in the European Union by the end of the projection period.

Biofuels meet 8% of world road-transport fuel demand in 2035 in the New Policies Scenario. Brazil continues to rank highest in the share of biofuels in total road-fuel consumption, reaching more than 40% in 2035 (Figure 12.3). The share of ethanol in biofuels consumption remains high in all countries. Although biodiesel continues to dominate biofuels use in the European Union, the share of ethanol rises from 27% in 2009 to 31% in 2035 in EU biofuels consumption. Over 2009-2035, biofuels meet about 20% of global incremental growth in road-transport fuel demand, the result of policy-driven increases in biofuels supply and demand-side efficiency measures to reduce oil consumption from road transport.

Advanced biofuels, such as those produced from ligno-cellulosic feedstocks, are assumed to be commercialised by 2020 in the New Policies Scenario. By 2035, advanced biofuels account for some 36% of total biofuels demand in OECD countries. The costs of advanced biofuels decline faster than those of conventional biofuels, on the assumption that investment in research and development in advanced biofuel technologies increases significantly. The large biomass demand requirements for a commercial advanced biofuel plant of up to 600 000 tonnes per year require complex logistical systems and good infrastructure in order to deliver the biomass at an economically competitive cost. Successful production of advanced biofuels can, therefore, be a particular challenge in rural areas of developing countries, where poor infrastructure and a complex pattern of land-ownership in small land holdings increase the complexity of feedstock logistics. Consequently, in the New Policies Scenario, advanced biofuels meet only about 5% of biofuels demand at the end of the projection period in non-OECD countries, mostly in China and India.

Figure 12.3 ● *Share of biofuels in total road-fuel consumption in selected regions by type in the New Policies Scenario*

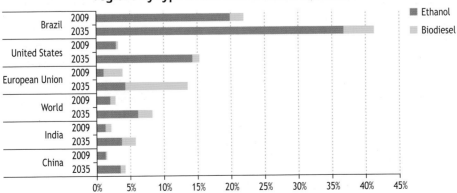

How green is your aircraft?

Compared with the road transport sector, aviation has fewer options to replace conventional fuels. The energy density of jet fuel is critical for providing adequate aircraft flying range, so shifting to gaseous fuels or electricity is impractical in the short term. Liquid hydrogen would require major changes in aircraft design. Ethanol is not a suitable alternative source of energy, due to its relatively low energy content and weight: an aircraft would need to be entirely redesigned in order to be operated with ethanol-based biofuels. Jet fuel is a form of kerosene, not radically different from diesel fuel for road vehicles, so high-quality, high energy-density biodiesel is the closest substitute. But at normal cruising altitudes, low air temperatures lead to problems with FAME biodiesel gelling in the fuel lines and tanks (Biofuels International, 2010). The term bio-derived synthetic paraffinic kerosene (Bio-SPK) refers to those biofuels that are suitable for use as aviation biofuels and closely resemble conventional jet fuels.

There are several promising technologies for making Bio-SPK from a variety of feedstocks and technology routes. Biomass-to-liquids (BTL) conversion of ligno-cellulosic feedstocks via Fischer-Tropsch synthesis is an interesting option for the medium term. Another option, which is very similar to conventional jet fuel and which has received considerable attention by airlines, is hydrogenated vegetable oil (HVO). Potential feedstocks are palm oil and waste vegetable oil, jatropha and camelina. Jatropha is a plant that can be grown in various soil conditions, including many that are not suitable for traditional agriculture, even though commercially attractive yields cannot be achieved on marginal land and cultivation is difficult. Camelina has similar characteristics to jatropha and is typically grown in temperate climates. Like jatropha, it is a crop that contains a lot of lipid, which can be extracted and converted to biofuels for aviation use. It is unclear as yet whether camelina offers any advantages over established crops (Schlumberger, 2010).

Algae are considered another promising feedstock for the large-scale production of biofuels for aviation. Algae are microscopic plants that grow suspended in water, undergoing a photosynthesis process that converts water, CO_2 and sunlight into oxygen and biomass. However, there is still uncertainty about the economics of algae-based biofuels and the availability of suitable locations to produce larger volumes. Considerable research and development needs to be carried out before algae can be commercialised.

The airline industry has shown great interest in testing and demonstrating the feasibility of using alternative fuels. A new aviation fuel specification, which will facilitate the use of alternative fuels, has been passed by the American Society for Testing and Materials (ASTM) International, the organisation which oversees international standards and specifications for jet fuel. Air New Zealand, Japan Airlines and Continental Airlines have carried out successful test flights using a blend of jatropha and traditional jet fuel. In 2009, Air France-KLM became the first airline to test biofuel in a passenger aircraft. The airline aims undertake to commercial flights that use biofuel from 2011.

Lufthansa announced that, in 2012, it will start running engines on some flights on a mixture of biofuel and kerosene. British Airways recently announced plans to build an organic waste BTL plant near London. The US Air Force has also undertaken extensive research on aviation biofuels.

Over the *Outlook* period, biofuels are expected to start to be regularly used for aviation by around 2020. But the pace of market growth is expected to depend on the vigour of government intervention.

In the New Policies Scenario, projected consumption calls for cumulative investment in biofuels production capacity of $335 billion (in year-2009 dollars) over the projection period (Figure 12.4). More than half of this investment, or about $180 billion, is for conventional production of ethanol, 10% for conventional biodiesel and the remainder for advanced biofuels. Around 60% of the total is invested in OECD countries. Over 50% of that investment goes to advanced biofuels technologies. Investment of more than $120 billion is required in non-OECD countries, nearly all of it in China and Brazil.

Figure 12.4 ● **Cumulative investment in biofuel production facilities in the New Policies Scenario by technology, 2010-2035** (in year-2009 dollars)

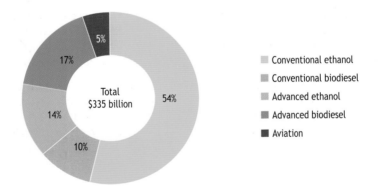

12

In general, with rapid technological progress, financing the construction of advanced biofuels plants should not be particularly difficult in countries like Brazil, China, India, South Africa and Thailand, since it will generally be possible to attract foreign direct investment in addition to domestic funding. However, for less-developed countries, the sheer scale of the investment needs could be a barrier for biofuels, as for other energy investments, since domestic funding possibilities are limited and significant administrative and governance risks may deter foreign companies from undertaking large investments (IEA, 2009a).

Government policies to support biofuels

In many countries, the biofuels industry depends heavily on government intervention, primarily in the form of obligations to blend biofuels into conventional fuels (blending mandates), production subsidies or both (Table 12.3). In such countries, these measures will continue to drive the expansion of biofuel production where it is otherwise not competitive with oil-based fuel production. Until recently, many biofuels programmes were conceived as part of farm-support policies, but a growing number of governments is now expanding or introducing such programmes for energy-security, economic and environmental reasons. Biofuels targets in the European Union, for example, can be seen as part of its commitment to reducing overall greenhouse-gas emissions.

Table 12.3 ● Current government support measures for biofuels in selected countries

	Targets/mandates*	Support measures
Argentina	E5 (2010)	Blending mandate; tax exemption
	B7 (2010)	Blending mandate; tax exemption
Australia (New South Wales)	E6 (2011)	Blending mandate; tax reduction
	B2	
Brazil	B5 (2010); E20-25	Blending mandate; tax reduction
Canada	E5 (2010)	Blending mandate
	B2 (2012)	Blending mandate
China (nine provinces)	E10	Biofuel mandate (50 Mt/y); fixed subsidy
Colombia	E10; B10 (2010); B20 (2012)	
France	7% by energy content	Obligation for fuel suppliers, under a tax for not complying with biofuel incorporation (Taxe Generale sur les Activites Polluantes)
Germany	6.75% by energy content (2010)	Blending mandate; tax reduction
India	E10; B5 (2012); 20% biofuels (2017)	
Italy	B3.5 (2010)	Tax reduction
Japan	500 million litres by 2010	
Korea	B2.5 (2011); B3 (2012)	Blending mandate
Paraguay	E24; B5	
Peru	E7.8 (2010); B5 (2010)	Blending mandate
South Africa	E2**; B2	Blending mandate; tax exemption
Spain	5.83% by energy content (2010)	Blending mandate; tax reduction
Sweden	5.75% by energy content (2010)	
Thailand	B2; B5 (2012)	
United Kingdom	3.6% (2010); 4.2% (2011); 4.7% (2012); 5.3% (thereafter)	Blending target; tax reduction
United States	49 billion litres (2010, of which 0.02 cellulosic ethanol); 78 billion litres (2015, of which 11.4 cellulosic); 136 billion litres (2022, of which 60 cellulosic)	Loan guarantees, production tax credit for cellulosic biofuels, VTEEC blending tax credit; blending target
Zambia	E5 (2011); B10 (2011)	Blending target

* Share of biofuels in total road-fuel consumption by volume (unless otherwise specified); E = Ethanol, and E5 represents a 5% share of ethanol in the final product fuel mix, similarly B = Biodiesel, and B7 represents a 7% share of biodiesel. Policies written in blue are mandatory.

** Use of corn as a feedstock is prohibited.

Source: IEA databases and analysis.

United States

Under the Energy Independence and Security Act of 2007, the Renewable Fuel Standard (RFS) in the United States requires that 9 billion gallons of renewable fuels (34 billion litres) are to be consumed annually by 2008, rising progressively to 36 billion gallons (136 billion litres) by 2022. The Act specifies that 21 billion gallons of the 2022 target must be advanced biofuels, defined as fuels that, on a life-cycle basis, must emit 50% less greenhouse gases than the gasoline or diesel fuel it replaces. The US Environmental Protection Agency (EPA) is investigating the possibility of approving a 15% blend, up from a maximum of 10% today (Box 12.4).

Box 12.4 ● Raising ethanol blend levels in the United States

In the United States, legislation allows for ethanol blends to gasoline of up to 10% (E10), but not beyond. This 10% "blend wall" is seen by many as consistent with the technical limit on how much ethanol can be blended into gasoline without causing problems for conventional vehicles; but it represents a major barrier to achieving biofuels targets. This issue is controversial. Extensive testing is being undertaken, with a view to allowing higher blend shares. The Environmental Protection Agency (EPA) decided in October 2010 to allow an increase in the blend rate of ethanol in gasoline to 15% from 10% for cars and light trucks built since 2007. Vehicles sold between 2001 and 2006 are subject to further testing.

The problem is how to impose a volumetric obligation — almost 140 billion litres (Table 12.3) — on a market which may be unable to absorb it. Who is to carry the risk? Farmers and ethanol producers would be delighted to see the required volumes, but fear that the market will be over-supplied and the obligation will fade away, leaving them exposed. The manufacturers have already suffered widespread bankruptcies, as prices for corn, the main feedstock for ethanol production in the United States, spiked in 2009. They are, therefore, pushing for approval of the use of E15 in older cars. Car manufacturers, however, fear they could be sued if owners of older cars buy fuel not suitable for their vehicles.

European Union

The Renewable Energy Directive 2009/28/EC mandates a share of renewable energy in total transport demand in EU member countries of at least 10% by 2020. This directive requires that, from the end of 2010, biofuels must generate greenhouse-gas emissions savings of at least 35%, compared with fossil fuels, if they are to count towards the renewables target; these savings rise to 50% in 2017 and 60% in 2018. The current EU interpretation of this policy classifies rapeseed biodiesel, which accounts for most European production, as meeting the 35% threshold; soyabean and palm-based biodiesel, primary sources for imports, fall below it. Although actual soya and palm biodiesel production may bring about higher emissions savings, this depends critically on the production process: emission savings are often well below 35%. Moreover, the challenge for foreign and domestic producers is likely to increase when indirect changes in land use are taken into account. The European Commission has recently funded several studies of indirect land use changes due to the use of biofuels. One showed that if domestic conventional biofuels were to be used to meet more than half of the 10% renewable fuels target by 2020, emissions from the indirect change in land use would be significant without substantial improvements in agricultural productivity, and would increase considerably as the share of domestically produced conventional biofuels increased (IFPRI, 2010). Advanced biofuels would be needed to reduce these emissions. The Commission has not yet issued requirements relating to the sustainability of the crops grown for biofuels production: the criteria are expected to be promulgated in November 2010. Germany has already placed limits on the origin of biofuels, an initiative which could shape EU policy more widely (FO Lichts, 2010a). The present basic requirements are laid out in several different directives, such as Directive 1998/70, which includes the EU Low Carbon Fuel Standard, the Renewable Energy Directive and the Fuel Quality Directive.

Brazil

Brazil is the world's largest producer of ethanol from sugar cane. Brazil's national ethanol programme, ProAlcool, was launched in response to the oil crises in the 1970s. Lead in gasoline was phased out completely in 1991 and limits on carbon monoxide, unburned hydrocarbons and sulphur emissions were tightened, boosting the attractiveness of biofuels. In 2003, car manufacturers, beginning with Volkswagen, introduced "flex-fuel" vehicles (FFVs), which are capable of running on any combination of ethanol and gasoline. Such vehicles allow consumers to choose the cheapest fuel, whatever the type. These vehicles accounted for 40% of the car fleet in Brazil in 2009, compared with only about 4% in the United States.[3] In Brazil, where ambient temperatures allow for higher blend shares than in the United States and the European Union, ethanol use is partly driven by mandatory ethanol blends and tax reductions for pure ethanol. From July 2007 to February 2010, the mandatory blend of ethanol in ethanol/gasoline blends was 25%. It was then reduced to 20% in an attempt to ease pressure on the sugar market, but was revised back up to 25% in May 2010. An obligation to blend 5% of biodiesel (B-5) into diesel fuels came into effect in January 2010.

Quantifying the value of government support to biofuels

Biofuels are generally not competitive with gasoline and diesel at market prices, so their production and use are encouraged by fiscal measures or other instruments. This year's *Outlook* analyses biofuels support schemes in 20 countries,[4] covering approximately 94% of total global biofuels consumption. The most common forms

---

3. There were some 8.3 million flex-fuel vehicles on the road in the United States in 2009. The number actually switching between the fuels is probably lower, however, as a recent survey in the United States found that 68% of E85 flex-fuel vehicle owners were not aware that they owned a flex-fuel vehicle.

4. United States, Germany, Spain, India, China, United Kingdom, Denmark, Portugal, France, Italy, Netherlands, Australia, Canada, Japan, Ireland, Greece, New Zealand, Austria, Brazil, Poland.

of support are tax credits and tax exemptions, import tariffs on foreign biofuels and blending mandates. Blending mandates have played an increasing role in recent biofuel support policies, many countries adding blending mandates to existing fiscal incentives or entirely replacing fiscal incentives by mandates. Some, but not all, of these measures can be considered subsidies (see Chapter 19).

In order to quantify the monetary value of government support to biofuels in this analysis, the tax advantage to biofuels, relative to the oil-based equivalent fuel, has been multiplied by the volume of biofuels consumed. Where blending mandates exist, tax reductions and biofuels prices were used for quantifying the implicit support through the blending mandate, which, in some cases is carried by the consumer (at least partially). Therefore, the value of what is called government support here represents a monetary value of all government interventions currently in place, irrespective of whether the cost is finally carried by the *government* or the *consumer*.

Our analysis finds that biofuels worldwide receive more financial support than any other renewable technology (see Chapter 9). In 2009, global support for biofuels was almost as high as that for solar photovoltaics (PV) and wind combined, reaching $20 billion; this is compared with about $23 billion for solar PV and wind. Biofuels support increased by 40% in the two years to 2009. The production of ethanol received most of the support, roughly $9 billion in 2007, rising to more than $13 billion in 2009 (Figure 12.5).

Figure 12.5 ● **Value of annual global government support to biofuels by type**

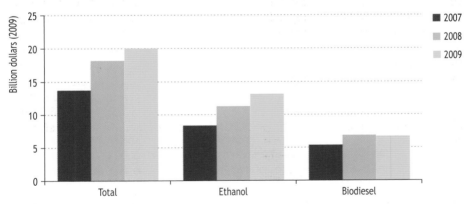

The United States provide the highest level of support to biofuels. In 2009, the value of support for biofuels production – the bulk of total support directed at ethanol – reached $8.1 billion (Table 12.4). This was largely the result of a $0.45 per gallon excise tax credit and a blending mandate. EU support cost $7.9 billion in 2009, of which the largest share is taken by biodiesel in Germany. Support in Brazil, with the third-largest support programme, takes the form of tax credits on pure ethanol and blending mandates. China ranks fourth.

Table 12.4 • Value of government support to biofuels in selected countries
   (billion $)

	2007	2008	2009
United States	4.9	6.6	8.1
Ethanol	*4.6*	*6.2*	*7.7*
Biodiesel	*0.3*	*0.4*	*0.4*
European Union	6.3	8.0	7.9
Ethanol	*1.3*	*2.0*	*2.1*
Biodiesel	*5.0*	*6.0*	*5.8*
Brazil	2.3	2.5	2.6
Ethanol	*2.3*	*2.5*	*2.6*
Biodiesel	*0.0*	*0.0*	*0.1*
China	0.3	0.6	0.5
Ethanol	*0.2*	*0.5*	*0.4*
Biodiesel	*0.1*	*0.1*	*0.1*

Government support to biofuels, as defined here, is not entirely paid for by governments. This is the case, for example, in Germany and the United States. The policy framework introduced in Germany in 2007 is something of a hybrid system, whereby only biofuel production above the level required by the national mandate attracts tax credits. In the case of biodiesel, Germany regularly far exceeded its own blending mandate between 2007 and 2009, with consumption reaching twice the level of the quota; as a result, the German government financed almost 55% of the cumulative $8.6 billion of support to biodiesel during those years, in the form of reduced tax revenues on road-transport fuel sales. The rest was paid by the consumer. For ethanol, the share of the government in total spending was lower, at about 20% between 2007 and 2009, as German ethanol consumption only slightly exceeded the mandate. In the United States, import tariffs on biofuels, in combination with blending mandates, increase the price of ethanol to consumers. Direct US government support takes the form of tax credits.

Brazil is a somewhat special case, as no pure gasoline is available to the consumer. Rather, Brazilian consumers can choose between pure ethanol or gasoline with a 25% ethanol blend (E25). Two types of regulations exist. One is a tax exemption on pure ethanol (hydrated ethanol), which, if compared with the tax on E25, accounted for $800 million of governmental support in 2007, increasing to $950 million in 2008 and $1 250 million in 2009. The rest is the impact of the blending mandate, where the ethanol part is taxed at a lower level than the gasoline part. According to Brazilian government officials, this difference in taxation does not represent a loss in tax revenues as taxes on the gasoline component have been increased to keep revenues constant.

Government support has played an important role in facilitating the growth in biofuels supply in recent years and is likely to continue to do so over the *Outlook* period. To estimate the amount of support that would be required (in monetary terms) in the New Policies Scenario, biofuels prices have been calculated using biofuel conversion costs and efficiencies, and biomass feedstock prices, projected to 2035.[5] These biofuels prices were then compared with gasoline and diesel prices before taxes by region over the projection period, and the increment was multiplied by the amount of biofuels consumed. Where biofuels prices break even with projected fossil-fuel costs over the projection period, such as in Brazil, support is assumed to be phased out. Using this approach, we calculate the average value of annual support for biofuels between 2010 and 2020 at $45 billion, increasing to about $65 billion between 2021 and 2035 (Figure 12.6). Ethanol absorbs most of the support, 60% on average, driven mainly by consumption in the United States. Biodiesel receives 40%, the European Union providing more support for biodiesel than other regions. Cumulatively, the support to biofuels in monetary terms is $1.5 trillion over the projection period.

Figure 12.6 ● Global average annual government support to biofuels in the New Policies Scenario

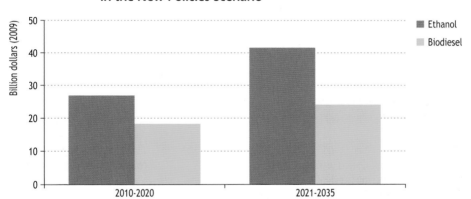

The cost of producing (or importing) biofuels is currently often significantly higher than the cost of imported oil. Consequently, government measures to encourage their production and use typically raise costs to motorists and the economy as a whole. There can be additional costs too, including the impact on food prices of devoting more land to biofuels production in the case of conventional biofuels. But the benefits can be large. These include a reduction in CO_2 emissions (the overall savings vary enormously among the different types of biofuels, technologies and location), benefits to the agriculture sector, especially in developing countries, and the energy-security benefit of reduced imports of oil. All other things being equal, the use of biofuels at the levels of the New Policies Scenario results in a volumetric reduction of oil consumption between 2009 and 2035 sufficient to meet twice the total demand for oil in the OECD in the year 2035.

---

5. See (IEA Bioenergy, 2009) for biofuel conversion costs and efficiencies, and OECD (2010) for the biomass feedstock prices used in this analysis.

Biofuels technologies

Conventional biofuels

Ethanol from sugar and starchy crops, and biodiesel from oilseed crops and animal fat, use well-established and simple conversion technologies. All current commercial biofuels production falls into these two categories. The main non-economic barriers to expanding the use of conventional biofuels are the demand for land and water, and the resulting competition with food and fibre production, as well as the threat to biodiversity. While these factors are potentially the same for any type of conventional biofuel, there are important differences according to the feedstock type and the region of production.

Ethanol

Ethanol is produced through a process of fermentation and distillation from sugar crops, such as sugar cane, sugar beet and sweet sorghum, or starch crops, such as corn, wheat and cassava. The basic production process from both types of crop is similar. But the energy requirement for the conversion of starch-based ethanol is much higher than that of sugar-based ethanol due to the additional process steps involved in converting starches into sugar. Ethanol can be used in blends of up to 10% in conventional spark ignition engines, or in blends of up to 100% in modified engines, although there is debate in many countries as to whether the 10% limit could be increased for newer vehicles. Though the energy content of ethanol is about two-thirds that of gasoline, when mixed with gasoline it has a higher octane rating, improves vehicle performance and can reduce CO_2 emissions.

Biodiesel

Biodiesel is produced from vegetable oil and animal fat through a process known as esterification. Major feedstocks are rapeseed, soyabean, palm and sunflower, but about 11% of the feedstock is estimated to be animal fat and used cooking oil (FO Lichts, 2010b). The production process provides additional co-products, typically bean cake, animal feed and glycerine, which can be used in several industries. Biodiesel can be blended with diesel or used in pure form in compression ignition engines with little or no modification to the engine. Its energy content is only about 90-95% that of diesel, but the overall fuel economy of the two fuels is generally comparable and biodiesel raises the cetane level and improves lubricity.[6] Biodiesel use can reduce emissions of CO_2 and particulate matter from the vehicle, compared with pure diesel, though the overall picture is complex. The use of palm oil as the feedstock is particularly controversial as it can irreversibly damage the environment if not grown sustainably, resulting in very high life-cycle greenhouse-gas emissions.

Advanced biofuels

In the production of conventional biofuels, only the starchy or sugary part of the plant is used for the production of fuel. These components represent a fairly small

---

6. The cetane number is a measurement of the combustion quality of diesel fuel during compression ignition. Lubricity is a measure of the reduction in friction of a lubricant.

percentage of the total plant mass, leaving large quantities of fibrous remains, such as seed husks and stalks. Much current research is focused on innovative processes to use these materials, of which 20% to 45% by weight is cellulose, to create fermentable sugars.

Successful conversion of such materials would make available a much broader range of biomass feedstocks. These include ligno-cellulosic feedstocks, such as wood, and agricultural residues such as straw, as well as perennial "woody" crops. When the conversion process is efficient, use of such residues and crops as feedstock can significantly reduce the area of land needed for growing crops for biofuels production, achieving higher biomass yields per hectare than biomass for many conventional biofuels. Other novel crops are being developed that may offer even higher productivity in the longer term.

Production from cellulose is technologically challenging and the cost of enzymes to break down the cellulose feedstock into fermentable sugars is high. A good deal of progress has been made at the research level in various processes, including biochemical and thermal processes, but no commercial scale conversion facilities have, as yet, been built.

One approach under development is to use a process similar to that used for coal-to-liquid (CTL) and gas-to-liquid (GTL) fuels, *i.e.* gasification, combined with Fischer-Tropsch (FT) synthesis (see Chapter 4). In this method, biomass must first be converted into a syngas through a two-step process involving thermal degradation of the biomass and cleaning of the derived gas. Then, FT synthesis is used to convert the syngas into biofuels. The products are of a similarly high quality to those derived from other fuel-synthesis processes. Biofuel-to-liquid (BTL)-diesel can be used in any given blend in conventional engines without modifications, which could be particularly interesting for the aviation industry. The BTL approach has advantages, such as reliance on non-food biomass.

Though no fully commercial conversion facilities have yet been built, developments in cellulosic ethanol and Fischer-Tropsch biodiesel are expected to drive the penetration of advanced biofuels in the New Policies Scenario and, more importantly, in the 450 Scenario. Demonstration projects have been successfully undertaken, such as DONG Energy's 5 million litres per year (Ml/year) straw-fed cellulosic ethanol plant in Denmark. POET, a large ethanol producer in the United States, developing a 95 Ml/year plant, recently announced it had reduced cellulosic ethanol costs to only $1/gallon ($0.26/litre) higher than corn ethanol costs. Choren, a German company, completed a 17 Ml/year BTL plant in Germany in 2008, but it still has not commenced commercial production (IEA, 2010).

Another interesting concept appears to be sugar-to-biodiesel conversion using yeast fermentation. AMYRIS, a US company, opened a pilot plant in California in 2008. Using bacteria for producing biodiesel from cellulosic materials is another concept under development by a research team of the US Joint BioEnergy Institute and the company LS9. This process might be able to produce a renewable fuel that can use existing distribution facilities.

Algae are now being intensively researched as a potential biofuel feedstock. In addition to their potentially high yields per unit land area, algae can grow in places unsuitable for agriculture, including industrial areas. Thus, their exploitation offers the prospect of a source of biofuel that avoids damage to ecosystems and competition with agriculture associated with other biomass resources. Although many testing and start-up companies are in operation in 12 countries, cost information is scarce. Biofuels from algae are, in any case, still at the research and development stage and face numerous obstacles related to energy and water needs, and productivity.

A successful transition to advanced biofuels will depend on several factors:

- Continuing strong public and private support for research and development, with particular emphasis on developing the links between industry, universities and government.

- Demonstration and pre-commercial testing, to reduce the risks to investors and make participation attractive to financial institutions.

- Development of widely-respected measures of performance, including life-cycle assessment tools to assess the net effects on the energy balance and on greenhouse-gas emissions and the impacts on water and ecosystems.

- Greater understanding of biomass resources through global mapping, in order to identify optimal growing areas and promising non-crop sources and to avoid unsustainable use.

Biofuels emissions

Biofuels are derived from renewable biomass feedstocks, but biofuels are not emission-free on a life-cycle basis. There is keen debate about the level of emissions savings that can be attributed to the use of biofuels and, more generally, to biomass (see the spotlight in Chapter 9).

Greenhouse-gas emissions can occur at any step of the biofuels supply chain.[7] Besides emissions at the combustion stage, greenhouse-gas emissions arise from fossil-energy use in the construction and operation of the biofuels conversion plant. In addition, the cultivation of biomass requires fertilisers, the use of machinery and irrigation, all of which also generate emissions.

To quantify the net greenhouse-gas emission savings relative to petroleum-based fuels, it is necessary to calculate the extent to which total emissions are offset by the uptake of CO_2 from the atmosphere during the growth of the biomass. If appropriate feedstocks and process conditions are chosen, biofuels can offer significant net

---

7. In IEA Statistics, biofuels (and biomass more generally) are not included in the data for CO_2 emissions from fuel combustion. This is because CO_2 emissions from biomass consumption for fuel production are assumed to be offset by CO_2 savings through biomass re-growth. This methodology is in line with the *1996 IPCC Guidelines*. Any departures from this assumption are counted within emissions from land use, land use change and forestry.

greenhouse-gas emissions savings over conventional fossil fuels. This is particularly the case with sugar cane ethanol, as much less energy is required to convert the biomass to ethanol. But variations are large and calculating average emissions savings is complex (Figure 12.7). Expectations are high that advanced biofuels will be produced from ligno-cellulosic biomass and will offer excellent greenhouse-gas savings, using non-food crop feedstocks.

Figure 12.7 ● Ranges of well-to-wheels emission savings relative to gasoline and diesel

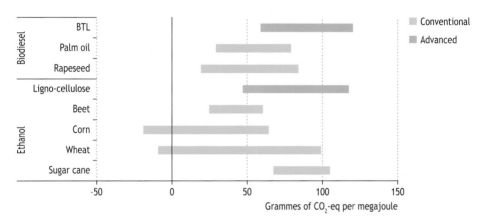

Grammes of CO_2-eq per megajoule

Note: Excludes land-use changes.
Source: Adapted from IEA (2009b).

The greenhouse-gas benefits of biofuels use can be reduced or even become negative if emissions arising from the associated change of land use are significant.[8] Such land use changes can be *direct*, as when feedstocks are grown on land that was previously forest or *indirect*, as when feedstock production for biofuels displaces other types of agricultural production undertaken on land elsewhere. So, for example, increasing the area dedicated to sugar cane or corn production for biofuels could displace cattle or soya production, which could lead to the conversion of forest land elsewhere for grazing cattle or producing soya. Such effects can be avoided when waste and residues are used as feedstock, or the feedstock is produced so as to give a higher yield per hectare or on land that is not otherwise used intensively to produce food or other products.

Using land for biofuels production that was previously covered with carbon-rich forest or where the soil carbon content is high can release considerable amounts of greenhouse gases, and even lead to a "carbon debt". In the worst cases, this debt could take hundreds or even thousands of years to recover via the savings in emissions

---

8. See, for example, UNEP (2009) and IEA Bioenergy (2009).

by substituting biofuels for fossil fuels. However, establishing perennial energy plantations on land previously used to produce annual crops or on impoverished or under-productive lands can lead to improvements in carbon stocks and enhance the overall greenhouse-gas emissions savings.

Indirect land-use changes are difficult to identify and model explicitly in greenhouse-gas balances. Several approaches are being developed to allow for such indirect effects. In some certification schemes (for example, the California Low-Carbon Fuel Standard), a specific greenhouse-gas penalty is added into calculations of overall balances to offset such effects; however, reaching consensus on what the penalties should be is difficult. In Brazil, a zoning approach has been developed that puts constraints on the areas in which sugar cane production can be expanded and the extent to which livestock intensity may be increased to compensate for displaced cattle production.

Biofuels costs

Outside of Brazil, biofuels generally cost much more to produce than conventional gasoline or diesel. Further cost reductions are achievable, even using existing technologies, through upscaling and improving efficiency and feedstock logistics (Figure 12.8). Oil-price variations, however, have numerous simultaneous effects. While biofuels could potentially become competitive at certain oil price levels, increasing use of biofuels lowers oil demand and, therefore, oil prices. In addition, increasing competition for fossil-fuel substitutes might create upward pressure on feedstock prices. Similarly, competition for land uses could increase feedstock prices, especially in the case of current biofuels, which use food crops.

Advanced biofuels, like BTL biodiesel or ligno-cellulosic ethanol, are currently not competitive with conventional fuels and are mostly in the demonstration phase. They are expected to be commercialised by 2020 in the New Policies Scenario. In the case of some advanced biodiesel, it may also be possible to produce better quality biofuels, with more favourable performance characteristics. Higher quality biofuels could have a wider range of uses, in particular for aircraft.

Producing advanced biofuels generally involves higher capital expenditures than current biofuels. However, ligno-cellulosic feedstocks, as used for the production of advanced biofuels, are generally cheaper than the feedstocks used for conventional biofuels, highlighting the potential for advanced biofuels to become cost-competitive in the long-run. Whether or not this comes about, however, depends on the interplay between a complex series of factors: the judgements reached on the sustainability of biofuels and the consequences for biofuels deployment; oil price levels and their impact on biomass feedstock prices, especially for food crops; land-use competition and its implications for feedstock prices; and the extent of cost reductions that can be achieved through upscaling and technological learning in the market place. Some biofuels could become competitive under the oil price assumptions in this *Outlook*.

Figure 12.8 ● Indicative cost ranges of selected biofuels versus gasoline and diesel prices

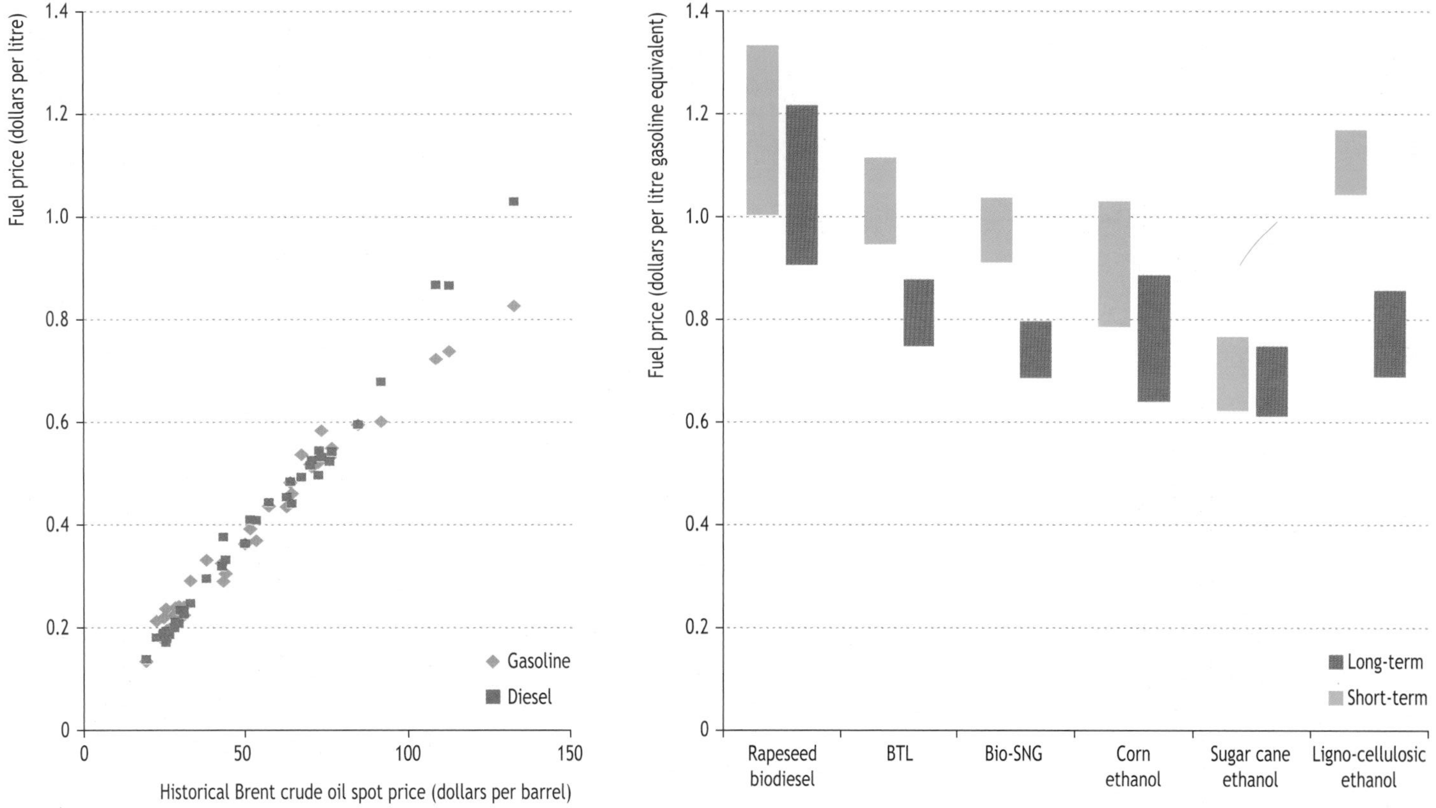

Source: Adapted from the Mobility Model of the IEA.

Note: Gasoline and diesel prices are quarterly from January 2000 to April 2010. Costs exclude subsidies. Feedstock costs are averages and can vary significantly by region.

12

ACHIEVING THE 450 SCENARIO AFTER COPENHAGEN

PREFACE

The climate change analysis in this year's *Outlook*, discussed in Part C, details the consequences of the United Nations Framework Convention on Climate Change conference in Copenhagen in December 2009 and its implications for achieving the 450 Scenario. In this scenario, government policies are assumed to be introduced that put the world on track for long-term stabilisation of the atmospheric concentration of greenhouse gases at 450 parts per million (ppm) of CO_2-equivalent. This is the level that would give us a reasonable chance of limiting the increase in global average temperature to 2° Celsius — the goal set in the Copenhagen Accord.

Chapter 13 sets out the policies and measures assumed in 450 Scenario. They are consistent with the national commitments that have so far been announced under the Accord, albeit on the assumption that they are implemented fully. The chapter also analyses the additional costs associated with achieving this scenario and the benefits that would accrue.

Chapter 14 looks at the implications of the 450 Scenario for the energy sector in detail, sector by sector, including the additional spending on low-carbon energy technologies necessary to achieve this transformation.

Chapter 15 quantifies the impact on oil markets of the policies assumed to be adopted in the 450 Scenario, including the outlook for demand, oil prices, production, trade and investment.

The 450 Scenario presented here is just one possible way of achieving emissions compatible with the climate goal. The trajectory of emissions assumed in this scenario is constrained by the outcome of the Copenhagen meeting and by the assumptions we have made in interpreting the national commitments that have been made. It is entirely possible — indeed, likely — that climate policy will change over the coming years. The scenario presented in the following pages is indicative of the level of action that would be needed globally to put the world on a more sustainable footing given the current policy context.

CHAPTER 13

ENERGY AND THE ULTIMATE CLIMATE CHANGE TARGET

How do we get there now?

H I G H L I G H T S

- Even in the most environmentally-ambitious interpretation of the Copenhagen Accord, as assumed in the 450 Scenario, energy-related CO_2 emissions reach 31.9 Gt in 2020 — a cumulative 17.5 Gt higher from 2008 to 2020 than in the trajectory estimated in *WEO-2009*, which assumed more intensive action earlier in the period. This means that to limit energy-related emissions to 21.7 Gt in 2035 dramatic emissions cuts are needed after 2020, involving a near-doubling of the annual average CO_2 intensity improvements achieved in the earlier period.

- Implementation of the Accord, however, could turn out to be less ambitious than we assume. The uncertainty surrounding the interpretation of non-Annex I country pledges could easily offset the maximum 3.1 Gt of reductions expected from Annex I countries (also uncertain). If this were the case, the 450 Scenario would likely be out of reach.

- Emissions savings in the 450 Scenario relative to the Current Policies Scenario are 3.5 Gt in 2020 and 20.9 Gt in 2035, or a 49% reduction. Just ten actions across five regions — the United States, the European Union, Japan, China and India — account for around half the emissions reductions. China alone accounts for 35% of abatement in 2035. While support for renewables and pricing of CO_2 in the power and industrial sectors are at the heart of emissions reductions in OECD countries (and China in the longer term), the phase-out of fossil-fuel subsidies is a crucial pillar of mitigation in the Middle East.

- Oil demand peaks around 2020 at 88 mb/d, and declines to 81 mb/d in 2035 in the 450 Scenario. Coal demand peaks before 2020, returning to 2003 levels by 2035. Gas is the least affected of the fossil fuels, increasing by 15% relative to today's level by 2035. Nuclear power and renewables make significant inroads in the energy mix, doubling their current share to 38% in 2035.

- To achieve the 450 Scenario, additional spending in the period 2010-2035 amounts to $18 trillion compared with the Current Policies Scenario, and $13.5 trillion compared with the New Policies Scenario. Cautious action before 2020, and the faster, deeper cuts required after 2020 as a result mean that achieving this year's 450 Scenario requires $1 trillion more spending than last year's 450 Scenario between 2010 and 2030. Global GDP would be reduced in 2030 by 1.9%, again more than the corresponding estimate made last year. Even on an ambitious interpretation, the targets and actions announced at Copenhagen do not represent the most efficient first steps towards a sustainable energy future.

Introduction

The Copenhagen Accord[1] sets a goal of limiting the long-term average increase in the global temperature to $2°C$ above pre-industrial levels. This is widely acknowledged to mean that the concentration of greenhouse gases in the atmosphere must be stabilised at a level no higher than 450 parts per million of carbon dioxide equivalent (ppm CO_2-eq).[2] This chapter and Chapters 14 and 15 examine the implications for the energy sector of achieving that target — the 450 Scenario.

This scenario differs from that presented in the *World Energy Outlook 2009*. The target last year was the same; but the path to it, though plausible, depended on early and vigorous action. We suggested the commitments which might be made at the 15th Conference of Parties to the United Nations Framework Convention on Climate Change (UNFCCC) in December 2009 in order to set the world on that path.

The situation this year demands a new starting point. On the one hand, the baseline has shifted, as the global economy is growing more strongly than was expected last year, meaning energy demand projections are higher; on the other, there are new specific national pledges, made at Copenhagen or since, but they fall short of what is necessary to follow the trajectory outlined last year. So the 450 Scenario set out here starts from these new realities: the goal is unchanged and determines what must be done, but the trajectory is no longer as efficient. The cost of that departure from the more efficient path can be loosely called the "cost of Copenhagen"; and the calculation of that cost forms one part of this chapter, arriving at the formidable figure of $1 trillion. More positively, the chapter shows in some detail where we need to go from here.

There is much scope for interpretation of the new pledges which nations have made. The New Policies Scenario, elaborated in Parts A & B of this *Outlook* adopts a relatively cautious interpretation. By contrast, the 450 Scenario discussed here assumes that countries will interpret their commitments ambitiously, taking more vigorous action between now and 2020 (see Table 13.1). Nonetheless, the achievement by 2020 leaves more to be accomplished after that date than was envisaged last year. The consequence is that much more demanding commitments are necessary in relation to the period 2020-2035. The carbon intensity of the energy sector needs to be reduced between 2020 and 2035 at a rate of 5.3% per year — four times the rate achieved between 1990 and 2008. The implications of this are explored below.

Scenarios require a foundation of assumptions. Those we have chosen for the 450 Scenario are already ambitious for the period up to 2020, as an interpretation of the commitments so far made. That does not mean that we believe no additional commitments relating to that period are possible. The Copenhagen Accord is due to be reviewed in 2015 and that review could result in a global decision to do more without delay. That would allow a more efficient trajectory to the ultimate goal to be followed, at lower cost. We have no basis now to assume significant new commitments in that

---

1. The Copenhagen Accord was the product of the 15th Conference of the Parties to the United Nations Framework Convention on Climate Change, December 2009. As of September 2010, 85 of the countries which have associated themselves with the Accord, accounting for 80% of global greenhouse-gas emissions, have registered emission reduction targets or commitments as to the actions they will take by 2020 (see Chapter 1).

2. See www.worldenergyoutlook.org for a summary of recent scientific findings that reinforce this conclusion.

timescale; but if the results we present here persuade governments to improve on the present sub-optimal path, the assumptions we have felt constrained to adopt for the 450 Scenario will have served a useful purpose.

The 450 trajectory in the new global context

The trajectory that might now be followed by world energy-related carbon-dioxide (CO_2) emissions, en route to long-term atmospheric stabilisation of greenhouse gases at 450 ppm CO_2-eq, has to be determined in two stages. First, there is the path to 2020, which we have taken as being set by the outcome of the Copenhagen negotiations. Second, the path beyond 2020 must be chosen, limited by the need to bring global annual emissions to an early peak so that, allowing for the slow dispersal of accumulated greenhouse gases from the atmosphere, the concentration of emissions can be brought down to the required level in a reasonable timescale. The trajectory of the 450 Scenario has been determined in this way.

The first difficulty is to interpret the commitments associated with the Copenhagen Accord (Box 13.1). The New Policies Scenario already goes beyond the Current Policies Scenario by making allowance for actions yet to be taken in pursuit of these commitments, albeit on a relatively cautious basis. The 450 Scenario, by contrast, interprets these commitments as rigorously as possible, and assumes that they are implemented with full vigour (see Table 1.1 in Chapter 1). For example, we have taken the upper figure of the 20-30% range of emissions reductions pledged by the European Union. The 450 Scenario also assumes the rapid implementation of the removal of fossil-fuel subsidies agreed by the G20. Beyond 2020, our analysis assumes that all countries contribute to the necessary action. Very stringent emissions targets are set in the OECD+ countries and Other Major Economies, with Other Countries selling emissions reduction credits in international carbon markets and receiving direct financing for mitigation.[3]

Box 13.1 ● **Uncertainties around the interpretation of Copenhagen Accord Pledges**

The Copenhagen Accord sets the goal of limiting global average temperature increase to 2°C, but it does not set out a path to reach this goal beyond 2020, and leaves many questions unresolved. Although pledges for 2020 have been made by 85 countries, many of these lack transparency, and there remain very substantial uncertainties about the interpretation of some of these targets in terms of their impact on global greenhouse-gas emissions. A number of countries, both Annex I and non-Annex I, have entered ranges rather than specific pledges. There are also a number of open questions relating to the provisions of the Accord and the future evolution of the Kyoto Protocol mechanisms, including the Clean Development Mechanism (CDM) and banking for future use of Assigned Amounts Units (AAUs).

---

3. See Annex C for regional definitions. The Copenhagen Accord envisages a "Green Climate Fund" to support actions taken in developing countries for adaptation and mitigation purposes.

Some of the issues are expected to be discussed and possibly settled at the Conference of the Parties in Cancun in December 2010. Of course, progress is uncertain. Even at the high end of the Annex I pledges, which assumes that all Annex I countries implement the most ambitious version of the pledges they have made in the context of the Copenhagen Accord — the uncertainty around non-Annex I countries' energy-related emissions in 2020 exceeds the maximum energy-related abatement attributable to the pledges of the Annex I countries in total (Figure 13.1). Of the total uncertainty around the non-Annex I countries' figures of 3.2 gigatonnes (Gt CO_2), we estimate uncertainty for Brazil may be over 350 million tonnes (Mt) CO_2, related to uncertainty about the baseline. Uncertainty about the Chinese pledge is estimated to be at least 2 Gt CO_2, while uncertainty surrounding the Indian pledge amounts to over 600 Mt CO_2, on the basis of different gross domestic product (GDP) estimates. All of these figures could be higher, depending on the assumptions made in calculating them. It is also unclear what level of emissions will result from the targets announced by Annex I countries, as many have entered ranges rather than specific targets. Although this uncertainty — at some 700 Mt CO_2 for all Annex I countries together — is less than that for non-Annex I countries, and is quantifiable and independent of assumptions, in that the targets are expressed against fixed baselines, it nonetheless adds to the difficulty of saying with any certainty what is the absolute emissions level associated with the Copenhagen Accord, and results in estimated total uncertainty of 3.9 Gt.

- **Uncertain baselines:** Where pledges are defined as deviations from a business-as-usual (BAU) baseline without a clear, or with more than one official projection, the absolute level of emissions implied by fulfilment of the pledge is not clear. For instance, BAU projections for energy-related CO_2 emissions from government sources in Brazil vary between 550 Mt CO_2 in 2019 to 900 Mt CO_2 in 2020. Similar problems exist for other countries.

- **Uncertain components of finance:** The Annex I countries pledged that they would "mobilise" finance of $100 billion per year by 2020 to fund mitigation and adaptation in developing countries. However, it is not clear from the Accord how much of this finance will be in the form of direct financial transfers to governments, and how much will come through carbon finance mechanisms, nor what the split might be between mitigation and adaptation. This is particularly relevant where pledges of action by non-Annex I countries are conditional on finance, as it is possible that the two conceptions of finance do not match.

- **Uncertainty around carbon market regulation:** The form that carbon markets and, therefore, carbon finance will take in the future remains very uncertain. No extension of the CDM has been agreed, nor has any linking between markets in Annex I countries. The accounting rules for offset credits generated in countries with targets that are not expressed in terms of absolute limits on emissions remain unelaborated, leading to the possibility of double-counting of reductions towards Annex I targets (in Mt CO_2 reductions) and non-Annex I targets (in, for example, carbon-intensity reductions).

- **Land use, land-use change and forestry (LULUCF):** There remains, as has historically been the case, uncertainty regarding not only the interpretation of pledges of abatement of emissions from land use, land-use change and forestry and what accounting method should be used for these emissions, but also around measurement of these emissions in the first place.

Figure 13.1 ● **Energy-related CO_2 emissions in Annex I and non-Annex I countries under the Copenhagen Accord in 2020[4]**

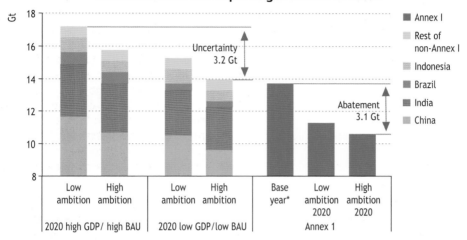

*According to country, base year is either 1990 or 2005.

The emissions pledges currently on the table in the climate negotiations — even in their most ambitious form — lead to higher energy-related CO_2 emissions in 2020 than those estimated in the *WEO-2009* analysis. Cumulative emissions to 2020 are 17.5 Gt higher and annual emissions reach 31.9 Gt CO_2 in 2020, compared with 30.7 Gt in *WEO-2009*. Therefore, to put the world on track for a 450 ppm stabilisation, action after 2020 will have to be much more stringent in order to keep total cumulative emissions to the same level by 2050, and the path that emissions follow over time in this year's scenario is quite different from the trajectory of *WEO-2009* (Figure 13.2).

Emissions in the Current Policies Scenario are still higher, reaching 35.4 Gt in 2020, and continuing to increase across the period, reaching a long-term level consistent with a temperature rise in excess of 6°C. Half of the increase in CO_2 emissions in 2020 since *WEO-2009* is due to a stronger than expected economic performance in major non-OECD countries, which meant that the fall in emissions in 2009 due to the financial crisis was limited to only 1%, rather than the nearly 3% that our earlier projections indicated.[5] Another 500 Mt of energy-related CO_2 emissions in 2020 have been added to our Current Policies Scenario due to the improved economic outlook up to 2020, compared with last year. The

4. Details of assumptions underpinning this figure can be found at www.worldenergyoutlook.org.
5. In fact, if China is excluded, global emissions did fall by nearly 3%, but unexpected growth in coal consumption in China offset most of this decline.

few newly-enacted policies now included in the Current Policies Scenario are insufficient to offset fully the increase in emissions due to the more dynamic economic prospect. These changes are taken into account in projecting the level of emissions in 2020 in our other two scenarios, but the levels reached are largely determined by the two different interpretations we have made of the intensity of the targets and measures associated with the Copenhagen Accord. The higher underlying level of emissions does, of course, have an effect on the choice of measures needed in the subsequent period to achieve these scenarios, and on the cost involved. This is further discussed below and in Chapter 14.

Figure 13.2 ● World energy-related CO₂ emissions by scenario

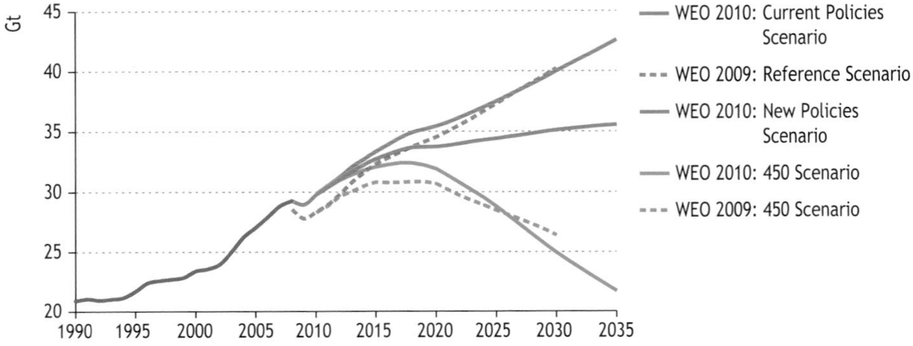

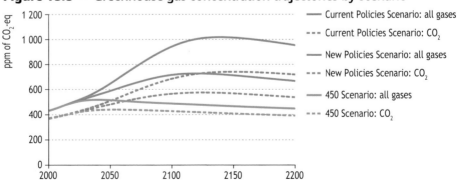

From the trajectories and emissions figures, it can be seen that the New Policies Scenario, while an improvement in environmental terms over the Current Policies Scenario, nonetheless puts us on a long-term path that is consistent with the atmospheric concentration of carbon dioxide equivalent stabilising at around 650 parts per million (Figure 13.3). According to the Intergovernmental Panel on Climate Change (IPCC), this level of concentration is likely to lead to a temperature increase of above 3.5°C (IPCC, 2007).[6]

Figure 13.3 ● Greenhouse-gas concentration trajectories by scenario

Source: IEA analysis using the MAGICC (version 5.3v2) and ENV-Linkages models.

---

6. More recent studies suggest that the IPCC estimates may be too optimistic. See for example Meinshausen *et al.* (2009).

Assumptions and methodology

The 450 analysis assumes that the pledges made in Copenhagen by Annex I countries, alongside the other abatement commitments of OECD+ countries,[7] are met using emission caps for the power and industry sector and carbon pricing. Different cap-and-trade systems in the OECD+ countries converge into a single system by 2020.[8] We assume a rule limiting the purchase of carbon offset credits by OECD+ countries to an amount no greater than one-third of their abatement commitment. This would permit some 940 Mt of abatement to be financed in Other Major Economies and Other Countries using an international offset mechanism, such as the CDM or its successor, at an estimated cost of $28 billion to the purchasing countries. It will be particularly important to ensure that mechanisms exist to prevent offset credits being counted towards the targets of more than one country; this analysis makes the simplifying assumption that there is no double counting. We have also assumed that there is no banking of unused emissions allowances from earlier periods for later use. To achieve the overall emissions target in Annex I countries in 2020, measures are also taken in the transport sector (stringent fuel-economy standards and incentives for biofuels) and in the buildings sector (implementation of energy standards and subsidies for renewables), reflecting the current political debate in each country. Non-Annex I countries take their mitigation action partly on the basis of co-financing, and are assumed to reach their stated goals. Domestic action in Other Major Economies and Other Countries, taking account of the various announced targets,[9] amounts to 840 Mt of abatement in 2020, compared with the Current Policies Scenario. Direct financial transfers, either bilaterally between countries or through a multilateral funding mechanism, are assumed to secure a further 420 Mt of abatement in these countries in 2020. Achieving this abatement requires some $250 billion in investment between 2010 and 2020. Annual expenditure increases across the period, reaching around $46 billion in 2020. If the cost to developed countries of financing this 420 Mt of abatement in developing countries is taken to be equivalent to the purchase of offset credits (that is, calculated based on the prevailing price of CO_2 and the amount of abatement achieved), it would add around $13 billion in direct transfers to the expenditure of $28 billion on offsets in 2020.

It is not clear whether the purchase of offset credits as well as direct transfers will be regarded as falling within the scope of the $100 billion of finance pledged by developed countries, nor what the split of expenditure might be between mitigation and adaptation. It is also unclear whether direct financial transfers will cover only

---

7. Annex C of this *WEO* contains regional definitions. Annex I countries and OECD+ countries broadly refer to the same group, with some exceptions. Annex I countries not in OECD+ are Belarus, Croatia, Monaco, Russia and Ukraine. OECD+ countries not in Annex I are Cyprus, Korea, Malta and Mexico.

8. While this may be seen as a bold assumption given the political context in late 2010, it allows us to model the energy sector without having to make specific assumptions about the most politically likely way in which carbon will be limited in countries which are now looking less likely to introduce carbon markets. While the same abatement could be achieved by other means, without a single linked carbon market, it is likely that costs would be higher.

9. Many non-Annex I countries announced that they would take actions to reduce emissions, but not in terms of quantative or intensity improvement targets. These are available at http://unfccc.int/home/items/5265.php.

13

marginal abatement costs – that is, be equivalent to purchase of offset credits in their calculation – or whether transfers will cover all investment over time to achieve that abatement. Since this is a matter for international negotiations, we have not taken a position on these questions.

The analytical framework applied to the period after 2020 assumes that the global community adopts a plausible combination of policy instruments to achieve the trajectory leading to a long-term concentration of greenhouse gases in the atmosphere no higher than 450 ppm CO_2-eq. These include: cap-and-trade systems; international agreements with sectoral targets for the iron, steel and cement industries; international agreements setting fuel-economy standards for passenger light-duty vehicles (PDLVs), aviation and shipping; and national policies and measures, such as building efficiency standards, labelling of appliances etc. Though the measures need to be applied more stringently, this policy framework is the same as that used for last year's analysis.[10] Policies and targets by region can be seen below (Table 13.1). Greater detail as to the specific policy assumptions by sector used for this year's analysis can be found in Annex B.

In the 450 Scenario, prices per tonne of CO_2 in OECD+ reach $45 in 2020, and climb to $120 in 2035. The carbon markets of the OECD+ and Other Major Economies are not directly linked, but both markets are assumed to allow access to offsets in other countries. This assumption is made to avoid a price slump in the OECD+, which would be a risk if linkage were to take place at very different price levels. By 2035 emissions in OECD+ countries are just over half their 1990 level, the price of CO_2 resulting in 90% of electricity generation coming from low-carbon technologies and ensuring widespread deployment of carbon capture and storage (CCS) in industry. Prices of CO_2 in Other Major Economies rise more steeply than in OECD+ countries, increasing from near zero in 2020 to $90 per tonne of CO_2 in only 15 years. After 2020, mitigation costs are also higher for other sectors than estimated last year.

In order to inform the 450 Scenario, we use a carbon-flow sub-model. It allows quantification of international emission trading and financing under different assumptions, estimating the price of permits, the volume and value of primary market trading, and the overall cost of abatement. The model uses country- and sector-specific marginal abatement curves derived from the World Energy Model. These are summed for all prices to build a global abatement curve. The global emissions level in the 450 Scenario determines the international equilibrium price for credits along this supply curve, and trade can be determined depending on a country's marginal abatement costs – a country with costs that are higher than the market price will purchase credits from those with costs below the market price. Subject to the constraints imposed on the model, such as a requirement to undertake a proportion of abatement domestically, marginal abatement costs are equalised, allowing the global abatement target to be met at minimum cost.

---

10. See www.worldenergyoutlook.org for a full description of the methodology used.

Table 13.1 ● Principal policy assumptions in the 450 Scenario by region

	To 2020	After 2020
OECD+	1. Implementation of a domestic carbon market from 2012. 2. A single OECD carbon market is created by 2020. 3. One-third of OECD emissions reductions (870 Mt from the baseline) are met via an international offset mechanism. 4. $100 billion in financing flows from the OECD to non-OECD countries by 2020.	• OECD-wide emissions trading scheme with cap on energy-related emissions of 5.8 Gt CO_2 in 2035. • Global sectoral agreements for light-duty vehicles, cement, iron and steel, aviation and shipping. • Buildings efficiency standards. • Sectoral target for passenger on-road car emissions of 75 g CO_2/km by 2035.
United States	17% reduction in GHG emissions compared with 2005.	
Japan	25% reduction in GHG emissions compared with 1990.	
European Union	30% reduction in GHG emissions compared with 1990.	
Australia	25% reduction in GHG emissions compared with 2000.	
New Zealand	20% reduction in GHG emissions compared with 1990.	
Mexico	30% reduction in GHG emissions compared with business-as-usual.	
Korea	30% reduction in GHG emissions compared with business-as-usual.	
Other Major Economies (OME) and Other Countries (OC)	Recipients of finance through an international offset mechanism and/or direct financial transfers to support domestic mitigation action, either through a multilateral mechanism or bilaterally.	• Emissions trading scheme introduced 2021, with emissions cap of 8.6 Gt CO_2 for 2035 for OME. • Domestic emissions trading schemes, sectoral agreements and international trading for cement, iron, steel, aviation and shipping. • Buildings efficiency standards. • Sectoral targets for on-road passenger car emissions of 85 g CO_2/km (OME) and 110 g CO_2/km (OC) by 2035 • Removal of fossil fuel subsidies (except in Middle East, where fossil fuel subsidisation rate is reduced to 20% by 2035).
Russia	25% reduction in GHG emissions compared with 1990.	
China	45% reduction in CO_2 intensity compared with 2005.	
India	25% reduction in CO_2 intensity compared with 2005.	
Brazil	39% reduction in GHG emissions compared with business-as-usual.	
Indonesia	41% reduction in GHG emissions compared with business-as-usual.	
South Africa	34% reduction in GHG emissions compared with business-as-usual.	

Note: GHG = greenhouse gas.

Total greenhouse-gas emissions and their energy-related component

All gases

In the 450 Scenario, greenhouse-gas emissions from all sources reach 46.2 Gt CO_2-eq by 2010, remain broadly flat for the next ten years, and then begin to fall rapidly, reaching a total of 21.4 Gt CO_2-eq in 2050, 40% lower than 1990 levels.[11] This trend is in sharp contrast with the Current Policies Scenario, where global emissions reach 71 Gt CO_2-eq in 2050. Emissions in the New Policies Scenario stabilise at around 50 Gt CO_2-eq, more than twice as high as in the 450 Scenario in 2050 (Figure 13.4).

Figure 13.4 ● **World anthropogenic greenhouse-gas emissions by type in the 450 Scenario**

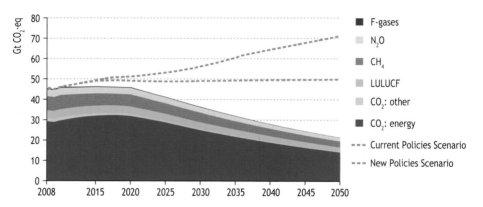

Note: F-gases include hydrofluorocarbons (HFCs), perfluorocarbons (PFCs) and sulphur hexafluoride (SF_6) from several sectors, mainly industry.

Sources: IEA-OECD analysis using MAGICC (version 5.3v2) and OECD Env-Linkages models.

In 2050, abatement of energy-related CO_2 emissions makes up 72% of the total reduction in emissions compared with the Current Policies Scenario. Emissions from methane (CH_4), nitrous oxide (N_2O) and F-gases peak before energy-related CO_2 emissions, due to the fact that the emission of these gases can readily be cut, at low cost, early in·the period. This means that their share of total emissions falls to 20% by 2020 but, as cheap abatement options for these gases are

---

11. The OECD ENV-Linkages model has been used to estimate the greenhouse-gas emissions trajectory compatible with the long-term target of stabilising the atmospheric concentration of greenhouse gases at 450 ppm CO_2-eq. The University Corporation for Atmospheric Research Model for the Assessment of Greenhouse-gas Induced Climate Change (MAGICC version 5.3v2) was used to confirm this result.

exhausted, their share in total greenhouse gases increases slightly and is just over 22% in 2050.

In the Copenhagen meeting and beyond, countries have also expressed strong interest in cutting emissions from forestry. The most ambitious target in the Copenhagen Accord could imply emissions from land use, land use change and forestry (LULUCF) peaking this year (2010). Most of the reduction in emissions would take place in Indonesia and Brazil, with some abatement also taking place in African countries. These emissions are exogenous to the ENV-Linkages model, and are assumed to halve between the beginning of the period and 2050, when they reach 2 Gt.

The atmospheric concentration of greenhouse gases in the 450 Scenario follows an overshoot trajectory — that is to say, it reaches a peak at some 520 ppm CO_2-eq around 2040 before falling back to 450 ppm CO_2-eq by around 2150. Although targeting a 450 ppm concentration is often treated as equivalent to a 2°C target, it is important to be clear that long-term stabilisation at 450 ppm by no means guarantees that the temperature increase will be limited to 2°C. There remains very substantial uncertainty around the sensitivity of the climate to greenhouse-gas emissions, as well as around the interplay of different factors and possible feedback effects. The IPCC 4th Assessment report (2007) pointed to 2°C as the mid-point of warming likely to be associated with stabilisation at 450 ppm, but more recent research suggests that the chances of limiting the temperature increase to 2°C at 450 ppm may be much lower than this. In addition, overshoot trajectories lead to much greater risk. If the temperature increases by more than 2°C in the period before concentrations fall back, there is a risk that the higher temperature reached could set in motion feedback loops. One example is melting permafrost, which leads to emissions of methane, and in turn, to a higher atmospheric concentration and greater warming. These risks and uncertainties strengthen the argument for taking even stronger action to curb emissions early in the period.

13

Energy-related CO_2 emissions

Energy-related CO_2 emissions continue to form the greatest part of global anthropogenic greenhouse gases emitted in the 450 Scenario, reaching 31.9 Gt CO_2 in 2020, or about 70% of total emissions. In order to move from this point to a trajectory that is compatible with long-term stabilisation of the atmospheric concentration at 450 ppm CO_2-eq, energy-related emissions need to fall to 21.7 Gt CO_2 by 2035. This is 3.5 Gt CO_2 lower than in the Current Policies Scenario in 2020, and 20.9 Gt CO_2 lower in 2035. Global emissions decline by an average of 680 Mt per year from 2020 to 2035. Emissions from OECD countries decline steadily from before 2015 and are 55% lower than 2005 levels in 2035 (or 48% lower than 1990). Emissions in non-OECD countries peak in 2018 at 19.8 Gt and decline thereafter, driven by large reductions in China (Figure 13.5). Nonetheless China is still the largest emitter in 2035, at 5.2 Gt, followed by India and the United States, each at 2.3 Gt, and the European Union at 1.8 Gt.

Figure 13.5 • Energy-related CO$_2$ emissions by region in the 450 Scenario

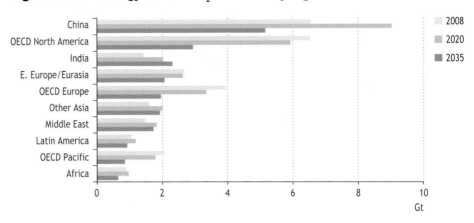

Global average energy-related CO$_2$ per-capita emissions decline gradually over the projection period, masking divergent underlying trends (Figure 13.6). Per-capita emissions in the United States, 18 tonnes CO$_2$ per person in 2008, decline to 15 tonnes per capita in 2020 and then begin to fall more steeply, to 6 tonnes CO$_2$ per person in 2035, an extremely dramatic and rapid change. China, meanwhile, sees its per-capita emissions exceed those of the European Union around 2020, as EU per-capita emissions fall. By this time, however, Chinese per-capita emissions have already peaked and they begin to fall back at a similar rate to those in the European Union across the second half of the projection period, just edging below the EU level by the end of the period. Per-capita emissions in India remain comparatively low across the period, though increasing slowly. By 2035, India is still emitting only 1.6 tonnes CO$_2$ per person.

Figure 13.6 • Energy-related CO$_2$ emissions per capita by region
in the 450 Scenario

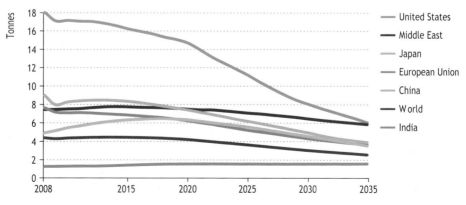

To decarbonise the energy economy to this extent, a doubling of the carbon intensity improvements achieved from 1990-2008 is required from 2008 to 2020 and twice that is required thereafter. The contrast with the Current Policies Scenario is notable, where, in the absence of any compelling force for change, the improvement post 2020 is barely higher than that seen between 1990 and 2008 (Figure 13.7). To put the improvements required in the 450 Scenario into perspective, the oil price shock in 1973 resulted in a 2.5% improvement in energy intensity between 1973 and 1974 — more than twice this improvement is needed in the 450 Scenario, sustained in each and every year from 2020 to 2035.

Figure 13.7 ● **Average annual change in CO$_2$ intensity by scenario**

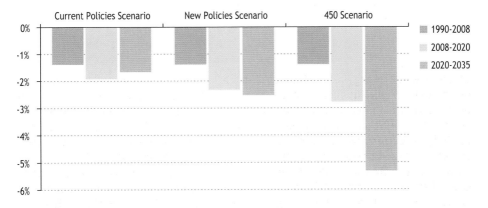

Where and how are the savings to be made?

Abatement by region

As we have seen, emissions in the 450 Scenario reach 31.9 Gt in 2020 and decline to 21.7 Gt in 2035, 20.9 Gt or 49% lower than in the Current Policies Scenario. To achieve those levels of abatement the engagement of all countries to impose stringent abatement measures is necessary as of 2020. In the OECD+ group, emissions are expected to rebound as these economies recover from the financial crisis, but, in the 450 Scenario, to fall steadily from before 2015. By 2035 OECD+ emissions collectively reach 5.9 Gt, just over half 1990 levels, a level similar to emissions from the United States today. Though emissions in Other Major Economies continue to grow until around 2020, they fall to current levels by 2030 and decline to 8.6 Gt by 2035. In Other Countries, growth in emissions continues through 2023, with a peak at 6.4 Gt and a slight decrease thereafter. In 2035, emissions are 6.1 Gt, 18% higher than in 2008 and 75% higher than in 1990.

Abatement in just six countries/regions, accounts for the bulk of the global CO$_2$ reductions, the share of these countries in the abatement, relative to the Current

Policies Scenario, growing from 66% in 2020 to 74% in 2035 (Figure 13.8). China's abatement is greater than that in the whole of the OECD+, at 7.4 Gt CO_2 or 35% of total abatement, compared with the OECD+'s 6.4 Gt CO_2. By contrast, India sees growth in emissions from 2020 to 2035, even in the 450 Scenario, although this growth is lower than in the Current Policies Scenario; and absolute emissions are 43% lower by 2035 in the 450 Scenario than in the Current Policies Scenario, but still nearly four times the level of 1990.

S P O T L I G H T

What role for phasing out fossil-fuel subsidies in climate change mitigation?

In 2009, the G20 agreed to "rationalize and phase out over the medium term inefficient fossil fuel subsidies that encourage wasteful consumption" and called upon the IEA, World Bank, OPEC and OECD to work together to produce a joint report analysing the scope of fossil fuel consumption subsidies and advising how the initiative should be taken forward. The results of the analysis can be seen in Chapters 19 and 20 of *WEO-2010*.

Fossil-fuel subsidies are estimated to have amounted to about $312 billion in 2009. Subsidised energy prices dampen the incentive for consumers to use energy efficiently, resulting in higher energy consumption and energy-related CO_2 emissions than would emerge if consumers were to pay for the full cost of energy. While in the Current Policies Scenario we assume that only countries with already, enacted policies will phase out subsidies, such as Russia or Indonesia, in the 450 Scenario, we assume much more ambitious action. By 2035, the only subsidies that are assumed to remain are in the Middle East, where the average subsidisation rate falls from current rates, in many cases well above 70%, depending on fuel and sector, to 20%.

The removal and reduction of subsidies in the 450 Scenario accounts for 1.4 Gt of CO_2 emissions reductions in 2035, compared with the Current Policies Scenario, or 7% of the global reductions. In the Middle East, it accounts for 29% (or around 280 Mt) of the abatement vis-à-vis the Current Policies Scenario by 2035. Phase out of subsidies is also important in North Africa, where it accounts for 33% of the abatement.

Chapter 19 discusses the effects of universal phase-out by 2020, a highly ambitious outcome, given the domestic difficulties the corresponding price increase could create. This would reduce CO_2 emissions by 1.5 Gt in 2020, roughly equivalent to the current emissions of Germany, France and Italy combined. The 450 Scenario assumes less ambitious (but perhaps more achievable) changes to fossil-fuel subsidies and the resulting abatement is correspondingly lower.

Figure 13.8 ● World energy-related CO_2 emission savings by region in the 450 Scenario

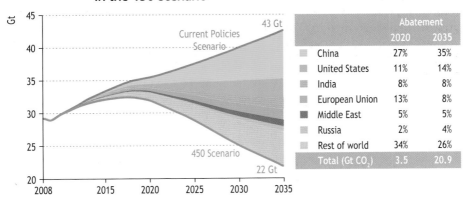

Abatement	2020	2035
China	27%	35%
United States	11%	14%
India	8%	8%
European Union	13%	8%
Middle East	5%	5%
Russia	2%	4%
Rest of world	34%	26%
Total (Gt CO$_2$)	3.5	20.9

Selecting the measures

The contribution made by different abatement measures to the 450 Scenario changes over time, as cheaper options are exhausted and more expensive options have to be taken up (Figure 13.9). End-use efficiency accounts for 67% of the 3.5 Gt abated in 2020, vis-à-vis the Current Policies Scenario, but its share declines to 47% by 2035, when total abatement is 20.9 Gt. Over time, the contribution made by energy efficiency is evenly split between abatement achieved through greater efficiency in direct combustion of fossil fuels (*e.g.* through the increased efficiency of coal furnaces) and abatement achieved as a result of lower electricity demand attributable to greater efficiency in end use (*e.g.* more efficient appliances) which reduces the combustion of fossil fuels in the power generation sector. Cheap end-use efficiency measures are quickly exploited in OECD+ countries, where consumers react to a price of CO_2 by putting in place efficiency measures in electricity use. The price of CO_2 is also instrumental in achieving energy efficiency improvements in direct use of fossil fuels in industry, while fuel economy standards are the key instrument for transport. Efficiency measures are also of more weight early in the period because other abatement measures, such as CCS, have longer lead-times.

Renewables, including biofuels, account for a slightly increasing share of CO_2 savings over time, provided that support policies are in place that go beyond the impact of the price of CO_2, their share growing from 19% in 2020 to 24% in 2035. The cost of those policies increases from some $60 billion in 2009 to more than $300 billion by 2035. Faster deployment of renewables, which reduces their capital costs, and higher electricity prices due to rising prices of CO_2 mean either that renewables become competitive earlier in the projection period, or that they require a lower level of support per unit of energy — for example, onshore wind in the United States becomes competitive in 2020 in the 450 Scenario, ten years

13

earlier than in the Current Policies Scenario. Nonetheless, the total amount of the support increases throughout the period, due to the rapid expansion in the use of renewable sources.

CCS becomes a key abatement technology by the end of the projection period, accounting for nearly 4 Gt of abatement by 2035. CCS is used in new coal (and gas-fired) power plants after 2020 in OECD+ and Other Major Economies and is also widely used as a retrofit measure (see Box 14.1 in Chapter 14). CCS becomes a key abatement option in certain industrial applications, as well as in energy transformation (*e.g.* coal-to-liquids). Nuclear power accounts for a fairly constant share of abatement across the period, increasing in absolute terms to 1.7 Gt by 2035.

Figure 13.9 ● *World energy-related CO$_2$ emission savings by policy measure in the 450 Scenario*

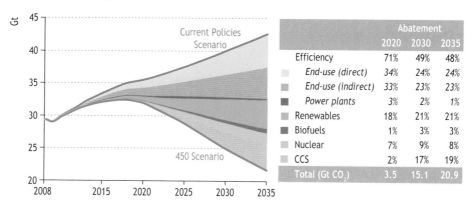

Abatement	2020	2030	2035
Efficiency	71%	49%	48%
End-use (direct)	34%	24%	24%
End-use (indirect)	33%	23%	23%
Power plants	3%	2%	1%
Renewables	18%	21%	21%
Biofuels	1%	3%	3%
Nuclear	7%	9%	8%
CCS	2%	17%	19%
Total (Gt CO$_2$)	3.5	15.1	20.9

Ten policies account for 45% of emissions reductions in 2020 and 54% in 2035 (Table 13.2). These policies are implemented in just five countries/regions – China, the European Union, the United States, India and Japan. The prompt implementation of policies in these countries is essential to the success of the 450 Scenario. Due to its sheer size and reliance on coal, Chinese industry is the largest source of abatement in 2020, with savings of nearly 350 Mt CO$_2$, closely followed by power generation in China, at just over 320 Mt CO$_2$. Those two sectors combined account for 19% of global savings by 2020 and 27% by 2035. These figures reflect the policy of the Chinese government to rebalance the economy, as well as action to realise renewables and nuclear capacity targets. Abatement in the building sector in China is also important. Pricing of CO$_2$ in the power sectors in European Union, United States and Japan unsurprisingly also plays a very important role by 2020, achieved both through emissions reductions in those countries and through the purchase of international offset credits, mainly from India. Overall, the composition of abatement in 2035, in terms of countries and sectors, is fairly similar to that in 2020.

Table 13.2 ● Key abatement* by policy area

2020				2035			
Country	Sector	Measures	Abatement (Mt CO$_2$)	Abatement (Mt CO$_2$)	Measures	Sector	Country
China	Industry	Rebalancing of the economy and efficiency improvements in iron, steel, cement and others.	348	3 434	Other Major Economies emissions trading scheme; extended support to renewables, nuclear, efficient coal.	Power generation	China
China	Power generation	Government capacity targets including wind 150 GW, nuclear 70 GW and hydro 300 GW.	321	2 237	Rebalancing of the economy, increasing use of CCS and efficiency improvements.	Industry	China
European Union	Power generation	Emissions trading scheme and expansion of renewable energy sources.	218	1 615	OECD+ emissions trading scheme; extended support to renewables and nuclear.	Power generation	United States
China	Buildings	50% of building stock has improved insulation to reduce energy consumption per unit area by 65% vs. 1980 level. 50% of appliances stock is highest efficiency standard.	168	970	100% of building stock has improved insulation to reduce energy consumption per unit area by 65% vs. 1980 level. 80% of appliances stock is highest efficiency standard. Penetration in the buildings sector of integrated solar energy systems.	Buildings	China
United States	Buildings	50% of existing buildings adopt retrofit measures to reduce energy consumption by 40% from 2008 level. Energy efficiency grants, use of renewables.	140	814	Support to renewables, nuclear and efficient coal; international offset projects.	Power generation	India
United States	Power generation	Target of 15 % from renewable sources; OECD+ emission trading scheme.	96	702	OECD+ emissions trading scheme; extended support to renewables.	Power generation	European Union

* In the 450 Scenario compared with the Current Policies Scenario.

13

Table 13.2 ● Key abatement* by policy area (continued)

2020				2035			
Country	Sector	Measures	Abatement (Mt CO$_2$)	Country	Sector	Measures	Abatement (Mt CO$_2$)
India	Power generation	International offset projects and support for solar PV.	79	India	Industry	Efficiency improvements in iron, steel, cement and others increasing use of CCS due to offset projects.	421
India	Industry	Efficiency improvements in iron, steel, cement and others due to international offset projects.	76	United States	Buildings	80% of new buildings are zero-energy buildings, 80% of existing buildings adopt retrofit measures to reduce energy consumption by 40% from 2008 level.	408
United States	Industry	Efficiency improvements and fuel-switching driven by OECD+ carbon price, implemented from 2013.	71	United States	Transport (road)	Passenger light-duty vehicle fuel economy standards, biofuels incentives and incentives for natural gas use in trucks.	398
Japan	Power generation	Increasing share of renewable in power generation to 15% by 2020; OECD+ emissions trading scheme.	53	European Union	Buildings	80% new buildings are zero-carbon footprint, 80% of existing buildings adopt retrofit measures to reduce energy consumption by 40% from 2008 level.	349
Total			1 571				11 348
Share of global abatement			45%				54%

* In the 450 Scenario compared with the Current Policies Scenario.

Implications for energy demand

In the 450 Scenario, total growth in both primary and final energy demand is restrained, compared with both the Current Policies Scenario and the New Policies Scenario, by the implementation of environmentally-ambitious policies and measures. World primary energy demand reaches 14 900 million tonnes of oil equivalent (Mtoe) in 2035, representing an annual average growth rate of less than half that seen from 1990 to 2008. Demand for all fuels is higher than today's levels by 2020, but by 2035 demand for both coal and oil has fallen below the level in 2008. Fossil fuels continue to be the major component of primary demand, although their share falls from more than 80% in 2008 to just over 60% in 2035. By contrast the share of nuclear and renewables in global primary demand increases to almost 40% in 2035, from less than one-fifth in 2008 (Figure 13.10).

The most dramatic change in energy demand growth over the period is seen in China, where, from 2000 to 2008, the growth in energy demand has been very steep, at around 9% per year on average. This growth begins to slacken off as early as 2012. From 2020 to the end of the projection period, energy demand in China remains almost flat. The United States also sees a change. Historically, its energy demand has grown at an average rate of around 1% per year. In the 450 Scenario, demand remains flat from 2008 to 2020, but then falls until 2030, when it once again stabilises. As a result of these trends, total primary energy demand in China, around 150 Mtoe lower than that of the United States in 2008, exceeds demand in the United States by more than 1 000 Mtoe by 2035.

Global demand for oil peaks just before 2020 at 88 million barrels per day (mb/d) in the 450 Scenario, after which it begins to fall, reaching 81 mb/d or around 3 800 Mtoe in 2035. By 2035, the share of oil in total primary energy demand has fallen to 26%, seven percentage points lower than in 2008. The implications for oil are further discussed in Chapter 15.

13

Figure 13.10 ● World primary energy demand by fuel in the 450 Scenario

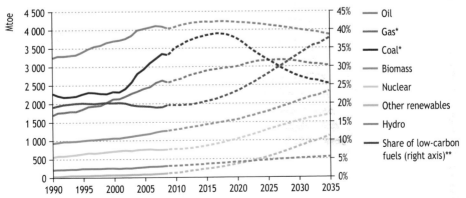

*Includes CCS. **Excludes CCS.

Primary natural gas consumption is projected to climb to 3.8 trillion cubic metres (tcm) in 2030, at an average annual growth rate of 0.8%, after which demand begins to fall. The slow rise in global gas demand to 2030 and fall thereafter masks very divergent trends in different regions. For example, gas demand in the United States rises sharply from 2020 through 2025, as the power sector shifts from coal to gas, but by 2035 gas demand in the United States has declined well below current levels, due to fuel-switching in power generation to nuclear and renewables. China and India both see steady growth in gas demand across the period, quadrupling their demand compared with 2008 levels by 2035. In Europe, demand for natural gas falls more or less steadily across the period. Despite these regional deviations from the global trend, the overall share of gas in the global primary energy mix remains at around 21% across the projection period.

Coal demand is the most affected in volume terms, peaking before 2020 at just over 5 500 million tonnes of coal equivalent (Mtce). Coal demand declines steeply in every year from 2020, returning to 2003 levels by 2035. Coal demand is by then some 3 600 Mtce, around 25% lower than today. The OECD+ coal market is significantly affected, with demand for coal in 2035 falling to less than half of the 2008 level (Figure 13.11).

Figure 13.11 ● **Primary energy demand by fuel and region in the 450 Scenario**

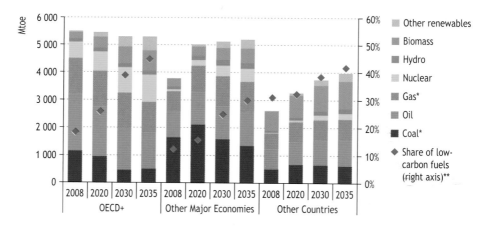

*Includes CCS. **Excludes CCS.

As a result of policies and measures implemented in the 450 Scenario, in particular price of CO_2 signals, demand for nuclear power and renewables combined reaches just over 5 600 Mtoe in 2035, almost two-and-a-half times the 2008 level. Demand for modern renewable energy (that is, renewables excluding traditional biomass[12]) nearly quadruples over the projection period, growing from some 843 Mtoe in 2008 to around 1 500 Mtoe in 2020 and, much more substantially, to 3 250 Mtoe in 2035 — representing an increase in share of total primary energy demand from 7% in 2008 to 11% in 2020 and 22% in 2035.

All regions see increases in demand for renewable energy, with some seeing dramatic growth. Renewable energy demand in India increases more than four-and-a-half times (Figure 13.12) and in China by nine times to more than 530 Mtoe by 2035. The United States also sees very substantial increase in demand for modern renewable energy by 2035, with demand reaching 550 Mtoe and accounting for 26% of total primary energy demand by 2035. Brazil remains (as is the case in all scenarios) the country with the largest share of renewables in total primary energy demand, 55% of energy coming from modern renewables in 2035.

Figure 13.12 ● Modern renewables primary energy demand by selected country/region in the 450 Scenario

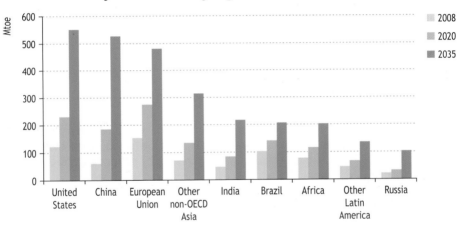

World electricity demand increases over time in all end-use sectors in the 450 Scenario, although by less than in the Current Policies Scenario. In the transport sector, demand for electricity remains more or less flat in the Current Policies Scenario, but increases in the 450 Scenario (Figure 13.13), reaching almost 1 500 terawatt-hours (TWh) by 2035, five-and-a-half times the 2008 level. This is driven by a major shift to electric vehicles. The share of nuclear in power generation increases by about 50% relative to current levels by 2035. Renewable-based generation increases to 45% of the global generation mix, almost two-and-a-half times higher than today, with wind power

---

12. Modern renewables encompass all renewable energy sources other than traditional biomass, which is defined as biomass consumption in the residential sector in developing countries and refers to the use of wood, charcoal, agricultural residues and animal dung for cooking and heating. All other biomass use is defined as modern.

increasing to almost 13% and solar photovoltaics (PV) and concentrating solar power (CSP) together to more than 6%. Overall, low-carbon fuels (nuclear, renewables and fossil-fuel power plants fitted with CCS) make up over three-quarters of electricity generation by 2035, up from less than one-third today.

Figure 13.13 ● *World electricity demand by sector in the 450 Scenario compared with the Current Policies Scenario*

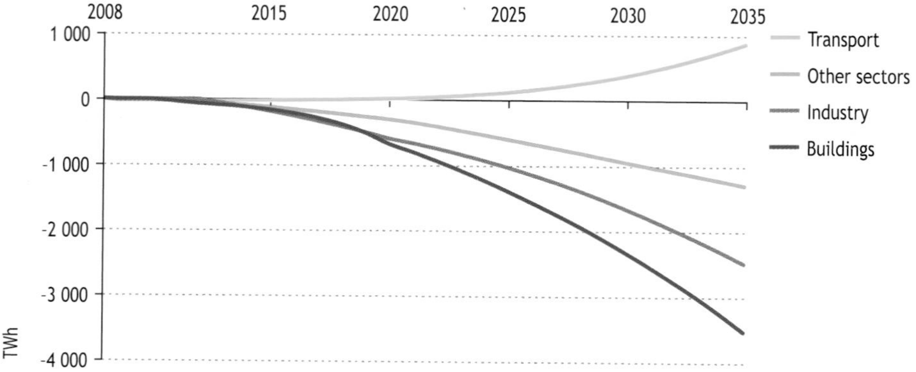

The cost of achieving the 450 Scenario

The global transformation of the energy sector to achieve the necessary reduction in CO_2 emissions requires very substantial spending on low-carbon technologies and energy efficiency. This spending includes capital spending by businesses, and consumer spending on cars, equipment and appliances (but not on their operation — meaning that the investment figures are gross, taking no account of savings in running costs attributable to more efficient appliances and cars).[13] The investment discussed here is additional to that incurred in the Current Policies Scenario. In the 450 Scenario, it amounts to $18 trillion in the period 2010 to 2035. Of this investment, only 12% (or $2.2 trillion) is incurred before 2020, more than half (or $9.4 trillion) in the decade from 2020 to 2030, and the remaining third (or $6.4 trillion) during the last five years of the projection period. This pattern is partly due to the fact that the abatement achieved in the period up to 2020, even with relatively vigorous action arising from the Copenhagen Accord, leaves much to be accomplished in the later period and at a higher capital cost per unit of CO_2 saved.

The greatest increase in investment is needed in the transport sector, where additional investment over the period, compared with the Current Policies Scenario, reaches $7.2 trillion (Figure 13.14). Almost 40% of this is incurred in the OECD+ countries, around one-quarter in Other Major Economies, around 20% in Other Countries and the remainder in international bunker fuels. The buildings sector is the second-largest area

---

13. See *WEO-2009* pp 260-1 for further details.

of cumulative additional investment, amounting to $5.6 trillion. About one-half of this is required in OECD+ countries. Of the cumulative investment needed in the power generation sector ($2.4 trillion) and the industry sector ($2 trillion), around 40% is incurred in the OECD+, 42% in Other Major Economies and the remaining 17% in Other Countries. Additional investment needs for biofuels are largest in the OECD+ countries, where around 70% of the total $0.7 trillion is invested.

Figure 13.14 ● Cumulative additional spending on low-carbon energy technologies in the 450 Scenario relative to the Current Policies Scenario

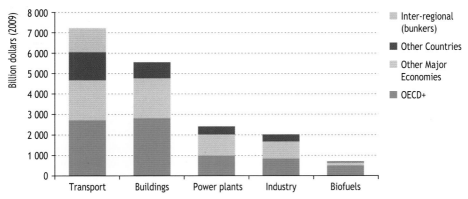

In 2020, the largest share of additional investment is needed in the European Union, with 23% of the total, just above China (Figure 13.15). By 2035, additional investment needs are greatest in China, at around one-quarter of the total, and second in the United States, at around 20%, while the European Union's share declines to just above 10%.

Figure 13.15 ● Annual additional spending on low-carbon energy technologies in the 450 Scenario relative to the Current Policies Scenario

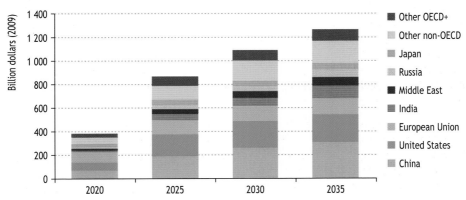

Note: Total additional spending differs from Figure 13.14 as inter-regional spending related to bunker fuels is excluded.

While the country in which investment takes place is not necessarily the country in which the cost of the investment is incurred – since some of the abatement achieved may be sold to other countries in the form of offset credits, or, if the abatement takes place in developing countries, may be financed by developed countries – it is nonetheless striking to note the dominance of a small number of countries in terms of the location of investment, notably China, the United States and the European Union, which together require more than half the additional investment across the period. In the case of China, the share of investment is lower than the corresponding share of abatement because China has lower investment costs per unit of abatement; for the United States, with higher abatement costs, the share of investment is higher than the share of abatement.

Box 13.2 ● Impact on government revenues

The impact on government revenues of policy interventions to address climate change is complicated due to the interactions with taxation and the economy as a whole. It is difficult, therefore, to establish the net impact on revenues. The reduction in demand for fossil fuels may mean that governments see a loss of revenues from value-added tax (VAT) and other fuel duties. However, in the transport sector, if fuel prices are allowed to fall in line with demand, the rebound effect (*i.e.* an increase in demand due to lower prices) could offset the impact of efficiency gains, at least to some extent, and reduce the emissions savings. To maximise the fuel economy gain (and CO_2 savings), consumer prices will need to remain the same, which can be achieved by raising VAT or other taxes on transport fuels.

Further, the implementation of policies to address climate change also brings with it other opportunities to raise revenues. Where a government puts a price on CO_2, whether through taxation or creating the conditions for a carbon market, government revenues can be raised if the price paid by polluters accrues to the government. We have assumed that all OECD+ countries enter into a single linked carbon market from 2013, in order to achieve the emissions reductions needed to meet the Copenhagen Accord pledges. If countries were to auction all emissions permits, revenues raised in 2020 would amount to a total of around $250 billion (equivalent to Portugal's 2009 GDP), decreasing to $185 billion in 2035, as OECD+ emissions fall. Of this 2035 figure, around $65 billion could be raised by the United States and $54 billion in the European Union. Even if only a percentage of permits are auctioned, the revenue raised could still be substantial, and could exceed the VAT lost in some countries.

The potential value of auctioning carbon permits in Other Major Economies is striking by the end of the period. We assume a linked carbon market by 2021 and if all permits were auctioned in the first year, the Other Major Economies could collectively raise around $120 billion. In 2035 this figure reaches $415 billion. China accounts for the greatest part of this, with a potential revenue stream of $90 billion in 2021 and as much as $270 billion in 2035.

In the context of global economic recovery and with many governments seeking to reduce debt accumulation and deficits, auctioning revenues could even assist in fiscal consolidation, rather than being used either for new or old spending commitments. Of course, in many countries it may not be politically feasible to auction 100% of permits, particularly early in the scheme. Even where there is auctioning, some governments may choose to hypothecate the *new* revenues from auctioning to specific uses, in order to garner political support for the introduction of such schemes. This could mean that the revenues cannot be seen as directly available to offset any revenue losses from the reduction in VAT and other tax receipts. Finance ministries will wish to take into account the interactive effects of climate policies on the public finances when taking decisions about the appropriate policy instruments to tackle climate change.

The cost of Copenhagen

Last year's *World Energy Outlook* assumed that Copenhagen would deliver a binding global agreement that would set in motion deep cuts in emissions by 2020. The actual outcomes of Copenhagen, even on an ambitious interpretation, result in emissions around 1.2 Gt CO_2 higher in 2020 than in last year's 450 Scenario. Achieving a 450 trajectory becomes that much more difficult. To compensate for the cumulative excess of 17.5 Gt CO_2 before 2020, rapid innovation is required after 2020 in all sectors, and the speed of the necessary transformation of the economy means that some investment decisions could be classed as economically irrational, for example, retiring power plants before their initial investment has been recouped (see Chapter 14 for further details). This results in costs — both in terms of macroeconomic impacts, and investment costs — that are higher than those seen last year.

Macroeconomic costs

The changes in supply and demand implied by the transformation of the way in which we produce and consume energy in the 450 Scenario, and the accompanying transformation of industrial processes and agricultural and forestry practices, mean that a new equilibrium is reached, and this affects the prices of a number of goods. Taking all this into account, we estimate that global GDP would be reduced in 2020 by the equivalent of 0.1%, compared with the Current Policies Scenario; in 2030 by the equivalent of 1.9%; and in 2035 by the equivalent of 3.2%.[14] This compares to last year's estimate, assuming earlier action, of a cost to GDP of 0.1% to 0.2% in 2020,[15] and 0.9% to 1.6% by 2030. While the loss of even 3.2% of global GDP in 2035 must be seen in the light of the prospective doubling of global GDP between 2008 and 2035 — and is therefore roughly equivalent to the loss of one year's growth — it is nonetheless noteworthy that it represents an impact on GDP that has more than doubled, compared with last year's

14. This does not take into account energy cost savings.

15. *WEO-2009* we presented a range of estimates, reflecting differing assumptions about the allocation of emissions permits and participation in global efforts to address climate change. This year's estimate corresponds to the lower end of last year's estimates.

estimates (Figure 13.16). This is driven partly by a change in economic expectations, which move the baseline (in this case, GDP in the Current Policies Scenario) up. The increase in the macroeconomic impact of mitigation, compared with last year, also reflects the non-optimal trajectory that is emerging from the current negotiations, although in order to present figures which are comparable to those presented last year, we have made the calculation (for this purpose alone) on the basis of the assumption that all countries participate in measures to curb emissions from 2013.

As with last year's estimates, the macroeconomic impact of achieving the 450 Scenario would be offset by the benefits of climate policy, including reduced energy demand. Moreover, GDP in the scenarios where no strong climate action is taken would be likely to be affected in the longer term by the unconstrained climate change entailed by those scenarios. The net impact of these opposing forces is difficult to quantify. For this reason, for modelling purposes the net level of global GDP is assumed to be unchanged in the 450, New Policies and Current Policies Scenarios.[16]

Figure 13.16 ● **Estimates of the percentage change in world GDP implied by the 450 Scenarios in *WEO-2009* and *WEO-2010***

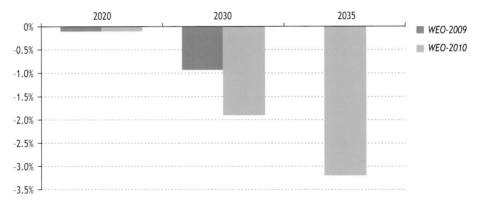

Source: OECD-IEA analysis, using OECD Env-Linkages model.

Implications for spending on low-carbon energy technologies

As discussed, the additional spending on low-carbon energy technologies needed to achieve the long-term stabilisation of atmospheric greenhouse gases has risen because of the failure to reach a more ambitious agreement at Copenhagen. A binding agreement to earlier and more stringent commitments is still possible before 2020 but, on the present basis and our assumptions about the arrangements beyond 2020, by 2030 the energy sector will have spent nearly $1 trillion more than we had estimated last year for an unchanged final result. Spending from 2010 to 2030 has risen from $10.6 trillion[17] to $11.6 trillion.

---

16. The estimated changes to GDP are taken from joint work with the OECD, using the OECD-ENV-Linkages model.

17. The figure specified in *WEO-2009* was $10.5 trillion in year-2008 dollars; this equates to $10.6 trillion in year-2009 dollars.

Higher emissions than seen in last year's 450 Scenario prior to 2020 are reflected in lower investment in the same period. As action to reduce emissions becomes more intense after 2020, so does investment. The transformation needed in the energy sector is no different in scale to that outlined in *WEO-2009* but has to occur much more rapidly — and more expensively — because it occurs later, leading to the $1 trillion excess overall by 2030.

In total, compared with last year's 450 Scenario, the investment needed to meet a 450 trajectory is higher in all sectors, other than transport (Figure 13.17). The buildings sector requires investment 23% higher than last year, while investment in industry increases by 31%. The power sector does not see a large increase — approximately 5% compared with last year's investment level. This is because demand is lower, meaning that the net investment needs for power generation are not very different from last year's 450 Scenario. Additional investment in transport is slightly lower than last year. This is largely due to a change in the Current Policies Scenario, where more of the cheap abatement options are already absorbed due to higher oil price assumptions, meaning that additional abatement becomes more costly, and therefore is delayed to the end of the period, meaning that much of it falls outside the time-frame analysed in *WEO-2009*.

Figure 13.17 ● Change in additional cumulative investment in *WEO-2010* 450 Scenario relative to *WEO-2009* 450 Scenario, 2010-2030

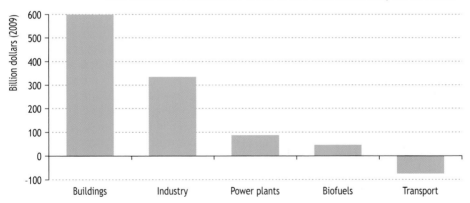

Benefits

The increased cost of reaching a 450 Scenario, based on the Copenhagen Accord, inevitably worsens any cost/benefit analysis of action on climate change. But this should not be allowed to obscure the scale of benefits, both in terms of the avoidance of climate change and the associated impacts and adaptation costs and in terms of other co-benefits. These include reduced local pollution and improved health outcomes as a result, which is quantified as a reduction in years of life lost, since these emissions are detrimental to human health. Energy security benefits, particularly in relation to oil, are discussed in Chapter 15.

Reduced local pollution

One of the benefits associated with moving to a low-carbon future is the associated reduction in the emissions of gases other than CO_2. Sulphur dioxide (SO_2), nitrogen oxides (NO_x) and particulate matter ($PM_{2.5}$) all have negative effects, both on human health and on the environment. Like climate change, the effects of these gases are not limited to the country or region in which they are emitted, but are felt beyond national borders. The policies aimed at reducing CO_2 emissions in the 450 Scenario also have the effect of reducing emissions of these air pollutants (Table 13.3). By 2035, SO_2 emissions are 61 Mt, or one-third lower than in the Current Policies Scenario. The majority of the decrease (27 Mt) takes place outside the OECD+, as in most OECD+ countries sulphur control measures are already in place, while non-OECD+ countries benefit from the reductions in SO_2 emissions primarily due to lower fossil-fuel consumption. NO_x emissions are 27% lower. $PM_{2.5}$ emissions are 8%, or 3.3 Mt, lower globally, though it should be noted that OECD+ emissions of particulates in the 450 Scenario are 17% higher in 2035 than in the Current Policies Scenario, due to greater use of biomass in the residential sector there. Emissions of particulates in non-OECD+ countries decrease by nearly 4 Mt. China and India benefit most. Due to their high reliance on coal, the paucity of pollution control mechanisms, and the expected exponential growth in car use, energy diversification measures have a particularly high value in these countries. Otherwise environmental costs could be high enough to pose a threat to future growth. A further benefit is a 23% global reduction in the costs of pollution control, compared with the Current Policies Scenario.

While reducing these pollutants has a positive impact on human health, insufficient data are available to allow for a quantitative global assessment of this impact. Estimates for European countries, China, India and the European part of Russia suggest that exposure to the concentrations of fine particles in ambient air which prevailed in 2005 will cause a loss of about 1.9 billion life-years, the vast majority of which, 1.6 billion life-years, in India and China, translating into a shortening of life-expectancy of more than one year.[18] Compared with the numbers presented in the last year's *Outlook*, the current estimates are lower. They take into account more conservative assumptions about relative risk factors for developing countries (China and India), resulting from recent findings of the Global Burden of Disease Study (forthcoming). In China the external costs of pollution — such as health costs, loss in labour productivity and loss in land productivity — amounted to 3.8% of the GDP in 2005 (World Bank, 2007). The 450 Scenario saves at least 750 million life-years compared with the Current Policies Scenario (Table 13.4), the vast majority of them in China and India. If the data were available, these figures would certainly be higher on a global basis.

---

18. By the statistical convention governing the measurement of the health impacts of (outdoor) air pollution, only the population above the age of 30 is taken into account in calculating the average effect on life-expectancy.

Table 13.3 ● Emissions of major air pollutants by region in the 450 Scenario (thousand tonnes)

	2005	2008	2020	2030	2035	Change versus Current Policies Scenario 2020	Change versus Current Policies Scenario 2035
Sulphur dioxide (SO_2)							
OECD+	29 553	22 765	12 083	9 520	9 463	-9%	-23%
United States	*13 793*	*9 985*	*4 179*	*2 790*	*2 835*	*-7%*	*-24%*
European Union	*7 839*	*5 551*	*2 499*	*2 066*	*2 038*	*-8%*	*-14%*
Japan	*753*	*637*	*505*	*465*	*446*	*-4%*	*-14%*
OME	45 767	48 590	42 213	30 367	28 862	-8%	-33%
Russia	*6 268*	*6 309*	*4 087*	*4 206*	*4 521*	*-3%*	*-7%*
China	*31 567*	*34 606*	*32 132*	*21 312*	*19 714*	*-9%*	*-38%*
Other Countries	20 248	22 475	22 646	22 803	22 252	-13%	-36%
India	*5 908*	*7 396*	*10 005*	*10 510*	*10 507*	*-11%*	*-40%*
World	95 569	93 830	76 942	62 690	60 577	-10%	-33%
Nitrogen oxides (NO_x)							
OECD+	37 337	32 402	17 961	13 945	13 657	-6%	-21%
United States	*17 203*	*14 379*	*7 232*	*5 347*	*5 229*	*-5%*	*-20%*
European Union	*11 054*	*9 536*	*5 167*	*3 860*	*3 794*	*-7%*	*-17%*
Japan	*2 289*	*1 899*	*912*	*635*	*570*	*-7%*	*-26%*
OME	28 637	31 866	31 917	27 933	27 947	-7%	-31%
Russia	*5 047*	*4 938*	*3 453*	*2 814*	*2 774*	*-4%*	*-21%*
China	*15 770*	*18 923*	*20 636*	*17 256*	*17 294*	*-8%*	*-35%*
Other Countries	19 559	20 551	20 574	22 943	24 506	-9%	-27%
India	*3 946*	*4 518*	*5 590*	*7 197*	*8 490*	*-10%*	*-30%*
World	85 533	84 820	70 453	64 821	66 110	-7%	-27%
Particulate matter ($PM_{2.5}$)							
OECD+	4 230	4 006	3 505	3 688	4 046	3%	17%
United States	*1 111*	*990*	*896*	*1 028*	*1 240*	*12%*	*57%*
European Union	*1 608*	*1 500*	*1 203*	*1 227*	*1 304*	*2%*	*10%*
Japan	*195*	*169*	*120*	*103*	*99*	*-5%*	*-14%*
OME	15 812	17 368	15 484	12 686	12 273	-3%	-15%
Russia	*1 332*	*1 372*	*1 312*	*1 246*	*1 223*	*-2%*	*-14%*
China	*12 463*	*13 883*	*12 026*	*9 524*	*9 129*	*-3%*	*-17%*
Other Countries	18 173	19 239	20 417	20 989	21 144	-2%	-8%
India	*5 066*	*5 488*	*5 624*	*5 720*	*5 800*	*-4%*	*-14%*
World	38 215	40 614	39 406	37 363	37 463	-2%	-8%

Note: The base year of these projections is 2005; 2008 is estimated by IIASA.
Source: IIASA (2010).

Table 13.4 ● Estimated life-years lost due to exposure to anthropogenic emissions of $PM_{2.5}$ (million life-years)

		Current Policies Scenario		450 Scenario	
	2005	2020	2035	2020	2035
China	1 163	1 565	1 573	1 491	1 215
India	432	854	1 466	792	1 085
Russia*	53	49	49	47	46
European Union	234	146	119	138	108

*European part only.
Source: IIASA (2010).

Avoided mitigation and adaptation costs

A valuation of the benefits of avoiding climate change is beyond the scope of our analysis. Estimates vary widely. One major variable is the discount factor used, an important consideration because the costs of unabated climate change would be incurred in the future, while the costs of mitigation are incurred now, meaning the former must be "discounted" to reflect the higher value society places on spending (or cost-saving) now. Of course the fact that emissions trajectories are uncertain and that the temperature increases associated with specific emissions trajectories can be calculated only probabilistically, make it even more difficult to assess the costs of unabated climate change. Yet, estimates have been made.

The UNFCCC (2007) has estimated that adaptation, in the absence of mitigation measures, would cost around $49-$101 billion dollars per year globally by 2030, which is well before the full impacts of climate change could be expected to be felt. A subsequent review of estimates of the cost of adaptation (Parry *et al.* 2009) concluded that the UNFCCC results were "likely to be substantial under-estimates", and placed the global estimated annual cost of adaptation in 2030 at two or three times the UNFCCC estimates for the sectors covered, and far higher again if other sectors are included (*e.g.* mining, manufacturing, retail, tourism). Including ecosystems protection alone could add up to around $300 billion per year to the estimates. These estimates of adaptation costs do not include any allowance for those economic impacts of climate change which cannot be avoided through adaptation measures due to technical or economic constraints (such as sea defences beyond a certain limit of sea level rise) and as such are only a partial estimate of costs which might be avoided through mitigation. Garnaut (2008), while focusing mainly on Australia, is emphatic that the costs of action are lower than the costs of inaction, reporting a net positive impact on Gross National Product (GNP) after 2050 with mitigation action. Ackerman and Stanton (2008) estimate that in the United States, the costs of unmitigated impacts in terms of hurricane damages, real-estate losses, energy-sector costs and water costs will amount to $1.8 trillion in 2100.

Moving from the New Policies Scenario to the 450 Scenario

This chapter examines the challenge of achieving the 450 Scenario. Where comparison against a baseline has been necessary, it has focused primarily on the policies and actions needed to move from the Current Policies Scenario to the 450 Scenario. However, if a comparison is made between the New Policies Scenario and the 450 Scenario, some of these policies and measures would already have been implemented to reach the New Policies Scenario and, therefore, a different set of measures and mix of technologies is needed.

The abatement needed to reach the 450 Scenario compared with the New Policies Scenario is 1.8 Gt in 2020, but reaches 13.7 Gt by 2035 (Figure 13.18). As cheaper abatement options are generally the first to be exploited, going beyond the New Policies Scenario requires greater use of the more expensive options. Therefore, biofuels and CCS both assume greater importance in moving between the New Policies Scenario and the 450 Scenario than in achieving the transition from the Current Policies Scenario to the New Policies Scenario. Renewables have a consistently lower share in moving beyond the New Policies Scenario, as they are already widely used to reach the New Policies Scenario, securing around 28% of the abatement from the Current Policies Scenario.

Nuclear power also plays a relatively smaller role in moving from the New Policies Scenario to the 450 Scenario in 2020 than it does in moving from the Current Policies Scenario to the 450 Scenario. Again, this is because government support and policy in the New Policies Scenario make nuclear power a relatively more important source of abatement in that scenario. Most of this effect is attributable to the extension of nuclear plant lifetimes in the European Union and promotion of nuclear in China. However, later in the period, nuclear power's share of abatement beyond the New Policies Scenario increases and becomes almost the same as the proportion of total abatement against either baseline.

Figure 13.18 ● **World energy-related CO$_2$ emission savings by policy measure in the 450 Scenario compared with the New Policies Scenario**

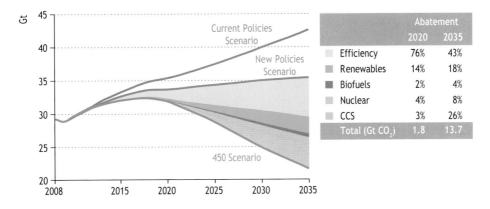

Energy efficiency policies and measures account for the largest share of emissions abatement both in achieving the New Policies Scenario and in moving beyond it, due to the large amount of cost-effective abatement potential which exists in this area.

In 2020, the support for renewables and nuclear already given in the New Policies Scenario means that going further relies heavily on higher energy efficiency. This share, however, becomes relatively less important towards the end of the projection period, as the scale of abatement required means that other technologies are more and more called into play.

Investment to go beyond the New Policies Scenario

Cumulative additional investment to reach the 450 Scenario, compared with the New Policies Scenario, amounts to $13.5 trillion dollars across the projection period, or three-quarters of the total cumulative additional investment needed to move from the Current Policies Scenario to the 450 Scenario. Emissions reductions in 2035 in the 450 Scenario compared with the New Policies Scenario are 13.7 Gt, or two-thirds of the total abatement with respect to the Current Policies Scenario (Figure 13.19). This higher proportionate investment compared with abatement reflects the fact that the abatement measures adopted to reach the New Policies Scenario are broadly less expensive on average than those necessary to move beyond it and reach a level of emissions compatible with a 450 trajectory, particularly towards the end of the period.

Figure 13.19 ● *Additional annual investment and abatement by scenario*

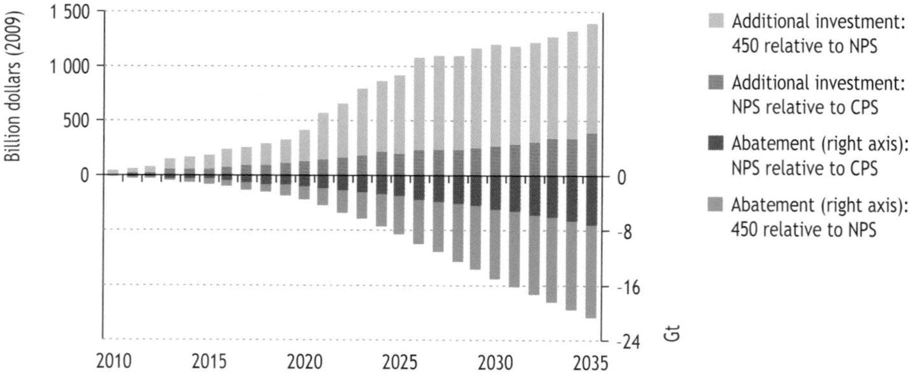

NPS = New Policies Scenario; CPS = Current Policies Senario; 450 = 450 Scenario.

Where is the abatement taking place?

In the 450 Scenario, compared with the New Policies Scenario, as is the case in comparison to the Current Policies Scenario, abatement in just six countries/regions accounts for the bulk of the global CO_2 reductions, growing from 60% in 2020 to 75% in 2035 (Figure 13.20), thus highlighting the key role of these countries in moving beyond the New Policies Scenario. By 2035, these countries' shares of abatement are almost the same for both the move from the Current Policies Scenario to the New Policies Scenario, and from the New Policies Scenario to the 450 Scenario. Earlier in the period however, these are some interesting differences. China and the European Union do more to reach the New Policies Scenario; the United States does less.

Figure 13.20 ● World energy-related CO_2 emissions savings by region/country in the 450 Scenario compared with the New Policies Scenario

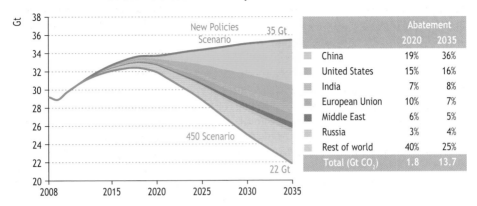

	Abatement	
	2020	2035
China	19%	36%
United States	15%	16%
India	7%	8%
European Union	10%	7%
Middle East	6%	5%
Russia	3%	4%
Rest of world	40%	25%
Total (Gt CO_2)	1.8	13.7

United States

In the United States, energy-related CO_2 emissions, having fallen in 2008 and 2009, rebound as of 2010, before falling again as policies are implemented to achieve the Copenhagen pledge of a 17% cut in emissions below 2005 levels by 2020. Domestic energy-related emissions fall to 5 Gt CO_2 by 2020, representing three-quarters of the pledged reductions; the remaining 260 Mt CO_2 reductions come through the purchase of international emissions-reduction credits. To achieve the 450 Scenario, total energy-related emissions fall by 60% between 2005 and 2035, to 2.3 Gt — in line with the over 80% reduction compared with 1990 by 2050 stated in the American Power Act. Most of this reduction is effected after 2020. The key enabling policies needed to achieve the reductions are:

■ The implementation of pricing CO_2 in the power sector — coupled with incentives for renewable energies and CCS. These policies account for 1.4 Gt (or 65% of the savings) in 2035 (Figure 13.21). The share of fossil-fuel generation drops from 71% today to 37% in 2035. In the short term, the natural gas share increases significantly, thanks to the price of CO_2, while that of coal drops. With widespread use of CCS, coal use rebounds after 2030. Generation from renewables quadruples from 2008 to 2035 to 1 800 TWh with economic support for renewables amounting to almost $500 billion over the *Outlook* period, an average of $19 billion per year.

■ Passenger light-duty vehicle (PLDV) fuel economy standards, biofuels incentives and incentives for the use of natural gas in trucks account for just over 30 Mt of savings in 2020 (11% of total abatement) and around 380 Mt in 2035. By 2035, biofuels account for one-quarter of fuel use in transport (from 3% today), and electric vehicles and plug-in hybrids account for 50% of PLDV sales in 2035.

■ The buildings and industry sectors account for 13% of the total emissions savings in 2035 (some 290 Mt).

13

European Union

We assume that the European Union will implement its target to make a 30% cut in emissions in 2020 compared with 1990, with a little less than 30% of the abatement bought as offsets (some 350 Mt). Emissions in 2020 are around 180 Mt lower, compared with the New Policies Scenario, in which we assume a 25% target for emissions reductions. The EU emissions trading system — comprising power generation, key industries and aviation — is expected to account for two-thirds of this reduction. To be on a 450 track, emissions have to reach less than 2 Gt in 2035, more than 50% below 2005 levels. The emissions trading scheme in the EU has to be strengthened significantly after 2020, as do fuel economy standards for vehicles. In 2035, 95% of electricity generation is low or zero carbon (including nuclear and coal and gas power plants with CCS). Renewables are responsible for a fairly stable proportion of abatement beyond the New Policies Scenario across the period, averaging 20%. This is not very different from their share of the burden in reaching the New Policies Scenario. CCS on the other hand is responsible for only 3% of abatement in 2020 — being an expensive abatement option, and slow to deploy — but 27% by 2035, making it the second most important abatement measure, after efficiency measures, which account for a falling share of abatement across the period, but are still achieving almost 340 Mt, or just over a third, of abatement compared with the New Policies Scenario by 2035.

Japan

Japan announced a pledge of emissions reductions of 25% below 1990 levels, on the premise that a fair and effective international framework will be established, in which all major economies participate, and that agreement will be reached by those economies on ambitious targets. In the 450 Scenario, this pledge is met through a combination of a reduction in domestic emissions to just under 920 Mt CO_2 by 2020 and purchase of certified emissions reductions on the international market. We assume a higher contribution of carbon offset credits for Japan compared with OECD+ countries as a whole as room for further efficiency improvements is limited and marginal abatement costs are generally higher than in other regions. Rapid decarbonisation of the Japanese economy continues to 2035, resulting in total energy-related emissions of nearly 445 Mt CO_2 by 2035, more than 60% below current levels. Total primary energy demand falls from around 500 Mtoe in 2008 to 440 Mtoe in 2035, a fall of 11%. In moving from the New Policies Scenario to the 450 Scenario, efficiency measures and renewables dominate early in the period, with an 84% share of abatement between them in 2020. Later in the period, CCS becomes much more important, accounting for some 75 Mt CO_2, or nearly a quarter, of abatement in 2035, against a negligible share in 2020. Renewables remain important across the period, accounting for slightly less than a quarter of all abatement, while nuclear accounts for 14% in 2020, and 16% in 2035, driven by strong government support and the rising CO_2 price experienced by all OECD+ countries.

China

Chinese emissions peak around 2020 at 9 Gt CO_2, in the 450 Scenario, and fall to 5.2 Gt CO_2 by 2035, 1.4 Gt CO_2 below 2008 levels. China accounts for 19% of global abatement in 2020, compared with the New Policies Scenario, and 36% in 2035. Even in 2020, China is still the most important single country in achieving the 450 Scenario.

In 2020, policies currently under discussion in China, to support the implementation of the carbon intensity target, account for 96 Mt or 27% of total abatement in China (Table 13.5). An additional 35 Mt of abatement annually results from a rebalancing of the economy towards services. China's emissions intensity improves by 48% compared with 2005 levels by 2020, exceeding its target.[19]

The further transformation required by the 450 Scenario will bring both challenges and opportunities to China, which is expected to maintain its leadership in the green growth race, remaining the world leader in wind installation (some 380 GW by 2035, 59% above the United States) and solar PV (nearly 190 GW in 2035, double the level of the United States). Thanks to demanding standards in the transport sector, China will become the largest world market for electric vehicles around 2020 — to the advantage of Chinese car manufacturers. Similarly, in the buildings sector, by 2035, the entire building stock will benefit from improved insulation (a 65% improvement in energy consumption per unit area compared with the 1980 levels); and 80% of appliances stock are expected to meet the highest efficiency standard currently applicable in the OECD. In the longer term, CCS development in China is expected to have a key role in the global deployment of this technology.

Table 13.5 ● **Abatement measures in China in the 450 Scenario compared with the New Policies Scenario in 2020**

Measures	Annual CO_2 savings (Mt)	Target* (WEO-2010 450 Scenario)
Economy-wide		
CO_2 intensity improvement target by 2020 relative to 2005		40-45% (48%)**
Share of non-fossil fuel in primary energy consumption by 2020		15% (16%)
Power generation		
More efficient fossil-fuel plants	12	n.a.
Nuclear	12	70 GW (70 GW)
Hydro	1	300 GW (298 GW)
Biomass	11	30 GW (11 GW)
Wind	34	100 GW (178 GW)
Solar	7	20 GW (20 GW)
Industry		
Efficiency improvement in iron, steel and cement	20	
Total	**96**	

*Economy-wide measures are announced government targets. Power generation measures are proposed targets.
**Includes offset credits.

---

19. This includes emissions reduction credits sold abroad.

India

India's emissions continue to grow across the projection period in the 450 Scenario, reaching 2 Gt CO_2 in 2020 and 2.3 Gt CO_2 by 2035. This growth, however, is much slower than that seen in the New Policies Scenario — emissions in the 450 Scenario are 6% and 31% lower in 2020 and 2035 respectively than emissions in the New Policies Scenario. India's Copenhagen Accord target of emissions intensity improvements of 20% to 25% by 2020 compared with 2005 is exceeded in the 450 Scenario, with the improvement reaching 42% by 2020.[20] The major contributors to abatement in India in 2035 are efficiency and renewables, at 47% and 24% of abatement respectively compared with the New Policies Scenario. CCS, which is not used to any significant degree in the New Policies Scenario, does play a strong role in moving beyond it to the 450 Scenario.

Russia

In the 450 Scenario, Russia's energy-related emissions remain more or less flat from 2008 to 2020 and are 1.6 Gt CO_2 in 2020, comfortably meeting Russia's Copenhagen Accord target of a 25% reduction in emissions, compared with 1990. However, more stringent cuts are needed after 2020. From that point, a steady decline in emissions begins and by 2035 emissions have fallen to 1.2 Gt CO_2.

Greater efficiency is the biggest component of emissions reductions, contributing 89% of abatement in 2020, compared with the New Policies Scenario, and 61% in 2035. By 2035, CCS is contributing 21% of abatement, compared with the New Policies Scenario, while renewables has the third-largest share, at 16%. The increase in renewables share in electricity generation from 16% in 2008 to 47% in 2035 represents a very large and rapid change; and a deep transformation is also achieved in the heat production system — CHP plants become much more efficient, with better optimised production of heat and power, and decreased losses in heat distribution.

Brazil

In associating themselves with the Copenhagen Accord, Brazil, along with other emerging markets, announced abatement targets for the first time. Brazil announced approximately 1 000 Mt CO_2 of abatement compared with business-as-usual (BAU), at least 800 Mt CO_2 of which will come from land use, land use change and forestry, primarily from reduced deforestation. However, substantial abatement is also necessary in the energy sector, in order to reach an emissions path that is compatible with a global 450 trajectory. Compared with the New Policies Scenario, energy-related emissions from Brazil are 12%, or 60 Mt CO_2 lower, at 440 Mt CO_2.

South Africa

South Africa has pledged to reduce emissions by 35% below BAU by 2020, and in the 450 Scenario South African energy-related emissions peak just before that, at 350 Mt CO_2. Meeting a 450 trajectory means that very stringent cuts are needed after

---

20. As in the case of China, emissions reduction credits sold abroad are included in this calculation.

Figure 13.21 ● Abatement by major region in the 450 Scenario compared with the New Policies Scenario

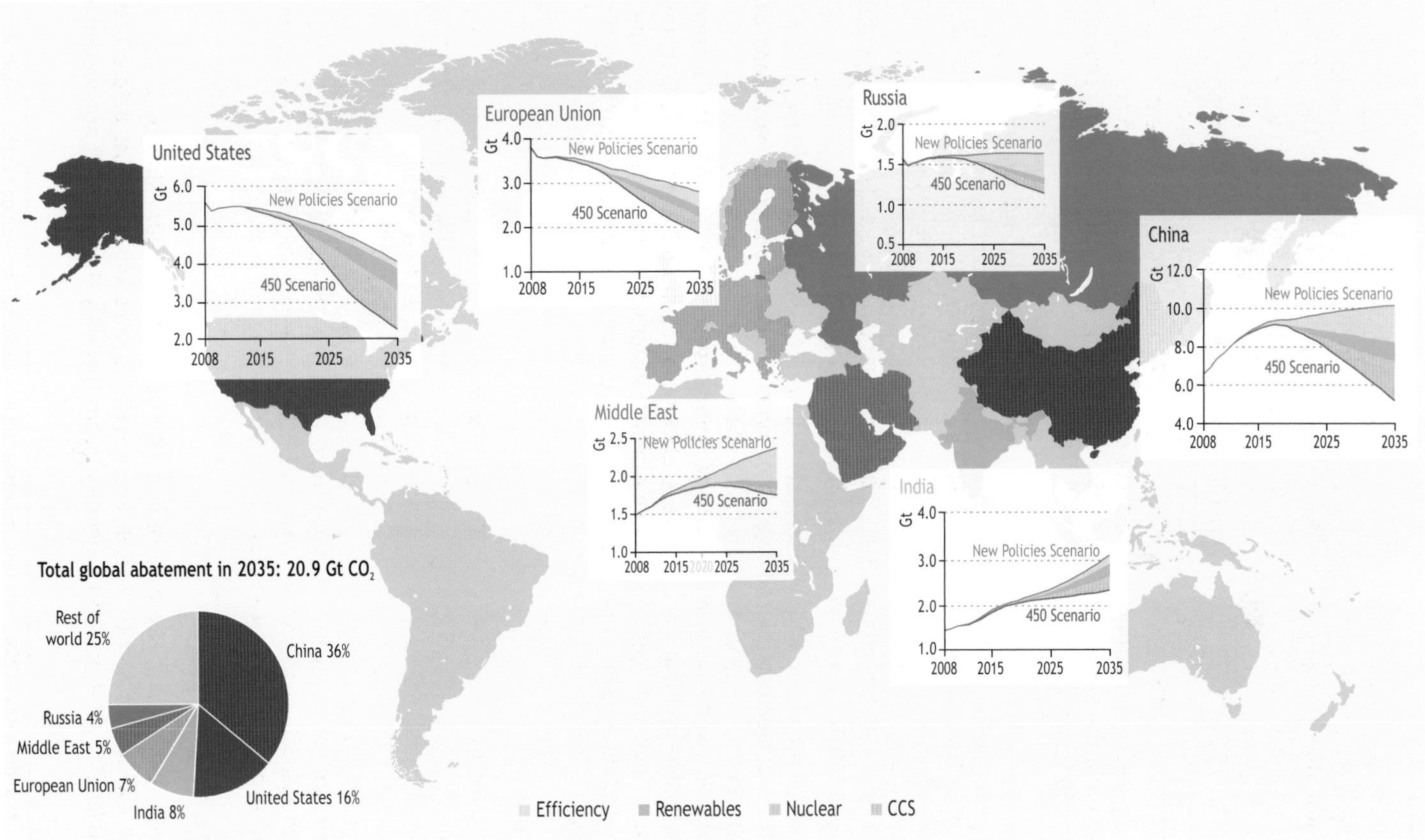

United States

Gt
- 6.0
- 5.0
- 4.0
- 3.0
- 2.0

New Policies Scenario

450 Scenario

2008 2015 2025 2035

European Union

Gt
- 4.0
- 3.0
- 2.0
- 1.0

New Policies Scenario

450 Scenario

2008 2015 2025 2035

Russia

Gt
- 2.0
- 1.5
- 1.0
- 0.5

New Policies Scenario

450 Scenario

2008 2015 2025 2035

China

Gt
- 12.0
- 10.0
- 8.0
- 6.0
- 4.0

New Policies Scenario

450 Scenario

2008 2015 2025 2035

Middle East

Gt
- 2.5
- 2.0
- 1.5
- 1.0

New Policies Scenario

450 Scenario

2008 2015 2025 2035

India

Gt
- 4.0
- 3.0
- 2.0
- 1.0

New Policies Scenario

450 Scenario

2008 2015 2025 2035

■ Efficiency   ■ Renewables   ■ Nuclear   ■ CCS

Total global abatement in 2035: 20.9 Gt CO$_2$

Rest of world 25%

Russia 4%

Middle East 5%

European Union 7%

India 8%

United States 16%

China 36%

The boundaries and names shown and the designations used on maps included in this publication do not imply official endorsement or acceptance by the IEA.

13

2020, and from 2025 emissions begin to fall rapidly, reaching 160 Mt CO_2 in 2035, 53% lower than 2008 emissions. This huge change is driven primarily by the widespread deployment of CCS technologies. In 2035, CCS accounts for 48% of South African abatement, compared with the New Policies Scenario, up from only 6% in 2020. This very substantial share is because, with a very rich coal resource and a CO_2 price of $90 per tonne in Other Major Economies, the application of CCS to coal-fired generation makes better economic sense for South Africa than a move to other sources of power. Energy-efficiency measures, which contribute 65% of abatement compared with the New Policies Scenario in 2020, have fallen to 20%, or just over 40 Mt CO_2 by 2035.

Indonesia

Indonesia's pledge associated with the Copenhagen Accord is to reduce emissions by 26% compared with BAU — though the government had earlier announced its willingness to cut emissions by 41% compared with BAU if funding were provided by the international community, and this is the assumed target in the 450 Scenario. Much of the abatement to achieve this target will come from reductions in emissions from deforestation, which dominate Indonesia's greenhouse-gas emissions, meaning that the trend in energy-related emissions in Indonesia (like Brazil) is likely to be rather different from the overall emissions trend. In the 450 Scenario, Indonesia's energy-related emissions increase steadily to 2023 and then level off, at around 530 Mt CO_2, where they remain until around 2030 before falling to just over 510 Mt CO_2 by 2035. Efficiency measures are the biggest contributor to Indonesian energy-related emissions savings compared with the New Policies Scenario, at 66%, or some 160 Mt CO_2 in 2035. Renewables and CCS together account for nearly a quarter of Indonesian energy-sector abatement by 2035.

CHAPTER 14

THE ENERGY TRANSFORMATION BY SECTOR

Where will the emissions cuts come from?

H I G H L I G H T S

- Abatement of energy-related CO_2 emissions in the 450 Scenario means that by 2035, the power sector is largely decarbonised, particularly in the developed countries, and the transport sector becomes the biggest emitter. Power-sector emissions in the 450 Scenario are more than halved, from nearly 12 Gt in 2008 to less than 5.3 Gt. By 2035, over three-quarters of global electricity generation is low-carbon. Given the limited efforts to 2020, rapid decarbonisation is needed thereafter, and over 90% of capacity additions are renewables (67%), nuclear (9%) and CCS-equipped (14%). Globally, renewables for power generation will require government support of some $3 trillion (in year-2009 dollars) to 2035. In the OECD+, this and a price of CO_2, reaching $120 per tonne in 2035 are the main drivers of an eight-fold increase in renewables generation (excluding hydro), compared with current levels.

- In 2010-2035, power-sector investment amounts to $11.1 trillion, a net increase of $2.4 trillion compared with the Current Policies Scenario. Almost 90% of the investment is in low-carbon technologies. Around one-third of new coal and gas CCGT plants (some 300 GW) are retired before the end of their technical lifetimes, over 100 GW of which do not fully recover their investment.

- By 2035, about 70% of global car sales are advanced vehicles (electric, plug-in hybrids and hybrids). China becomes the world's largest electric vehicle market just before 2020, accounting for 40% of global sales by 2035. Support for biofuels grows from $20 billion in 2009 to $125 billion in 2035 – most of it in the United States and the European Union. Additional investment in the transport sector, compared with the Current Policies Scenario, totals $7.2 trillion. As the Copenhagen Accord does not provide a policy framework to incentivise early deployment of alternative vehicles and aircraft, almost 90% of the investment takes place after 2020.

- As efficiency improvements approach saturation, CCS technology is expected to play a key role in reducing emissions from industry, accounting for some 40% of abatement in 2035. Additional investment, relative to the Current Policies Scenario, in industry amounts to $2 trillion in the period 2010-2035.

- Achieving the 450 Scenario requires large improvements in the energy efficiency of the buildings sector. Additional investment, relative to the Current Policies Scenario, to achieve this change amounts to $5.6 trillion from 2010 to 2035, some 90% of it after 2020.

Overview

This chapter examines in more detail the sectoral changes effected by the policies discussed in Chapter 13. The power sector is the largest source of global energy-related CO_2 emissions today, contributing just over 40% of emissions, and accordingly has been the focus of the greatest abatement effort. Transport is the second-largest emitter, followed by industry and buildings. In the Current Policies Scenario, as emissions from all sectors continue to grow, the sectoral shares of emissions remain largely unchanged across the period, with power generation increasing its share slightly to 44%. In the New Policies Scenario, there is little change, though abatement from the power sector leads to a reduction in its share to 39% of all energy-related emissions by 2035 (Figure 14.1).

Figure 14.1 ● *Share of total energy-related CO_2 emissions by sector and scenario*

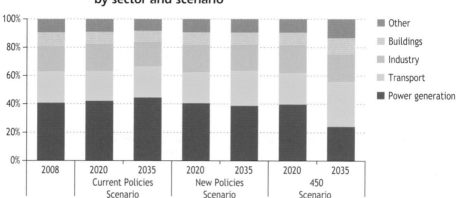

The 450 Scenario, however, leads to a major change. By 2035, the power sector achieves such significant abatement — two-thirds of the overall total, *vis-à-vis* the Current Policies Scenario — that the transport sector (where emissions reductions are more expensive) becomes the largest source of energy-related emissions (Figure 14.2). More than one-third of the power sector abatement (4.9 gigatonnes [Gt]), or 23% of total abatement) comes from reduced demand for electricity. This abatement is driven, in OECD+ countries (from 2013) and in Other Major Economies (from 2021), by implementation of a cap-and-trade system, with its accompanying rising price of CO_2, together with policies to support the deployment of renewable energy.[1] All these measures substantially decarbonise the power sector, reducing its share in total energy-related emissions to below one-quarter. In OECD+ countries together the change is even more pronounced, with emissions from power generation accounting for only 15% of total energy-related emissions by 2035. Such a transformation of the power sector, though dramatic, is achievable — the policy instruments and technologies needed to

---

1. See Annex B for policies and measures by region. Annex C contains regional definitions.

achieve it are known. Even when the abatement potential of the power sector has been taken fully into account, to bring global emissions to a long-term sustainable level will require substantial further emissions reductions in other sectors. This highlights that a long-term strategy to reduce greenhouse-gas emissions needs to address all sectors, identifying policies and investing in R&D for low-carbon technologies across the board.

Since abatement in the transport sector is more costly, by 2035 it is the largest source of energy-related CO_2 emissions, with a share of almost one-third of global emissions, despite deployment of more efficient vehicles, both hybrids and electric vehicles. Urgent action is needed to tackle trucks and other modes of transport and to deploy widely the (still immature) technologies in these areas. Careful consideration should be given to end-use prices, as in the more relaxed oil market associated with the 450 Scenario, there is a risk of a "rebound effect" as end-use prices fall — meaning that emissions savings from efficiency could be eroded by the increased demand associated with lower end-user prices for fossil fuels (see Chapter 15 and the section on transport in this chapter).

Industry is the third-largest emitter in 2035, with its share growing marginally to below one-fifth. Efficiency improvements are important to achieve emission reductions, but technology changes will be necessary to achieve CO_2 savings. Carbon capture and storage (CCS) in industrial processes becomes important during the projection period, as do new methods of manufacturing, such as the production of clinker at lower temperatures.

The buildings sector (which includes the residential and services sub-sectors), the fourth-largest emitter in 2035, also needs to undergo a low-carbon revolution. Currently, little attention is paid to this sector, and as a result, the options available now are limited. Retrofitting buildings in the OECD is very costly. New building in most developing countries does not at present prioritise CO_2 emissions reductions. This will need to change in the future if a long-term sustainable path is to be achieved.

14

Figure 14.2 ● Energy-related CO_2 emissions abatement by sector in the 450 Scenario compared with the Current Policies Scenario

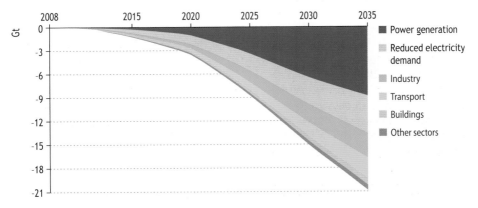

Power generation

Fuel mix and generating technologies

A radical transformation of the power generation sector is necessary to move to a low-carbon future. This requires a concerted push to low-carbon technologies[2] that not only displace inefficient thermal plants, but meet the relentless growth in electricity demand, while maintaining an affordable and reliable service to consumers. Action is required on the supply side through a different technology mix and energy efficiency improvements. As well, growth in electricity demand must be reduced as part of this far-reaching change. This is achieved through the adoption of cap-and-trade systems, and CO_2 prices rising to $120 per tonne in OECD+ countries in 2035 and $90 per tonne in Other Major Economies in 2035, as well as support to renewables and changes in regulation (see Annex B). In the 450 Scenario, electricity demand grows at an average annual rate of 1.9% between 2008 and 2035, compared with 2.5% in the Current Policies Scenario. This represents a drop of 5 300 terawatt-hours (TWh), or around 16%, by 2035, corresponding to the combined total current production of OECD North America.

In the 450 Scenario, the installation of thermal plants without CCS is significantly lower, in favour of renewables and nuclear technology (Table 14.1). Nuclear also plays an important role in providing baseload operation and, in some cases, has the potential to provide backup capacity. More than 500 gigawatts (GW) of new nuclear capacity is installed globally by 2035, while a change of policies in several countries favours the lifetime extension of nuclear plants. The net result is that the overall nuclear capacity operating in 2035 more than doubles relative to today.

Table 14.1 ● Capacity additions by fuel and region in the 450 Scenario (GW)

	2010-2020				2021-2035			
	World	OECD+	OME**	OC**	World	OECD+	OME**	OC**
Coal	575	91	356	127	438	140	236	62
*CCS-equipped**	*13*	*9*	*3*	*1*	*408*	*188*	*213*	*8*
Oil	31	6	17	8	35	9	13	12
Gas	434	148	186	100	480	168	215	97
*CCS-equipped**	*4*	*4*	*1*	*0*	*173*	*104*	*69*	*1*
Nuclear	137	46	75	16	387	165	145	77
Hydro	364	60	192	113	497	65	167	265
Biomass	73	44	17	12	234	80	89	65
Wind – onshore	430	245	150	35	840	381	302	157
Wind – offshore	67	42	20	4	298	170	91	36
Solar PV	123	81	21	21	652	238	233	181
Concentrating solar power	39	18	11	9	185	75	68	42
Geothermal	12	5	1	5	35	13	7	15
Marine	1	1	-	0	19	17	1	1
Total	2 285	787	1 047	451	4 100	1 522	1 568	1 010

*Note: CCS-equipped capacity additions in the table may exceed the overall additions of the corresponding fuel as this figure includes plant retrofit.

**OME = Other Major Economies. OC = Other Countries. Regional definitions can be found in Annex C.

---

2. Low-emission technologies refer to fossil-fuel plants fitted with carbon capture and storage technologies, nuclear plants, and renewable generating technologies, including hydropower.

Renewable plant additions account for over 60% of the global additions between 2010 and 2035, more than 40% of these being new wind installations. By comparison, renewables make up about one-third of the total global additions in the Current Policies Scenario over the projection period.

This shift towards low-carbon technologies occurs through a combination of policies to promote their use and discourage the use of fossil fuels, in particular coal plants without CCS. Rising prices of CO_2 change the cost ranking of new plants to the benefit of low-carbon technologies, as well as changing the merit order of existing plants at the expense of older inefficient fossil-fuel plants. Driven by growth in non-OECD countries, global installed coal capacity continues to increase in the period to 2020 (although not at the pace of the Current Policies Scenario), even as energy-efficiency measures reduce the need for new capacity and renewables installations increase. After 2020, mainly thanks to the introduction of cap-and-trade systems in OECD+ and Other Major Economies, older inefficient coal plants are rapidly retired, with most of the existing installed capacity being taken out of service – often before the end of its technical lifetime – within the projection period (Figure 14.3). By then, CCS-fitted plants increase significantly, with many existing plants being retrofitted in order to remain economic and extend their lifetime. Due to the rising price of CO_2, some 300 GW (or around one-third) of new coal and gas CCGT plants built between now and 2035 will be retired well before the end of their technical lifetime and in several cases even before they have achieved a commercial return on the capital invested. Around 100 GW fall into this category, representing a net loss of around $70 billion or 28% of the investment cost.

Figure 14.3 ● World installed coal-fired generation capacity in the 450 Scenario relative to the Current Policies Scenario

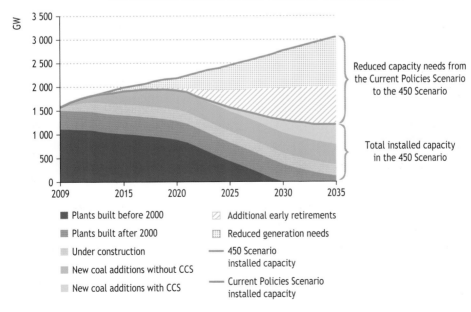

14

As a result of this transformation, electricity generation from coal-fired plants without CCS in 2035 is set to fall by more than two-thirds, compared with today. Most of the drop occurs in OECD+ countries, with coal generation increasing only in countries that do not introduce cap-and-trade systems by the end of the projection period (e.g. India). By comparison, in the Current Policies Scenario, coal generation without CCS doubles to 16 300 TWh. Strikingly, by 2035 in the 450 Scenario, coal generation from plants fitted with CCS reaches more than 3 000 TWh, which exceeds that from coal plants not equipped with CCS and represents about three-quarters of the total generation from all CCS-fitted plants (Figure 14.4).

Figure 14.4 ● **Incremental world electricity generation by fuel and scenario, 2008-2035**

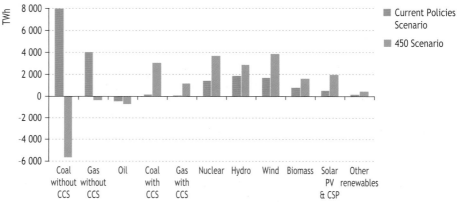

During this radical transformation, the flexible operational nature of gas-fired generation and its lower CO_2 content makes it an attractive "bridging" fuel. Consequently, gas-fired generation increases through to the late 2020s, to 45% above current levels, then reduces to about 20% more than today's levels, as a result of the rise in installed capacity of low-carbon generating plants (Figure 14.5).

Figure 14.5 ● **World electricity generation by type and scenario**

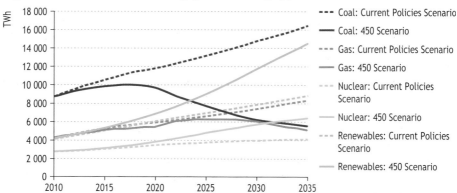

Box 14.1 ● Carbon capture and storage

Carbon capture and storage (CCS) plays a key role in the 450 Scenario greenhouse-gas mitigation portfolio. By 2035, the contribution of CCS to global emissions reductions, compared with the Current Policies Scenario, amounts to nearly 4 Gt of CO_2 (or 19%), up from less than 90 Mt in 2020 (or 2%). This is a very ambitious CCS growth path that requires CCS to be applied beyond coal in the power sector, where it accounts for 2 Gt of avoided emissions in 2035. CCS technologies are also adopted for gas power plants (700 Mt) and in emissions-intensive industrial sectors, such as cement, iron and steel, chemicals, and pulp and paper (1.3 Gt). By 2030, in order to compensate for the higher emissions to 2020, CCS plays a more important role in mitigation than in last year's 450 Scenario, to the extent of 1.2 Gt more. In particular, CCS retrofit plays a larger role, especially in China and the United States.

This level of application of CCS requires investment of $1.3 trillion in excess of that in the Current Policies Scenario from 2010 to 2035, which is about 8% of the overall investment needed to achieve the 450 trajectory. Most of the CCS projects occur in OECD+ countries, where the price of CO_2 in the power and industry sectors makes it a viable option after a phase between 2010 and 2020 during which government intervention to fund CCS demonstration projects runs at an average annual level of $3.5 to $4 billion (IEA, 2009a).

Although OECD+ countries are expected to take the lead in CCS deployment in the next decade, CCS technology spreads rapidly to Other Major Economies soon after 2020, as a price of CO_2 is introduced. By 2035, CCS technologies account for 21% of abatement in Other Major Economies (2 Gt of CO_2) and 25% of abatement in OECD+ (1.6 Gt of CO_2). This level of abatement requires expanded international collaboration and financing for CCS demonstration in developing countries, possibly including through the Clean Development Mechanism or an alternative financing mechanism generating offset credits. It will also require effective development of legal and regulatory frameworks and systematic mapping of storage sites (IEA, 2010).

14

In the 450 Scenario, global renewables-based generation is set to grow almost four-fold to 14 500 TWh by 2035, more than 60% higher than in the Current Policies Scenario. This increase is driven primarily by wind power, which doubles its output, compared with the Current Policies Scenario, to 4 100 TWh in 2035. Solar photovoltaics (PV) and concentrating solar power (CSP) respectively treble and more than quadruple their contribution by 2035 and collectively provide 2 000 TWh. Significant growth is also observed from hydro, with over 900 TWh, and biomass generation, with more than 800 TWh. More than 70% of the growth with respect to the Current Policies Scenario occurs in non-OECD countries, notably China and India, their collective share of world renewable electricity generation jumping from 19% today to 32% by 2035. In the United States, the share of renewables-based generation in total generation increases from below 10% today to more than one-third in 2035, mainly attributable to the rapid roll out of wind power.

CO₂ emissions

The 450 Scenario necessitates a rapid decarbonisation of the power generation sector since it currently generates more than 40% of global energy-related CO_2 emissions. Through targeted policies and incentives for the deployment of new capacity additions with low emissions, the CO_2 intensity (defined as the CO_2 emission content per unit of generation) drops by 2035 to a quarter of today's level, at just above 130 grammes of CO_2 per kilowatt-hour (gCO_2/kWh) (Figure 14.6).

Figure 14.6 ● **Change in world CO₂ emissions from power generation in the 450 Scenario compared with the Current Policies Scenario**

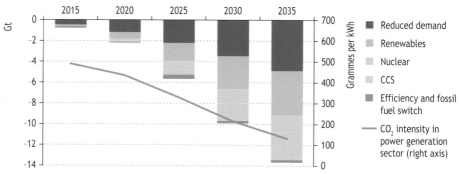

In 2035, compared with the Current Policies Scenario, emissions reductions from reduced demand represent more than one-third of the sectoral savings, closely followed by savings from generation by renewables at just over 30%, plants fitted with carbon capture and storage technology just under 20%, and nuclear plants 13%. Other emissions-saving measures, such as more efficient gas and coal plants and, in several countries, coal-to-gas switching, provide the remainder of the savings. The options available to decarbonise the market vary markedly across regions and reflect the distinct nature of those markets. Globally, total CO_2 emissions in the 450 Scenario are more than halved, from nearly 12 Gt in 2008 to around 5.3 Gt in 2035 (Figure 14.7).

Figure 14.7 ● **Change in world CO₂ emissions from power generation in the 450 Scenario compared with 2008**

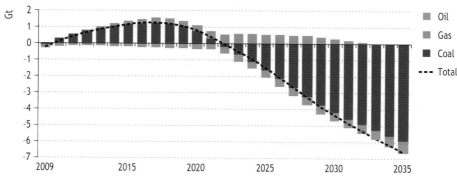

Investment in generating capacity[3]

Investments in the power sector in the 450 Scenario amount to $11.1 trillion over the period 2010-2035, of which more than 60% is absorbed by renewable plants, 17% nuclear, 7% CCS and 14% fossil-fuel plants without CCS. Investment in low-carbon technologies accelerates sharply after 2020, when these technologies account for over 90% of total investment.

In the 450 Scenario, some $100 billion per year on average is invested by China between 2010 and 2020. This is almost 60% above the European Union, and almost two-and-a-half times the expenditure in the United States (Figure 14.8). In 2010-2020, 75% of the Chinese investment goes into low-carbon technologies, with substantial investment in hydro, wind and nuclear technologies. In the European Union, three-quarters of investments go to renewables alone, particularly wind and solar PV during the same decade.

Figure 14.8 ● *Share of average annual global investment by technology type in the 450 Scenario*

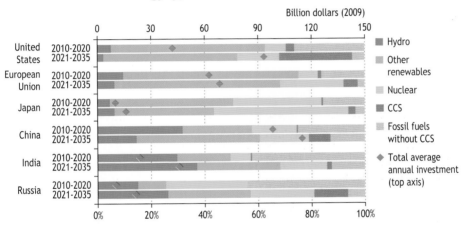

Note: Excludes investment in rooftop PV.

After 2020, greater annual expenditure on power generation is required by all countries in order to move towards a 450 ppm world. Significant reduction of emissions in the United States requires investment to more than double compared with the period 2010-2020, to more than $90 billion annually, with a strong push towards CCS technology, nuclear and non-hydro renewables. By comparison, higher expenditure on low-emission technologies in the previous decade by the European Union means that average annual investment there rises by around 10% to just under $70 billion per year. Investments in fossil-fuel plants without CCS technology in China drop from 25% of the total in the previous decade to 13% post 2020, with overall expenditure rising to $115 billion per year in 2021-2035.

---

3. This section focuses on power sector investment, which excludes investment in rooftop photovoltaics. This is reported under investment in buildings.

Over the projection period, net additional investments in the power sector in the 450 Scenario with respect to the Current Policies Scenario amount to $2.4 trillion. Investments for fossil-fuel plants without CCS are reduced by $2.1 trillion, while investments in additional low-carbon technologies cost a further $4.5 trillion, of which $1 trillion is spent on wind, $0.9 trillion on solar PV and CSP, $0.8 trillion on hydro plants, $0.4 trillion on other renewables, and $0.7 trillion each on nuclear plants and fossil-fuel plants fitted with CCS.

Government support for renewables

Electricity generated from most renewable technologies in the majority of countries is not yet competitive with electricity from non-renewables plants. Renewable technologies, therefore, require support if their share in electricity generation is to increase. The mechanisms involved can take many forms, such as feed-in tariffs and producer tax credits (see Chapters 9 and 10). In order to become competitive, most renewable technologies need to reduce their costs or to see rising costs for alternative fuels and technologies and, therefore, rising wholesale electricity prices. A combination of these two factors is likely. The unit costs of renewable technologies are likely to fall with technological development and wider deployment. Wholesale prices depend on several factors, with the cost of fossil fuels and the eventual price of CO_2 being the main determinant.

In the Current Policies Scenario, wholesale prices rise throughout the projection period, mainly due to rising fossil-fuel prices. The wholesale prices in the OECD+ countries double by 2035 with respect to 2009 reaching an average of about $90 per megawatt-hour (MWh). In the 450 Scenario, they increase even further in OECD+ and Other Major Economies, despite lower gas prices and falling coal prices, mainly due to the introduction of prices of CO_2 (Figure 14.9). In the OECD+ countries, the wholesale prices increase to two-and-a-half times the 2009 levels in real terms, to almost $110 per MWh.

Figure 14.9 ● **Additional price impact of the cost increase to the electricity producer in selected OECD+ countries resulting from the CO_2 price in the 450 Scenario**

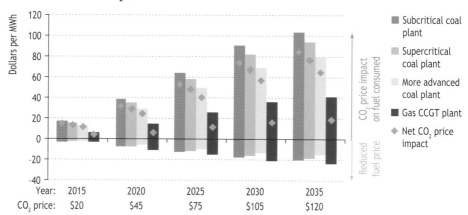

Over the projection period, generation from renewables (excluding large hydro, which is commercially competitive in the majority of countries) grows from around 840 TWh in 2009 to more than 4 000 TWh by 2035 in the Current Policies Scenario and to 8 800 TWh in the 450 Scenario. This corresponds to an increase in the share in global generation from 4% today to 11% by 2035 in the Current Policies Scenario and to 28% in the 450 Scenario.

The increases in renewable electricity generation and in the wholesale price have an impact on the support that renewable technologies need in order to be competitive. This impact is complex, and can best be explained by looking at support for renewables by three indicators: first, the necessary support per unit of electricity generated; second, the total value of support; and third, the support in relation to the electricity wholesale price.

In the 450 Scenario, the level of government support per unit of renewable generation (averaged across all renewable sources, except large hydro) decreases from $55 per MWh today to $20 per MWh by 2035. This fall is due to the increasing wholesale prices, which make renewables more competitive with other fuels, and cost reductions through increasing learning and deployment for all renewable sources, which bring down their unit cost, further enhancing competitiveness.

Indeed, the combination of falling technology costs and increasing wholesale prices driven by the rising price of CO_2 means that some technologies become fully cost-competitive during the *Outlook* period in some regions. For example, in the United States, onshore wind becomes competitive by 2030 in the Current Policies Scenario, and by 2020 in the 450 Scenario. On a global level, support for onshore wind falls to below $5 per MWh by 2035 in the Current Policies Scenario and by 2030 in the 450 Scenario. The support needed per unit of electricity produced for solar PV and CSP falls markedly in the 450 Scenario, to one-seventh and under one-quarter of the support needed today, respectively.

While support per unit of renewable electricity generated falls over the period, the increase in renewables deployment necessary to achieve the 450 Scenario means that cumulative global financial support for renewable electricity generation grows, reaching $3 trillion over the period 2010-2035, almost $1 trillion higher than in the Current Policies Scenario. In that scenario, the support grows over the decade to 2020, but after this point, the rate of growth in support slows and annual average support sees very little increase (Figure 14.10). In the 450 Scenario, by contrast, renewable support needs to increase substantially after 2020. In this scenario, support continues to grow through the period 2020-2035, reaching almost $180 billion by 2035, about 90% higher than in the Current Policies Scenario.

Support for renewables can also be expressed as a percentage of wholesale prices. In 2009, support for renewable generation in the OECD+ countries ranged from $2 to $8 per MWh, equivalent to an average increase over and above the wholesale prices of 9%. Over the entire projection period, the average amount of the financial support for renewable generation per unit of total electricity produced (that is, electricity from both non-renewable and renewable sources) is almost 30% higher

14

than current levels in the 450 Scenario. Despite the increase in the absolute level, the share over and above the wholesale price declines to just 6% on average in the OECD+ countries. In the European Union it is equivalent to 8% of the wholesale price, in the United States to 5% and in Japan to 3% (Figure 14.11). In the 450 Scenario, the overall amount of financial support for electricity generation from renewable sources in the OECD+ countries in the period 2010-2035 increases by only 15% with respect to the Current Policies Scenario, despite additional cumulative generation from renewable sources (excluding large hydro) of 35%, or about 14 000 TWh.

Figure 14.10 ● Average annual global support for renewable electricity by scenario

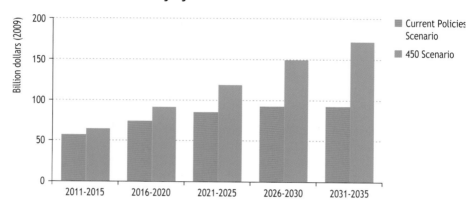

Figure 14.11 ● Average wholesale electricity prices and renewable support costs by scenario and major region, 2010-2035

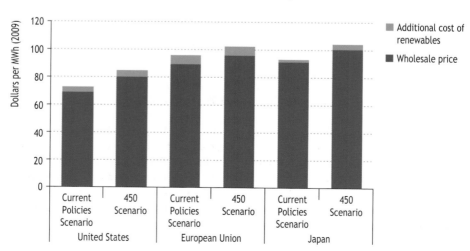

Transport

Transport fuel demand

Total global transport oil consumption in the 450 Scenario grows only slightly from 2 150 million tonnes of oil equivalent (Mtoe) to some 2 300 Mtoe in 2020, levels out thereafter and decreases to about 2 200 Mtoe in 2035. Nevertheless, oil remains the dominant fuel in the transport sector even in the 450 Scenario, with a share of 77% in all transportation fuels, down from 94% in 2008. Most of the oil savings in the 450 Scenario occur in road transport, which accounts for more than 80% of all oil savings by 2035. Among road vehicles, passenger light-duty vehicles (PLDVs) account for more than three-quarters of the oil savings, some 560 Mtoe by 2035. Savings in aviation account for an additional 135 Mtoe, or 15% of total oil savings in 2035.

Energy consumption in transport in the 450 Scenario becomes more diversified over the projection period, with biofuels, natural gas and electricity playing more important roles. Biofuels reach almost 400 Mtoe by 2035 in the 450 Scenario, a share of 14%; natural gas consumption increases to about 130 Mtoe in 2035 and electricity to almost the same level. Most of the growth in the use of alternative fuels occurs in road transport, where the potential for fuel switching is the greatest. The increasing use of electricity in the transport sector as a whole is largely a result of electrification in road transport, which accounts for almost 90% of the increase in electricity demand in the transport sector by 2035 (Figure 14.12).

Figure 14.12 ● **World fuel consumption in the transport sector in the 450 Scenario**

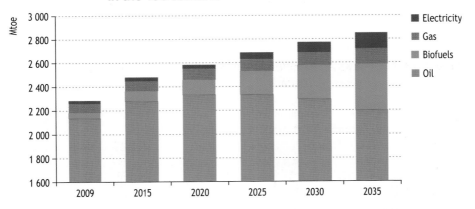

14

This transformation is facilitated by the policy framework adopted in the 450 Scenario (Box 14.2). Many of the cheap efficiency measures with short pay-back periods are often cost-effective for the first owner of a new car. For this reason, some of the potential to increase efficiency is already taken up in the Current Policies Scenario and, more significantly, in the New Policies Scenario (see Chapter 3). Without measures to promote behavioural changes, such as the purchase of smaller cars or modal shifts to mass transport systems, very substantial deployment of alternative cars and fuels is necessary to achieve the 450 Scenario.

Box 14.2 ● The policy framework for the transport sector in the 450 Scenario

The policy framework assumed in *WEO-2010* is specified in Annex B. It includes five key pillars :
- International sectoral agreements in the passenger light-duty vehicles (PLDV) sector and aviation (both domestic and international), which provide CO_2 emission limits for new cars and aircraft in all countries.
- Full technology spill-over from PLDVs to light commercial vehicles (LCVs).
- Alternative fuel support policies.
- National policies and measures in other segments of the transport sector.
- Retail fuel prices are kept (through taxation in OECD countries and subsidy removal in non-OECD countries) at a level similar to that reached in the Current Policies Scenario. This scheme is necessary to offset the rebound effect that could occur due to lower oil prices.

Sectoral targets for PLDVs are used or are under discussion in several countries. Their use is justified by the fact that the PLDV (and aviation) sectors are global, dominated by several international companies using homogenous technology. Use of a common sectoral target allows for long-term planning and security in investment and technology development. Further, it harmonises technology across countries and allows for cost reductions through centralised manufacturing. The sectoral targets for PLDVs relate to the sales of new vehicles. They are on-road targets for new sales, taking account of both efficiency improvements and deployment of alternative fuels, and do not assume significant behavioural changes by consumers. The CO_2 targets in 2035 for OECD+ (75 gCO_2/km in the 450 Scenario), Other Major Economies (85 gCO_2/km in the 450 Scenario) and Other Countries (105 gCO_2/km in the 450 Scenario) are averages for each region. For aviation, the sectoral agreement assumed requires the global aviation fleet to improve its average fuel consumption by 45% over today's level, to 2.5 litres per 100 revenue passenger kilometres (RPK) in 2035.[4]

The sectoral target for PLDVs and the assumed technology spill-over to LCVs lead to an improvement of more than 50% in the average fuel economy of new cars in both segments in 2035, compared with today, in line with the targets of the Global Fuel Economy Initiative.[5] For medium- and heavy-freight traffic, the possibility for spill-over is significantly lower given the maturity especially of diesel engines and the fact that cost-effectiveness is already an important criterion for decisions in this segment. This leads to the assumption of an additional 5% efficiency improvement in 2035, compared with the average efficiency in the Current Policies Scenario, for this road transport segment.

As an example of the impact of the assumption on retail fuel prices, gasoline prices reach $ 3.60/gallon in the United States in 2035 (an increase of more than 50% over 2009 levels), $2.10/litre in the EU (some 20% above 2009), $1.70/litre in Japan (almost 35% above 2009) and $1.40/litre in China (more than 65% above 2009).

---

4. Revenue passenger kilometres is a common aviation industry measure of demand.
5. See www.50by50campaign.org for details.

The transformation required in the 450 Scenario leads to significant changes in global vehicle sales. By 2035, about 70% of PLDV sales are advanced vehicles (electric cars, hybrids and plug-in hybrids) (Figure 14.13). Almost 60% of vehicles sold still primarily use internal combustion engines, but either in hybrid vehicles, with battery backup, or in highly-efficient flex-fuel vehicles, able to use any combination of oil-based fuels or biofuels. Natural-gas-vehicle sales make up another 2% of sales by 2035, while fuel-cell vehicles are commercialised only towards the end of the *Outlook* period. China overtakes the United States as the largest market for electric cars by 2018, and remains by far the largest market for electric cars and plug-in hybrids throughout the rest of the projection period, accounting for one fifth of global electric car sales by 2020, and 39% by 2035.

The technology spill-over from PLDV to light commercial vehicles (LCVs) carries similar changes into this segment. Pure internal combustion engine vehicles, which account for 98% of sales today, make up only about 22% of sales by 2035, while hybrid, plug-in hybrid and electric vehicles constitute 70% of total LCV sales. The remaining sales are natural-gas vehicles and fuel-cell vehicles, the latter deployed in commercial fleet demonstration projects.

Figure 14.13 ● Vehicle sales by type and scenario, 2035

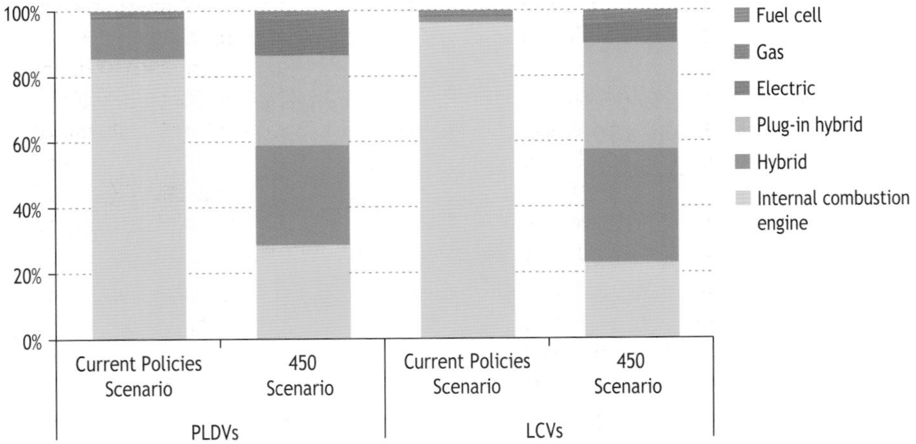

Note: PLDVs = passenger light-duty vehicles; LCVs = light commercial vehicles.

The transformation we describe takes place in a different time frame from that projected last year, and sales of electric cars in 2020 are lower. In the absence of a more ambitious global climate policy agreement, there is no global framework supporting technological change in transport, leaving the deployment of electric cars up to industry and national governments. This has the effect of reducing global electric car sales by the year 2020 to 2% of total PLDV sales, down from 4% in last year's *Outlook*. Similarly, sales of plug-in hybrids reach only 5% of total sales by 2020, down from 12% in last year's *Outlook*.

CO_2 emissions

In the 450 Scenario, global emissions from the transport sector reach 7.2 Gt CO_2 in 2020, but fall to 6.9 Gt by 2035 (around 300 million tonnes [Mt] above the 2008 level), having peaked soon after 2020. This differs from the emissions trajectory projected in last year's 450 Scenario, in which global transport emissions continued to rise across the projection period, reaching 7.7 Gt CO_2 in 2030. This is partly due to a downward revision of the emissions trajectory in the Current Policies Scenario, largely as a result of higher oil price assumptions in this year's *Outlook*. These induce higher efficiency improvements to the global car fleet. The biggest change from last year's 450 Scenario is seen in non-OECD countries. While last year's projections see emissions from transport in non-OECD countries rising across the projection period, the new 450 path entails a much slower rise and levelling off towards the end of the period. Some non-OECD countries even begin to see a drop in their emissions at the end of the period.

The largest contributor to emission savings is increasing end-use efficiency. However, the share of abatement achieved through efficiency falls over the period, because a large proportion of the possible efficiency gains are deployed before 2020, limiting the additional abatement that can be achieved from this source thereafter. Fuel switching, the biggest component of which is to biofuels in road transport and aviation, is responsible for more than 40% of abatement in the transport sector by 2035 — up from just 16% in 2020 (Figure 14.14).

Figure 14.14 ● **World transport-related CO_2 emission abatement in the 450 Scenario**

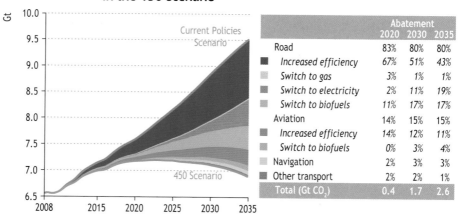

Driven by the adoption of a CO_2 cap-and-trade system in the power sector, and the resulting price of CO_2 levels, average emissions per kWh of electricity are substantially reduced, thus increasing the amount of carbon saved through the adoption of electric cars and decreasing the marginal costs of abatement (Figure 14.15). With increasing decarbonisation, well-to-wheels emissions from electric cars are significantly lower than those from vehicles using oil-based transportation fuels.

Figure 14.15 ● Sales of electric and plug-in hybrid vehicles in the 450 Scenario and CO$_2$ intensity in the power sector by scenario

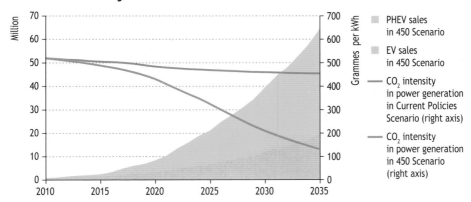

Note: Includes passenger light-duty and light commercial vehicles. PHEV = plug-in hybrid electric vehicles. EV = electric vehicles.

········· S P O T L I G H T ·········

Can e-bikes make a difference?

Two-wheelers, whether bicycles or powered two-wheelers such as mopeds, scooters and motorcycles, are an important means of mobility around the world. In many developing countries, two-wheelers are a first affordable step towards individual mobility. More than 95% of all powered two-wheelers are produced in China, Southeast Asia and Japan. Although powered two-wheelers are generally very fuel efficient, they contribute disproportionally to pollutant emissions and noise (IEA 2009b).

There could be a niche for electric two-wheelers (e-bikes), which generate no emissions and little noise during operation. Electric bikes are bicycles powered by normal human effort, but able to use electrical assistance from a modest battery and motor. The scooter-style e-bike is a more sophisticated machine which does not require human effort. It offers typically a range of about 40 kilometres.

Electric bikes could become a very important means of transport, especially in urban areas, where problems with local pollution, noise and congestion are pressing. Today, e-bikes are particularly popular in China, partly due to the ban on gasoline-fuelled scooters in several big cities, such as Beijing and Shanghai.

How far e-bikes will substitute for other forms of transport in the future is unknown. In the 450 Scenario, e-bikes are assumed to replace other motorised two-wheelers and are projected to make up around 20% of two-wheeler sales by 2035.

14

Changes in the aviation sector are responsible for some 15% of CO_2 abatement in the transport sector by 2035, as a result of a prevailing sectoral agreement that encourages increased biofuels consumption and additional technical, operational and infrastructure measures. Technical and equipment measures include installation of new wingtips, measures to reduce drag, early aircraft retirements, engine retrofits and upgrades. Operational measures cover fuel-management techniques, other pilot techniques and weight reductions. Improvements in infrastructure involve redesigned flight paths and more efficient traffic control. Together, such efficiency measures are responsible for three-quarters of emission savings. The rest is due to the adoption of biofuels (see Spotlight in Chapter 12).

International shipping and domestic navigation are increasingly important in climate discussions, but shipping contributes only 3% to global transport-related emissions reductions in the 450 Scenario, compared with the Current Policies Scenario. This is because improving hydrodynamics and increasing motor efficiency, the installation of sails and speed reductions are occurring as measures to reduce oil consumption already under the oil price assumptions in the Current Policies Scenario. Nevertheless, given the importance of the sector for future climate negotiations, and the fact that emissions from this sector will need to be addressed, much effort will be expended to improve the data and to inform decision making.

Investment in transport

In the 450 Scenario, the assumed sectoral agreements in the passenger light-duty vehicle and aviation sectors, as well as national policies in other transport sectors, lead to additional investment, compared with the Current Policies Scenario, of $7.2 trillion for the entire transport sector. Almost 90% of the additional investment over the *Outlook* period takes place after 2020. Passenger cars account for almost 60% of the investment, followed by aviation, with just under a quarter (Figure 14.16).

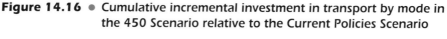

Figure 14.16 ● Cumulative incremental investment in transport by mode in the 450 Scenario relative to the Current Policies Scenario

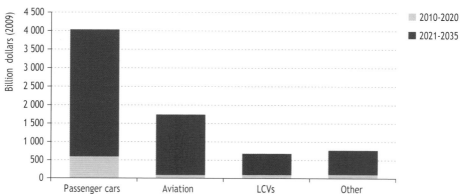

About 38% of the additional investment occurs in OECD+ countries, mostly due to high investment needs in road transport for electrification. Other Major Economies account for about 27% and Other Countries 19%. The remaining 17% is needed in international shipping and aviation.

The electrification of road transport is probably the biggest challenge and entails significant costs. In the 450 Scenario, the marginal cost of CO_2 abatement in the PLDV segment of road transport ranges from $95 per tonne of CO_2 in the United States, to $180 in the European Union, and up to values well in excess of $200 per tonne of CO_2 in Japan.

Financial support to biofuels for transport in the 450 Scenario totals $2.2 trillion across the period, growing from $20 billion in 2009 to $125 billion in 2035. The United States is responsible for more than 60% of global spending on ethanol and is therefore the clear leader, whereas the European Union leads the global support to biodiesel, responsible for about 40% of the total. The developing world has significantly more access to low-cost biomass and, with global technology spill-over, makes significantly less financial support available to biofuels over the projection period.

Industry[6]

Industrial energy demand

In the 450 Scenario, the compound annual average growth rate of final energy demand in the industry sector between 2008 and 2035 falls to 1% per year, from 1.7% in the Current Policies Scenario. This is lower than the growth rate of 1.5% which was seen between 1990 and 2008 (Figure 14.17). This slowing in demand growth is driven, particularly in the second half of the projection period, by the stabilisation of production in the emerging economies of energy-intensive basic materials, such as iron and steel and cement, and by improvements in energy efficiency.

Energy demand in this sector in the OECD+ begins to decline before 2020, while demand in Other Major Economies levels off around 2020. In Other Countries demand continues to grow throughout the projection period, though the growth rate is lower than in the Current Policies Scenario. As a result the share in demand of the OECD+ falls to 27% in 2035, from 37% in 2008. Other Major Economies and Other Countries expand their shares, from 42% to 44% and from 21% to 28%, respectively.

14

Global coal and oil demand in the sector peak before 2020, then begin to decline, due to relatively rapid price increases because of their high carbon content and the phase out of subsidies. By contrast, gas demand grows slowly but constantly, and electricity demand also grows, backed by fuel switching, increased sophistication in manufacturing processes and more recycling (though the rate of growth is slower than in the Current Policies Scenario). The share in industrial energy demand of fossil fuels

---

6. Industry sector energy demand and CO_2 emission are calculated in accordance with IEA energy balances *i.e.* including neither demand/emissions from coke ovens, blast furnaces and petrochemical feedstocks (which appear in the "other energy sector" or "non-energy use sector"), nor process-related CO_2 emissions (which are outside the energy sector, and fall under the "CO_2 : other" category in Figure 13.4 in Chapter 13).

drops to 53% in 2035, from 61% in 2008. The share of electricity expands to 33%, from 26%, and electricity becomes the largest energy source used in the sector, as in the buildings sector in all three scenarios.

Figure 14.17 ● **Industrial energy demand by scenario**

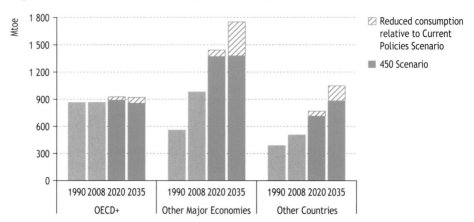

Box 14.3 ● **The policy framework for the industry sector in the 450 Scenario**

The policy framework assumed in *WEO-2010* is specified in Annex B. It includes three key pillars:

● A cap-and-trade system in OECD+ countries as of 2013, and in Other Major Economies as of 2021.

● International sectoral agreements for the iron and steel, and non-metallic minerals sectors for all countries.

● National policies and measures for other industries.

Industry joins the power sector as part of a cap-and-trade system from 2013 in OECD+ countries and from 2021 in Other Major Economies. The prices of CO_2 in OECD+ are $45 per tonne and $120 per tonne in 2020 and 2035, respectively. In Other Major Economies, the price of carbon reaches $90 per tonne in 2035. This cap-and-trade system promotes improvements in energy efficiency, fuel switching from carbon-rich energy sources (such as coal) to low-carbon fuels and deployment of carbon capture and storage technology.

In addition, international sectoral agreements are assumed for the iron and steel, and non-metallic minerals sectors. The international sectoral agreements function as a complement to domestic and regional cap-and-trade systems and national policies by limiting carbon leakage. These agreements help accelerate improvement in energy efficiency in these industrial sub-sectors.

Many countries implement national policies to improve energy efficiency in industrial sectors, in the form of government R&D and preferential tax and credit policies for the deployment of more efficient equipment. In the 450 Scenario, the introduction of equipment with the best available technology in its class in terms of efficiency, recycling and adoption of new materials are assumed, together with promotion of fuel switching to lower-carbon fuels (IEA, 2008; 2009c and 2010). Importantly, China is assumed to rebalance its economy by promoting the growth of the services and light industry sectors to a greater extent than it does in the Current Policies Scenario.

CO_2 emissions

Energy-related direct CO_2 emissions[7] from industry begin to decline around 2020 and are lower than present levels by around 2030, in contrast to continued growth in the Current Policies Scenario (Figure 14.18). The reduction of CO_2 emissions from industry, despite the growth of energy demand, comes from fuel switching to lower-carbon and carbon-free energy and electricity. Even when indirect CO_2 emissions from electricity and heat are included, the reduction is sustained.

Figure 14.18 ● *Change in industrial energy-related CO_2 emissions by scenario and region, 2008-2035*

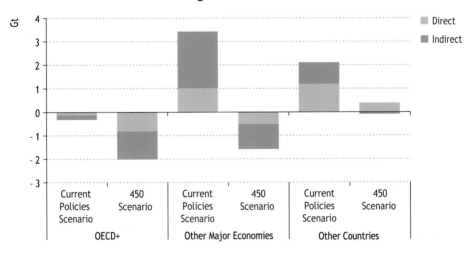

14

The world economy — especially in developing countries — is recovering from the financial crisis faster than expected. This faster recovery puts upward pressure on energy demand and CO_2 emissions in both the Current Policies and 450 Scenarios, compared with last year's projections. The change in CO_2 emissions (including indirect emissions) can be decomposed into the "activity effect" (that is, the change in emissions due to the change in activity in the industry sector, in terms of value-added), the "energy intensity effect" (that is, the change in emissions due to the reduction in

7. CO_2 emissions from fossil-fuel combustion, not including process-related emissions.

the amount of energy needed to carry out each unit of those activities) and the "CO_2 content effect" (that is the change in emissions due to the different rate of emissions from energy sources used). This decomposition shows that the total fall in emissions over the period, of almost 1 Gt, is driven, at the global level, by the energy intensity effect and the CO_2 content effect. Taken alone, the activity effect actually increases emissions — but the CO_2 content and energy intensity effects compensate for this, as emissions are reduced by fuel switching to cleaner fuels and through the use of CCS. A greater role for CCS is expected than was seen in last year's analysis. Although CCS requires additional energy consumption for separation and capture of CO_2, the reduction in emissions due to CCS use in the industry sector is 1.3 Gt in 2035, accounting for 42% of the total reduction compared with the Current Policies Scenario.

Investment in more energy-efficient industrial equipment

Additional investment for industrial efficiency improvements over the projection period in the 450 Scenario, relative to the Current Policies Scenario, amounts to some $1.4 trillion in 2009 prices. The projection period is longer by five years (or one-quarter) than in last year's analysis; but investment across this longer period increases by more than 50%, as more stringent measures to achieve greater reduction in emissions are required beyond 2030 technology. Additionally, some $640 billion of investment for CCS technology is required. Total investment required in the industry sector to meet the 450 Scenario is equivalent to 0.3% of the cumulative value added in the industry sector in the same period.

Although the investment taking place in Other Major Economies marginally exceeds that in other regions, it does not automatically follow that Other Major Economies are the largest investor as at least part of the finance will come from outside the region. Investment in the OECD+ region generates much lower energy and CO_2 savings per dollar (Figure 14.19), due to the high investment unit cost there.

Figure 14.19 ● Share in additional investment, CO_2 reduction and energy savings in industry by region in the 450 Scenario

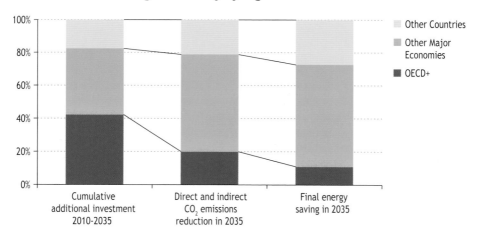

Buildings

Energy use in buildings

Energy used in buildings grows at an average annual rate of 0.6% in this scenario, from 2 850 Mtoe in 2008 to over 3 360 Mtoe in 2035. This represents a substantial slow down, compared with the Current Policies Scenario, where energy demand grows at an average annual rate 1.2% between 2008 and 2035. In 2035, energy savings, compared with the Current Policies Scenario, amount to around 530 Mtoe in 2035, meaning that the buildings sector contributes 30% of the savings in total final energy consumption in 2035. Similarly to other sectors, energy demand in the buildings sector is higher in 2020 than in last year's 450 Scenario, making the path from 2020-2035 much more difficult, as savings in that period need to be greater to compensate for the additional emissions up to 2020. Energy savings, compared with the Current Policies Scenario, and corresponding investment are concentrated in the later years, with 13% taking place before 2020, 46% between 2021 and 2030, and 41% between 2031 and 2035.

Box 14.4 ● The policy framework for the buildings sector in the 450 Scenario

Achieving the 450 Scenario requires strong policy intervention to reduce emissions from the buildings sector, especially after 2020. This includes a wide range of policies and measures in all majors economies, from net zero-energy buildings in Japan and zero-carbon footprint buildings in the European Union applicable to new buildings constructed in the next decades, to mandatory building code standards and labelling requirements for equipment and appliances in Russia, China and India (see Annex B). The implementation of those policies and measures is responsible for about two-thirds of the energy saving in the building sector. Further savings are achieved by the higher electricity prices in the 450 Scenario compared with the Current Policies Scenario. The higher electricity prices, resulting from the assumed increase in prices of CO_2, play an important role in promoting energy efficiency measures installations, ensuring that energy costs become a key purchasing criteria for consumers in the building sector, and also pave the way for the greater switch towards the use of renewable building materials.

14

The use of fossil fuels in the buildings sector peaks around 2020 and declines thereafter, by 2035 fossil fuel use is 8% lower than in 2008. Consumption of coal and oil is reduced by 34% and 23% respectively from 2008 to 2035, while gas use increases by 6%. Electricity use, despite significant energy efficiency savings, grows at 1.5% per year, driven by electricity demand for appliances in non-OECD countries (where electricity demand grows at 2.8% per year) and by fuel switching. Of all fuels used in buildings, modern biomass and renewables experience the fastest growth, with an average annual growth rate of 3.8% and 9.2% respectively over the projection period. Solar thermal accounts for 75% of this increase, as it is widely used for space and water heating. Solar heating meets 23% of the space heating and water heating demand in Japan and 8% in the United States in 2035, increasing from 2% and less than 1% respectively today. Solar is particularly important in China, which is by far the largest market for solar thermal collectors worldwide: solar thermal meets 9% of the total commercial energy demand for buildings in China in 2035. In India, as part of the national solar mission, 20 million square metres of solar collectors are installed by 2020, so solar thermal grows at an annual rate of 20% between 2009 and 2020.

Achieving the 450 Scenario requires very ambitious improvements in the energy efficiency of the buildings sector, especially in Other Major Economies. For example, in the services sector in Other Major Economies, energy intensity (energy consumed per unit of value added) needs to decline at an average annual rate of 2.2% between 2008 and 2020 and at 3.5% between 2020 and 2035. This represents a substantial change compared with the trend to date, with minimal change in energy intensity between 2000 and 2008.

CO$_2$ emissions

Global CO$_2$ emissions from the building sector decrease at an annual rate of 0.5% over the projection period in the 450 Scenario. Due to the higher energy consumption level in 2020 compared with last year's 450 Scenario, emissions from the buildings sector in 2020 are over 120 Mt higher than projected last year. Nonetheless, if the CO$_2$ emissions from the generation of electricity used in buildings are attributed to this sector (rather than being attributed to the power generation sector), incremental CO$_2$ emissions from the buildings sector are reduced by 3.5 Gt worldwide from 2008 to 2035 (Figure 14.20).

Figure 14.20 ● **Change in energy-related CO$_2$ emissions in the buildings sector by scenario and region, 2008-2035**

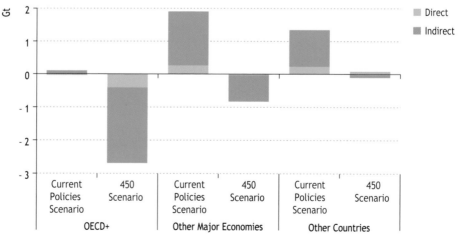

CO$_2$ emissions in Other Major Economies in the Current Policies Scenario increase by 1.9 Gt, while in the 450 Scenario they fall by 0.8 Gt. This means that the biggest reduction in incremental CO$_2$ emissions from the buildings sector between the two scenarios occurs in Other Major Economies, due to the combination of a slowdown in growth in electricity consumption, from 4% per year on average in the Current Policies Scenario to 2.7% per year in the 450 Scenario, a less carbon-intensive power sector, and a switch to renewable energy.

Investment in energy-related equipment in buildings

In the 450 Scenario, additional spending on energy-related equipment in buildings, including heating and cooling equipment, insulation, office equipment and household appliances, over the period 2010-2035 – over and above the Current Policies Scenario – is $5.6 trillion. Around 62% of the additional investment goes into more efficient electricity and heat use – including appliances, heat pumps, space heating and cooling – and 21% into renewables, notably into solar water heaters and decentralised PV. Cumulative additional investment to 2030 is 23% higher than in last year's 450 Scenario. Because most of the energy savings occur after 2020, almost 89% of the investment is needed in the period 2021-2035, with 43% being spent in the period 2031-2035. Investment needed between 2031 and 2035 almost equals that needed between 2021-2030. More than half of the incremental investment is needed in OECD+ countries, where the additional investment needs to reduce energy consumption further are high because energy use is already relatively efficient in these countries (Figure 14.21). Spending on more efficient appliances and office equipment is substantial (although the incremental cost is modest, relative to the large savings in electricity consumption achieved). Other Major Economies need to invest an additional $2 trillion, compared with the Current Policies Scenario, mainly in electrical appliances. Around 57% of this investment is needed in China. The Other Countries group need an extra $775 billion, of which just over $140 billion is invested in India.

Figure 14.21 ● Investment by region and fuel in the buildings sector

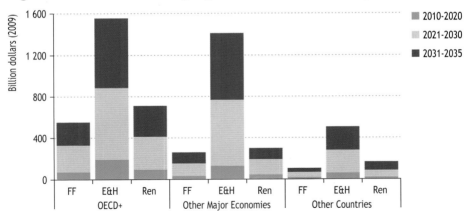

Note: FF = Fossil fuels (refers to investment in equipment in buildings powered by fossil fuels, *e.g.* gas-fired central heating).

E&H = Electricity and heat (refers to investment in appliances and lighting equipment powered by electricity generated elsewhere).

Ren = Renewables (refers to investment in renewable generation in buildings and renewable heat sources, such as solar photovoltaic generation in buildings and geothermal heat).

CHAPTER 15

IMPLICATIONS FOR OIL MARKETS
Who benefits as oil demand growth declines?

H I G H L I G H T S

- The policies to respond to the challenges of climate change and energy security that are assumed in the 450 Scenario lead to lower fuel prices. The resulting economic benefits to consuming countries offset part of the costs associated with transforming the energy sector. Nonetheless, the main oil exporters continue to benefit from growing oil exports and increasing oil-export revenues.

- Global oil demand in the 450 Scenario peaks at 88 mb/d before 2020, only 4 mb/d higher than in 2009, and then falls to 81 mb/d in 2035 (18 mb/d lower than in the New Policies Scenario). There are stark differences in demand trends between regions: OECD oil demand plummets, from 42 mb/d in 2009 to 28 mb/d in 2035, while non-OECD demand continues to rise, from 36 mb/d to 46 mb/d.

- In the 450 Scenario, global oil production peaks at just under 86 mb/d before 2020 and falls to 78 mb/d in 2035. Non-OPEC production steadily declines to under 37 mb/d in 2035, a net loss of almost 11 mb/d. OPEC production, in contrast, rises to almost 42 mb/d in 2035, an increase of 8 mb/d, thanks to its lower production costs and increased output of NGLs.

- In the 450 Scenario, there is still a need to build 50 mb/d of new capacity to compensate for falling production from existing fields. Nonetheless, the volume of oil which has to be found and developed from new sources by 2035 is only two-thirds that in the New Policies Scenario, allowing the oil industry to discard some of the more costly and more environmentally sensitive prospective projects. The 450 Scenario implies investment along the oil-supply chain of $6.4 trillion in 2010-2035, 21% less than in the New Policies Scenario.

- Energy security is enhanced in the 450 Scenario by the greater diversity of the energy mix. By 2035, the world relies on oil for about a quarter of its energy needs, seven percentage points less than today. There is greater substitution of biofuels, electricity and natural gas for oil in transport, and less reliance on oil supplies from, and transported through, politically sensitive regions.

- In the 450 Scenario, annual spending on oil imports in 2035 by the five largest importers — China, the European Union, the United States, India and Japan — is around $560 billion, or one-third, lower than in the New Policies Scenario. OPEC's cumulative oil revenues in 2010-2035 amount to $27 trillion or about $1 trillion per year. While this is 16% lower than in the New Policies Scenario, it is more than a three-fold increase compared with the last quarter century.

Introduction

Turning the 450 Scenario into reality would require an unprecedented mobilisation of finance and technology in all types of oil-consuming capital stock from cars to boilers, and aircraft to petrochemical plants. The policy measures that would drive these actions would have important repercussions on the oil market. For consuming countries, the economic benefits that would accrue from policy-driven reductions in demand and prices would help offset part of the significant costs associated with achieving the 450 Scenario (Table 15.1). Oil exporters, despite understandable concerns about lower global oil demand, would see continued growth in the demand for their oil, a rising oil price and a tripling of their revenues, compared to the last 25 years. They, too, would enjoy environmental benefits. This chapter quantifies the implications for oil demand of the policies that are assumed to be adopted in the 450 Scenario and discusses the consequences for oil prices, production, trade and investment.

Table 15.1 ● Key oil market indicators by scenario

	Current Policies Scenario (CPS)	New Policies Scenario	450 Scenario
Oil prices in 2035 ($ 2009 per barrel)	135	113	90
Oil demand in 2035 (mb/d)	107.4	99.0	81.0
OPEC production in 2035 (mb/d)	54	50	42
Peak oil demand (year)	after 2035	after 2035	2018
Conventional crude oil supply in 2035	Slightly increasing	Plateau at 68-69 mb/d	Declining
Remaining conventional recoverable oil resources in 2035 (billion barrels)	1 619	1 647	1 702
Investment in oil supply 2010-2035 ($ 2009 billion)	8 852	8 053	6 380
Additional investment in biofuels 2010-2035 compared with CPS ($ 2009 billion)	-	94	720
Additional investment in road transport 2010-2035 compared with CPS ($ 2009 billion)	-	1 770	5 492
Long-term concentration of greenhouse gases (ppm CO_2-eq)	1 000 ppm	650 ppm	450 ppm
Eventual likely temperature increase* (°C)	>6	3.5	2

*Mean of the range, from IPCC, 2007.

Demand

Primary oil demand trends

Oil demand in the 450 Scenario peaks before 2020 at slightly over 88 million barrels per day (mb/d) and declines steadily thereafter to 81 mb/d in 2035, 3 mb/d below 2009 levels (Table 15.2). This is in sharp contrast with the trends projected in the Current Policies Scenario, in which demand continues to increase to 107 mb/d in 2035, and in the New Policies Scenario, where demand grows, though less rapidly, to 99 mb/d in 2035 (see Chapter 3). The earlier peak in demand in the 450 Scenario is driven by policies that are assumed to be put in place to meet a stringent interpretation of the greenhouse-gas emission-reduction targets that have already been adopted for 2020 and to set the energy sector on a long-term trajectory that would ensure that the goal of limiting the concentration of greenhouse gases in the atmosphere to 450 parts per million (ppm) of carbon-dioxide equivalent (CO_2-eq) is achieved. Oil's role in the world primary energy mix is reduced significantly in the 450 Scenario; the global economy relies on oil for around a quarter of its energy needs in 2035, two percentage points less than in the New Policies Scenario and seven points less than today.

Policies in the transport sector account for more than three-quarters of the reduction in oil demand in 2035, relative to the New Policies Scenario; around 80% of these transport-related oil savings come from road transport. The main measures that drive these reductions in transport oil demand are international sectoral agreements that set very ambitious CO_2 emissions limits per vehicle, and gasoline and diesel pricing and tax policies (see Chapter 14).

Table 15.2 ● **Primary oil demand* by region in the 450 Scenario** (mb/d)

	1980	2009	2020	2030	2035	2009-2035**	Change vs NPS in 2035	Change vs CPS in 2035
OECD	**41.3**	**41.7**	**38.2**	**31.9**	**28.0**	**-1.5%**	**-21%**	**-28%**
North America	20.8	22.0	20.6	17.1	14.7	-1.5%	-24%	-29%
United States	*17.4*	*17.8*	*16.6*	*13.5*	*11.4*	*-1.7%*	*-24%*	*-29%*
Europe	14.4	12.7	11.5	9.7	8.7	-1.4%	-16%	-27%
Pacific	6.1	7.0	6.1	5.1	4.6	-1.6%	-18%	-23%
Japan	*4.8*	*4.1*	*3.3*	*2.7*	*2.4*	*-2.1%*	*-18%*	*-25%*
Non-OECD	**20.0**	**35.8**	**42.2**	**44.9**	**45.6**	**0.9%**	**-16%**	**-23%**
E. Europe/Eurasia	9.1	4.6	4.8	4.8	4.7	0.1%	-12%	-19%
Russia	*n.a*	*2.8*	*2.8*	*2.7*	*2.7*	*-0.1%*	*-11%*	*-14%*
Asia	4.4	16.4	21.1	24.6	25.9	1.8%	-14%	-19%
China	*1.9*	*8.1*	*11.4*	*12.8*	*13.1*	*1.9%*	*-14%*	*-19%*
India	*0.7*	*3.0*	*4.1*	*5.7*	*6.6*	*3.1%*	*-12%*	*-20%*
ASEAN	*1.1*	*3.7*	*4.0*	*4.2*	*4.4*	*0.7%*	*-16%*	*-21%*
Middle East	2.0	6.5	7.7	7.4	7.2	0.4%	-22%	-31%
Africa	1.2	3.0	3.0	2.9	2.9	-0.1%	-24%	-30%
Latin America	3.4	5.3	5.6	5.2	5.0	-0.3%	-21%	-28%
World***	**64.8**	**84.0**	**87.7**	**84.1**	**81.0**	**-0.1%**	**-18%**	**-25%**
European Union	*n.a*	*12.2*	*10.9*	*9.1*	*8.1*	*-1.6%*	*-16%*	*-28%*

*Excludes biofuels demand, which is projected to rise from 1.1 mb/d (in energy-equivalent volumes of gasoline and diesel) in 2009 to 2.6 mb/d in 2020 and to 8.1 mb/d in 2035. See Chapter 3 for a precise definition of oil in this *WEO*.
** Compound average annual growth rate. *** Includes international marine and aviation fuel.
Note: NPS = New Policies Scenario; CPS = Current Policies Scenario.

15

Regional trends

The global oil demand trends in the 450 Scenario mask stark differences between regions. Oil demand in OECD countries declines steadily, from around 42 mb/d in 2009 to 28 mb/d in 2035, while demand in non-OECD countries increases, from 36 mb/d to 46 mb/d, over the same period (Figure 15.1). China, where demand grows by 5.0 mb/d; India, 3.6 mb/d; the Middle East, 0.7 mb/d; and ASEAN countries, 0.7 mb/d, account for most of the global increase. Despite the measures introduced in the transport sector, the spectacular growth in the vehicle stock in those countries continues to push up their oil use. China becomes the largest oil consumer soon after 2030, surpassing the United States, where demand is in decline. Non-OECD Asia gains 13 percentage points in market share over the *Outlook* period, accounting for almost a third of global oil demand by 2035.

Figure 15.1 ● **Change in oil demand by region in the 450 Scenario compared with 2008**

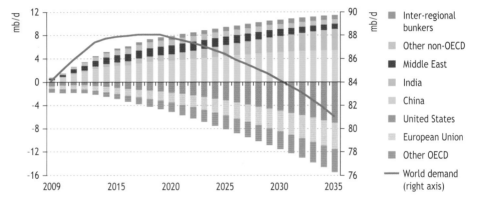

Sectoral trends

In the 450 Scenario, with the exception of transport and industry, global oil demand declines in all sectors between 2009 and 2020 (Figure 15.2). This is in contrast to the New Policies Scenario, in which only the power sector sees a reduction. After 2020, global oil demand falls even in transport and industry, as increasingly stringent policies take effect. During that period, demand in the transport sector declines most in absolute terms, due to its magnitude and the fact that the limited remaining oil use in other sectors (for example, diesel generators in rural areas and oil used as feedstock for petrochemicals and chemicals) is the most costly and difficult to displace. Although the share of oil use declines steeply in all sectors after 2020, oil remains the dominant fuel in the transport sector and in non-energy use (Table 15.3).

Figure 15.2 ● Annual average change in world oil demand by sector in the 450 Scenario

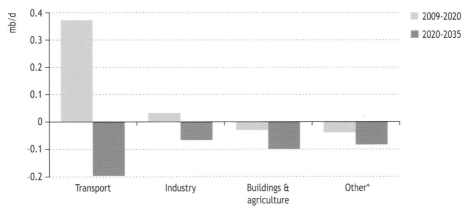

*Includes power generation, other energy sectors and non-energy use.

Table 15.3 ● World oil demand by sector in the 450 Scenario (Mtoe)

| | 2009 | 2020 | 2035 | Oil share of total sectoral fuel consumption | | |
				2009	2020	2035
Power generation	238	161	99	5%	3%	2%
Other energy sector	281	263	195	22%	19%	14%
Industry	333	351	301	14%	12%	10%
Transport	2 138	2 336	2 202	93%	90%	77%
Buildings and agriculture	450	433	362	15%	13%	10%
Non-energy use	572	632	657	75%	73%	71%
Total	4 012	4 175	3 816	33%	30%	26%

Mtoe = million tonnes of oil equivalent.

15

Impact of lower oil demand on oil prices

In the 450 Scenario, crude oil import prices increase more slowly than in the other scenarios in *WEO-2010*, reflecting lower demand. In real terms, the price needed to balance supply and demand reaches $90/barrel (in 2009 dollars) in 2020 and remains stable at that level thereafter (Figure 15.3). The IEA crude oil import price in 2025 is, on average, $15/barrel lower than in the New Policies Scenario; in 2035, it is $23/barrel lower. Compared with the Current Policies Scenario, prices are $30/barrel lower in 2025 and $45/barrel lower in 2035. Nonetheless, in the 450 Scenario, there is a price increase between 2009 and 2035 of almost $30/barrel, or 49% in real terms.

Figure 15.3 ● *Average IEA crude oil import price by scenario*

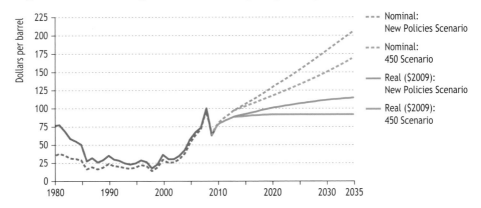

Oil-related CO$_2$ emissions

In the 450 Scenario, the global economy relies on oil for around a quarter of its energy needs in 2035, seven percentage points less than today. However, despite the policy-driven fall in demand for oil in the 450 Scenario, the share of oil in global CO$_2$ emissions actually increases from 37% in 2008 to 46% in 2035 (Figure 15.4). By contrast, the share falls slightly in the New Policies Scenario. In the 450 Scenario, soon after 2025, oil overtakes coal to become the leading source of emissions from fossil-fuel consumption, as demand for coal falls even more sharply. The message is twofold: first, climate mitigation strategies that do not tackle oil use (and oil in transport in particular) will fail in the longer term; second, climate change mitigation and energy diversification away from oil are closely interlinked.

Figure 15.4 ● *Share of world energy-related CO$_2$ emissions by fuel and scenario*

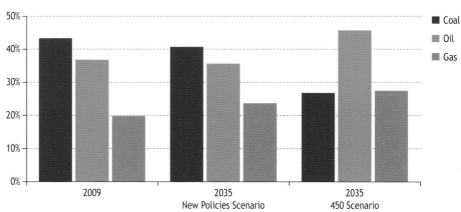

Production

Lower oil demand growth in the 450 Scenario obviously means that oil production grows less too. The strong policies to reduce oil demand that are assumed to be adopted to respond to the challenge of climate change result in a peak in global oil production of just under 86 mb/d before 2020 (Figure 15.5), production following an undulating plateau for much of the 2010s.[1] From around 2020, global oil production gradually declines, reaching 78 mb/d in 2035.

Figure 15.5 ● **World oil production by source in the 450 Scenario**

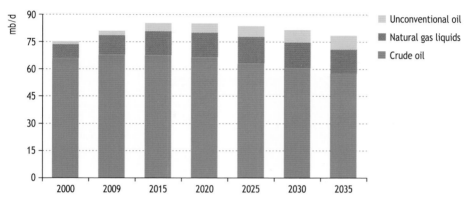

In the 450 Scenario, the breakdown of oil production, both between the different types of oil supply, and between OPEC and non-OPEC, changes notably over the projection period (Table 15.4). Global production of conventional crude oil declines continuously over the next quarter century, from 68 mb/d in 2009 to 58 mb/d in 2035. In contrast, production of natural gas liquids (NGLs) rises, from around 11 mb/d today to over 13 mb/d in 2035, and their share of total production increases from 13% to 17%. The rising share of NGLs results from the quicker growth in production of natural gas relative to oil in the 450 Scenario and because an increasing share of gas production occurs in regions with "wet gas", *i.e.* gas that contains a significant amount of NGLs.

The role of unconventional oil in world oil production also expands, albeit to a lesser extent than in the New Policies Scenario. Production increases from 2.3 mb/d in 2009 to 7.4 mb/d in 2035 (2.1 mb/d less than in the New Policies Scenario). By 2035, unconventional oil represents 9% of global production, compared with 3% in 2009. Growth is fastest in the current decade and then tapers off, with declining world oil demand, causing oil prices to level off, which reduces the attractiveness of investing in projects to develop these higher-cost resources. Although the production of unconventional sources of oil generally emits significantly more greenhouse gases than most conventional sources, growth in output is assumed to be made possible by the introduction of new

15

---

1. Production is total supply (which equals demand) less volumetric processing gains.

technologies which reduce emissions. Canadian oil sands remain the main source of unconventional supply, with just over 3 mb/d of production in 2035. Venezuelan extra-heavy oil also continues to play a significant role, together with coal-to-liquids (CTL), gas-to-liquids (GTL) and, to a lesser extent, oil shales (see Chapter 4).

Table 15.4 ● *Oil supply by source in the 450 Scenario* (mb/d)

	1980	2009	2015	2020	2025	2030	2035	2009-2035*
OPEC	25.5	33.4	38.9	40.1	41.1	41.5	41.7	0.9%
Crude oil	24.7	28.3	30.6	31.4	31.0	31.4	31.8	0.5%
Natural gas liquids	0.9	4.6	6.9	7.1	8.3	8.0	7.6	1.9%
Unconventional	0.0	0.5	1.4	1.6	1.8	2.0	2.3	6.1%
Non-OPEC	37.1	47.7	46.4	45.1	42.7	40.1	36.7	-1.0%
Crude oil	34.1	39.6	37.0	35.1	32.2	29.2	25.9	-1.6%
Natural gas liquids	2.8	6.2	6.5	6.5	6.5	6.3	5.7	-0.3%
Unconventional	0.2	1.8	3.0	3.4	4.0	4.6	5.1	4.0%
World production	62.6	81.0	85.3	85.2	83.8	81.6	78.5	-0.1%
Crude oil	58.8	67.9	67.6	66.5	63.2	60.6	57.7	-0.6%
Natural gas liquids	3.7	10.8	13.3	13.6	14.8	14.3	13.3	0.8%
Unconventional	0.2	2.3	4.4	5.0	5.8	6.6	7.4	4.5%
Processing gains	1.2	2.3	2.5	2.5	2.5	2.5	2.5	0.3%
World supply	63.8	83.3	87.8	87.7	86.3	84.1	81.0	-0.1%
*World liquids supply**	*63.9*	*84.4*	*89.6*	*90.3*	*90.6*	*90.1*	*89.1*	*0.2%*

* Compound average annual growth rate.
** Includes biofuels (see Chapter 12 for details of biofuels projections).

Non-OPEC oil production in the 450 Scenario declines steadily to less than 37 mb/d in 2035, a net loss of production of almost 11 mb/d compared with today and 9 mb/d less than in the New Policies Scenario (Figure 15.6). Lower oil prices reduce the profitability of new investment in the relatively high-cost resources in non-OPEC regions, which become increasingly expensive to produce over time. The resultant fall in investment accentuates the decline in mature basins in non-OPEC regions. The fall in non-OPEC output accelerates through the *Outlook* period, reaching an average of 700 kb/d per year in the first half of the 2030s. OPEC production, in contrast, rises to over 40 mb/d in 2020 and almost 42 mb/d in 2035, an increase of 8 mb/d, thanks to its lower production costs, which leave it less affected by the drop in oil prices, and increased output of NGLs. Although the increase in OPEC production over the 25-year period is 8 mb/d less than in the New Policies Scenario, it is still bigger than the increase in OPEC production in 1980-2009. OPEC's share of world production rises considerably in the 450 Scenario, from 41% in 2009 to 53% in 2035.

Even though global oil production drops by 2.5 mb/d between 2009 and 2035 in the 450 Scenario, there is still a need to develop some 50 mb/d of new capacity in order to compensate for the decline in production at existing fields as they pass

Figure 15.6 ● Change in oil production by source and scenario, 2009-2035

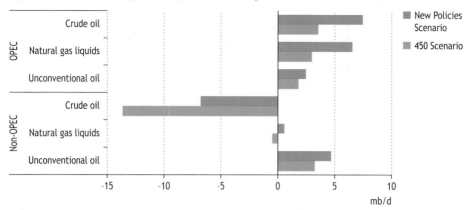

their peak level of production and flow-rates begin to drop (Figure 15.7). This is just over four times the current production capacity of Saudi Arabia. However, the need for exploration to find and then develop reservoirs that are as yet unknown is only two-thirds of that in the New Policies Scenario, a difference of almost 60 billion barrels. This reduction is equivalent to two-thirds of the estimated volume of oil that is thought to remain to be found in the Arctic and is comparable to the total volume of oil discovered during the past five years. As the oil industry typically develops easy-to-find oil first, this reduced need to bring on new capacity allows the industry to dispense with some of the more costly and more environmentally sensitive projects.

OPEC's cumulative production of conventional oil (crude and NGLs) in the 450 Scenario is some 18 billion barrels lower in the period 2009-2035 than in the New Policies Scenario. This amounts to 1.5 years of output at current rates of production that would remain in the ground to be produced when conditions make this economically advantageous. Particularly in OECD countries, where oil demand falls most rapidly, the fall in demand for oil products projected in the 450 Scenario is likely to speed up the closure of smaller, less profitable refineries.

Figure 15.7 ● World oil production by type in the 450 Scenario

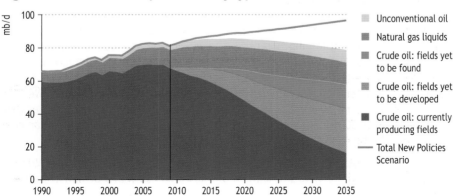

Investment

The oil production trends in the 450 Scenario imply a need for cumulative investment along the oil-supply chain of over $6.4 trillion (in 2009 dollars) in 2010-2035. Capital spending amounts on average to $245 billion per year, but it falls through the *Outlook* period as global oil demand drops, production shifts increasingly towards less costly regions and technology drives down unit costs. Almost three-quarters of projected oil-sector investment is needed in non-OECD regions (Figure 15.8). Investment in OECD countries is high relative to OECD production capacity, because of higher unit costs.

Figure 15.8 ● **Cumulative oil sector investment* by region and activity in the 450 Scenario, 2010-2035**

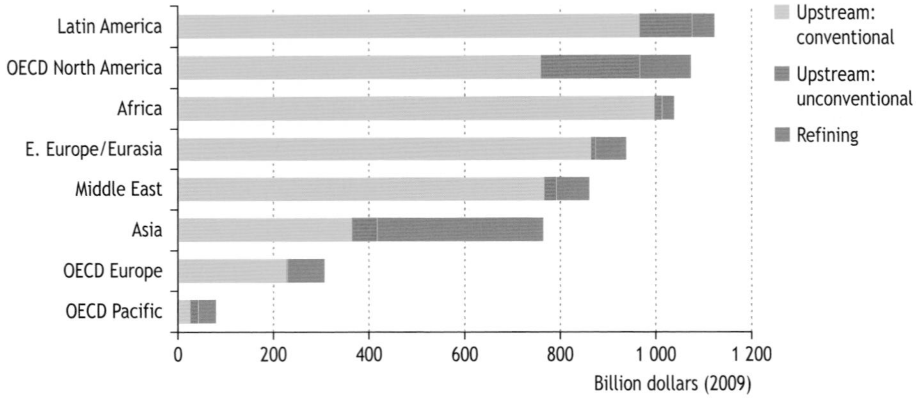

*Excludes an additional $210 billion of investment in oil tankers and oil pipelines for international trade.

Capital spending on upstream exploration and development dominates oil-sector investment, accounting for 85% of the total. Approximately $5 trillion is invested in conventional oil developments and some $440 billion in unconventional oil projects. Almost 12% of total spending is directed to investments in oil refining, where it increasingly goes towards improving conversion and quality-treatment capability to meet ever more stringent fuel-quality standards. Investment in oil tankers and oil pipelines for international trade amounts to $210 billion in 2010-2035.

Investment in oil supply in the 450 Scenario is 21% lower than in the New Policies Scenario, with the bulk of the reduction coming after 2020. This drop results from the reduced need to bring on new production capacity, including the most costly deepwater offshore oil projects. Upstream investment by OPEC countries, which are together responsible for the bulk of the projected increase in supply, is $310 billion, or 15% lower than in the New Policies Scenario.

What role can biofuels play in a carbon-constrained world?

In the 450 Scenario, biofuels production increases from 1.1 mb/d in 2009 to 8.1 mb/d in 2035 — 3.7 mb/d (or 85%) more than in the New Policies Scenario. The 7 mb/d increase in biofuels production over the *Outlook* period is the largest among all the different sources of liquid fuels, making the contribution of biofuels to global liquid supply as large as unconventional oil by 2035, each contributing around 8-9%. Biofuels are not included in oil in our definition, but are a substitute for it, so policies to promote biofuels can have a major impact on oil demand and the need to develop new oil supplies.

In the 450 Scenario, the global output of advanced biofuels, such as ligno-cellulosic ethanol, reaches 5.3 mb/d in 2035, equal to roughly two-thirds of total biofuels production. Because of worries about food security, the potential impact on greenhouse-gas emissions of land use changes (see Chapter 12), water resources and biodiversity, the potential for increasing the output of conventional biofuels, with the exception of sugar cane ethanol, is limited. Though not yet produced commercially, advanced biofuels, characterised by the use of non-food biomass feedstocks, such as woody and cellulosic plants and waste material, promise to overcome those issues. The potential is large. If around 10% of global agricultural and forestry residues was diverted to this purpose, it would be enough to cover all of the biomass needs for advanced biofuels production in the 450 Scenario.

Development of advanced biofuels is seen as essential to displace middle distillate petroleum fuels in trucks, ships and aircraft beyond 2035. Reaching the levels of production described in the 450 Scenario will be very difficult; the need for investment in research and development is estimated to be on the order of $100-120 billion to 2030 (IEA, 2009). Key short-term objectives include:

- Cutting the production costs of ligno-cellulosic ethanol by 2020 to $0.60 per litre of gasoline equivalent, mainly via improved enzymes.

- Cutting the production costs of biomass-to-liquids by 2020 to $0.70 per litre of gasoline equivalent, by optimising biomass gasification and synthesis-gas production.

The IEA's *Energy Technology Perspectives 2010* study shows very real signs that some of the necessary changes are starting to occur, in part due to recent implementation of "green" stimulus package funding for clean energy technologies (IEA, 2010). Current support policies for advanced biofuels in the United States and, to a lesser extent, in Europe, are an essential first step in the right direction, but will be insufficient to mobilise the full resources needed for the 450 Scenario. The engagement of those developing countries with large biomass resources and with well-developed infrastructure, such as Brazil and China, will be key to the successful roll-out of advanced biofuels technology.

15

Implications for oil-importing countries

Oil trade

At the global level, the volume of inter-regional oil trade in the 450 Scenario expands until around 2020 before starting to decline. By 2035 it reaches 39.5 mb/d, compared with 36.7 mb/d in 2009. Oil imports into the OECD drop sharply over the *Outlook* period, but this is more than offset by an increase in demand for imports from other regions (Table 15.5). In the United States, oil imports drop by 45%, from 10.4 mb/d in 2009 to 5.7 mb/d in 2035 — a level last seen in the mid-1980s. All other OECD countries also see a decline in their oil-import requirements, compared with current levels, ranging from a 15% cut in OECD Europe to a 42% cut in Japan. The savings are significant *vis-à-vis* the New Policies Scenario. For the OECD in aggregate, oil net imports in 2035 in the 450 Scenario are 3.7 mb/d lower than in the New Policies Scenario.

Table 15.5 ● **Oil net imports in key regions in the 450 Scenario** (mb/d)

	2009	2020	2035	Change vs. NPS		Change vs. CPS	
				2020	2035	2020	2035
OECD	23.0	22.4	14.1	-2%	-21%	-4%	-30%
North America	8.4	8.2	3.0	2%	-33%	1%	-39%
United States	*10.4*	*10.2*	*5.7*	*-1%*	*-27%*	*-1%*	*-34%*
Europe	8.2	8.6	7.0	-3%	-16%	-7%	-29%
Pacific	6.4	5.5	4.1	-5%	-19%	-7%	-24%
Japan	*4.0*	*3.3*	*2.3*	*-5%*	*-18%*	*-8%*	*-25%*
Korea	*2.0*	*1.9*	*1.5*	*-5%*	*-13%*	*-6%*	*-15%*
China	4.3	7.9	10.7	-2%	-17%	-3%	-21%
India	2.2	3.4	6.0	-2%	-11%	-8%	-19%
Indonesia	0.3	0.3	0.5	-14%	-40%	-21%	-45%
World*	36.7	40.9	39.5	-3%	-18%	-5%	-25%
European Union	*10.0*	*9.7*	*7.5*	*-4%*	*-16%*	*-8%*	*-29%*

* Total net imports for all *WEO* regions/countries (some of which are not shown in this table), not including trade within *WEO* regions.

In contrast with the OECD, non-OECD Asian countries see an increase in imports in the 450 Scenario, albeit not to the extent projected in the New Policies Scenario. Growth in demand from increasing vehicle ownership and industrial activity more than offsets the impact of strong demand-side efficiency and fuel diversification policies. China and India experience the biggest jump in absolute terms. China's net imports grow from 4 mb/d in 2009 to 11 mb/d in 2035 — but this is still a reduction of over 2 mb/d, compared with the New Policies Scenario.

The fall in oil trade seen in the 450 Scenario, compared with the New Policies Scenario, would have several other important implications. The volume of oil transiting key choke-points (such as the Strait of Hormuz, Strait of Bab el-Mandab and the Suez Canal) would be lower than in the New Policies Scenario. Furthermore, the absolute volumes of oil stocks IEA countries are obliged to hold to meet their membership obligations (equivalent to 90 days of oil net imports) would be lower in 2035 than today, with a commensurate reduction in the cost of maintaining oil storage. For China and India — both of which are now developing strategic oil storage facilities — cumulative spending on oil storage in the 450 Scenario to maintain the same level of emergency preparedness would be much less onerous than in the New Policies Scenario.

Oil-import bills and intensity

Lower oil-import requirements and lower international oil prices significantly reduce oil import bills in the 450 Scenario, compared with the New Policies Scenario. In 2035, the five-largest importers — China, the European Union, the United States, India and Japan — collectively spend around $560 billion, or one-third, less than in the New Policies Scenario. These savings increase over time as the impact of efficiency and diversification measures grows and as the difference between oil prices in the different scenarios increases.

In some OECD importing countries, oil-import bills are lower in 2035 than in 2009. The oil-import bill in the United States peaks in 2015, at around $350 billion, and declines to some $190 billion in 2035, 19% below 2009 levels and less than half the peak value reached in 2008. The savings for the United States are also very large compared to the import bill in the New Policies Scenario, almost $135 billion in 2035. Among OECD countries, the proportionate impact on the import bill is highest in the United States, but the reduction in other countries is also marked (Figure 15.9). In the European Union, import bills peak around 2015, at $320 billion, and decline steadily to $250 billion in 2035. This level is slightly higher than the 2009 level, but 33% lower than the peak value reached in 2008.

Figure 15.9 ● Oil-import bills in selected countries by scenario

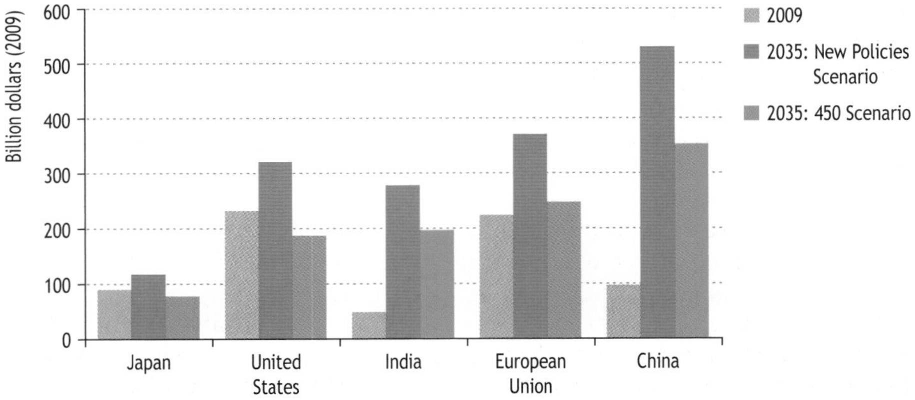

Note: Calculated as the value of net imports at prevailing average international prices. The split between crude/refined products is not taken into account.

Spending on oil imports by China and India increases in the 450 Scenario, compared with current levels, but is significantly lower than in the New Policies Scenario. In 2035, China's spending on oil imports is almost $180 billion (or 34%) and India's $80 billion (or 29%) lower than in the New Policies Scenario. Nonetheless, at around $350 billion and

15

$190 billion respectively in 2035, China's oil-import bill overtakes that of the United States around 2025; and India's overtakes that of the United States, to take second position, by 2035.

The 450 Scenario projections imply a declining level of spending on oil imports as a share of GDP in all major importing countries (Figure 15.10). This share spiked in 2008, following the run-up in oil prices and the global economic slow-down. In 2035, oil-import spending represents less than 1% of GDP in the United States and the European Union, down from 2.8% and 2.2% respectively in 2008. As a share of GDP, oil-import bills in China and India are lower in 2035 than in 2009. They represent a higher percentage of GDP in the New Policies Scenario.

Figure 15.10 ● Oil-import bills as a share of GDP at market exchange rates in selected countries by scenario

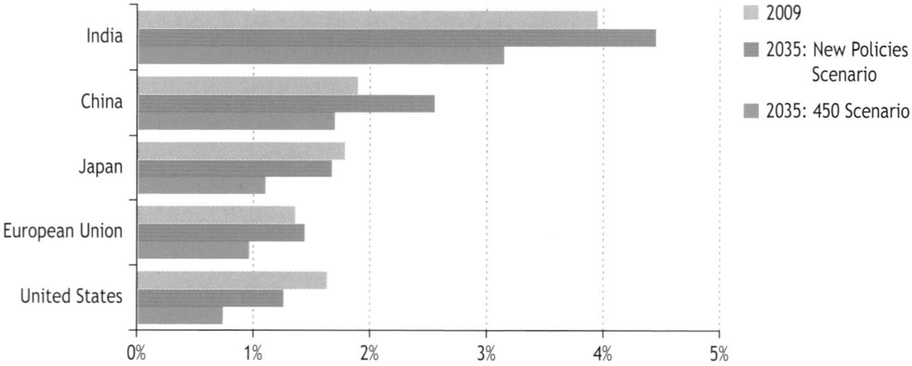

Note: Calculated as the value of net imports at prevailing average international prices. The split between crude/refined products is not taken into account.

The policies that are assumed to be adopted in the 450 Scenario improve the efficiency of oil use and diversify the energy mix, in favour of lower carbon sources. This leads to a significant reduction in oil intensity — measured as oil use per dollar of GDP — over the *Outlook* period, reducing the vulnerability of oil-consuming countries to price volatility (Table 15.6).

Table 15.6 ● Oil intensity by region in the 450 Scenario
(toe per thousand $ of GDP at market exchange rates)

	2009	2015	2020	2025	2030	2035	Change 2009-2035
United States	0.058	0.049	0.041	0.034	0.027	0.020	-65%
European Union	0.036	0.030	0.026	0.022	0.018	0.015	-58%
Japan	0.038	0.032	0.026	0.022	0.019	0.016	-58%
China	0.077	0.057	0.046	0.040	0.035	0.030	-61%
India	0.117	0.088	0.073	0.064	0.058	0.051	-56%
Middle East	0.198	0.172	0.144	0.116	0.092	0.073	-63%

Implications for oil-producing countries

Domestic energy use and related emissions

Although the major oil producers would export less oil in the 450 Scenario than in the other two scenarios, there are at least partially offsetting economic and environmental benefits. In the 450 Scenario, primary energy demand in the Middle East rises much less than in the New Policies Scenario, by 41% between 2008 and 2035, against 69% in the New Policies Scenario. Per-capita energy consumption in the Middle East in the 450 Scenario declines at 0.2% per year on average over the projection period, reaching about 2.8 tonnes of oil equivalent (toe) in 2035 (Figure 15.11). This is still high compared with the global average (as a result of the region's hot climate, which necessitates considerable air-conditioning, the importance of energy-intensive industries in the economy and relatively inefficient energy production and consumption practices), but is 17% lower than the level reached in the New Policies Scenario. The fall in energy intensity in the 450 Scenario gathers pace towards the end of the *Outlook* period.

Figure 15.11 ● *Energy intensity and per-capita consumption in the Middle East by scenario*

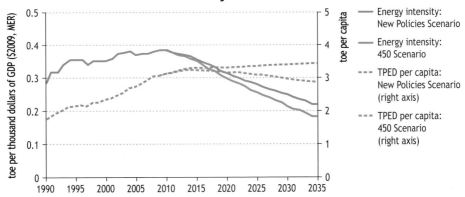

Note: MER = market exchange rate. TPED = total primary energy demand.

15

Reduced use of energy — mainly oil and gas — would bring important environmental benefits. Though the Middle East region's energy-related CO_2 emissions in the 450 Scenario continue to increase until soon after 2020, from just under 1.5 gigatonnes (Gt) in 2008 to a peak of 1.9 Gt, they then fall to around 1.7 Gt in 2035, the region making a growing contribution to the global abatement of CO_2 (Table 15.7). By the end of the *Outlook* period, the Middle East's CO_2 emissions are more than a quarter below the level reached in the New Policies Scenario. The policies assumed to be adopted in the 450 Scenario in the Middle East also lead to a big reduction in the emission of local air pollutants. By 2035, the region's sulphur-dioxide emissions are 25% lower than in the New Policies Scenario. Nitrogen-oxide emissions are reduced by 21% and emissions of particulates ($PM_{2.5}$) also drop.

Table 15.7 ● *Emissions of energy-related CO₂ and major air pollutants in the Middle East by scenario* (million tonnes)

		New Policies Scenario		450 Scenario		2008-2035*	Change vs. NPS in 2035
	2008	2020	2035	2020	2035		
CO₂	1 476	1 934	2 354	1 833	1 735	0.6%	-26%
Sulphur dioxide (SO₂)	4.56	3.80	3.52	3.72	2.65	-2.0%	-25%
Nitrogen oxides (NOₓ)	4.26	4.47	5.38	4.32	4.26	0.0%	-21%
Particulate matter (PM₂.₅)	0.76	0.82	0.75	0.82	0.71	-0.2%	-6%

* Compound average annual growth rate in the 450 Scenario.
Source: Information on local pollutants from IIASA (2010).

Oil exports and revenues

Despite lower global demand for oil in the 450 Scenario, oil exports by OPEC producers increase from 26 mb/d in 2009 to 34 mb/d in 2035. An increasing share of this oil production is directed towards exports. This results from the decline in the rate of growth in domestic demand, thanks to the assumed reduction in subsidies and the introduction of more efficient cars and trucks, as OPEC countries benefit from technology spill-over from the faster deployment of advanced vehicles in global markets.

OPEC's cumulative oil revenues in the 450 Scenario in 2010-2035 are projected to amount to $27 trillion in 2009 dollars (Figure 15.12). While this is 16% lower than earnings in the New Policies Scenario, it is still three times more in real terms than their earnings over the last quarter century. Moreover, as a result of the assumed reduction in subsidies, in the Middle East state revenues from domestic sales of oil products increase by $430 billion, compared with the New Policies Scenario.

Figure 15.12 ● *Cumulative OPEC oil-export revenues by scenario*

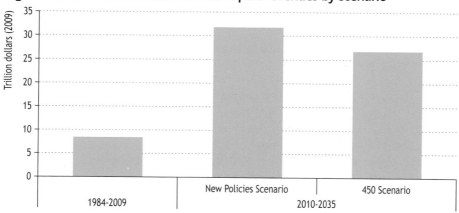

Note: Calculated as the value of net exports at prevailing average international prices. The split between crude/refined products is not taken into account.

PREFACE

In recognition of the growing importance of Caspian countries in the global energy market, Part D of this *Outlook* takes a detailed look at the internal energy markets in the region and its oil and gas supply potential. The regional supply and demand balance is assessed together with the prospects for the production and export of oil and gas, the factors affecting their prospects (including potential barriers to investment), and the implications of energy developments in the region for global energy security and environmental sustainability.

Chapter 16 provides an overview of the current status of and recent trends in Caspian energy markets, describes the macroeconomic and political context and the assumptions underpinning the analysis. This chapter assesses the main drivers of demand and offers detailed demand projections for the major countries and the region as a whole, under different scenarios, with a particular focus on the New Policies Scenario. Chapter 17 details, country-by-country, the hydrocarbon and coal resources of the region and the prospects for their development, assessing the difficulty of extraction and the related costs, the potential for expanding the transportation infrastructure for domestic markets and exports, the trends in current investment and future investment needs. Chapter 18 integrates the regional outlook into the global picture, quantifying the implications for global energy security and climate change.

These chapters use the terms "Caspian" and "Caspian region" as shorthand for a diverse group of countries in the South Caucasus and in Central Asia: Armenia, Azerbaijan and Georgia in the South Caucasus; and Kazakhstan, the Kyrgyz Republic, Tajikistan, Turkmenistan and Uzbekistan on the eastern side of the Caspian Sea (these latter five countries are collectively described as Central Asia in the text). The main focus is on Azerbaijan, Kazakhstan, Turkmenistan and Uzbekistan, the largest consumers and producers of energy in the region. This approach excludes two countries, Russia and Iran, which, as littoral states, are undeniably Caspian. However, including these two countries fully within this analysis would also have the effect of broadening its scope far beyond the Caspian Sea, particularly since the bulk of the huge energy resources of both Russia and Iran lie outside the Caspian basin. The Russian and Iranian role in the Caspian region is therefore a part of the discussion, particularly as it relates to oil and gas exploration and transportation, but these countries are not included in the regional projections and detailed analysis.

CHAPTER 16

CASPIAN DOMESTIC ENERGY PROSPECTS

How much energy will Caspian countries need?

H I G H L I G H T S

- Although Caspian countries account for only 1.4% of global primary energy use, policies and trends in their domestic energy use, beyond being critical to the region's social and economic development, have an influence on world prospects by determining the volumes available for export.

- Recent trends in Caspian energy use have been shaped by the economic upheavals and, in some cases, political instability and conflict that followed the collapse of the Soviet Union. An economic revival throughout the region began in the late 1990s and reversed the overall decline in energy use, though the region's primary energy demand in 2008 was still only 85% of that in the early 1990s.

- Yet the region remains highly energy-intensive, reflecting continuing gross inefficiencies in the way energy is used, as well as climatic and structural economic factors. If the region were to use energy as efficiently as OECD countries, consumption of primary energy in the Caspian as a whole would be cut by 80 Mtoe, or one-half. How quickly this energy-efficiency potential might be exploited hinges largely on government policies, especially on energy pricing (all main Caspian countries subsidise at least one form of energy), market reform and improved access to financing for energy projects.

- Caspian energy demand in aggregate expands progressively between 2008 and 2035 in all three scenarios. In the New Policies Scenario, total Caspian primary energy demand grows at an average rate of 1.4% per year, reflecting an absence of strong policy intervention to curb the growth in energy use; by 2035, demand is about 50% higher than in 2008. Primary energy intensity intensity falls by 47% between 2008 and 2035, approaching the current average level in the rest of the world. It falls 56% in the 450 Scenario and 43% in the Current Policies Scenario.

- The primary energy mix changes little over the *Outlook* period. Fossil fuels account for 95% of Caspian primary energy mix in 2035 in the New Policies Scenario, down only slightly from today and implying little progress towards more sustainable energy use. Natural gas sees the biggest increase in absolute terms, though its share of total demand rises only slightly, from 60% to 62%. Oil's share stays flat, as rising demand for transport is partially offset by lower use in power plants. The contribution of modern renewables remains marginal.

- Kazakhstan and Turkmenistan see the fastest rates of growth in energy use, reflecting relatively rapid economic growth and, in the case of Turkmenistan, persistently high subsidies assumed. Their economies remain highly energy-intensive. The more modest increase in energy needs in Azerbaijan and Uzbekistan are met primarily by gas and oil.

Figure 16.1 ● Key energy features of Caspian countries

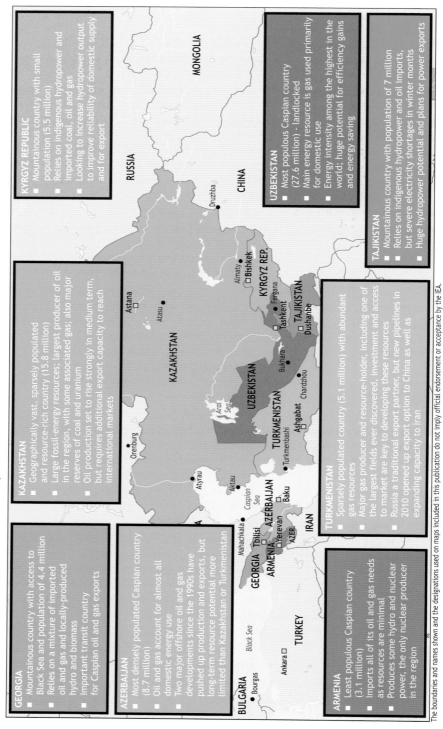

GEORGIA
- Mountainous country with access to Black Sea and population of 4.4 million
- Relies on a mixture of imported oil and gas and locally-produced hydro and biomass
- Important transit country for Caspian oil and gas exports

AZERBAIJAN
- Most densely populated Caspian country (8.7 million)
- Oil and gas account for almost all domestic energy use
- Two major offshore oil and gas developments since the 1990s have pushed up production and exports, but long-term resource potential more limited than Kazakhstan or Turkmenistan

ARMENIA
- Least populous Caspian country (3.1 million)
- Imports all of its oil and gas needs as resources are minimal
- Produces some hydro and nuclear power; the only nuclear producer in the region

KAZAKHSTAN
- Geographically vast, sparsely populated and resource-rich country (15.8 million)
- Large fossil-energy resources; largest producer of oil in the region, with some associated gas; also major reserves of coal and uranium
- Oil production set to rise strongly in medium term, but requires additional export capacity to reach international markets

KYRGYZ REPUBLIC
- Mountainous country with small population (5.5 million)
- Relies on indigenous hydropower and imported coal, oil and gas
- Looking to increase hydropower output to improve reliability of domestic supply and for export

UZBEKISTAN
- Most populous Caspian country (27.6 million) - landlocked
- Main energy resource is gas used primarily for domestic use
- Energy intensity among the highest in the world; huge potential for efficiency gains and energy saving

TAJIKISTAN
- Mountainous country with population of 7 million
- Relies on indigenous hydropower and oil imports, but severe electricity shortages in winter months
- Huge hydropower potential and plans for power exports

TURKMENISTAN
- Sparsely populated country (5.1 million) with abundant gas resources
- Major gas producer and resource-holder, including one of the largest fields ever discovered, investment and access to market are key to developing these resources
- Russia a traditional export partner, but new pipelines in 2010 opened up export option to China as well as expanding capacity to Iran

The boundaries and names shown and the designations used on maps included in this publication do not imply official endorsement or acceptance by the IEA.

Overview of Caspian energy

The Caspian region, endowed with abundant hydrocarbon and other energy resources, is set to emerge as an important contributor to global energy supplies and, therefore, to world energy security (Figure 16.1). The region has significantly expanded its oil and gas exports to international markets since the beginning of the 1990s (Table 16.1) and Azerbaijan, Kazakhstan, Turkmenistan and, to a lesser extent, Uzbekistan all have the potential to increase hydrocarbon production in the coming years. The increases in output so far have been associated with, and encouraged by, an emerging diversity of export routes and markets, first for oil and more recently for gas, diminishing reliance on export routes through Russia (see Chapter 17 for a detailed discussion).

Table 16.1 ● Key energy indicators for the Caspian*

	Unit	1990	2000	2008	2000-2008**
GDP (MER)	$2009 billion	157	105	225	10.0%
GDP (PPP)	$2009 billion	299	201	427	9.9%
Population	million	66	71	76	0.9%
Primary energy demand	Mtoe	198	128	169	3.6%
Primary energy demand per capita	toe	2.99	1.80	2.22	2.6%
Energy intensity	toe per $1 000 (2009, MER)	1.27	1.22	0.75	-5.9%
Oil net trade***	mb/d	-0.24	0.76	2.10	13.5%
Natural gas net trade***	bcm	40.77	33.52	63.42	8.3%
Energy-related CO_2 emissions	Mt	547.9	320.9	412.0	3.2%

* Armenia, Azerbaijan, Georgia, Kazakhstan, Kyrgyz Republic, Tajikistan, Turkmenistan and Uzbekistan.
** Compound average annual growth rate. *** Negative value indicates imports.
Note: MER = market exchange rate. PPP = purchasing power parity.

The eight countries included in this analysis cover a large geographic area, a total of over 4 million square kilometres across the heart of Eurasia, from the Black Sea in the west to the Chinese border in the east, with Kazakhstan accounting for almost two-thirds of this. But the combined population of these eight countries, at 76 million in 2008, is relatively small, only slightly more than that of Turkey, at 75 million. With the exception of Uzbekistan, Caspian energy producers do not have large domestic markets for their hydrocarbon production and so have to look outside the region for opportunities to monetise their resources. In doing so, they usually have to rely on transit routes through neighbouring countries; the problem of energy transit, reflecting the large overland distance between resources and the main demand centres, is a recurring challenge for Caspian producers.

The policies and trends in Caspian countries' domestic energy use are critical to the region's social and economic development and, although the share of Caspian countries in global demand is small, are also significant for global supply, since they have an

16

impact on the volumes available for export. There is huge potential in the region to use energy more efficiently; and the share of energy-related greenhouse-gas emissions in the Caspian region is also much higher than population and gross domestic product (GDP) levels would imply. The comparison with Turkey is useful because population levels and climatic conditions are broadly similar: total energy-related carbon dioxide (CO_2) emissions for the eight Caspian countries were over 410 million tonnes in 2008, more than 50% higher than the emissions figure for Turkey, even though total Caspian output (GDP in purchasing power parity [PPP] terms) is only half as large.[1]

Figure 16.2 ● Total primary energy demand in the Caspian by country

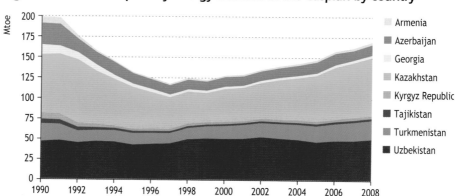

Recent trends in energy demand in the Caspian region (Figure 16.2) have been shaped by the economic upheavals and, in some cases, political instability and conflict that followed the collapse of the Soviet Union at the beginning of the 1990s. Many countries saw a slump in industrial production and general economic activity in the early part of the 1990s and a concomitant fall in energy use, although this is less visible in Uzbekistan and Turkmenistan. An economic revival throughout the region that began in the late 1990s halted, and then reversed, the overall decline in energy use, at least until the onset of the global financial and economic crisis in 2008-2009. However, the region's primary energy demand in 2008, *i.e.* the total energy input into the eight regional economies, was still only 85% of the equivalent figure at the end of the Soviet period.

The fact that Caspian countries used less energy in 2008 than they did in 1990, in particular that electricity consumption was 207 terawatt-hours (TWh) in 2008 compared to 249 TWh in 1990, might suggest that these countries now have reserve

---

1. In 2008, Turkey's GDP (in PPP terms) amounted to $924 billion; its total primary energy demand was 98.5 Mtoe and total energy-related CO_2 emissions were 263 million tonnes. This analysis does not include energy projections to 2035 for Turkey as a single country.

capacity available to support future economic growth. The reality is quite different. Many of the energy assets now used in the region were built in Soviet times and the technical deterioration of this capacity was accelerated after 1991 by a shortage of funds for maintenance and upgrading. As a result, many Caspian countries are already facing energy shortages, of electricity in particular, and have a major task ahead to attract investment in new infrastructure and generation capacity. The nature and scale of this challenge is explored later in this chapter.

Discussion of the Caspian region as a whole implies a homogeneity among the countries that does not exist in practice. In Soviet times, the oil and gas industries in Azerbaijan and Uzbekistan, for example, were developed much more intensively than those in Turkmenistan and Kazakhstan. Since independence in 1991, although these countries received similar inheritances in terms of technology, infrastructure and institutions from the Soviet Union, and some similarities remain, the Caspian countries have followed divergent paths based on their own policy choices and models of development, distinctive demographic, political and economic circumstances, and a wide variety in the size and type of resource endowments. The analysis below highlights how these different factors, and policy intentions for the future, affect patterns of energy use across the region.

Trends in energy production and investment

Energy production in the Caspian region is dominated by the four main fossil-fuel resource-holders (Figure 16.3): Azerbaijan, Kazakhstan, Turkmenistan and Uzbekistan. The decline in production in the 1990s reflects primarily a fall in gas production in Turkmenistan, where output collapsed from 84 billion cubic metres (bcm) in 1991 to as little as 13 bcm in 1998 (because of a gas dispute with Russia). After 1999, Turkmenistan gas production recovered to a range of 60-70 bcm per year, before the economic crisis and the decline in European gas demand in 2009 precipitated another lengthy dispute with Russia over export volumes and prices.

From the late 1990s, the region's energy production was bolstered by increased oil output from Kazakhstan, which jumped from 450 thousand barrels per day (kb/d) in 1996 to about 1 million barrels per day (mb/d) by 2002 and then to 1.6 mb/d in 2009. Since 2004 there has also been a rapid rise in Azerbaijani oil output, which also more than tripled, from around 300 kb/d to 1.1 mb/d, by 2010.

16

Kazakhstan has received the bulk of international direct investment in the Caspian region since 1991. As of 2008, the total level of foreign investment since 1991 in the eight countries amounted to $102 billion; around $69 billion of this, *i.e.* over two-thirds, was in Kazakhstan, well above its 40% share of regional GDP. Azerbaijan has also been a major recipient of international investment, with inflows of over $3 billion in both 2003 and 2004, during a major expansion of the offshore Azeri-Chirag-Guneshli project. Although investments in Uzbekistan and Turkmenistan have picked up since 2006 and 2007, their share of cumulative investment in 1991-2008 was only 4% and 6% respectively (UNCTAD, 2009).

Figure 16.3 ● Total energy production in the Caspian by country

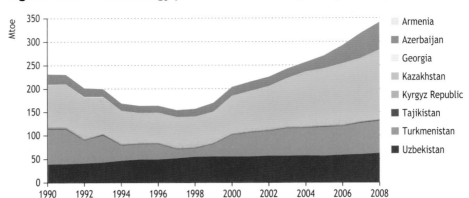

Trends in politics and governance

All the Caspian countries have taken important steps to consolidate their position as independent states since 1991, but the record of building democratic political institutions and moving away from a centrally planned economy has been mixed. Most leaders recoiled from the idea of economic "shock therapy" in the early years after 1991 and ideas of "guided democracy" and "gradual reform" have tended to prevail but, beyond this, there has not been any regional consensus on a model for national development and there are strong variations across the region. Some countries have loosened central control over both the political and economic spheres, as for example in Georgia and, to a lesser extent, in Armenia and the Kyrgyz Republic. In some other countries, such as Azerbaijan and Kazakhstan, economic liberalisation has often proceeded faster than political change. In Turkmenistan and Uzbekistan there has been very little, or only very gradual, economic or political change.

Some countries have also been seriously affected by conflict, in particular the Nagorno-Karabakh war between Armenia and Azerbaijan from 1988-1994 and the civil war in Tajikistan from 1992-1997. Georgia experienced prolonged periods of civil unrest in the early 1990s as well as the war with Russia in 2008. A legacy of conflict and mistrust is one of the reasons why efforts to promote regional co-operation, whether in the South Caucasus or in Central Asia, have met with only limited success. Overall, there have been relatively few examples in the region since 1991 of power being transferred following freely contested elections; more often, political transitions have either been managed within an existing ruling elite or via upheaval and unrest (as, for example, in the Kyrgyz Republic in 2010).

Concerns about the quality of national governance and institutions throughout the region remain. Survey data show that all Caspian countries are still widely perceived as prone to serious domestic and public-sector corruption, although this varies by country (Transparency International, 2009). Doubts also exist about the commitment of some countries to respect property rights and contract stability, factors which discourage foreign and private investment and significantly raise the costs of doing business.

Key assumptions

GDP and population

GDP for the Caspian region as a whole fell sharply for several years after 1991, reflecting the difficult adjustment for regional economies away from the previous economic system and the decline in trade between former Soviet republics. Declines were strongest in those countries affected by unrest and conflict, notably Tajikistan and the countries of the South Caucasus; national output in Uzbekistan was affected least. Growth rates picked up again from the latter part of the 1990s. The period 2000-2008 saw a vigorous expansion of 9.9% per year, driven in particular by impressive growth in Kazakhstan, (the region's largest economy) which averaged 9.3% growth from 2000-2008; in Turkmenistan (14.3% per year over the same period); and in Azerbaijan (16.1%), which was the fastest growing economy in the world from 2005 to 2007.

The global economic crisis did not spare the Caspian region: net importers were affected by the run-up in energy and food prices to 2008, while energy exporters felt some impact from the subsequent declines in commodity prices; all countries experienced a fall in remittances sent home by workers abroad (many working in Russia) and a sharp fall in capital inflows. Uzbekistan, whose economy has only limited interaction with international markets, was again the least affected, while Kazakhstan was among the hardest hit. The relatively open Kazakhstan economy, its banking sector in particular, was vulnerable to tightening credit markets and the burst of a real estate bubble, which contributed to a slowdown in the rate of GDP growth to 1.2% in 2009, the worst performance since 1998. The use of accumulated public savings, notably the National Fund of the Republic of Kazakhstan, which holds a portion of national oil revenues, helped to mitigate the worst effects of the crisis.

Caspian GDP growth assumptions for the period to 2015 are based on the latest IMF projections (IMF, 2010). Most Caspian countries are set to grow relatively strongly over this period. Beyond 2015, rates of growth are adjusted upwards in periods where countries have large projected increases in hydrocarbon exports but, overall, are assumed to fall progressively in the longer term as the economies mature and, in some cases, hydrocarbon production levels off. The region as a whole is assumed to grow by 3.8% per year on average over the full projection period (above the worldwide average of 3.2%), with Turkmenistan seeing the fastest rate, over 5% per year, the result of rapid expansion in gas production and exports, which is expected to drive economic development (Table 16.2). Caspian GDP per capita (at market exchange rates) is projected to more than double from an average of $2 900 in 2008 to over $6 700 in 2035, but remains well below that of OECD Europe, at $48 000 in 2035. Among Caspian countries, Kazakhstan's per-capita GDP remains the highest, reaching $18 000 in 2035 (up from $6 900 in 2008).

The total population of the eight countries covered by this study is assumed to grow from 76 million in 2008 to 93 million in 2035, an annual growth rate of 0.7%. In 2009, Uzbekistan was the most populous country, with 28 million inhabitants, followed by

16

Kazakhstan and Azerbaijan. Over time the share of the regional population living in urban areas is also expected to increase, from a current 44% to 54% in 2035, pushing up demand for modern energy services, as they are more readily available in towns and cities.

Table 16.2 ● **Indicators and assumptions for population and GDP in the Caspian**

	Population			GDP ($2009, PPP)			GDP per capita ($2009, PPP)		
	2008 (million)	1990-2008*	2008-2035*	2008 (billion)	1990-2008*	2008-2035*	2008	1990-2008*	2008-2035*
Azerbaijan	9	1.1%	0.7%	78	3.7%	2.8%	9 042	2.6%	2.1%
Kazakhstan	16	-0.2%	0.4%	180	1.9%	4.1%	11 480	2.2%	3.6%
Turkmenistan	5	1.8%	0.9%	31	4.5%	5.4%	6 104	2.7%	4.4%
Uzbekistan	27	1.6%	0.9%	72	2.6%	4.3%	2 653	1.0%	3.4%
Other Caspian**	20	0.2%	0.7%	66	-0.4%	2.6%	3 361	-0.7%	1.9%
Total	76	0.8%	0.7%	427	2.0%	3.8%	5 604	1.2%	3.1%
World	6 692	1.3%	0.9%	70 395	3.3%	3.2%	10 519	2.0%	2.3%
Russia	142	-0.2%	-0.4%	2 291	0.6%	3.0%	16 155	0.8%	3.5%
OECD Europe	543	0.5%	0.2%	16 351	2.2%	1.6%	30 094	1.7%	1.4%

* Compound average annual growth rate.
** Armenia, Georgia, Kyrgyz Republic and Tajikistan.
Note: GDP and population assumptions are the same for all three scenarios (see Chapter 1).

Energy and climate policies

As elsewhere in this *Outlook*, the Caspian analysis is based on three scenarios (see Chapter 1 for a full description). Detailed results for the New Policies Scenario are presented here for the four main Caspian countries (Azerbaijan, Kazakhstan, Turkmenistan and Uzbekistan), for the other four countries as a group (Armenia, Georgia, the Kyrgyz Republic and Tajikistan) and for the Caspian region as a whole. The New Policies Scenario assumes the implementation, albeit in a relatively cautious manner, of policy commitments on matters of energy and environment that have been announced but not yet implemented. In many regions of the world, the pledges made under the Copenhagen Accord are a central pillar of these commitments, but this is less the case for the Caspian region. Kazakhstan is the only Caspian country to have submitted to the UN Framework Convention on Climate Change (UNFCCC) a quantified economy-wide emissions target under the Copenhagen Accord; its stated goal is to reduce emissions by 15% by 2020, compared to the base year of 1992.[2] Of the other non-Annex I countries in the region, only Armenia and Georgia have provided information to the UNFCCC on nationally appropriate mitigation actions and these do not include quantitative targets.

---

2. Kazakhstan ratified the Kyoto Protocol on 19 June 2009 and, therefore, is considered an Annex I Party for the purposes of the Protocol, but it remains a non-Annex I Party for the purposes of the Convention.

The New Policies Scenario, therefore, takes into account the broad policy intentions and, where available, targets on energy and environment that are set out in national strategy documents and sectoral programmes. Each country has, to a greater or lesser extent, set out medium-term policy aims in these areas, for example Kazakhstan's Strategic Development Plan to 2020 (President of Kazakhstan, 2010), has specific targets in such areas as reductions in energy intensity, increasing the share of renewable energy and upgrading power generation capacity.[3] Where new policies affect energy and carbon intensity over a specified period, e.g. to 2015 or 2020, it is assumed that additional measures will be introduced to maintain the pace of decline in intensity through to 2035, including, where appropriate, pricing reforms. The general assumption for the New Policies Scenario, that announced policy commitments are implemented only cautiously, is particularly pertinent to the Caspian region, where policy announcements and targets have tended to be declaratory, lacking both the administrative mechanisms and the budgetary or financial support necessary for implementation. For the purposes of this scenario, given the high degree of uncertainty, we have assumed in most cases that the policies actually implemented will not be strong enough to reach the stated targets within the intended timeframe. This assumption varies from country to country, based on an assessment of the quality of governance and, in particular, the country's capacity to formulate and implement policy.[4]

Regional demand outlook

Overview

In aggregate, the Caspian region accounts for only 1.4% of global total primary energy use, reflecting its low population and correspondingly modest share of global economic activity. By contrast, the region's energy intensity, measured by its energy use per dollar of GDP, remains well above the average of the rest of the world: the energy intensity of GDP (in PPP terms) of the region as a whole in 2008 was more than 30% higher than that of Russia, more than double the global average and over three times that of Europe. This results mainly from the relatively inefficient ways in which energy is used, a legacy of the Soviet era, and from climatic factors that boost heating needs in winter. Despite important advances in recent years (IEA, 2009), there is considerable scope remaining to improve energy efficiency; much of the fall in energy intensity in recent years was the result of declining industrial production rather than implementation of specific policies or measures to promote efficiency. Political stability and large investments — and continued market reforms to stimulate them — will be needed for this potential to be realised. More efficient energy use could largely offset the impetus to demand growth that is likely to come from robust rates of economic growth. Together with the rate of economic growth, this issue of energy intensity is a major source of uncertainty surrounding future energy demand in the region.

16

---

3. The State Programme for the Accelerated Industrial and Innovative Development of the Economy 2010-2014 was launched by Kazakhstan in 2010. It includes a target of 1 TWh of electricity production from renewables by 2014, among a range of measures that have implications for energy sector development.

4. Assessments were based on the World Bank's World Governance Indicators (World Bank, 2009a).

Pricing reforms, involving reductions in subsidies to various fuels and energy services, will play a particularly important role in stimulating investments in more efficient ways of using energy and promoting energy conservation. All main Caspian countries subsidise at least one form of energy, in some cases extremely heavily (Figure 16.4). Natural gas and electricity are generally the most heavily subsidised, especially to households. In Turkmenistan, for example, there is no charge to residential users for electricity and gas supplied up to a certain threshold. In most cases, prices are set below the levels that would prevail in a truly competitive market ("reference prices") for social and industrial policy reasons (see Chapter 19). These subsidies are assumed to be reduced progressively over the projection period in the 450 Scenario (where they are eliminated by 2035), while, in the absence of any firm plans to change current pricing and subsidy policies, a continuation of those policies is assumed for the other two scenarios.

Figure 16.4 ● **Energy subsidies in selected Caspian countries, 2009**

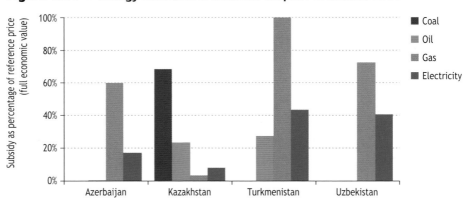

Note: See Chapters 19 and 20 for a detailed analysis of energy subsidies.

Primary energy demand and fuel mix

Caspian energy demand, in aggregate, expands progressively between 2008 and 2035 in all three scenarios (Table 16.3). Annual demand growth is only slightly slower in the New Policies Scenario compared with the Current Policies Scenario — 1.4% versus 1.7% on average in 2008-2035 — reflecting the absence of strong plans to address rising energy use and related emissions, compared with most other regions (especially the OECD). This analysis suggests that the region's policy ambitions, as they stand, do not move it substantially away from a "business-as-usual" scenario towards more efficient or sustainable patterns of energy use. In the 450 Scenario, however, demand grows much more slowly — by only 0.8% per year — on the assumption that Caspian countries do adopt strong measures to exploit much of their considerable potential for improving energy efficiency and for switching to low-carbon fuels and technologies.

In the New Policies Scenario, total Caspian primary energy demand in 2035 is about 46% higher than in 2008. Kazakhstan and Uzbekistan remain the largest consumers in the region in all scenarios, their combined share holding steady at about 71% at the end of the projection period. Turkmenistan sees the fastest rate of energy demand growth but the overall size of its energy market remains modest, accounting for less than 15% of total Caspian energy use by 2035.

Table 16.3 ● Primary energy demand by country in the Caspian by scenario (Mtoe)

	1990	2008	2015	2020	2025	2030	2035	2008-2035*
New Policies	198.4	169.4	205.2	219.8	234.5	240.7	246.5	1.4%
Azerbaijan	26.0	12.9	14.8	15.8	16.7	17.3	17.8	1.2%
Kazakhstan	71.3	71.0	86.2	92.5	101.9	104.1	106.6	1.5%
Turkmenistan	22.1	23.7	30.7	32.6	34.2	35.6	36.7	1.6%
Uzbekistan	46.6	50.5	60.2	63.4	65.6	67.2	68.8	1.1%
Other Caspian	32.5	11.3	13.3	15.4	16.1	16.4	16.6	1.4%
Current Policies	198.4	169.4	205.6	223.8	244.2	256.0	267.4	1.7%
450	198.4	169.4	196.5	205.2	213.2	211.0	207.6	0.8%

* Compound average annual growth rate.

For the region as a whole, the New Policies Scenario does not result in more than a marginal change to the primary energy mix in the period to 2035. The share of natural gas in total Caspian primary energy use rises slightly, from 60% in 2008 to 62% in 2035, and gas accounts for more than 65% of the total incremental primary energy demand over the projection period; natural gas remains the single largest fuel in Azerbaijan, Turkmenistan and Uzbekistan and overtakes coal to become the most important fuel in Kazakhstan. Oil demand expands in line with primary energy demand, averaging 1.5% per year, as higher consumption in the transportation sector is tempered partially by less oil-burning for power generation. Nuclear power grows at the fastest rate among primary energy fuels, averaging 4.9% per year, on the assumption that new nuclear power plants are commissioned during the projection period, one each in Armenia and Kazakhstan (Figure 16.5).

Figure 16.5 ● Primary energy demand in the Caspian by fuel in the New Policies Scenario

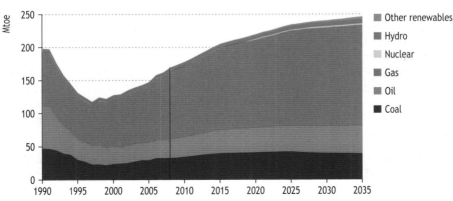

Legend: Other renewables, Hydro, Nuclear, Gas, Oil, Coal

16

The fuel mix in *final* energy consumption in the Caspian is set to continue recent trends, characterised, as in the rest of world, by a growing share of electricity. At present, electricity accounts for 12% of total final energy use in the region, though the share varies markedly across countries; it is highest in the Kyrgyz Republic and Tajikistan, which have low-cost hydropower resources, and lowest in Kazakhstan, Turkmenistan and Uzbekistan, where industry relies much more on fossil fuels for process energy. In the New Policies Scenario, the share of electricity is projected to rise to 14%, converging on, but far from reaching, the average level worldwide of 23% in 2035. The share of heat in final energy consumption is projected to drop from 10% in 2008 to 7% in 2035, as improved efficiency and conservation reduce heating needs.

All Caspian countries have taken steps over the past two decades to address some of the extremely inefficient and wasteful energy practices and technologies that had become entrenched during the Soviet era. But these efforts have only gone so far and considerable potential remains for improving energy efficiency and curbing emissions of greenhouses gases and other pollutants. The technical potential for improving energy efficiency using current technology is thought to be much greater than in the rest of the world. The potential for reducing the amount of energy used for district heating is particularly large (Box 16.1).

Box 16.1 ● Caspian potential for saving energy in district heating

In all Caspian countries, a significant share of energy used in buildings takes the form of district heat. In many cases, heat is produced, distributed and consumed very inefficiently. Modernising district heating plants and rehabilitating or replacing inefficient combined heat and power (CHP) plants alone could reduce overall primary energy consumption in Eastern Europe and Central Asia (including Russia) by an estimated 17% by 2030 (World Bank, 2010). Further energy savings could be realised by reducing heat-distribution losses, by insulating buildings and by installing metering and thermostats in buildings to discourage waste.

Heat is priced at well-below the true cost of supply in most Caspian countries, but the inefficient use of district heat is only partly due to low prices. Another reason is that, especially in the residential sector, end-users are often not billed for the actual amount of heat they use because supplies to individual dwellings are not metered. Thus, there is little incentive to use heat efficiently or conserve it. In Kazakhstan, heating tariffs for residential buildings are often based on the size of the apartment, so there is no incentive to limit consumption (UNECE, 2008). In addition, in large housing blocks, it is often not possible to adjust the amount of heat supplied to each apartment. As a result, simply raising prices for heat would make no difference to consumption; people would still need to heat their apartments and so higher prices would simply result in many households being unable or unwilling to pay, a common problem in many parts of the region in recent years. Experience has shown that policies to remove heat subsidies are generally effective only when accompanied by investments in metering and heat- control systems, and by the introduction of billing systems based on individual households' actual consumption (Von Moltke *et al.*, 2003).

Based on year-2008 energy intensities, we estimate that, were the region to use energy as efficiently as OECD countries, consumption of primary energy in the Caspian region, as a whole, could be cut by more than 80 million tonnes oil equivalent (Mtoe), which is one-half of the level in 2008 (Figure 16.6).[5] The energy-savings potential in both absolute and percentage terms is greatest in Kazakhstan, where energy use could in principle be lowered by more than one-half, mainly in the industrial, residential and commercial sectors. The savings potential is also very large in Turkmenistan — especially in the services and residential sectors, as well as in distribution systems — and in Uzbekistan, where about half of the over-consumption is related to inefficient industrial energy use. By contrast, much of the potential that existed until recently in Azerbaijan has now been exploited, thanks in large part to pricing reforms.

Figure 16.6 ● Energy savings potential in the main Caspian countries, 2008

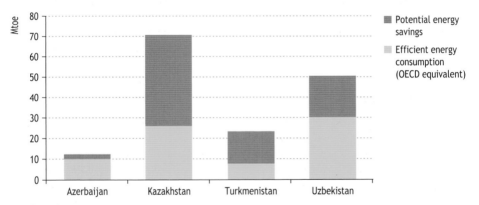

Note: The potential savings are calculated sector-by-sector based on the actual energy intensity of OECD countries, adjusted for structural economic differences and climatic factors.

How quickly the region's energy-efficiency potential will actually be exploited will be largely driven by government policies, in particular with respect to energy pricing, market reform and the financing of energy investments (see Spotlight). In the Current Policies Scenario, in which no change in policy is assumed, progress is naturally slowest, with efficiency gains coming solely from past measures and from the progressive replacement of inefficient capital stock by plants using more advanced technologies. The New Policies Scenario does not produce a large additional gain in efficiency, compared to a continuation of current policies. However, the strong policies assumed in the 450 Scenario, including pricing reforms and standards, do result in a significantly faster improvement.

16

5. This estimate is broadly in line with the results of other recent studies. For example, a 2008 UN study estimated the energy savings potential across all sectors through improved efficiency and conservation in Central Asia at 35-40% of current consumption (UNESCAP, 2008). An earlier study put the potential even higher, at around one-half, some 60% of it in industry (including energy production), up to one-quarter in housing, 7-8% in transport and 6-7% in agriculture (SPECA, 2007).

What policies can unlock the Caspian's energy savings potential?

Energy demand growth is normally correlated quite closely with economic growth, but in countries with very high energy-efficiency potential — such as those in the Caspian region — this energy savings potential can be a very important and cost-effective source of additional energy 'supply'. Achieving these gains can be politically difficult, but evidence from the Caspian region, as in all former Soviet Union's countries, suggests that the policy formula for success rests on three vital foundations:

- A move towards market-based energy pricing, reliably regulated, as a means of triggering the investments needed to replace obsolete and inefficient technologies.

- Metering of energy so that consumption can be attributed to individual consumers.

- A governance structure that can ensure that energy is regularly and fully paid for, as well as to provide targeted incentives and support for vulnerable social groups (to replace broad subsidies).

Where governments have managed to take action in all these areas, there has been a significant impact. In Georgia, reform of the electricity sector since the late 1990s and the resulting reduction in commercial and technical losses meant that, for the period from 1999-2008, electricity consumption increased by only 10% while the economy nearly quadrupled in size. In Azerbaijan, too, price rises in 2007 were accompanied by a metering programme and efforts to improve collection rates, with the result that electricity consumption fell from almost 21 TWh in 2006 to around 17 TWh in 2007 and 2008, at a time when the Azerbaijani economy enjoyed double-digit growth.

There are also examples where partial or erratic implementation of policies has not produced the same gains. In Uzbekistan, for example, there has been a major drive to install electricity and gas meters and promote payment discipline, but tariff increases have not been sufficient to have a large impact on consumer behaviour. As a result, the link between energy demand growth and GDP remains broadly intact. In the Kyrgyz Republic, events in early 2010 have increased the perception of political risk associated with electricity market reform in Central Asia. The government tried to make up for postponing previous tariff increases with a precipitous jump in electricity prices. This step was widely seen as contributing to the unrest that brought down the government later in the spring.

While the combination of pricing, metering and better sector governance can reduce waste, these are best seen as one-time gains related to the manifold inefficiencies inherited from the Soviet period. They are not a medium-term substitute for more sophisticated efficiency policies and measures, for example, the development of new building codes or appliance standards, or the development of energy efficiency strategies and institutions. But they are an essential first step.

Figure 16.7 ● Primary energy intensity in the Caspian and Russia in the New Policies Scenario

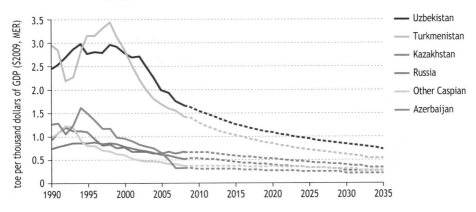

The projected efficiency gains, together with an overall shift in the structure of economic activity to less energy-intensive activities (notably services), is expected to result in a continuing decline in the energy intensity of the Caspian region in all scenarios (Figure 16.7). In the New Policies Scenario, primary energy intensity falls by 47% between 2008 and 2035, approaching but not reaching the current average level for the world as a whole. Intensity falls by 43% in the Current Policies Scenario and by 56% in the 450 Scenario.

Figure 16.8 ● Comparison of per-capita primary energy demand to GDP per capita in the New Policies Scenario (1990, 2000, 2008, 2020, 2035)

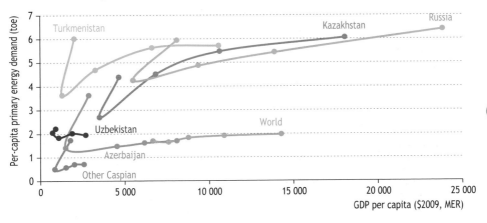

16

Per-capita energy consumption in the Caspian region is set to continue to expand at a higher rate than in the rest of the world. Today, each person in the region consumes on average 2.2 tonnes of oil equivalent (toe), compared with 1.8 toe worldwide.

Among the four Caspian countries studied in detail, per-capita consumption is highest in Kazakhstan and Turkmenistan, reflecting climatic factors and, in the case of Kazakhstan, the large industrial component in total consumption (Figure 16.8). It is lowest in Azerbaijan, due to a relatively low level of industrial activity outside of the oil sector (oil extraction accounted for 95% of industrial activity in 2007), a relatively buoyant services sector, reflecting Azerbaijan's important (and traditional) regional role in trade and transport, as well as energy-efficiency improvements introduced in recent years. In the New Policies Scenario, Caspian per-capita consumption is projected to rise to about 2.7 toe by 2035, still well above the world average of 2.0 toe.

Electricity generation and other sectoral trends

In the Caspian region as a whole, the buildings and agriculture sectors (including residential energy uses and services) account for the largest share of overall energy consumption. The relative importance of this and other sectors varies widely across the region: industry, for example, accounts for close to one-half of all energy consumed in Kazakhstan, but less than a fifth in Azerbaijan.[6] Power generation accounts for about 40% of primary energy use in Azerbaijan and about 30% in both Kazakhstan and Uzbekistan, mainly reflecting differences in the extent to which industry relies on electricity for its energy needs.

Figure 16.9 ● Incremental energy demand in the Caspian by sector and fuel in the New Policies Scenario, 2008-2035

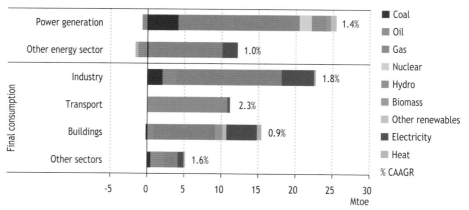

Note: CAAGR = compound average annual growth rate.

In general, the importance of electricity in final energy uses grows, matching the trend in the rest of the world. Nonetheless, in the New Policies Scenario, the share of power generation (including CHP plants) in total primary energy demand remains fairly stable

---

6. The quality of data on the breakdown of energy use by sector is very poor for some countries and a large proportion of final energy consumption in some cases is not specified in official statistics. Improving the quality and reliability of energy data will be important to informing policy choices in the future.

at 32% throughout the projection period, thanks to the energy savings achieved through increased efficiency in power generation and reduced losses in distribution. Overall energy use for power and heat generation is projected to expand at an average rate of 1.4% per year. Energy demand in the industry sector grows on average at 1.8% per year, with gas accounting for most of the increase (Figure 16.9).

In the New Policies Scenario, the use of energy in the transport sector is projected to grow at the fastest rate, averaging 2.3% per year. A planned expansion of regional rail links will absorb some of the increased demand for transport, but most of this growth will be for road transport, as rising incomes boost demand for mobility and road infrastructure improves; the vast distances and landlocked geographical situation of Caspian countries will encourage more driving. Oil consumption in the transport sector is projected to double over the projection period. Yet, despite this growth, the transport sector's share of total final consumption remains relatively small, increasing from 11% in 2008 to 14% by 2035 in the New Policies Scenario. At 130 cars per thousand people, projected car ownership in 2035 remains low relative to Russia and the world average (Figure 16.10).

Figure 16.10 ● Road oil consumption and passenger light-duty vehicle ownership in the Caspian in the New Policies Scenario

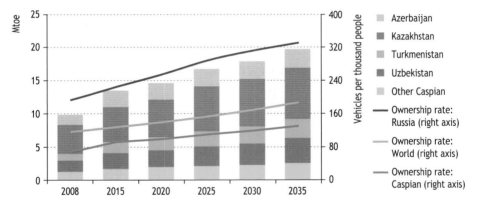

The Caspian region relies heavily on fossil fuels to generate electricity, with natural gas accounting for 38% of the electricity produced and coal for 31% in 2008. Hydropower accounts for 27% of the total, mainly in Tajikistan, the Kyrgyz Republic, Georgia and Kazakhstan. The fuel mix mirrors the geographical distribution of natural resource endowments: the four energy exporting countries (Azerbaijan, Kazakhstan, Turkmenistan and Uzbekistan) rely almost exclusively on fossil fuels (Figure 16.11). Coal is the principal fuel in Kazakhstan, contributing to 79% of total generation, while natural gas is the most important fuel for electricity generation in the other energy-exporting countries. Hydropower generates almost all electricity in Tajikistan and the Kyrgyz Republic and is the largest contributor to Georgia's domestic electricity production. Armenia is currently the only country in the region with nuclear power production.

16

Figure 16.11 ● Electricity generation in the Caspian by country and fuel, 2008

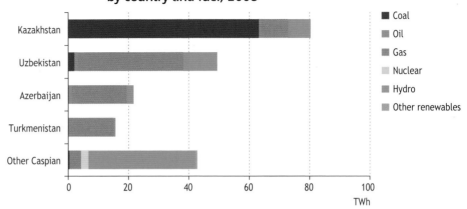

The Caspian region's considerable potential to generate electricity from modern renewable energy sources, notably wind, solar and biomass, remains largely untapped so far, due to the abundance of domestic fossil-fuel resources, energy subsidies (which hold down fossil-fuel input prices and electricity prices) and the lack of regulatory and policy incentives. The contribution of renewables to the region's current generation mix, excluding large hydro, is negligible.

In the New Policies Scenario, electricity generation in the Caspian region is projected to grow by 1.9% per year between 2008 and 2035, reaching 344 TWh by 2035 (Figure 16.12). Caspian electricity demand per capita grows by 1.2% per year over the projection period – in line with the world average – with Turkmenistan and Kazakhstan registering the fastest growth (2.5% and 1.7% per year, respectively). Kazakhstan is projected to continue to rank well ahead of the other Caspian countries for per-capita electricity demand by 2035, when it is about double the world average.

Natural gas, coal and hydropower are expected to remain the principal primary fuels, contributing respectively 48%, 23% and 23% to total generation in 2035. Electricity production from natural gas increases at 2.7% per year on average, with growth primarily concentrated in the fossil fuel-rich countries, where natural gas plants progressively replace old oil-fired power plants and, to some extent, coal plants, particularly in Kazakhstan. Hydropower shows the second-fastest growth among energy sources for electricity production, at 1.3% per year over the projection period, while oil use for power generation declines steeply, on average by over 4% per year, as old power plants are replaced with gas- and coal-fired plants. Other renewables represent 2% of generation by 2035, stimulated by feed-in tariffs and other incentives put in place in some countries of the region, with wind providing the largest contribution. However, despite a growing interest in the development and deployment of modern low-carbon energy sources, mainly in the other Caspian countries and, to a lesser degree, in Azerbaijan and Kazakhstan, the contribution of other renewables in meeting energy needs in all countries is expected to remain marginal in all three scenarios. Kazakhstan is assumed to bring one nuclear reactor into production towards the end of the *Outlook* period; Armenia is also assumed to commission a new reactor by 2020, to replace the existing one that will soon be retired.

Figure 16.12 ● Electricity generation in the Caspian by fuel in the New Policies Scenario

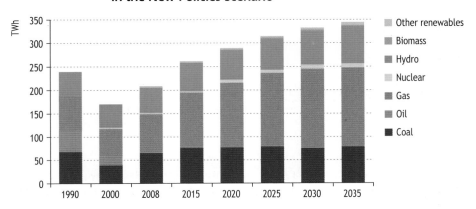

Installed generating capacity in the Caspian region amounted to 55 gigawatts (GW) at the end of 2008. Although this is well in excess of current demand, much of this capacity is obsolete: all but one-fifth was built during the Soviet era and more than one-third was built over 40 years ago (Figure 16.13). We assume that 32 GW, about 60% of current installed capacity, will be retired as it comes to the end of its operating life. Most of this will be coal- and gas-fired capacity. Some Caspian countries, for example Azerbaijan and Turkmenistan have already embarked on a process of modernisation of their national power systems through the replacement of old and inefficient power plants.

Figure 16.13 ● *Age profile of installed thermal and nuclear capacity in the Caspian, 2008*

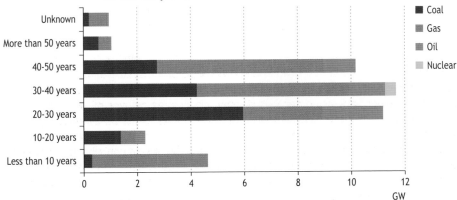

Source: Platts World Electric Power Plants Database, December 2009 version.

16

In the New Policies Scenario, total installed capacity reaches 86 GW by 2035, with the share of natural gas increasing sharply, from 41% in 2008 to 52% by 2035. The average thermal efficiency of gas-fired power plants, including CHP plants, rises from 43% in 2008 to 50% by 2035. By contrast, the shares of oil and coal generating capacity fall heavily, from 6% and 28% in 2008 to 1% and 20% respectively in 2035.

In the New Policies Scenario, the power sector across the Caspian region requires $152 billion of investment in the period from 2010 to 2035, of which $72 billion is needed for power generation and $80 billion for modernisation and expansion of transmission and distribution systems. Gas-fired capacity will absorb about 36% of total power generation investments, coal 30% and hydro 18%. Kazakhstan accounts for most of the investment in coal-fired plants, while in Turkmenistan and Uzbekistan almost all generation investment goes to gas-fired plant (Figure 16.14). More than half of the total investments implemented in hydropower capacity are expected to take place in the energy-importing countries, in particular Tajikistan and the Kyrgyz Republic, where significant untapped potential remains.

Figure 16.14 ● **Cumulative power sector investment in the Caspian by country and type in the New Policies Scenario, 2010-2035**

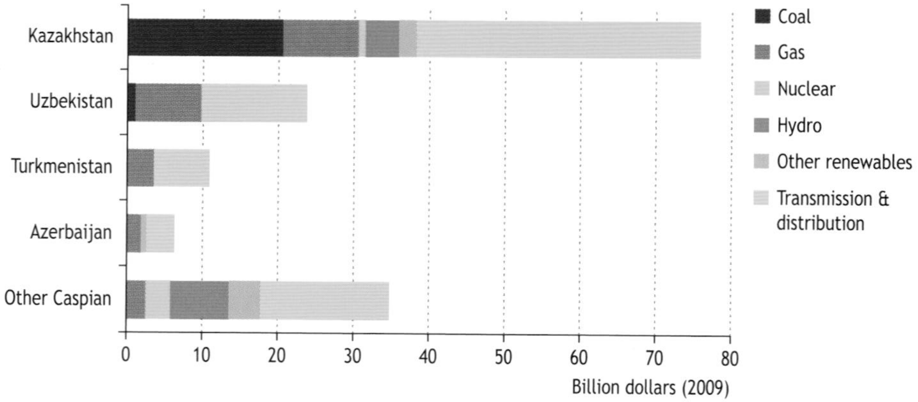

Analysis by country

The remainder of this chapter examines the demand profiles of Azerbaijan, Kazakhstan, Turkmenistan and Uzbekistan, individually, and then the collective profile of Armenia, Georgia, the Kyrgyz Republic and Tajikistan. The projected energy demand trends across the Caspian countries show some similarities. All of them are characterised by a continuing heavy dependence on fossil fuels, typically natural gas for producing heat in the household and industrial sectors, as well as oil for transport. There are also several differences. Kazakhstan is the only country that uses coal in any significant volume, while Armenia is currently the sole producer of nuclear power. Hydropower continues to dominate electricity production in the Kyrgyz Republic, Tajikistan and Georgia, but remains small in all other countries, where gas is generally the dominant fuel for power generation. Access to modern energy by households varies somewhat across the region, largely reflecting income levels, urbanisation rates and the extent of governmental measures (Box 16.2).

Box 16.2 ● Access to energy in the Caspian

A standard indicator of access to energy in many parts of the world is the electrification rate, *i.e.* the percentage of the population that has a connection to electricity networks. This standard approach does not hold for the Caspian region; official data show rates of access to electricity are very high, above 99% in all of the countries studied (see Chapter 8), reflecting the importance attached to electrification in Soviet economic planning. But despite these data, there is strong evidence to suggest that, in reality, access to modern energy services is limited in some Caspian countries, particularly outside the large cities and in remote regions, where incomes are generally lowest. Average electricity consumption per capita in the residential sector in the Caspian region is low, less than two-thirds of that in the rest of the world. There are regular incidences, particularly in Central Asia, of load-shedding and brownouts: a striking example was the near-collapse of Tajikistan's electricity system over the winter of 2007-2008, when low reservoir levels and poor hydrological conditions provoked electricity shortages and cut-offs.

There are two major barriers to energy access; reliability of energy supply and affordability. Poorly maintained Soviet-era infrastructure is the major constraint on reliability of supply, and funds for investment are limited in many cases by prices that are below cost-recovery levels. Yet even these subsidised prices can create difficulties for consumers, resulting in increased non-payment for electricity: collection rates dropped to 74% in the Kyrgyz Republic and only 54% in Uzbekistan in 2006 (ADB, 2009). Kazakhstan is the richest country in the Caspian region by GDP per capita, but, even here, 40% of the population is classified as poor or near-poor (Ramani, 2009). Further evidence of the affordability issue across the Caspian region comes from the high level of "commercial losses" in most national electricity statistics, which suggests a relatively high number of illegal connections to the grid.

Many low-income rural households across the region also lack access to clean fuels for cooking and heating. In Azerbaijan, over 20% of rural households rely on straw, wood or coal, with many of them cooking over an open fire. Three-quarters of the rural households in Armenia rely on wood for cooking and heating using open fires. Poverty in rural areas of Tajikistan forces households to use traditional biomass for cooking and heating, thus leading to degradation of local resources and less food for livestock. In some cases, decentralised deployment of renewable energy technologies could be a way forward; for example, micro hydropower has a lot of potential in isolated mountainous communities in Tajikistan and the Kyrgyz Republic. But tackling the problem of energy access across the Caspian region as a whole hinges on broader efforts to alleviate poverty and to support the provision of modern energy infrastructure.

16

Azerbaijan

Azerbaijan is the least energy-intensive of the four main Caspian countries, partly due to the relatively small contribution of industry to GDP and a fairly dynamic (albeit small) services sector, and partly because of gains in energy efficiency and reduced waste in recent years. Energy use rose steadily through the early to mid-2000s, but fell sharply in 2007, seemingly as a result of a switch back to gas for heating (as domestic gas supply from the Shah Deniz field became available), which cut electricity use and, therefore, the need to generate power in the most inefficient plants. A large tariff increase (electricity prices tripled in January 2007) together with the implementation of a large-scale metering programme and a drive to improve collection rates also contributed. This downward adjustment is expected to prove a one-off phenomenon. Continuing high rates of GDP growth — driven largely by rising oil and gas production — are expected to continue to push up domestic energy use in Azerbaijan over the projection period, especially in the period to 2020, though the intensity of policy action will affect the pace of growth to some degree. In the New Policies Scenario, Azerbaijani primary energy demand increases by 38% between 2008 and 2035, an average rate of increase of 1.2% per year (Table 16.4).

Table 16.4 ● Primary energy demand in Azerbaijan by fuel in the New Policies Scenario (Mtoe)

	1990	2008	2015	2020	2025	2030	2035	2008-2035*
Oil	11.6	4.1	4.8	4.9	5.0	5.0	5.1	0.8%
Gas	14.2	8.6	9.8	10.7	11.4	12.0	12.2	1.3%
Other**	0.2	0.2	0.2	0.3	0.3	0.4	0.5	3.5%
Total	26.0	12.9	14.8	15.8	16.7	17.3	17.8	1.2%

* Compound average annual growth rate.
** Includes coal, hydro, biomass and other renewables.

Natural gas, sourced from abundant indigenous resources, dominates energy demand in Azerbaijan, accounting for close to two-thirds of primary energy use, with oil products making up almost all the rest. The picture changes only marginally over the projection period; although oil use in the transport sector grows by 2% per year, the share of oil in total primary energy demand falls from 32% in 2008 to 29% in 2035 in the New Policies Scenario, mainly due to increased use of gas and electricity in end-use sectors; oil-burning to generate power and heat had already dropped to very low levels by 2008 (Figure 16.15). The share of gas increases correspondingly, from 67% to 69%. The contribution of other fuels, mainly hydropower, barely changes (at around 2%) as Azerbaijan has limited unexploited hydropower resources. Even though Azerbaijan is now looking more seriously at other renewables, such as the potential for wind power on the Absheron peninsula, these are not assumed to make a visible contribution to the energy mix before 2035 in the New Policies Scenario.

In final uses, natural gas remains the dominant fuel, with its share rising marginally from 42% in 2008 to 44% 2035. Gas is expected to remain the primary fuel for heating and, although new housing units being built are expected to be considerably more energy efficient than the old housing stock, the sheer increase in living space is expected to boost heating needs. Gas also remains the leading fuel for power generation and its share of total output grows from 88% in 2008 to 93% in 2035. Electricity sees its share of final energy demand decline marginally, from 17% in 2008 to 16% in 2035, although this remains a higher share than in other Caspian countries. Increasing prosperity is expected to boost demand for electric appliances, while the continuing long-term shift in economic activity to services encourages the use of electrical equipment. Air-conditioning demand is set to grow too, evening out the seasonal differences between summer and winter load. The process of electricity-sector reform is far from complete and further efficiency improvements are expected, spurred by stronger policy efforts, continued tariff reform and the ongoing implementation of a large-scale metering programme.

Figure 16.15 ● **Incremental energy demand in Azerbaijan by sector and fuel in the New Policies Scenario, 2008-2035**

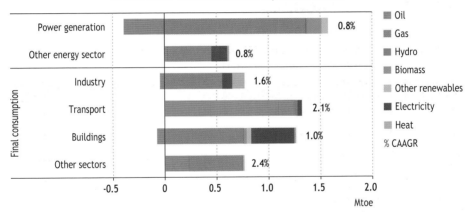

Power-generation fuel input needs are projected to grow by 0.8% per year in the New Policies Scenario, more slowly than overall electricity demand, as inefficient gas-fired plants and (the few remaining) oil-fired plants are replaced by much more efficient combined-cycle gas turbines (CCGTs). The Azerbaijan authorities have already embarked on a large-scale modernisation programme of the country's electricity generation capacity: the country's first CCGT unit, with a capacity of 400 megawatts (MW) was commissioned in 2002, a second 520 MW unit was inaugurated in 2009 and a third 760 MW plant is due to be completed in 2011-2012. As of 2009, natural gas had almost completely backed out fuel oil in power generation (Figure 16.16). Improvements in generation and transmission have meant a dramatic improvement in the reliability of electricity supply since the late 1990s, although there are still occasional power outages. In the New Policies Scenario, the Azerbaijani power sector requires about $6.5.billion of investment in the period to 2035, the lowest level among the main Caspian countries, of which about $4 billion is needed for modernisation and expansion of transmission and distribution systems (Figure 16.14).

16

Figure 16.16 ● Electricity generation in Azerbaijan by fuel in the New Policies Scenario

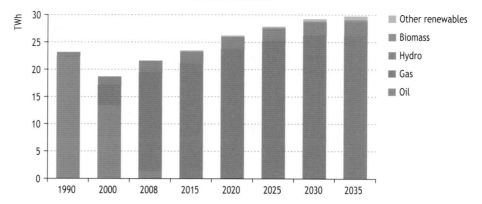

Industrial energy use is projected to grow by 1.6% per year between 2008 and 2035 in the New Policies Scenario. A new methanol plant under construction is assumed to use 0.5 bcm of gas per year from 2012. The development of energy-intensive industries, such as petrochemicals, is expected to continue, underpinned by relatively low-cost local energy supplies.

Kazakhstan

In contrast with other Caspian countries, Kazakhstan's energy mix is dominated by coal. This is mainly used in power generation and accounts for 44% of total primary energy demand. The country's economy is the third most energy-intensive among the resource-owning Caspian countries in PPP terms (behind Uzbekistan and Turkmenistan), but still among the most intensive in the world. Kazakhstan also has the highest per-capita energy consumption in the Caspian region. The prospects for energy use in Kazakhstan hinge on the implementation of efforts to improve energy intensity and efficiency, central pillars of the country's overall plan for strategic development (President of Kazakhstan, 2010) and the key to curbing the growth in greenhouse-gas emissions. In developing this scenario, we have assumed partial achievement of Kazakhstan's target, submitted to the UN Framework Convention on Climate Change, to reduce emissions by 15% by 2020, compared with 1992. Energy-related CO_2 emissions do decline by 2020 (relative to 1992) but only by 2%, rather than 15%. Our findings suggest that full implementation of this target would require additional policy measures to improve efficiency and reduce intensity, and also a much faster move away from coal use in power generation.[7]

---

7. We have tried to strike a balance in the New Policies Scenario between known initiatives for electricity generation, including, for example, the coal-fired 1.2 GW Balkhash thermal plant that is due to be commissioned by 2015, and the policy goals for emissions that imply a move towards a lower-emission power mix. A system of national emissions allowances has been established for large enterprises and power generation plants and the Kazakh Ministry of Environmental Protection is considering the introduction of a national emissions trading scheme, with a pilot phase possible in 2011-2012. It is not yet clear how this scheme would be structured and implemented.

In the New Policies Scenario, primary energy demand increases by 50% between 2008 and 2035 — an average rate of increase of 1.5% per year (Table 16.5). Demand growth averages 2.9% per year in 2009-2015, but slows towards the latter part of the *Outlook* period as rates of economic growth are tempered by a levelling off of oil exports (see Chapter 17). Coal use is expected to grow at the lowest rate of any fuel, at 0.6% per year to 2035, because of a lower assumed share in power generation and because electricity and gas meet most of the incremental demand in the industrial sector.

Table 16.5 ● **Primary energy demand in Kazakhstan by fuel in the New Policies Scenario** (Mtoe)

	1990	2008	2015	2020	2025	2030	2035	2008 -2035*
Coal	40.0	30.9	37.7	38.6	39.6	37.1	36.0	0.6%
Oil	19.9	11.8	13.9	14.7	16.6	16.8	17.5	1.5%
Gas	10.7	27.5	33.6	38.0	44.3	47.8	50.1	2.2%
Nuclear	-	-	-	-	-	0.6	0.6	n.a.
Hydro	0.6	0.6	0.7	0.8	0.9	1.1	1.3	2.5%
Other**	0.1	0.2	0.3	0.4	0.5	0.7	1.1	7.2%
Total	71.3	71.0	86.2	92.5	101.9	104.1	106.6	1.5%

* Compound average annual growth rate.
** Includes biomass and other renewables.

Figure 16.17 ● **Incremental energy demand in Kazakhstan by sector and fuel in the New Policies Scenario, 2008-2035**

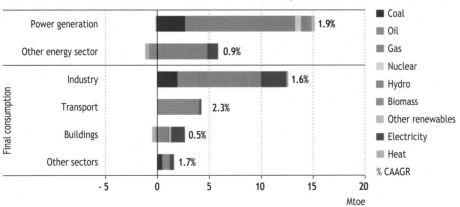

Among final sectors, transport sees the fastest rate of increase, 2.3% per year on average, between 2008 and 2035. Increased economic activity combined with Kazakhstan's singular geographic characteristics — a vast land area, landlocked location and the long distances between the centres of energy and industrial production and demand — drive this increase in transport demand. The government is planning to modernise and expand the road infrastructure, which will facilitate rapid growth in freight and, with it, demand for diesel. Industrial demand is expected to grow at a slower pace, averaging 1.6% per year, due to limited development of the most energy-intensive sub-sectors, such as iron and steel and metallurgy, together with improved energy efficiency in heat production and the use of process energy. Natural gas takes the largest share of incremental industrial energy demand. Demand in the buildings sector is set to grow at the slowest pace, by 0.5% per year over 2008-2035, reflecting the impact of higher prices and measures to promote more efficient energy use and conservation. In this sector, gas is expected to take market share from district heat, as gas is preferred for new buildings (Figure 16.17).

The power sector is expected to be the single biggest contributor to increased primary energy demand in Kazakhstan, accounting for 42% of the total increase in the New Policies Scenario. Electricity demand is set to grow faster than any primary fuel, at an annual rate of 2.4% per year; even though much of the country's existing generation capacity is expected to be replaced over the projection period, rising demand more than offsets the impact of the higher thermal efficiency of new centralised power stations and co-generation plants. Coal will remain the dominant fuel in the power-generation mix, but its share will decline steeply from 79% in 2008 to 51% in 2035 as part of the efforts that are assumed to be made to meet Kazakhstan's target under the Copenhagen Accord. As a result, natural gas is set to make a growing contribution, as the supply of gas produced in association with oil in the west of the country expands and domestic infrastructure links are expanded. The share of gas-fired power in total generation rises from 11% in 2008 to 34% in 2035, as it meets over 60% of the total incremental electricity production. Hydropower, mainly in the south of the country, also sees its share rise, albeit modestly, from 9% to 10% (Figure 16.18).

Figure 16.18 ● Electricity generation in Kazakhstan by fuel in the New Policies Scenario

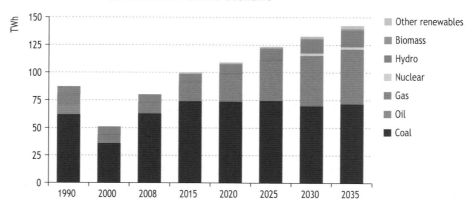

The New Policies Scenario also takes into account the government's plan to build a new nuclear power plant in the Mangistau region: we assume that a 300 MW reactor is commissioned by the end of the projection period, accounting for 2% of total electricity generation in 2035. Total investment needs in the power sector amount to $76 billion over 2010-2035, about one-half of overall energy investment needs in the region. This amount is almost equally divided between power generation and the transmission and distribution sector (Figure 16.14 above). The electricity sector currently suffers from large losses in transmission and distribution and, although the commissioning of a second north-south power transmission line in 2009 has eased the problem, parts of southern Kazakhstan still have to rely on occasional intermittent electricity imports from Tajikistan and the Kyrgyz Republic.

The potential for policy-driven energy savings in Kazakhstan is enormous (Figure 16.6, above), but only a small proportion of this potential is exploited in the New Policies Scenario. The biggest contributor to these energy savings is the power and heat sector, as a result of increased efficiency in generation and in end-uses. A law on renewable energy sources was passed in 2009 and legislation on energy efficiency is currently under development and is assumed to be passed; but these new policies fall a long way short of what would be required to meet the country's climate target in the absence of a radical shift in the fuel mix. Estimates of the country's renewables potential suggest that the market could absorb an estimated 3 billion kilowatt-hours (kWh) of electricity generated from renewable sources already today, rising to 10 billion kWh, or 6% of Kazakhstan's total electricity needs, by 2024 (REEEP, 2009). However, lacking evidence as yet of implementation of the 2009 legislation, we assume in the New Policies Scenario that the development of renewable resources will continue to be limited by competition from relatively low-cost fossil fuels. Although modern renewables grow relatively rapidly in this scenario, starting from a very low base they still account for barely 1% of total primary energy demand by 2035. Access to international mechanisms, such as the multi-donor Clean Technology Fund, could help to accelerate the realisation of Kazakhstan's energy savings and renewables potential; Kazakhstan has presented an investment plan to this fund and could receive up to $200 million for projects, such as 200 MW of small hydro and up to 100 MW of wind power.

Turkmenistan

According to official statistics, Turkmenistan relies on gas and oil for all of its energy needs: gas accounted for 78% of primary demand and oil for the remaining 22% in 2008. It is likely that small amounts of biomass are also used by poor households in remote areas, but this is not reported. In the New Policies Scenario, total primary energy demand is projected to grow at an annual rate of 1.6% per year from 2008 to 2035. This is well below the assumed rate of GDP growth of 5.4% per year (that is underpinned by a big rise in gas exports), implying a significant decline in energy intensity. However, this decline reflects the likelihood of structural changes in the economy and the gradual replacement of inefficient capital stock, rather than the implementation of specific policies on energy efficiency. Among primary fuels, oil consumption grows most rapidly. Natural gas, the indigenous resources of which are very large, remains the dominant fuel, growing almost as fast as oil demand and continuing to provide 78% of domestic needs in 2035 (Table 16.6). Approximately one-third of this gas is used for power generation, 20% in the residential sector, about 15% in industry and the rest (about one-third) for the energy needs of the gas industry itself (upstream activities, compression for transportation and losses).[8]

16

---

8. IEA estimates based on industry sources, as no precise breakdown of final consumption is available.

Table 16.6 ● Primary energy demand in Turkmenistan by fuel in the New Policies Scenario (Mtoe)

	1990	2008	2015	2020	2025	2030	2035	2008-2035*
Oil	7.5	5.2	7.1	7.4	7.8	8.0	8.1	1.6%
Gas	14.3	18.5	23.6	25.2	26.4	27.6	28.6	1.6%
Other**	0.36	0.00	0.00	0.00	0.01	0.02	0.05	21.3%
Total	22.1	23.7	30.7	32.6	34.2	35.6	36.7	1.6%

* Compound average annual growth rate.
** Includes coal, hydro, biomass and other renewables.

Total final consumption in Turkmenistan is expected to almost double over the projection period, rising on average by 2.3% annually. The transport sector sees the fastest growth at 4% per year, driven by rising incomes and increased trade within the region. Industrial demand grows by 2.8% per year, with consumption coming mainly from the cotton and metallurgical industries (Figure 16.19). Among final fuels, electricity is set to continue to grow most rapidly, at 3.7% per year, as a result of relatively high rates of GDP growth and assumed continuation of policies that heavily subsidise electricity. Households are currently supplied with electricity up to 35 kWh per month free of charge and only a small charge is made per kWh consumed over and above that threshold. There are no plans to change this policy.

Figure 16.19 ● Incremental energy demand in Turkmenistan by sector and fuel in the New Policies Scenario, 2008-2035

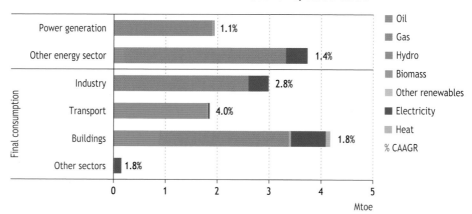

Electricity generation in the country remains exclusively gas-fired over the projection period, increasing from 15 TWh in 2008 to over 34 TWh in 2035, an annual growth rate of 3.1%. The Turkmenistan government has plans to modernise and expand power generation, transmission and distribution capacity, as well as the gas distribution grid. Both the electricity system and gas pipeline networks are characterised by old and inefficient equipment. The average thermal efficiency of the country's power plants is estimated at 25%, far below the world average of over 45%. Electricity and gas distribution losses are also very high. Two power stations at Mary and Turkmenbashi are being upgraded, involving the installation of CCGTs, and three more such plants are to be built, at a total cost of $2.2 billion. By 2035, the average efficiency of gas-fired power plants is projected to increase to 42%, still significantly below current best-practice standards. Total investment needs in the electricity sector amount to $11 billion in 2010-2035, most of which is to expand and modernise transmission and distribution lines (Figure 16.14).

Uzbekistan

Uzbekistan has the second-largest energy market in the Caspian region, by dint of its large population (which makes up 36% of the total for the region) and of the extremely high energy intensity of its economy. Oil and gas account for virtually all the primary energy used in the country, reflecting its sizeable resources. The outlook for Uzbek energy use largely depends on the extent to which the considerable potential for using energy more efficiently is exploited, thereby countering the underlying upward pressure on demand from economic and population growth. In the New Policies Scenario, primary energy demand in 2035 reaches a level 36% higher than in 2008, an average rate of growth of 1.1% per year. This occurs in spite of progressive improvements in efficiency as the economy matures, with energy intensity dropping by an average of 3% per year. Gas use accounts for the bulk of this increase and oil for most of the rest (Table 16.7).

Table 16.7 ● **Primary energy demand in Uzbekistan by fuel in the New Policies Scenario** (Mtoe)

	1990	2008	2015	2020	2025	2030	2035	2008 -2035*
Coal	3.4	1.2	1.5	1.8	2.0	2.3	2.7	2.9%
Oil	10.1	4.7	5.9	6.4	6.1	6.6	7.3	1.6%
Gas	32.5	43.6	51.7	54.2	56.4	57.2	57.3	1.0%
Hydro	0.6	1.0	1.0	1.0	1.0	1.0	1.0	0.0%
Other**	0.0	0.0	0.0	0.0	0.1	0.2	0.5	31.9%
Total	46.6	50.5	60.2	63.4	65.6	67.2	68.8	1.1%

* Compound average annual growth rate.
** Includes biomass and other renewables.

The New Policies Scenario assumes relatively modest policy action to tackle the country's energy problems, which include gross inefficiencies in the production and use of energy and ageing infrastructure. Since the collapse of the Soviet Union, the government's approach has been to avoid the perceived social challenges of a rapid transition to a market economy; very gradual introduction of market principles into the economy is assumed to continue. This implies, for example, that the current patterns of tariff increases for end-users will continue to outpace inflation by a small margin, but rules out any prospect that subsidies for energy use will be removed.

Gas is expected to remain the bedrock of the energy system. The bulk of the gas consumed in Uzbekistan is used for power and heat generation and for direct heating in the residential sector. Gas is the main fuel for power and heat generation, accounting

16

for 70% of the country's total output of electricity and over 90% of heat production in 2008. There are an estimated 7 500 centralised boiler stations (with 25 000 boilers in total), dating back to Soviet times, that generate heat for distribution to industry and collective residential buildings. In principle, replacement of existing gas-fired power and heat plants with modern CCGTs to co-generate heat and power would yield up to 30% in energy savings (CAREC, 2009); but, since 1991, only two new power generation projects have been completed, both designed to expand capacity rather than replace old, inefficient plants. By 2015, however, the government aims to replace 570 MW of low-efficiency generation capacity and install three CCGT power plants, totalling 1 600 MW. Construction of two such projects, in Tashkent and Navoi, began in 2009 and the go-ahead has been given for a third project for new 800 MW CCGT units at the Talimarjan plant in the south of the country.

There is considerable unmet demand for gas and electricity in Uzbekistan: most electricity generation capacity is in the north of the country, whereas gas production and the main population centres are in southern regions. Limited north-south transmission capacity results in regular shortages of electricity and there are also reports that gas supplies to the domestic market are restricted in order to free up gas for export during the winter months, when demand is at its peak. These problems are exacerbated by high losses in both the electricity and gas distribution systems. The government aims to increase exports of natural gas in the coming years but, at the same time, there are plans to increase gas supplies to new petrochemical projects and to a new 35 kb/d gas-to-liquids plant in Karchi, which will consume around 3.5 bcm of gas per year when operational. This plant, which will be built by Uzbekneftegaz (the state oil company), Petronas and Sasol, using Sasol technology, is assumed to come on stream by 2020, reducing the country's need to import oil products (see Chapter 17). The commitment of gas for export and for specific domestic industrial projects may lead to restraints on gas use by other domestic consumers, implemented either through curtailed supply and shortages or alternatively through efficiency gains. In total, gas use is projected to reach 67 bcm in 2020 and 70 bcm in 2035 in the New Policies Scenario (Figure 16.20).

Figure 16.20 ● Primary natural gas demand in Uzbekistan by sector in the New Policies Scenario

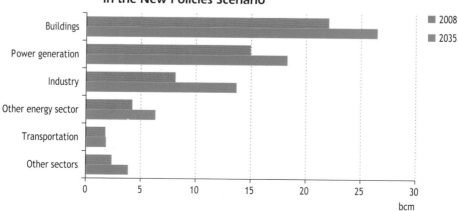

Oil consumption has declined significantly in recent years and is currently running at only one-half the level of the early 1990s, the result of sharp increases in prices, improved efficiency and conservation and fuel switching by end-users (especially in industry). The country is a small net importer of crude oil and a minor exporter of refined products, with indigenous production dwindling more or less in parallel with consumption in recent years. The scope for further reductions in demand is now limited and rising incomes are expected to drive up oil demand, especially for transport fuels, in the longer term. In the New Policies Scenario, demand reaches 127 kb/d in 2020 and 161 kb/d in 2035, up from about 100 kb/d in 2009.

Among final fuels, electricity consumption is expected to continue to grow rapidly. This projection hinges on the success of efforts to attract much-needed investment in expanding power generation and transmission capacity and replacing obsolete plants. Very low tariffs have discouraged investment in efficiency and conservation. The state power company, Uzbekenergo, inherited the assets of the Soviet era without debt, enabling the cost-recovery tariff to be set at a level that covered only operating and maintenance costs. Subsidised gas prices have also kept operating costs down artificially. The need to invest heavily in the coming years to replace existing infrastructure (most plants are 30 to over 50 years old) will increase the capital base and drive up the level of tariffs needed to recover costs. This is likely to spur investments on the demand side, curbing the growth of demand. We project electricity generation to increase from 49 TWh in 2008 to 74 TWh in 2035 (Figure 16.21), with total investments in the power sector amounting to about $24 billion, of which about 60% goes to transmission and distribution networks. Natural gas remains the leading fuel for electricity generation, its share of output rising from 70% in 2008 to 73% in 2035. Government policy aims to increase the share of coal in the fuel mix and coal registers the fastest growth rate, 5% per year, boosting its share of total generation to over 10% in 2035.

Figure 16.21 ● Electricity generation in Uzbekistan by fuel in the New Policies Scenario

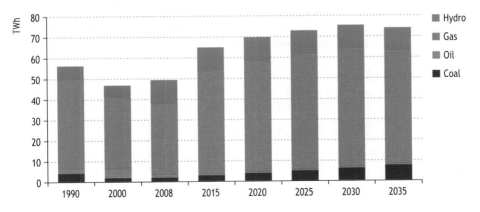

Alone among the countries of the Caspian region, Uzbekistan has put in place a functioning framework for Clean Development Mechanism (CDM) projects under the Kyoto Protocol. Seven CDM projects have been approved for implementation, with many others at different stages of preparation. Financial transfers under other international support mechanisms could also assist in developing Uzbekistan's large renewables potential, notably in solar energy, but renewable sources of electricity generation are assumed to struggle against subsidised gas-based electricity. The contribution of renewable energy sources to overall electricity generation therefore remains very small, at less than 1% in 2035 in the New Policies Scenario.

Other countries

The other countries in the Caspian region, Armenia, Georgia, the Kyrgyz Republic and Tajikistan, accounted for less than 7% of the region's total primary energy demand in 2008, reflecting their small populations and their relatively modest shares of global economic activity. All four countries suffered steep economic declines in the 1990s, with their combined GDP contracting by more than 60% from 1990 to 1995. By 2008, GDP had recovered to $66 billion (in PPP), but was still below the 1990 level. Energy use fell even more. Today, the four countries' combined primary energy demand per capita is far lower than that of the other Caspian countries, and among the lowest in the world. It amounted to 0.6 toe in 2008, raising concerns about the levels of access to energy (Box 16.2).

Armenia, Georgia, the Kyrgyz Republic and Tajikistan have only limited proven indigenous fossil resources and therefore have a heavy dependence on fossil-fuel imports from neighbouring countries, notably oil for transport and natural gas for the industry and power generation sectors. Armenia is the only country in the Caspian region with nuclear power production, from a single 376 MW reactor that provided about 40% of the country's total electricity production in 2008. The Armenian authorities announced in 2009 the construction of a new 1 060 MW nuclear plant, to be completed by 2017. This plant will replace the operating one, which is close to its 30 year design life and is situated in a seismic zone. We assume that this new nuclear power plant is commissioned by 2020.

Georgia, the Kyrgyz Republic and Tajikistan are the main producers of hydropower in the Caspian region and each country depends heavily on hydropower for its domestic electricity needs. Hydropower meets over 98% of domestic electricity demand in Tajikistan (90% for the Kyrgyz Republic, 85% for Georgia) although production is subject to seasonal variations that do not coincide with the demand profile. There is large scope for increasing hydropower production, particularly in the mountainous regions of Central Asia, where only a small part of the technically available hydropower potential has been exploited: current estimates of the extent of this exploitation to date are 5% in Tajikistan (Government of the Republic of Tajikistan, 2008) and 13% in the Kyrgyz Republic (CASE, 2008). Major projects are planned in both of these countries, such as the huge Rogun project (with installed capacity of 3.6 GW) in Tajikistan and Kambarata (1.9 GW) in the Kyrgyz Republic. In Georgia, the government estimates that the domestic installed hydropower generation capacity is about 20% of the country's potential (Ministry of Energy of Georgia, 2010); it has launched a programme to make more than 80 small and large green-field sites available to private investors for hydropower development. Together with new high-voltage interconnections, this initiative is designed to increase the country's export capacity. In the New Policies Scenario, we assume that total installed power capacity in Armenia, Georgia, the Kyrgyz Republic and Tajikistan rises from 12.2 GW in 2008 to 18 GW in 2035, an increase of almost 50% over the projection period, with hydropower capacity increasing from 7.9 GW to 11.3 GW. However, there is much uncertainty about the pace of new hydropower developments, in particular the major Central Asian projects that are extremely capital-intensive and face opposition from countries further downriver, including Uzbekistan in the case of the Rogun plant (see section on regional electricity co-operation in Chapter 18).

Total energy demand in aggregate in the four countries expands progressively between 2008 and 2035 in all three scenarios, though at different speeds. In the New Policies Scenario,[9] total primary energy demand is projected to increase at an annual average rate of 1.4% from 2008 to 2035 (Table 16.8). The share of natural gas in the four countries' energy mix declines from 35% in 2008 to 26% in 2035.

Table 16.8 ● Primary energy demand in Armenia, Georgia, Kyrgyz Republic and Tajikistan by fuel in the New Policies Scenario (Mtoe)

	1990	2008	2015	2020	2025	2030	2035	2008-2035*
Coal	4.0	0.7	0.7	0.7	0.7	0.6	0.6	-0.6%
Oil	13.9	2.6	3.1	3.4	3.6	3.8	4.0	1.6%
Gas	11.1	3.9	4.7	5.1	5.3	4.9	4.3	0.3%
Nuclear	0.0	0.6	0.7	1.7	1.7	1.7	1.7	3.7%
Hydro	3.1	3.1	3.5	3.6	3.7	4.0	4.4	1.3%
Other**	0.5	0.4	0.6	0.9	1.2	1.4	1.7	5.6%
Total	32.5	11.3	13.3	15.4	16.1	16.4	16.6	1.4%

* Compound average annual growth rate.
** Includes biomass and other renewables.

Among fossil fuels, oil use in the four countries grows most strongly, at 1.6% per year, reaching 4 Mtoe by 2035, while coal use is expected to contract at an annual rate of 0.6%, mainly due to the phasing out of old coal-fired power plants. Among end-use sectors, transport demand grows most briskly, at 2% per year on average over the *Outlook* period. Industrial energy demand grows by only 1% per year, thanks to improvements in energy efficiency (Figure 16.22). Among primary fuels, gas accounts for virtually all of the increase in industrial energy demand, due to rising consumption in Armenia and Georgia. Gas availability in Georgia is boosted by its position along the main transit route through the South Caucasus. In end-use sectors, the use of electricity in industry is projected to grow most quickly, averaging 1.5% per year, primarily because of aluminium production in Tajikistan. With the exception of

16

9. Only Armenia and Georgia have submitted information to the UNFCCC on nationally appropriate mitigation actions, along with general policy indications, however, with no specific quantitative targets.

Figure 16.22 ● Incremental energy demand in Armenia, Georgia, Kyrgyz Republic and Tajikistan by sector and fuel in the New Policies Scenario, 2008-2035

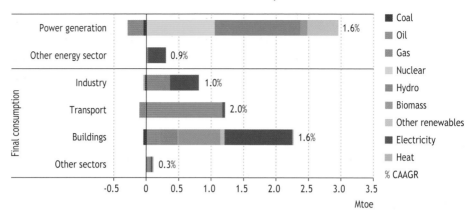

Georgia, where gas-fired power is competitive, government policies encourage the use of low-cost hydroelectricity in the industry sector in order to reduce the dependence on imported gas. Hydropower remains the main fuel for electricity generation, covering 80% of total production at the end of the projection period. Nuclear power also grows significantly, as a result of the new plant that is assumed to start operation in Armenia in 2020. For the region as a whole, electricity generation is projected to increase by 50% over the projection period, implying collective investment needs in the power sector of about $35 billion between 2010 and 2035, of which just over one-half is for generation (Figure 16.14, above).

CHAPTER 17

HYDROCARBON RESOURCES AND SUPPLY POTENTIAL

What will it take to unlock the Caspian's energy riches?

H I G H L I G H T S

- The Caspian region contains substantial resources of oil and natural gas, which could underpin a sizeable increase in production and exports. But potential barriers to their development, notably the complexities of financing and constructing transportation infrastructure across several countries, the investment climate and uncertainty over demand for Caspian gas in key export markets, are expected to constrain this expansion.

- In the New Policies Scenario, Caspian oil production grows strongly, mainly as a result of step increases in the capacity of three super-giant fields already under development. Production in the region jumps from 2.9 mb/d in 2009 to a peak of around 5.4 mb/d between 2025 and 2030, before falling back to 5.2 mb/d by 2035. Kazakhstan is the main driver of oil production growth.

- Most of the incremental oil output goes to exports, which double to a peak of 4.6 mb/d soon after 2025. This will require a sizeable increase in export capacity, in particular from Kazakhstan. Investment decisions on new export infrastructure are expected soon and must balance various commercial and strategic considerations.

- Caspian gas production jumps from an estimated 159 bcm in 2009 to nearly 260 bcm by 2020 and over 310 bcm in 2035 in the New Policies Scenario. Turkmenistan and, to a lesser extent, Azerbaijan and Kazakhstan drive this expansion. As with oil, exports are projected to grow rapidly, reaching nearly 100 bcm in 2020 and 130 bcm in 2035, up from 63 bcm in 2008.

- While Russia will remain a purchaser of Caspian gas, there will be greater diversity in Caspian gas trade as the region expands its access to new markets. The further development of a southern corridor from Azerbaijan to Turkey and other European markets paves the way for larger volumes of Azerbaijan gas to move westwards; Azerbaijan exports reach 35 bcm by 2035 – up from 5 bcm in 2009.

- The commissioning of the Turkmenistan-China pipeline has shifted the centre of gravity of Central Asia's gas sector eastwards. We project Chinese imports from the Caspian region to reach about 60 bcm by 2035 in the New Policies Scenario, although concerns about over-reliance on this supply route may limit the rate at which imports grow. Export to Russia will be constrained by the development of Russia's own resources, its more efficient gas use and the evolution of demand for Russian exports to Europe and to new markets in the east.

Overview

The Caspian region contains substantial resources of both oil and gas, though they are unevenly distributed geographically. These resources could underpin a sizable increase in production and exports over the next two decades or so. But there are major potential barriers to their development. Chief among these are the sheer scale of the investments needed, uncertainties over the quality of governance and the investment climate, the complexities of financing and constructing transportation infrastructure that passes through several countries before reaching export markets, technical difficulties associated with some of the upstream developments, competition from other resources (including, in the case of gas, liquefied natural gas and unconventional gas) and — crucially — uncertainty over the level of demand for Caspian gas in key export markets. All these factors will combine to constrain, at least to some degree, the growth of supply of Caspian hydrocarbons.

In the New Policies Scenario, Caspian oil production grows markedly — especially over the first 15 years of the projection period — mainly as a result of step increases in the capacity of major fields already under development in Kazakhstan and Azerbaijan. Production from Azerbaijan, Kazakhstan, Turkmenistan and Uzbekistan rises from an estimated 2.9 million barrels per day (mb/d) in 2009 to 4.4 mb/d in 2020 and to a peak of around 5.4 mb/d between 2025 and 2030, before falling back to 5.2 mb/d by 2035 (Figure 17.1). Although oil demand across the region continues to grow with economic expansion, total production remains much higher, freeing up oil for export. The volume of exports peaks at 4.6 mb/d soon after 2025 and falls back to about 4.3 mb/d in 2035, up from about 2.3 mb/d in 2009. Azerbaijan and Kazakhstan remain the only significant exporters of oil.

Figure 17.1 ● **Caspian* oil balance in the New Policies Scenario**

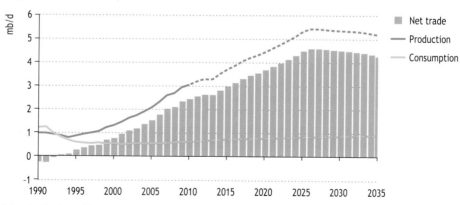

* See the Preface of Part D for the definition of the Caspian region in this *Outlook*.

Natural gas production is also set to expand substantially over the projection period. In the New Policies Scenario, marketed gas production in the Caspian countries in aggregate jumps from 188 billion cubic metres (bcm) in 2008 (and an estimated 159 bcm in 2009) to nearly 260 bcm by 2020 and 315 bcm in 2035 (Figure 17.2). As with oil, gas demand is set to grow less than production in volume terms, yielding a significant expansion of the region's net exports. By 2035, total net exports are projected to reach nearly 100 bcm in 2020 and 130 bcm in 2035, up from only 63 bcm in 2008. The biggest contributors to this increase in exports are Turkmenistan and Azerbaijan.

Figure 17.2 ● Caspian gas balance in the New Policies Scenario

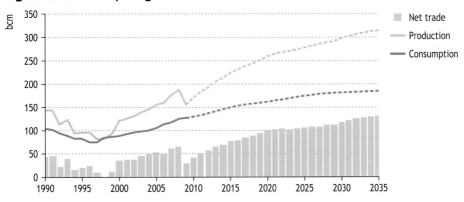

The oil and gas landscape of the Caspian region has been transformed in the years since 1991. From a starting position as constituent republics within a tightly integrated Soviet economic system, the Caspian resource-owners have become autonomous actors on the international energy stage, asserting national authority over management of their resources and creating new links with export markets. Diversification of routes and markets has contributed to more reliable market-based pricing of exports and, thereby, created stronger incentives to develop resources. This process has been quicker for oil than for natural gas; although more oil-export capacity is required to accommodate future production growth, the relative ease and flexibility of transportation of oil meant that it was traded to multiple destinations and with reference to international prices quite soon after 1991. By contrast, natural gas exports from Turkmenistan, which has been the region's largest gas exporter, were characterised for much of the 1990s and early 2000s by non-cash and barter payments well below the international value of the gas. It was only after 2005 that Gazprom, the major purchaser of Central Asian gas, was ready to concede cash payments and higher export prices, reflecting the importance that Central Asian supplies had come to assume in the Russian gas balance (at least until the economic crisis in 2008-2009), as well as increased competition for Caspian gas resources from China and also from other potential consumers in Europe and southern Asia.

Aside from the distance to export markets, oil and gas exploration and production in the Caspian region has to cope with some distinctive challenges. While average upstream exploration and development costs are reasonable by international standards (Box 17.1), developments in the Caspian Sea operate in a very difficult and fragile natural environment, with the shallower northern Caspian waters habitually freezing

17

from November until March. The Russian Volga-Don canal system from the Black Sea is the only maritime route into the Caspian region, creating logistical difficulties for companies bringing in drilling and other large equipment. Upstream developments since 1991 have also had to deal with rapidly shifting legal and regulatory frameworks and shifting balances between state and commercial influence over the sector, as the Caspian countries established their national systems of resource management and then, in some cases, toughened the conditions for upstream operators from the mid-2000s onwards.

Box 17.1 ● How do Caspian upstream costs compare?

For all the technical challenges associated with Caspian production, the costs of getting Caspian resources out of the ground compare favourably with those in most other regions. The actual and planned capital expenditure and estimated production profiles for the six main Caspian oil and gas fields (the Azeri-Chirag-Guneshli complex, Karachaganak, Kashagan, Shah Deniz, South Yolotan and Tengiz) show a wide variety of costs. Overall, we calculate that the capital cost of developing these fields averages around $8 per barrel of oil and $55 per thousand cubic metres of gas (around $8.5 per barrel of oil equivalent [boe], or $1.5/MBtu). These are at the lower end of the estimated range of costs for the Eastern Europe/Eurasia region of $7-19/boe and below the global average (see Figure 3.28 in Chapter 3). These estimates, combined with the size of Caspian resources and — for much of the region — their relative accessibility to outside investors, help to explain the continued interest of national and international oil and gas companies in the Caspian upstream sector. But these figures tell only a part of the story and need to be considered alongside a wider range of risks and costs, including regulatory and fiscal requirements, operating expenditures (lifting costs) and, most crucially for the Caspian, the distance, expense and complexity involved in bringing resources to international markets.

Foreign investment has been central to the development of oil and gas production in Azerbaijan and Kazakhstan, but much less so elsewhere in the region. Investment by privately-owned international companies was the dominant element of this story in the 1990s, but since 2000 an increasing share of foreign investment has come from national oil and gas companies in Asia, including Korea, Malaysia, India and, in particular, China. State-owned Chinese companies have become heavily involved in various upstream and mid-stream projects as investors, service providers, operators and as purchasers of Caspian hydrocarbons. New export infrastructure both for oil (since 2006) and gas (since late 2009) now connects Central Asia to the fast-growing Chinese market. Both as a source of investment capital and as a major export market, China will continue to have a strong influence on trends in Caspian production and trade through the projection period and beyond.

China's growing role in the region is challenging the traditional predominance of Russia in Central Asia and also provides stiff competition for other international companies seeking investment opportunities in the region – all to the benefit of the Caspian countries themselves. But, even as a greater share of Caspian resources is exported to the east, it is worth keeping China's current investment position in perspective. As of 2009, we estimate the share of Chinese companies in the oil and gas production of the four main Caspian producers at 7%, resulting largely from a 19% share in Kazakhstan oil output (Figure 17.3). This is well behind the 38% share of privately-owned companies in Caspian output, including the international oil companies involved in the Azeri-Chirag-Guneshli field development in Azerbaijan and the Tengiz and Karachaganak projects in Kazakhstan. Production from Chinese-led projects is set to increase in the coming years, for example from the China National Petroleum Corporation (CNPC) Bagtyyarlyk contract area in eastern Turkmenistan. However, at least in the period to 2025, the overall share of Chinese companies in Caspian oil and gas production is likely to go down rather than up – especially once output from the Kashagan field, in which international companies currently have an 83% share, starts and then builds up.

Figure 17.3 ● **Estimated Caspian oil and gas production by type of company, 2009**

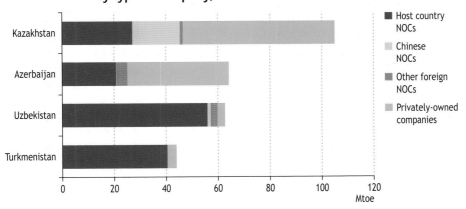

Note: Host country NOC refers to national oil and gas companies operating in their own country; NOCs (whether China or foreign) refers to state-owned or state-controlled companies; shares of production in projects governed by production-sharing agreements (PSA) are allocated according to the ownership of the PSA.

17

Oil

Overview and market context

There are 48 billion barrels of proven oil reserves in the Caspian region, equal to about 3.5% of the world total. Remaining recoverable resources amount to almost 109 billion barrels, or 5% of global oil resources (Table 17.1). Proven reserves alone would be sufficient to sustain current rates of production for about 45 years (though production will certainly increase), about the same as the global average of 46 years. The bulk of the region's reserves and resources are in Kazakhstan. Azerbaijan and Turkmenistan also hold significant volumes of recoverable oil (though proven reserves in the latter are small). Resources in the rest of the region are small. Oil has been produced in parts of the region for over a hundred years but the Caspian as a whole — particularly offshore — is still relatively under-explored and exploited. The region contains three super-giant fields (containing more than 5 billion barrels of proven and probable reserves) as well as at least a dozen giant fields. Two super-giant fields are already producing: the Tengiz field in Kazakhstan and the Azeri-Chirag-Guneshli (ACG) complex in Azerbaijan; the Kashagan field in Kazakhstan is one of the few remaining super-giant fields not yet in production. The three fields collectively hold nearly one-fifth of the region's initial endowment of recoverable oil. To date only 19% of the total volume of ultimately recoverable oil in the Caspian region has been produced, compared with a global average of 33%.

Table 17.1 ● Conventional oil resources in the Caspian by country, end-2009 (billion barrels)

	Proven reserves	Ultimately recoverable resources	Cumulative production	Remaining recoverable resources
Azerbaijan	7.0	29.9	11.7	18.2
Kazakhstan	39.8	78.2	9.2	68.9
Turkmenistan	0.6	19.5	3.6	15.9
Uzbekistan	0.6	5.5	1.1	4.3
Other Caspian*	–	1.4	0.2	1.3
Total	48.0	134.4	25.8	108.6
Share of world	*3.5%*	*3.9%*	*2.3%*	*4.7%*

* Armenia, Georgia, Kyrgyz Republic and Tajikistan.
Sources: O&GJ (2009); BGR (2009); data provided to the IEA by the US Geological Survey; IEA databases and analysis.

In the New Policies Scenario, the three super-giant fields account for 93% of the increase in production between 2009 and 2020 (Figure 17.4).These three fields collectively account for over half of Caspian oil production at their peak after 2020 (and about 3% of total world oil production), after which time a growing share of output comes from smaller fields awaiting development and fields that have yet to be found. Development of the three projects and the related export infrastructure will stimulate further developments in deposits nearby.

Kazakhstan will be the main driver of Caspian oil production, contributing all of the net growth in the region's oil output in the New Policies Scenario. Azerbaijan's oil production levels off by around 2014 and starts to fall gradually after 2020. Output growth in the rest of the Caspian is minimal (Table 17.2).

Figure 17.4 ● Oil production in the Caspian by major field*
in the New Policies Scenario

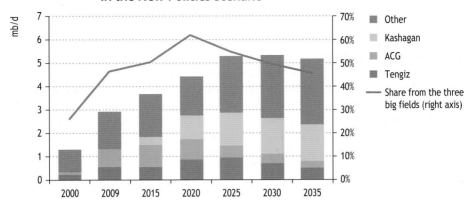

* Kashagan and Tengiz (Kazakhstan); Azeri-Chirag-Guneshli (Azerbaijan).

Table 17.2 ● Oil production in the Caspian by country
in the New Policies Scenario (mb/d)

	1990	2000	2009	2015	2020	2025	2030	2035
Azerbaijan	0.3	0.3	1.1	1.3	1.3	1.2	1.1	0.9
Kazakhstan	0.5	0.7	1.6	2.0	2.8	3.8	3.9	3.9
Turkmenistan	0.1	0.1	0.2	0.3	0.3	0.3	0.3	0.2
Uzbekistan	0.1	0.2	0.1	0.1	0.1	0.1	0.1	0.1
Other Caspian	0.0	0.0	0.0	0.0	0.0	0.0	0.0	0.0
Total	1.0	1.3	2.9	3.7	4.4	5.3	5.4	5.2
Share of world	*1.5%*	*1.7%*	*3.5%*	*4.1%*	*4.9%*	*5.7%*	*5.6%*	*5.2%*

Detailed analysis of the production history of the region's largest oil fields demonstrates the critical role played in the profile of production and the rate at which reserves are developed by access to export infrastructure and by short-term fluctuations in export demand. Applying a approach similar to that used in our global field-by-field study of oil fields in *WEO-2008* and gas fields in *WEO-2009*, we analysed the 44 largest oil and gas fields in the Caspian to determine regional resource management characteristics.[1] As the super-giant oil fields in the region are at an early stage of development, the results are best compared to those of giant fields elsewhere (Table 17.3).

17

---

1. Results for the gas fields are included below in the gas section.

Table 17.3 ● Production-weighted average annual observed decline rates of oilfields by region (%)

	Post-peak			Post-plateau		
	Giant	Large	Total	Giant	Large	Total
Caspian	3.4	6.3	3.6	2.9	5.8	3.0
E. Europe/Eurasia*	5.0	12.1	5.1	5.1	12.4	5.3
World*	6.5	10.4	5.1	6.6	10.7	5.8

* Decline rates from IEA (2008a).

Overall the decline rate trends follow those of previous studies, with the larger fields declining more slowly, but the Caspian rates are lower than the global averages as they are heavily influenced by the extended profiles of fields where production has been curtailed due to export and investment restrictions. Furthermore, exploitation of the largest fields in the region has typically been done in multiple phases, often with periods of very low activity between them, further extending the time required to deplete the resources. Average flow rates during plateau production were 3.7% of proven and probable (2P) reserves per year, lower than the global averages for fields of similar sizes.

Most of the increase in Caspian oil production between 2008 and 2035 is exported. In the New Policies Scenario, total exports are projected to almost double, from 2.35 mb/d in 2009 to a peak of about 4.5 mb/d in 2030, before falling back to 4.3 mb/d in 2035. Kazakhstan accounts for all the growth in the region's exports; Azerbaijan's exports level off by the middle of the 2010s and begin to decline by around 2020 in line with the downward trend in its oil production (Table 17.4). The prospects for Caspian oil flows are explored later in this chapter.

Table 17.4 ● Oil net exports in the Caspian by country in the New Policies Scenario (mb/d)

	1990	2000	2009	2015	2020	2025	2030	2035
Azerbaijan	0.05	0.16	0.97	1.23	1.19	1.06	0.96	0.77
Kazakhstan	0.16	0.55	1.33	1.69	2.47	3.42	3.57	3.55
Turkmenistan	-0.03	0.07	0.09	0.15	0.14	0.11	0.09	0.07
Uzbekistan	-0.14	0.01	0.01	-0.02	-0.03	-0.02	-0.03	-0.05
Other Caspian	-0.28	-0.03	-0.05	-0.07	-0.07	-0.08	-0.08	-0.09
Total	-0.24	0.76	2.35	2.98	3.69	4.49	4.51	4.27

Note: Negative numbers indicate net imports.

Azerbaijan

Azerbaijan was quick to open its doors to international investors after independence in 1991 and has maintained an open stance on upstream investment since then.

The development of the Azeri-Chirag-Guneshli (ACG) group of offshore fields and the opening of the Baku-Tbilisi-Ceyhan oil export pipeline in 2006 were visible and successful results of this policy, making Azerbaijan one of the very few countries outside OPEC that has increased its conventional oil output since 2000. However, a number of other international companies that partnered with the national oil company, SOCAR, to explore offshore blocks did not discover hydrocarbons in commercial quantities, tempering some of the early optimism about the size of Azerbaijan's oil wealth.

With no other fields under development in Azerbaijan that are of comparable size to ACG, output from that project will have a large impact on overall Azerbaijani production. The remaining production will come from a variety of small onshore and offshore fields, developed both independently by SOCAR and with local and international partners. SOCAR is adding offshore expertise to its long experience onshore; as well as operating the shallow-water Guneshli field, the national company is completing a first test well in deeper water at the Umid field in 2010. There is also liquids production from the Shah Deniz gas condensate field; this was around 35 thousand barrels per day (kb/d) in 2009 but could rise to over 100 kb/d during Phase II development (see Azerbaijan gas section for full discussion of this field).

Overall, Azerbaijani output is projected to rise from an estimated 1.1 mb/d in 2009 to a plateau of 1.3 mb/d by 2012 (Figure 17.5). Compared with production, domestic oil consumption is relatively low at under 80 kb/d in 2008; demand for oil in Azerbaijan has fallen by more than half since 1990 due to economic contraction in the 1990s and then the replacement of fuel oil by natural gas in power generation. Even though domestic oil consumption is projected to recover through the projection period, driven by increased demand for oil in the transportation sector, the bulk of Azerbaijan's production is still available for export.

Longer-term projections are driven by overall resource potential: new discoveries, as well as new projects and enhanced recovery at existing fields, such as ACG, are expected to mitigate a decline in Azerbaijani output after 2020. We project total production to decline gradually after 2020, falling to 0.9 mb/d in 2035 in the New Policies Scenario. Despite the mixed exploration record of the 1990s, there are some signs of a second wave of investor interest in the Azerbaijani offshore, with France's Total re-starting exploration in 2010 at Absheron (it had been part of a previous production-sharing agreement, or PSA, consortium with the US company, Chevron, at this field) and Germany's RWE is also looking to revisit a previously explored structure at Nakhichevan. BP has an agreement with SOCAR on the Shafag and Asiman prospects, which have not yet been explored. The longer-term production potential would be bolstered by the resolution of disputes over Azerbaijan's maritime borders. This includes the Azerbaijan claim to the mid-Caspian Serdar/Kyapaz field (see also the section on Turkmenistan) and also the promising areas not currently open to exploration, near the Azerbaijani border zone with Iran. Exploration of the Alov field (called Alborz in Iran) by a BP-led consortium was postponed in 2001 after the Iranian military turned survey vessels back from the contract area. Political decisions could open up this area during the projection period, with a potentially material impact on the region's oil balance.

17

Figure 17.5 ● Azerbaijan's oil balance in the New Policies Scenario

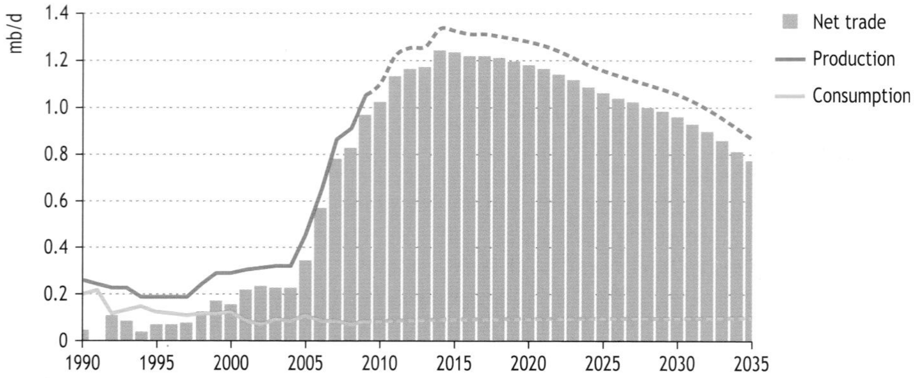

Azeri-Chirag-Guneshli (ACG)

The ACG group of fields in the Caspian Sea (Figure 17.6) is estimated to contain up to 9 billion barrels of recoverable resources and accounted for about three-quarters of Azerbaijan's output in 2009. The complex was originally discovered during the Soviet era, but only the shallow part of the Guneshli field was developed. This area is now operated independently by SOCAR, accounting for more than two-thirds of the company's operated oil and gas production.

The rest of ACG, which now has five production platforms, is operated by BP on behalf of the Azerbaijan International Operating Company (AIOC), a consortium of BP and eight other companies, including SOCAR, Chevron, Statoil and ExxonMobil. Following signature of the PSA in 1994, production at the Chirag field started in 1997 and then increased in three subsequent phases, boosting total ACG output from an average of around 130 kb/d in 2004 to 670 kb/d in 2007 and then to around 800 kb/d by early 2010. Production was held back in 2008 by an explosion on the Baku-Tbilisi-Ceyhan pipeline, the main export route for Azerbaijani oil, and then from late 2008 and into 2009 by a gas leak at the Central Azeri Platform.

Further production growth in the coming years is expected to come from the Chirag Oil Project, scheduled to start production from a sixth platform in late 2013. The Chirag Project, a $6 billion investment approved in March 2010, will push output levels from ACG towards the 1 mb/d mark. With this investment in place, production from ACG is set to remain above 900 kb/d until 2018 or 2019, before declining. Remaining ACG reserves would be sufficient to justify additional investment in production, but it is not certain when, or whether, this will materialise. A constraint is the expiry of the current PSA at the end of 2024, reducing incentives for the existing consortium to sanction any new projects after the Chirag Oil Project. A decision on the extension of the agreement will be important in determining the medium-term trajectory of ACG production.

Figure 17.6 ● Main oil deposits and export routes in the South Caucasus

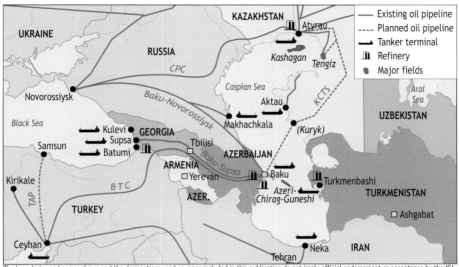

The boundaries and names shown and the designations used on maps included in this publication do not imply official endorsement or acceptance by the IEA.

Azerbaijan oil export and transit routes

The pipeline network in the South Caucasus (Table 17.5) has grown to accommodate increases in output from the ACG complex and there is already enough oil transportation capacity in place to accommodate Azerbaijan's total projected export levels for the period to 2035. A strategic question for Azerbaijan and the region is how the export infrastructure in the South Caucasus will be used when domestic production in Azerbaijan starts to decline, and whether – and when – new transportation capacity will be required in order to bring additional oil from Kazakhstan to market. Azerbaijan has already become a transit country for Kazakhstan oil – an estimated 65 kb/d was shipped across the Caspian Sea to Baku in 2009 – and the South Caucasus is an important strand in Kazakhstan's own vision of multiple export routes for oil (see the section Kazakhstan Caspian Transportation System (KCTS) below).

Table 17.5 ● Azerbaijan's oil export routes

	Route	Current capacity (kb/d)	Date commissioned, length	Possible capacity increases
Northern	Baku-Novorossiysk (Russia)	100	1983, 1 330 km	None planned
Western	Baku-Supsa (Georgia)	100	1999, 833 km	Possible new Baku-Black Sea coast pipeline
Baku- Ceyhan	Baku-Tiblisi-Ceyhan (Turkey)	1 200	2006, 1 768 km	Up to 1.6 mb/d with new pumping stations
Rail	Baku- Batumi / Kulevi (Georgia)	Around 220	n.a.	n.a.

17

The main export pipeline in the 1990s was from Baku to the Russian Black Sea port of Novorossiysk (this became known as the Northern Export Route), which has been used by SOCAR and initially also by AIOC. The advantage of competitive transportation tariffs along this route has, though, been offset by periodic instability in the North Caucasus and the absence of a quality bank within the Russian pipeline system operated by Transneft, to compensate Azerbaijani exporters for the loss of value for Azeri Light crude oil when it is mixed with lower quality crudes. In planning additional export routes, the Azerbaijani authorities and ACG partners have also been wary about the strategic implications of giving Russia a hold over Azerbaijani exports.

A second export option was completed in 1999, the so-called Western Export Route running from Baku to Supsa on the Georgian Black Sea Coast. Capacity along this route rose to around 150 kb/d by 2003, but it is now operating at around 100 kb/d. This became the preferred route for export of early oil and, but for a period of repair from 2006-2008, has been used consistently for ACG oil exports.

Since 2006, the main export route for ACG oil has been the Baku-Tbilisi-Ceyhan pipeline, which runs across Azerbaijan and Georgia to the Turkish Mediterranean coast. The capacity of this route reached 1 mb/d in 2008 and 1.2 mb/d in March 2009, with the addition of facilities to use drag-reducing agents in the pipelines. Exports along this route averaged only 780 kb/d in 2009. Under-utilisation of capacity was linked in part to the gas leak at the Central Azeri platform, which constrained ACG output; this had the temporary effect of raising the transportation tariff, which is calculated on the basis of a guaranteed rate of return to investors. One consequence of this tariff increase was that exports of Kazakhstan oil via the BTC pipeline, which began in October 2008, stopped in 2009 and switched back to rail routes to the Georgian oil terminals of Batumi and Kulevi. However, as of 2010, some crude from Turkmenistan is being exported through the BTC pipeline.

The routes to market via Azerbaijan and Georgia — as part of the Kazakhstan Caspian Transportation System (KCTS) — are important export options for Kazakhstan's oil producers as their production capacity increases over the coming two decades. There are plans to build a fourth oil terminal south of Baku at Garadagh, which would add new receiving capacity alongside the existing terminals at Dubendi, Sangachal and Baku itself. We assume that transit of Kazakhstan oil through the South Caucasus increases rapidly, starting in 2020, from current levels of under 100 kb/d to over 1 mb/d by 2025. Given still significant, though declining, Azeri exports, this will require more than 500 kb/d in additional export capacity through the South Caucasus in the early 2020s (Figure 17.7). No decisions have yet been taken about whether this would mean expansion of the existing Baku-Tbilisi-Ceyhan pipeline and/or construction of a new pipeline from Baku to the Black Sea.

Kazakhstan

With 40 billion barrels of proven oil reserves, Kazakhstan has the largest share of Caspian reserves and has the potential to become one of the world's leading oil producers. Production of oil amounted to 1.6 mb/d in 2009 but is projected to increase

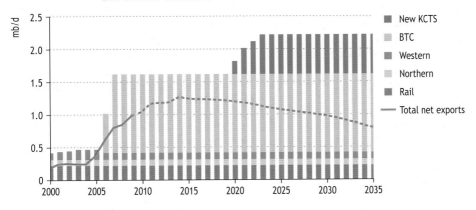

Figure 17.7 ● Azerbaijan's oil net exports and transit capacity by source in the South Caucasus in the New Policies Scenario

rapidly in the coming years as existing projects are expanded and the huge, but much-delayed, Kashagan field begins production. In the New Policies Scenario, Kazakhstan output rises to 2 mb/d in 2015 and then climbs to 2.8 mb/d in 2020 and to a plateau of 4 mb/d soon after 2025 (Figure 17.8). Kazakhstan is the only country among the Caspian producers where production grows throughout the projection period. The main uncertainties relate not to hydrocarbon resource availability but rather to the climate for investment and operation of upstream projects, and to the availability of sufficient export capacity.

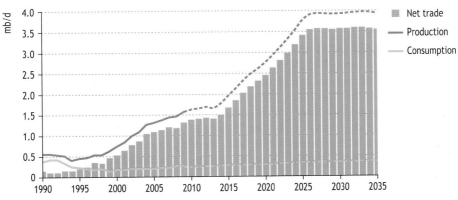

Figure 17.8 ● Kazakhstan's oil balance in the New Policies Scenario

The prospects for Kazakhstan's oil production to 2035 rest to a large extent on three fields: Karachaganak (a gas/condensate field), Tengiz and Kashagan, all of which are being developed by major international consortia (and which are described in more detail below). There are also more mature production areas and multiple smaller prospects being developed by domestic and foreign companies — increasingly by companies from China. After the Tengiz and Karachaganak operating companies and the national oil company, KazMunaiGaz, the next largest oil producers in Kazakhstan in 2009 were AktobeMunaiGaz, majority owned by CNPC since 2003, and MangistauMunaiGaz, in which CNPC bought a 50% stake in 2009. As of 2009, Chinese companies accounted for an estimated 19% share of Kazakhstan oil output.

The hydrocarbons sector is the engine for Kazakhstan's national economic development and the government in recent years has been seeking to increase the use of local labour, goods and services by oil and gas companies as well as seeking higher revenues from the industry as a whole. Examples of the latter include the introduction of an oil-export duty in mid-2008 (suspended when prices fell in 2009, but re-imposed in August 2010) and a new tax code in 2009. Existing PSA projects with tax stability clauses have not been immune from fiscal pressures; on the contrary, there has been a general drive to review and re-balance agreements concluded in the 1990s, when oil prices were much lower. A related priority has been to ensure a prominent role for KazMunaiGaz in major oil and gas projects. The government's twin concerns over fiscal conditions and national ownership — alongside frustration over project delays and cost over-runs — came together during the extended negotiations over the Kashagan project. Although these negotiations were concluded in 2008, similarly protracted discussions over new or revised commercial terms could affect other upstream developments in Kazakhstan, holding back the projected increases in oil output.

Kashagan

The super-giant Kashagan field lies under the shallow-water Kazakhstan sector of the Caspian Sea (Figure 17.9) and is estimated to contain around 35 billion barrels of oil in place, of which between 7 and 9 billion barrels is deemed recoverable, or up to 12.5 billion barrels with gas re-injection.[2] The development timetable has been delayed several times since appraisal work began in the 1990s. We assume that first oil comes at the end of 2013, one year behind the current official schedule. The composition of the Kashagan consortium has changed several times since the PSA was signed in 1997 (see Table 17.6 for current ownership shares), with KazMunaiGaz acquiring its current 16.8% stake both through the government's pre-emption rights (an 8.3% share in 2005) and during a later dispute, fuelled by project delays, which gave the national company a much greater role in project management.[3] The North Caspian Sea PSA runs until 2041 and, given that first oil was originally foreseen for 2005, project delays mean that the productive life of the PSA will be at least eight years shorter than originally envisaged. The contract area includes Kashagan itself and three smaller discoveries — Kalamkas More, Aktote and Kairan — that are also scheduled for development during this period.

---

2. See Chapter 3 for definitions of terms relating to reserves and resources.

3. Within the consortium, Eni is responsible for the execution of the initial phase of the project, but operational responsibilities for Phase II were restructured in 2008: Shell has responsibility for overall management of production operations, together with KazMunaiGaz, and leads the offshore development work; ExxonMobil takes charge of drilling; and Eni retains responsibility for the onshore plant.

Table 17.6 ● Ownership of the main Caspian upstream and midstream oil projects

	Karachaganak (KPO)	Kashagan (NCOC)	Tengiz (TCO)	CPC pipeline	ACG (AIOC)	BTC pipeline
BG Group	32.5%			2.0%		
BP					37.4%	30.1%
Chevron	20.0%		50.0%	15.0%	11.3%	8.9%
ConocoPhillips		8.4%				2.5%
Eni	32.5%	16.8%		2.0%		5.0%
ExxonMobil		16.8%	25.0%	7.5%	8.0%	
Inpex		7.6%			11.0%	2.5%
Lukoil	15.0%		5.0%	12.5%		
Shell		16.8%		3.7%		
StatoilHydro					8.6%	8.7%
Total		16.8%				5.0%
TPAO					6.8%	6.5%
KazMunaiGaz		16.8%	20.0%	19.0%		
Russia (Transneft)				31.0%		
SOCAR					10.0%	25.0%
Other				7.3%	6.9%	5.8%

Note: The ownership structure of the various components of the Kazakhstan Caspian Transportation System has yet to be determined. KPO = Karachaganak Petroleum Operating; NCOC = North Caspian Operating Company; CPC = Caspian Pipeline Consortium; TCO = Tengizchevroil; and AIOC = Azerbaijan International Operating Company; BTC = Baku-Tbilisi-Ceyhan.

The north Caspian, with its shallow waters freezing over every winter, presents a very challenging setting for conventional offshore platforms. The solution found by the consortium is to build artificial islands for all offshore activities. Over the course of full-field development, six major island hubs are envisaged to collect production from numerous drilling islands and more than 200 wells. The reservoir is deep and has very high pressure, with large volumes of associated gas, which has high non-hydrocarbon content (see section on Kazakhstan gas for a discussion of Kashagan gas use). The north Caspian is also an environmentally sensitive area and the consortium has to adhere to strict standards including the "zero discharge" principle for all platform-generated wastes, which have to be transported and processed onshore.

To date, only Phase I — also called the "experimental" programme — has been approved and production from this phase, from first oil in 2013, is expected to rise to 300 kb/d by 2015. Full Phase I production of 450 kb/d is likely from 2016, assuming that the construction of some additional sour gas re-injection facilities, originally foreseen as part of Phase II development, is brought forward and synchronised with Phase I.

The second phase of the Kashagan project, which will incorporate the development of the Kalamkas field, is still in conceptual planning and the start-up date is a major uncertainty for the long-term production profile. In the New Policies Scenario, we assume that production from Phase II starts in 2019, with peak production from the first two phases combined (including Kalamkas) reaching around 1.1 to 1.2 mb/d in the early 2020s. Adding a third phase then brings projected Kashagan output to the anticipated

17

plateau level of 1.5 mb/d by 2025, a level that is expected to be maintained until around 2030. It will become clear only once production begins whether the anticipated plateau production of 1.5 mb/d is a conservative or optimistic figure. Potential constraints are by no means all below the ground; production could also be held back by a lack of export capacity (see Kazakhstan oil export routes below).

Figure 17.9 ● **Oil fields and infrastructure in the North Caspian**

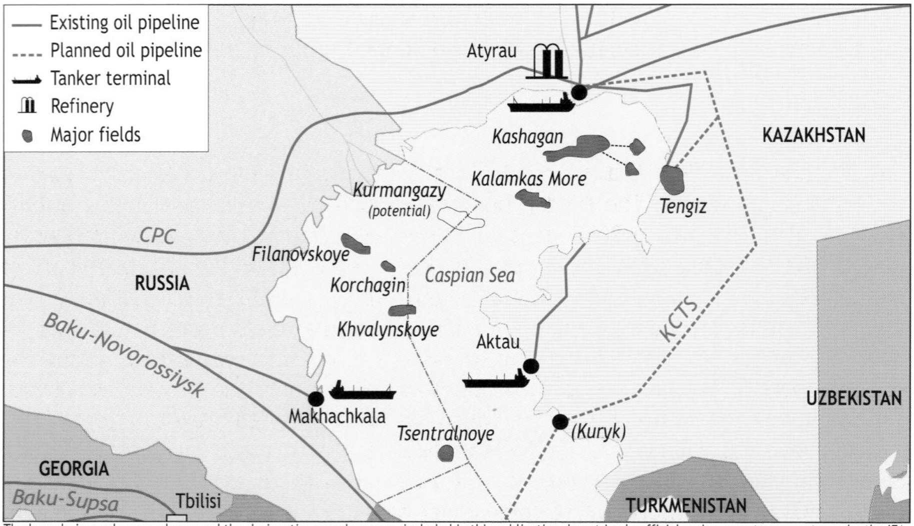

The boundaries and names shown and the designations used on maps included in this publication do not imply official endorsement or acceptance by the IEA.

Tengiz

The Tengiz field in western Kazakhstan is currently the largest producing field in Kazakhstan, with output of 450 kb/d in 2009, just under 30% of the country's total output (Figure 17.4 above). Estimates of recoverable resources range between 6 and 9 billion barrels, from initial oil in place of around 26 billion barrels. Production planning for the Tengiz field dates back to the 1980s and commercial output started in 1991, but field development began in earnest only with the signature of a 40-year partnership agreement in 1993 with the Tengizchevroil (TCO) consortium, led by Chevron (Table 17.6). As the largest single contributor to Kazakhstan oil production growth since 1991, the TCO consortium has so far been at the forefront of efforts to expand the country's export capacity; this mantle will fall to the Kashagan consortium over the next 15-year period.

Although Tengiz is onshore, some of the technical challenges at the field are similar to those at Kashagan, including a deep, high-temperature and high-pressure reservoir and an elevated hydrogen sulphide content in the associated gas, which has made sulphur management a major issue for the operator. The completion of new processing and reinjection facilities at the site in 2008 (the Sour Gas Injection and Second Generation Project) brought production up to its current capacity of 540 kb/d. However, there are significant possibilities to expand output beyond this level. Once sanctioned by the Kazakhstan government, a Future Growth Project, including new re-injection capabilities for sour gas, could bring total capacity to between 800-900 kb/d after 2015. This project would not exhaust the field's potential; we project production to reach 850 kb/d towards 2020 and to peak at around 950 kb/d in 2025, before entering a period of gradual decline. Since 2001, the main export route for Tengiz has been the Caspian Pipeline Consortium (CPC) oil pipeline, although around 150 kb/d still leaves the field by rail (just under half of which is then shipped from Aktau across the Caspian for transit via Azerbaijan). The much-delayed expansion of the CPC pipeline is critical to accommodate the anticipated increases in Tengiz output to 2020.

Karachaganak

The Karachaganak field is a gas condensate field in northwest Kazakhstan, close to the border with Russia. The field is being developed by a consortium under a 1997 PSA that runs through to 2038. It is currently the second-largest liquids-producing field in Kazakhstan, with output of 270 kb/d representing around one-sixth of Kazakhstan's total production. Initial hydrocarbons in place are estimated at 9 billion barrels of condensate and 1.4 trillion cubic metres (tcm) of natural gas, of which 2.4 billion barrels and 450 bcm of gas are considered recoverable during the contract period. Despite the large volumes of natural gas, the project economics are driven by condensate production, which has been easier to bring to international markets (see Kazakhstan gas section for discussion of natural gas). Karachaganak is the only major field development in Kazakhstan without any participation from KazMunaiGaz. The consortium has come under regulatory and fiscal pressure in recent years, including the levy of around $1 billion in export duties that is disputed by the consortium, which claims an exemption under their PSA. The persistent problems facing the Karachaganak development have led to speculation that a resolution may involve KazMunaiGaz acquiring a stake in the project.

Phase II development of the field was completed in 2004 and construction of a fourth stabilisation train for condensate production was sanctioned in 2006. This will add around 55 kb/d of processing capacity in 2011, bringing total production capacity above 300 kb/d. Phase III of the project was originally envisaged for 2012, but it is still under review and has not yet been sanctioned; we assume that this project will be implemented in stages from 2014. The impact of Phase III would be primarily on natural gas — exports would double to 16 bcm per year — but it would also result in an increase in liquids output to above 350 kb/d and enable these levels to be maintained from the late 2010s until around 2025. Most current liquids export from Karachaganak is through the Caspian Pipeline Consortium oil pipeline to Novorossiysk, although the Karachaganak partners will need to seek additional routes to market for the expanded output under Phase III development.

Other fields

Future oil-production potential in Kazakhstan is concentrated in the offshore Caspian Sea and in the west of the country. There are a number of other prospective

17

developments, particularly offshore, that could contribute to production growth during the projection period. In 2009, Shell completed a successful appraisal well and production test at the offshore Pearls project (under a PSA with KazManaiGaz and Oman Oil) and the same year ConocoPhillips, together with KazMunaiGaz and Abu-Dhabi-based Mubadala, signed project agreements for the Nursultan (or 'N') block further south. Among the other offshore prospects, there have been high expectations of the Kurmangazy field that Russia's Rosneft and KazMunaiGaz are jointly developing, but two exploratory wells, in 2006 and 2009, failed to find hydrocarbons. Apart from Tengiz and Karachaganak, the main onshore producing areas are in the Mangistau region around Aktau (fields owned by MangistauMunaiGaz, in which China's CNPC has a 50% stake, and the Uzen fields), Aktobe in the northwest of the country (where China's CNPC has an 85% stake in the main production company) and the Turgay basin in Central Kazakhstan around Kumkol (where, likewise, Chinese companies have a significant share in the main fields). Most of this existing onshore production is mature but new investments, small discoveries and field work-overs are likely to mitigate the rate of decline in output.

Kazakhstan oil export routes

Access to international markets has been a key strategic dilemma for Kazakhstan and its upstream operators and this will remain the case, particularly in the period to 2025, when production is expected to grow rapidly. Existing export capacity is estimated at just over 1.5 mb/d, compared with actual exports of around 1.3 mb/d in 2009. We project that current infrastructure (Figure 17.10) will again be insufficient for projected export levels by 2015 and that by 2025 Kazakhstan will require an additional 2 mb/d of export capacity for the projected increase in oil output to materialise.

Figure 17.10 ● Main oil deposits and export routes in Central Asia

The boundaries and names shown and the designations used on maps included in this publication do not imply official endorsement or acceptance by the IEA.

In this *Outlook*, we assume expansion along all the main export routes (Table 17.7), with the largest increases in capacity during the period to 2020 coming from the Caspian Pipeline Consortium (CPC) pipeline and then, in the period to 2025, from the KCTS for trans-Caspian oil shipments (Figure 17.11). The fact that the CPC pipeline expansion is the first of the envisaged capacity increases will, in the short term, increase Kazakhstan's reliance on export routes through Russia, which already account for more than three-quarters of Kazakhstan oil exports. There are risks associated with such a high level of dependence on a single transit country, although any sustained attempt to use this leverage would, in the longer term, encourage the development of alternative routes, such as KCTS or further expansion of the pipeline to China.

Table 17.7 ● Kazakhstan's oil export routes

	Route	Current capacity (kb/d)	Date commissioned, length	Possible capacity increases
Atyrau-Samara	*Uzen*-Atyrau-Samara (Russia)	300	1970s, 1 500 km	Discussions about expansion to 400 or 500 kb/d; no decision
Caspian Pipeline Consortium (CPC)	Tengiz-Novorossiysk (Russia)	650	2001, 1 510 km	Decision likely 2010, expansion to 1.34 mb/d by 2015*
Kazakhstan-China	Atyrau- Alashankou (China)	200	open in stages from 2004, total 2 163 km	Expansion likely to 400 kb/d by mid 2010s; further expansion possible
Trans-Caspian (+ KCTS)	Exports via port of Aktau (Kuryk for KCTS)	300	various segments	Planned KCTS could add up to 1.1 mb/d capacity by 2025
Rail	Various	Up to 250	n/a	Up to 500 kb/d with new rail export option from Kashagan

* CPC expansion is expected to bring total throughput capacity to 67 million tonnes per year (1.34 mb/d using the standard conversion factor, but up to 1.5 mb/d in practice for a light blend of oil).

Figure 17.11 ● Kazakhstan's oil net exports and transit capacity in the New Policies Scenario

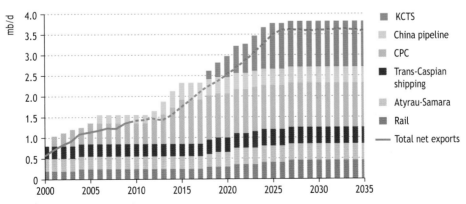

Note: For the purposes of this calculation, it is assumed that 80% of CPC capacity is available for Kazakhstan producers.

We assume that the expansion of trans-Caspian shipments (via the KCTS) takes place only from 2018-2019, to coincide with the assumed timing for Phase II development of the Kashagan field. This means that Phase I Kashagan production (450 kb/d by 2016) has to reach markets through a combination of expanded capacity in the CPC pipeline, the pipeline from Atyrau to Samara, the existing infrastructure for trans-Caspian shipments from Aktau, the Kazakhstan-China pipeline or by rail.

Synchronising upstream developments with decision-making on export capacity among multiple state and commercial partners, all of whom have their own interests and preferences, will be a hugely complex task. For this reason, the picture presented above is subject to a large degree of uncertainty. Aside from political and policy considerations, such as KazMunaiGaz's wish to have the operatorship of future transportation routes, a complicating factor is that the companies involved have different stakes in the various upstream and midstream projects (Table 17.5 above), with each company bearing its own responsibility for getting its share of production to market. For example, partners in the CPC oil pipeline are more likely to have access to capacity in this line, while others have a greater incentive to push for new capacity and alternative routes.

Atyrau-Samara

This connection to the Transneft-operated Russian system, with actual current capacity estimated at 300 kb/d, was the major export route for Kazakhstan oil in the 1990s, prior to the start of operation of the CPC line. A disadvantage of this route (as with the Baku-Novorossiysk line for Azerbaijan) is that Kazakhstan oil is mixed with heavier Russian Urals blend; the spread between the respective prices has narrowed over the past ten years, but Urals blend still trades at a discount of around 1-2% to Tengiz crude. A quality bank within the Transneft system would mitigate this risk, but this has yet to be implemented. KazTransOil and Transneft have discussed possible expansion of this line to 400-500 kb/d, but no decision has yet been taken. This route is mainly used by KazMunaiGaz, MangistauMunaiGaz and other smaller companies. We assume a modest capacity increase to 400 kb/d by 2020.

Caspian Pipeline Consortium

The Caspian Pipeline Consortium oil pipeline links the Tengiz field to a Black Sea export terminal near Novorossiysk. It is the only export pipeline with partial private ownership operating on Russian territory. First oil was transported in 2001. For many years the pipeline has been operating at levels well above its nominal capacity of 565 kb/d, through the use of drag-reducing agents; actual average throughput in 2009 was closer to 700 kb/d. After years of delay, the consortium partners approved in 2008 a plan to increase the current pipeline capacity to 1.34 million b/d at an estimated cost of $4.5 billion. A final investment decision on this expansion is expected by the end of 2010, and we assume that capacity is expanded in stages to over 800 kb/d in 2012, 1.2 mb/d in 2013 and 1.34 mb/d when completed in 2015. The expansion will result in additional volumes of oil being exported through the Black Sea and therefore increases the incentive to develop a new pipeline route to bypass the Turkish Straits (Box 17.2).

Box 17.2 ● **By-passing the Turkish Straits**

The projected rise in crude exports from the Caspian region is heightening concerns about the possibility of a tanker accident in the congested Turkish Straits. Although free passage is guaranteed under the 1936 Montreux Convention, safety and environmental considerations along the narrow waterway can mean significant delays. Around 2 mb/d of oil passed through the Straits in 2009. Volumes have fallen since the mid-2000s, as Russia switched some oil export to ports in the Baltics and, more recently, to the ESPO pipeline that runs eastwards, but the planned expansion of deliveries along the Caspian Pipeline Consortium (CPC) line to Novorossiysk will again increase the amount of crude oil requiring passage from the Black Sea. The Turkish government is considering ways both to encourage the use of oil pipelines to bypass the waterway and to apply severe sanctions in the event of an accident.

There are various pipeline options that could ease the pressure on the Straits, including routes through Ukraine, Romania, Bulgaria and Turkey itself. No final decisions have been taken, but a route across Turkey from Samsun on the Black Sea coast to the existing oil terminal in Ceyhan appears well placed to make progress, as support for the longstanding proposal to link Bourgas in Bulgaria with Alexandroupolis in Greece has waned. The 555 km Samsun-Ceyhan project, also called the trans-Anatolian pipeline, would have an initial capacity of 600 kb/d. Since Turkey already hosts another export pipeline that avoids the Black Sea and Turkish Straits – the Baku-Tiblisi-Ceyhan line – this new link would have the effect of increasing regional dependence on Turkey for oil exports. But the main challenge facing this project, as with all the bypass proposals, is to guarantee throughput and finance construction when shippers want to keep open the – cheaper – option of passage through the Straits, if it is available.

Trans-Caspian shipments and the Kazakhstan Caspian Transportation System (KCTS)

Around 220 kb/d, or 17%, of the oil exported from Kazakhstan in 2009 was transported by barge from the Kazakhstan port of Aktau across the Caspian to Baku in Azerbaijan, Makhachkala in Russia and Neka in Iran. There is potential to modernise and upgrade these existing flows through Aktau. In addition, development of the planned Kazakhstan Caspian Transportation System (KCTS) would result in a major expansion of trans-Caspian trade and a step change in transit flows through Azerbaijan and on to the Black Sea and/or Mediterranean oil terminals. Kazakhstan and Azerbaijan have put in place a basic framework to facilitate the development of KCTS, including a 2006 inter-governmental accord and a series of subsequent understandings between KazMunaiGaz and SOCAR. However, many key issues regarding ownership, financing and operation of the new transportation system remain open.

17

The KCTS project involves a 700-km pipeline from the Kashagan onshore facility at Eskene to a new port facility at Kuryk in Kazakhstan (south of Aktau), probably with a tie-in to the Tengiz export infrastructure en route; a tanker fleet (likely to be of new 60 thousand tonne "Caspian-class" vessels, in place of the presently available small and ageing craft)[4] and unloading facilities in Azerbaijan; and an expansion of transportation capacity through Azerbaijan and the South Caucasus, either to Ceyhan via the Baku-Tbilisi-Ceyhan route or to the Black Sea Coast. The decision to go for tanker transportation across the Caspian, rather than a trans-Caspian oil pipeline, has been shaped in part by political considerations, including opposition from Russia to such a proposal (Box 17.3).

Box 17.3 ● Caspian Sea legal issues

Is the Caspian Sea indeed a "sea"? Or a "lake"? Or something else entirely? A long-running legal debate on this issue might appear arcane, but it has already affected the development of energy investment and trade in the Caspian region and could continue to do so (Janusz, 2005). At the root of the problem is that the Caspian Sea does not fit easily into any of the existing categories offered by international law. It is not easily recognisable as a "sea", subject to the UN Convention on the Law of the Sea (and only Russia of the five littoral states has ratified the Convention); nor can it persuasively be shown that the Caspian Sea should be considered as a "lake" or governed as a condominium, i.e. for common use or equal share among all littoral states.

Existing treaties on the Caspian — the main one being a 1940 Treaty between the Soviet Union and Iran — do not offer much guidance, since they do not clarify rights related to the energy sector, for example, oil and gas exploration, and do not define seabed boundaries. The validity of the treaty was in any event challenged after 1991 by Azerbaijan, Kazakhstan and Turkmenistan. The resulting legal uncertainty has had implications for the development of offshore oil and gas resources. It was not immediately evident after 1991 to what extent and in which areas the littoral states could claim sovereignty over sub-soil resources. This led to disputes over exploration in areas claimed by more than one state. The clearest examples are the mid-Caspian Serdar/Kyapaz field between Azerbaijan and Turkmenistan, and fields in the south Caspian between Azerbaijan and Iran. Turkmenistan has also claimed that parts of the Azeri-Chirag-Guneshli complex lie in its territorial waters.

---

4. For the moment the only shipyards in the Caspian equipped to build 60 000 deadweight tonne vessels are in Astrakhan, Russia and Neka, Iran, although Azerbaijan, among others, is planning to develop such capacity. It remains to be seen whether Kazakhstan will be able to act as an equal partner on the trans-Caspian segment of KCTS alongside the established Azerbaijani shipping interests in building and operating a new Caspian fleet.

Since the Caspian Sea appears to be a specific case, the clearest way to resolve questions about its legal status would have been a comprehensive agreement among the five littoral states. But, even though negotiations began in the early 1990s, such a broad agreement has proved elusive. In the meantime, there has been progress in more limited negotiations. Between 1998 and 2003, Russia, Azerbaijan and Kazakhstan signed bilateral treaties settling delimitation of the seabed and subsoil. But this still leaves key issues unresolved between Turkmenistan and Azerbaijan, and between both of these countries and Iran.

State practice since 1991 has strengthened the right of Caspian states to develop oil and gas resources in their national sectors. The same is not yet true of sub-sea pipelines between national sectors. Although the Caspian Sea contains thousands of kilometres of pipeline, none of these connects different coastal states and all current trans-Caspian energy trade is by tanker. Russia and Iran have in the past taken the view that any international Caspian sub-sea pipelines must have the approval of all littoral states and have also raised concerns about the environmental impact of pipeline developments; the opposing view is that, if two coastal states have agreed to connect their pipeline systems, there should be no obstacle to prevent them doing so. For as long as the Caspian Sea legal framework remains unclear, decisions on the direction and nature of future oil and gas flows across the Caspian, while continuing to be driven by commercial considerations, will be subject to a strong measure of political calculation and, therefore, uncertainty.

The KCTS project has been developed to anticipate the large expected increases in output from Phase II of the Kashagan field and we therefore assume that additional capacity along the KCTS route starts to become available in 2018-19, later than originally scheduled, and reaches 1.1 mb/d in 2025. This increase is based on our projections of Kazakhstan's additional oil transportation capacity needs in the New Policies Scenario. Even with additional time, the risk remains that delayed implementation of this costly and complex multi-stage project will become a constraint on realising Kazakhstan's upstream potential.

Kazakhstan-China

The idea for a Kazakhstan-China pipeline from the oil hub Atyrau on the north Caspian shore to Alashankou in China's north-western Xinjiang region dates back to 1997 and the project has been implemented in stages. China's CNPC has taken responsibility for pipeline construction and financing, and construction of the first (western-most) stage of the project was completed in 2004. Since then, this section has been bringing oil from the Aktobe region (where CNPC has upstream production) west to Atyrau. These flows will need to be reversed to complete an eastward link between Atyrau and the Chinese market.

17

The second stage of construction was the section between Atasu in north-eastern Kazakhstan and Alashankou in China. This was commissioned in 2006, and brings oil to China primarily from the Kumkol fields in which CNPC has an interest, as well as some Russian oil exports from western Siberian fields that enter Kazakhstan via the Omsk-Pavlodar pipeline. The final, central, stage of the project, completed in 2009, connects the first two parts between Kenkiyak and Kumkol. The capacity of the overall pipeline is around 200 kb/d and we assume that this is increased to 400 kb/d by 2015. Even though netbacks from western Kazakhstan to China are currently less attractive than along other routes because of the huge distances involved, further expansion of eastward export capacity after 2015 should not be ruled out — particularly if there are delays in implementation of the KCTS (see the analysis of netbacks below) and if price reform proceeds in China (see Chapter 20).

Rail

Exports by rail to multiple destinations in Europe and China accounted for around 80 kb/d of Kazakhstan export in 2009, with this option also used to deliver oil to the Aktau terminal for trans-Caspian shipment. The rail system offers an important element of flexibility when other, cheaper options are close to capacity. With this in mind, the Kashagan partners have initiated planning for an Eskene West Rail Project, which could add a 300 kb/d rail-export facility at the Kashagan field. A new rail link between Kazakhstan and Iran, via Turkmenistan, is also planned to be complete in 2011.

Other Caspian oil producers

Turkmenistan

Turkmenistan is a small net exporter of oil, with production of around 200 kb/d and domestic consumption of about 100 kb/d in 2009. Turkmenistan's gas output, mainly in the east, is fairly 'dry' and so there are only small volumes of condensate produced; oil output is concentrated in western areas and offshore in the Caspian Sea. The state oil company, Turkmenneft, is the main oil producer, but most of the increase in production in recent years has come from two PSA projects with foreign companies: the offshore Cheleken block being developed by the United Arab Emirates' Dragon Oil and Eni's onshore Burun field. These projects produced around 40 kb/d and 25 kb/d respectively in 2009.

Although the Turkmenistan authorities have a very ambitious national target for oil production of over 2 mb/d by 2030, we project only modest growth in Turkmenistan's oil output in the period to 2035, with output reaching 290 kb/d in 2020 and then falling back to 250 kb/d in 2035. Additional oil is likely to come from the Petronas-operated Block 1 fields, but the main potential increment comes from another offshore field lying in a disputed mid-Caspian area between Turkmenistan and Azerbaijan (called Serdar in Turkmenistan, Kyapaz in Azerbaijan). Turkmenistan signed a PSA for field development with Buried Hill, a Canadian Company, in 2007, but exploration has been held back pending the outcome of discussions with Azerbaijan on the maritime border.

Uzbekistan

Uzbekistan is a mature oil (and gas) region and, given the country's relatively limited resource potential, the country is unlikely to be able to significantly expand its crude oil and nature gas liquids (NGL) production in the longer term. Falling reservoir pressure and rising water at the country's main producing fields brought output in 2009 down to little more than 110 kb/d, well down on the peak of 175 kb/d reached in 1997. As a result, Uzbekistan's oil industry is struggling to meet domestic demand. Today, roughly half of the country's oil output is in the form of condensate produced from gas fields. A priority for the government is to limit imports of oil, in line with the general policy of self-sufficiency, and, if possible, to become a net oil exporter. In the New Policies Scenario, total oil production is projected to continue to stagnate in the medium term, but rebounds at the start of the 2020s with the assumed commissioning of a gas-to-liquids project that is currently under development as a joint project between the state oil and gas company Uzbekneftegaz, Petronas and South Africa's Sasol. Oil imports edge higher in the period to 2020, with the gas-to-liquids project only temporarily reversing the long-term trend towards reliance on imports.

The main production area is the Bukhara-Khiva region in the south-west of the country towards the border with Turkmenistan, which accounts for around 70% of national production and contains the Kokdumalak field, the country's largest. As part of its efforts to stem the decline in production, the government is seeking to encourage upstream investment by foreign companies: for the moment, the main investors are national oil companies from Asia and Russia, with CNPC the largest foreign investor in 2009 and 50% owner of the Mingbulak field, one of the few with significant growth potential. All the crude oil produced in Uzbekistan is processed at three refineries at Ferghana, Altyarik and Bukhara, with total capacity of around 220 kb/d.

Russia

Although exploration began in the Russian sector of the Caspian Sea in 1995, commercial production started only in 2010 from Lukoil's Yuri Korchagin field; this region is set to play a small but increasing role in Russia's overall oil production.[5] Output from the Caspian basin and Caucasus regions of Russia (including onshore) was around 90 kb/d in 2008, but Russia's latest energy strategy forecasts that this could reach 420 to 440 kb/d by 2030 (Government of Russia, 2009).

The main contribution to offshore production in the period to 2020 will come from the Korchagin field and also from the nearby Filanovskoye field discovered in 2005, whose probable and possible reserves of 600 million barrels made it one of the largest oil discoveries of the decade in Russia. However, Lukoil is making the case to the Russian fiscal authorities that tax breaks will be essential to permit viable development of these fields, proposing that they be exempt from oil export duty at least until they reach their planned production capacities of 50 kb/d for Korchagin (in 2011) and 180 kb/d for Filanovskoye (in 2018). With appropriate fiscal incentives, Lukoil estimates that production from the Russian sector of the Caspian could reach 320 kb/d in 2020.

17

---

5. Russia's Caspian production was not modelled separately for this *Outlook*.

Agreement with Kazakhstan on their Caspian maritime border in 2002 opened up some prospective areas for exploration, with the parties agreeing on joint development of fields straddling the border (a model used successfully elsewhere that could yet be the key to unlocking resources in other disputed areas of the Caspian). Since these are understood to be gas condensate fields, they are covered in more detail in the Russia gas section below.

Russia's influence over Caspian oil developments extends well beyond its national sector of the Caspian Sea. The infrastructure inherited from the Soviet era bound the Caspian countries closely to their former fellow republic. Despite the commissioning of the Baku-Tbilisi-Ceyhan and Kazakhstan-China pipelines in 2006, over half of the total export flows from Azerbaijan, Kazakhstan and Turkmenistan were still transported along Russian routes in 2009. Russian companies also have large investments in the Caspian region; Lukoil has stakes in the Tengiz and Karachaganak projects and in the CPC pipeline (see Table 17.5 above) as well as significant upstream projects of its own in Kazakhstan and Uzbekistan.

Iran

There is no commercial oil production in the Iranian sector of the Caspian Sea and Iran has only recently moved to start exploration. Activities were held back in the past by an insistence from the Iranian side that a comprehensive agreement on the legal status of the Caspian Sea should precede any decisions on exploration and production. Iran has also given priority to developing more accessible reserves and so has only limited experience with deeper offshore activities. In recent years, access to international offshore technology has also been restricted by sanctions.

However, Iran's position has evolved since the mid-2000s. A crucial step was the construction, in Iran's Neka shipyard, of the country's first semi-submersible rig (the 'Íran Alborz'), which was inaugurated in 2009. The Oil Ministry has identified 46 prospective structures for exploration in Iranian territorial waters[6] and drilling of the first of three exploratory wells started in February 2010. In advance of the results of new exploration, there are no indications as yet of commercial quantities of oil in the Iranian sector. Any commercial discoveries, though, would raise questions about how Iran would manage their development and to what extent any prospective investor would require legal certainty over ownership of the subsea resources in order to proceed with a project; the latter could provide an incentive to resolve the southern Caspian maritime borders. Under any circumstances, given the time necessary to develop new prospects in the Caspian region, commercial production from the Iranian section of the Caspian Sea is unlikely before the mid-2020s.

In the meantime, Iran's direct role in the Caspian oil industry is limited to the oil swaps arrangement that sees volumes of oil delivered to Neka, mainly from Turkmenistan and Kazakhstan, exchanged for equivalent volumes from Iran's oil terminals in the Gulf. An average of 90 kb/d was exported from the Caspian region in this way in 2009 and

---

6. The Iranian definition of its territorial waters overlaps considerably with areas claimed by Azerbaijan and Turkmenistan, respectively; for the moment, though, there are no reports to suggest that Iran is conducting exploration in the disputed areas.

Iranian officials claim that this could easily be expanded to 300 kb/d. But political risk and sanctions continue to limit use of this route, as well as to inhibit progress on the various proposals to build export pipelines from the Caspian region south through Iran. The oil swaps arrangement broke down in mid-2010, with exporters switching to Baku, reportedly because of an increase in the Iranian swaps tariff.

The prospects for Caspian oil export flows

Gaining access to transportation infrastructure to facilitate exports has been a perennial problem for Caspian oil producers; building pipelines and putting in place other means of moving oil to regional and international markets has been made more difficult by a complex web of logistical, regulatory and political constraints. Kazakhstan and Azerbaijan have had some success in developing oil export capacities, notably with the construction of the CPC and BTC pipelines that today provide the principal routes along which the two countries export oil (Figure 17.12). Yet our projections for the New Policies Scenario show a doubling of net oil exports from the Caspian region to around 4.5 mb/d in the period between 2025 and 2035. This will call for a sizeable expansion of export capacity. Kazakhstan, in particular, now requires another big increase in capacity in the period to 2025 if it is to realise its production potential.

Figure 17.12 ● **Caspian oil export flows, 2009** (million tonnes)

The boundaries and names shown and the designations used on maps included in this publication do not imply official endorsement or acceptance by the IEA.

17

How export flows evolve over the coming decade and a half as production increases will depend on near-term investment decisions that must balance a range of commercial and strategic considerations. Our analysis of current netback values suggests that, for Azerbaijan, pipeline routes to the Mediterranean and to the Georgian Black Sea coast are the most competitive solution, even when taking into account the cost, for the latter, of an additional pipeline bypassing the Turkish straits (Figure 17.13). Estimated netbacks for Kazakhstan are lower across the board because of the longer distance to market but, likewise, routes to the Black Sea (via the CPC pipeline and via the South Caucasus) and Mediterranean (via the South Caucasus) are among the most attractive. The worsening of the commercial terms for swap arrangements through Iran in mid-2010, with a reported rise in the swap tariff, is reflected in these calculations. Netbacks for routes through the Transneft system in Russia (but not through the CPC pipeline) are affected by a loss of value as Azerbaijan and Kazakhstan crudes are mixed with Urals blend. Finally, the least attractive export option for the moment is the Kazakhstan-China pipeline, where netbacks suffer because of the long distance to market.

Figure 17.13 ● Estimated Caspian oil export netbacks*

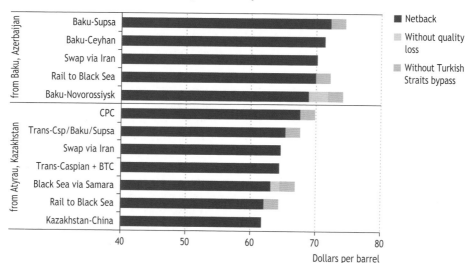

* Estimates based on information available in mid-2010. The Kazakhstan data do not include the effect of the oil export duty, re-introduced in August 2010 at a level of $20 per tonne (equivalent to around $2.7 per barrel).

In developing export strategies and routes, companies will continue to seek the maximum value for their exports allowing for the risks of instability in transit countries and of monopoly control. Against this background, it is clear that the Kazakhstan Caspian Transportation System is a key project for the next phase of Caspian oil development. There are strong commercial and strategic arguments for this project but, even so, many different political and commercial interests will need to be balanced and aligned for this complex route to proceed. If new routes to the west such as the KCTS are not realised in time, there would be an opportunity for China to increase its imports of Caspian oil.

Black swans and wild cards: what could change the pattern of Caspian resource development?

Projections are always subject to a large degree of uncertainty, but a projected trajectory for Caspian resource development is, perhaps, more vulnerable than most to unexpected twists and turns. Turkmenistan's gas production history in the years since 1991 is a case in point: on two occasions, in 1997-1998 and again in 2009, Turkmenistan gas production and export suffered an abrupt and sharp collapse because of disputes over prices and volumes with Russia, which at the time held a near-monopoly position for Turkmenistan export. It could be argued that this sort of episode will become less common as Caspian producers develop more diverse and resilient export systems. Nonetheless, the potential for disputes, accidents, policy reversals, new discoveries and other high-impact events to affect patterns of Caspian resource development remains significant.

A major cause of this uncertainty is the existence of long, overland, multi-country routes to market, which make the reliability of export flows contingent on a long chain of political relationships and circumstances. The Russia-Georgia conflict in August 2008 itself resulted in only a minor interruption to transit flows of oil and gas (IEA, 2008b), but the hostilities — and the unrelated explosion and fire on the Baku-Tbilisi-Ceyhan pipeline that preceded it — were a reminder of the potential vulnerabilities of Caspian export routes. Events in Georgia have not curtailed plans to expand transit flows through the South Caucasus, but perceptions of political risk continue to play a large part in shaping patterns of export. The limited role played by Iran in Caspian hydrocarbon exports provides a case in point. These considerations have also hindered the development of a southern export route for Central Asian gas through Afghanistan to markets in the sub-continent. Current perceptions of risk will undoubtedly shift many times over the projection period, with important consequences for upstream and midstream decision-making.

A second consideration is that the Caspian countries are still relatively new independent states and are, in many cases, still in the process of defining national strategies for development. This creates a relatively large potential for policy shifts on issues such as resource development, upstream access, export routes or domestic energy use. This potential is amplified by the fact that decision-making in many parts of the Caspian, and certainly in all of the four main resource-owning countries, is heavily concentrated at the top of the respective political systems. Changes at the top could presage significant changes to existing policies. Attitudes towards offshore drilling and trans-Caspian hydrocarbon trade could also be transformed by any threat to the fragile enclosed environment of the Caspian Sea, a risk brought home by the Macondo disaster in the Gulf of Mexico (see Chapter 3).

17

Finally, parts of the Caspian region are still relatively unexplored, including some offshore areas — for example, those disputed areas in the southern Caspian where exploration has not yet been possible — and parts of Turkmenistan, and so the potential for major new discoveries remains high. Undiscovered resource potential is not limited to the existing producers: among the "wild cards" are the possibility of new hydrocarbon discoveries in neighbouring countries, notably in Afghanistan and in Tajikistan, where the geology is said to be promising but where exploratory work is at an early stage.

Natural gas

Overview and market context

The Caspian region's proven reserves of conventional natural gas amount to 13 trillion cubic metres (tcm), or 7% of the world total (Table 17.8). Remaining recoverable resources are much larger, at an estimated 26 tcm. The region's largest gas field — the super-giant South Yolotan field — is still being appraised, but is large enough to support production well in excess of 100 bcm/year. However, supply from this field and from other parts of the region will be constrained by above-ground factors: the sheer scale of the investment required (heightened by the fact that much of the region's gas is sour, (*i.e.* it contains a high percentage of hydrogen sulphide) and the need for export infrastructure, often crossing multiple borders, to distant international markets. In total, proven reserves in the Caspian region are sufficient to sustain current rates of production for 70 years, compared to the global average of about 60 years.

Table 17.8 ● Conventional natural gas resources in the Caspian by country, end-2009 (tcm)

	Proven reserves	Ultimately recoverable resources	Cumulative production	Remaining recoverable resources
Azerbaijan	1.4	4.4	0.3	4.1
Kazakhstan	2.0	6.1	0.4	5.8
Turkmenistan	7.9	14.2	2.3	11.9
Uzbekistan	1.7	5.2	1.5	3.7
Other Caspian	0.2	0.3	0.0	0.3
Total	13.2	30.3	4.5	25.8
Share of world	7.2%	6.5%	5.0%	6.9%

Sources: Cedigaz (2009); USGS (2000 and 2008) and information provided to the IEA.

Our analysis of the region's largest gas fields shows production characteristics that are comparable to those of the giant fields included in our worldwide study in *WEO-2009* (Table 17.9). Like the oil fields studied in the previous section, many fields exhibit interruptions or significant changes in production rates, caused by export or investment restrictions. The analysis has been made based on the results of giant fields worldwide, as this is the average size of fields produced historically in the Caspian.

Table 17.9 ● **Plateau production characteristics and production-weighted average annual decline rates for gas fields**

	Time to reach plateau (years)	Plateau length (years)	Plateau production rate (% of reserves per year)	Post-peak decline (%)	Post-plateau decline (%)
Caspian	7.9	8.8	4.4	8.4	6.1
World*	9.7	7.6	4.3	8.2	9.4

* Results for giant gas fields from IEA (2009).

The Dauletabad field in Turkmenistan is the only super-giant field with sufficient historical production to be included in the decline analysis but, as its production profile includes distinct phases of exploitation from high initial production rates through a rapid decline and a later plateau (until the export issues of 2009 again curtailed production), calculated decline rates alter greatly depending on the period of time considered. The increasing significance of export volumes for Azerbaijan and Turkmenistan gas production will continue to affect future field production profiles in the region, in particular that of the super-giant South Yolotan field in Turkmenistan.

Table 17.10 ● **Natural gas production in the Caspian by country in the New Policies Scenario** (bcm)

	1990	2000	2009	2015	2020	2025	2030	2035
Azerbaijan	10	6	17	20	36	43	49	49
Kazakhstan	7	12	36	47	49	55	61	68
Turkmenistan	85	47	41	85	104	110	119	128
Uzbekistan	41	56	66	72	70	70	69	69
Other Caspian	0.3	0.1	0.1	0.1	0.1	0.1	0.1	0.1
Total	143	121	159	224	260	278	298	315
Share of world	*6.9%*	*4.8%*	*5.1%*	*6.3%*	*6.8%*	*6.9%*	*6.9%*	*6.9%*

Turkmenistan and, to a lesser extent, Azerbaijan and Kazakhstan, will drive most of the increase in Caspian gas production over the next quarter of a century. In the New Policies Scenario, total Caspian production doubles, from an estimated 159 bcm in 2009[7] to 315 bcm in 2035 (Table 17.10). Turkmenistan contributes 55% of this output growth but accounts for three-quarters of the growth in exports. Azerbaijan is the other major contributor to export growth (Table 17.11).

17

7. 2009 Caspian production figures are anomalous because of the sharp decline in output from Turkmenistan; for comparison, total Caspian production in 2008 was 188 bcm, of which Turkmenistan contributed 71 bcm.

Table 17.11 ● Natural gas net exports in the Caspian by country in the New Policies Scenario (bcm)

	1990	2000	2009	2015	2020	2025	2030	2035
Azerbaijan	-8	0	8	9	23	29	35	35
Kazakhstan	-6	1	4	7	4	2	4	8
Turkmenistan	67	32	17	56	73	78	85	92
Uzbekistan	1	5	15	8	3	1	-1	-1
Other Caspian	-13	-4	-3	-6	-6	-6	-6	-5
Total	41	34	40	74	97	103	117	129

Note: Negative numbers indicate net imports.

In Kazakhstan, production increases are largely absorbed by growing domestic demand. Uzbekistan, which today accounts for more than one-third of the region's gas production, is expected barely to contribute to the growth in output to 2035. The region's three largest gas fields contribute much of the increase in output, their share of total Caspian production rising from less than 10% in 2009 to almost 40% by 2035 (Figure 17.14).

Figure 17.14 ● Natural gas production in the Caspian by major field* in the New Policies Scenario

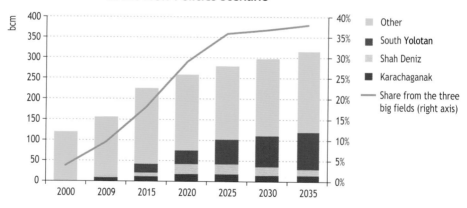

* South Yolotan (Turkmenistan), Karachaganak (Kazakhstan) and Shah Deniz (Azerbaijan).

Azerbaijan

The discovery of natural gas at the offshore Shah Deniz field in 1999 ushered in a period of increasing interest in Azerbaijan's potential as a gas producer and supplier. In 2006 Azerbaijan ceased importing gas from Russia and the following year became a net exporter for the first time. The country's current proven reserves are 1.4 tcm, a major part of which are in the Shah Deniz field. Gas production potential has been held back by uncertainties over the marketing and transit arrangements for gas trade with

Europe, but there are signs in 2010 that these issues are being resolved: this opens up the prospect of a significant expansion in production and export, starting in the second half of the 2010s.

Azerbaijan natural gas production was 16.7 bcm in 2009, with over one-third of the total coming from Shah Deniz. Associated gas from the ACG group of oilfields contributed a further quarter of output, with the rest coming from a number of smaller fields. Most of the associated gas from the ACG complex is re-injected to maintain reservoir pressure and maximise oil production, but SOCAR holds the rights to any excess gas not required for operations or re-injection. The other main gas producing field is the shallow water Guneshli field, operated separately by SOCAR.

In the New Policies Scenario, the country's gas production is projected to increase modestly to 2015 to around 20 bcm, before climbing from 2017 as Phase II of the Shah Deniz field development gets underway. This is projected to bring total production to 36 bcm by 2020, of which about 23 bcm will be available for export (Figure 17.15). Beyond 2020, remaining gas resources are expected to keep output moving steadily upwards, reaching 43 bcm in 2025 and just under 50 bcm in 2035.

Figure 17.15 ● Azerbaijan's natural gas balance in the New Policies Scenario

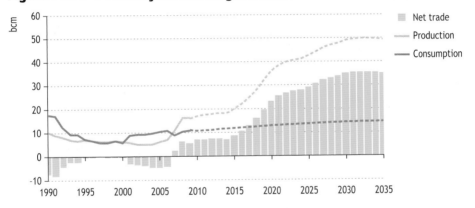

The potential for incremental gas production after 2020 comes again from Shah Deniz, where there are significant volumes that could be developed after Phase II. There are also preliminary discussions regarding the deep gas resources underlying the giant ACG complex, mainly in the Guneshli field at a depth of more than 6 000 metres. However, these resources are not covered explicitly by the 1994 PSA, so a new agreement will be needed if they are to be developed. The improved understanding of the complex geology of the Azerbaijani south Caspian that has been gained in recent years has raised hopes of deeper gas prospects elsewhere. Total is exploring the Absheron field, where a first well is being drilled in 2010, while Germany's RWE has proposed drilling at Nakhichevan. In addition, the gas that is being re-injected at the ACG complex will, at some point, become available for commercial production, the timing depending on decisions about oil recovery. If this gas is indeed produced, it is unlikely to make any significant contribution to Azerbaijan's overall production until towards the very end of the projection period.

17

Shah Deniz

The Shah Deniz gas condensate field has estimated recoverable resources of up to 1.2 tcm and is being developed by a consortium including BP, as the operator, SOCAR, Statoil, Total, Lukoil, OIEC (Iran) and TPAO (Turkey). The field was brought into production in 2006; the four wells currently producing at Shah Deniz are deep, at more than 5 000 metres, and are highly productive, with an average flow rate per well in excess of 4 million cubic metres per day. Gas is delivered to the Sangachal terminal and then into the South Caucasus Pipeline. Production is expected to reach the plateau level for Phase I of 8.6 bcm/year in 2011, which should be maintained until at least 2020. Three-quarters of Phase I output is contracted to the Turkish market.

With Phase I set to produce a total of around 100 to 120 bcm, *i.e.* only around 10% of recoverable resources, the Shah Deniz field has sufficient volumes to support a large increase in production in the medium term. Deep and complex wells present considerable technical challenges and high costs for Phase II development. Nonetheless, the main uncertainty concerns who will buy the gas and how it will be transported to market. Three main European pipeline projects, described in more detail below, have been competing to secure Azerbaijani supplies for markets beyond Turkey in southeast Europe. Turkey itself has been seeking additional Azerbaijani gas supplies to meet fast-growing domestic demand. Russia and Iran have also expressed interest in importing gas from Shah Deniz.[8] Lengthy commercial negotiations on gas sales and transit arrangements held up a decision on Phase II development, but progress on these issues between Azerbaijan and Turkey in 2010 means that we assume first gas will be produced in Phase II towards the end of 2016, with exports to Turkey and beyond from 2017 and plateau production of 16 bcm from around 2019.

Looking further ahead, there are large additional gas resources that could be exploited in future phases: an appraisal well drilled in 2007 to a depth of more than 7 300 metres discovered a new high-pressure reservoir in a deeper structure than those currently being exploited. The current PSA for Shah Deniz development runs to 2030 but the field has the potential to remain the mainstay of Azerbaijani gas production beyond this date.

Azerbaijan natural gas export routes and markets

The delays with Shah Deniz Phase II have highlighted the importance of transparent transportation arrangements for the realisation of Caspian natural gas potential. With such arrangements in place, Azerbaijani natural gas, potentially joined by other sources of gas from the Caspian and Middle East, is set to open up a new southern corridor for gas supply to Europe over the next few years.[9] With the exception of volumes supplied to Russia — assumed to be up to 3 bcm/year — and small amounts supplied across the southern border to Iran, we assume that all gas exported from Azerbaijan will be transported along this corridor, for sale to Georgia, Turkey and to markets in southeast

---

8. Russia concluded a gas sales agreement with SOCAR in 2009, but this gas is supplied by SOCAR from its operated production and it does not involve volumes from the Shah Deniz field.
9. A southern corridor refers here to the possibility of a supply route or routes linking new sources of gas supply in the Caspian and/or the Middle East to European markets.

Europe. An open question, discussed in the Turkmenistan gas section below, is whether any gas from the eastern side of the Caspian Sea will join the volumes exported from Azerbaijan.

Figure 17.16 ● **Natural gas export routes in the South Caucasus**

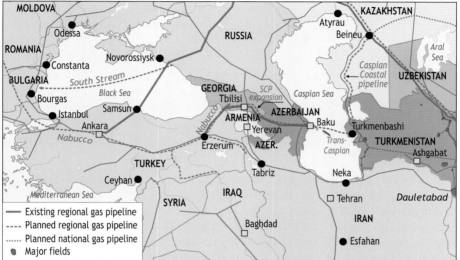

The boundaries and names shown and the designations used on maps included in this publication do not imply official endorsement or acceptance by the IEA.

The June 2010 agreement between Azerbaijan and Turkey allocates some 6 bcm/year from plateau production of Shah Deniz Phase II to the Turkish market and specifies the transit conditions for a further 8.5 bcm/year for onward export to the rest of Europe; around 1.5 bcm will be used within Azerbaijan and Georgia. This agreement has allowed firm negotiations to begin on onward gas deliveries to Greece and elsewhere in southeast Europe between the Shah Deniz consortium and partners in the main pipeline projects that are competing for gas supply: Nabucco, the Trans-Adriatic Pipeline and the Interconnector Greece-Italy (Table 17.12). Alongside these pipeline projects, various ideas remain on the table to bring gas directly across the Black Sea by pipeline or in the form of liquefied or compressed natural gas to Romania or Bulgaria. Although they are more costly, these proposals provide potential back-up in case the pipeline options run into difficulties.

Given uncertainties about the timing and availability of alternative gas supply from elsewhere in the Caspian and Middle East, volumes from Shah Deniz Phase II (at least 8.5 bcm/year or up to 14.5 bcm/year if Turkey chooses to re-export its allocation) are critical to the immediate prospects for each of the three main pipeline projects. If Shah Deniz is perceived to be the only source of gas available from the region to 2020, this will disadvantage the pipeline proposal with the largest capacity — Nabucco — since it would mean that initial Shah Deniz gas would have to shoulder the high up-front costs

17

Table 17.12 ● Azerbaijan's main westward gas-export pipeline projects

	Route	Length (km)	Start date	Annual capacity	Estimated cost ($ billion)
South Caucasus (existing)	Azerbaijan-Georgia-Turkey	692	2006	Current 8 bcm, expansion up to 20 bcm	Initial 1.7
Nabucco	Turkey-Bulgaria-Romania-Hungary-Austria	4 030 incl. feeder lines	From 2016	Initial 8 bcm, expansion up to 25-31 bcm	10.1
Trans-Adriatic	Greece-Albania-Italy	405 + 115 offshore	From 2016	Initial 10 bcm, expansion up to 20 bcm	1.9
Interconnector Greece-Italy (IGI)	Greece-Italy (+ connection Greece-Bulgaria)	600 + 207 offshore	From 2016	8 bcm (to Italy)	1.4

Note: Cost estimates are provided by the project consortia (converted from euros); other industry and analyst estimates are considerably higher.

of a large-capacity pipeline on its own until additional supplies became available.[10] On the other hand, although the IGI (via Greece to Italy) and Trans-Adriatic pipeline proposals might appear better attuned to short-term market conditions, neither of these proposals provides, as Nabucco does, for the construction of new transportation capacity across Turkey. Given rising domestic demand, the Turkish network will require substantial investment if capacity is not to become a constraint on transit flows from the Caspian and Middle East. It may be that a hybrid option, combining aspects of the different proposals, will emerge as a favoured choice but, in this *Outlook*, we take no view as to which project in southeast Europe is most likely to be built. However, we do conclude that, whichever project is built, there is room for only one of these proposals (as currently conceived) to make headway, at least in the period to 2025.

South Caucasus Pipeline

The South Caucasus Pipeline provides the conduit for gas from Baku via Georgia to the Turkish border, where it feeds into the BOTAS-operated pipeline network to the Erzurum gas hub in eastern Turkey. The pipeline was built by a BP-led consortium to accommodate gas export from Phase I of the Shah Deniz development, using the same rights of way as the Baku-Tbilisi-Ceyhan oil pipeline for much of its route. Gas transportation began in December 2006. The current capacity of the pipeline is just under 8 bcm/year. A decision on expansion of the pipeline to around 20 bcm/year will be synchronised with Phase II production from Shah Deniz, which is assumed to start towards the end of 2016. Any additional westward gas export from Azerbaijan beyond Shah Deniz Phase II, or transit of gas from Turkmenistan, would require new capacity beyond the 20 bcm/year — possibly another pipeline running alongside the existing link to Turkey.

---

10. The cost of additional compression to bring the Nabucco pipeline from its initial 8 bcm capacity to the targeted levels of 25 to 31 bcm/year is estimated at only around 15% of the total capital expenditure (officially $10.1 billion).

Kazakhstan

Kazakhstan has 2 tcm of proven gas reserves, much of it associated with oil. According to the Ministry of Oil and Gas, three-quarters of Kazakhstan's gas reserves are in three fields, Karachaganak, Tengiz and Kashagan. There is a strong commercial logic behind the decision of the upstream operators of Tengiz and Kashagan to prioritise production of liquids over gas. Aside from the long distance to international markets, much of Kazakhstan's natural gas is sour and, therefore, requires expensive processing. Even Kazakhstan's major gas-producing field, Karachaganak, is driven by condensate production rather than the natural gas itself, which is sold at a low price across the nearby border to Russia for processing at the Orenburg facility. IEA preliminary data show gas production in 2009 of 36 bcm.[11]

A strategic problem for Kazakhstan is that the gas reserves and main production areas, in the west and northwest of the country, are geographically distant from the major demand centres in the south and there is no north-south supply link between the two. For the moment, Kazakhstan manages this through a swap arrangement with Gazprom, whereby some of the volumes exported from Karachaganak to Russia in the northwest (around 4.6 bcm in 2009) are exchanged for equivalent amounts of gas at the southern border with Uzbekistan. Kazakhstan also purchases small volumes of gas directly from Uzbekistan.

The Kazakhstan government is keen to reduce its dependence on gas imports. A first strand in this drive for gas self-sufficiency is development of more proximate domestic gas resources, in particular the Amangeldy group of fields in the southern Zhambyl region that is among the few non-associated gas prospects in Kazakhstan, with the potential to produce up to 0.3 bcm/year. Secondly, the government is proposing to build a new pipeline link from Beineu in the northwest to the southern city of Shymkent. This project has enjoyed support from China, since it would also allow for Kazakhstan gas to join the main Turkmenistan-China pipeline running across southern Kazakhstan. A China-Kazakhstan joint venture is set to start project construction, with the link planned to be operational in 2014 and providing up to 10 bcm of annual capacity, which could be expanded up to 15 bcm/year at a later stage.

In the New Policies Scenario, Kazakhstan's production of gas outpaces domestic consumption in the next few years, so that the country remains a net exporter (Figure 17.17). Production is projected to reach 47 bcm in 2015 and 55 bcm in 2025. The country emerges as a major gas producer towards the end of the projection period, with output reaching almost 70 bcm by 2035. After 2030, gas output could be boosted by the production of re-injected gas at some of the major oil fields. The timing of this so-called 'gas blowdown' is not certain and may well be pushed beyond 2035 by

17

---

11. The Ministry of Oil and Gas, KazMunaiGaz and the national statistical agency cite different data for natural gas production and domestic consumption. The variations are assumed to be linked to differences in accounting for the gas that is flared or used for different technical purposes in oil production and also the raw gas sent from Karachaganak to Russia for processing in Orenburg, over half of which is currently *imported* back into Kazakhstan. According to the Ministry of Oil and Gas, total gas production in 2009 was 36 bcm but the volume of marketable gas in 2009 available for distribution was less than half of this figure, at 15.6 bcm, with 8.6 bcm of this used domestically and a net 7 bcm of export.

continued possibilities for enhanced oil recovery. The evolution of the Kazakhstan gas balance will be highly contingent on government policy decisions on the position of gas in the domestic fuel mix: we have assumed that gas partially replaces domestic coal in power generation, as a way to meet Kazakhstan's emissions goals (see Chapter 16).

Figure 17.17 ● *Kazakhstan's natural gas balance in the New Policies Scenario*

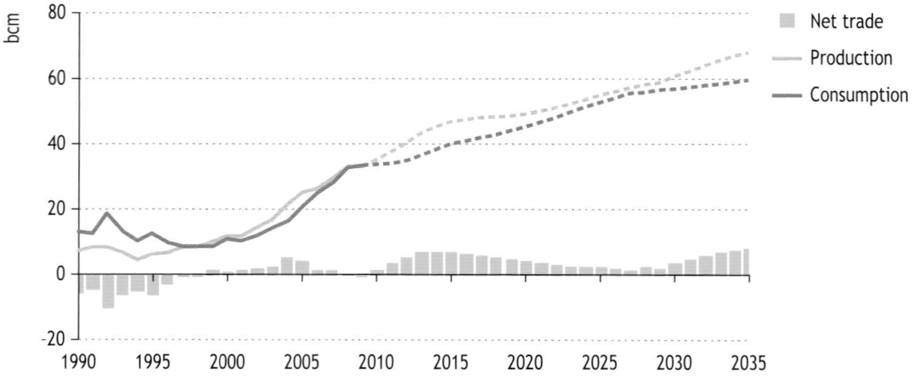

Karachaganak

Karachaganak, near the border with Russia, is Kazakhstan's largest gas producing field. Net of gas-re-injection, around 8 to 9 bcm/year of untreated gas is sold from the field to KazRosGas, which manages the transfer to Orenburg for processing and the marketing of the treated gas. In the first six months of 2010, KazRosGas — a joint venture between KazMunaiGaz and Gazprom — received around 3.6 bcm of processed Karachaganak gas from Orenburg (from 4.25 bcm of raw gas from the field), three-quarters of which was then marketed in Kazakhstan itself.

Phase III development of the field would allow for a doubling of natural gas exports to 16 bcm/year. A decision on this project has been delayed; we now assume that it is implemented in stages from 2014. At various times, Kazakhstan has sought to encourage processing of at least part of Karachaganak output within Kazakhstan itself, so as to capture a greater share of the value of the gas domestically.

Tengiz and Kashagan

The Tengiz oil field is Kazakhstan's second-largest gas-producing field, with a record gas output of 7 bcm in 2009. The large volumes of associated gas in the field have been a challenge for Tengizchevroil (TCO). There are no high-value markets for the gas within easy reach, and the gas requires expensive processing at the field to remove non-hydrocarbons, including a large share of hydrogen sulphide. The latter contributed to a build-up of sulphur in storage at the Tengiz facilities, which in turn caused environmental and regulatory problems. TCO, like other upstream companies in Kazakhstan, has also been under pressure from the government to reduce gas flaring (Box 17.4).

Box 17.4 ● Gas flaring in the Caspian

Natural gas produced in association with oil is flared in all the four main energy-producing Caspian countries. According to satellite data, Kazakhstan flared 5.2 bcm in 2008 (making it the sixth-largest gas-flaring country in the world), Uzbekistan 2.7 bcm, Turkmenistan 1.5 bcm and Azerbaijan 0.4 bcm (NOAA, 2009). The total amount of Caspian gas flared, 9.8 bcm, was about 7% of the global total. Official data is, in some cases, at odds with these satellite readings: the Kazakhstan Ministry of Oil and Gas, for example, estimates that only 1.7 bcm was flared in 2009, down from 3.1 bcm in 2006. Whatever the true state of gas flaring in the region, it is clear that major efforts are being made to reduce this wasteful practice. In Kazakhstan, increasing regulatory pressure culminated in 2010 in new legislation that prohibits routine gas flaring (*i.e.* all but flaring to ensure safe operation of the plant). Gas flared at the Tengiz field, for example, has declined by 95% since 1.8 bcm was flared in 2000. However, the problem persists at many smaller fields in the region, where there is no infrastructure to use or treat the associated gas.

Compared to the situation in the early 2000s, the Tengiz partners are now re-injecting much larger quantities of gas and marketing more of the treated gas (as well as more of the sulphur). Gas production from Tengiz is projected to remain at around current levels, *i.e.* 6 to 8 bcm/year, until the late 2020s. The Kashagan project is facing many of the same technical challenges as Tengiz and the partners are planning to resolve them in similar ways. A first priority is to maximise reinjection of sour gas (up to 15% hydrogen sulphide and 4% carbon dioxide), which will begin from the start of Phase I production. During Phases II and III, when oil production is due to reach 1.5 mb/d, total gas production from the field will rise rapidly to as much as 30 bcm/year, of which up to 25 bcm/year will be re-injected.

The volumes of initial and re-injected gas in Tengiz and Kashagan mean that these have the potential, at some point, to become major gas-producing fields; the combined incremental production potential for Kazakhstan amounts to some 30 to 50 bcm/year. A decision to start producing this re-injected gas is likely only once the bulk of the oil in these fields has been recovered, since once the gas blowdown begins and the field is de-pressurised, the output of liquids will fall sharply. Given the volumes of recoverable oil in Tengiz and Kashagan, large-scale gas production is likely only towards the end of the projection period. However, this coincides with the end of the current PSA arrangements for both fields, so the question of who would pay for the additional gas-processing facilities would need to be resolved.

Kazakhstan natural gas export routes and markets

Kazakhstan's traditional export route has been to Russia, both directly from the Karachaganak field and also via the Central Asia-Centre (CAC) pipeline system that has been the main export artery for Central Asia. The CAC pipelines have the advantage

17

of running close to the main gas-producing regions in western Kazakhstan. There have been various proposals to upgrade and expand this system, including the addition of a new Caspian Coastal pipeline from the Turkmenistan Caspian coast (described in more detail in the Turkmenistan gas section below). However, the commissioning of the Turkmenistan-China gas pipeline in 2009 has opened up a second export option for Kazakhstan, which could be used from 2012 once the Beineu-Shymkent domestic pipeline link starts operation. We assume that, from the mid 2010s, Kazakhstan will export gas to both Russia and China and that this will allow upstream operators to get more value from their gas sales. Gas is currently sold on the domestic market and to KazRosGas at prices that are well below the international netback level.

Kazakhstan is a major transit country for exports from Turkmenistan and Uzbekistan (via the CAC, Bukhara-Urals and Turkmenistan-China pipelines) and also for Russia (some Russian pipelines, for example from Orenburg, cross northern Kazakhstan). Transit of gas from Turkmenistan to Russia collapsed in 2009 to below 12 bcm, from previous levels above 40 bcm/year, and transit of Uzbekistan exports was 13 bcm. Due to weaker demand from Russia for Central Asian gas, reduced transit volumes are expected to persist at least until 2013-2015, meaning that there will be ample spare capacity in the CAC system and little incentive in the short term for Kazakhstan to invest in upgraded or new capacity on export pipelines to Russia.

Turkmenistan

With 7.9 tcm of proven reserves, Turkmenistan is the largest gas resource holder in the Caspian region. The reserves figures quoted by Turkmenistan officials are higher still, with estimates in excess of 20 tcm, putting Turkmenistan closer to the range of proven reserves in Iran or Qatar (Government of Turkmenistan, 2006). Different methodologies and classification systems account for some of the difference between international and Turkmenistan estimates, but the main reason is a lack of exploration and verifiable appraisal, particularly of fields that have been discovered but not developed. This issue is now being addressed, particularly in relation to Turkmenistan's major discovery at South Yolotan.

Despite high per-capita gas consumption, Turkmenistan's relatively small population has in most years made Turkmenistan the leading gas-exporting country in the Caspian region. However, exports declined dramatically in 2009 as a result of an explosion in April on the main CAC export pipeline and Gazprom's reluctance to take Turkmenistan volumes at a time of sharply reduced demand for its own gas. For Ashgabat, the only alternative route to market at the time was a relatively small-capacity connection southwards along the Caspian Sea to Iran. The breakdown of Turkmenistan's main export route had a major impact on production, which we estimate to have declined to around 40 bcm in 2009 from 71 bcm in 2008. Even though it is distant from the main demand centres, Turkmenistan was arguably the country most affected by the global gas glut.

The outlook for Turkmenistan gas has improved since mid-2009, although short-term export opportunities are still well below the country's production potential. The major development in late 2009 was the commissioning of the 7 000 km Turkmenistan-China pipeline, which puts Turkmenistan (and other Central Asian producers) in the unique position of having overland connections to both the western and eastern Eurasian markets. In addition, the Turkmenistan authorities concluded contracts for the development of the super-giant South Yolotan field in the southeast of the country (see below), which is the cornerstone of Turkmenistan's plans to increase output over the coming years. The official production target is 250 bcm per year by 2030. While the resource base could conceivably support an expansion of this magnitude, there is considerable uncertainty about whether sufficient investment will be forthcoming and whether markets for such large incremental gas volumes will be available. In the New Policies Scenario, we project output rising to 85 bcm by 2015, to over 110 bcm by 2025 and to almost 130 bcm by 2035. As a result, exports rebound from a low of less than 20 bcm in 2009 to over 55 bcm by 2015, 80 bcm by 2025 and 90 bcm by 2035 (Figure 17.18).

Figure 17.18 ● **Turkmenistan's gas balance in the New Policies Scenario**

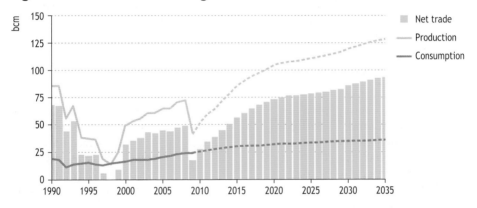

Turkmenistan's policy on upstream developments is that international companies are welcome to invest in offshore developments, but that onshore, where the bulk of gas reserves are located, their role is limited to providing assistance on a contractual basis to the state-owned company, Turkmengaz. Offshore gas output in the Turkmenistan section of the Caspian Sea, primarily associated gas produced by Malaysia's Petronas, is projected to reach 10 bcm per year by 2015, although it is not clear where and how this gas will be marketed. Russia's Itera and RWE both took the first steps towards joining the offshore producers in 2009, signing PSAs on different offshore blocks. A less promising sign was the decision by Wintershall/Maersk/OMEL to surrender rights to other blocks, after concluding that the results of initial drilling were not encouraging enough to proceed with further exploration.

A number of companies are looking at the larger resources available onshore, but the only international company that has been granted direct access to the major inland resources is China's CNPC, which has a PSA for the Bagtyyarlyk contract area on the right (east) bank of the Amu Darya river, near the border with Uzbekistan. Given the

17

reluctance of the Turkmenistan authorities to countenance other onshore PSAs, some international oil companies are looking at how onshore service contracts might best be structured to include an element of longer-term risk and reward. It is not yet clear whether this approach will find favour with the Turkmenistan government.

One way or another, though, the Turkmenistan authorities may need to harness international expertise and technology to support new gas developments. The next generation of Turkmenistan onshore gas fields will be more expensive and complex to develop than those brought into production up to now, because they are deeper, have higher pressure and temperature, and have higher concentrations of hydrogen sulphide and carbon dioxide. For the moment, as the example of the South Yolotan field suggests, the Turkmenistan authorities believe that their policy of relying on classical service contracts for onshore gas developments will deliver the necessary results. We assume that a continuation of the current policy of restricting onshore access, alongside the difficulties of gaining access to major export markets, constrains the pace of future gas developments.

South Yolotan

The development of the South Yolotan field in southeast Turkmenistan, discovered in 2006, will be the main driver of export growth over the *Outlook* period, both for Turkmenistan and for the region as a whole (Figure 17.19). The Turkmenistan government commissioned an international audit of the field's resources to lend credibility to claims about its size. The mean estimate of gas-in-place was 6 tcm, within a range of 4 to 14 tcm. Further appraisal wells drilled since the audit have apparently indicated that the deposit is even more extensive than originally thought, indicating that South Yolotan is likely to be the mainstay of Turkmenistan production for many years to come, as production from the main existing fields, Shatlyk and Dauletabad, declines. The current mean resource estimates would allow for, at minimum, five additional phases of development of comparable size to the 30 bcm/year output planned for the initial phase.

With the assistance of at least $3 billion in loans from China, the Turkmenistan authorities awarded a series of contracts in 2009 for the first phase of field development at South Yolotan. Official expectations are that first gas will be produced as soon as 2012. The contracts, totalling $9.7 billion, were awarded to the United Arab Emirates-based companies, Petrofac and Gulf Oil & Gas, South Korea's LG and Hyundai, and CNPC. These contracts cover drilling, sub-surface development and surface handling and construction, with Turkmengaz retaining overall project management. Developing such a large and complex field, with an estimated non-hydrocarbon gas content of around 8%, will involve huge technical challenges. Turkmengaz does not yet have a track record of bringing large new fields into production and, given the pattern of delays experienced by other large investment projects in the Caspian region, the announced schedule for field development could well slip. We assume first gas from South Yolotan in 2013, rising to 30 bcm/year by 2020, with most of this gas being exported to China via the recently completed Turkmenistan-China pipeline.

Once the initial technical hurdles of the first phase have been overcome, increased production from the field will depend on new investments in both upstream developments and the availability of export infrastructure and associated transit and sales agreements. None of these can be taken for granted. We assume a further expansion of South Yolotan production in the early 2020s, bringing total output from the field to 60 bcm/year in 2025, with demand and infrastructure being sufficient to justify a third development phase pushing production levels up to 90 bcm/year by the end of the projection period.

Figure 17.19 ● Main natural gas deposits and pipeline routes in Central Asia

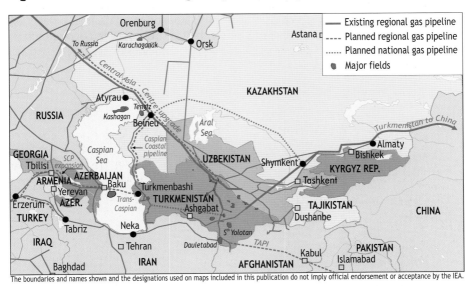

The boundaries and names shown and the designations used on maps included in this publication do not imply official endorsement or acceptance by the IEA.

Turkmenistan natural gas export routes and markets

Turkmenistan's gas-export strategy aims to maximise the value of the country's gas reserves by increasing the number of export options and also the domestic flexibility to bring large gas volumes to any export point on Turkmenistan's border. The benefits of diversification became engrained in Turkmenistan's strategic thinking during the 1990s and early 2000s, when the country's reliance on a single export route to Russia meant having to accept disadvantageous terms for gas trade: two gas disputes with Russia in 1997-1998 and 2009 led to a shut-off of exports and of a large part of the country's production. However, Turkmenistan's ability to implement its strategy has been limited by its reluctance to get involved in any midstream projects beyond its borders. The government's insistence upon sales at the Turkmenistan border means reliance upon Turkmenistan's customers to bring the necessary export infrastructure to the country's frontier. Leaving aside pipelines to Iran, Turkmenistan's southern neighbour, only China has proved capable of constructing a new long-distance

17

pipeline to Turkmenistan. It is not yet clear that other potential customers for Turkmenistan gas in south Asia and in Europe will be able or willing to emulate the Chinese example.

The proposal for an East-West domestic gas pipeline is designed to give Turkmenistan the ability to send gas to multiple export points on its external borders. A tender for this pipeline was announced in 2009, but in 2010 Turkmengaz announced that it was proceeding with the $2 billion, 900 km pipeline on its own. It is not clear if and when this link will be completed, in view of the cost and the original intention to synchronise it with the stalled Caspian Coastal Pipeline project. Nonetheless, the project has some potential strategic value to Turkmenistan, since the country's ability to transport up to 30 bcm/year in either direction − from the main gas producing regions in the east to the Caspian coastal area, or from offshore Caspian production towards the east − could provide a crucial element of flexibility in negotiations on gas sales and pricing.

Box 17.5 ● **Putting a price on Caspian natural gas exports**

Securing an export price related to the international market value of natural gas has been a long-standing goal for Caspian exporters. These efforts have started to bear fruit since the mid-2000s, as competition for gas exports from the region has intensified. Agreements reached between Azerbaijan and Turkey in 2010, which pave the way for further development of a southern gas corridor, should enable Azerbaijan to link its export prices to those prevailing in Europe. In theory, similar access is also available to all Central Asian gas producers after Gazprom agreed that, from 2009, export prices would be linked to the price in Gazprom's main continental European markets, minus transportation and other costs. But, since 2009, the reduced need for Russia to call on Central Asian gas supplies has reduced the incentive for Gazprom to buy this gas at such prices. With the start of gas export to China from Turkmenistan in 2010, the outlook for gas-export pricing in Central Asia now depends on a complex interaction between energy-sector developments in Russia, China and, to a lesser extent, in Iran. The price that any of these countries will be willing to pay for Caspian gas will depend on the cost and availability of alternative domestic supplies (including competing fuels) and, in the case of China, the price of imported LNG and coal, and of potential gas imports from Russia.

Russia: Central-Asia Centre system and Caspian Coastal Pipeline

Russia signed a long-term framework agreement with Turkmenistan in 2003 that sets out annual target volumes for gas purchases for the period to 2028, but actual sales have been well below these indicative levels. According to the original agreement, annual gas exports from Turkmenistan to Russia should have been of the order of 70-80 bcm/year from 2009 onwards, whereas in practice they did not move beyond a 40-50 bcm/year range, before falling to under 12 bcm in 2009. Russia and Turkmenistan revised their target export schedule down sharply at the end of 2009, specifying that Russia would buy up to 30 bcm/year for 2010 and subsequent years; actual Turkmenistan exports to Russia in 2010 are expected to remain at around 10 to 12 bcm.

The medium-term Russian call on Central Asian gas, and Turkmenistan gas in particular, is uncertain. The upper limit is the figure for overall gas imports included in the *Energy Strategy of Russia for the Period to 2030* (Government of Russia, 2009), which envisages that imported gas will contribute 88-94 bcm/year to the Russian gas balance.[12] Our projections in the New Policies Scenario suggest that the call on Central Asian gas may be considerably lower than this. As energy-efficiency improvements in Russia drive down Russian gas demand growth, the surplus of Russian production over domestic consumption rises from 209 bcm in 2008 to 311 bcm in 2035; if Russia were to keep its share of the EU gas market steady at around one-quarter of total EU-27 gas consumption (the level in 2008), then the Russian capacity to export gas to non-EU markets would increase from 75 bcm in 2008 to almost 150 bcm in 2035 in the New Policies Scenario, even without any gas imports from Central Asia. Ultimately, decisions on gas imports from the Caspian will result from a mixture of cost and strategic considerations: imports may increase if Central Asian gas becomes available at lower prices (Box 17.5); the import requirement could also rise if there are delays to Gazprom's upstream developments in Russia, such as the gas projects on the Yamal peninsula (although non-Gazprom production within Russia would in all probability be a cheaper way to fill any gap). Russian imports could also be driven up by a strategic decision to forestall the development of alternative export routes from Central Asia to Europe. However, our projections do not suggest a systemic need for imports from Central Asia for Russia to meet its export obligations or ambitions.

If indeed Russia's call on Turkmenistan gas remains limited in the medium term to less than 30 bcm,[13] as we assume, this will continue to undermine the long-standing plans to upgrade and extend the export infrastructure between Central Asia and Russia. These include the plans to rehabilitate, and raise capacity on, the main Soviet-era CAC pipeline route connecting the gas-producing areas in Turkmenistan and Uzbekistan to the Russian gas network via Kazakhstan, as well as the 2007 proposal for a new Caspian Coastal (or Prikaspiskiy) Pipeline running up the Caspian coast from the Turkmenistan coast via Kazakhstan to Russia[14]. The CAC system has suffered from a lack of investment in maintenance and capacity has declined considerably as a result, from nominal capacity of close to 100 bcm/year (in total) to current estimates of 45-55 bcm/year (Fredholm, 2008). Gazprom's stated intention has been to restore it to a level of about 80 bcm/year. However, future Turkmenistan exports, plus up to 15 bcm/year from

17

---

12. The only current source of gas import to Russia is the Caspian region. The Russian energy strategy to 2030 states that gas import levels will be based on economic conditions on international gas markets and the state of Russia's fuel and energy balance, but does not refer to contractual obligations to purchase defined amounts.

13. Published Gazprom projections do not foresee more than 38 bcm of imports from the whole of Central Asia (*i.e.* similar to 2010 levels) until at least 2013.

14. There is an existing line, the third string of the Central-Asia Centre system, that was commissioned in 1972 and runs along this coastal route, but the available capacity is assumed to be very low.

Uzbekistan, are not expected to exceed current capacity. Under these circumstances, it will be very difficult for Gazprom to make the case, both internally and to Kazakhstan and Turkmenistan, for investment in new transportation capacity.

China: Turkmenistan-China pipeline

The commissioning of the Turkmenistan-China pipeline in December 2009 marked a major shift in the politics and economics of east Caspian gas, signalling the end of Russia's hold on large-volume gas purchases not only from Turkmenistan, but also from the transit countries, Uzbekistan and Kazakhstan. The backbone of this project is a 30-year gas sales agreement for 30 bcm/year of gas, concluded in July 2007, which was later extended to 40 bcm/year. The capacity of the Turkmenistan-China link is scheduled to rise from the current 10 bcm to around 40 bcm/year by 2012, with the completion of a second string and additional compression; in mid-2010, there were reports that China was interested in raising the ultimate capacity of this eastward corridor to 60 bcm/year.

The Turkmenistan-China pipeline project was implemented with impressive speed. The initial framework agreement on gas co-operation between Turkmenistan and China was concluded only in 2006 and construction began in the latter part of 2007. The entire pipeline route — the longest in the world — stretches for close to 7 000 km, consisting of under 200 km within Turkmenistan itself, around 500 km through Uzbekistan, 1 300 km through Kazakhstan, and then the remainder within China (including a second West-East pipeline that is scheduled for completion in 2011) to bring gas to the main gas consumption areas. The line was financed by Chinese government loans and much of it built by the China National Petroleum Corporation (CNPC), in joint ventures with local companies in Uzbekistan and Kazakhstan.

Initial gas supplies for the pipeline are coming from Turkmengaz fields in southeast Turkmenistan and from CNPC fields in the Bagtyyarlyk contract area near the Uzbekistan border, where production began in 2010. CNPC expects its Bagtyyarlyk operations to produce up to 13 bcm/year from 2012, leaving Turkmengaz with the obligation to find the remaining export volumes from existing fields and from South Yolotan. Gas exports will be drawn east by the imperative for Turkmenistan to repay at least $3 billion in loans taken from China in 2008-2009.

Iran: Turkmenistan-Iran pipelines

Turkmenistan is connected to the Iranian domestic gas network by two pipelines: the link between Korpedzhe and Kurt-Kui in the west, which runs along the Caspian coastline and which opened in 1997, and a short cross-border pipeline, which was commissioned in late 2009 between the major south-eastern field of Dauletabad and the Iranian grid at Khangiran. Exports to Iran were in the range of 5 to 7 bcm/year for much of the 2000s, but these are likely to increase in 2010 as new capacity is available and while exports via other routes are limited. Total export capacity is now 14 bcm/year and if, as anticipated, exports in 2010 are close to these levels then Iran would be Turkmenistan's main export partner for this year. There are plans to raise total export potential to 20 bcm/year by doubling the capacity of the Dauletabad-Khangiran line.

Since Iran's domestic production and consumption are almost in balance, imports from Turkmenistan have the effect of freeing up a roughly equivalent amount of gas for export from Iran to Turkey. The link between Turkmenistan imports and Iran's ability to export was highlighted in December 2007, when an interruption to Turkmenistan supply — caused by a dispute over pricing and exceptionally cold weather — meant a suspension of Iran pipeline exports to Turkey. The expansion of Turkmenistan exports to Iran in 2010 may allow for greater volumes to be supplied to Turkey. They may also facilitate a new Iranian export pipeline project, which we assume to go ahead, from southern Iran to Pakistan. Given the difficulties facing a direct infrastructure link from Turkmenistan to Pakistan via Afghanistan, an indirect link via Iran may be the most likely way in the medium term for Turkmenistan to have an impact on the gas balance in south Asia.

South Asia: Turkmenistan-Afghanistan-Pakistan-India pipeline

The idea for a pipeline linking Turkmenistan to the fast-growing gas markets of south Asia initially gained momentum when Turkmenistan gas was assumed to be available relatively cheaply. However, higher price expectations on the Turkmenistan side since the mid-2000s have complicated the economics of this project. Despite renewed political interest in this project in 2010, continued concerns over the security situation along the pipeline route in Afghanistan have made it difficult to attract commercial interest. Gas demand growth in south Asia is projected to be strong, with India's import requirement rising from 10 bcm in 2008 to 75 bcm in 2035 in the New Policies Scenario. However, given the large uncertainties, we assume that this project is not implemented during the projection period.

Europe via a southern gas corridor: trans-Caspian gas transportation

The possibility that supplies from Turkmenistan might pass through a southern corridor to Europe is widely discussed, but there are a number of important difficulties that still need to be resolved. The most likely source of supply in the near term is the offshore associated gas being developed in the Turkmenistan section of the Caspian Sea. We assume that there will be around 10 bcm of associated gas available from offshore Turkmenistan fields by 2015, and this gas does not, for the moment, have an obvious route to market. Delays in constructing the proposed Caspian Coastal Pipeline to Russia are likely to be prolonged (as discussed above) and this creates an opportunity for European companies; RWE and Austria's OMV are particularly interested in this offshore gas as a means of securing additional supplies of gas for the Nabucco pipeline project.[15]

A main obstacle is the lack of export infrastructure to bring this gas westward across or around the Caspian Sea. Among the different technical solutions, a sub-sea pipeline could be built from shore-to-shore or, alternatively, there could be a mid-Caspian interconnector tying offshore Turkmenistan production platforms to the existing Azerbaijani offshore infrastructure; such a link would involve only around 60 km of subsea pipeline. There are different views as to the political feasibility of these pipeline ideas, because of expected opposition from Iran and from Russia; what is certain is

17

---

15. RWE was awarded an offshore exploration licence for Turkmenistan offshore Block 23 in the Caspian Sea in 2009.

that any such pipeline link will, at the very least, require the committed support of both Turkmenistan and Azerbaijan, who, thus far, have not managed to agree on their maritime border. A second option is the technically unproven possibility of exporting compressed natural gas by tanker across the Caspian (Box 17.6). Finally, there is the possibility of expanding exports to Iran, thereby allowing Iran to increase its gas exports to Turkey, or, alternatively, of constructing a new pipeline link around the southern Caspian through Iran; the route through Iran, while potentially competitive on cost, is ruled out for the moment by political considerations.

Box 17.6 ● LNG and CNG as options for Caspian gas exports

Exports by pipeline may appear the obvious choice for landlocked Caspian gas exporters, but when pipeline proposals run into difficulties, whether for commercial or political reasons, the search for alternatives ways to market intensifies. This has led a variety of interested Caspian parties to consider the viability of transporting gas either as liquefied natural gas (LNG) or compressed natural gas (CNG) across both the Caspian and the Black Seas. The technology for LNG trade is well established and the potential costs for any Caspian project are relatively easy to estimate; unfortunately, for the short distances that would be involved, this option would be prohibitively expensive. Assessing the cost of CNG trade, however, is more difficult because there are, for the moment, very few comparable examples of sea-borne CNG transportation. Cost estimates are therefore subject to large uncertainties; but a CNG system, whether of multiple small barges or of larger tankers, would be likely to cost less than LNG over a relatively short distance, though still considerably more than a subsea pipeline.

Our baseline assumption for capital expenditure on a generic 500 km subsea pipeline with capacity of 12 bcm/year is around $2 billion. This implies, with operating expenditure, a gas transportation cost of $0.70 to $0.80/MBtu. By contrast, the construction of a typical LNG liquefaction, shipping and regasification chain with capacity of 5 million tonnes per year might involve capital expenditure of around $5 billion, leading to a gas transportation cost of over $3/MBtu – more than four times the estimated cost of a sub-sea pipeline. There are different technical possibilities for CNG transportation, but while the loading and unloading facilities would be considerably cheaper than for LNG liquefaction and regasification, many more vessels would be required for transportation. We estimate that gas transportation costs for a 5 bcm/year, 500 km CNG system would be in the range of $1.40 to $2.00/MBtu.

Despite these difficulties, we assume that, since Turkmenistan has few other export options, up to 10 bcm of offshore Turkmenistan gas will, one way or another, be made available to supplement volumes available for export to Turkey and/or other European markets. We assume that these flows start in the latter part of the 2010s, but are not joined by gas from onshore Turkmenistan fields. The validity of these assumptions will depend on many factors, including not only market developments and costs but also regional political relationships and European policy on energy supply diversity. Alongside the different transportation options considered by individual companies, the European Commission has proposed a Caspian Development Corporation, designed to facilitate trans-Caspian gas trade (Box 17.7).

Box 17.7 ● The Caspian Development Corporation

Diversification of Europe's sources of gas supply has become a strategic priority for the European Union, particularly after the disputes between Russia and Ukraine, in 2006 and again in 2009, which interrupted European gas supply. Caspian gas resources could help to provide such diversity but, looking beyond Azerbaijan, it has proved difficult for individual European companies to secure gas supplies from the east of the Caspian — not least because of the challenge of putting in place the necessary infrastructure for trade across or under the Caspian Sea. With this in mind, the European Commission launched, in 2009, the idea of a Caspian Development Corporation (CDC) to co-ordinate gas purchasing and infrastructure development, with the aim of kick-starting Turkmenistan's gas trade with Europe (EC, 2009).

Different commercial models for CDC are being discussed and no decision has yet been taken to implement the concept. The basic idea is to aggregate demand from different potential purchasers of Turkmenistan gas, so as to present Turkmenistan with the offer of a single, large long-term gas contract. Such a contract could help to underpin the financing of new transportation infrastructure to the east of Baku, trigger the development of otherwise untapped gas reserves in Turkmenistan and bolster gas availability along the southern corridor.

There are a number of challenges facing CDC, including uncertainties over future European import needs, the compatibility of the concept with European competition legislation and, crucially, the political willingness of Turkmenistan and Azerbaijan to pursue a trans-Caspian gas link. There are also questions about upstream aspects. The volumes being discussed for CDC of up to 30 bcm/year are larger than those likely to be available from the Turkmenistan offshore areas and would need to be met in large part from the onshore resources. Yet development of these onshore fields, for the moment at least, is off-limits to international companies (with the exception of the PSA awarded to China's CNPC). Under these circumstances, it is not clear how Turkmenistan would be able to guarantee the availability of sufficient gas output for CDC and whether these guarantees would be enough to justify taking the risks involved in financing a new gas pipeline across the Caspian.

Uzbekistan

Uzbekistan is well endowed with gas resources, though their size and quality is expected to limit the extent to which production can be further increased in the years to come. Proven reserves are estimated at 1.7 tcm and Uzbekistan is currently the world's 14th largest gas producer. Marketed production amounted to an estimated 66 bcm in 2009, only slightly below the all-time peak of 67 bcm in 2008 and about two-thirds higher than in 1990. In the New Policies Scenario, production is projected to continue to climb only marginally, reaching a plateau just above 70 bcm in 2015 (Figure 17.20) before tailing off slightly to 69 bcm in 2035. This reflects our judgement that new investments in gas production will not be enough to raise production levels significantly. With consumption projected to rise steadily from 54 bcm in 2008 to 67 bcm in 2020 and 70 bcm in 2035, the margin for gas export progressively falls such that, by 2030, Uzbekistan's supply and demand for gas are in balance and the country even becomes a small net importer.

Natural gas production is currently concentrated in the southwest of the country, near the border with Turkmenistan, and this area includes the two largest gas-producing fields, Shurtan and Kokdumalak. Output at many of the existing fields operated by Uzbekneftegaz, the state oil and gas company, is at plateau or in decline. The authorities hope that foreign investment in new gas developments can bolster production levels in the future and new fields are now being developed under PSAs by companies from Russia, China and other Asian countries, often in partnership with Uzbekneftegaz.

A key project for Uzbekistan's future gas output involves the Kandym group of fields in the southwest. This is being developed by Lukoil and is estimated to hold 250 bcm of gas. Production began in 2007 and could increase up to 8 bcm/year, but Lukoil has postponed plans to build a gas processing plant and we assume that the field will now reach plateau production only towards the mid-2010s. Gas production from a second Lukoil PSA further south (Southwest Gissar) is currently small, but could rise to 5 bcm/year at a later stage.

Figure 17.20 ● **Uzbekistan's gas balance in the New Policies Scenario**

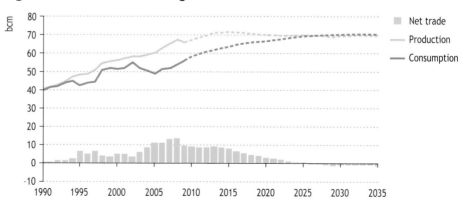

Aside from Lukoil, the main companies with an interest in Uzbekistan gas production are Gazprom, Petronas, CNPC, the Korean National Oil Company (KNOC) and Kogas. Of these, Gazprom is the principal company with productive assets, notably its stake in the Gissarneftegaz joint venture that is the country's second-largest gas producer (after Uzbekneftegaz) from fields in the southwest. Petronas and CNPC also have exploration projects in this southwestern region, the traditional oil and gas province of Uzbekistan where most of the remaining gas reserves are located.

In recent years, the attention of foreign investors has also shifted to the North Ustyurt plateau and areas in the dried-up bed of the Uzbekistan sector of the Aral Sea. Commercial production from these regions is limited for the moment and the exploration record has been mixed: Gazprom, for example, gave up licenses for three of its blocks there in 2009 to Uzbekneftegaz. However, among the more positive prospects are the Surgil gas field, which is being developed by a group of Korean companies, and the Aral Sea exploration block, where two test wells drilled by a large multi-national consortium (including Lukoil, CNPC, Petronas and KNOC) showed encouraging results in 2010.

These upstream projects operated by foreign companies are expected to contribute an ever larger proportion of Uzbekistan's gas production over the projection period, but it is not clear that they will be sufficient to offset declines at existing fields. While the country's gas resources are considerable, there are doubts about how quickly they can be developed in view of technical factors, including the small average size of deposits and high content of hydrogen sulphide. There are also administrative barriers to investment, such as a new excise tax, introduced in 2009, which the government would like to apply to the existing PSAs.

Uzbekistan natural gas export routes and markets

Russia is currently Uzbekistan's main gas export market, with small volumes also being supplied to Kazakhstan, the Kyrgyz Republic and Tajikistan. Uzbekistan is also an important transit country for exports from Turkmenistan, both to Russia and also, since the end of 2009, to China. Total Uzbekistan gas exports amounted to 14 bcm in 2008 and an estimated 15 bcm in 2009 (according to preliminary data). There are plans to export gas to China. In June 2010, CNPC signed a framework agreement with Uzbekneftegaz to purchase 10 bcm/year of gas from Uzbekistan, though details of timing and pricing were not provided. It is not yet clear which fields have been identified as the source of export flows to China, although the logical choice would be the main southern fields that are close to the Turkmenistan border and the pipeline route.

The projections in the New Policies Scenario suggest that gas availability for export will hold steady at around 10 bcm/year to 2015, but then diminish gradually to 2035 as domestic demand outpaces production. This projection casts doubt on the official ambition to maintain or expand gas exports to Russia, while also starting gas sales to China. Over time, a shrinking export surplus would also make it more difficult for the government to attract foreign investment in the Uzbekistan upstream, since domestic wholesale gas prices may well not be high enough to attract such investment. However, this export projection is heavily contingent upon Uzbekistan's domestic needs for gas in power generation and in the residential and industrial sectors. There is considerable scope for improving the efficiency of gas use in all sectors, which could offset to a large

17

degree the rise in demand as incomes and economic activity grow (see Chapter 16). Relatively small changes in the prospects for demand and production would have a large impact on the availability of gas for export.

Russia

Fields being developed in the Russian offshore sector of the Caspian have promising gas prospects. The main projects are the Tsentralnoye field, which is being explored by Lukoil and Gazprom in partnership with KazMunaiGaz, and the Khvalynskoye field, originally a 50/50 venture between Lukoil and KazMunaiGaz, before the Kazakhstan side sold half of its stake in 2009 to Total and GDF Suez. The Russian energy strategy (Government of Russia, 2009) foresees that gas production from offshore Caspian Sea fields will rise gradually to reach 21 to 22 bcm/year by 2030.

Prospects for natural gas export flows

Overall Caspian gas exports are set to expand considerably over the projection period, though the direction and routes over which they will flow will hinge on a complex set of geopolitical and economic factors. The period from 2009-2010 may turn out to have been a pivotal one for Caspian gas exports. Whereas in 2008, over 80% of export from the Caspian countries was destined for Russia the corresponding figure in 2010 is likely to be closer to 55%. While this still leaves Russia as the main purchaser of Caspian gas, the trend towards greater diversity in Caspian gas trade is unlikely to be reversed. This change reflects the sudden drop in Russia's purchases of Turkmenistan gas in 2009-2010. But it presages an expected long-term shift in gas trade away from primary reliance on Russia and towards new markets in the west and, crucially, in the east. Agreements on the development of a southern corridor from Azerbaijan to Turkey and other European markets pave the way for larger volumes of Azerbaijan gas to move along this route, with at least the possibility to attract additional Turkmenistan volumes. At the same time, the commissioning of the Turkmenistan-China pipeline — and the coincidence of this development with sharply reduced demand from Russia for Central Asia gas — has shifted the centre of gravity of Central Asia's gas sector sharply eastwards.

How much gas Russia will need or want to import from Central Asia in the longer term is uncertain. As the world's largest gas resource-holder and exporter — Russia has felt few incentives to facilitate Caspian countries' direct access to international markets, where Caspian exporters could compete with Russian gas. Instead, the emerging pattern is that Caspian gas (primarily gas from Central Asia) acts as a backstop to the Russian gas balance, to be called upon as necessary if there is a shortfall in Russia due to under-investment or unexpectedly strong demand, or bought as a trading opportunity if available at a discount to international prices.

There may also be limits to China's appetite for Central Asian gas: 40 bcm/year — the current capacity of the Turkmenistan-China pipeline — represents around half of China's projected import requirement in 2020 or about 18% of total Chinese gas demand. This effect may be mitigated by delays with field development, but such a high level of dependence on a single and potentially vulnerable supply route may see

China placing a ceiling on imports from Turkmenistan and Uzbekistan. Although we project an expansion of Central Asian exports to China to around 60 bcm by the end of the projection period (coming almost entirely from Turkmenistan), China may want to regulate the speed at which exports from Central Asia increase to this level on the grounds of energy security (see Chapter 18).

17

CHAPTER 18

REGIONAL AND GLOBAL IMPLICATIONS
Why does the future of Caspian energy matter?

H I G H L I G H T S

- Caspian energy resources could help drive social and economic development across the region. Exports of oil and gas have accounted for a growing share of GDP in Azerbaijan, Kazakhstan and Turkmenistan in recent years. In the New Policies Scenario, the share continues to rise in the short to medium term, but falls in all countries towards the end of the projection period as export volumes dwindle and economic activity in the non-hydrocarbon sector takes over as the main driver of growth.

- The variety of resource endowments across the Caspian creates incentives for intra-regional energy trade and co-operation, particularly in the South Caucasus and Central Asia. But the main energy-trading arrangements, centred on electricity, collapsed with the break-up of the Soviet Union and have since only partially been rebuilt, engendering a big loss of economic efficiency. Progress in re-establishing links will require more market-reflective pricing and is likely to be slow.

- The Caspian has the potential to make a significant contribution to global energy security, by reducing the need to develop more expensive sources of hydrocarbons and by increasing the diversity of sources of supply in importing regions. In all three scenarios, the share of the Caspian in world inter-regional trade of both oil and gas increases between now and 2035, from 6% to 9% for oil and from 4% to 11% for gas in the New Policies Scenario. Creating a diverse and flexible system of export routes will enable the Caspian region to gain access to international market prices for its resources and contribute fully to global oil and gas security.

- Oil transit flows through Russia and Georgia are set to grow strongly, while Azerbaijan will become a major transit country for Kazakhstan oil exports after 2020. Gas transit will also expand substantially, with Kazakhstan, Uzbekistan and Georgia the main conduits. How best to mitigate transit risks is a crucial challenge for Caspian producers.

- The Caspian accounts for only 1.4% of global energy-related CO_2 emissions, reflecting its low population and relatively small GDP. Yet the region's CO_2 intensity is extremely high, mainly because of heavy reliance on fossil fuels and high energy intensity. Uzbekistan has the most CO_2-intensive economy in the world after Iraq, and Kazakhstan is not far behind. The potential for cutting emissions is big: in the 450 Scenario, emissions are lowered by a quarter, vis-à-vis the New Policies Scenario, by 2035. The global impact would be small, but these countries still have much to gain from moving to a more sustainable development model.

Energy in national and regional economic development

Caspian energy resources have the potential to become the motor for much-needed social and economic development across the region, both as an input to national and regional economic activity and as a source of export revenue and government income. Using oil and gas wealth to support more diversified and sustainable economic growth in non-energy sectors of the economy is a central strand in the national economic strategies of all four major oil and gas exporters in the region. These countries are also likely to become an important source of outward investment within the region: Kazakhstan and Azerbaijan, for example, are already among the main sources of foreign investment in Georgia, with Kazakhstan's KazMunaiGaz and Azerbaijan's SOCAR owning oil export terminals on the Black Sea coast at Batumi and Kulevi respectively.

Exports of oil and gas have accounted for a growing share of gross domestic product (GDP) in Azerbaijan, Kazakhstan and Turkmenistan in recent years and will remain a major driver of economic activity and source of government revenues. Uzbekistan differs in this respect from the region's other energy producers as it is not a large exporter — and projected domestic gas consumption exceeds production by 2030 in the New Policies Scenario, making the country a gas net importer (see Chapters 16 and 17). In 2008, oil and gas exports accounted for about two-thirds of the value of total exports from Azerbaijan, Kazakhstan and Turkmenistan combined. In the three countries taken together, the share of oil and gas export revenue in GDP has increased rapidly since 2000, because of higher output in Azerbaijan and Kazakhstan, the sharp rise in international oil and gas prices to 2008, and the convergence between the price offered for Turkmenistan gas exports and international market-based prices. Although the share of energy export revenue in Turkmenistan's GDP suffered a sharp decline in 2009, it is expected to pick up again in the coming year as export levels recover.

In the New Policies Scenario, the importance of export revenues to GDP continues to rise in the short to medium term, but falls in all countries towards the end of the projection period, as export volumes dwindle and economic activity in the non-hydrocarbon sector — boosted to a significant degree by the infrastructure investment financed out of oil and gas revenues earlier in the projection period — takes over as the main driver of growth. Nonetheless, total oil and gas export revenues in the producing countries increase four-fold over the projection period, from about $51 billion in 2009 to over $200 billion in 2035 (Figure 18.1).[1]

Yet the high dependence on energy-export revenues also creates risks, notably the so-called "natural resource curse". There is evidence from some resource-owning countries that an abundance of natural resources can actually hinder economic growth and human development in the longer term. Three reasons are commonly given: difficulties with macroeconomic management, when high resource export earnings strengthen the exchange rate and in turn discourage production in other sectors of

---

1. The opportunity to generate substantial revenue from energy exports is not limited to the hydrocarbons sector; both Tajikistan and the Kyrgyz Republic are expected to increase their exports of electricity through an expansion of hydropower capacity.

the economy, also known as "Dutch Disease"; the volatility of resource prices and therefore of earnings, which can complicate national fiscal and debt management; and the negative effects of large natural resource revenues on the development of national institutions and governance.

Figure 18.1 ● Oil and gas export revenues in selected Caspian countries in the New Policies Scenario

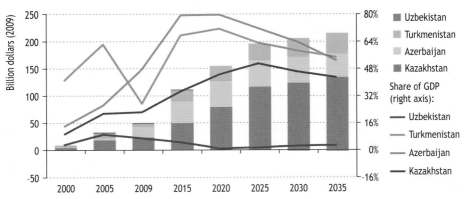

Note: Calculated as the value of net exports net of transportation costs at prevailing average international prices. The split between crude/refined products is not taken into account. GDP is at market exchange rates.

There are few signs as yet that the resource curse is afflicting the Caspian region: energy exports appear to be positively linked to GDP growth (CASE, 2008). However, as Figure 18.1 demonstrates, the Caspian countries are only at the start of a period of exceptionally large revenue receipts from oil and gas exports. Each of the main energy-exporting countries has established a national fund that sets aside a proportion of revenues for future development needs (Kalyuzhnova, 2008): Azerbaijan set up such a fund in 1999, Kazakhstan in 2000 and Turkmenistan in 2008. As of early-2010, an estimated total of $83 billion was held in these three funds. However, growing revenues may increase the temptation to spend a higher share of this income on short-term political priorities to the detriment of long-term economic development.

Regional energy co-operation

The variety of resource endowments across the Caspian region creates incentives for energy trade and co-operation within the region, particularly since the South Caucasus and Central Asia, respectively, are linked by a common energy infrastructure, inherited from Soviet times. The Soviet-era energy systems were designed to exploit the differences in resource endowment across the republics, in particular the synergies between countries with fossil fuels, those with hydropower and, in the South Caucasus, the nuclear generation capacity built in Armenia. However, the main regional energy-

18

trading arrangements collapsed after 1991 and have since been only partially rebuilt, leaving cross-border trade at a fraction of former levels. In Central Asia, for example, regional electricity trade collapsed from 25 gigawatt-hours (GWh) in 1990 to only 4 GWh in 2008 (ADB, 2010).

The multiple inter-dependencies that characterise the Caspian region (Table 18.1) create an opportunity for states to meet their energy needs at lower cost and with greater reliability. Some examples exist of productive bilateral energy co-operation, as between Azerbaijan and Georgia in the South Caucasus, and between Kazakhstan and the Kyrgyz Republic in Central Asia. There are also multiple regional initiatives that rely on or promote co-operation (Box 18.1). But, despite these, frameworks for regional energy trade remain relatively weak, at considerable economic cost.

Box 18.1 ● Towards a common energy space?

There has been a proliferation of initiatives to promote regional co-operation in the energy sector, particularly in Central Asia: this is a priority for many international organisations and donors working in the region. The results so far have tended to be more declaratory than tangible; but there are two regional institutions that could have an important influence on energy developments during the period to 2035. The first of these is the Shanghai Cooperation Organisation, a regional political and security forum of which all the Central Asian countries, except Turkmenistan, are members. The participation of both Russia and China is crucial and the organisation has the potential to contribute significantly to the management of political risks in Central Asia, with implications for energy projects.

The second is the proposal for a customs union, initially involving Russia, Kazakhstan and Belarus, which came into being in mid-2010 and which may evolve into a fully-fledged "common economic space". This initiative is the latest in a long line of proposals among the former Soviet states for economic integration, all of which, up until now, have failed to make much headway. For the moment, the significance of the customs union for the energy sector is limited. Russia, for example, continues to levy oil-export duties at its own border (and Kazakhstan has resumed doing so from August 2010) even though the logic of a customs union would suggest a common levy on a common external border.[2] Likewise, though there are provisions on freedom of transit across the territory of participating countries for oil and oil products, natural gas is excluded. For the projection period, one key issue in relation to this initiative is whether price distortions within and between countries will lessen or even disappear. A widespread move towards market-reflective prices for energy, notably for natural gas, would facilitate the creation of a genuine common economic space — and would also have the effect of making the Russian market a much more attractive destination for Caspian gas producers.

---

2. This was a key element in the oil-transit dispute between Russia and Belarus in early 2010.

Table 18.1 ● Main energy and water relationships in the Caspian

	Provides	Relies on
Azerbaijan	Oil / gas to Georgia Transit services to Kazakhstan	Georgia for oil/gas transit Iran for gas supply (to Nakichevan)
Armenia		Russia for oil/gas/nuclear fuel supply Iran for gas supply
Georgia	Transit services to Azerbaijan / Kazakhstan	Azerbaijan for oil/gas supply
Kazakhstan	Gas, electricity to Russia Coal, gas to Kyrgyz Republic Transit services to Turkmenistan, Uzbekistan	Russia, (Azerbaijan, Georgia) for oil transit Uzbekistan for gas/electricity supply (to south) Kyrgyz Republic for water services
Kyrgyz Republic	Water services to Kazakhstan	Uzbekistan, Kazakhstan for fossil fuels
Turkmenistan	Gas to Russia, Iran	Russia, Uzbekistan, Kazakhstan for gas transit Tajikistan for water services
Tajikistan	Water services to Uzbekistan	Uzbekistan for gas
Uzbekistan	Gas to Russia, Kazakhstan, Tajikistan, Kyrgyz Republic Transit services to Turkmenistan	Tajikistan for water services

These regional relationships matter all the more because Caspian energy resources do not have easy routes to external markets. The main hydrocarbon producers are all landlocked and Uzbekistan is even "double landlocked" in that none of its neighbours has access to international waters either. In many cases, the route to market runs through other producer countries, which may be reluctant to facilitate transit for potential competitors. Mutually beneficial arrangements between countries along the main export routes remain crucial to the development of the region's resources and to realising the region's potential contribution to Eurasian and global energy security.

Electricity trade and the electricity-water nexus

Regional electricity grids used to operate both in Central Asia — including southern Kazakhstan — and in the South Caucasus. The unified South Caucasus grid ceased to operate soon after 1991, amid tensions and then conflict in the region between Armenia and Azerbaijan. A Central Asian energy grid has functioned fitfully until the present day. There is significant potential to increase cross-border electricity trade but there are few signs of any real commitment to a regional approach to energy rationalisation and security. The tendency in recent years has rather been towards the autonomous and, where possible, self-sufficient operation of national electricity systems, along with specific bilateral arrangements for export, as for example between Uzbekistan and Afghanistan.

The development, over time, of a regional concept of rational energy use and energy security would avoid the current very large loss of economic efficiency associated with unnecessary investments in new generation and transmission capacity and inefficiencies in the operation of existing capacity. The electricity generation mix in the South Caucasus, for example, offers scope for exploiting the synergies between Azerbaijan's

18

gas-fired power generation, Georgia's hydropower and Armenia's nuclear capacity. On an even larger scale, there is huge scope in Central Asia for more co-ordinated seasonal exchanges of energy and water between the two mountainous countries with hydropower potential, Tajikistan and the Kyrgyz Republic, and their hydrocarbon-rich neighbours.

A connection between energy and water use along the Amu Darya and Syr Darya rivers was established in the 1960s, when hydropower facilities built in the mountainous republics of Tajikistan and Kyrgyzstan released water in the summer months in order to satisfy the water needs of republics further downstream, mainly for irrigation of the cotton crop. In return, Tajikistan and Kyrgyzstan received fossil fuels to run their thermal power plants during the winter months, during which time (even though electricity demand was high) the upriver countries would cut back hydropower generation.

These regional arrangements no longer operate effectively. A main reason for this is the continued existence of subsidised power prices in Tajikistan and the Kyrgyz Republic, which now have to pay market-reflective prices for their fossil-fuels imports. This has constrained the possibilities for thermal power plants in these countries to generate electricity and heat competitively during the winter months (because of subsidised power in those countries). As a consequence, the operation of the key reservoirs — Nurek in Tajikistan and Toktogul in the Kyrgyz Republic — switched away from summer irrigation and gave priority instead to generating electricity in the winter months (Figure 18.2). This change in operation has had a significant impact on regional energy supply and on water distribution along the entire river system. For downstream countries it has often meant water shortages in the summer months, affecting agricultural production, and flooding in the winter. For Tajikistan and the Kyrgyz Republic, it has resulted in regular shortages of power, which reached their nadir in Tajikistan in the winter of 2007-2008 (UNDP, 2009). These shortages have fuelled social unrest in the Kyrgyz Republic and Tajikistan.

Figure 18.2 ● **Water releases from the Toktogul reservoir by season in the Kyrgyz Republic**

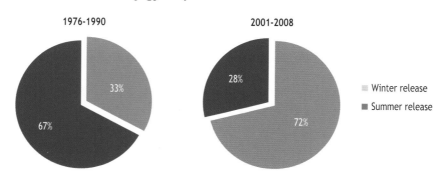

Source: Data provided to the IEA by the Ministry of Industry, Energy and Mineral Resources of the Kyrgyz Republic.

Although there are some functional water-sharing arrangements in place at a bilateral level, as for example between Kazakhstan and the Kyrgyz Republic, it has proved extremely difficult to resurrect a stable overall framework for cross-border energy and water transfers. The Kyrgyz Republic and Tajikistan both intend to complete large new hydropower plants that could improve the reliability of electricity supply and bring revenue from export (see also Chapter 16), but these projects — in particular the huge 3.6 gigawatt (GW) Rogun dam project in Tajikistan — are extremely capital intensive, face opposition from downstream countries and will take many years to complete. In the near term, alongside continued disagreement about these projects, there is likely to be continued tension about the seasonal patterns of operation of the existing hydro plants, about the pricing of internationally traded hydrocarbons and, more controversially, whether and how to compensate for water services provided by the hydro-rich countries. A prolonged failure to find sustainable bilateral or regional solutions to cross-border energy and water issues will hold back economic and social development, and pose threats to the environment throughout the projection period.

Oil and gas transit

As oil and gas production increases in the Caspian region in the coming years, so too will the reliance of regional producers on energy transit. Of the main flows of oil out of the region (around 80% of total production was exported in 2008), only the direct deliveries from Kazakhstan to China do not involve transit. Russia and Georgia are set to see the largest oil transit flows throughout the period to 2035 (Figure 18.3). Completion of the Caspian Transportation System (see Chapter 17) would make Azerbaijan a major transit country for Kazakhstan oil exports after 2020. For natural gas, the Caspian region will soon have some of Eurasia's most important transit relationships by volume, with Kazakhstan and Uzbekistan joined, again, by Georgia as the main transit countries.

Figure 18.3 ● Oil and gas transit in selected Caspian countries in the New Policies Scenario

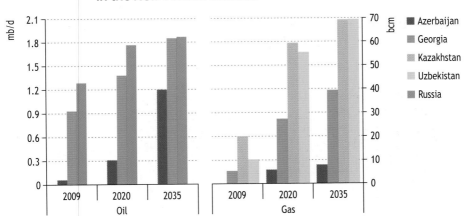

Note: Gas transit flows in Central Asia in 2009 were strongly affected by the sharp decline in exports from Turkmenistan to Russia: Russia itself is not included as a gas transit country since it purchases and re-sells Caspian gas rather than offering transit services.

The stability of these transit relationships and the ability of countries along the energy supply chain to mitigate transit risks is a crucial challenge for Caspian producers (Box 18.2). If transit risks remain — or are perceived to remain — relatively high, this would adversely affect the competitive position of Caspian gas supplies through the projection period, particularly if improvements in technology bring liquefied natural gas (LNG) transportation costs down and/or supplies of unconventional gas become available in, or closer to, the main demand centres.

Box 18.2 ● Mitigating transit risks in the Caspian

The notion of transit risk covers the possibility that transit flows are interrupted, or become prohibitively expensive, or that requests to expand transit capacity or gain access to existing capacity are turned down. All of these issues are constant preoccupations for Caspian energy policy makers, as well as midstream and upstream operators. The optimal solution combines resilient, flexible export options and strong international transit disciplines. Flexibility implies that any country or company not offering competitive commercial conditions for transit ultimately faces the prospect that resources will find an alternative way to market. But the natural advantages of certain geographical routes to market and the high capital costs of building transportation infrastructure make flexibility difficult to achieve in practice. And pipelines, once built, are very inflexible. A set of rules and mechanisms to settle disputes is needed to ensure that pipelines continue to operate reliably over long periods and cope with changing political and economic circumstances.

The main international rules on energy transit are in the 1994 Energy Charter, which has been signed and ratified by all the Caspian countries (as defined in this *WEO*). The Treaty provisions build on principles from the World Trade Organisation. They oblige participating states to take the necessary measures to facilitate transit of energy and prohibit the interruption or reduction of flows, even in the event of a dispute (Energy Charter, 2002). But an attempt among Energy Charter members to fill in greater operational detail on issues such as transit tariffs, the creation of new transit infrastructure and access to transit networks has yet to reach a conclusion after more than ten years of talks. As a result, the multilateral legal and institutional guarantees for energy transit remain relatively weak, all the more so since Russia — which had not in any case ratified the 1994 Treaty — announced in 2009 that it would no longer apply the Treaty provisions, even on a provisional basis. With or without the Treaty, no Caspian gas producer has yet succeeded in negotiating transit rights through the Russian gas network. In the absence of strong international mechanisms, bilateral and selective multilateral political relationships become that much more important in defining and protecting exports routes from the Caspian; in the case of the Turkmenistan-China pipeline, the importance of the transit countries' relationship with China provides the strongest incentive for them to ensure reliable and uninterrupted flows.

Ultimately, the more Caspian gas appears "stranded" *i.e.* lacks an export market outlet, the slower the pace of resource development and the more likely it is that other local options to commercialise the gas will be pursued. The petrochemical industry, which has been expanding in some Caspian countries, provides one option for countries to use more gas on the domestic market, with intermediate or final chemical products then exported instead of the natural gas itself. Another option would be to convert natural gas into oil products using gas-to-liquids technology, again either for domestic needs or for export. A 35 thousand barrels per day (kb/d) project, requiring 3.5 billion cubic metres (bcm) per year of natural gas, is already planned in Uzbekistan (see Chapter 16).

Implications of Caspian resource development for global energy security

Energy security is a global concern. In a global energy market, changes in the supply/demand balance and fuel mix in any one country or region inevitably affect all other market participants through international trade. With abundant fossil and other energy resources, the Caspian has the potential to make a significant contribution to energy security in the rest of the world. Broadly defined (Box 18.3), the potential global energy-security benefits of Caspian energy developments are two-fold:

- Higher exports of oil and gas from the region will add to global supplies; while Caspian oil and gas should not be considered "low-cost" by international standards, particularly once transportation costs are taken into account, their development would nonetheless reduce the need to develop other, more expensive sources of hydrocarbons and would result in lower prices than would otherwise be the case.

- They will also increase the diversity of the sources of oil and gas supplies into importing regions.

In all three scenarios presented in this *Outlook*, the share of the Caspian region in world production and exports of both oil and gas increases through the projection period.

There are also important energy-security benefits to the rest of the world to be gained from exploiting the enormous potential for energy savings within the Caspian region, in addition to the accompanying economic benefits to those countries and the broader environmental benefits (see the concluding section). Today, over-consumption of energy in the four leading Caspian countries (calculated on a sector-by-sector analysis of the efficiency of energy use in the region, see Chapter 16) accounts for roughly half of all the energy consumed. This reduces in aggregate the amount of oil available for export by about 320 kb/d and the amount of exportable gas exports by about 50 bcm per year.

Oil security

In the case of oil, the share of the Caspian in world production grows from 3.5% in 2009 to more than 5% in 2035 in each scenario. The region's share of global exports also increases in all scenarios, from 6% to about 9% in the New Policies and Current Policies Scenario, and to just under 8% in the 450 Scenario (Figure 18.4). To put this in perspective, a 9% share of global oil exports in 2035 is equivalent to projected oil net exports from Latin America. This would be enough to meet almost all the projected import requirements of North America in 2035 or, alternatively, to cover one-third of China's oil imports.

Figure 18.4 ● **Share of the Caspian in world oil supply by scenario**

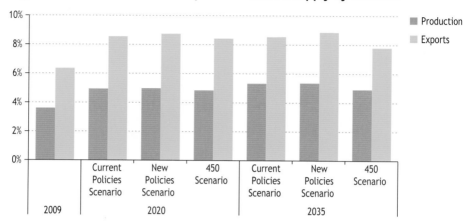

As described in detail in Chapter 16, Kazakhstan accounts for the bulk of Caspian production and exports. Azerbaijan is the other major producer, but peak production of 1.3 million barrels per day (mb/d) is reached as early as 2014, before declining to under 900 kb/d in 2035 (Figure 18.5). By contrast, Kazakhstan is projected to join the small elite of countries with output of more than 2 mb/d in 2016 and then to become one of the top ten oil producers in the world in 2022, a ranking that it maintains until the end of the *Outlook* period. Kazakhstan's output in 2035 is expected to be around 3.9 mb/d, only marginally down from the peak reached earlier in the 2030s. As a result, Kazakhstan is the fourth-largest contributor to global oil-production growth over the *Outlook* period, second only to Brazil among non-OPEC countries (Figure 18.6).

Figure 18.5 ● **Oil production in the Caspian by country in the New Policies Scenario**

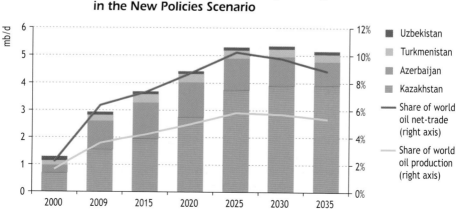

Figure 18.6 ● Incremental oil production by selected country in the New Policies Scenario, 2009-2035

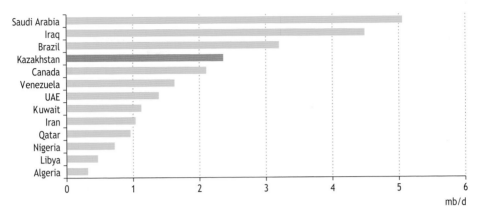

We assume that the projected additional exports of oil will flow along multiple routes to international markets, but the share that will flow along each route remains highly uncertain (see Chapter 17). There is a particular need for Kazakhstan to develop new export pipeline capacity in the period to 2025 as it develops its vast oil resources. From a global oil-security perspective, the important consideration is not whether any individual market is importing more or less Caspian oil, but whether Caspian oil is able to reach multiple markets on a commercial basis – without any single transit route, country or company having such a dominant role in transportation as to create effective monopoly control over these flows of oil. Reliable access to market will create the necessary commercial incentives to encourage continued upstream investment in the region. Moreover, dependable supplies of Caspian oil to both European or Asian markets would reduce the need of those regions to import oil from the Middle East by tanker via vulnerable routes, such as the Straits of Hormuz, the Suez Canal and the Straits of Malacca (see Chapter 3).

Box 18.3 ● Defining energy security

Energy security, broadly defined, means adequate, affordable and reliable supplies of energy. It matters because energy is essential to economic growth and human development. No energy system can be entirely secure in the short term, because disruptions or shortages can arise unexpectedly, whether through sabotage, political intervention, strikes, technical failures, accidents or natural disasters. In the longer term, under-investment in energy production or transportation capacity can lead to shortages and consequently unacceptably high prices. So energy security, in practice, is best seen as a problem of risk management, that is reducing to an acceptable level the risks and consequences of disruptions and adverse long-term market trends.

Secure energy supply is a public good, as the benefit derived from it by one consumer does not reduce the benefit to everyone else. Markets alone do not reflect the cost to society of a supply failure because it is beyond the power of an individual supplier or consumer to guarantee security. Correspondingly, all market players benefit from action to safeguard energy security, whether or not they have contributed to it. That is why governments must take ultimate responsibility for ensuring an adequate degree of security within the framework of open, competitive markets, and why international collaboration through the International Energy Agency makes sense.

Short-term threats to security concern unexpected disruptions, whether of a political, technical, accidental or malevolent nature. Long-term threats relate to a lack of deliverability, caused by deliberate or unintentional under-investment in capacity. Both short-term disruptions and under-investment result in higher prices, causing hardship to consumers and harming economic prospects. The two are linked: under-investment also renders the energy system more vulnerable to sudden supply disruptions, accentuating their impact on prices, while experience of short-term disruptions shakes market confidence in supply, increasing the risk of under-investment in production.

Concerns about energy security have evolved over time with changes in the global energy system and perceptions about the risks and potential costs of supply disruptions. In the 1970s and 1980s, the focus was on oil and the dangers associated with over-dependence on oil imports. Today, worries about energy security extend to natural gas, which is increasingly traded internationally, and the reliability of electricity supply. There are growing concerns about whether competitive markets for electricity and gas, as they currently operate, provide sufficient incentive for building capacity. And concerns about energy security must be reconciled with worries about the environmental impact of energy production and use, including their contribution to climate change. Action to improve energy security will not be effective if it provokes an environmental revolt.

Most governments have developed policies designed to ensure adequate investment in energy supply infrastructure to meet projected needs, to promote more efficient energy use in order to reduce the risk of demand running ahead of deliverability, to encourage diversity in the fuel mix, geographic sources and supply routes, and to improve market transparency, in order to help suppliers and consumers make economically efficient investment and trading decisions, and governments to take informed policy decisions. Many have put in place measures to respond to short-term disruptions, including co-ordinated use of emergency oil stocks, plans to redirect supply flows and demand-side management. Oil emergency stocks and co-ordinated responses to a supply disruption form a central pillar of the energy-security policies of IEA countries.

Gas security

The overall picture for gas is similar to that for oil, with the Caspian's share of global production increasing from 5% in 2009 to around 7% in 2035, with the result fairly consistent across all the scenarios (Figure 18.7). As with oil, the region's share in global exports is higher than its share of global production, with the Caspian region accounting for about 11% of global gas export in the New Policies Scenario; this falls slightly in the 450 Scenario, in which global gas demand is projected to fall after 2025 as the market penetration of renewables and nuclear power increases (see Chapter 5).

Figure 18.7 ● Share of the Caspian in world natural gas supply by scenario

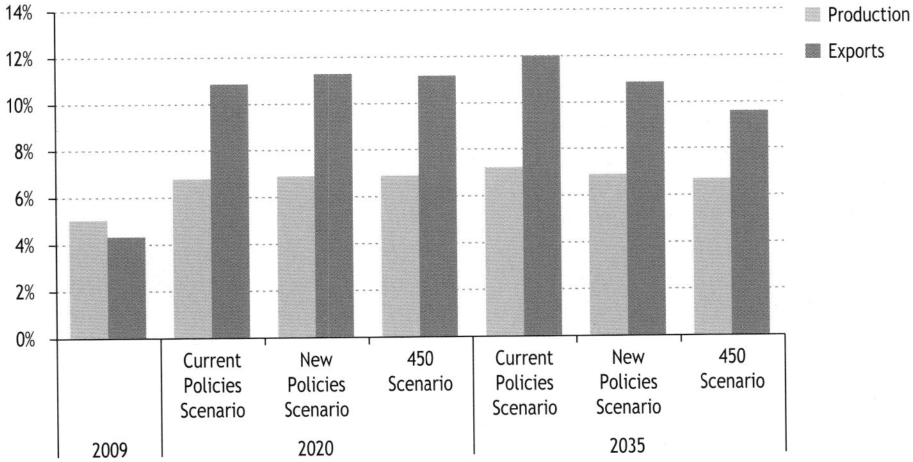

Whereas the major contribution to Caspian oil output growth comes from Kazakhstan, the analogous role for natural gas is played by Turkmenistan, which becomes, from 2015, one of the ten largest gas producers in the world in the New Policies Scenario. All the Caspian energy resource-owners, with the exception of Uzbekistan, see a substantial expansion in projected natural gas production in the period to 2035 (Figure 18.8). The bulk of total Caspian gas production will still be consumed within the region, but in Turkmenistan and Azerbaijan (with relatively small populations and lower gas demand) additional production also translates into large increases in exports.

The projected increase in total Caspian gas exports, from a low point in 2009 of under 30 bcm to almost 130 bcm/year by the end of the projection period in the New Policies Scenario, represents an important addition to the world's supply. But, in contrast to the oil market, it is difficult to assess the energy-security repercussions of Caspian gas exports in global terms since there is not yet a global gas market, but rather a series of regional gas markets loosely connected by inter-regional trade. In the case of landlocked Caspian gas reserves, decisions on the direction of exports carry strategic weight, both for the importing country and also for the producers themselves, since gas trade implies a set of long-term relationships essential to the development of domestic economies.

18

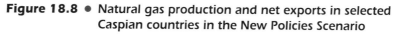

Figure 18.8 ● Natural gas production and net exports in selected Caspian countries in the New Policies Scenario

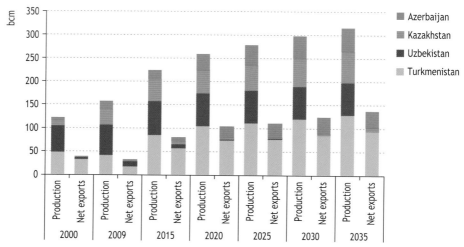

Russia was the main purchaser of Caspian gas until recently, notwithstanding the sharp drop in Turkmen gas imports in 2009, but this is set to change with the take-off of Turkmen exports to China and the prospect of rising exports from Azerbaijan and possibly Turkmenistan to Turkey and other European markets through a southern corridor. How much gas Russia will need or want to import from Central Asia in the longer term remains highly uncertain (see Chapter 17). Russia is a pillar of global gas supply and, insofar as Central Asian gas contributes to the reliability of that supply, this gas contributes indirectly to Eurasian gas security. However, Russian control over Central Asian gas flows deprives Eurasia of an important element of market diversity.

Similar considerations apply to Iran as a market and route to market, in that Iranian demand for gas from the Caspian region is likely to be contingent on shifts in the Iranian gas balance. At present, with Iran in most years a marginal net importer of gas, Turkmenistan plays an important role in meeting Iran's domestic needs, while freeing Iran to export small quantities to Turkey and, possibly, also to Pakistan in the future. But in the medium term, if Iran (as projected in this *Outlook*) has a larger surplus of its own gas to export, the incentives for Iran to offer Central Asia favourable conditions for gas trade become less compelling.

For Europe and for China, direct gas trade with the Caspian region promises diversity of supply, but it is not axiomatic that this equates to increased *security* of supply if this diversity comes with additional political or economic risks. The routes from the Caspian gas reserves to these markets are long and complex, and it will be a critical challenge to mitigate transportation and transit risks in order to unlock the energy-security benefits that Caspian gas could potentially provide (see Box 18.2).

In the case of Europe, the southern corridor for the supply of gas from the Caspian and Middle East is seen as having the potential to supply a significant part of the European

Union's future gas needs (EC, 2009). We project that the share of Caspian gas in OECD Europe's gas supply to increase from around 2% today to more than 8% by 2035 in the New Policies Scenario. The benefits of market diversity from an open and transparent energy corridor to the Caspian will be felt more widely across southeast and central Europe, where dependence on Russia as a single gas supplier is very high.

For China, gas from the Caspian region assumes even greater importance as a share of domestic supply and imports: our projections suggest that one-half of China's gas imports may come via the Turkmenistan-China pipeline by 2020 (see Chapter 17). This may lead China to place a ceiling on imports from Central Asia, on the grounds of energy security. Chinese perceptions about the reliability of this route will play a part in this calculation, as will the evolution of their domestic gas balance (Figure 18.9).

Figure 18.9 ● **Caspian share of markets and imports in OECD Europe and China in the New Policies Scenario**

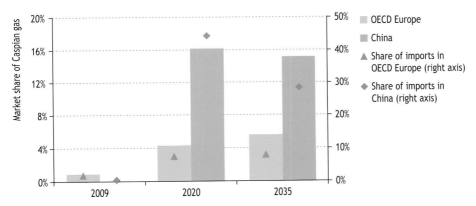

Implications for climate change

The Caspian region today accounts for only 1.4% of global energy-related carbon-dioxide (CO_2) emissions, reflecting its low population and relatively small GDP. Kazakhstan is the main emitter of CO_2, accounting for half of the region's total emissions; in the New Policies scenario, total Caspian emissions are projected to rise by almost half between 2008 and 2035 (Table 18.2). Yet the region's carbon intensity (measured as CO_2 emissions per dollar of GDP) is extremely high, mainly because of the region's heavy reliance on fossil energy and its high *energy* intensity. This is a consequence of climatic factors, the region's dependence on heavy industry, a lack of energy-efficiency regulations and standards, large subsidies and obsolete technologies (see Chapter 16). Uzbekistan and Turkmenistan have the highest carbon intensities among the Caspian countries. Indeed, Uzbekistan has the most carbon-intensive economy in the world after Iraq on a purchasing-power parity basis.

18

Table 18.2 ● Energy-related CO_2 emissions in the Caspian by country in the New Policies Scenario (million tonnes)

	1990	2008	2015	2020	2025	2030	2035	2008-2035*
Azerbaijan	63.2	29.3	31.9	34.2	35.9	37.0	37.9	1.0%
Kazakhstan	236.4	201.6	249.9	266.3	290.0	289.7	293.5	1.4%
Turkmenistan	46.6	47.3	60.3	64.0	67.2	69.8	71.9	1.6%
Uzbekistan	119.8	114.9	137.9	145.9	151.3	155.5	159.2	1.2%
Other Caspian**	63.2	29.3	31.9	34.2	35.9	37.0	37.9	1.0%
Total	529.3	422.4	511.9	544.7	580.3	589.1	600.4	1.3%

* Compound average annual growth rate.
** Armenia, Georgia, Kyrgyz Republic and Tajikistan.

A significant reduction in carbon intensity for the Caspian region is projected in the New Policies Scenario (Figure 18.10), but this is related mainly to the expected rapid expansion of GDP growth, rather than to the introduction of specific policies aimed at addressing some of the extremely inefficient energy practices and technologies that are commonplace in the region. Emission levels in 2035 remain high by global standards in Uzbekistan and Turkmenistan and also in Kazakhstan, with its continued reliance on coal for power generation despite a large increase in gas-fired electricity production.

Figure 18.10 ● Carbon intensity in Caspian countries and selected other countries in the New Policies Scenario

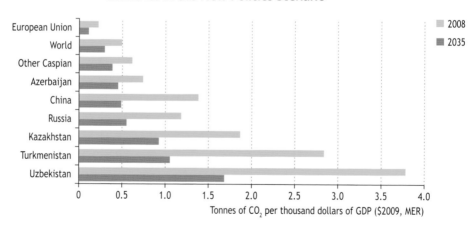

Note: MER = market exchange rate.

As discussed in Chapter 16, the policies that are currently envisaged for implementation in the Caspian region (as considered in the New Policies Scenario), do not move the region far away from their current, business-as-usual, course. In aggregate, the Caspian region is therefore not projected to take much advantage of the many opportunities that exist to move to more efficient and sustainable patterns of energy use.

By contrast, in the 450 Scenario, which assumes Caspian countries implement strong policy actions to improve energy efficiency and to deploy low-carbon fuels and technologies, the region's energy-related CO_2 emissions in 2035 are 140 million tonnes (Mt), or about 25%, lower than in New Polices Scenario (Figure 18.11). Energy-efficiency improvements in end-use sectors, notably buildings, deliver about three-quarters of the emissions savings in this scenario. The retirement of older, inefficient fossil-fuel-fired plants and their replacement with more efficient technologies accounts for an additional 17% in 2020. The power sector's contribution to overall emissions abatement declines over the projection period, as the gap in the thermal efficiency of electricity generation between the two scenarios narrows. The increased deployment of renewables, mainly hydropower, accounts for most of the remaining emissions savings. Even with these large efficiency gains, the Caspian region's contribution to the global reduction in emissions needed to achieve the 450 Scenario, compared with the New Policies Scenario, is about 1%, reflecting the region's relatively small share of global energy demand. However, it is also relevant to note that gas exported to China has a very beneficial effect on China's own emissions (Box 18.4).

Figure 18.11 ● Energy-related CO_2 emissions abatement in the Caspian by source in the 450 Scenario compared with the New Policies Scenario

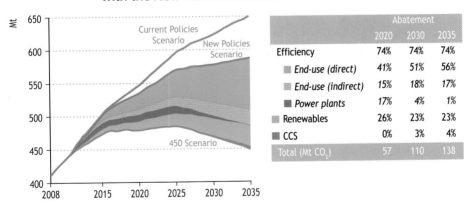

	Abatement		
	2020	2030	2035
Efficiency	74%	74%	74%
■ End-use (direct)	41%	51%	56%
▨ End-use (indirect)	15%	18%	17%
■ Power plants	17%	4%	1%
Renewables	26%	23%	23%
■ CCS	0%	3%	4%
Total (Mt CO_2)	57	110	138

Even if action by Caspian countries has at best only a small global impact on climate change, these countries still have much to gain from moving to a more sustainable development model, not least in order to tackle the many serious local environmental issues facing the region. Moreover, while there are various opinions about the impact of climate change on the region, some of the risks are already becoming apparent, particularly in Central Asia; for example, there is already an observed decrease in glacier coverage in the Tian Shen and Pamir mountains — the sources of much of the region's water supply (EDB, 2009). Over the projection period, larger and earlier snow melt, heavier winter rainfall and dryer, hotter summers add to the discrepancy between water availability and the region's irrigation needs. The links between energy security, water, climate change and human development in the Caspian region are likely to become even closer in the years to come.

18

Box 18.4 ● How big are the climate benefits of Caspian gas going east?

As an important new source of supply to the Chinese market, gas from Central Asia provides energy that would otherwise have to be supplied from domestic sources or from imports from other sources. The main competing fuel for gas in China is domestic coal for use in power generation and industry, and, to a degree, oil for industry and buildings. Combustion of natural gas produces significantly less greenhouse-gas emissions than coal or oil, so the substitution of these fuels by imported gas brings a significant gain in terms of reduced emissions. The Caspian region, primarily Turkmenistan, is projected to increase gas exports to China by up to 60 bcm per year by the end of the projection period in the New Policies Scenario. If all of this gas were to displace coal from China's energy consumption, then the cumulative savings in terms of CO_2 emissions from 2009-2035 would amount to 1.35 gigatonnes of CO_2, an average of 50 Mt of CO_2 savings per year. This is close to 80% of Turkmenistan's average annual emissions over the same period.

PREFACE

The focus on energy subsidies in this *Outlook* demonstrates the impact of fossil-fuel subsidy removal for energy markets, climate change and government budgets. The analysis builds on work undertaken by the IEA in response to a call from G-20 Leaders to the IEA, OECD, OPEC and World Bank at the Pittsburgh Summit in 2009 to prepare a joint report to support the G-20 commitment to phase out and rationalise inefficient fossil-fuel subsidies.

Chapter 19 focuses on energy subsidies that encourage over-consumption by reducing the price of fossil fuels below the full cost of supply. It assesses quantitatively the extent of such subsidies and provides the results of modelling undertaken to determine the potential energy savings and carbon-dioxide emissions reduction from their removal. The chapter discusses the importance of ensuring that subsidy-reform programmes are carefully designed, as low-income households, spending a higher percentage of their household income on energy, can be disproportionately affected by the elimination of subsidies. The chapter closes with a summary of plans being pursued around the world to move towards market-based pricing.

Chapter 20 entails a more detailed discussion on energy subsidies in five countries that have plans gradually to introduce market-based pricing: China, India, Indonesia, Iran and Russia. Key drivers behind the polices being implemented have varied from country to country, as have expectations over the likelihood that lasting reform will take hold, in view of the economic, political and social barriers that must be overcome.

CHAPTER 19

ANALYSING FOSSIL-FUEL SUBSIDIES
What would getting prices right entail?

H I G H L I G H T S

- Fossil-fuel subsidies result in an economically inefficient allocation of resources and market distortions, while often failing to meet their intended objectives. Subsidies that artificially lower energy prices encourage wasteful consumption, exacerbate energy-price volatility by blurring market signals, incentivise fuel adulteration and smuggling, and undermine the competitiveness of renewables and more efficient energy technologies. For importing countries, subsidies often impose a significant fiscal burden on state budgets, while for producers they quicken the depletion of resources and can thereby reduce export earnings over the long term.

- Fossil-fuel consumption subsidies worldwide amounted to $312 billion in 2009. The annual level fluctuates widely with changes in international energy prices, domestic pricing policy, exchange rates and demand. The vast majority were provided in non-OECD countries, which are projected to contribute 93% of incremental global energy demand to 2035 in the New Policies Scenario.

- Considerable momentum is building globally to cut fossil-fuel subsidies. In September 2009, G-20 leaders committed to phase out and rationalise inefficient fossil-fuel subsidies, a move that was closely mirrored in November 2009 by APEC leaders. Many countries are now pursuing reforms, but steep economic, political and social hurdles will need to be overcome to realise lasting gains.

- Reforming inefficient energy subsidies would have a dramatic effect on supply and demand balances in global energy markets. A universal phase-out of all fossil-fuel consumption subsidies by 2020 would cut global primary energy demand by 5%, compared with a baseline in which subsidies remain unchanged. This amounts to the current consumption of Japan, Korea and New Zealand combined. Oil demand would be cut by 4.7 mb/d by 2020, or around one-quarter of current US demand.

- Phasing out fossil-fuel consumption subsidies could represent an integral building block for tackling climate change. A complete phase-out would reduce carbon-dioxide emissions by 5.8%, or 2 Gt, by 2020. This amounts to a significant share of the abatement needed to be on track by 2020 to limit the global temperature increase to 2°C.

- In countries with low levels of modern energy access, subsidies in the residential sector for kerosene, electricity and LPG — fuels that often support the basic needs of the poor — represented just 15% of fossil-fuel consumption subsidies in 2009. Nonetheless, subsidy-reform programmes need to be carefully designed as low-income households are likely to be disproportionately affected.

Defining energy subsidies[1]

The IEA defines an energy subsidy as any government action directed primarily at the energy sector that lowers the cost of energy production, raises the price received by energy producers or lowers the price paid by energy consumers. Many energy subsidies are difficult to measure, so for practical reasons much narrower definitions are often adopted that include only those subsidies that can be quantified and for which data are readily available. The broad definition used by the IEA is designed to capture all of the diverse and obscure types of energy subsidy that commonly exist.

Energy subsidies are frequently differentiated according to whether they confer a benefit to producers or consumers, or whether they support traditional fossil fuels or cleaner forms of energy. In this chapter, the focus is on fossil fuels and, predominately, on consumption subsides.[2] Subsides can be further distinguished according to the channels through which they are administered; these include budgetary payments, regulations, taxes and trade instruments (Table 19.1). They can be grouped as either direct transfers, such as grants to expedite the deployment of fledgling energy technologies, or indirect transfers, such as the regulation of end-use prices.

Fossil-fuel consumption subsidies, which lower prices to end-users are now rare in most OECD countries, but are still present in many other regions. Production subsidies, by contrast, involve measures that seek to expand domestic supply. They remain an important form of subsidisation in both OECD and non-OECD countries, though many subsidies in this category have also been phased out, with the shift towards more market-oriented economic and energy policies and liberalisation of international trade. Both production and consumption subsidies, by encouraging excessive production or consumption, can lead to an inefficient allocation of resources and market distortions.

OECD countries and a number of emerging economies have been introducing production subsidies and/or support programmes to aid the diffusion of renewable energy, nuclear power and carbon capture and storage (CCS). In many cases the introduction of transitional incentives to move cleaner and more efficient technologies quickly towards market competitiveness can help to reduce greenhouse-gas emissions and pollution, and lead to a more diverse energy mix. Such assistance for renewable energy is covered in Chapter 9.

Rationale for energy subsidies and the need for reform

Historically, the rationale for the introduction of energy subsidies has been to advance particular political, economic, social and environmental objectives, or to address

---

1. This chapter builds on analysis undertaken in response to a call from G-20 Leaders to the IEA, OECD, OPEC and World Bank at the Pittsburgh Summit in 2009 to prepare a Joint Report on energy subsidies (IEA/OECD/OPEC/World Bank, 2010).

2. To increase the availability and transparency of energy subsidy data, the IEA has established an energy subsidy online database: www.worldenergyoutlook.org/subsidy.asp.

problems in the way markets operate. In practice, subsidies have often proved to be an unsuccessful or inefficient means of achieving their stated goal. But, once introduced, they tend to command the political support of a particular section of society and become difficult to reform or eliminate.

Table 19.1 ● Common types of energy subsidies

	Description	Examples
Trade instruments	Quotas Technical restrictions Tariffs	Tariffs on imported ethanol
Regulations	Price controls Demand guarantees and mandated deployment rates Market-access restrictions Preferential planning consent Preferential resource access	Gasoline price regulated at $0.03 per litre in Venezuela Regulations that prioritise use of domestic coal for power generation
Tax breaks	Rebates or exemptions on royalties, duties, producer levies and general consumption taxes Tax credits Accelerated depreciation allowances on equipment	Favourable tax deduction on oil and gas fields and coal deposits Excise exemptions for fuel used in international air, rail or water transport
Credit	Low-interest or preferential rates on loans to producers	Loan guarantees to finance new nuclear power plants
Direct financial transfer	Grants to producers or consumers	Home heating assistance programmes for the elderly and low income earners
Risk transfer	Limitation of financial liability	Limits on the energy industry's financial liability in the event of an accident
Energy-related services provided by government at less than full cost	Direct investment in energy infrastructure Public research and development	Provision of seismic data for oil and gas exploration

Source: Adapted from UNEP and IEA, 2002.

The most common justifications for the introduction of energy subsidies are:

■ **Alleviating energy poverty:** Fossil-fuel subsidies have been seen as a means of helping to improve the living conditions of the poor by making cleaner and more efficient fuels affordable and accessible; for example, liquefied petroleum gas (LPG) in place of traditional biomass.

■ **Boosting domestic supply:** Subsidies have been introduced to support indigenous fuel production in a bid to reduce import dependency. They have also been used at times to support a country's foreign and strategic economic policies by helping the overseas activities of national energy companies.

■ **Redistributing national resource wealth:** In major energy-producing countries, subsidies in the form of artificially low energy prices are often seen as a means of

19

sharing the value of indigenous natural resources. They are also used in an effort to encourage economic diversification and employment by improving the competitiveness of energy-intensive industries, such as petrochemicals and aluminium.

■ **Protecting employment:** Energy subsidies, usually in the form of tariffs or trade restrictions, are often used to maintain regional employment, especially in periods of economic downturn or transition.

■ **Protecting the environment:** Energy subsidies are increasingly being introduced to promote the use of clean energy sources and new technologies to combat climate change. The underlying rationale is that the market, alone, fails to capture all the costs of producing and using some fuels, so support may be required to offset these external costs.

In recent years there has been growing momentum to phase out fossil-fuel subsidies as many were seen to be failing to serve effectively the aforementioned objectives. While also, in a period of persistently high prices, imposing unsupportable financial burdens on countries importing energy at world prices and selling it domestically at lower, regulated prices. As a share of GDP at market exchange rates, spending on oil and gas imports in many emerging economies spiked in 2008, reaching levels well above those seen during the first and second oil shocks. For example, spending on oil and gas imports reached 6.9% of GDP in India and 3% in China. Many countries seized the opportunity presented by the fall in prices after mid-2008 to reduce subsidies without having a major impact on inflation (since the fall in world prices cushioned consumers from the upward pressure on prices resulting from subsidy removal) and without provoking consumer wrath.

A related motivation for phasing out fossil-fuel subsidies stems from their adverse impact on investment resources. Where fossil-fuel consumption is subsidised through consumer price controls, the effect, in the absence of offsetting compensation payments to companies, is to reduce energy companies' revenues, which limits their ability to invest in, maintain and expand the energy infrastructure. This problem is particularly prevalent within the electricity sector of many developing countries (leading to rolling blackouts or low levels of electricity access), but also exists in the oil, natural gas and coal sectors.

There are many other good reasons to phase out subsidies (Figure 19.1). They can encourage wasteful consumption, thereby leading to faster depletion of finite energy resources, and can also discourage rationalisation and efficiency improvements in energy-intensive industries. There is a strong empirical link between low energy prices and excessive consumption. Extremely high rates of electricity consumption in parts of the Middle East and North Africa, for instance, can be shown to derive from cheap electricity tariffs (or even free electricity in some cases) rather than demography or healthy economic growth. The resulting subsidy, in certain cases, has over-burdened government resources at the expense of social and economic expenditures (Khatib, 2010).

Fossil-fuel subsidies exacerbate energy price-volatility on global markets by dampening normal demand responses to changes in international prices. For example, many market analysts were surprised by the robustness of global oil demand, despite the dramatic increases in crude-oil prices, during the first half of 2008. This has now been attributed in part to artificially low energy prices in many countries, which blunted market signals. A survey of 131 countries carried out by the International Monetary Fund (IMF) found that in 2008 around two-thirds of countries failed to fully pass through the sharp rise in international prices for gasoline and one-half failed to pass through the full increase in the cost of diesel (Coady *et al.*, 2010). Cutting subsidies, by shifting the burden of high prices from government budgets to individual consumers, would lead to a much faster and stronger demand response to future changes in energy prices and free up government revenues for other urgent needs.

Figure 19.1 ● **Potential unintended effects of fossil-fuel consumption subsidies**

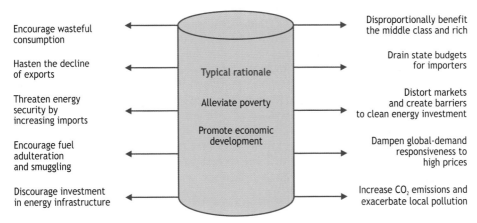

Encourage wasteful consumption

Hasten the decline of exports

Threaten energy security by increasing imports

Encourage fuel adulteration and smuggling

Discourage investment in energy infrastructure

Typical rationale

Alleviate poverty

Promote economic development

Disproportionally benefit the middle class and rich

Drain state budgets for importers

Distort markets and create barriers to clean energy investment

Dampen global-demand responsiveness to high prices

Increase CO_2 emissions and exacerbate local pollution

Energy subsidies can encourage fuel adulteration,[3] and the substitution of subsidised fuels for more expensive fuels (Shenoy, 2010). In some countries, subsidised kerosene intended for household cooking and lighting is diverted for unauthorised use as diesel fuel due to wide price differentials. Fuel smuggling can also arise, since an incentive is created to sell subsidised products in neighbouring countries where prices are unsubsidised and, therefore, higher. This has been an issue for years in many parts of the world, particularly in southeast Asia, Africa and the Middle East. The effect in subsidising countries is a substantial financial transfer to smugglers, while recipient countries experience losses from uncollected taxes and excise duties, due to reduced sales in the legitimate market. Removing subsidies would eliminate incentives both to adulterate fuels and to smuggle them across borders.

Although energy subsidies are often intended to help redistribute income to the poor, the greatest benefit typically goes to those who consume the most energy, *i.e.* who can afford to own motor vehicles, electrical appliances, etc.. The IMF has estimated that

19

---

3. Fuel adulteration typically involves illegally mixing cheaper fuels into particular categories of fuel, often leading to problems with engine performance, increased environmental pollution and reduced fuel-excise tax revenues.

80% of the total benefits from petroleum subsidies in 2009 accrued to the richest 40% of households (Coady et al., 2010). Nonetheless, the removal of even poorly targeted energy subsidies needs to be carefully implemented, since the adverse impact on poor households can be disproportionally large (see the section on subsidies and energy poverty below).

Importing countries have clear incentives to remove energy subsidies to consumers because of their associated direct financial burden. But a number of energy-rich exporting countries have also moved to phase out subsidies, or expressed interest in doing so, concerned not only by the high cost of the subsidies but also the resulting low efficiency in domestic energy use: the consequences can be sharp domestic demand growth and reduced availabilities for export (Birol et al., 1995). Over time, such subsidies may even threaten to curtail the exports that earn vital state revenue streams. In May 2010, top officials from Saudi Aramco warned that Saudi oil export capacity would be restricted to less than 7 million barrels per day (mb/d) by 2028 if domestic energy demand continued to rise at its current pace and that changes to energy prices may be needed to slow its growth. Furthermore, a number of major oil exporters, including Iran, Nigeria and Kazakhstan, rely on imports of refined petroleum products, principally because low regulated prices preserve artificially high demand and undermine investment in adequate refining capacity. This problem is particularly acute if refiners are not reimbursed by governments for their losses. The potential significance of energy-subsidy reform in producing regions is highlighted by the results of the Current Policies Scenario, in which no new measures to phase out subsidies are assumed: in Middle East countries the increase in domestic oil demand to 2020 is projected to absorb 24% of the growth in crude oil production.

Energy subsidies can have varying environmental effects. In some instances, for example where subsidies enable poor communities to switch from the traditional use of biomass to modern fuels, they can have positive implications for the local environment by minimising deforestation and household air pollution. In the vast majority of cases, however, fossil-fuel subsides are counterproductive in reaching local and global environmental goals. Subsidised energy prices dampen incentives for consumers to use energy more efficiently, resulting in higher consumption and greenhouse-gas emissions than would otherwise occur. Furthermore, fossil-fuel subsidies undermine the development and commercialisation of renewable energy and other technologies that could become more economically attractive. Even marginal shifts from fossil fuels to renewable energy could help to accelerate the learning effect for renewables and cause unit production costs to decline.

Measuring fossil-fuel consumption subsidies

Measuring energy subsidies is a complex undertaking due to the varying definitions of what constitutes a subsidy and the availability of adequate data. In this section, we quantify subsidies to fossil-fuel consumption. These subsidies were chosen for measurement because they have a particularly important impact on global energy trends affecting economic growth, energy security and the environment, as recognised by the political decision taken in Pittsburgh in September 2009 (Box 19.1). Subsidies to energy production are not quantified in our analysis. Estimating subsidies to energy production is a complicated endeavour, due to the different sources, recipients and categories of producer support, and as the data are in many cases of poor quality or nonexistent.

Box 19.1 ● **The G-20 and APEC commitments to phase out fossil-fuel subsidies**

A key step towards reforming energy subsidies was taken in September 2009 when G-20 leaders met in Pittsburgh, United States, and committed to "rationalize and phase out over the medium term inefficient fossil fuel subsidies that encourage wasteful consumption", a move that was closely mirrored by Asia-Pacific Economic Cooperation (APEC) leaders in November 2009. These commitments were made in recognition that such subsidies distort markets, impede investment in clean energy sources and undermine efforts to deal with climate change. Recognising the importance of providing those in need with essential energy services, including through the use of "targeted cash transfers and other appropriate mechanisms", the G-20 called on:

● Individual countries to "phase out such subsidies" and for energy and finance ministers of G-20 countries to develop their country implementation strategies and timeframes and report back to the next G-20 Summit;

● The IEA, the Organisation of the Petroleum Exporting Countries (OPEC), the Organisation for Economic Co-operation and Development (OECD) and the World Bank to provide an analysis of the scope of energy subsidies and suggestions for the implementation of this initiative.

A Joint Report (IEA/OECD/OPEC/World Bank, 2010) was prepared by the four organisations in response to this request. It draws on their relevant expertise and work, as well as on input and comments from other organisations and experts.[4] At the G-20 Toronto Summit in late June 2010, leaders welcomed the work of the finance and energy ministers to fulfil the Pittsburgh fossil-fuel subsidies pledge and encouraged continued and full implementation of country-specific strategies. The G-20 also committed to review progress towards this pledge at upcoming leaders' summits.

The price-gap approach

The IEA measures energy consumption subsidies using a price-gap approach. This compares final consumer prices with reference prices, which correspond to the full cost of supply or, where appropriate, the international market price, adjusted for the

19

---

4. The report is available at www.worldenergyoutlook.org/docs/G20_Subsidy_Joint_Report.pdf. Details of subsidy reforms currently being undertaken or proposed by countries are available at www.g20.org/Documents2010/expert/Annexes_of_Report_to_Leaders_G20_Inefficient_Fossil_Fuel_Subsidies.pdf.

costs of transportation and distribution. The estimates cover subsidies to fossil fuels consumed by end-users and subsidies to fossil-fuel inputs to electric power generation. Simple as the approach may be conceptually, compiling the necessary price data across different fuels and sectors and computing reference prices are formidable tasks.

The price-gap approach is the most commonly applied method for quantifying consumer subsidies.[5] It is designed to capture the net effect of all subsidies that reduce final prices below those that would prevail in a competitive market. However, estimates produced using the price-gap approach do not capture all types of intervention known to exist. They, therefore, tend to be understated as a basis for assessing the impact of subsidies on economic efficiency and trade. For example, the method does not take account of revenue losses in countries where under-collection of energy bills (particularly for electricity) is prevalent, or where energy theft is rife. Despite these limitations, the price-gap approach is a valuable tool for estimating subsides and for undertaking comparative analysis of subsidy levels across countries to support policy development (Koplow, 2009).

For countries that import a given product, subsidy estimates derived through the price-gap approach are explicit. That is, they represent net expenditures resulting from the domestic sale of imported energy (purchased at world prices in hard currency), at lower, regulated prices. In contrast, for countries that export a given product — and therefore do not pay world prices — subsidy estimates are implicit and have no direct budgetary impact. Rather, they represent the opportunity cost of pricing domestic energy below market levels, i.e. the rent that could be recovered if consumers paid world prices. For countries that produce a portion of their consumption themselves and import the remainder (such as Iran), the estimates presented here represent a combination of opportunity costs and direct government expenditures.

Reference prices

For net importing countries, reference prices have been calculated based on the import parity price: the price of a product at the nearest international hub, adjusted for quality differences, plus the cost of freight and insurance to the importing country, plus the cost of internal distribution and marketing and any value-added tax (VAT). VAT was added to the reference price where the tax is levied on final energy sales, as a proxy for the tax on economic activities levied across the country in question. Other taxes, including excise duties, are not included in the reference price. Therefore, in the case of gasoline, even if the pre-tax pump price in a given country is set by the government below the reference price, there would be no net subsidy if an excise duty large enough to make up the difference is levied. For net exporting countries, reference prices were based on the export parity price: the price of a product at the nearest international hub adjusted for quality difference, minus the cost of freight and insurance back to the exporting country, plus the cost of internal distribution and marketing and any VAT.

---

5. Kosmo (1987), Larsen and Shah (1992) and Coady et al., (2010), for example, have used this approach.

For oil products, average distribution and marketing costs for all countries were based on costs in the United States ($0.08 per litre for gasoline and diesel). The assumed costs for shipping refined products, in contrast, vary according to the distance of the country from its nearest hub and have been taken from average costs as reported in industry data. For gas and coal, transportation and internal distribution costs have been estimated based on available shipping data. All calculations have been carried out using local prices and the results have been converted to dollars at market exchange rates.

Reference prices have been adjusted for quality differences, which affect the market value of a fuel. For example, for countries that rely heavily on relatively low-quality domestic coal but also import small volumes of higher quality coal, such as India and China, reference prices are set below observed import prices.

Unlike oil, gas and coal, electricity is not extensively traded over national borders, so there is no reliable international benchmark price. Therefore, electricity reference prices were based on annual average-cost pricing for electricity in each country (weighted according to output levels from each generating option). In other words, electricity reference prices were set to account for the cost of production, transmission and distribution, but no other costs, such as allowances for building new capacity, were included. They were determined using reference prices for fossil fuels and annual average fuel efficiencies for power generation. An allowance of $15 per megawatt-hour (MWh) and $40/MWh was added to account for transmission and distribution costs for industrial and residential uses, respectively. To avoid over-estimation, electricity reference prices were capped at the levelised cost of a combined-cycle gas turbine (CCGT) plant.

Some authorities regard the above method of determining reference prices as inappropriate. In particular, a number of energy resource-rich countries are of the opinion that the reference price in their markets should be based on their cost of production, rather than prices on international markets as applied within this analysis. The basis for this view typically is that these countries are using their natural resources in a way that effectively promotes their general economic development, and that this approach more than offsets the notional loss of value by selling the resource internally at a price below the international price. The counter-argument is that such an approach results in an economically inefficient allocation of resources and reduces economic growth in the longer term.

Cross-subsidies between sectors, *i.e.* where some consumers are charged a price above cost so as to offset lower prices for other consumers, have not been taken into account in this analysis. For example, in many countries commercial and industrial consumers often pay a price above cost so as to finance lower prices for the agriculture and residential sectors, while the opposite situation can also be found in other countries (for example, where aluminium producers are able to negotiate special low electricity rates). Furthermore, as the price-gap method measures an average variance in prices, it does not capture the variability in prices by time-of-day or region that are often vitally important in giving new technologies entry points into energy markets. Similarly, it does not pick up direct subsidies to consumers that are tied to fuel purchases, such as the discounted fuel coupons used by some developing countries or heating rebate schemes.

19

Box 19.2 ● *Sample calculation: estimating gasoline subsidies in Venezuela*

The first step is to calculate the appropriate reference price. Venezuela was a net exporter of gasoline in 2009 and therefore we start with the free-on-board (FOB) price, or the price of a product at the border. Taking the average spot price of gasoline in 2009 at the nearest hub, the United States, the fob price is calculated by subtracting the average cost of freight and insurance to transport gasoline between Venezuela and the United States. Given a spot price of 0.89 bolívares fuertes (VEF) ($0.41) per litre and a shipping cost of VEF 0.02 ($0.01) per litre, the FOB price is VEF 0.87 per litre. To complete the calculation of reference prices and arrive at the price consumers would see at their local pump, retail and distribution cost are added as well as any VAT. Assuming distribution and retail costs equal to those in the United States, VEF 0.17 ($0.08) per litre, the final reference price for gasoline in 2009 was VEF 1.04 ($0.48) per litre. No VAT is applied to gasoline sales in Venezuela.

As average end-use prices for gasoline in 2009 were reported as VEF 0.06 ($0.03) per litre, the price gap then amounts to VEF 0.98 per litre. To estimate the total value of the subsidy to gasoline, we take the price gap multiplied by total final consumption (estimated at 15.9 billion litres), arriving at a gasoline subsidy of approximately VEF 15.6 billion ($7.3 billion).

Subsidy estimates

The IEA estimates that the value of fossil-fuel consumption subsidies (including subsidies to electricity generated from fossil fuels) amounted to $312 billion in 2009 (Figure 19.2). This finding is based on the price-gap method outlined above and an extensive survey to identify those countries that subsidise fossil-fuel consumption. In total, 37 such countries were identified, estimated to represent over 95% of global subsidised fossil-fuel consumption. Remaining subsidised consumption occurs in countries where reliable data on energy consumption and prices are unavailable.[6]

The $312 billion estimate comprises subsidies to fossil fuels used in final consumption and to fossil-fuel inputs to electric power generation. In 2009, oil products and natural gas were the most heavily subsidised fuels, attracting subsidies totalling $126 billion and $85 billion, respectively. Subsidies to electricity consumption were also significant, reaching $95 billion in 2009. At only $6 billion, coal subsidies were comparatively small.

Almost all of the consumption subsidies identified were found in non-OECD countries (Figure 19.3). In absolute terms, the biggest subsidies are in those countries with the largest resource endowments. For a given fuel, net-exporting countries do not incur

---

6. All but two of the countries identified as subsidising energy consumption were outside the OECD; production subsidies rather than consumption subsidies are by far the most prevalent form of subsidisation in OECD countries.

Figure 19.2 ● Economic value of fossil-fuel consumption subsidies by type

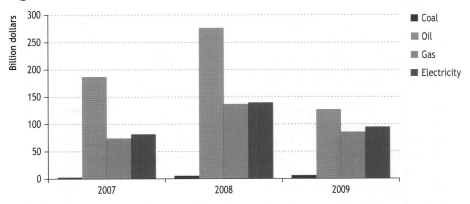

Figure 19.3 ● Economic value of fossil-fuel consumption subsidies by country and type, 2009

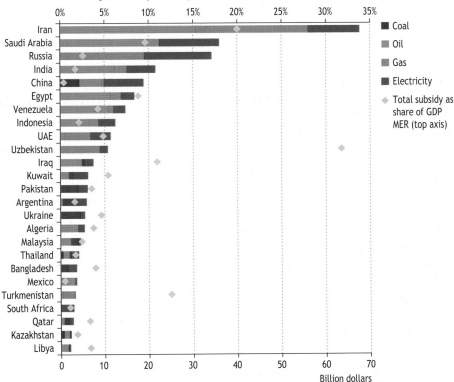

Note: MER = market exchange rates.

hard-currency expenditures by pricing domestic energy products below their value in international markets, as long as prices are set above the cost of production. Iran's subsidies reached $66 billion (the highest of any country), most of this sum going to oil products and natural gas. The next highest level of subsidisation identified was in Saudi Arabia, at $35 billion, followed by Russia, at $34 billion. Of the net-importing countries, India had the biggest subsidies in 2009, at $21 billion, followed by China with $19 billion. However, when viewed on a per-capita basis or as a percentage of GDP, subsidies in China and India were low compared with most other countries in the dataset.

Fossil-fuel subsidisation rates, expressed as a proportion of the full cost of supply, vary considerably by fuel and also by country. For the countries surveyed here, fossil fuels were subsidised at a weighted-average rate of 22%, meaning consumers paid roughly 78% of competitive market reference prices (Figure 19.4). Natural gas was the most highly subsidised fuel, at an average rate of 51% in 2009. Subsidisation rates for natural gas are comparatively high since many supplies are still priced within limited domestic markets, even as the global market for liquefied natural gas continues to grow. Oil products were subsidised at an average rate of 19%, electricity at 18% and coal at 7%.

The magnitude of energy subsides fluctuates from year-to-year with changes in world prices, domestic pricing policy, exchange rates and demand. Of these factors, movements in world prices typically have by far the greatest impact on variations in subsidy levels. In 2008, when fossil-fuel prices surged in international markets during the first half of the year, the value of energy consumption subsidies was estimated at $558 billion, a dramatic increase from 2007, when the total was $343 billion. Declining world prices were the main reason for the sharp drop in the value of subsidies between 2008 and 2009. However, some of the observed drop can also be attributed to deliberate interventions to raise consumer prices (thereby, shrinking the price-gap) in order to reduce the burden on government finances.

Some countries, including China and Mexico, manage price volatility by regulating domestic prices for certain energy products. Although the intent may not be to hold average prices over a period below market levels, rising international energy prices can inadvertently lead to market transfers to consumers (an effect picked-up by the price-gap approach). Conversely, when world prices fall, the situation can lead to unexpected revenues. For example, when oil prices were high in 2008, Mexico's fuel-excise mechanism resulted in estimated oil-product subsidies of $22 billion. However, these all but vanished with the fall in prices in 2009. Experience has shown that governments often find it hard to increase domestic prices when international prices are increasing and not to immediately pass through the full extent of any subsequent price falls. During the rapid run-up in world oil prices in early 2008, many countries abandoned automatic price adjustments in order to shield consumers, but they subsequently faced criticism for being slow to adjust downward after prices fell sharply later in the year.

In contrast to consumption subsidies, less work has been done to date to quantify subsidies to fossil-fuel production. These subsidies are often administered via indirect mechanisms, such as complex tax concessions, that make them difficult to identify and challenging to estimate. Nonetheless, the Global Subsidies Initiative (GSI), a Geneva-based programme of the International Institute for Sustainable Development (IISD),

Figure 19.4 ● Fossil-fuel consumption subsidy rates as a proportion of the full cost of supply, 2009

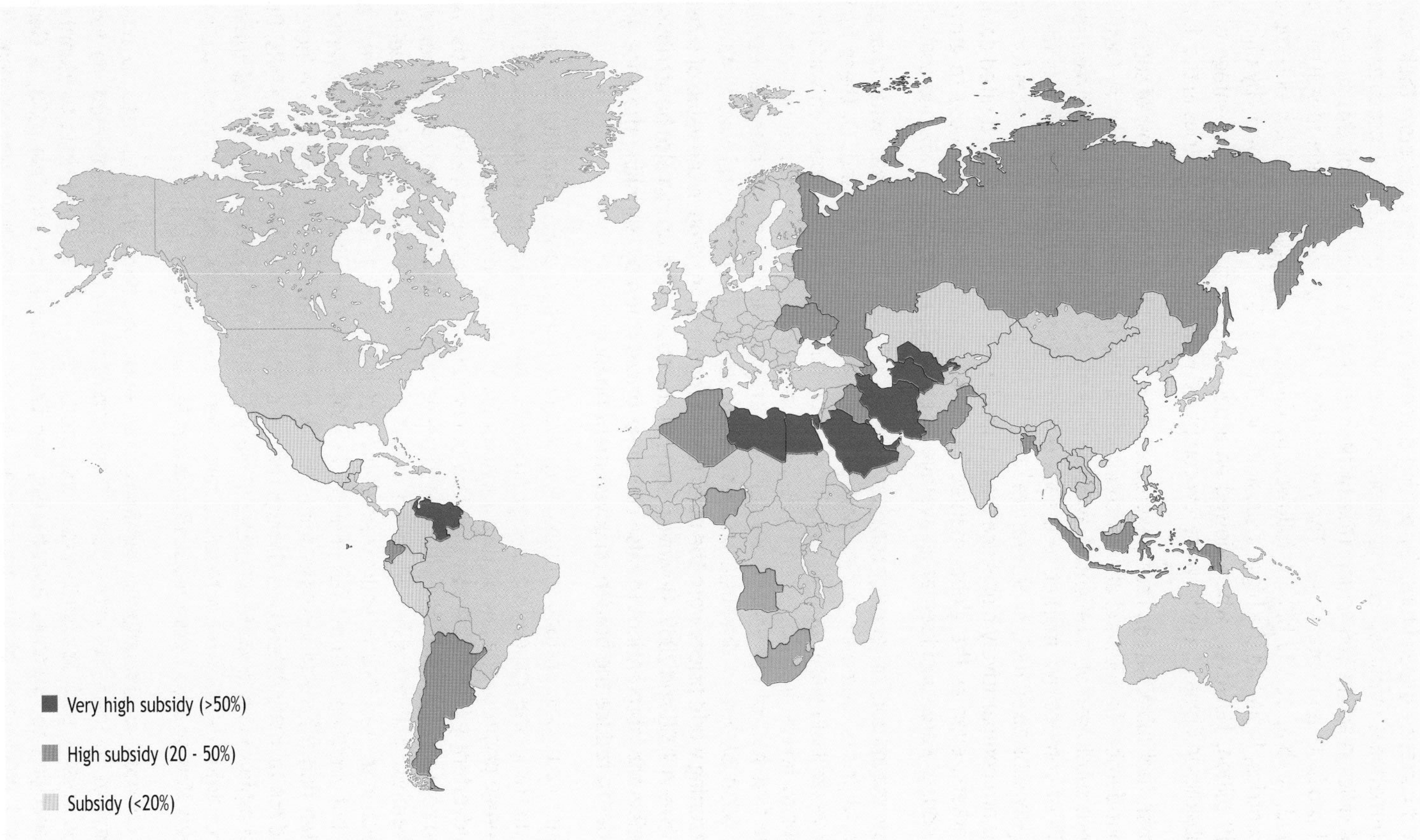

■ Very high subsidy (>50%)

▨ High subsidy (20 - 50%)

▨ Subsidy (<20%)

The boundaries and names shown and the designations used on maps included in this publication do not imply official endorsement or acceptance by the IEA.

has estimated that worldwide fossil-fuel production subsidies may be of the order of $100 billion per year (GSI, 2010). The GSI has also mapped information sources on producer subsidies in certain countries and identified methods for deriving estimates.[7]

As one means of attracting greater investment in renewable forms of energy, many countries around the world — especially those in the OECD, but also a number of emerging economies such as Brazil, China, India, Indonesia, Argentina and Philippines — are introducing support programmes. Such support, if well designed, can be beneficial for both energy security and the environment. Globally, we estimate that support for renewables totalled $57 billion in 2009 (see Chapter 9). As a basis for comparison, total support to renewables in 2009 was equivalent to only 18% of the value of fossil-fuel consumption subsidies in 2009. It is worth noting, however, that support for cleaner forms of energy are generally direct payments, whereas a high proportion of fossil-fuel subsidies are opportunity costs.

Implications of phasing out fossil-fuel consumption subsidies

Method and assumptions

This section quantifies the energy savings that would result from the phase-out of fossil-fuel consumption subsidies and the implications for CO_2 emissions. The comparison is with a baseline case in which subsidy rates from 2010 remain unchanged relative to their average level in 2007-2009.[8] Because subsidies tend to fluctuate as a result of market volatility, this provides a reasonable basis for estimating the impact of the subsidy phase-out, even though the magnitude of subsidies may rise or fall sharply in a given year. The analysis is based on the premise that subsidies to consumers lower the end-user prices of energy products and thus lead to higher levels of consumption than would occur in their absence. The unsubsidised, or reference, prices are calculated using the price-gap analysis described above.

To illustrate the magnitude of the gains possible by eliminating subsidies, the analysis assumes a gradual phase-out of all subsidies to fossil-fuel consumption, globally, over the period 2011-2020. The assumption of universal phase-out by 2020 is frankly optimistic, in view of the steep economic, political and social hurdles that would first need to be overcome. On the other hand, a growing number of countries have already announced plans that, if fully implemented, would eliminate or reduce their subsidies well before 2020 (see below: announced plans to phase out subsidies). Generally, it

---

7. Details are available at www.globalsubsidies.org/files/assets/mapping_ffs.pdf and www.globalsubsidies.org/files/assets/pb7_ffs_measuring.pdf.

8. Although beyond the scope of this analysis, social and equity impacts resulting from energy subsidy removal also need to be a central consideration in the design of any phase-out programme (see, for example, IEA/ OECD/OPEC/World Bank, 2010; Ellis, 2010).

can be expected that phase-out programmes will progress more rapidly in importing countries than in exporting countries, as direct fiscal costs are usually a more urgent spur to action than forgone revenue.

The price elasticity of demand — the percentage change in the quantity demanded in response to a percentage change in price — determines the extent to which consumption can be expected to fall in response to the removal of subsidies to each fuel and in each sector and country. Since the price elasticity of demand for energy is negative, increases in price lead to a reduction in consumption. Energy demand is more elastic in the long-run as people find additional ways to curb demand over time, *e.g.* as they did following the oil shocks of 1973 and 1979.

As the reduction in energy demand that would result from a phase-out programme would act to lower world prices for fossil fuels, the modelling takes account of the rebound effect (also known as the "take-back effect"), which refers to the increase in the demand for energy services (transport, heating, refrigeration, lighting, etc.) which occurs when the overall cost of the service declines. Consumers and businesses change their behaviour as a consequence of lower costs, *e.g.* raise thermostat levels in the winter; cool their buildings more in the summer; buy more appliances or operate them more frequently; or drive their vehicles more. These behavioural changes erode some of the energy savings that arise from higher prices or technical energy-efficiency improvements. The rebound effect is seen predominately in those countries without consumption subsidies, as they experience an overall decrease in end-use prices as subsidy removal leads to a fall in world prices for fossil fuels.

The inter-fuel substitution effects that would result from subsidy phase-out are also taken into account. For instance, phasing out subsides for natural gas in a certain market may lead to an increase in the use of coal. Energy-related carbon-dioxide (CO_2) savings linked to the abolition of subsidies are then determined, based on the CO_2 emission factor. This calculation is performed for each fuel and the results added to determine the total change.

Energy demand

Compared with a baseline case in which subsidy rates remain unchanged, the complete phase-out of consumption-related fossil-fuel subsidies between 2011 and 2020 would cut global primary energy demand by 5%, or 738 million tonnes of oil equivalent (Mtoe), by 2020 (Figure 19.5). This reduction is equivalent to the current energy consumption of Japan, Korea, and New Zealand combined. Furthermore, reductions in energy demand (relative to the baseline) would continue to be realised after 2020 as consumers continue to change their behaviour over time. For example, a power company burning oil to produce electricity may not have the choice of switching to a less costly alternative overnight, but could decide to build new, non-oil capacity if it expects higher input prices to persist as a permanent feature of the market. Similarly, a rise in the price of gasoline might encourage a motorist to buy a more fuel-efficient car, but only when the existing vehicle is traded or scrapped.

19

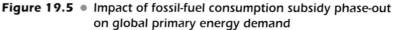

Figure 19.5 ● Impact of fossil-fuel consumption subsidy phase-out on global primary energy demand

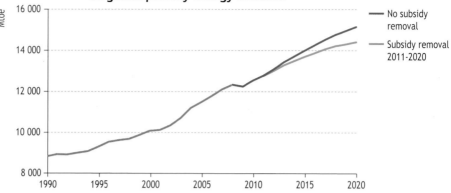

Phasing out energy subsidies would cut global oil demand by 4.7 million barrels per day (mb/d) by 2020, with savings predominately in the transport sector (Figure 19.6). Demand for transport fuels is relatively inelastic in the short term, but the prospect of higher prices over a longer period can influence consumer behaviour, resulting in a significant reduction in demand. Demand responses vary markedly by country, according to the magnitude of subsidies and types of oil products subsidised.

Figure 19.6 ● Oil savings resulting from consumption subsidy phase-out, 2020

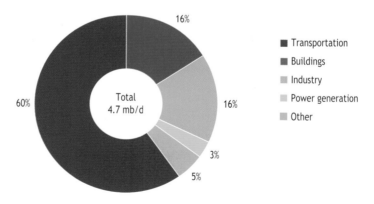

Where consumption is subsidised, eliminating energy subsidies would reduce dependence on imports and lead to an immediate improvement in the fiscal position of many governments. Moreover, exposing consumers to market-driven price signals would strengthen and accelerate the demand response, which in turn would contribute to reducing volatility in global markets. The phase-out of energy subsidies would have a number of other positive effects on long-term energy security by encouraging diversification of the energy mix and slowing down the depletion of finite fossil-fuel resources.

CO$_2$ emissions

The phase-out of fossil-fuel consumption subsidies over 2011-2020 would reduce global energy-related CO$_2$ emissions by 5.8% by 2020 compared with a baseline case in which subsidy rates remain unchanged (Figure 19.7). This amounts to savings of 2 gigatonnes (Gt) of CO$_2$ by 2020, equivalent to the current combined emissions of Germany, France, United Kingdom and Italy. Reduced demand growth for fossil fuels would also lead to lower emissions of particulate matter and other air pollutants.

Figure 19.7 ● Impact of fossil-fuel consumption subsidy phase-out on global energy-related CO$_2$ emissions

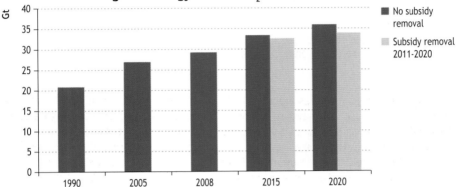

The analysis in this chapter, together with the climate analysis presented in Part C, illustrates the importance of the G-20 commitment to phase out inefficient fossil-fuel subsidies in tackling climate change and the role it could play in implementing the commitments under the Copenhagen Accord (Figure 19.8). The complete phase-out of fossil-fuel subsidies by 2020 would reduce CO$_2$ emissions by 1.5 Gt with respect to the Current Policies Scenario, which itself already assumes a certain degree of subsidy

Figure 19.8 ● Impact of fossil-fuel consumption subsidy phase-out on global energy-related CO$_2$ emissions compared with the Current Policies and 450 Scenarios

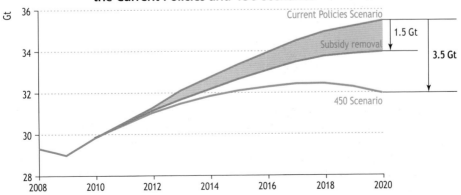

phase-out, based on currently announced plans. These savings amount to over 40% of the abatement required between the Current Policies Scenario and the 450 Scenario in 2020.

Do subsidies to energy production encourage wasteful consumption?

It is generally accepted that fossil-fuel consumption subsidies encourage wasteful consumption as the reduction of consumer prices below the full cost of supply leads to excessive use, which is detrimental to energy security and to the fight against climate change. But is there any empirical evidence to support the case that subsidies for production also lead to higher consumption to any significant degree?

The answer depends on how subsidies are administered and the extent to which lower production in a country that removes its subsidies is displaced by production elsewhere. As stated, the rationale for the introduction of production subsidies is typically to encourage domestic production of energy so as to support local industry and employment and improve energy security. One of the most prominent examples of production subsidy can be seen in Germany, which spent almost €1.8 billion in 2009 subsidising hard coal mining (these subsidies are due to be completely eliminated by 2018). But a large share of this total was allocated to cover the cost of closing down mines and compensating workers who had lost their jobs, so it was unlikely to have directly encouraged wasteful consumption of coal. However, in other cases, subsidies maintain production that would otherwise be uneconomic, for example, some countries provide subsidies that enable high-cost local coal producers to compete against imports. Many countries also offer subsidies for oil and gas production such as reduced royalties for leases in certain areas.

Removing fossil-fuel production subsidies would typically have the effect of making domestic production less competitive compared with imports and would, therefore, tend to lower indigenous production. The extent to which investment and production would be shifted to other parts of the world, and the extent to which prices would rise or fall as a result, would depend on the shape of the supply curve. In practice, the global effect might be small, if the volumes involved were small.

Nonetheless, there are other reasons that support a close review of the efficiency and effectiveness of fossil-fuel production subsidies. For example, if the removal of coal subsidies in an economy leads to greater imports of higher-quality coal, it would clearly benefit the environment (Steenblik and Coroyannakis, 1995). Furthermore, by propping up less efficient producers, production subsidies can create barriers to the introduction of cleaner technologies and fuels and discourage the uptake of more efficient production practices.

Subsidy reform could free up budgetary resources that could be better used elsewhere in the economy. For example, estimated fossil-fuel consumption subsidies in 2009 amounted to 45% of the additional yearly investment in low-carbon technologies and energy efficiency (around $700 billion on average) that is required to meet the 450 Scenario (see Chapter 13). However, a portion of the liberated funds would need to be directed towards the costs involved with subsidy removal, such as creating a comprehensive social welfare net.

Subsidies and energy poverty

Although the intent of many energy-consumption subsidies is to make energy services more affordable and accessible for the poor, studies have repeatedly shown them to be an inefficient and often ineffective means of doing so. The cost of these subsidies falls on the entire economy, but benefits are conditional upon the purchase of subsidised goods and thus tend to accrue disproportionately to middle and higher-income groups. Poor households may be unable to afford even subsidised energy or related services, or may have no physical access to them (for example, rural communities lacking a public transport network or a connection to an electricity grid). In general, subsidies for liquid fuels are particularly difficult to target, given the ease with which such fuels can be sold on the black market. In comparison, the distribution of electricity and piped natural gas is more easily monitored and controlled.

We estimate that subsidies in the residential sector to kerosene, LPG and electricity in countries with limited household access to modern energy (defined as countries with electrification rates of under 90% or modern fuels access under 75%) represented just 15% of the $312 billion of consumption subsidies in 2009 (Table 19.2). There is considerable evidence that most of these subsidies in any case go to richer households. The Co-ordinating Ministry of Economic Affairs of Indonesia, for example, reported that the top 40% of high-income families absorb 70% of energy subsidies, while the bottom 40% of low-income families reap only 15% of the benefits (IEA, 2008).

Although low-income households only benefit from a small proportion of energy subsidies they are still likely to be disproportionately affected by their removal, as they spend a higher percentage of their household income on energy. Similarly, subsidies can bring considerable benefits to the poor when they encourage switching to cleaner and more efficient fuels or enhance access to electricity (see Chapter 8). Therefore, any moves to phase-out subsidies must be carefully designed so as not to restrict access to essential energy services or increase poverty. Providing financial support for economic restructuring or poverty alleviation is essential to smoothing the path for fossil-fuel subsidy reform. In most successful cases of energy-subsidy reform, support has been well-targeted, temporary and transparent.[9] In undertaking major changes, assessments should be made regarding the extent to which the economy and society can absorb the impacts of the reform. Furthermore, the phase-out of fossil-fuel subsidies should be

19

---

9. See, for example, Laan et al., (2010), and IMF (2008).

considered as a package, particularly if broader structural reforms are underway or being contemplated. Pre-announcing a strategy and timeframe for phasing in subsidy reform can help households and businesses to adjust to these reforms (UNEP, 2008).

Table 19.2 ● Subsidies in the residential sector for electricity, kerosene and LPG in countries with low levels of modern energy access, 2009

	Energy poverty indicators		Presence of subsidies in residential sector			Electricity, LPG & kerosene subsidies as a share of total subsidies
	Electrification rate	Modern fuels access	Electricity	LPG	Kerosene	
Angola	26%	52%	yes	no	no	25%
Bangladesh	41%	10%	yes	yes	yes	27%
China	99%	68%	yes	yes	no	74%
India	66%	25%	yes	yes	yes	87%
Indonesia	65%	46%	yes	yes	yes	56%
Nigeria	51%	31%	yes	no	no	100%
Pakistan	62%	31%	yes	yes	no	35%
Philippines	90%	52%	no	yes	no	17%
South Africa	75%	74%	no	yes	no	4%
Sri Lanka	77%	20%	yes	no	no	63%
Thailand	99%	65%	yes	yes	no	61%
Vietnam	98%	41%	yes	yes	no	70%

Announced plans to phase out subsidies[10]

In recent years, many countries have implemented or proposed reforms to bring their domestic energy prices into line with the levels that would prevail in an undistorted market (Table 19.3). These efforts contributed to a small but important extent to the reduction in our estimates for energy subsidies in 2009 relative to 2008. The key drivers behind the moves have varied from country to country, as have expectations over the likelihood that lasting reform will take hold, in view of the political and social barriers that first need to be overcome.

In *Russia*, natural gas and electricity prices were not reformed during the earlier economic crisis of the 1990s and have remained government-regulated, leading to subsidies. Following the priorities set forth in *Russia's Energy Strategy to 2020*, substantial progress has now been made to introduce more market-based gas and electricity pricing, especially in the industrial sector. Domestic natural-gas tariffs for

---

10. See Chapter 20 for a detailed discussion on plans to reform subsidies in China, India, Indonesia, Iran and Russia.

industry have risen consistently since 2000, with plans for tariffs to converge with export prices (excluding transport costs and export duties) through 2014 based on the balancing of revenues from domestic and export sales. In the electricity sector, prices in the wholesale market are scheduled to be fully liberalised in 2011. Electricity tariffs to the residential sector are planned to be gradually raised via three-year vesting contracts, which set out a pre-defined schedule of price increases, starting in January 2011.

Table 19.3 ● Plans to reform energy subsidies in selected countries

	Description of announced plans
China	Oil product prices were indexed to a weighted basket of international crude prices in 2008. Natural gas prices rose by 25% in May 2010. Plans exist to remove preferential power tariffs for energy-intensive industries and to extend tiered pricing for households.
Egypt	Plans to eliminate energy subsidies to all industries by the end of 2011.
India	Abolished gasoline price regulation in June 2010 and plans to do the same for diesel. The price of natural gas paid to producers under the regulated price regime was increased by 230% in May 2010.
Indonesia	Set goal to reduce spending on energy subsidies 40% by 2013 and fully eliminate fuel subsidies by 2014.
Iran	Plans to replace subsidised energy pricing with targeted assistance to low-income groups over the period 2010-2015. Reforms call for the prices of oil products, natural gas and electricity to rise to market-based levels.
Malaysia	Announced reductions in subsidies for petrol, diesel and LPG as the first step in a gradual subsidy-reform programme.
Mexico	Intends to phase out subsidies to gasoline and diesel by the end of 2010.
Russia	Natural gas prices for industrial users are to continue increasing toward international levels through 2014 based on the balancing of revenues from domestic and export sales. Pricing in the wholesale electricity market is scheduled to be fully liberalised in 2011.
Saudi Arabia	The Electricity and Co-Generation Regulatory Authority (ECRA) plans minor electricity tariff increases for industrial and commercial users.
South Africa	Plans to increase electricity tariffs by approximately 25% per year over 2010-2013.

In *South Africa*, subsidised electricity pricing, coupled with non-payment by customers and an inability of utilities to enforce property rights, has led to a lack of investment and a shortage of electricity capacity. Rolling blackouts have provided strong impetus for recent price increases and plans to further raise tariffs in coming years. In 2010, the National Energy Regulator of South Africa (NERSA) granted Eskom, the state utility, permission to raise average rates by approximately 25% per year over 2010-2013. Through cross-subsidies, it will maintain its Free Basic Electricity programme, which provides targeted subsidies to the poor through a minimum amount of free electricity for essential services.

In *Iran*, subsidised under-pricing of domestic energy has strained the economy, forced reliance on refined product imports and led to widespread energy inefficiency. To reduce the fiscal burden of subsidies and lessen exposure to international economic

19

sanctions, Iran enacted a subsidy reform law in 2010. The subsidy reform law would increase prices to market-based levels over 2010-2015 and use the savings to replace price subsidies with targeted assistance to low-income groups. Current plans call for oil-product prices to rise to at least 90% of the average Persian Gulf export price; for natural gas prices to be raised to 75% of the export price; and for average domestic electricity prices to be determined according to the full cost of production.

Currently, *China* is active in implementing energy price and tax reforms to optimise its energy consumption and reduce energy intensity. In 2008, the National Development and Reform Commission (NDRC) introduced a new pricing regime for oil products that pegs domestic prices to a weighted average of a basket of international crude-oil prices. In the first half of 2010, the NDRC announced changes to the mechanism for pricing natural gas and increased the transmission tariff, effectively bringing natural gas prices closer to import-parity levels. The NDRC implemented electricity retail price increases for non-residential users in 2009, and in mid-2010 it abolished preferential electricity tariffs for certain energy-intensive industries and announced that progressive tariffs for electricity would be implemented nationwide for households, *i.e.* tariffs that increase in line with consumption levels. Since 2009, China has experimented with pilot schemes that allow direct power sales from large generators to end-users.

Although earnings from energy taxes in *India*, which go predominately to the state governments, far outweigh the cost of subsidies, which is borne by the central government, the country is in the process of energy price and tax reform (Government of India, 2010). In June 2010, the federal government announced that gasoline prices would henceforth be market-driven and the intention to later apply market-driven pricing for diesel. It also announced immediate price increases for diesel, LPG and kerosene. Natural gas pricing reform was also implemented in mid-2010, allowing state-run Oil & Natural Gas Corp. (ONGC) and Oil India Ltd. (OIL) to sell gas from new fields at market rates instead of regulated prices. Furthermore, the price of natural gas more than doubled under the regulated price regime in 2010.

In 2010, *Indonesia* announced plans to eliminate energy subsidies by 2014. According to the May 2010 revised state budget, 12.8% of government expenditure in 2010 will be devoted to energy-consumption subsidies, compared with 32% in 2008. The 2010 state budget allows the government to raise domestic fuel prices if crude oil prices rise more than 10% above the budgeted level of $65 per barrel. In June 2010, power tariffs were raised by an average of 10% in a bid to reduce the fiscal burden on the state budget and boost revenues for Indonesia's state power company. Indonesia has an ongoing programme to phase out the use of kerosene in favour of LPG. The energy ministry is considering a new plan to restrict the use of subsidised fuel to public transportation vehicles and cars purchased before 2005.

Despite *Mexico* being the world's seventh-largest crude-oil producer, subsidised energy prices have represented a serious economic strain on the government budget and contributed to increasing reliance on refined product imports. Mexico is currently reforming its excise arrangements for refined products with the intention of eliminating gasoline and diesel subsidies by late 2010 and those for LPG by late 2012. As part of the process, retail prices for gasoline, diesel and LPG – set by the government – have been increasing on a monthly basis since December 2009.

Box 19.3 ● **The IEA energy-subsidy online database**

As highlighted by the G-20, increasing the availability and transparency of energy subsidy data is an essential step in building momentum for global fossil-fuel subsidy reform. Improved access to data on fossil-fuel subsidies will raise awareness about their magnitude and incidence and encourage informed debate on whether the subsidy represents an economically efficient allocation of resources or whether it would be possible to achieve the same objectives by alternative means. Transparency of subsidy data can also encourage consistent presentation and provide a useful baseline from which progress to phase out subsides can be monitored (Hale, 2008; Laan, 2010).

As a contribution to the process of increasing transparency of energy-subsidy data, the IEA has established an online database which allows public access to data on fossil-fuel subsidies, including breakdowns by country, by fuel and by year. This new database represents an extension of the systematic analysis of energy subsidies that the IEA has been undertaking through the *World Energy Outlook* series since 1999. It will be updated annually as a means of tracking the progress being made by countries to phase-out fossil fuel subsidies. The database is available at www.worldenergyoutlook.org/subsidies.asp.

The database has been constructed following an extensive survey of end-use price data. A key source of data was the IEA's quarterly publication, *Energy Prices and Taxes*. Other sources include official statistics, international and national energy companies, consulting firms and investment banks' research reports. The IEA's network of energy and country experts and their local energy contacts have also contributed substantially to the identification and verification of end-user prices. Additional data were extracted from databases, reports and personal communications with various organisations, including the Asian Development Bank, OPEC, IMF, Latin American Energy Organization (OLADE) and the European Bank for Reconstruction and Development.

19

CHAPTER 20

COUNTRY SUBSIDY PROFILES
Iran, Russia, China, India and Indonesia

H I G H L I G H T S

- Iran, Russia, China, India and Indonesia each have a long history of subsidising the consumption of fossil fuels. Today their subsidies remain sizeable, collectively amounting to $152 billion in 2009. All five countries have plans to gradually introduce market-based pricing. To date, progress varies significantly from country to country. If lasting reforms are achieved, they will have a noticeable effect on supply and demand balances in each country's domestic market, as well as important implications for global energy and emissions trends.

- The enormous subsidies in Iran continue to burden the economy and contribute to deep inefficiencies in the energy sector. In 2010, a far-reaching subsidy reform law was enacted. It seeks to introduce market-based pricing of oil products, gas and electricity over 2010-2015, and replace subsidies by targeted assistance to low-income groups. We estimate that Iran had the highest bill for fossil-fuel consumption subsidies of any country in 2009, at $66 billion.

- In Russia, controls over gas and electricity tariffs continue to result in subsidies that contribute to wasteful energy use, hinder competitiveness and limit investment in the energy sector. But gas and electricity prices have risen steadily in recent years, reducing subsidies, and further moves toward more market-based pricing are planned. We estimate that fossil-fuel consumption subsidies in Russia totalled $34 billion in 2009.

- China has made significant progress over recent decades in bringing its domestic energy prices closer to global market levels, and is continuing to push ahead with reforms. Today, however, some energy prices are still set or guided by the central government in pursuit of various socio-economic goals. We estimate that fossil-fuel consumption subsidies in China in 2009 amounted to $19 billion.

- India has been actively reforming its energy-pricing policy to reduce the fiscal burden on the state budget. In 2010, the government implemented a landmark natural gas pricing reform and made major changes to pricing arrangements for refined oil products, with a focus on those used disproportionately by wealthier consumers. We estimate that India's fossil-fuel consumption subsidies totalled $21 billion in 2009.

- Indonesia has set a goal of a 40% reduction in spending on energy subsidies by 2013 and eliminating them entirely by 2014. To lessen the adverse impact of these reforms on the poor, the government plans to increase targeted assistance to low-income groups. In 2009, we estimate that Indonesia's fossil-fuel consumption subsidies were $12 billion.

Iran

Energy sector overview

With vast reserves, Iran is one of the world's largest oil and natural gas producers. Iran has a population of 72 million and an expanding economy, which grew at an estimated rate of 5.5% per year between 2000 and 2008 (Table 20.1). Energy demand increased at 6.8% annually during that period. Iran's primary energy mix is dominated by oil and gas, which comprised 44% and 54%, respectively, of the total in 2008. Together, these fuels also account for about 90% of electricity generation. Natural gas is favoured for domestic use in order to free up oil for export, leading to a doubling in gas demand since 2000 and Iran becoming the third-largest gas consumer in the world. Heavily subsidised energy consumption has left a legacy of inefficient energy use, environmental degradation, inadequate investment and fuel import dependence. Energy intensity has risen since 2000 and, in 2008, stood 40% above the global average.

Table 20.1 ● **Key economic and energy indicators for Iran**

	Unit	1990	2008	1990-2008*
Population	million	54	72	1.6%
GDP (PPP) per capita	$ (2009)	6 767	11 299	2.9%
Energy demand	Mtoe	68	202	6.2%
Energy demand per capita	toe	1.26	2.81	4.6%
Energy intensity	toe per $1 000 GDP (PPP, 2009)	0.19	0.25	1.6%
Oil net exports	mb/d	2.3	2.6	0.8%
Natural gas net imports**	bcm	-2.0	2.4	n.a.
Electricity consumption	TWh	49	164	6.9%

*Compound average annual growth rate. **Negative values indicate net exports.
Note: Mtoe = million tonnes of oil equivalent. PPP = purchasing power parity. TWh = terawatt-hours.

Oil and gas activities play a central role in supporting Iran's economy, generating about 80% of its export revenues in 2008. At the end of 2009, proven oil reserves totalled 138 billion barrels, ranking third in the world, with production averaging 4.3 million barrels per day (mb/d). Iran held the second-largest gas reserves, at 30 trillion cubic metres (tcm) (O&GJ, 2009), with production standing at 144 billion cubic metres (bcm) in 2009. Despite this rich endowment of hydrocarbon resources and its place as a major oil and gas producer, Iran imports about one-third of the gasoline and 10% of the diesel it consumes; it is also a net importer of a small amount of natural gas.

Iran's lack of self-sufficiency in certain energy products results from subsidised energy consumption and the slow rate of energy sector development and modernisation. Substantial investment capital and new technology are required to stem or reverse the decline of major oil fields, expand refining capacity, modernise ageing electricity generation and distribution infrastructure, and develop new upstream oil and gas projects. With limited domestic capacity to undertake these projects, Iran must seek to attract investment and technology from outside its borders. This has proven challenging, however, given state control over the energy sector and international sanctions in response to Iran's nuclear programme.

Energy pricing and subsidy policy

The chronic under-pricing of domestic energy in Iran represents a large subsidy that burdens the economy and contributes to deep inefficiencies in the energy sector. As part of its long-standing welfare policy, Iran has regulated prices across nearly all fuels – including oil products, natural gas and electricity – and sectors at well below the full economic cost of supply. These energy subsidies cause myriad problems: they lead to excessive and inefficient energy use, boost fuel imports, discourage much-needed investment in the energy sector, incentivise fuel smuggling and increase pollution. Iran's fuel import dependence and energy sector investment shortfalls are further compounded by international economic sanctions in response to its nuclear programme. Moreover, the current system of energy subsidies is highly inequitable, with the country's richest 30% of the population estimated by government officials to receive 70% of the subsidies (Harris, 2010).

Recognising these problems and the unsustainable nature of such high levels of energy subsidies, Iran has made several attempts to reduce energy subsidies or target them better. Iran's Third Five-Year Development Plan (2000-2005) called for more market-based pricing of energy products; but price hikes were subsequently scaled back after strong public opposition.

In more recent years, international economic sanctions targeting fuel imports have created additional impetus for Iran to rein in consumption. This led to implementation of the current gasoline rationing system. Launched in 2007, this system initially limited drivers of private vehicles to 120 litres per month of gasoline at the highly subsidised price of $0.10 per litre, with additional gasoline available for purchase at higher prices (approximately $0.40 per litre). Rations at the subsidised price have since decreased to 100 litres per month (in 2008), 80 litres per month (in 2009) and 60 litres per month (in 2010). Thus, some percentage of consumed gasoline is bought at the most highly subsidised price, while the rest is paid for at prices closer to international levels. Iranian officials believe that the system has avoided additional imports and resulted in savings estimated at $11 billion (Gonn, 2010). The gasoline rationing system is, however, planned to come to an end with the introduction of new subsidy reforms in the second half of 2010.

The Fifth Five-Year Development Plan (2010-2015) called for ambitious economic reforms and once more focused on overhauling energy subsidies. In early 2010, after more than a year of debate in the Iranian Parliament (*Majlis*), culminating in endorsement from the Guardian Council, a law outlining far-reaching subsidy reform was enacted. The subsidy reform law calls for gradual implementation of market-based energy pricing and the replacement of subsidies by targeted assistance to lower income groups. The following are Iran's key objectives for domestic energy pricing under the reform law between 2010 and 2015:

20

- The prices for gasoline, diesel, kerosene, liquefied petroleum gas (LPG) and other oil derivatives are to rise to at least 90% of the Persian Gulf export free-on board (FOB) price.

- Gas tariffs for households are to rise to 75% of the Persian Gulf export price. Industrial gas consumers may receive preferential rates.

- Average electricity prices are to be determined according to the full cost of production.

To compensate for higher prices and the impact on low-income groups, 50% of the fiscal benefit resulting from increased prices would be redistributed to low-income consumers via direct cash and non-cash payments. Another 30% would be allocated for raising energy efficiency in key sectors of the economy and improving public transportation. The remaining 20% would be used to offset government expenses associated with higher energy costs or as an additional safety net. Funds generated by the reform are to be distributed by a newly formed organisation called the Subsidy Reform Organisation, which will be overseen by a multi-ministerial board. The level of the government's year-to-year fiscal benefit from higher prices is to be authorised by the Iranian Parliament through the annual budget process (Amuzegar, 2010). The target to be generated in the first year is $20 billion, with implementation of the reforms planned to start in the second half of the 2010/11 Iranian calendar year (which began in March 2010).

Even though the Iranian government is now legally obliged to complete energy subsidy reforms, the path to full implementation is still unclear. Among points lacking clarity (as of September 2010) are: the level and frequency of price adjustments for fuels each year; the definition of those eligible for compensatory payments; and the amount and duration of those payments. How to minimise inflation, which already stood above 10% in 2009, while increasing prices for energy products is a major concern, as is the overall effect of the reforms on the competitiveness of domestic industry, already at a disadvantage due to the ongoing sanctions (Harris, 2010). The redistribution of revenues generated and administration of the Subsidy Reform Organisation will be key to implementation; however, the revenue stream from the sale of energy products at market-based prices will depend on consumer reaction to higher prices, which is difficult to predict. Moreover, as with previous reforms, social unrest could test the strength of political will behind the subsidy reform law.

Should the process to implement subsidy reforms falter, energy in Iran will remain significantly under-priced. With consumers paying only a fraction of international prices for oil products, as low as $0.10 per litre of gasoline and $0.02 per litre of diesel, refining capacity is insufficient to meet demand and Iran imports about one-third of its gasoline and 10% of its diesel consumption. The bill incurred for purchasing these fuels at world prices is substantial, and the costs are not recovered when diesel and gasoline are under-priced in the domestic market (Figure 20.1). Additional subsidies for gasoline and diesel exist as opportunity costs, when domestically produced supplies are sold below the price they could command in international markets (where they would also earn hard currency).

Without price reforms, gas too will remain highly subsidised. Low-income households and industries in need of financial support benefit from cheap gas prices. Since the majority of gas subsidies in Iran result from artificially low pricing of domestically produced gas, they are largely implicit and are not identified as expenditures in the central budget. They nonetheless represent an opportunity cost through forgone capture of rents. Even without the constraints of rising domestic gas demand, Iran would not currently be able to capitalise on revenues from gas exports since it lacks the export capacity, via pipeline or liquefied natural gas (LNG), necessary to ship large quantities of gas to international markets.

Figure 20.1 ● **Estimated gasoline and diesel import bill of Iran**

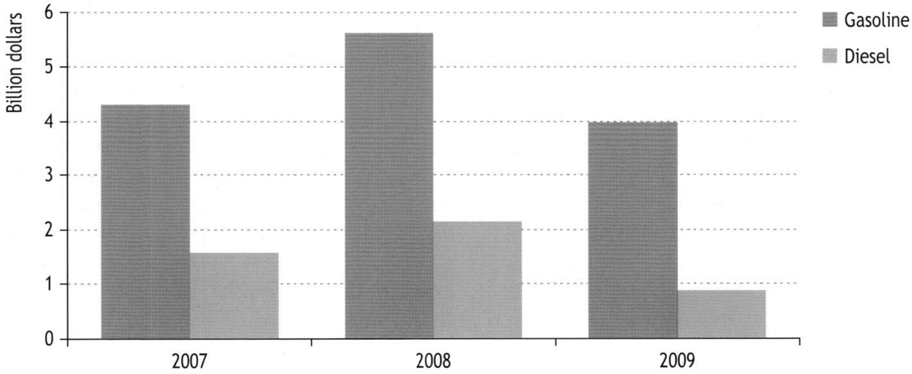

Note: Import bill is estimated using net imports of gasoline and diesel, and reference prices for each fuel.
Source: IEA analysis.

Electricity tariffs to consumers in Iran are also under-priced, failing to recover even full operating costs. It follows that prices do not provide for the investment costs needed to modernise and expand Iran's ageing electricity generation and grid infrastructure. Thus, Iran's subsidised electricity prices limit the resources available for investment, since no profit is made to reinvest, while also encouraging wasteful consumption and thereby increasing the need for investment in new power plants.

Subsidy estimates

Based on the methodology outlined in Chapter 19, we estimate that the economic value of fossil-fuel consumption subsidies in Iran was some $66 billion in 2009 (Table 20.2), the highest amount of any country in the world. Since only limited reforms were then underway, the decline in the estimate from 2008 ($98 billion) to 2009 was almost entirely the result of falling prices for fuels on world markets.

On a per-capita basis and as a share of GDP, subsidies were $895 and 20%, respectively, in 2009. With energy subsidised at an average rate of 89%, consumers paid only 11% of the competitive market price for energy products. Oil products, mostly transport fuels, accounted for almost half of the country's subsidies in 2009, at a cost of about $30 billion. Much of the subsidised gasoline and diesel had to be imported and paid for

20

with hard currency. Consumption subsidies for natural gas were almost $25 billion in 2009 (mostly in the form of revenue forgone). Electricity subsidies in Iran are estimated to have exceeded $11 billion.

Table 20.2 ● Fossil-fuel consumption subsidies in Iran

		Unit	2007	2008	2009
Total	Subsidies	$ billion	64.6	97.7	66.4
	per capita	*$*	*909*	*1 358*	*895*
	as a share of GDP (MER)	*%*	*22.6*	*29.3*	*20.1*
	Rate of subsidisation	%	86	90	89
By fuel	Oil products	$ billion	36.6	52.0	30.1
	rate of subsidisation	*%*	*86*	*90*	*88*
	Natural gas	$ billion	18.8	30.5	24.8
	rate of subsidisation	*%*	*93*	*94*	*95*
	Electricity	$ billion	9.2	15.2	11.4
	rate of subsidisation	*%*	*77*	*84*	*82*

Russia

Energy sector overview

Given Russia's role as a major producer, exporter and consumer of fossil fuels, its energy sector is of major economic and political importance. After falling steeply in the 1990s, its demand grew at an average rate of 1.3% per year from 2000, making it the third-largest consumer of primary energy in the world in 2008. The principal component of the primary energy mix in Russia is natural gas (53%), spurred on by a country-wide gasification policy to support development and utilise an abundant resource. Oil (21%) and coal (17%) account for most of the remainder. Domestic heat and electricity production account for around 60% of Russian gas consumption. Widespread inefficiencies have long plagued Russia's energy sector; although energy intensity has been declining since the late 1990s (Table 20.3), it was still 72% above the global average in 2008 (and around 50% higher than the energy intensity of Canada).

Table 20.3 ● Key economic and energy indicators for Russia

	Unit	1990	2008	1990-2008*
Population	million	148	142	-0.2%
GDP (PPP) per capita	$ (2009)	13 937	16 155	0.8%
Energy demand	Mtoe	880	688	-1.4%
Energy demand per capita	toe	5.95	4.85	-1.1%
Energy intensity	toe per $1 000 GDP (PPP, 2009)	0.43	0.30	-1.9%
Oil net exports	mb/d	5.3	7.3	1.7%
Natural gas net exports	bcm	185.0	208.6	0.7%
Coal net exports	Mtce	6.2	64.8	13.9%
Electricity consumption	TWh	1 074	1 021	-0.3%

*Compound average annual growth rate.

With 48 tcm of gas reserves and 60 billion barrels of oil reserves at the end of 2009, Russia controls a large share of global hydrocarbon resources (O&GJ, 2009). With production of 589 bcm in 2009, Russia was the world's second-largest gas producer after the United States. As a key exporter of gas, it accounts for more than one-quarter of European gas supply. Due to the curtailing of oil output in Saudi Arabia, in 2009 Russia became the world's leading oil producer, at 10.2 mb/d. Many of its largest fields are in decline, however, and sustained development of Russia's oil and gas resources depends on very large-scale investment. Conditions for investment have fluctuated over the past decade, becoming more onerous for external investors in recent years. Continued state involvement in the energy sector, including the frequent renegotiation of project terms to favour Rosneft and Gazprom (the two large national oil and gas companies), has hindered competitiveness and further reduced the flow of foreign capital. Overcoming the challenge of opening east Siberian oil and gas fields will require less burdensome and more stable fiscal terms.

Energy pricing and subsidy policy

Domestic prices for oil products and coal in Russia are market-driven. Regulated under-pricing of natural gas and electricity, however, persists from the Soviet era and was not reformed during the economic crisis of the 1990s when Russian consumers, both industrial and residential, were in no position to pay more for energy. Price controls result in a high level of under-pricing to some sectors, particularly households, which have continued to be cross-subsidised by industry. Although natural gas and electricity prices still fall below the full economic cost of supply, the government has made substantial progress in introducing more market-based gas and electricity pricing, especially in the industrial sector.

The under-pricing of domestic gas and electricity has fostered the systemic problems that the Russian government is trying to address. Subsidies have bred inefficiency throughout the Russian economy, which ranks as one of the least efficient in the world in terms of GDP per unit of energy consumed. Furthermore, artificially low prices have resulted in revenue losses for natural gas producers and electricity generators, leading to low availability of capital for investment and reducing the confidence of investors in the viability of new investment.

The *Russian Energy Strategy to 2020,* approved by the federal government in 2003, gave priority to improving energy sector competitiveness in domestic and export markets. One important feature was a commitment to the gradual liberalisation of domestic prices for energy. In line with the strategy, Russia has steadily increased natural gas tariffs for industry, with a view toward convergence with export prices by 2014 (Gazprom, 2010). Moreover, electricity market reform is now well advanced, with liberalisation of wholesale electricity market scheduled for completion in 2011. These target dates have been subject to revision with changes in economic circumstances or other factors.

The Federal Tariff Service (FTS) sets wholesale tariffs for natural gas destined for industrial and power sector use; tariffs to residential and municipal customers are established on a local basis by regional energy commissions for 60 price zones. Gazprom is required by law to supply pre-negotiated volumes of gas to customers at regulated

20

prices, regardless of profitability; additional gas can be purchased from Gazprom or independent producers at higher prices. These subsidised prices entail an opportunity cost, borne by Gazprom and supported by its export earnings.

Gas tariffs for Russian industry (in rouble terms) have been consistently increased since 2000, by approximately 15% to 25% each year. In 2007, the government adopted the goal of achieving equal profitability from sales to domestic and export markets by 2011 (Russian Federation Decree #333, 2007), signalling that industrial tariffs would continue to converge with average netback prices for exported gas (excluding export duties and transportation costs). The target date for full parity was extended to 2014, following the surge in oil prices during 2008 and the subsequent economic downturn. Gradual price hikes nonetheless have narrowed the gap between domestic prices and prices in European markets (Figure 20.2). Recent market developments, such as the shale gas boom and increasing LNG availability, have applied downward pressure on long-term contracts and European spot prices, and are contributing toward this convergence. Another notable result of the Russian government's resolve to continue the increase in domestic gas prices was that Gazprom recorded its first-ever profit from domestic sales in 2009.

Figure 20.2 ● **Natural gas prices for industry in Russia compared with average European netbacks**

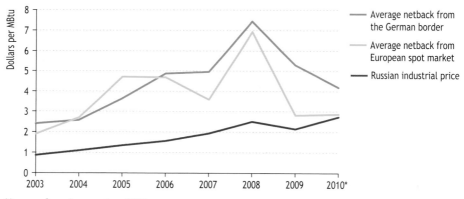

*Average from January-June 2010.
Note: Netback is the price at the Russian border, which excludes export duties and transportation costs.
Source: IEA databases and analysis.

The Russian government has also taken steps to liberalise the electricity sector in order to create more competition and attract needed investment. Restructuring has occurred in two major stages. First, the state-controlled utility, RAO UES of Russia, was split into separate generation (privatised) and transmission and distribution (state-owned) entities, a process completed in 2008. Following this reorganisation, several foreign utilities entered the electricity sector. Second, Russia is restructuring its electricity markets to achieve more cost-reflective pricing. This process is partially complete. Electricity market restructuring began with the establishment in 2006 of a two-tiered wholesale market. Most electricity was initially bought and sold under government-regulated bilateral contracts, set by the FTS; the rest was freely traded. The plan was to increase the share of electricity traded in the free segment by 5% to 15% every six months, with a corresponding reduction in the volume of electricity available at controlled rates (Russian Federation Resolution #529, 2006). Reflecting this wholesale market reform, average industrial electricity prices have risen more than 50% between 2006, when the process began, and 2009. Full liberalisation of the wholesale market is currently scheduled for 2011.

The pricing of electricity for residential consumers remains government-controlled, rapid price increases to households in the past having given rise to social and political difficulties. Non-payment for electricity and heating was widespread during and after Russia's economic crisis in the early 1990s. Regional energy commissions establish tariffs for regulated electricity sold in the retail market, within bounds set by the FTS. For prices of unregulated electricity purchased by retail suppliers from the wholesale market, the FTS imposes limits determined by the average cost of electricity in the wholesale market during the previous month, a regulated tariff for transmission and infrastructure services, and a profit margin. Starting in 2011 and lasting up to three years, electricity prices for the residential sector will be set by contracts at fixed prices to suppliers and consumers (Russian Federation Law #36-FZ, 2003). Those prices will continue to rise through the contract period, gradually scaling back price subsidies.

Subsidy estimates

Based on the price-gap approach, we estimate that the cost of natural gas and electricity consumption subsidies in Russia in 2009 was almost $34 billion (Table 20.4). The reduction in the subsidy bill between 2008 and 2009 was the combined result of the fall in fossil-fuel prices in international markets and the rise of domestic tariffs for gas and electricity, stemming from efforts to bring about more cost-reflective prices.

In 2009, subsidies were $238 per person and 2.7% of GDP. Fossil-fuel consumption in Russia was subsidised at an average rate of 23%, meaning that consumers paid 77% of the full economic cost for energy products. Subsidies for natural gas were the highest, estimated at almost $19 billion in 2009. Although prices in electricity markets have been gradually moving towards market levels, the under-pricing of electricity still resulted in large subsidies in 2009, estimated at just under $15 billion.

Table 20.4 ● Fossil-fuel consumption subsidies in Russia

		Unit	2007	2008	2009
Total	Subsidies	$ billion	33.3	53.8	33.6
	per capita	$	235	380	238
	as a share of GDP (MER)	%	2.6	3.2	2.7
	Rate of subsidisation	%	23	25	23
By fuel	Natural gas	$ billion	18.4	30.7	18.7
	rate of subsidisation	%	50	54	50
	Electricity	$ billion	14.9	23.2	14.9
	rate of subsidisation	%	29	32	27

20

China

Energy sector overview

In less than a generation, China, which was a largely self-sufficient energy consumer, has become the world's fastest-growing energy consumer (and importer) and a major player in global energy markets. The country's primary energy demand stood at 2 131 Mtoe in 2008, or equivalent to 93% of the energy demand of the United States in that year (Table 20.5). Despite the global financial and economic crisis, China's economy has remained resilient, growing at 9.1% in 2009 (NBS, 2010). Preliminary data suggest that China overtook the United States in 2009 to become the world's largest energy user. Coal plays a very prominent role in China's energy mix, accounting for 66% of total primary energy consumption in 2008, though the demand for other fuels is also growing rapidly. China's energy intensity, measured as the ratio of energy demand to GDP, is about 1.5 times the global average but has declined substantially over the last several decades as a result of efficiency improvements and structural change, particularly the growth of the service sector.

Oil accounts for 17% of total primary energy demand in China. At 8.1 mb/d in 2009, Chinese oil demand is close to 50% of oil demand in the United States. China, which was a net oil exporter in the early 1990s, has become the world's second-largest oil importer, with net imports of 4.3 mb/d in 2009. The country's proven oil reserves amount to 20.4 billion barrels, enough to sustain current production levels for 15 years (O&GJ, 2009). China's gas demand has been growing at over 10% per year for the past few years and the country started to import LNG in 2006. Current gas consumption accounts for 3.3% of total primary energy demand, a share set to increase significantly in the next decade and beyond. China's proven gas reserves are around 3 tcm (O&GJ, 2009), or enough to sustain current production levels for 35 years.

Table 20.5 ● **Key economic and energy indicators for China**

	Unit	1990	2008	1990-2008*
Population	million	1 141	1 333	0.9%
GDP (PPP) per capita	$ (2009)	1 330	6 282	9.0%
Energy demand	Mtoe	872	2 131	5.1%
Energy demand per capita	toe	0.76	1.60	4.2%
Energy intensity	toe per $1 000 GDP (PPP, 2009)	0.57	0.25	-4.4%
Oil net imports**	mb/d	-0.4	3.9	n.a
Natural gas net imports	bcm	0	4.5	n.a
Coal net exports	Mtce	7.9	13.1	2.8%
Electricity consumption	TWh	650	3 490	9.8%

*Compound average annual growth rate. **Negative values indicate net exports.

China is both the largest consumer and producer of coal in the world and, until recently, was an important net exporter. In 2009, it was a net importer of coal, as domestic supply struggled to keep up with high demand growth and logistical constraints meant that imports were often the cheaper option. However, coal net imports totalled only 1% of demand (nonetheless, volumetrically large by the standards of global steam coal trade). Chinese coal mines, mostly in the north and west, are located far from the main demand centres along the south-eastern coastline. China's installed power generation capacity totalled 874 GW in 2009 and around 70% of it is coal-fired. Despite the global economic crisis, electricity generation grew by 6.7% in 2009 to reach 3 681 TWh and has continued to grow strongly in 2010 (China Electricity Council, 2010).

Energy pricing and subsidy policy

China has made significant progress in bringing domestic energy prices closer to global market levels and is continuing to push ahead with new reforms. These efforts have contributed to the reduction in energy intensity experienced since 1980. However, many energy prices in China are still set or guided by the central government, in pursuit of various socio-economic goals. Furthermore, in many provinces, local government can also still influence retail energy prices.

Prices for crude oil produced in China are already determined on the basis of the price for comparable grades of oil sold in international markets. Prices for most refined oil products also now generally match the international levels. Indeed, in 2009, the average retail prices for gasoline and diesel in China were respectively 40% and 32% above those in the United States (due to higher taxes). Prices of oil products are now determined in accordance with a set of Administration Measures for Petroleum Prices ("Measures"), promulgated by the National Development and Reform Commission (NDRC). The objective is to create a price-adjustment mechanism that allows domestic oil product prices to track international market levels while insulating the domestic economy from the excessive volatility of petroleum prices on world markets. The Measures state that when a change in the average price of crude oil on the international market lasts for 22 consecutive days and exceeds a level of 4%, prices of domestic refined oil products may be adjusted accordingly. In accordance with these reforms, the NDRC has raised maximum retail prices for gasoline and diesel six times since January 2009, with the latest increase of 5% taking place in April 2010 (Figure 20.3). This type of price regulation can lead to temporary subsidies if adjustments are too slow or are insufficient to match rapid price increases in international markets. But the situation can be reversed, with the government making revenue gains if controlled prices are too slow to readjust for falling world prices.

In 2007, faced with increased demand and increasing imports, China lifted its remaining price controls for coal and began to introduce a market-based pricing system. Coal prices for power generation are now largely set by direct negotiations between coal producers and power companies, without any direct price control. However, state intervention does still occur; in the interest of taming inflation, in June 2010, the NDRC issued a price-freeze directive to its domestic producers. Coking-coal prices now largely reflect international prices. In this study, the reference prices for China's coal have been adjusted down from the average prices paid by China for the foreign coal it imports to reflect the lower quality of China's domestic coal.

20

Figure 20.3 ● Petroleum product prices in China compared to Singapore spot prices

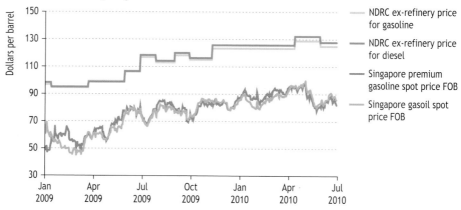

Sources: FACTS (2010); IEA analysis.

Compared with the action taken on oil and coal, moves to establish market-based pricing for gas in China have been slow and prices have remained relatively low compared to those on international markets. The government has maintained a cost-plus price regime that comprises three elements: (A) ex-plant price; (B) transportation tariff; and (C) end-user price. Both (A) and (B) are under the control of the central government, while (C) is under the control of the local government of each province (Ni, 2009). While China was self-sufficient in gas, its gas-price regime did not result in any explicit subsidies. However, the recent increase in imports has made gas subsidisation more explicit. With gas imports set to grow significantly, it is already accepted that higher prices will be needed, particularly to encourage producers to increase indigenous supply, including from unconventional sources such as coalbed methane and shale gas. In 2010, the NDRC announced a 25% rise in the price for domestically produced onshore gas and, in June, finally abolished the previous pricing system while expanding the limited scope for price negotiation between buyer and seller. The increase will have the effect of narrowing the price gap between domestic supplies and imported gas (such as LNG) and between domestic gas and rival energy sources (such as liquefied petroleum gas [LPG]). Many analysts view this as an important step towards an eventual market-based pricing regime. The increase should induce consumers to use gas more efficiently, and should accelerate investment by China's national oil companies in domestic exploration and production and development of LNG and long-distance pipeline gas import projects.

Prices for electricity in China vary from province to province and are set by NDRC on the basis of surveys of generating costs and recommendations from local pricing bureaus. Electricity prices in the residential sector — which accounted for just 19% of the country's total power consumption in 2008 — are lower than in other sectors. The most recent tariff increase occurred in November 2009, when NDRC increased the prices for non-residential use by 2.8 fen ($0.004) per kilowatt-hour (kWh). NDRC has been considering a proposal to introduce a tiered electricity pricing mechanism for residents (under which prices would increase with consumption) and the elimination of preferential tariff arrangements for certain industries is under way.

Since 2009, the Chinese government has undertaken pilot schemes, in four provinces and one municipality, which allow direct sales of electricity by power generators to end users. In October 2009, the first direct purchase programme between 15 aluminium manufacturers and power generators was established. This allows generators to sell power from an over-supplied region to industrial users. At present, there is no separate grid pricing and the government sets the reference transmission and distribution price as well as the retail price for each province. The absence of separate grid pricing distorts generation pricing and hinders efficient and adequate grid investments. Power shortages frequently occur (especially during the summer and winter peak-demand seasons) as the sector struggles to keep up with rapid economic growth, which is expected to continue into the foreseeable future. More efficient and cost-reflective pricing would enhance investment in and the efficiency of China's power sector.

Subsidy estimates

Based on the price-gap approach outlined in Chapter 19, we estimate that the economic value of fossil-fuel consumption subsidies in China in 2009 was well over $18 billion (Table 20.6). When viewed on a per-capita basis and as a share of GDP, subsidies were $14 and 0.4%, respectively. The reduction in China's subsidy bill between 2008 and 2009 was partly the result of the fall in fossil-fuel prices in international markets, but the ongoing price reforms, in particular of oil products, also played a role. In 2009, fossil fuels in China were subsidised at an average rate of 4%, meaning consumers paid 96% of competitive market reference prices.

Table 20.6 ● Fossil-fuel consumption subsidies in China

		Unit	2007	2008	2009
Total	Subsidies	$ billion	17.2	45.4	18.6
	per capita	$	*13*	*34*	*14*
	as a share of GDP (MER)	%	*0.5*	*1.0*	*0.4*
	Rate of subsidisation	%	4.5	8.5	3.9
By fuel	Oil products	$ billion	11.8	24.6	5.0
	rate of subsidisation	%	*6*	*9*	*3*
	Natural gas	$ billion	0.0	7.1	0.5
	rate of subsidisation	%	*0*	*26*	*2*
	Coal	$ billion	1.0	3.2	4.3
	rate of subsidisation	%	*3*	*6*	*7*
	Electricity	$ billion	4.4	10.4	8.8
	rate of subsidisation	%	*3*	*6*	*4*

India

Energy sector overview

In India is the world's fourth-largest energy consumer with a total primary energy demand of 621 Mtoe in 2008, or equivalent to the primary demand of Brazil, Indonesia and Saudi Arabia combined. Between 2000 and 2008, average annual GDP grew at

20

7.3% while primary energy demand grew robustly at 3.8% per year. Given its rapidly expanding population and emerging economy, India has significant potential for further energy demand growth. Coal is the key fuel in the current energy mix, accounting for 42% of the total in 2008. Oil and gas comprised 23% and 6%, respectively. Insufficient indigenous hydrocarbon resources have led India to import a growing share of its energy (Table 20.7). Oil and gas reserves, in particular, are limited, but even with extensive steam-coal reserves, India has had to import coal to satisfy domestic consumption requirements. Expanding energy infrastructure — in both the oil and gas and electricity sectors — is a major priority for India. Some 400 million people are currently without access to electricity; more than twice that number rely on the traditional use of biomass for cooking (see Table 8.1 in Chapter 8).

Oil consumption in India was around 3 mb/d in 2009. Due to increasing domestic demand and a recent expansion of refining capacity, crude oil imports have risen to around 2.6 mb/d. India is a net exporter of most refined petroleum products, although it remains a net importer of kerosene and LPG, which are used mainly by households for lighting and cooking. India's proven oil reserves amounted to 5.6 billion barrels at the end of 2009, or enough to sustain current production levels for 19 years (O&GJ, 2009). India's natural gas consumption has more than tripled in the last two decades, with LNG imports starting in 2004. In 2009, net imports of natural gas reached 15 bcm, about 30% of total consumption. At the end of 2009, India's proven natural gas reserves were estimated to be 1.1 tcm, or enough to sustain current production levels for 30 years. India has the world's fourth-largest hard coal endowment, totalling 239 billion tonnes in 2007 (BGR, 2009).

Table 20.7 ● Key economic and energy indicators for India

	Unit	1990	2008	1990-2008*
Population	million	850	1 140	1.6%
GDP (PPP) per capita	$ (2009)	1 296	2 927	4.6%
Energy demand	Mtoe	319	620	3.8%
Energy demand per capita	toe	0.38	0.54	2.1%
Energy intensity	toe per $1 000 GDP (PPP, 2009)	0.29	0.19	-2.4%
Oil net imports	mb/d	0.5	2.2	8.3%
Natural gas net imports	bcm	0.0	10.5	n.a.
Coal net imports	Mtce	5.9	52.3	12.9%
Electricity consumption	TWh	291	839	6.1%

*Compound average annual growth rate.

Energy pricing and subsidy policy

India has been actively reviewing its energy-pricing policy over the last several years, as price regulation has led to a significant burden on the federal budget. In 2009, the Kirit Parikh Committee, led by a former member of the nation's Planning Commission, was set up to suggest a viable and sustainable new policy for the pricing of diesel, petrol, LPG and kerosene. In June 2010, after reviewing the recommendations stemming from the Parikh Report (Government of India, 2010), the federal government announced that the price for gasoline, which is used in India primarily by wealthier car owners and accounts for only 8% of the country's refined product demand, would no longer be regulated, thereby empowering state-owned refiners to set retail prices (leading to an average $0.07 per litre, or 7% increase). The government announced that the same arrangements would be adopted later for diesel, which accounts for 33% of refined product demand. The government also announced an immediate increase in the regulated prices of diesel (by $0.04 per litre), LPG (by $0.76 per cylinder) and kerosene (by $0.07 per litre). There are no current plans to abolish state control of LPG and kerosene pricing.

These recent moves towards market-based prices will reduce the federal government's spending on subsidies and are very positive for India's oil sector. Prior to the new pricing arrangements, India's government-owned oil marketing companies (OMCs) were incurring substantial losses on sales as the regulated prices at which they were obliged to sell gasoline, diesel, LPG and kerosene were often below cost. The bulk of these losses were, however, reimbursed through a combination of oil bonds[1] and direct payments from the state. The government also required state-controlled oil producers to discount the price of crude oil sold to state-owned refiners, so as to reduce their retail losses and therefore the amount the state had to reimburse to the marketing companies (Clarke and Graczyk, 2010). The producers' losses were not similarly reimbursed by the state. The recent reforms have created a more balanced playing field between state-owned and private-sector fuel retailers, whose retail prices were previously higher than those of their state-run rivals, because they were not eligible for the state subsidies. In some cases, this had led to privately run service station chains being moth-balled,[2] but now many are being significantly expanded. State-controlled oil producers are no longer required to provide discounts on crude sales, so they can begin to function more like private companies and will be able to change market-based crude prices.

Despite the significant subsidies that still exist in India's oil sector, consumers often do not benefit at the pump. End-use prices for gasoline (and to a lesser extent diesel) remain high relative to many parts of the world because of significant rates of tax that are levied by different levels of government (Figure 20.4). For example, the retail price of unleaded gasoline was $0.91 per litre in India in 2009, compared with $0.62 in the United States. Taxes on transportation fuels in India represent an important source of revenue for both state governments and the federal government. In June 2010, the federal government called upon the states to lower the sales taxes (or value-added tax, VAT) they levy on refined products in a bid to cushion the impact of recent price hikes. In response, the Delhi state government rolled back an increase in sales tax on diesel that it had announced in the state budget. It appears that other states will soon follow. VAT rates vary dramatically across the individual states but are generally high. Andhra Pradesh, a state on the south-eastern coast of India, has the highest VAT on petrol, at 33%.

---

1. Oil bonds are debt securities issued by the government to OMCs to be traded by these companies for liquid cash, or to be used as collateral for borrowing in financial markets.

2. In 2008, Reliance Industries Limited closed all of its petrol stations as the subsidies that were being paid to the state-owned oil marketing companies were making operations economically unviable.

20

In contrast to gasoline and diesel, prices for kerosene and LPG cylinders for domestic use remain heavily subsidised and there are no existing plans for price controls to be removed. Kerosene is widely used in India by poorer segments of the population for home lighting. The subsidies on LPG cylinders created an incentive to illegally divert the gas into the non-subsidised commercial sector. Subsidised kerosene has been similarly diverted and mixed with diesel. To overcome these problems, the Petroleum Ministry has been considering plans to limit the number of gas cylinders each consumer may purchase at the subsidised price, with any additional cylinders being charged at the market price.

Figure 20.4 ● Average refined product prices and taxes in India, 2009

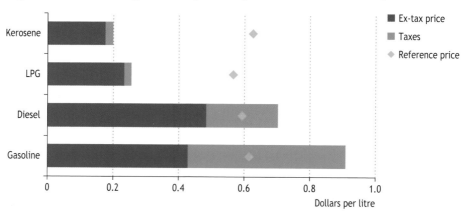

Note: Prices and taxes of products are based on Mumbai area. The calculation of reference prices is discussed in Chapter 19.
Sources: Ministry of Petroleum and Natural Gas (2009); IEA databases and analysis.

India's hydrocarbon vision statement envisages the share of natural gas in the energy mix increasing to about 20% by 2025. Although much of this will be met by imports, the government wants to encourage an expansion of domestic production and reforms to current pricing arrangements are viewed as pivotal to encouraging the necessary investment. Low gas prices to producers currently deter investment in mature fields and in exploration of deep and ultra-deep waters, where almost all recent gas discoveries have been made.

Currently, there are two main pricing regimes for natural gas in India. The first is the administered pricing mechanism (APM), under which the government sets prices for gas produced by state-run oil companies (ONGC and OIL) from fields previously allocated to them at no cost. APM gas currently accounts for about 60% of the gas sold in India. The second applies to non-APM or free-market gas, which is gas produced domestically from joint-venture fields or gas imported as LNG. For pricing purposes, there are two categories within this non-APM gas. The pricing of gas produced from joint-venture gas fields licensed before the New Exploration Licensing Policy (NELP) came into effect is determined by the production sharing contract (PSC) provisions signed by the consortium with the government. The pricing of the substantial quantity of gas expected to be derived from the gas fields awarded by the government under the NELP will be based on a pricing formula linked to the crude oil price. The price of LNG imported under long-term contracts is governed by the sale and purchase agreements (SPAs) between the LNG seller and the buyer, while spot cargoes are purchased on mutually agreeable commercial terms (Mercados, 2010).

In May 2010, India increased by 230% the price received by producers for APM gas to $4.2 per million British thermal units (MBtu). This sharp increase was deemed necessary as the price previously paid to producers was insufficient to cover their costs, let alone to provide for new investment. The new price is equivalent to the price realised by Reliance Industries Limited (RIL) for gas from its new offshore fields in the Krishna-Godavari basin. Furthermore, the government has announced that any new gas discovered by ONGC and OIL in the blocks given to them under the APM regime can be sold at market rates.

The sharp rise in the APM price will increase the cost of fertiliser production and power generation: these sectors consume around 90% of total APM gas production. However, end-user prices for fertiliser are protected by separate subsidies, so that they are not expected to increase. Similarly, the increase in end-use electricity prices will be only marginal, as the bulk of electricity is generated from coal and APM gas represents only a small share of the natural gas used in the power sector. Other smaller users of APM gas, such as compressed natural gas (CNG) and piped gas distributors, are expected to pass through the increase to end users.

All issues relating to coal production, supply and pricing in India are controlled by the Ministry of Coal. Although the bulk of the country's electricity production is fuelled by domestic coal, the sector has increasingly been turning to coal imports as local output proved unable to keep pace with rapid demand growth and the quality of Indian coal proved insufficient for modern, energy efficient, low-emissions power plants. Reforms in India's steam coal industry are expected to slowly bring domestic coal prices in line with import parity levels, with due allowance for quality differences. In June 2010, state-owned Coal India Ltd, which is responsible for almost 90% of the county's coal production, announced that it would move to price its premium grades on an import parity basis. Currently, coal prices in India are as much as 50% cheaper than imports, partly due to quality differences. Indian coal is generally of poor quality and its low calorific value means that more coal is needed to generate the same amount of electricity. However, it is also due to discounts provided in order to support low electricity prices. As more than 80% of India's electricity is generated from coal, the implementation of the coal pricing reforms can be expected to lead to much higher power prices.

The Indian government continues to pursue reform of electricity markets, in order to address chronic problems of under-investment and poor quality of service. With the enactment of the Electricity Act of 2003, India initiated a much-needed overhaul of its power sector. The Act consolidates the laws relating to generation, transmission, distribution, trading and use of electricity. It also lays out plans to rationalise electricity tariffs. The Act does not, however, specify the time frame for elimination of subsidies, which remain very large. Individual states, rather than the central government of India,

20

are primarily responsible for setting electricity tariffs under the Act. The states have the largest share of generation and transmission assets, and almost all distribution is under their control.

Several state electricity boards (SEBs) provide electricity at heavily subsidised rates (or, in certain cases, for free) to some sections of the community. In several cases the subsidies to the agriculture and household sectors are funded by means of above-cost tariffs for commercial and industrial customers and railways. The generally poor cost-recovery rate, the very low price and the widespread non-payment of electricity all deter private investors. These conditions have also resulted in many SEBs becoming financially weak, harming their capacity to invest in building new generating plant and in maintaining and extending the network. Electricity subsidies in the agriculture sector lead to the excessive use of electricity in pumping, for example, which may lead to wasteful use of water and land degeneration.

A more cost-reflective electricity pricing system, together with measures to reduce thefts and system losses, would enhance the attractiveness of new investment in supply and improve the efficiency. Where over-pricing of industrial electricity currently occurs, its elimination would improve industrial competitiveness. Across India, in 2009 the average residential electricity tariff was $0.07/kWh, or 57% of the OECD average (Figure 20.5). Industry tariffs, at $0.09/kWh, were roughly the same as the OECD average level.

Figure 20.5 ● **Electricity prices with India compared with selected countries, 2009**

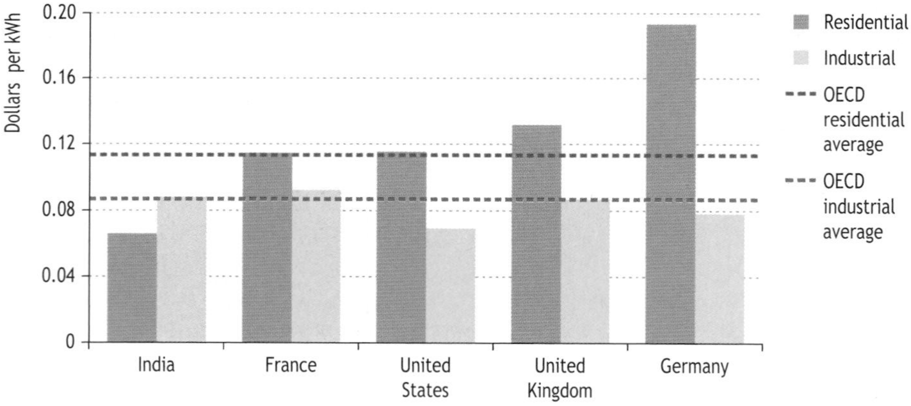

Source: IEA analysis.

Subsidy estimates

Based on the price-gap methodology, we estimate that the economic value of fossil-fuel consumption subsidies in India in 2009 was just over $21 billion (Table 20.8), with the largest amount ($12 billion) going to oil products, such as kerosene and LPG. On a per-capita basis and as a share of GDP, subsidies amounted to $18 per person and 1.7%, respectively. In 2009, fossil fuels in India were subsidised at an average rate of 15%, with consumers paying 85% of market-based reference prices. The reduction in India's

World Energy Outlook 2010 - FOCUS ON ENERGY SUBSIDIES

subsidy bill between 2008 and 2009 resulted primarily from the fall in fossil-fuel prices in international markets. With new energy pricing reforms in 2010, we would expect to see further reductions in India's subsidy bill in the years ahead, but much will depend on the success of reform implementation and on movements in international prices.

Table 20.8 ● Fossil-fuel consumption subsidies in India

		Unit	2007	2008	2009
Total	Subsidies	$ billion	24.6	41.0	21.1
	per capita	*$*	*22*	*36*	*18*
	as a share of GDP (MER)	*%*	*2.2*	*3.4*	*1.7*
	Rate of subsidisation	%	18	24	15
By fuel	Oil products	$ billion	17.7	32.2	12.1
	rate of subsidisation	*%*	*20*	*31*	*18*
	Natural gas	$ billion	2.1	2.9	2.7
	rate of subsidisation	*%*	*18*	*80*	*77*
	Electricity	$ billion	4.9	5.9	6.3
	rate of subsidisation	*%*	*13*	*11*	*12*

Indonesia

Energy sector overview

Indonesia is rich in natural resources. It is the world's leading steam-coal exporter, a leading LNG exporter and, until recently, has been an oil exporter. Indonesia is the fourth most populous nation and has a quickly developing economy, with annual growth of 5.1% from 2000 to 2009. Indonesia's primary energy demand almost doubled between 1990 and 2008 (Table 20.9). Fossil fuels account for 66% of Indonesia's primary energy demand. Coal demand, mainly for power generation and industry, grew fastest at 14% per year from 1990 to 2008.

Table 20.9 ● Key economic and energy indicators for Indonesia

	Unit	1990	2008	1990-2008*
Population	million	178	228	1.4%
GDP (PPP) per capita	$ (2009)	2 339	4 033	3.1%
Energy demand	Mtoe	104	199	3.7%
Energy demand per capita	toe	0.58	0.87	2.3%
Energy intensity	toe per $1 000 GDP (PPP, 2009)	0.25	0.22	-0.8%
Oil net imports**	mb/d	-0.9	0.3	n.a.
Natural gas net exports	bcm	27.7	37.1	1.6%
Coal net exports	Mtce	4.3	180.7	23.1%
Electricity consumption	TWh	33	149	8.7%

*Compound average annual growth rate. **Negative values indicate net exports.

Indonesian proven oil reserves stood at 4 billion barrels at the end of 2009 (O&GJ, 2009). At 2009 production levels of 982 thousand barrels per day (kb/d), these reserves would sustain production for another 11 years. The further development of Indonesia's oil reserves largely depends on the ability to attract investment. At the end of 2009,

20

Indonesia's proven gas reserves were estimated to be 3 tcm while its gas production stood at 75 bcm. In 2008 and 2009, almost two-thirds of Indonesia's gas production was exported in the form of LNG. Most future production will come from large gas projects, such as the deep-water Natuna D-Alphas gas field. Domestic demand for gas is expected to grow rapidly, which will limit gas exports. The Indonesian government has already implemented several programmes, such as the National City Gas Network Program for Households, designed to promote the use of gas in order to reduce oil consumption. Insufficient power capacity investment in Indonesia has caused frequent supply failures, particularly on the islands of Java and Bali, and further investment in this sector is also needed to improve the electrification rate, which is estimated to be about 65%.[3]

Energy pricing and subsidy policy

Indonesia has a long history of directly subsidising energy as a means of supporting the incomes of poor households. Previously, subsidies were available for industry and all segments of the population, but coverage has become increasingly targeted and the number of subsidised fuels has declined. Today, the state regulates the prices of selected fuels (kerosene, LPG in small containers, and some grades of diesel and gasoline) and sets electricity rates, all below market levels. The prices of non-regulated products fluctuate with global prices. The subsidised grade of gasoline (Premium 88 RON)[4] was priced at $0.42 per litre in 2009 (or 96% of the mean of Platts Singapore price for gasoline 92 RON), while the non-subsidised grade (Premium 92 RON) averaged $0.57 per litre. The government has been attempting to cut back further on this energy subsidisation, as it is a large burden (directly and indirectly) on the state budget and has led to increasing reliance on imported energy. To mitigate the social impacts and the political unpopularity of price rises, the government has introduced additional social support programmes (some involving unconditional direct cash transfers) as well as a new programme of health insurance for the poor and a school operational assistance programme.

Regulated energy prices are also a serious deterrent to much-needed investment in energy efficiency, new infrastructure for electricity and developing Indonesia's significant remaining oil and gas resources. They also reduce incentives to exploit the country's significant renewable energy potential. Indonesia's progress in reforming subsidies has been slow: attempts to raise fuel prices have sparked riots and created political divisions. As a result of high international energy prices, subsidies for oil products and electricity in Indonesia peaked in 2008 at 3.5% of GDP or 20% of the total national budget.

In the Medium-Term Plan (RPMN) published in 2010, the government has set a goal of reducing the direct and indirect cost of subsidies by 40% by 2013, and eliminating fuel subsidies entirely by 2014. The gap between international and domestic prices is to be progressively reduced, in an effort to minimise the impact on the poor. In 2009, subsidies on higher grades of diesel and heating oil were completely phased

---

3. See www.worldenergyoutlook.org/database_electricity/electricity_access_database.htm
4. RON refers to the research octane number.

out. About 40 million households switched from subsidised kerosene to LPG as part of the ongoing "Zero Kero" programme. LPG is more efficient for cooking and less polluting for households than kerosene. Under this programme, the government plans to phase out subsidies to kerosene by 2012, but will first provide poor households that currently use kerosene for cooking with conversion packages, comprising a stove and a 3-kilogram LPG canister. The successful completion of the programme is expected to have a significant effect on the government subsidy bill, as kerosene is the most heavily subsidised oil product and accounted for about 32% of the estimated total oil subsidy in 2009.

Part of Indonesia's strategy to curb spending on energy subsidies is to improve the distribution mechanisms, so that subsidised fuels can be better directed towards low income earners and small-scale industries. The state no longer subsidises oil products for larger industrial consumers and, in July 2010, the government announced that it intended to eliminate subsidies for owners of private cars (except for those purchased before 2005), while motorcycle users and public transportation vehicles will remain eligible. The government also plans to enhance the monitoring of distribution channels and increase penalties for the illegal use of subsidised fuel. Another part of Indonesia's strategy is to maintain and expand those existing support mechanisms that promote the development of renewables (particularly biofuels, geothermal and solar), so as to diversify the energy mix away from fossil fuels.

In Indonesia, natural gas prices for the residential and commercial sectors (such as for restaurants and hotels) are regulated by the government through BPH Migas, the downstream oil and gas regulator. However, the gas prices paid by electricity generators and fertiliser plants are not regulated and have increased significantly in recent years. The prices of gas paid by power generators ranged from $5/MBtu to $6/MBtu in 2009, while the industrial sector paid an average of $5.48 per MBtu.

Electricity prices in Indonesia are regulated by the government. For many years, the regulated price has been below the cost of production, with the difference being made up by a direct government subsidy to the state-owned electricity enterprise, PT PLN (which levies the electricity tariff). Though theoretically providing an economic price to PT PLN, these arrangements, in practice, have severely limited the ability of the state-owned enterprise to expand capacity and undertake regular maintenance, thereby contributing to the increasing frequency of significant supply disruptions throughout the country. To address these capacity shortages, in 2008 the government launched the first phase of a rapid-build programme to add 10 gigawatts (GW) of new capacity and is now initiating the second phase to add another 10 GW. Progressive tariffs, which increase with rising consumption, are already being used by PT PLN to meet the double challenge of delivering its public service obligation to provide electricity to new consumers while also minimising demand growth. PT PLN has also introduced a system under which industrial consumers pay for the capacity they require as well as the electricity consumed.

In June 2010, the Indonesian government raised power tariffs by an average of 10%. Larger industrial users saw a bigger increase, but prices for low-income households remained unchanged. As the new tariff is still below PT PLN's cost of production,

20

it will not completely eliminate the need for a government subsidy, but will reduce the overall burden on the state budget. The government has also been reviewing the possibility of allowing PT PLN to charge market-based prices, complemented by direct government support to low-income households.

Subsidy estimates

We estimate that the economic value of fossil-fuel consumption subsidies in Indonesia in 2009 was over $12 billion (Table 20.10). When viewed on a per-capita basis and as a share of GDP, subsidies were $53 per person and 2.3% respectively. In 2009, energy in Indonesia was subsidised at an average rate of 25%, meaning consumers paid 75% of competitive market reference prices. The reduction in Indonesia's subsidy bill between 2008 and 2009 was primarily the result of the fall in fossil-fuel prices in international markets. Based on the reforms to energy pricing that Indonesia introduced in 2010, we would expect to see further reductions in the country's subsidy bill in the years ahead. The extent of this will depend on how successfully the reforms are implemented, and on movements in international prices for those fuels that remain subsidised. It is estimated that oil product subsidies in Indonesia in 2009 amounted to almost $9 billion. The subsidies went predominantly to kerosene, gasoline and diesel. The total value of the electricity subsidy was $3.6 billion, with the residential sector accounting for more than 80% of the total.

Table 20.10 ● Fossil-fuel consumption subsidies in Indonesia

		Unit	2007	2008	2009
Total	Subsidies	$ billion	13.2	17.9	12.2
	per capita	$	*58*	*78*	*53*
	as a share of GDP (MER)	%	*3.0*	*3.5*	*2.3*
	Rate of subsidisation	%	18	24	25
By fuel	Oil products	$ billion	11.3	13.2	8.6
	rate of subsidisation	%	*31*	*28*	*28*
	Electricity	$ billion	1.9	4.7	3.6
	rate of subsidisation	%	*21*	*35*	*31*

ANNEXES

TABLES FOR SCENARIO PROJECTIONS

General note to the tables

The tables detail projections for *energy demand*, gross *electricity generation* and *electrical capacity*, and *carbon dioxide (CO₂) emissions* from fuel combustion in the Current Policies, New Policies and 450 Scenarios. The following regions/countries are covered: World, OECD[1], OECD North America, the United States, OECD Europe, the European Union, OECD Pacific, Japan, non-OECD, Eastern Europe/Eurasia, the Caspian, Russia, non-OECD Asia, China, India, the Middle East, Africa, Latin America and Brazil. The definitions for regions, fuels and sectors can be found in Annex C. In the table headings *CPS* refers to the Current Policies Scenario and *450* refers to the 450 Scenario.

Data for *energy demand*, gross *electricity generation* and CO_2 *emissions* from fuel combustion up to 2008 are based on IEA statistics, published in *Energy Balances of OECD Countries, Energy Balances of Non-OECD Countries* and CO_2 *Emissions from Fuel Combustion*.

Both in the text of this *WEO* and in the tables, rounding may lead to minor differences between totals and the sum of their individual components. Growth rates are calculated on a compound average annual basis and are marked "n.a." when the base year is zero or the value exceeds 200%. Nil values are marked "-".

Definitional note to the tables

Total primary energy demand (TPED) is equivalent to *power generation* plus *other energy sector* excluding electricity and heat, plus *total final consumption (TFC)* excluding electricity and heat. *TPED* does not include ambient heat from heat pumps or electricity trade. Sectors comprising *TFC* include *industry, transport, buildings* (residential and services) and *other* (agriculture and non-energy use). Projected *electrical capacity* is the net result of existing capacity plus additions less retirements. *Total CO_2* includes emissions from *other energy sector* in addition to the *power generation* and *TFC* sectors shown in the tables. CO_2 *emissions* and *energy demand* from international marine and aviation *bunkers* are included only at the world *transport* level. CO_2 *emissions* do not include emissions from industrial waste and non-renewable municipal waste.

---

1. Chile, Israel and Slovenia joined the OECD in mid-2010, but as accession negotiations were continuing when our modelling work commenced, these countries are not included in the OECD in this *Outlook*.

World: New Policies Scenario

	Energy demand (Mtoe)							Shares (%)		CAAGR (%)
	1990	2008	2015	2020	2025	2030	2035	2008	2035	2008-2035
TPED	8 779	12 271	13 776	14 556	15 263	16 014	16 748	100	100	1.2
Coal	2 233	3 315	3 892	3 966	3 986	3 984	3 934	27	23	0.6
Oil	3 222	4 059	4 252	4 346	4 440	4 550	4 662	33	28	0.5
Gas	1 674	2 596	2 919	3 132	3 331	3 550	3 748	21	22	1.4
Nuclear	526	712	818	968	1 078	1 178	1 273	6	8	2.2
Hydro	184	276	331	376	417	450	476	2	3	2.0
Biomass and waste	904	1 225	1 385	1 501	1 627	1 780	1 957	10	12	1.7
Other renewables	36	89	178	268	384	521	699	1	4	7.9
Power generation	2 984	4 605	5 282	5 723	6 135	6 564	6 980	100	100	1.6
Coal	1 228	2 169	2 515	2 554	2 576	2 575	2 531	47	36	0.6
Oil	375	271	200	176	156	137	125	6	2	-2.8
Gas	581	1 016	1 137	1 250	1 344	1 454	1 547	22	22	1.6
Nuclear	526	712	818	968	1 078	1 178	1 273	15	18	2.2
Hydro	184	276	331	376	417	450	476	6	7	2.0
Biomass and waste	59	87	126	165	227	316	425	2	6	6.1
Other renewables	32	74	155	235	337	454	602	2	9	8.1
Other energy sector	899	1 295	1 423	1 480	1 530	1 573	1 613	100	100	0.8
Electricity	*182*	*290*	*332*	*360*	*385*	*408*	*430*	*22*	*27*	*1.5*
TFC	6 289	8 423	9 525	10 059	10 535	11 045	11 550	100	100	1.2
Coal	763	823	1 014	1 035	1 023	1 010	994	10	9	0.7
Oil	2 608	3 502	3 775	3 900	4 029	4 177	4 314	42	37	0.8
Gas	950	1 308	1 468	1 548	1 626	1 711	1 794	16	16	1.2
Electricity	835	1 446	1 776	1 993	2 196	2 404	2 608	17	23	2.2
Heat	333	258	283	289	293	295	295	3	3	0.5
Biomass and waste	796	1 070	1 186	1 259	1 321	1 383	1 449	13	13	1.1
Other renewables	4	15	24	34	47	67	96	0	1	7.2
Industry	1 808	2 351	2 882	3 069	3 188	3 304	3 409	100	100	1.4
Coal	470	646	819	845	841	839	837	27	25	1.0
Oil	326	332	367	365	359	350	337	14	10	0.1
Gas	366	466	547	582	610	639	666	20	20	1.3
Electricity	379	603	783	884	961	1 033	1 099	26	32	2.2
Heat	150	113	129	131	132	133	132	5	4	0.6
Biomass and waste	117	191	235	262	284	311	338	8	10	2.1
Other renewables	0	0	0	1	1	1	1	0	0	2.0
Transport	1 576	2 299	2 514	2 669	2 834	3 035	3 244	100	100	1.3
Oil	1 483	2 150	2 320	2 437	2 561	2 714	2 864	94	88	1.1
Bunkers	*199*	*335*	*355*	*380*	*404*	*433*	*463*	*15*	*14*	*1.2*
Electricity	21	23	30	35	41	48	57	1	2	3.4
Biofuels	6	45	81	109	136	167	204	2	6	5.7
Other fuels	67	81	83	88	95	107	120	4	4	1.5
Buildings	2 247	2 850	3 082	3 249	3 410	3 570	3 729	100	100	1.0
Coal	237	125	133	129	120	109	97	4	3	-0.9
Oil	331	344	350	349	346	338	327	12	9	-0.2
Gas	431	617	653	686	720	755	787	22	21	0.9
Electricity	404	781	914	1 020	1 134	1 256	1 379	27	37	2.1
Heat	173	142	151	156	158	159	159	5	4	0.4
Biomass and waste	668	827	861	878	888	891	891	29	24	0.3
Other renewables	4	14	22	31	43	61	89	0	2	7.0
Other	657	923	1 046	1 072	1 104	1 136	1 167	100	100	0.9

World: Current Policies and 450 Scenarios

	Energy demand (Mtoe)						Shares (%) 2035		CAAGR (%) 2008-2035	
	2020	2030	2035	2020	2030	2035	CPS	450	CPS	450
	Current Policies Scenario			450 Scenario						
TPED	14 896	16 941	18 048	14 127	14 584	14 920	100	100	1.4	0.7
Coal	4 307	4 932	5 281	3 743	2 714	2 496	29	17	1.7	-1.0
Oil	4 443	4 826	5 026	4 175	3 975	3 816	28	26	0.8	-0.2
Gas	3 166	3 722	4 039	2 960	3 106	2 985	22	20	1.7	0.5
Nuclear	915	1 040	1 081	1 003	1 495	1 676	6	11	1.6	3.2
Hydro	364	416	439	383	483	519	2	3	1.7	2.4
Biomass and waste	1 461	1 621	1 715	1 539	2 022	2 316	10	16	1.3	2.4
Other renewables	239	384	468	325	789	1 112	3	7	6.3	9.8
Power generation	5 930	7 087	7 747	5 552	5 843	6 138	100	100	1.9	1.1
Coal	2 843	3 349	3 638	2 381	1 443	1 259	47	21	1.9	-2.0
Oil	187	159	154	161	106	99	2	2	-2.1	-3.7
Gas	1 263	1 550	1 728	1 168	1 234	1 090	22	18	2.0	0.3
Nuclear	915	1 040	1 081	1 003	1 495	1 676	14	27	1.6	3.2
Hydro	364	416	439	383	483	519	6	8	1.7	2.4
Biomass and waste	151	243	308	178	398	541	4	9	4.8	7.0
Other renewables	208	330	399	278	684	953	5	16	6.4	9.9
Other energy sector	1 513	1 678	1 758	1 418	1 395	1 371	100	100	1.1	0.2
Electricity	370	437	472	346	364	373	27	27	1.8	0.9
TFC	10 224	11 544	12 239	9 779	10 257	10 460	100	100	1.4	0.8
Coal	1 072	1 123	1 152	997	904	868	9	8	1.3	0.2
Oil	3 985	4 439	4 662	3 751	3 650	3 521	38	34	1.1	0.0
Gas	1 565	1 765	1 875	1 486	1 563	1 596	15	15	1.3	0.7
Electricity	2 040	2 548	2 831	1 933	2 230	2 376	23	23	2.5	1.9
Heat	297	317	325	281	262	248	3	2	0.9	-0.1
Biomass and waste	1 234	1 299	1 325	1 285	1 543	1 692	11	16	0.8	1.7
Other renewables	31	54	69	46	105	159	1	2	5.9	9.2
Industry	3 132	3 512	3 716	2 967	3 076	3 110	100	100	1.7	1.0
Coal	876	937	972	811	750	730	26	23	1.5	0.5
Oil	380	384	382	351	321	301	10	10	0.5	-0.4
Gas	587	658	696	556	592	604	19	19	1.5	1.0
Electricity	908	1 115	1 227	857	971	1 012	33	33	2.7	1.9
Heat	134	141	144	128	120	114	4	4	0.9	0.0
Biomass and waste	246	276	294	264	321	347	8	11	1.6	2.2
Other renewables	1	1	1	1	1	1	0	0	2.0	2.2
Transport	2 710	3 182	3 433	2 588	2 770	2 850	100	100	1.5	0.8
Oil	2 483	2 891	3 102	2 336	2 292	2 202	90	77	1.4	0.1
Bunkers	381	441	475	368	374	374	14	13	1.3	0.4
Electricity	34	46	53	38	80	128	2	4	3.1	6.5
Biofuels	107	142	163	122	283	386	5	14	4.8	8.2
Other fuels	86	103	114	92	114	134	3	5	1.3	1.9
Buildings	3 296	3 690	3 893	3 160	3 298	3 364	100	100	1.2	0.6
Coal	133	122	115	125	97	82	3	2	-0.3	-1.5
Oil	363	373	374	327	286	264	10	8	0.3	-1.0
Gas	699	790	839	651	659	654	22	19	1.1	0.2
Electricity	1 041	1 314	1 468	983	1 114	1 165	38	35	2.4	1.5
Heat	160	172	178	151	138	131	5	4	0.8	-0.3
Biomass and waste	871	868	855	879	904	917	22	27	0.1	0.4
Other renewables	29	51	65	43	99	151	2	4	5.8	9.2
Other	1 087	1 160	1 197	1 064	1 113	1 137	100	100	1.0	0.8

World: New Policies Scenario

Electricity generation (TWh)	1990	2008	2015	2020	2025	2030	2035	Shares (%) 2008	Shares (%) 2035	CAAGR (%) 2008-2035
Total generation	11 821	20 183	24 513	27 373	30 016	32 696	35 336	100	100	2.1
Coal	4 427	8 273	10 195	10 630	10 931	11 160	11 241	41	32	1.1
Oil	1 338	1 104	791	689	606	529	480	5	1	-3.0
Gas	1 726	4 303	5 199	5 881	6 430	7 032	7 557	21	21	2.1
Nuclear	2 013	2 731	3 139	3 712	4 136	4 520	4 883	14	14	2.2
Hydro	2 145	3 208	3 844	4 367	4 848	5 232	5 533	16	16	2.0
Biomass and waste	131	267	405	547	768	1 087	1 476	1	4	6.5
Wind	4	219	756	1 229	1 749	2 278	2 851	1	8	10.0
Geothermal	36	65	96	131	177	225	279	0	1	5.6
Solar PV	0	12	58	130	264	428	632	0	2	15.8
CSP	1	1	29	56	102	185	340	0	1	24.6
Marine	1	1	1	2	6	20	63	0	0	19.2

Electrical capacity (GW)	2008	2015	2020	2025	2030	2035	Shares (%) 2008	Shares (%) 2035	CAAGR (%) 2008-2035
Total capacity	4 719	5 952	6 581	7 186	7 867	8 613	100	100	2.3
Coal	1 514	1 930	2 047	2 092	2 160	2 229	32	26	1.4
Oil	438	429	349	292	253	236	9	3	-2.3
Gas	1 230	1 527	1 629	1 743	1 901	2 064	26	24	1.9
Nuclear	391	431	502	555	602	646	8	8	1.9
Hydro	945	1 119	1 271	1 410	1 520	1 602	20	19	2.0
Biomass and waste	52	75	98	134	184	244	1	3	5.9
Wind	120	358	535	703	862	1 035	3	12	8.3
Geothermal	11	16	21	27	34	42	0	0	5.1
Solar PV	15	57	110	197	294	406	0	5	13.0
CSP	1	10	17	30	52	91	0	1	19.3
Marine	0	0	1	2	6	17	0	0	16.7

CO_2 emissions (Mt)	1990	2008	2015	2020	2025	2030	2035	Shares (%) 2008	Shares (%) 2035	CAAGR (%) 2008-2035
Total CO_2	20 924	29 260	32 741	33 739	34 395	35 053	35 442	100	100	0.7
Coal	8 296	12 579	14 865	15 084	14 999	14 818	14 416	43	41	0.5
Oil	8 805	10 805	11 289	11 580	11 886	12 257	12 624	37	36	0.6
Gas	3 823	5 875	6 586	7 075	7 510	7 979	8 402	20	24	1.3
Power generation	7 476	11 918	13 364	13 668	13 831	13 908	13 756	100	100	0.5
Coal	4 927	8 670	10 058	10 175	10 190	10 083	9 767	73	71	0.4
Oil	1 192	864	638	561	495	435	397	7	3	-2.8
Gas	1 357	2 384	2 668	2 933	3 146	3 391	3 593	20	26	1.5
TFC	12 435	15 852	17 752	18 368	18 826	19 359	19 851	100	100	0.8
Coal	3 231	3 629	4 498	4 583	4 510	4 440	4 362	23	22	0.7
Oil	7 053	9 266	9 958	10 309	10 663	11 076	11 460	58	58	0.8
Transport	4 393	6 403	6 911	7 262	7 633	8 089	8 539	40	43	1.1
Bunkers	613	1 033	1 096	1 172	1 245	1 332	1 422	7	7	1.2
Gas	2 152	2 958	3 295	3 476	3 653	3 842	4 029	19	20	1.2

World: Current Policies and 450 Scenarios

	Electricity generation (TWh)						Shares (%) 2035		CAAGR (%) 2008-2035	
	2020	2030	2035	2020	2030	2035	CPS	450	CPS	450
	Current Policies Scenario			450 Scenario						
Total generation	28 032	34 716	38 423	26 505	30 170	31 981	100	100	2.4	1.7
Coal	11 789	14 784	16 455	9 704	6 269	5 609	43	18	2.6	-1.4
Oil	736	625	606	622	391	361	2	1	-2.2	-4.1
Gas	5 907	7 419	8 342	5 446	6 012	5 071	22	16	2.5	0.6
Nuclear	3 510	3 992	4 147	3 848	5 737	6 433	11	20	1.6	3.2
Hydro	4 238	4 834	5 110	4 454	5 618	6 032	13	19	1.7	2.4
Biomass and waste	493	825	1 052	594	1 379	1 889	3	6	5.2	7.5
Wind	1 080	1 653	1 936	1 383	3 197	4 107	5	13	8.4	11.5
Geothermal	120	174	200	142	291	391	1	1	4.3	6.9
Solar PV	119	288	352	164	723	1 179	1	4	13.3	18.5
CSP	37	110	185	144	519	838	0	3	21.8	28.8
Marine	2	12	39	3	34	72	0	0	17.1	19.8

	Electrical capacity (GW)						Shares (%) 2035		CAAGR (%) 2008-2035	
	2020	2030	2035	2020	2030	2035	CPS	450	CPS	450
	Current Policies Scenario			450 Scenario						
Total capacity	6 611	8 056	8 875	6 529	7 689	8 605	100	100	2.4	2.3
Coal	2 184	2 748	3 056	1 932	1 282	1 225	34	14	2.6	-0.8
Oil	355	273	268	337	227	205	3	2	-1.8	-2.8
Gas	1 667	2 028	2 261	1 543	1 729	1 802	25	21	2.3	1.4
Nuclear	476	535	551	519	760	849	6	10	1.3	2.9
Hydro	1 230	1 398	1 473	1 297	1 634	1 750	17	20	1.7	2.3
Biomass and waste	89	144	181	106	230	307	2	4	4.7	6.8
Wind	477	662	751	592	1 148	1 423	8	17	7.0	9.6
Geothermal	20	28	32	23	43	56	0	1	4.0	6.2
Solar PV	101	206	242	138	485	748	3	9	10.8	15.5
CSP	12	31	50	42	141	221	1	3	16.7	23.3
Marine	1	3	11	1	9	19	0	0	14.5	17.2

	CO_2 emissions (Mt)						Shares (%) 2035		CAAGR (%) 2008-2035	
	2020	2030	2035	2020	2030	2035	CPS	450	CPS	450
	Current Policies Scenario			450 Scenario						
Total CO_2	35 437	40 009	42 589	31 908	24 937	21 724	100	100	1.4	-1.1
Coal	16 424	18 519	19 742	14 156	8 001	5 820	46	27	1.7	-2.8
Oil	11 861	13 127	13 782	11 080	10 425	9 944	32	46	0.9	-0.3
Gas	7 151	8 362	9 065	6 673	6 511	5 960	21	27	1.6	0.1
Power generation	14 903	17 416	18 931	12 676	7 480	5 257	100	100	1.7	-3.0
Coal	11 345	13 284	14 403	9 433	4 474	2 753	76	52	1.9	-4.2
Oil	594	503	486	511	337	314	3	6	-2.1	-3.7
Gas	2 965	3 629	4 042	2 732	2 669	2 190	21	42	2.0	-0.3
TFC	18 793	20 712	21 699	17 637	15 947	15 018	100	100	1.2	-0.2
Coal	4 734	4 908	5 006	4 415	3 277	2 836	23	19	1.2	-0.9
Oil	10 543	11 836	12 475	9 886	9 426	8 985	57	60	1.1	-0.1
Transport	*7 398*	*8 617*	*9 248*	*6 962*	*6 841*	*6 579*	*43*	*44*	*1.4*	*0.1*
Bunkers	*1 175*	*1 357*	*1 460*	*1 136*	*1 157*	*1 158*	*7*	*8*	*1.3*	*0.4*
Gas	3 516	3 967	4 217	3 337	3 244	3 197	19	21	1.3	0.3

OCD: New Policies Scenario

	Energy demand (Mtoe)							Shares (%)		CAAGR (%)
	1990	2008	2015	2020	2025	2030	2035	2008	2035	2008-2035
TPED	4 477	5 421	5 468	5 516	5 543	5 578	5 594	100	100	0.1
Coal	1 068	1 128	1 093	1 017	936	845	715	21	13	-1.7
Oil	1 850	2 035	1 925	1 858	1 782	1 705	1 630	38	29	-0.8
Gas	840	1 271	1 293	1 341	1 374	1 413	1 450	23	26	0.5
Nuclear	450	592	633	671	708	747	790	11	14	1.1
Hydro	101	113	121	125	129	133	135	2	2	0.7
Biomass and waste	141	229	300	359	416	482	554	4	10	3.3
Other renewables	28	52	103	146	197	254	319	1	6	7.0
Power generation	1 702	2 241	2 312	2 356	2 411	2 467	2 492	100	100	0.4
Coal	749	904	880	811	744	663	535	40	21	-1.9
Oil	150	83	48	33	25	23	22	4	1	-4.8
Gas	175	436	445	474	496	524	557	19	22	0.9
Nuclear	450	592	633	671	708	747	790	26	32	1.1
Hydro	101	113	121	125	129	133	135	5	5	0.7
Biomass and waste	52	69	94	114	136	161	186	3	7	3.7
Other renewables	25	44	90	128	172	217	266	2	11	6.9
Other energy sector	394	475	460	457	451	445	441	100	100	-0.3
Electricity	*105*	*124*	*124*	*126*	*127*	*126*	*125*	*26*	*28*	*0.0*
TFC	3 080	3 696	3 743	3 793	3 811	3 833	3 854	100	100	0.2
Coal	231	135	130	125	113	104	95	4	2	-1.3
Oil	1 579	1 802	1 739	1 695	1 636	1 574	1 514	49	39	-0.6
Gas	590	737	747	759	765	769	768	20	20	0.2
Electricity	548	795	848	889	928	966	995	22	26	0.8
Heat	40	59	62	63	63	63	62	2	2	0.2
Biomass and waste	88	160	206	245	280	320	367	4	10	3.1
Other renewables	4	8	12	18	25	37	53	0	1	7.5
Industry	820	849	887	898	885	871	854	100	100	0.0
Coal	159	110	108	104	95	87	79	13	9	-1.2
Oil	166	124	119	112	104	95	86	15	10	-1.4
Gas	225	253	262	265	260	254	246	30	29	-0.1
Electricity	220	267	283	291	294	296	294	31	34	0.4
Heat	14	25	25	25	24	23	22	3	3	-0.4
Biomass and waste	36	71	89	100	107	115	125	8	15	2.1
Other renewables	0	0	0	1	1	1	1	0	0	2.0
Transport	934	1 191	1 178	1 174	1 153	1 134	1 123	100	100	-0.2
Oil	907	1 128	1 092	1 067	1 027	987	950	95	85	-0.6
Electricity	8	10	12	13	15	17	21	1	2	2.8
Biofuels	0	31	50	67	82	97	115	3	10	5.0
Other fuels	19	22	24	26	29	33	37	2	3	1.8
Buildings	975	1 229	1 256	1 304	1 357	1 414	1 466	100	100	0.7
Coal	68	21	18	17	15	13	12	2	1	-2.0
Oil	206	171	154	147	139	130	120	14	8	-1.3
Gas	307	428	428	435	443	450	452	35	31	0.2
Electricity	314	511	544	576	610	644	672	42	46	1.0
Heat	26	34	36	38	39	40	40	3	3	0.6
Biomass and waste	52	57	65	76	88	104	123	5	8	2.9
Other renewables	3	7	11	16	23	33	48	1	3	7.4
Other	351	427	422	418	416	414	411	100	100	-0.1

OECD: Current Policies and 450 Scenarios

	Energy demand (Mtoe)						Shares (%)		CAAGR (%)	
	2020	2030	2035	2020	2030	2035	2035		2008-2035	
	Current Policies Scenario			450 Scenario			CPS	450	CPS	450
TPED	5 595	5 787	5 877	5 365	5 220	5 215	100	100	0.3	-0.1
Coal	1 117	1 079	1 055	944	454	496	18	10	-0.2	-3.0
Oil	1 894	1 821	1 783	1 786	1 477	1 289	30	25	-0.5	-1.7
Gas	1 351	1 460	1 519	1 261	1 270	1 098	26	21	0.7	-0.5
Nuclear	631	669	683	694	913	965	12	19	0.5	1.8
Hydro	124	130	132	127	137	141	2	3	0.6	0.8
Biomass and waste	339	416	457	384	599	728	8	14	2.6	4.4
Other renewables	139	211	249	169	370	497	4	10	6.0	8.7
Power generation	2 400	2 565	2 639	2 294	2 290	2 362	100	100	0.6	0.2
Coal	903	867	835	750	275	318	32	13	-0.3	-3.8
Oil	36	29	30	25	14	13	1	1	-3.7	-6.5
Gas	474	540	579	434	465	315	22	13	1.1	-1.2
Nuclear	631	669	683	694	913	965	26	41	0.5	1.8
Hydro	124	130	132	127	137	141	5	6	0.6	0.8
Biomass and waste	109	148	168	117	174	205	6	9	3.4	4.1
Other renewables	122	182	211	147	313	405	8	17	6.0	8.6
Other energy sector	465	468	475	431	386	364	100	100	-0.0	-1.0
Electricity	128	132	134	122	118	117	28	32	0.3	-0.2
TFC	3 839	3 964	4 020	3 704	3 655	3 614	100	100	0.3	-0.1
Coal	130	120	114	117	102	94	3	3	-0.6	-1.3
Oil	1 727	1 683	1 658	1 636	1 371	1 204	41	33	-0.3	-1.5
Gas	768	798	811	733	719	702	20	19	0.4	-0.2
Electricity	902	997	1 041	868	924	943	26	26	1.0	0.6
Heat	65	69	71	61	59	56	2	2	0.7	-0.2
Biomass and waste	231	268	288	267	424	522	7	14	2.2	4.5
Other renewables	16	29	37	22	57	92	1	3	6.1	9.7
Industry	907	905	899	872	856	838	100	100	0.2	-0.1
Coal	108	101	96	97	86	81	11	10	-0.5	-1.1
Oil	116	105	98	111	96	87	11	10	-0.9	-1.3
Gas	266	262	259	253	247	240	29	29	0.1	-0.2
Electricity	298	309	311	286	286	281	35	34	0.6	0.2
Heat	26	25	25	24	22	21	3	3	0.1	-0.6
Biomass and waste	92	102	109	100	118	126	12	15	1.6	2.2
Other renewables	1	1	1	1	1	1	0	0	2.0	2.3
Transport	1 190	1 193	1 195	1 148	1 056	1 007	100	100	0.0	-0.6
Oil	1 084	1 065	1 056	1 032	833	703	88	70	-0.2	-1.7
Electricity	13	17	19	15	34	57	2	6	2.5	6.8
Biofuels	67	81	87	71	151	202	7	20	3.9	7.2
Other fuels	26	31	34	30	38	45	3	4	1.5	2.6
Buildings	1 319	1 446	1 510	1 268	1 331	1 362	100	100	0.8	0.4
Coal	17	14	13	16	12	10	1	1	-1.7	-2.7
Oil	153	145	141	135	100	82	9	6	-0.7	-2.7
Gas	444	472	485	418	404	388	32	28	0.5	-0.4
Electricity	582	662	702	559	596	597	46	44	1.2	0.6
Heat	39	43	45	37	36	35	3	3	1.1	0.1
Biomass and waste	69	82	89	83	131	164	6	12	1.7	4.0
Other renewables	15	27	35	20	53	86	2	6	6.1	9.8
Other	423	419	416	416	412	408	100	100	-0.1	-0.2

OECD: New Policies Scenario

	Electricity generation (TWh)							Shares (%)		CAAGR (%)
	1990	2008	2015	2020	2025	2030	2035	2008	2035	2008-2035
Total generation	7 560	10 673	11 290	11 794	12 262	12 694	13 018	100	100	0.7
Coal	3 057	3 882	3 798	3 596	3 375	3 059	2 536	36	19	-1.6
Oil	683	378	210	142	110	98	95	4	1	-5.0
Gas	769	2 365	2 479	2 674	2 778	2 890	3 052	22	23	0.9
Nuclear	1 725	2 272	2 431	2 574	2 718	2 866	3 033	21	23	1.1
Hydro	1 170	1 312	1 405	1 455	1 501	1 542	1 576	12	12	0.7
Biomass and waste	123	221	307	383	471	570	668	2	5	4.2
Wind	4	188	544	783	1 019	1 260	1 493	2	11	8.0
Geothermal	29	41	56	72	92	109	124	0	1	4.2
Solar PV	0	12	44	87	149	206	259	0	2	12.1
CSP	1	1	15	25	43	75	123	0	1	20.0
Marine	1	1	1	2	6	20	59	0	0	19.0

	Electrical capacity (GW)						Shares (%)		CAAGR (%)
	2008	2015	2020	2025	2030	2035	2008	2035	2008-2035
Total capacity	2 546	2 855	2 957	3 086	3 246	3 397	100	100	1.1
Coal	655	678	663	630	587	520	26	15	-0.9
Oil	217	191	125	95	83	81	9	2	-3.6
Gas	746	830	847	873	924	974	29	29	1.0
Nuclear	324	334	351	366	384	403	13	12	0.8
Hydro	444	465	481	494	505	514	17	15	0.5
Biomass and waste	40	53	65	79	95	110	2	3	3.8
Wind	97	245	331	408	481	552	4	16	6.7
Geothermal	7	9	12	14	17	18	0	1	3.6
Solar PV	15	44	74	113	145	175	1	5	9.6
CSP	1	5	8	12	21	34	0	1	15.8
Marine	0	0	1	2	5	16	0	0	16.4

	CO_2 emissions (Mt)							Shares (%)		CAAGR (%)
	1990	2008	2015	2020	2025	2030	2035	2008	2035	2008-2035
Total CO_2	11 004	12 544	12 169	11 793	11 285	10 732	9 981	100	100	-0.8
Coal	4 099	4 322	4 207	3 900	3 525	3 094	2 455	34	25	-2.1
Oil	4 977	5 272	4 961	4 783	4 578	4 379	4 195	42	42	-0.8
Gas	1 928	2 950	3 001	3 111	3 183	3 258	3 331	24	33	0.5
Power generation	3 908	4 935	4 751	4 481	4 191	3 852	3 316	100	100	-1.5
Coal	3 025	3 648	3 553	3 264	2 950	2 564	1 966	74	59	-2.3
Oil	474	264	153	105	81	73	71	5	2	-4.8
Gas	409	1 022	1 045	1 112	1 160	1 215	1 280	21	39	0.8
TFC	6 509	6 937	6 755	6 639	6 428	6 214	5 995	100	100	-0.5
Coal	1 012	590	572	551	499	457	416	9	7	-1.3
Oil	4 145	4 646	4 460	4 336	4 162	3 981	3 806	67	63	-0.7
Transport	2 661	3 327	3 221	3 147	3 028	2 909	2 802	48	47	-0.6
Gas	1 352	1 701	1 723	1 752	1 767	1 776	1 773	25	30	0.2

OECD: Current Policies and 450 Scenarios

	Electricity generation (TWh)						Shares (%)		CAAGR (%)	
	2020	2030	2035	2020	2030	2035	2035		2008-2035	
	Current Policies Scenario			450 Scenario			CPS	450	CPS	450
Total generation	11 971	13 120	13 650	11 513	12 109	12 323	100	100	0.9	0.5
Coal	3 995	4 048	4 012	3 317	1 294	1 520	29	12	0.1	-3.4
Oil	160	125	129	109	58	54	1	0	-3.9	-7.0
Gas	2 668	2 953	3 116	2 453	2 676	1 644	23	13	1.0	-1.3
Nuclear	2 420	2 568	2 621	2 663	3 502	3 703	19	30	0.5	1.8
Hydro	1 442	1 508	1 533	1 475	1 594	1 642	11	13	0.6	0.8
Biomass and waste	361	515	593	393	615	738	4	6	3.7	4.6
Wind	750	1 082	1 237	841	1 638	1 991	9	16	7.2	9.1
Geothermal	70	94	100	75	133	169	1	1	3.4	5.4
Solar PV	80	157	179	112	328	442	1	4	10.6	14.4
CSP	21	58	93	71	239	354	1	3	18.8	24.8
Marine	2	12	37	3	32	66	0	1	16.9	19.4

	Electrical capacity (GW)						Shares (%)		CAAGR (%)	
	2020	2030	2035	2020	2030	2035	2035		2008-2035	
	Current Policies Scenario			450 Scenario			CPS	450	CPS	450
Total capacity	2 936	3 226	3 350	2 919	3 221	3 496	100	100	1.0	1.2
Coal	678	668	632	608	295	295	19	8	-0.1	-2.9
Oil	128	87	86	115	59	50	3	1	-3.4	-5.3
Gas	856	968	1 031	802	858	869	31	25	1.2	0.6
Nuclear	330	347	350	361	466	490	10	14	0.3	1.5
Hydro	476	495	502	488	522	533	15	15	0.5	0.7
Biomass and waste	62	86	98	67	102	121	3	3	3.4	4.2
Wind	320	427	475	351	597	703	14	20	6.1	7.6
Geothermal	11	15	16	12	20	24	0	1	3.0	4.6
Solar PV	68	113	125	94	231	302	4	9	8.2	11.8
CSP	7	17	26	20	63	91	1	3	14.6	20.1
Marine	1	3	10	1	9	18	0	1	14.3	16.8

	CO_2 emissions (Mt)						Shares (%)		CAAGR (%)	
	2020	2030	2035	2020	2030	2035	2035		2008-2035	
	Current Policies Scenario			450 Scenario			CPS	450	CPS	450
Total CO_2	12 324	12 182	12 080	11 070	7 109	5 771	100	100	-0.1	-2.8
Coal	4 306	4 070	3 901	3 565	779	537	32	9	-0.4	-7.4
Oil	4 885	4 734	4 669	4 587	3 680	3 150	39	55	-0.4	-1.9
Gas	3 134	3 378	3 510	2 917	2 650	2 085	29	36	0.6	-1.3
Power generation	4 872	4 816	4 767	4 067	1 434	836	100	100	-0.1	-6.4
Coal	3 643	3 461	3 318	2 975	419	262	70	31	-0.4	-9.3
Oil	116	92	95	81	45	43	2	5	-3.7	-6.5
Gas	1 113	1 263	1 354	1 011	969	531	28	63	1.0	-2.4
TFC	6 766	6 668	6 601	6 384	5 135	4 440	100	100	-0.2	-1.6
Coal	572	527	499	512	291	205	8	5	-0.6	-3.8
Oil	4 421	4 296	4 229	4 179	3 356	2 858	64	64	-0.3	-1.8
Transport	_3 197_	_3 141_	_3 115_	_3 043_	_2 457_	_2 074_	_47_	_47_	_-0.2_	_-1.7_
Gas	1 773	1 844	1 873	1 692	1 489	1 377	28	31	0.4	-0.8

A

OECD North America: New Policies Scenario

	\multicolumn{7}{c	}{Energy demand (Mtoe)}	\multicolumn{2}{c	}{Shares (%)}	CAAGR (%)
	1990	2008	2015	2020	2025	2030	2035	2008	2035	2008-2035
TPED	2 245	2 731	2 759	2 789	2 812	2 836	2 846	100	100	0.2
Coal	488	580	579	552	518	477	417	21	15	-1.2
Oil	914	1 052	1 009	986	956	922	880	39	31	-0.7
Gas	516	669	670	692	708	727	749	24	26	0.4
Nuclear	180	245	256	266	283	299	312	9	11	0.9
Hydro	51	58	61	62	64	65	66	2	2	0.5
Biomass and waste	78	105	139	168	197	233	275	4	10	3.6
Other renewables	19	22	44	63	87	114	147	1	5	7.3
Power generation	850	1 096	1 135	1 155	1 181	1 206	1 213	100	100	0.4
Coal	419	524	528	504	473	433	368	48	30	-1.3
Oil	46	27	18	12	9	8	7	2	1	-5.0
Gas	95	199	199	214	225	239	259	18	21	1.0
Nuclear	180	245	256	266	283	299	312	22	26	0.9
Hydro	51	58	61	62	64	65	66	5	5	0.5
Biomass and waste	41	22	34	42	52	65	79	2	7	4.8
Other renewables	18	19	40	56	76	98	122	2	10	7.0
Other energy sector	190	241	237	238	237	238	240	100	100	-0.0
Electricity	*56*	*62*	*63*	*65*	*66*	*66*	*66*	*26*	*28*	*0.3*
TFC	1 535	1 860	1 875	1 905	1 923	1 941	1 956	100	100	0.2
Coal	60	36	33	32	27	24	22	2	1	-1.9
Oil	803	948	920	908	887	861	830	51	42	-0.5
Gas	360	393	392	393	394	394	391	21	20	-0.0
Electricity	271	390	412	432	452	471	486	21	25	0.8
Heat	3	8	7	7	6	6	6	0	0	-1.2
Biomass and waste	37	83	105	127	146	168	196	4	10	3.2
Other renewables	0	3	5	7	11	16	25	0	1	8.6
Industry	357	379	392	395	388	380	372	100	100	-0.1
Coal	50	34	31	30	26	23	21	9	6	-1.8
Oil	58	47	45	43	40	37	33	12	9	-1.3
Gas	138	146	148	148	145	141	137	38	37	-0.2
Electricity	94	105	110	112	112	112	110	28	30	0.2
Heat	1	6	6	6	6	5	5	2	1	-0.7
Biomass and waste	16	41	53	57	59	62	66	11	18	1.8
Other renewables	-	0	0	0	0	0	0	0	0	0.7
Transport	559	710	707	711	710	707	705	100	100	-0.0
Oil	540	669	656	647	632	614	593	94	84	-0.4
Electricity	1	1	2	2	2	3	4	0	1	4.9
Biofuels	-	21	29	41	53	65	80	3	11	5.1
Other fuels	18	19	19	20	22	25	28	3	4	1.5
Buildings	456	577	584	604	629	657	682	100	100	0.6
Coal	10	2	1	1	0	0	0	0	0	-8.1
Oil	64	56	47	44	40	35	30	10	4	-2.3
Gas	184	212	208	208	209	211	210	37	31	-0.0
Electricity	175	282	299	317	336	354	370	49	54	1.0
Heat	2	2	1	1	1	1	0	0	0	-4.3
Biomass and waste	21	21	23	28	33	40	49	4	7	3.3
Other renewables	0	3	4	6	10	15	23	0	3	8.4
Other	163	194	192	195	196	197	196	100	100	0.0

OECD North America: Current Policies and 450 Scenarios

	Energy demand (Mtoe)						Shares (%) 2035		CAAGR (%) 2008-2035	
	2020	2030	2035	2020	2030	2035	CPS	450	CPS	450
	Current Policies Scenario			450 Scenario						
TPED	2 820	2 918	2 966	2 703	2 604	2 606	100	100	0.3	-0.2
Coal	580	580	578	509	218	274	19	11	-0.0	-2.7
Oil	994	964	942	950	779	664	32	25	-0.4	-1.7
Gas	698	737	759	657	699	591	26	23	0.5	-0.5
Nuclear	265	279	289	270	362	378	10	14	0.6	1.6
Hydro	62	64	65	62	66	68	2	3	0.4	0.5
Biomass and waste	161	200	222	180	302	380	7	15	2.8	4.9
Other renewables	61	94	111	73	178	252	4	10	6.2	9.4
Power generation	1 178	1 254	1 288	1 116	1 093	1 140	100	100	0.6	0.1
Coal	528	523	508	463	172	221	39	19	-0.1	-3.1
Oil	12	8	8	9	5	4	1	0	-4.6	-6.4
Gas	215	238	254	204	269	179	20	16	0.9	-0.4
Nuclear	265	279	289	270	362	378	22	33	0.6	1.6
Hydro	62	64	65	62	66	68	5	6	0.4	0.5
Biomass and waste	42	61	72	42	68	85	6	7	4.4	5.0
Other renewables	55	81	93	65	151	206	7	18	6.0	9.1
Other energy sector	241	248	256	219	193	180	100	100	0.2	-1.1
Electricity	65	69	70	63	61	61	27	34	0.4	-0.0
TFC	1 917	1 979	2 007	1 862	1 840	1 818	100	100	0.3	-0.1
Coal	33	31	30	29	26	24	1	1	-0.7	-1.5
Oil	916	903	893	878	731	631	45	35	-0.2	-1.5
Gas	398	403	404	381	366	353	20	19	0.1	-0.4
Electricity	438	483	504	420	451	463	25	25	1.0	0.6
Heat	7	7	6	7	6	5	0	0	-0.7	-1.4
Biomass and waste	119	139	150	138	234	295	7	16	2.2	4.8
Other renewables	6	13	18	9	26	46	1	3	7.4	11.1
Industry	400	397	393	383	373	362	100	100	0.1	-0.2
Coal	31	30	28	28	25	23	7	6	-0.6	-1.3
Oil	45	42	40	43	39	36	10	10	-0.6	-1.0
Gas	149	145	141	141	134	129	36	36	-0.1	-0.5
Electricity	115	118	118	109	107	103	30	28	0.4	-0.1
Heat	6	6	6	6	5	5	1	1	-0.2	-0.9
Biomass and waste	54	57	60	57	63	66	15	18	1.4	1.8
Other renewables	0	0	0	0	0	0	0	0	0.7	1.5
Transport	715	724	728	699	654	627	100	100	0.1	-0.5
Oil	651	644	640	628	504	419	88	67	-0.2	-1.7
Electricity	2	3	4	3	13	29	1	5	4.6	12.8
Biofuels	41	54	59	45	105	143	8	23	4.0	7.4
Other fuels	20	23	25	24	31	36	3	6	1.1	2.4
Buildings	608	663	690	585	618	634	100	100	0.7	0.4
Coal	1	0	0	1	0	0	0	0	-4.7	-9.0
Oil	46	43	41	38	21	14	6	2	-1.2	-5.1
Gas	212	218	221	200	186	173	32	27	0.2	-0.8
Electricity	319	360	380	306	328	329	55	52	1.1	0.6
Heat	1	1	0	1	1	0	0	0	-4.2	-4.6
Biomass and waste	24	28	30	32	56	75	4	12	1.4	4.9
Other renewables	6	12	17	8	25	44	3	7	7.4	11.1
Other	194	196	195	194	196	195	100	100	0.0	0.0

A

OECD North America: New Policies Scenario

	Electricity generation (TWh)							Shares (%)		CAAGR (%)
	1990	2008	2015	2020	2025	2030	2035	2008	2035	2008-2035
Total generation	3 801	5 253	5 521	5 775	6 024	6 254	6 429	100	100	0.8
Coal	1 790	2 266	2 264	2 234	2 160	2 008	1 744	43	27	-1.0
Oil	209	117	78	53	39	34	30	2	0	-4.9
Gas	406	1 082	1 115	1 222	1 292	1 371	1 502	21	23	1.2
Nuclear	687	942	982	1 020	1 085	1 146	1 196	18	19	0.9
Hydro	593	678	710	725	742	757	771	13	12	0.5
Biomass and waste	90	82	121	156	201	261	323	2	5	5.2
Wind	3	60	202	287	383	500	614	1	10	9.0
Geothermal	21	24	34	46	58	68	76	0	1	4.4
Solar PV	0	2	10	24	48	78	110	0	2	16.9
CSP	1	1	4	8	16	29	56	0	1	16.6
Marine	0	0	0	0	0	3	6	0	0	21.3

	Electrical capacity (GW)						Shares (%)		CAAGR (%)
	2008	2015	2020	2025	2030	2035	2008	2035	2008-2035
Total capacity	1 252	1 333	1 383	1 455	1 524	1 581	100	100	0.9
Coal	355	356	373	377	351	306	28	19	-0.5
Oil	94	79	45	31	30	29	7	2	-4.3
Gas	448	461	471	484	501	523	36	33	0.6
Nuclear	121	126	131	139	147	153	10	10	0.9
Hydro	186	192	196	200	203	206	15	13	0.4
Biomass and waste	15	21	26	33	43	53	1	3	4.7
Wind	28	84	115	146	182	217	2	14	7.9
Geothermal	4	6	7	9	10	11	0	1	3.8
Solar PV	1	7	17	31	49	68	0	4	16.0
CSP	0	2	3	5	8	14	0	1	13.5
Marine	0	0	0	0	1	1	0	0	17.2

	CO_2 emissions (Mt)							Shares (%)		CAAGR (%)
	1990	2008	2015	2020	2025	2030	2035	2008	2035	2008-2035
Total CO_2	5 547	6 529	6 421	6 301	6 092	5 839	5 479	100	100	-0.6
Coal	1 910	2 229	2 235	2 130	1 963	1 757	1 448	34	26	-1.6
Oil	2 450	2 755	2 637	2 571	2 492	2 406	2 306	42	42	-0.7
Gas	1 187	1 545	1 549	1 601	1 638	1 676	1 724	24	31	0.4
Power generation	2 011	2 618	2 603	2 517	2 387	2 217	1 959	100	100	-1.1
Coal	1 640	2 060	2 078	1 976	1 832	1 638	1 338	79	68	-1.6
Oil	148	91	60	40	29	25	23	3	1	-5.0
Gas	222	466	465	500	526	554	598	18	31	0.9
TFC	3 192	3 528	3 431	3 386	3 304	3 216	3 107	100	100	-0.5
Coal	267	155	142	137	117	105	94	4	3	-1.9
Oil	2 098	2 465	2 384	2 341	2 278	2 202	2 110	70	68	-0.6
Transport	1 576	1 960	1 922	1 895	1 852	1 800	1 737	56	56	-0.4
Gas	827	907	905	908	909	909	903	26	29	-0.0

OECD North America: Current Policies and 450 Scenarios

	Electricity generation (TWh)						Shares (%)		CAAGR (%)	
	2020	2030	2035	2020	2030	2035	2035		2008-2035	
	Current Policies Scenario			450 Scenario			CPS	450	CPS	450
Total generation	5 850	6 420	6 676	5 610	5 955	6 097	100	100	0.9	0.6
Coal	2 331	2 432	2 413	2 045	857	1 104	36	18	0.2	-2.6
Oil	55	38	34	40	21	19	1	0	-4.5	-6.4
Gas	1 214	1 352	1 441	1 191	1 625	1 011	22	17	1.1	-0.3
Nuclear	1 016	1 069	1 108	1 038	1 390	1 449	17	24	0.6	1.6
Hydro	724	746	754	725	766	786	11	13	0.4	0.5
Biomass and waste	154	238	283	157	274	344	4	6	4.7	5.5
Wind	279	404	468	301	663	851	7	14	7.9	10.3
Geothermal	45	61	63	47	83	107	1	2	3.6	5.7
Solar PV	25	59	72	29	118	181	1	3	15.1	19.1
CSP	7	20	35	38	152	238	1	4	14.6	23.1
Marine	0	2	5	0	5	8	0	0	20.3	22.5

	Electrical capacity (GW)						Shares (%)		CAAGR (%)	
	2020	2030	2035	2020	2030	2035	2035		2008-2035	
	Current Policies Scenario			450 Scenario			CPS	450	CPS	450
Total capacity	1 387	1 521	1 578	1 348	1 449	1 618	100	100	0.9	1.0
Coal	374	389	369	344	140	166	23	10	0.1	-2.8
Oil	48	32	30	44	25	21	2	1	-4.1	-5.4
Gas	475	520	553	447	498	503	35	31	0.8	0.4
Nuclear	130	137	142	133	177	184	9	11	0.6	1.6
Hydro	196	201	202	196	206	210	13	13	0.3	0.5
Biomass and waste	26	39	46	27	45	56	3	3	4.2	5.0
Wind	111	151	171	119	233	289	11	18	7.0	9.1
Geothermal	7	9	10	7	12	15	1	1	3.3	4.9
Solar PV	17	37	44	20	76	115	3	7	14.2	18.3
CSP	2	6	9	10	38	57	1	4	11.6	19.5
Marine	0	0	1	0	1	2	0	0	16.2	18.5

	CO_2 emissions (Mt)						Shares (%)		CAAGR (%)	
	2020	2030	2035	2020	2030	2035	2035		2008-2035	
	Current Policies Scenario			450 Scenario			CPS	450	CPS	450
Total CO_2	6 450	6 432	6 380	5 930	3 689	2 951	100	100	-0.1	-2.9
Coal	2 237	2 187	2 117	1 940	247	200	33	7	-0.2	-8.5
Oil	2 600	2 543	2 512	2 473	1 972	1 654	39	56	-0.3	-1.9
Gas	1 613	1 702	1 751	1 517	1 469	1 098	27	37	0.5	-1.3
Power generation	2 618	2 620	2 585	2 304	763	448	100	100	-0.0	-6.3
Coal	2 074	2 036	1 967	1 800	164	134	76	30	-0.2	-9.6
Oil	42	28	25	30	16	15	1	3	-4.6	-6.4
Gas	502	556	593	475	583	299	23	67	0.9	-1.6
TFC	3 428	3 391	3 361	3 267	2 619	2 229	100	100	-0.2	-1.7
Coal	144	134	128	125	71	50	4	2	-0.7	-4.1
Oil	2 367	2 327	2 300	2 263	1 808	1 512	68	68	-0.3	-1.8
Transport	1 908	1 886	1 874	1 838	1 478	1 227	56	55	-0.2	-1.7
Gas	918	930	933	879	740	667	28	30	0.1	-1.1

A

United States: New Policies Scenario

	Energy demand (Mtoe)							Shares (%)		CAAGR (%)
	1990	2008	2015	2020	2025	2030	2035	2008	2035	2008-2035
TPED	1 915	2 281	2 280	2 290	2 291	2 288	2 272	100	100	-0.0
Coal	460	546	544	523	494	454	403	24	18	-1.1
Oil	757	852	814	790	760	723	676	37	30	-0.8
Gas	438	543	526	529	530	537	545	24	24	0.0
Nuclear	159	218	225	233	247	259	269	10	12	0.8
Hydro	23	22	24	25	26	26	27	1	1	0.7
Biomass and waste	62	85	113	140	167	198	235	4	10	3.8
Other renewables	14	16	33	49	68	90	117	1	5	7.8
Power generation	750	949	976	987	1 002	1 013	1 009	100	100	0.2
Coal	396	495	500	481	456	417	361	52	36	-1.2
Oil	27	13	9	4	3	4	4	1	0	-4.6
Gas	90	168	159	165	167	177	189	18	19	0.4
Nuclear	159	218	225	233	247	259	269	23	27	0.8
Hydro	23	22	24	25	26	26	27	2	3	0.7
Biomass and waste	40	19	29	36	44	55	65	2	6	4.6
Other renewables	14	13	29	42	58	75	95	1	9	7.6
Other energy sector	150	178	168	162	155	149	144	100	100	-0.8
Electricity	49	48	48	49	50	50	49	27	34	0.0
TFC	1 294	1 542	1 541	1 561	1 568	1 574	1 574	100	100	0.1
Coal	56	30	26	25	21	18	16	2	1	-2.3
Oil	683	782	754	740	717	687	651	51	41	-0.7
Gas	303	328	321	320	319	317	312	21	20	-0.2
Electricity	226	328	344	359	374	388	398	21	25	0.7
Heat	2	7	6	6	6	5	5	0	0	-1.5
Biomass and waste	23	65	84	104	122	143	169	4	11	3.6
Other renewables	0	3	4	6	10	15	22	0	1	8.4
Industry	284	295	296	296	287	278	269	100	100	-0.3
Coal	46	28	25	24	21	18	16	10	6	-2.1
Oil	44	34	31	30	27	24	21	12	8	-1.8
Gas	110	116	114	112	109	105	100	39	37	-0.6
Electricity	75	79	79	79	78	76	73	27	27	-0.3
Heat	-	6	5	5	5	5	4	2	2	-0.9
Biomass and waste	9	32	42	46	48	51	55	11	20	2.0
Other renewables	-	0	0	0	0	0	0	0	0	0.7
Transport	488	601	598	601	597	590	580	100	100	-0.1
Oil	472	565	554	544	527	505	477	94	82	-0.6
Electricity	0	1	1	1	1	2	3	0	1	6.0
Biofuels	-	20	27	39	50	62	76	3	13	5.1
Other fuels	15	16	16	17	19	21	24	3	4	1.5
Buildings	389	491	497	513	533	555	575	100	100	0.6
Coal	10	2	1	1	0	0	0	0	0	-9.1
Oil	48	38	30	26	23	18	14	8	2	-3.7
Gas	164	186	182	181	181	181	179	38	31	-0.1
Electricity	152	249	264	279	295	310	322	51	56	1.0
Heat	2	1	1	1	1	1	0	0	0	-5.5
Biomass and waste	14	13	15	19	24	30	38	3	7	4.1
Other renewables	0	2	4	6	9	14	22	0	4	8.5
Other	133	155	149	150	151	150	149	100	100	-0.1

United States: Current Policies and 450 Scenarios

	Energy demand (Mtoe)						Shares (%)		CAAGR (%)	
	2020	2030	2035	2020	2030	2035	2035		2008-2035	
	Current Policies Scenario			450 Scenario			CPS	450	CPS	450
TPED	2 313	2 353	2 366	2 224	2 101	2 091	100	100	0.1	-0.3
Coal	543	543	542	487	204	260	23	12	-0.0	-2.7
Oil	796	757	727	767	616	513	31	25	-0.6	-1.9
Gas	535	542	546	506	541	436	23	21	0.0	-0.8
Nuclear	232	241	248	235	317	332	11	16	0.5	1.6
Hydro	25	26	26	25	27	28	1	1	0.6	0.8
Biomass and waste	134	168	187	147	249	312	8	15	3.0	5.0
Other renewables	48	75	90	58	147	211	4	10	6.7	10.1
Power generation	1 004	1 054	1 073	952	907	942	100	100	0.5	-0.0
Coal	499	494	481	447	165	214	45	23	-0.1	-3.1
Oil	5	4	4	3	3	2	0	0	-4.0	-6.3
Gas	166	173	178	156	216	129	17	14	0.2	-1.0
Nuclear	232	241	248	235	317	332	23	35	0.5	1.6
Hydro	25	26	26	25	27	28	2	3	0.6	0.8
Biomass and waste	36	53	62	36	56	68	6	7	4.4	4.7
Other renewables	42	63	73	50	124	170	7	18	6.6	10.0
Other energy sector	164	157	158	153	126	115	100	100	-0.4	-1.6
Electricity	50	51	52	47	45	44	33	38	0.2	-0.3
TFC	1 570	1 600	1 608	1 526	1 491	1 464	100	100	0.2	-0.2
Coal	26	24	23	23	20	19	1	1	-1.0	-1.8
Oil	746	722	703	719	586	497	44	34	-0.4	-1.7
Gas	324	324	323	311	294	280	20	19	-0.1	-0.6
Electricity	364	397	411	349	370	378	26	26	0.8	0.5
Heat	6	6	5	6	5	4	0	0	-1.0	-1.7
Biomass and waste	97	115	125	111	193	245	8	17	2.4	5.0
Other renewables	6	12	17	8	24	41	1	3	7.4	10.9
Industry	300	290	284	287	273	261	100	100	-0.1	-0.5
Coal	25	24	22	22	20	18	8	7	-0.9	-1.6
Oil	31	29	27	29	26	24	9	9	-0.9	-1.3
Gas	114	107	103	107	98	92	36	35	-0.4	-0.9
Electricity	81	80	78	77	72	68	28	26	-0.0	-0.5
Heat	5	5	5	5	4	4	2	2	-0.3	-1.0
Biomass and waste	43	45	49	46	52	55	17	21	1.6	2.0
Other renewables	0	0	0	0	0	0	0	0	0.7	1.5
Transport	604	602	597	592	547	520	100	100	-0.0	-0.5
Oil	547	530	517	530	418	340	87	65	-0.3	-1.9
Electricity	1	2	3	2	11	24	0	5	5.6	14.2
Biofuels	39	51	56	39	91	124	9	24	3.9	7.0
Other fuels	17	19	21	21	27	32	3	6	1.0	2.6
Buildings	516	558	579	497	522	534	100	100	0.6	0.3
Coal	1	0	0	1	0	0	0	0	-5.1	-9.4
Oil	28	24	22	22	8	2	4	0	-2.0	-10.4
Gas	184	188	189	174	161	148	33	28	0.1	-0.8
Electricity	281	315	330	270	287	286	57	54	1.1	0.5
Heat	1	1	0	1	1	0	0	0	-5.4	-5.8
Biomass and waste	15	19	20	22	42	57	3	11	1.6	5.6
Other renewables	6	12	17	7	23	40	3	8	7.5	11.0
Other	150	149	148	150	150	149	100	100	-0.2	-0.2

United States: New Policies Scenario

	Electricity generation (TWh)							Shares (%)		CAAGR (%)
	1990	2008	2015	2020	2025	2030	2035	2008	2035	2008-2035
Total generation	3 203	4 343	4 529	4 716	4 896	5 057	5 169	100	100	0.6
Coal	1 700	2 133	2 134	2 130	2 077	1 933	1 712	49	33	-0.8
Oil	131	58	41	20	16	17	17	1	0	-4.5
Gas	382	911	889	935	952	1 010	1 098	21	21	0.7
Nuclear	612	838	864	895	949	994	1 032	19	20	0.8
Hydro	273	257	280	289	298	304	310	6	6	0.7
Biomass and waste	86	72	107	138	177	227	277	2	5	5.1
Wind	3	56	178	247	327	424	515	1	10	8.6
Geothermal	16	17	25	34	44	51	57	0	1	4.6
Solar PV	0	2	9	23	44	71	99	0	2	16.6
CSP	1	1	3	6	12	24	50	0	1	16.1
Marine	-	-	-	-	-	2	4	-	0	n.a.

	Electrical capacity (GW)						Shares (%)		CAAGR (%)
	2008	2015	2020	2025	2030	2035	2008	2035	2008-2035
Total capacity	1 062	1 113	1 145	1 204	1 250	1 283	100	100	0.7
Coal	334	334	352	360	335	295	31	23	-0.5
Oil	71	58	26	18	20	21	7	2	-4.4
Gas	409	406	405	408	412	420	39	33	0.1
Nuclear	106	110	113	120	126	130	10	10	0.8
Hydro	100	104	107	109	111	112	9	9	0.4
Biomass and waste	12	17	22	28	36	44	1	3	5.0
Wind	25	73	97	123	153	181	2	14	7.5
Geothermal	3	4	5	7	8	8	0	1	3.9
Solar PV	1	6	15	28	43	58	0	5	15.6
CSP	0	1	2	3	6	12	0	1	12.8
Marine	0	0	0	0	0	1	0	0	18.2

	CO_2 emissions (Mt)							Shares (%)		CAAGR (%)
	1990	2008	2015	2020	2025	2030	2035	2008	2035	2008-2035
Total CO_2	4 850	5 571	5 445	5 307	5 088	4 812	4 442	100	100	-0.8
Coal	1 797	2 086	2 092	2 010	1 865	1 669	1 397	37	31	-1.5
Oil	2 042	2 227	2 133	2 068	1 991	1 899	1 787	40	40	-0.8
Gas	1 011	1 257	1 220	1 229	1 231	1 244	1 259	23	28	0.0
Power generation	1 848	2 385	2 370	2 287	2 165	1 997	1 759	100	100	-1.1
Coal	1 550	1 946	1 965	1 888	1 763	1 577	1 312	82	75	-1.4
Oil	88	45	32	15	12	12	13	2	1	-4.6
Gas	210	393	372	385	390	408	434	16	25	0.4
TFC	2 730	2 917	2 811	2 758	2 669	2 566	2 441	100	100	-0.7
Coal	245	127	112	106	89	78	68	4	3	-2.3
Oil	1 788	2 032	1 954	1 911	1 842	1 755	1 650	70	68	-0.8
Transport	1 376	1 654	1 622	1 593	1 543	1 479	1 399	57	57	-0.6
Gas	697	759	744	741	738	734	724	26	30	-0.2

United States: Current Policies and 450 Scenarios

	Electricity generation (TWh)						Shares (%)		CAAGR (%)	
	2020	2030	2035	2020	2030	2035	2035		2008-2035	
	Current Policies Scenario			450 Scenario			CPS	450	CPS	450
Total generation	4 772	5 181	5 351	4 572	4 790	4 876	100	100	0.8	0.4
Coal	2 196	2 292	2 277	1 970	822	1 068	43	22	0.2	-2.5
Oil	21	20	20	16	12	10	0	0	-3.9	-6.2
Gas	932	969	1 002	905	1 317	735	19	15	0.4	-0.8
Nuclear	891	925	954	903	1 217	1 273	18	26	0.5	1.6
Hydro	289	299	302	289	308	320	6	7	0.6	0.8
Biomass and waste	137	213	252	139	234	288	5	6	4.7	5.3
Wind	244	343	392	255	560	706	7	14	7.5	9.9
Geothermal	33	47	49	35	64	84	1	2	4.0	6.1
Solar PV	24	55	69	27	108	160	1	3	15.0	18.7
CSP	6	17	31	34	145	226	1	5	14.2	22.8
Marine	-	1	3	0	3	5	0	0	n.a.	n.a.

	Electrical capacity (GW)						Shares (%)		CAAGR (%)	
	2020	2030	2035	2020	2030	2035	2035		2008-2035	
	Current Policies Scenario			450 Scenario			CPS	450	CPS	450
Total capacity	1 151	1 252	1 287	1 117	1 182	1 321	100	100	0.7	0.8
Coal	353	366	348	326	133	159	27	12	0.1	-2.7
Oil	28	22	22	24	13	11	2	1	-4.2	-6.7
Gas	410	429	447	392	427	428	35	32	0.3	0.2
Nuclear	113	117	120	114	153	160	9	12	0.5	1.5
Hydro	107	110	110	107	112	115	9	9	0.4	0.5
Biomass and waste	21	33	39	22	37	46	3	3	4.7	5.2
Wind	96	128	143	100	195	239	11	18	6.6	8.7
Geothermal	5	7	8	6	9	12	1	1	3.5	5.2
Solar PV	16	34	41	18	67	97	3	7	14.1	17.8
CSP	2	5	8	10	36	55	1	4	11.0	19.3
Marine	0	0	1	0	1	1	0	0	17.2	19.3

	CO$_2$ emissions (Mt)						Shares (%)		CAAGR (%)	
	2020	2030	2035	2020	2030	2035	2035		2008-2035	
	Current Policies Scenario			450 Scenario			CPS	450	CPS	450
Total CO$_2$	5 420	5 310	5 202	5 030	2 942	2 273	100	100	-0.3	-3.3
Coal	2 089	2 040	1 978	1 852	214	170	38	7	-0.2	-8.9
Oil	2 089	2 011	1 956	2 004	1 566	1 283	38	56	-0.5	-2.0
Gas	1 242	1 258	1 268	1 174	1 162	820	24	36	0.0	-1.6
Power generation	2 362	2 338	2 291	2 114	640	338	100	100	-0.1	-7.0
Coal	1 958	1 922	1 860	1 739	149	120	81	35	-0.2	-9.8
Oil	16	14	15	12	9	8	1	2	-4.0	-6.3
Gas	387	402	415	363	482	211	18	62	0.2	-2.3
TFC	2 792	2 712	2 651	2 673	2 104	1 762	100	100	-0.4	-1.9
Coal	112	102	97	98	52	35	4	2	-1.0	-4.7
Oil	1 929	1 858	1 806	1 855	1 446	1 181	68	67	-0.4	-2.0
Transport	*1 603*	*1 552*	*1 514*	*1 553*	*1 225*	997	*57*	*57*	*-0.3*	*-1.9*
Gas	751	752	748	720	606	546	28	31	-0.1	-1.2

OECD Europe: New Policies Scenario

	Energy demand (Mtoe)							Shares (%)		CAAGR (%)
	1990	2008	2015	2020	2025	2030	2035	2008	2035	2008-2035
TPED	1 601	1 820	1 802	1 813	1 817	1 826	1 843	100	100	0.0
Coal	442	313	275	242	218	194	158	17	9	-2.5
Oil	601	634	594	573	544	516	497	35	27	-0.9
Gas	258	457	463	480	496	511	518	25	28	0.5
Nuclear	204	240	236	237	230	227	241	13	13	0.0
Hydro	38	45	48	51	53	55	56	2	3	0.8
Biomass and waste	53	108	138	163	187	212	237	6	13	3.0
Other renewables	5	23	47	67	88	111	136	1	7	6.8
Power generation	611	752	745	754	770	788	806	100	100	0.3
Coal	270	226	193	164	147	128	96	30	12	-3.1
Oil	47	24	14	11	7	6	6	3	1	-5.1
Gas	40	159	161	171	183	196	202	21	25	0.9
Nuclear	204	240	236	237	230	227	241	32	30	0.0
Hydro	38	45	48	51	53	55	56	6	7	0.8
Biomass and waste	8	40	51	61	72	81	89	5	11	3.0
Other renewables	3	19	41	59	77	96	115	3	14	6.9
Other energy sector	147	151	142	140	136	133	130	100	100	-0.5
Electricity	38	45	42	42	42	41	41	30	31	-0.4
TFC	1 114	1 280	1 293	1 314	1 320	1 327	1 340	100	100	0.2
Coal	122	56	52	50	45	42	38	4	3	-1.4
Oil	515	562	535	518	493	467	449	44	34	-0.8
Gas	204	280	284	292	295	298	298	22	22	0.2
Electricity	190	266	281	295	309	324	336	21	25	0.9
Heat	37	46	49	50	50	51	51	4	4	0.4
Biomass and waste	44	67	86	102	115	130	147	5	11	2.9
Other renewables	2	4	6	8	11	15	21	0	2	6.6
Industry	319	310	318	322	318	314	310	100	100	-0.0
Coal	70	36	35	33	30	28	25	12	8	-1.3
Oil	57	43	39	36	33	29	26	14	8	-1.8
Gas	78	86	89	90	88	86	82	28	27	-0.2
Electricity	86	108	112	116	118	119	120	35	39	0.4
Heat	13	15	16	15	15	15	14	5	5	-0.3
Biomass and waste	14	22	26	30	34	38	41	7	13	2.4
Other renewables	0	0	0	0	0	0	0	0	0	3.5
Transport	264	342	342	343	331	319	316	100	100	-0.3
Oil	258	323	312	307	290	273	265	95	84	-0.7
Electricity	5	6	8	9	10	11	12	2	4	2.3
Biofuels	0	10	19	25	27	30	34	3	11	4.7
Other fuels	1	2	3	3	4	5	6	1	2	3.5
Buildings	400	482	493	515	539	563	586	100	100	0.7
Coal	48	17	15	14	13	12	11	4	2	-1.8
Oil	95	74	68	66	63	59	55	15	9	-1.1
Gas	108	176	177	183	189	193	195	37	33	0.4
Electricity	95	146	155	166	177	190	200	30	34	1.2
Heat	24	30	33	34	35	36	36	6	6	0.7
Biomass and waste	29	34	39	45	52	60	69	7	12	2.6
Other renewables	1	3	5	7	10	14	19	1	3	6.5
Other	131	147	140	133	132	130	129	100	100	-0.5

OECD Europe: Current Policies and 450 Scenarios

	Energy demand (Mtoe)						Shares (%) 2035		CAAGR (%) 2008-2035	
	2020	2030	2035	2020	2030	2035	CPS	450	CPS	450
	Current Policies Scenario			450 Scenario						
TPED	1 843	1 913	1 949	1 774	1 753	1 759	100	100	0.3	-0.1
Coal	292	265	256	234	125	132	13	8	-0.7	-3.1
Oil	594	578	570	550	464	417	29	24	-0.4	-1.5
Gas	486	542	569	440	410	362	29	21	0.8	-0.9
Nuclear	203	194	185	250	306	323	10	18	-1.0	1.1
Hydro	50	53	55	52	56	57	3	3	0.7	0.9
Biomass and waste	155	186	202	172	244	284	10	16	2.4	3.6
Other renewables	63	95	112	76	148	184	6	10	6.1	8.0
Power generation	762	815	842	745	760	784	100	100	0.4	0.2
Coal	212	193	188	160	60	72	22	9	-0.7	-4.1
Oil	11	8	8	9	4	4	1	0	-4.0	-6.6
Gas	173	210	228	144	119	76	27	10	1.3	-2.7
Nuclear	203	194	185	250	306	323	22	41	-1.0	1.1
Hydro	50	53	55	52	56	57	6	7	0.7	0.9
Biomass and waste	57	74	82	63	88	99	10	13	2.7	3.4
Other renewables	55	82	97	66	127	152	12	19	6.2	8.0
Other energy sector	142	141	140	136	124	118	100	100	-0.3	-0.9
Electricity	*43*	*44*	*44*	*41*	*39*	*39*	*31*	*33*	*-0.1*	*-0.5*
TFC	1 339	1 398	1 426	1 281	1 273	1 264	100	100	0.4	-0.0
Coal	51	46	43	47	40	36	3	3	-0.9	-1.6
Oil	536	524	517	498	421	377	36	30	-0.3	-1.5
Gas	296	314	323	279	276	272	23	22	0.5	-0.1
Electricity	299	333	350	291	310	318	25	25	1.0	0.7
Heat	52	56	58	49	47	45	4	4	0.9	-0.1
Biomass and waste	97	112	120	108	157	184	8	15	2.1	3.8
Other renewables	8	12	15	9	21	31	1	2	5.2	8.2
Industry	324	324	323	313	310	306	100	100	0.2	-0.0
Coal	34	31	30	31	28	26	9	8	-0.7	-1.2
Oil	37	32	29	35	30	26	9	9	-1.4	-1.8
Gas	90	90	89	86	83	81	27	26	0.1	-0.2
Electricity	117	122	124	114	116	116	38	38	0.5	0.3
Heat	16	16	16	15	14	13	5	4	0.1	-0.5
Biomass and waste	28	33	35	31	39	43	11	14	1.8	2.5
Other renewables	0	0	0	0	0	0	0	0	3.5	3.7
Transport	352	356	358	333	304	289	100	100	0.2	-0.6
Oil	317	316	316	296	246	216	88	75	-0.1	-1.5
Electricity	8	10	11	9	15	20	3	7	2.1	4.3
Biofuels	24	25	26	24	39	47	7	16	3.6	6.0
Other fuels	3	5	6	3	4	6	2	2	3.6	3.6
Buildings	522	581	609	504	529	541	100	100	0.9	0.4
Coal	15	13	11	13	10	9	2	2	-1.5	-2.5
Oil	68	64	62	61	47	39	10	7	-0.7	-2.4
Gas	187	205	214	176	175	173	35	32	0.7	-0.1
Electricity	168	196	211	163	175	178	35	33	1.4	0.7
Heat	35	40	42	33	33	32	7	6	1.2	0.2
Biomass and waste	43	51	56	48	70	83	9	15	1.8	3.3
Other renewables	7	11	14	9	20	29	2	5	5.2	8.2
Other	140	137	135	132	129	128	100	100	-0.3	-0.5

OECD Europe: New Policies Scenario

	Electricity generation (TWh)							Shares (%)		CAAGR (%)
	1990	2008	2015	2020	2025	2030	2035	2008	2035	2008-2035
Total generation	2 632	3 600	3 745	3 914	4 078	4 239	4 370	100	100	0.7
Coal	1 011	934	823	710	642	576	444	26	10	-2.7
Oil	203	104	59	44	31	25	24	3	1	-5.3
Gas	167	869	879	942	990	1 033	1 039	24	24	0.7
Nuclear	782	922	906	909	883	871	923	26	21	0.0
Hydro	443	521	563	593	617	638	653	14	15	0.8
Biomass and waste	21	113	150	183	219	249	277	3	6	3.4
Wind	1	120	318	455	573	676	773	3	18	7.1
Geothermal	4	10	12	16	20	24	29	0	1	4.0
Solar PV	0	7	27	48	75	92	103	0	2	10.2
CSP	-	0	8	13	23	39	56	0	1	35.3
Marine	1	1	1	2	4	14	48	0	1	18.3

	Electrical capacity (GW)						Shares (%)		CAAGR (%)
	2008	2015	2020	2025	2030	2035	2008	2035	2008-2035
Total capacity	881	1 059	1 093	1 123	1 178	1 243	100	100	1.3
Coal	198	215	185	153	139	121	22	10	-1.8
Oil	66	57	42	31	22	22	7	2	-4.0
Gas	191	235	232	240	261	281	22	23	1.4
Nuclear	137	134	134	130	127	131	16	11	-0.2
Hydro	190	205	214	222	229	233	22	19	0.8
Biomass and waste	20	26	32	37	42	46	2	4	3.1
Wind	65	151	201	240	272	301	7	24	5.8
Geothermal	2	2	3	3	4	5	0	0	3.2
Solar PV	11	30	45	59	67	73	1	6	7.3
CSP	0	3	4	7	11	17	0	1	20.2
Marine	0	0	1	1	4	14	0	1	16.0

	CO_2 emissions (Mt)							Shares (%)		CAAGR (%)
	1990	2008	2015	2020	2025	2030	2035	2008	2035	2008-2035
Total CO_2	3 884	3 946	3 701	3 555	3 402	3 240	3 021	100	100	-1.0
Coal	1 667	1 213	1 061	929	826	716	543	31	18	-2.9
Oil	1 636	1 676	1 571	1 519	1 436	1 354	1 300	42	43	-0.9
Gas	581	1 057	1 069	1 108	1 140	1 170	1 178	27	39	0.4
Power generation	1 349	1 383	1 220	1 113	1 048	977	829	100	100	-1.9
Coal	1 105	936	799	680	599	508	352	68	42	-3.6
Oil	151	77	45	33	24	19	18	6	2	-5.1
Gas	93	371	376	400	425	450	458	27	55	0.8
TFC	2 360	2 373	2 302	2 267	2 184	2 097	2 028	100	100	-0.6
Coal	525	244	232	220	200	183	167	10	8	-1.4
Oil	1 370	1 484	1 415	1 374	1 302	1 226	1 173	63	58	-0.9
Transport	764	968	935	919	869	818	792	41	39	-0.7
Gas	465	645	655	673	682	687	687	27	34	0.2

OECD Europe: Current Policies and 450 Scenarios

	Electricity generation (TWh)						Shares (%)		CAAGR (%)	
	2020	2030	2035	2020	2030	2035	2035		2008-2035	
	Current Policies Scenario			450 Scenario			CPS	450	CPS	450
Total generation	3 967	4 377	4 576	3 856	4 060	4 138	100	100	0.9	0.5
Coal	932	911	920	696	232	298	20	7	-0.1	-4.1
Oil	48	34	34	39	16	15	1	0	-4.0	-7.0
Gas	948	1 095	1 155	780	619	307	25	7	1.1	-3.8
Nuclear	780	744	711	960	1 176	1 241	16	30	-1.0	1.1
Hydro	584	622	635	604	651	665	14	16	0.7	0.9
Biomass and waste	170	226	252	191	271	311	6	8	3.0	3.8
Wind	434	613	693	478	817	940	15	23	6.7	7.9
Geothermal	15	20	22	17	31	39	0	1	3.0	5.2
Solar PV	43	70	75	62	150	183	2	4	8.9	12.6
CSP	12	33	50	29	73	90	1	2	34.8	37.7
Marine	1	9	29	2	24	50	1	1	16.1	18.5

	Electrical capacity (GW)						Shares (%)		CAAGR (%)	
	2020	2030	2035	2020	2030	2035	2035		2008-2035	
	Current Policies Scenario			450 Scenario			CPS	450	CPS	450
Total capacity	1 072	1 173	1 219	1 098	1 241	1 327	100	100	1.2	1.5
Coal	197	175	162	160	91	87	13	7	-0.7	-3.0
Oil	43	25	25	40	16	14	2	1	-3.6	-5.6
Gas	235	280	300	230	233	237	25	18	1.7	0.8
Nuclear	114	108	101	140	167	174	8	13	-1.1	0.9
Hydro	211	224	228	218	235	239	19	18	0.7	0.8
Biomass and waste	30	38	42	33	45	51	3	4	2.7	3.5
Wind	195	254	279	209	314	351	23	26	5.5	6.4
Geothermal	3	3	4	3	5	6	0	0	2.4	4.1
Solar PV	41	54	56	56	108	127	5	10	6.3	9.5
CSP	4	10	15	8	22	27	1	2	19.7	22.5
Marine	0	3	8	1	7	14	1	1	13.8	16.1

	CO_2 emissions (Mt)						Shares (%)		CAAGR (%)	
	2020	2030	2035	2020	2030	2035	2035		2008-2035	
	Current Policies Scenario			450 Scenario			CPS	450	CPS	450
Total CO_2	3 823	3 792	3 797	3 350	2 314	1 964	100	100	-0.1	-2.6
Coal	1 132	1 016	978	883	271	205	26	10	-0.8	-6.4
Oil	1 569	1 528	1 510	1 456	1 186	1 045	40	53	-0.4	-1.7
Gas	1 122	1 248	1 309	1 011	857	714	34	36	0.8	-1.4
Power generation	1 317	1 302	1 318	1 013	375	233	100	100	-0.2	-6.4
Coal	877	787	763	650	125	93	58	40	-0.8	-8.2
Oil	36	26	26	30	13	12	2	5	-4.0	-6.6
Gas	404	488	530	334	238	128	40	55	1.3	-3.9
TFC	2 328	2 314	2 302	2 169	1 789	1 591	100	100	-0.1	-1.5
Coal	227	204	191	206	123	90	8	6	-0.9	-3.6
Oil	1 419	1 384	1 364	1 319	1 076	940	59	59	-0.3	-1.7
Transport	*947*	*945*	*944*	*885*	*736*	*647*	*41*	*41*	*-0.1*	*-1.5*
Gas	682	726	747	644	591	560	32	35	0.5	-0.5

European Union: New Policies Scenario

	Energy demand (Mtoe)							Shares (%)		CAAGR (%)
	1990	2008	2015	2020	2025	2030	2035	2008	2035	2008-2035
TPED	**1 632**	**1 749**	**1 722**	**1 723**	**1 719**	**1 719**	**1 732**	**100**	**100**	**-0.0**
Coal	455	304	262	220	194	168	135	17	8	-3.0
Oil	601	606	568	544	514	483	461	35	27	-1.0
Gas	295	440	444	459	472	486	491	25	28	0.4
Nuclear	207	244	238	244	240	237	251	14	14	0.1
Hydro	25	28	31	32	33	34	35	2	2	0.8
Biomass and waste	46	107	139	165	189	213	239	6	14	3.0
Other renewables	3	18	41	59	79	98	120	1	7	7.3
Power generation	**644**	**731**	**714**	**717**	**730**	**744**	**762**	**100**	**100**	**0.2**
Coal	286	229	192	155	135	115	87	31	11	-3.5
Oil	61	25	14	10	7	6	6	3	1	-5.3
Gas	54	149	150	160	171	183	189	20	25	0.9
Nuclear	207	244	238	244	240	237	251	33	33	0.1
Hydro	25	28	31	32	33	34	35	4	5	0.8
Biomass and waste	8	40	51	62	72	81	89	5	12	3.0
Other renewables	3	16	38	55	72	88	105	2	14	7.2
Other energy sector	**149**	**144**	**135**	**132**	**129**	**125**	**122**	**100**	**100**	**-0.6**
Electricity	*39*	*43*	*40*	*39*	*39*	*38*	*38*	*30*	*31*	*-0.5*
TFC	**1 124**	**1 219**	**1 230**	**1 243**	**1 242**	**1 242**	**1 249**	**100**	**100**	**0.1**
Coal	120	44	40	37	33	29	25	4	2	-2.0
Oil	500	536	510	491	464	435	413	44	33	-1.0
Gas	228	276	280	286	289	290	290	23	23	0.2
Electricity	185	246	257	269	280	292	302	20	24	0.8
Heat	54	49	52	53	53	54	54	4	4	0.4
Biomass and waste	37	67	87	103	116	132	150	6	12	3.0
Other renewables	1	2	3	5	7	10	15	0	1	8.1
Industry	**341**	**295**	**301**	**303**	**297**	**292**	**286**	**100**	**100**	**-0.1**
Coal	68	30	29	27	23	21	18	10	6	-1.8
Oil	57	42	38	35	31	28	25	14	9	-1.9
Gas	97	88	91	92	89	87	83	30	29	-0.2
Electricity	85	98	101	104	105	106	106	33	37	0.3
Heat	19	15	15	15	15	14	13	5	5	-0.5
Biomass and waste	14	22	26	30	33	37	41	7	14	2.4
Other renewables	-	0	0	0	0	0	0	0	0	3.6
Transport	**258**	**330**	**334**	**334**	**319**	**303**	**296**	**100**	**100**	**-0.4**
Oil	252	312	303	296	277	256	244	94	82	-0.9
Electricity	5	6	7	8	9	10	11	2	4	2.3
Biofuels	0	10	20	26	29	32	36	3	12	4.8
Other fuels	1	2	3	3	4	5	6	1	2	3.3
Buildings	**394**	**455**	**465**	**484**	**505**	**528**	**548**	**100**	**100**	**0.7**
Coal	48	11	9	8	7	6	5	2	1	-3.0
Oil	89	70	64	62	60	57	53	15	10	-1.0
Gas	110	168	169	174	179	182	184	37	33	0.3
Electricity	90	136	144	153	162	173	182	30	33	1.1
Heat	33	33	36	37	39	40	40	7	7	0.7
Biomass and waste	23	34	39	45	52	60	70	7	13	2.7
Other renewables	1	2	3	4	7	10	14	0	3	8.0
Other	**132**	**139**	**130**	**123**	**121**	**120**	**118**	**100**	**100**	**-0.6**

European Union: Current Policies and 450 Scenarios

	Energy demand (Mtoe)						Shares (%)		CAAGR (%)	
	2020	2030	2035	2020	2030	2035	2035		2008-2035	
	Current Policies Scenario			450 Scenario			CPS	450	CPS	450
TPED	1 753	1 802	1 831	1 690	1 663	1 665	100	100	0.2	-0.2
Coal	268	231	219	213	115	121	12	7	-1.2	-3.3
Oil	566	547	537	523	435	387	29	23	-0.5	-1.7
Gas	465	516	543	424	396	352	30	21	0.8	-0.8
Nuclear	211	204	196	257	307	325	11	20	-0.8	1.1
Hydro	31	33	33	32	35	36	2	2	0.7	0.9
Biomass and waste	156	189	205	174	246	285	11	17	2.4	3.7
Other renewables	55	83	97	67	129	160	5	10	6.5	8.5
Power generation	724	763	788	713	727	751	100	100	0.3	0.1
Coal	201	173	165	152	63	74	21	10	-1.2	-4.1
Oil	11	8	7	9	4	4	1	1	-4.5	-6.7
Gas	162	196	215	137	115	76	27	10	1.4	-2.5
Nuclear	211	204	196	257	307	325	25	43	-0.8	1.1
Hydro	31	33	33	32	35	36	4	5	0.7	0.9
Biomass and waste	58	74	82	64	87	98	10	13	2.7	3.4
Other renewables	51	76	88	62	115	137	11	18	6.5	8.3
Other energy sector	135	132	131	129	118	113	100	100	-0.3	-0.9
Electricity	*40*	*40*	*40*	*39*	*37*	*37*	*30*	*33*	*-0.3*	*-0.5*
TFC	1 269	1 314	1 335	1 214	1 194	1 182	100	100	0.3	-0.1
Coal	39	33	30	34	27	24	2	2	-1.5	-2.2
Oil	510	493	483	472	393	347	36	29	-0.4	-1.6
Gas	291	308	316	275	270	265	24	22	0.5	-0.1
Electricity	272	300	314	267	282	288	23	24	0.9	0.6
Heat	55	59	62	52	50	47	5	4	0.9	-0.1
Biomass and waste	98	114	122	110	158	186	9	16	2.2	3.8
Other renewables	4	8	9	5	14	23	1	2	6.3	9.9
Industry	305	302	300	294	287	282	100	100	0.1	-0.2
Coal	28	24	22	25	21	19	7	7	-1.2	-1.7
Oil	36	30	28	34	28	25	9	9	-1.5	-1.9
Gas	92	92	91	87	84	81	30	29	0.1	-0.3
Electricity	106	108	109	102	103	102	36	36	0.4	0.1
Heat	16	15	15	15	13	13	5	5	-0.0	-0.7
Biomass and waste	28	33	35	31	38	42	12	15	1.8	2.5
Other renewables	0	0	0	0	0	0	0	0	3.6	8.3
Transport	343	342	340	324	292	275	100	100	0.1	-0.7
Oil	306	301	297	286	233	201	87	73	-0.2	-1.6
Electricity	8	10	11	9	14	19	3	7	2.1	4.2
Biofuels	25	26	27	25	41	49	8	18	3.7	6.0
Other fuels	3	5	6	4	5	6	2	2	3.4	3.3
Buildings	491	544	570	475	496	508	100	100	0.8	0.4
Coal	8	6	5	8	4	3	1	1	-2.7	-4.4
Oil	64	61	60	58	44	37	10	7	-0.6	-2.4
Gas	178	194	202	168	166	163	35	32	0.7	-0.1
Electricity	154	178	191	151	162	164	33	32	1.2	0.7
Heat	39	44	46	36	36	35	8	7	1.2	0.1
Biomass and waste	43	52	57	49	70	84	10	16	2.0	3.4
Other renewables	4	7	9	5	14	23	2	4	6.3	9.9
Other	130	127	125	122	118	117	100	100	-0.4	-0.6

European Union: New Policies Scenario

Electricity generation (TWh)								Shares (%)		CAAGR (%)
	1990	2008	2015	2020	2025	2030	2035	2008	2035	2008-2035
Total generation	2 568	3 339	3 437	3 572	3 703	3 832	3 938	100	100	0.6
Coal	1 050	940	815	668	590	517	389	28	10	-3.2
Oil	221	105	58	42	29	24	23	3	1	-5.5
Gas	191	786	791	853	893	934	936	24	24	0.6
Nuclear	795	937	912	937	922	910	963	28	24	0.1
Hydro	286	327	355	369	380	392	402	10	10	0.8
Biomass and waste	20	110	148	183	217	247	274	3	7	3.4
Wind	1	119	313	446	557	647	723	4	18	6.9
Geothermal	3	6	8	10	14	18	21	0	1	4.9
Solar PV	0	7	27	48	74	91	102	0	3	10.2
CSP	-	0	8	13	23	39	56	0	1	35.3
Marine	1	1	1	2	4	14	48	0	1	18.3

Electrical capacity (GW)							Shares (%)		CAAGR (%)
	2008	2015	2020	2025	2030	2035	2008	2035	2008-2035
Total capacity	835	1 002	1 026	1 044	1 090	1 149	100	100	1.2
Coal	201	217	183	146	132	115	24	10	-2.0
Oil	71	60	43	31	23	22	8	2	-4.2
Gas	183	224	219	225	244	265	22	23	1.4
Nuclear	139	135	138	135	132	137	17	12	-0.1
Hydro	143	155	161	166	170	174	17	15	0.7
Biomass and waste	20	26	31	37	42	46	2	4	3.1
Wind	65	149	199	235	263	286	8	25	5.7
Geothermal	1	2	2	3	3	3	0	0	3.6
Solar PV	11	30	45	59	66	71	1	6	7.2
CSP	0	3	4	7	11	17	0	1	20.2
Marine	0	0	1	1	4	14	0	1	16.0

CO_2 emissions (Mt)								Shares (%)		CAAGR (%)
	1990	2008	2015	2020	2025	2030	2035	2008	2035	2008-2035
Total CO_2	4 037	3 808	3 542	3 348	3 173	2 996	2 780	100	100	-1.2
Coal	1 727	1 180	1 008	838	725	610	452	31	16	-3.5
Oil	1 643	1 613	1 513	1 454	1 364	1 274	1 210	42	44	-1.1
Gas	666	1 015	1 022	1 057	1 084	1 112	1 118	27	40	0.4
Power generation	1 491	1 377	1 195	1 049	972	899	765	100	100	-2.2
Coal	1 169	949	798	643	553	458	317	69	41	-4.0
Oil	195	79	46	33	23	19	18	6	2	-5.3
Gas	127	349	351	373	396	422	430	25	56	0.8
TFC	2 374	2 251	2 181	2 137	2 042	1 943	1 863	100	100	-0.7
Coal	521	199	182	169	148	131	115	9	6	-2.0
Oil	1 332	1 418	1 355	1 309	1 230	1 145	1 081	63	58	-1.0
Transport	745	934	908	887	829	767	729	41	39	-0.9
Gas	520	634	644	659	665	668	666	28	36	0.2

European Union: Current Policies and 450 Scenarios

	Electricity generation (TWh)						Shares (%)		CAAGR (%)	
	2020	2030	2035	2020	2030	2035	2035		2008-2035	
	Current Policies Scenario			450 Scenario			CPS	450	CPS	450
Total generation	3 614	3 934	4 094	3 540	3 706	3 771	100	100	0.8	0.5
Coal	878	810	799	660	237	299	20	8	-0.6	-4.2
Oil	44	31	30	38	16	14	1	0	-4.6	-7.2
Gas	860	993	1 057	719	583	293	26	8	1.1	-3.6
Nuclear	808	782	753	985	1 180	1 248	18	33	-0.8	1.1
Hydro	362	382	389	377	406	416	10	11	0.7	0.9
Biomass and waste	170	225	252	190	268	304	6	8	3.1	3.8
Wind	425	586	647	468	752	853	16	23	6.5	7.6
Geothermal	9	13	15	12	24	30	0	1	3.6	6.3
Solar PV	43	69	74	62	148	180	2	5	8.9	12.5
CSP	12	33	50	27	68	84	1	2	34.8	37.3
Marine	1	9	29	2	23	50	1	1	16.1	18.4

	Electrical capacity (GW)						Shares (%)		CAAGR (%)	
	2020	2030	2035	2020	2030	2035	2035		2008-2035	
	Current Policies Scenario			450 Scenario			CPS	450	CPS	450
Total capacity	1 003	1 076	1 114	1 036	1 148	1 219	100	100	1.1	1.4
Coal	194	162	147	162	96	90	13	7	-1.1	-2.9
Oil	43	24	23	41	16	14	2	1	-4.1	-5.8
Gas	220	259	281	218	215	216	25	18	1.6	0.6
Nuclear	118	113	106	143	168	176	10	14	-1.0	0.9
Hydro	158	166	169	165	177	181	15	15	0.6	0.9
Biomass and waste	29	38	42	33	45	50	4	4	2.8	3.5
Wind	192	245	264	206	294	324	24	27	5.4	6.2
Geothermal	2	2	3	2	4	5	0	0	2.7	4.7
Solar PV	41	53	55	56	106	124	5	10	6.2	9.4
CSP	4	10	15	8	20	25	1	2	19.7	22.2
Marine	0	3	8	1	7	14	1	1	13.8	16.0

	CO_2 emissions (Mt)						Shares (%)		CAAGR (%)	
	2020	2030	2035	2020	2030	2035	2035		2008-2035	
	Current Policies Scenario			450 Scenario			CPS	450	CPS	450
Total CO_2	3 612	3 518	3 498	3 167	2 192	1 837	100	100	-0.3	-2.7
Coal	1 035	880	828	799	238	171	24	9	-1.3	-6.9
Oil	1 505	1 453	1 425	1 395	1 119	974	41	53	-0.5	-1.9
Gas	1 072	1 186	1 245	974	834	692	36	38	0.8	-1.4
Power generation	1 245	1 187	1 190	966	393	246	100	100	-0.5	-6.2
Coal	833	709	671	618	140	104	56	42	-1.3	-7.9
Oil	35	24	23	30	13	12	2	5	-4.5	-6.7
Gas	378	454	495	318	239	130	42	53	1.3	-3.6
TFC	2 199	2 165	2 142	2 045	1 660	1 460	100	100	-0.2	-1.6
Coal	176	148	135	157	78	48	6	3	-1.4	-5.1
Oil	1 355	1 309	1 279	1 257	1 007	868	60	59	-0.4	-1.8
Transport	*917*	*902*	*890*	*856*	*698*	*603*	*42*	*41*	*-0.2*	*-1.6*
Gas	669	708	727	631	575	544	34	37	0.5	-0.6

A

OECD Pacific: New Policies Scenario

	Energy demand (Mtoe)							Shares (%)		CAAGR (%)
	1990	2008	2015	2020	2025	2030	2035	2008	2035	2008-2035
TPED	**631**	**870**	**908**	**914**	**913**	**916**	**905**	**100**	**100**	**0.1**
Coal	138	236	240	222	200	174	139	27	15	-1.9
Oil	335	349	321	299	281	267	254	40	28	-1.2
Gas	66	145	160	169	170	175	184	17	20	0.9
Nuclear	66	107	141	168	195	221	238	12	26	3.0
Hydro	11	10	11	12	12	13	13	1	1	1.1
Biomass and waste	10	17	23	27	32	37	41	2	5	3.4
Other renewables	4	7	11	16	22	29	36	1	4	6.4
Power generation	**241**	**392**	**431**	**447**	**460**	**473**	**473**	**100**	**100**	**0.7**
Coal	60	155	160	144	124	102	71	40	15	-2.8
Oil	56	31	15	10	9	9	10	8	2	-4.3
Gas	40	78	85	89	88	89	95	20	20	0.7
Nuclear	66	107	141	168	195	221	238	27	50	3.0
Hydro	11	10	11	12	12	13	13	2	3	1.1
Biomass and waste	3	6	9	11	13	15	17	2	4	3.7
Other renewables	3	6	9	13	19	23	28	1	6	6.2
Other energy sector	**57**	**84**	**81**	**79**	**77**	**74**	**71**	**100**	**100**	**-0.6**
Electricity	*11*	*17*	*19*	*19*	*19*	*19*	*18*	*20*	*26*	*0.3*
TFC	**431**	**555**	**575**	**574**	**568**	**564**	**558**	**100**	**100**	**0.0**
Coal	49	43	45	44	41	38	35	8	6	-0.8
Oil	261	292	283	269	256	245	234	53	42	-0.8
Gas	26	64	70	74	76	78	79	12	14	0.8
Electricity	86	139	155	162	167	170	172	25	31	0.8
Heat	0	5	6	6	6	6	6	1	1	0.4
Biomass and waste	7	10	14	16	19	21	24	2	4	3.2
Other renewables	2	1	2	3	4	5	8	0	1	7.3
Industry	**145**	**160**	**177**	**181**	**179**	**176**	**172**	**100**	**100**	**0.3**
Coal	39	40	42	41	38	36	33	25	19	-0.8
Oil	51	35	35	34	31	29	27	22	15	-1.0
Gas	10	21	25	27	27	27	27	13	16	0.9
Electricity	40	53	62	64	65	65	65	33	38	0.7
Heat	-	3	3	3	3	3	3	2	2	0.1
Biomass and waste	5	8	10	12	14	16	17	5	10	3.0
Other renewables	0	0	0	0	0	0	0	0	0	0.6
Transport	**110**	**139**	**129**	**120**	**112**	**107**	**102**	**100**	**100**	**-1.1**
Oil	109	136	124	113	105	99	93	97	91	-1.4
Electricity	2	2	3	3	3	4	5	2	4	2.9
Biofuels	-	0	1	1	1	2	2	0	2	7.1
Other fuels	0	1	2	2	3	3	3	1	3	3.2
Buildings	**120**	**170**	**179**	**185**	**189**	**194**	**198**	**100**	**100**	**0.6**
Coal	10	2	2	2	1	1	1	1	1	-1.0
Oil	47	40	39	38	37	36	35	24	17	-0.6
Gas	15	40	42	44	45	46	47	24	24	0.6
Electricity	44	83	90	94	98	100	102	49	51	0.8
Heat	0	2	2	3	3	3	3	1	1	0.9
Biomass and waste	2	2	2	3	3	4	5	1	2	3.2
Other renewables	1	1	1	2	3	4	6	1	3	7.2
Other	**56**	**85**	**89**	**89**	**88**	**87**	**86**	**100**	**100**	**0.0**

OECD Pacific: Current Policies and 450 Scenarios

	Energy demand (Mtoe)						Shares (%)		CAAGR (%)	
	2020	2030	2035	2020	2030	2035	2035		2008-2035	
	Current Policies Scenario			450 Scenario			CPS	450	CPS	450
TPED	932	956	963	888	863	850	100	100	0.4	-0.1
Coal	246	235	222	202	112	91	23	11	-0.2	-3.5
Oil	306	279	270	285	235	208	28	24	-1.0	-1.9
Gas	167	181	191	164	161	145	20	17	1.0	0.0
Nuclear	163	197	209	173	244	264	22	31	2.5	3.4
Hydro	12	12	12	13	15	16	1	2	0.9	1.9
Biomass and waste	24	30	33	32	52	64	3	8	2.6	5.2
Other renewables	15	22	26	20	44	61	3	7	5.1	8.5
Power generation	459	496	509	434	437	438	100	100	1.0	0.4
Coal	164	152	140	127	42	24	28	6	-0.4	-6.6
Oil	13	12	14	7	5	5	3	1	-2.9	-6.5
Gas	86	92	97	86	77	60	19	14	0.8	-0.9
Nuclear	163	197	209	173	244	264	41	60	2.5	3.4
Hydro	12	12	12	13	15	16	2	4	0.9	1.9
Biomass and waste	10	13	15	12	18	22	3	5	3.1	4.6
Other renewables	12	18	21	16	35	47	4	11	5.0	8.1
Other energy sector	81	80	79	76	69	65	100	100	-0.2	-0.9
Electricity	19	20	20	18	17	17	25	26	0.6	-0.0
TFC	583	586	587	560	542	532	100	100	0.2	-0.2
Coal	45	43	41	41	36	34	7	6	-0.2	-0.9
Oil	274	256	248	260	218	195	42	37	-0.6	-1.5
Gas	75	81	84	73	77	77	14	15	1.0	0.7
Electricity	166	180	186	157	163	163	32	31	1.1	0.6
Heat	6	6	6	6	6	6	1	1	0.7	0.2
Biomass and waste	15	17	18	20	34	43	3	8	2.2	5.5
Other renewables	2	4	4	3	9	15	1	3	5.1	9.9
Industry	184	184	183	176	173	170	100	100	0.5	0.2
Coal	43	40	38	38	34	31	21	19	-0.2	-0.9
Oil	35	31	29	33	28	25	16	15	-0.6	-1.2
Gas	26	28	29	26	29	30	16	18	1.2	1.3
Electricity	66	69	70	63	63	62	38	36	1.0	0.6
Heat	3	3	3	3	3	3	2	2	0.4	-0.1
Biomass and waste	11	12	13	13	16	18	7	11	2.0	3.1
Other renewables	0	0	0	0	0	0	0	0	0.6	0.6
Transport	123	113	109	116	98	91	100	100	-0.9	-1.6
Oil	116	105	101	109	82	68	92	75	-1.1	-2.5
Electricity	3	4	4	3	6	8	4	9	2.4	5.2
Biofuels	1	1	2	2	7	11	1	13	6.6	15.0
Other fuels	2	3	3	3	3	3	3	3	3.2	3.4
Buildings	188	203	210	179	184	186	100	100	0.8	0.3
Coal	2	1	1	2	1	1	1	1	-0.8	-1.4
Oil	39	38	38	36	32	29	18	16	-0.2	-1.1
Gas	44	48	50	42	43	42	24	23	0.8	0.2
Electricity	96	106	111	90	93	91	53	49	1.1	0.4
Heat	3	3	3	3	3	3	1	1	1.0	0.6
Biomass and waste	3	3	3	3	5	6	2	3	1.7	4.1
Other renewables	2	3	4	3	8	13	2	7	5.1	10.3
Other	89	86	85	89	87	86	100	100	-0.0	0.0

OECD Pacific: New Policies Scenario

	Electricity generation (TWh)						Shares (%)		CAAGR (%)
	1990	2008	2015	2020	2025	2030	2035	2008	2035	2008-2035
Total generation	1 127	1 820	2 023	2 106	2 160	2 201	2 219	100	100	0.7
Coal	257	682	711	652	573	475	348	38	16	-2.5
Oil	270	157	74	45	40	38	40	9	2	-4.9
Gas	197	414	485	510	495	487	510	23	23	0.8
Nuclear	255	409	542	646	750	849	915	22	41	3.0
Hydro	133	114	132	137	142	147	152	6	7	1.1
Biomass and waste	12	26	36	44	52	60	68	1	3	3.6
Wind	-	8	25	42	62	83	106	0	5	10.0
Geothermal	4	7	9	11	14	16	18	0	1	3.7
Solar PV	0	3	7	15	26	36	45	0	2	11.0
CSP	-	0	2	3	5	7	11	0	0	34.0
Marine	-	-	0	1	2	3	5	-	0	n.a.

	Electrical capacity (GW)						Shares (%)		CAAGR (%)
	2008	2015	2020	2025	2030	2035	2008	2035	2008-2035
Total capacity	412	462	481	508	545	573	100	100	1.2
Coal	102	107	105	100	97	92	25	16	-0.4
Oil	57	55	37	33	30	31	14	5	-2.2
Gas	107	134	144	149	162	170	26	30	1.7
Nuclear	66	74	86	98	110	119	16	21	2.2
Hydro	68	69	71	72	73	74	16	13	0.3
Biomass and waste	4	6	7	9	10	11	1	2	3.4
Wind	4	10	15	22	28	34	1	6	8.6
Geothermal	1	1	2	2	3	3	0	0	3.4
Solar PV	3	6	13	22	29	35	1	6	10.1
CSP	0	1	1	1	2	3	0	1	15.0
Marine	-	0	0	0	1	1	-	0	n.a.

	CO_2 emissions (Mt)						Shares (%)		CAAGR (%)
	1990	2008	2015	2020	2025	2030	2035	2008	2035	2008-2035
Total CO_2	1 573	2 069	2 047	1 937	1 791	1 653	1 482	100	100	-1.2
Coal	521	880	911	841	737	621	464	43	31	-2.3
Oil	892	841	753	693	650	619	588	41	40	-1.3
Gas	160	348	383	402	404	413	429	17	29	0.8
Power generation	548	934	927	851	756	657	529	100	100	-2.1
Coal	280	653	676	608	519	418	276	70	52	-3.1
Oil	174	96	48	31	28	28	30	10	6	-4.3
Gas	94	185	204	212	208	211	223	20	42	0.7
TFC	956	1 037	1 022	986	940	902	860	100	100	-0.7
Coal	220	191	198	194	182	169	155	18	18	-0.8
Oil	676	697	661	620	582	553	522	67	61	-1.1
Transport	*321*	*399*	*364*	*334*	*308*	*292*	*274*	*38*	*32*	*-1.4*
Gas	60	149	163	172	176	180	183	14	21	0.8

OECD Pacific: Current Policies and 450 Scenarios

	Electricity generation (TWh)						Shares (%)		CAAGR (%)	
	2020	2030	2035	2020	2030	2035	2035		2008-2035	
	Current Policies Scenario			450 Scenario			CPS	450	CPS	450
Total generation	2 153	2 322	2 398	2 047	2 094	2 088	100	100	1.0	0.5
Coal	732	704	679	577	205	118	28	6	-0.0	-6.3
Oil	58	53	61	31	20	20	3	1	-3.4	-7.4
Gas	506	506	521	483	431	327	22	16	0.9	-0.9
Nuclear	624	755	802	665	936	1 014	33	49	2.5	3.4
Hydro	134	140	144	146	178	191	6	9	0.9	1.9
Biomass and waste	37	51	57	46	70	83	2	4	3.0	4.4
Wind	37	65	75	62	158	200	3	10	8.6	12.6
Geothermal	11	14	15	12	19	23	1	1	2.9	4.6
Solar PV	12	28	32	21	59	78	1	4	9.6	13.3
CSP	2	5	7	4	14	26	0	1	32.2	38.5
Marine	0	1	3	1	4	8	0	0	n.a.	n.a.

	Electrical capacity (GW)						Shares (%)		CAAGR (%)	
	2020	2030	2035	2020	2030	2035	2035		2008-2035	
	Current Policies Scenario			450 Scenario			CPS	450	CPS	450
Total capacity	478	531	553	473	531	552	100	100	1.1	1.1
Coal	107	104	101	104	65	42	18	8	-0.1	-3.3
Oil	37	30	31	31	18	16	6	3	-2.3	-4.7
Gas	146	168	178	125	127	129	32	23	1.9	0.7
Nuclear	85	102	108	87	121	132	19	24	1.8	2.6
Hydro	69	71	71	74	82	84	13	15	0.2	0.8
Biomass and waste	6	9	10	8	12	14	2	2	2.8	4.2
Wind	14	22	25	23	51	63	5	12	7.3	11.0
Geothermal	2	2	2	2	3	4	0	1	2.7	4.3
Solar PV	11	22	25	18	47	60	4	11	8.7	12.3
CSP	1	1	2	1	4	7	0	1	13.4	18.7
Marine	0	0	1	0	1	2	0	0	n.a.	n.a.

	CO_2 emissions (Mt)						Shares (%)		CAAGR (%)	
	2020	2030	2035	2020	2030	2035	2035		2008-2035	
	Current Policies Scenario			450 Scenario			CPS	450	CPS	450
Total CO_2	2 051	1 958	1 903	1 790	1 106	856	100	100	-0.3	-3.2
Coal	936	867	807	742	260	132	42	15	-0.3	-6.8
Oil	716	663	647	659	522	452	34	53	-1.0	-2.3
Gas	398	428	449	389	324	273	24	32	0.9	-0.9
Power generation	937	895	863	749	295	155	100	100	-0.3	-6.4
Coal	692	638	588	525	131	35	68	23	-0.4	-10.2
Oil	39	38	44	22	16	16	5	10	-2.9	-6.5
Gas	206	219	231	203	148	104	27	67	0.8	-2.1
TFC	1 010	963	939	947	727	621	100	100	-0.4	-1.9
Coal	202	190	180	181	97	65	19	11	-0.2	-3.9
Oil	635	585	565	597	472	406	60	65	-0.8	-2.0
Transport	342	309	297	320	243	200	32	32	-1.1	-2.5
Gas	173	188	194	169	158	150	21	24	1.0	0.0

A

Japan: New Policies Scenario

	Energy demand (Mtoe)							Shares (%)		CAAGR (%)
	1990	2008	2015	2020	2025	2030	2035	2008	2035	2008-2035
TPED	**439**	**496**	**495**	**491**	**486**	**482**	**470**	**100**	**100**	**-0.2**
Coal	77	114	112	102	87	74	57	23	12	-2.5
Oil	250	214	181	164	153	144	135	43	29	-1.7
Gas	44	84	89	93	93	94	98	17	21	0.6
Nuclear	53	67	90	105	121	134	138	14	29	2.7
Hydro	8	7	8	8	8	8	8	1	2	1.0
Biomass and waste	5	7	10	12	13	14	15	1	3	2.9
Other renewables	3	3	5	8	11	14	17	1	4	6.1
Power generation	**174**	**220**	**237**	**246**	**254**	**261**	**261**	**100**	**100**	**0.6**
Coal	25	60	62	53	42	34	21	27	8	-3.8
Oil	51	26	11	7	6	6	7	12	3	-4.7
Gas	33	52	56	59	58	57	61	24	23	0.6
Nuclear	53	67	90	105	121	134	138	31	53	2.7
Hydro	8	7	8	8	8	8	8	3	3	1.0
Biomass and waste	2	4	6	8	9	10	11	2	4	3.3
Other renewables	1	3	4	7	10	12	14	1	5	6.2
Other energy sector	**38**	**50**	**45**	**42**	**38**	**34**	**29**	**100**	**100**	**-2.0**
Electricity	*7*	*9*	*10*	*10*	*10*	*10*	*10*	*19*	*33*	*0.1*
TFC	**300**	**319**	**313**	**308**	**302**	**296**	**288**	**100**	**100**	**-0.4**
Coal	32	28	29	29	28	26	24	9	8	-0.6
Oil	184	171	155	145	136	128	120	54	42	-1.3
Gas	15	33	34	35	36	36	37	10	13	0.5
Electricity	64	83	90	94	96	98	98	26	34	0.6
Heat	0	1	1	1	1	1	1	0	0	1.5
Biomass and waste	3	3	3	4	4	4	5	1	2	2.2
Other renewables	1	1	1	1	2	2	3	0	1	5.9
Industry	**103**	**87**	**94**	**95**	**92**	**89**	**85**	**100**	**100**	**-0.1**
Coal	31	27	28	28	26	25	23	32	27	-0.7
Oil	37	23	24	23	21	20	18	27	21	-0.9
Gas	4	7	9	10	11	11	12	9	14	1.7
Electricity	29	26	30	31	30	30	29	30	34	0.4
Heat	-	-	-	-	-	-	-	-	-	n.a.
Biomass and waste	3	3	3	3	4	4	4	3	4	1.5
Other renewables	-	-	-	-	-	-	-	-	-	n.a.
Transport	**72**	**78**	**63**	**56**	**51**	**47**	**43**	**100**	**100**	**-2.1**
Oil	70	76	61	54	48	44	39	98	90	-2.5
Electricity	1	2	2	2	2	3	3	2	8	2.9
Biofuels	-	-	0	0	1	1	1	-	2	n.a.
Other fuels	-	-	0	0	0	0	0	-	1	n.a.
Buildings	**84**	**113**	**115**	**118**	**120**	**122**	**124**	**100**	**100**	**0.3**
Coal	1	1	1	1	1	1	1	0	1	1.1
Oil	36	32	30	30	29	29	28	28	23	-0.5
Gas	11	25	24	24	25	25	25	22	20	0.0
Electricity	34	55	58	61	63	65	65	49	53	0.6
Heat	0	1	1	1	1	1	1	1	1	1.5
Biomass and waste	0	0	0	0	0	0	0	0	0	7.3
Other renewables	1	1	1	1	2	2	3	1	3	6.2
Other	**41**	**41**	**41**	**40**	**39**	**37**	**36**	**100**	**100**	**-0.4**

Japan: Current Policies and 450 Scenarios

	Energy demand (Mtoe)						Shares (%)		CAAGR (%)	
	2020	2030	2035	2020	2030	2035	2035		2008-2035	
	Current Policies Scenario			450 Scenario			CPS	450	CPS	450
TPED	497	497	495	476	454	440	100	100	-0.0	-0.4
Coal	109	97	89	90	52	37	18	8	-0.9	-4.1
Oil	169	153	149	156	126	111	30	25	-1.3	-2.4
Gas	93	98	102	89	74	65	21	15	0.7	-0.9
Nuclear	101	117	122	109	149	161	25	36	2.2	3.3
Hydro	8	8	8	8	10	11	2	2	1.0	1.8
Biomass and waste	10	12	13	13	19	21	3	5	2.4	4.2
Other renewables	8	11	12	11	25	34	2	8	4.7	8.8
Power generation	248	267	275	238	242	241	100	100	0.8	0.3
Coal	59	53	48	45	13	3	18	1	-0.8	-10.7
Oil	9	10	12	4	3	3	4	1	-2.9	-8.0
Gas	59	62	64	55	38	28	23	12	0.7	-2.3
Nuclear	101	117	122	109	149	161	45	67	2.2	3.3
Hydro	8	8	8	8	10	11	3	4	1.0	1.8
Biomass and waste	6	9	10	8	10	11	4	5	2.9	3.5
Other renewables	6	9	10	9	19	25	4	10	4.9	8.4
Other energy sector	42	35	31	40	31	27	100	100	-1.7	-2.3
Electricity	*11*	*10*	*10*	*10*	*9*	*9*	*32*	*33*	*0.3*	*-0.2*
TFC	312	307	304	300	282	272	100	100	-0.2	-0.6
Coal	30	29	27	27	24	22	9	8	-0.2	-0.9
Oil	148	134	128	140	115	102	42	38	-1.1	-1.9
Gas	35	37	38	35	37	37	13	14	0.6	0.5
Electricity	95	101	104	90	92	90	34	33	0.8	0.3
Heat	1	1	1	1	1	1	0	0	1.6	0.9
Biomass and waste	3	4	4	5	8	10	1	4	1.2	5.2
Other renewables	1	2	2	2	6	9	1	3	3.7	10.1
Industry	96	93	91	92	87	84	100	100	0.2	-0.1
Coal	29	28	26	26	23	21	29	25	-0.2	-0.9
Oil	23	21	19	22	18	16	21	20	-0.7	-1.3
Gas	10	11	12	11	13	15	13	18	1.6	2.5
Electricity	31	31	31	30	29	28	34	33	0.6	0.2
Heat	-	-	-	-	-	-	-	-	n.a.	n.a.
Biomass and waste	3	3	3	3	4	4	3	4	0.6	1.4
Other renewables	-	-	-	-	-	-	-	-	n.a.	n.a.
Transport	57	50	47	55	44	39	100	100	-1.9	-2.6
Oil	55	47	43	52	38	30	92	79	-2.1	-3.4
Electricity	2	3	3	2	4	5	6	14	2.1	4.5
Biofuels	0	0	1	1	2	3	1	7	n.a.	n.a.
Other fuels	0	0	0	0	0	0	1	1	n.a.	n.a.
Buildings	119	127	131	113	114	114	100	100	0.5	0.0
Coal	1	1	1	1	1	1	1	1	1.3	0.5
Oil	31	31	31	29	25	23	24	21	-0.1	-1.1
Gas	25	26	26	24	23	22	20	19	0.2	-0.4
Electricity	61	67	70	58	59	57	54	50	0.9	0.1
Heat	1	1	1	1	1	1	1	1	1.6	0.9
Biomass and waste	0	0	0	0	0	1	0	0	2.7	12.5
Other renewables	1	2	2	2	5	9	1	8	4.0	10.5
Other	40	37	36	40	37	36	100	100	-0.5	-0.4

A

Japan: New Policies Scenario

	Electricity generation (TWh)							Shares (%)		CAAGR (%)
	1990	2008	2015	2020	2025	2030	2035	2008	2035	2008-2035
Total generation	836	1 075	1 169	1 211	1 237	1 252	1 252	100	100	0.6
Coal	117	288	297	261	212	168	110	27	9	-3.5
Oil	248	139	59	33	29	29	32	13	3	-5.2
Gas	167	283	332	353	341	324	337	26	27	0.6
Nuclear	202	258	346	403	463	514	531	24	42	2.7
Hydro	89	76	87	89	91	95	99	7	8	1.0
Biomass and waste	11	22	30	35	40	44	47	2	4	2.8
Wind	-	3	10	20	34	45	57	0	5	12.0
Geothermal	2	3	4	5	6	7	7	0	1	3.8
Solar PV	0	2	5	11	20	25	29	0	2	10.0
CSP	-	-	-	-	-	-	-	-	-	n.a.
Marine	-	-	-	-	0	1	3	-	0	n.a.

	Electrical capacity (GW)						Shares (%)		CAAGR (%)
	2008	2015	2020	2025	2030	2035	2008	2035	2008-2035
Total capacity	268	290	296	311	333	346	100	100	0.9
Coal	46	46	44	40	39	37	17	11	-0.8
Oil	50	48	32	29	27	28	19	8	-2.2
Gas	68	83	91	94	103	107	25	31	1.7
Nuclear	48	50	56	63	69	71	18	21	1.5
Hydro	47	48	48	49	49	50	18	14	0.2
Biomass and waste	4	5	6	7	7	8	1	2	2.8
Wind	2	5	8	12	16	19	1	5	8.9
Geothermal	1	1	1	1	1	1	0	0	3.4
Solar PV	2	5	10	17	21	24	1	7	9.4
CSP	-	-	-	-	-	-	-	-	n.a.
Marine	-	-	-	0	0	1	-	0	n.a.

	CO_2 emissions (Mt)							Shares (%)		CAAGR (%)
	1990	2008	2015	2020	2025	2030	2035	2008	2035	2008-2035
Total CO_2	1 063	1 147	1 072	998	915	843	763	100	100	-1.5
Coal	293	414	424	387	331	280	212	36	28	-2.4
Oil	655	529	432	387	360	339	319	46	42	-1.9
Gas	115	204	215	224	224	224	233	18	31	0.5
Power generation	363	469	441	396	344	301	253	100	100	-2.3
Coal	128	263	271	234	186	144	87	56	34	-4.0
Oil	157	80	35	20	19	20	22	17	9	-4.7
Gas	78	126	135	141	139	137	144	27	57	0.5
TFC	655	637	592	566	537	510	481	100	100	-1.0
Coal	150	135	137	136	130	122	112	21	23	-0.7
Oil	470	426	377	349	324	303	282	67	59	-1.5
Transport	*208*	*225*	*179*	*158*	*142*	*129*	*115*	*35*	*24*	*-2.4*
Gas	35	76	78	81	83	85	87	12	18	0.5

Japan: Current Policies and 450 Scenarios

	Electricity generation (TWh)						Shares (%)		CAAGR (%)	
	2020	2030	2035	2020	2030	2035	2035		2008-2035	
	Current Policies Scenario			450 Scenario			CPS	450	CPS	450
Total generation	1 227	1 298	1 328	1 170	1 178	1 158	100	100	0.8	0.3
Coal	286	260	245	219	68	14	18	1	-0.6	-10.5
Oil	45	45	54	23	13	13	4	1	-3.5	-8.5
Gas	361	351	353	321	217	151	27	13	0.8	-2.3
Nuclear	386	449	469	420	571	616	35	53	2.2	3.3
Hydro	89	94	99	92	112	123	7	11	1.0	1.8
Biomass and waste	29	38	42	36	46	50	3	4	2.4	3.0
Wind	17	35	39	37	98	122	3	11	10.5	15.3
Geothermal	5	5	6	5	8	11	0	1	2.6	5.1
Solar PV	9	20	22	18	43	52	2	5	8.7	12.4
CSP	-	-	-	-	-	-	-	-	n.a.	n.a.
Marine	-	0	1	-	1	4	0	0	n.a.	n.a.

	Electrical capacity (GW)						Shares (%)		CAAGR (%)	
	2020	2030	2035	2020	2030	2035	2035		2008-2035	
	Current Policies Scenario			450 Scenario			CPS	450	CPS	450
Total capacity	292	320	330	285	312	309	100	100	0.8	0.5
Coal	44	40	37	43	24	5	11	2	-0.8	-7.7
Oil	32	27	28	26	14	12	8	4	-2.2	-5.2
Gas	92	105	110	73	68	62	33	20	1.8	-0.3
Nuclear	56	64	66	57	77	83	20	27	1.2	2.0
Hydro	48	49	50	49	53	54	15	18	0.2	0.5
Biomass and waste	5	6	7	6	8	8	2	3	2.3	3.0
Wind	7	12	14	14	32	39	4	13	7.6	11.9
Geothermal	1	1	1	1	1	2	0	1	2.4	4.7
Solar PV	8	16	18	15	36	43	5	14	8.2	11.7
CSP	-	-	-	-	-	-	-	-	n.a.	n.a.
Marine	-	0	0	-	0	1	0	0	n.a.	n.a.

	CO_2 emissions (Mt)						Shares (%)		CAAGR (%)	
	2020	2030	2035	2020	2030	2035	2035		2008-2035	
	Current Policies Scenario			450 Scenario			CPS	450	CPS	450
Total CO_2	1 044	986	958	917	580	444	100	100	-0.7	-3.5
Coal	416	382	355	337	133	61	37	14	-0.6	-6.8
Oil	402	368	358	366	282	244	37	55	-1.4	-2.8
Gas	226	237	245	214	165	139	26	31	0.7	-1.4
Power generation	428	409	403	341	143	70	100	100	-0.6	-6.8
Coal	259	231	213	196	51	3	53	4	-0.8	-15.8
Oil	27	30	36	13	8	9	9	12	-2.9	-8.0
Gas	143	148	154	132	84	59	38	84	0.7	-2.8
TFC	578	542	523	542	409	348	100	100	-0.7	-2.2
Coal	141	135	127	126	69	47	24	13	-0.2	-3.9
Oil	356	321	306	335	261	224	59	64	-1.2	-2.4
Transport	*161*	*138*	*128*	*153*	*111*	*90*	*24*	*26*	*-2.1*	*-3.4*
Gas	81	86	89	81	79	78	17	22	0.6	0.1

Non-OECD: New Policies Scenario

	Energy demand (Mtoe)							Shares (%)		CAAGR (%)
	1990	2008	2015	2020	2025	2030	2035	2008	2035	2008-2035
TPED	4 103	6 516	7 952	8 660	9 315	10 002	10 690	100	100	1.9
Coal	1 165	2 187	2 799	2 949	3 050	3 139	3 220	34	30	1.4
Oil	1 173	1 688	1 972	2 108	2 254	2 412	2 569	26	24	1.6
Gas	834	1 325	1 625	1 791	1 957	2 137	2 297	20	21	2.1
Nuclear	76	120	185	297	370	431	482	2	5	5.3
Hydro	84	163	210	250	288	317	340	3	3	2.8
Biomass and waste	763	995	1 085	1 142	1 210	1 298	1 402	15	13	1.3
Other renewables	8	37	75	122	187	267	379	1	4	9.0
Power generation	1 283	2 364	2 970	3 367	3 724	4 097	4 488	100	100	2.4
Coal	478	1 265	1 635	1 743	1 833	1 912	1 996	53	44	1.7
Oil	225	188	153	144	131	114	103	8	2	-2.2
Gas	406	581	691	776	847	930	991	25	22	2.0
Nuclear	76	120	185	297	370	431	482	5	11	5.3
Hydro	84	163	210	250	288	317	340	7	8	2.8
Biomass and waste	7	18	32	52	90	154	239	1	5	10.1
Other renewables	8	30	64	106	165	237	336	1	7	9.4
Other energy sector	505	820	963	1 023	1 079	1 128	1 172	100	100	1.3
Electricity	77	166	208	234	258	281	305	20	26	2.3
TFC	3 010	4 392	5 427	5 886	6 320	6 779	7 232	100	100	1.9
Coal	532	688	883	910	909	906	899	16	12	1.0
Oil	830	1 366	1 681	1 825	1 989	2 170	2 338	31	32	2.0
Gas	360	571	721	789	861	941	1 026	13	14	2.2
Electricity	287	651	928	1 104	1 268	1 438	1 614	15	22	3.4
Heat	293	199	221	227	230	232	232	5	3	0.6
Biomass and waste	708	910	980	1 014	1 041	1 062	1 080	21	15	0.6
Other renewables	0	7	11	16	22	30	43	0	1	6.8
Industry	988	1 502	1 995	2 171	2 303	2 434	2 556	100	100	2.0
Coal	311	536	712	741	746	752	758	36	30	1.3
Oil	159	207	248	252	255	255	251	14	10	0.7
Gas	140	213	285	317	350	385	419	14	16	2.5
Electricity	159	336	500	593	666	737	804	22	31	3.3
Heat	137	89	104	106	108	109	110	6	4	0.8
Biomass and waste	81	120	146	163	178	195	213	8	8	2.1
Other renewables	-	0	0	0	0	0	0	0	0	-4.0
Transport	444	774	982	1 115	1 277	1 468	1 658	100	100	2.9
Oil	377	687	873	990	1 130	1 294	1 451	89	88	2.8
Electricity	13	13	18	22	26	31	36	2	2	3.8
Biofuels	6	15	31	41	55	69	87	2	5	6.8
Other fuels	48	58	60	62	66	74	83	8	5	1.3
Buildings	1 272	1 621	1 826	1 945	2 053	2 156	2 263	100	100	1.2
Coal	170	104	115	113	106	96	85	6	4	-0.7
Oil	125	174	195	202	206	208	208	11	9	0.7
Gas	124	189	225	251	278	305	335	12	15	2.1
Electricity	90	270	369	444	524	611	707	17	31	3.6
Heat	147	108	114	118	120	120	119	7	5	0.4
Biomass and waste	616	770	796	803	800	787	768	47	34	-0.0
Other renewables	0	7	11	15	21	29	41	0	2	6.6
Other	306	496	624	654	687	722	756	100	100	1.6

Non-OECD: Current Policies and 450 Scenarios

	Energy demand (Mtoe)						Shares (%)		CAAGR (%)	
	2020	2030	2035	2020	2030	2035	2035		2008-2035	
	Current Policies Scenario			450 Scenario			CPS	450	CPS	450
TPED	**8 920**	**10 712**	**11 696**	**8 395**	**8 969**	**9 296**	**100**	**100**	**2.2**	**1.3**
Coal	3 190	3 852	4 226	2 799	2 260	1 999	36	22	2.5	-0.3
Oil	2 168	2 563	2 768	2 021	2 123	2 153	24	23	1.8	0.9
Gas	1 816	2 262	2 520	1 699	1 836	1 887	22	20	2.4	1.3
Nuclear	284	371	398	309	583	711	3	8	4.5	6.8
Hydro	240	286	308	256	346	377	3	4	2.4	3.2
Biomass and waste	1 121	1 205	1 257	1 156	1 402	1 553	11	17	0.9	1.7
Other renewables	100	173	219	156	419	615	2	7	6.8	10.9
Power generation	**3 531**	**4 521**	**5 108**	**3 258**	**3 553**	**3 775**	**100**	**100**	**2.9**	**1.7**
Coal	1 940	2 482	2 803	1 631	1 169	942	55	25	3.0	-1.1
Oil	151	130	124	136	92	86	2	2	-1.5	-2.9
Gas	789	1 010	1 149	733	769	775	22	21	2.6	1.1
Nuclear	284	371	398	309	583	711	8	19	4.5	6.8
Hydro	240	286	308	256	346	377	6	10	2.4	3.2
Biomass and waste	42	95	139	62	224	336	3	9	8.0	11.5
Other renewables	85	148	188	131	371	548	4	15	7.0	11.4
Other energy sector	**1 048**	**1 210**	**1 283**	**987**	**1 009**	**1 007**	**100**	**100**	**1.7**	**0.8**
Electricity	*242*	*305*	*339*	*224*	*246*	*256*	*26*	*25*	*2.7*	*1.6*
TFC	**6 004**	**7 139**	**7 744**	**5 707**	**6 207**	**6 437**	**100**	**100**	**2.1**	**1.4**
Coal	942	1 003	1 038	880	803	773	13	12	1.5	0.4
Oil	1 877	2 315	2 529	1 747	1 905	1 943	33	30	2.3	1.3
Gas	797	966	1 064	753	844	894	14	14	2.3	1.7
Electricity	1 138	1 551	1 791	1 064	1 306	1 433	23	22	3.8	3.0
Heat	232	248	254	220	203	192	3	3	0.9	-0.1
Biomass and waste	1 004	1 031	1 037	1 018	1 098	1 134	13	18	0.5	0.8
Other renewables	14	25	31	24	48	67	0	1	5.6	8.6
Industry	**2 225**	**2 607**	**2 818**	**2 095**	**2 220**	**2 272**	**100**	**100**	**2.4**	**1.5**
Coal	768	836	876	714	663	649	31	29	1.8	0.7
Oil	263	279	285	240	225	214	10	9	1.2	0.1
Gas	322	396	437	303	346	364	16	16	2.7	2.0
Electricity	610	806	916	571	685	731	33	32	3.8	2.9
Heat	108	115	119	103	98	93	4	4	1.1	0.2
Biomass and waste	154	174	185	164	203	221	7	10	1.6	2.3
Other renewables	0	0	0	0	0	0	0	0	-4.0	-4.0
Transport	**1 139**	**1 547**	**1 762**	**1 072**	**1 318**	**1 434**	**100**	**100**	**3.1**	**2.3**
Oil	1 017	1 385	1 571	936	1 085	1 125	89	78	3.1	1.8
Electricity	21	29	34	23	46	71	2	5	3.5	6.4
Biofuels	40	61	76	51	111	149	4	10	6.3	9.0
Other fuels	60	72	81	62	76	89	5	6	1.2	1.6
Buildings	**1 977**	**2 244**	**2 383**	**1 892**	**1 967**	**2 002**	**100**	**100**	**1.4**	**0.8**
Coal	116	108	102	110	86	72	4	4	-0.1	-1.4
Oil	210	228	233	192	186	183	10	9	1.1	0.2
Gas	255	318	354	233	255	266	15	13	2.3	1.3
Electricity	459	652	767	424	518	567	32	28	3.9	2.8
Heat	121	129	132	114	102	97	6	5	0.8	-0.4
Biomass and waste	802	786	766	796	773	753	32	38	-0.0	-0.1
Other renewables	14	24	30	24	47	65	1	3	5.4	8.5
Other	**663**	**741**	**781**	**648**	**701**	**729**	**100**	**100**	**1.7**	**1.4**

Non-OECD: New Policies Scenario

	Electricity generation (TWh)							Shares (%)		CAAGR (%)
	1990	2008	2015	2020	2025	2030	2035	2008	2035	2008-2035
Total generation	4 261	9 510	13 223	15 579	17 754	20 002	22 318	100	100	3.2
Coal	1 369	4 391	6 397	7 034	7 556	8 101	8 706	46	39	2.6
Oil	655	726	581	546	496	431	385	8	2	-2.3
Gas	957	1 939	2 720	3 206	3 652	4 141	4 505	20	20	3.2
Nuclear	288	458	709	1 137	1 418	1 655	1 850	5	8	5.3
Hydro	975	1 895	2 439	2 913	3 346	3 690	3 958	20	18	2.8
Biomass and waste	8	47	98	164	297	517	808	0	4	11.1
Wind	0	31	212	446	730	1 018	1 357	0	6	15.1
Geothermal	8	24	40	59	84	116	155	0	1	7.2
Solar PV	0	0	14	43	115	222	374	0	2	31.1
CSP	-	-	14	32	59	110	216	-	1	n.a.
Marine	-	-	-	0	0	1	4	-	0	n.a.

	Electrical capacity (GW)						Shares (%)		CAAGR (%)
	2008	2015	2020	2025	2030	2035	2008	2035	2008-2035
Total capacity	2 173	3 098	3 624	4 100	4 621	5 216	100	100	3.3
Coal	858	1 252	1 384	1 462	1 574	1 709	39	33	2.6
Oil	222	238	224	198	170	155	10	3	-1.3
Gas	485	697	782	870	977	1 090	22	21	3.0
Nuclear	66	97	152	188	219	244	3	5	4.9
Hydro	501	653	790	916	1 015	1 089	23	21	2.9
Biomass and waste	13	22	33	55	90	134	1	3	9.2
Wind	24	114	204	295	380	483	1	9	11.8
Geothermal	4	7	9	13	18	23	0	0	6.8
Solar PV	0	13	36	84	148	231	0	4	27.6
CSP	0	5	10	17	30	57	0	1	25.6
Marine	-	-	0	0	0	1	-	0	n.a.

	CO_2 emissions (Mt)							Shares (%)		CAAGR (%)
	1990	2008	2015	2020	2025	2030	2035	2008	2035	2008-2035
Total CO_2	9 307	15 682	19 477	20 774	21 865	22 990	24 039	100	100	1.6
Coal	4 198	8 256	10 659	11 184	11 474	11 723	11 961	53	50	1.4
Oil	3 215	4 500	5 233	5 626	6 064	6 546	7 007	29	29	1.7
Gas	1 894	2 926	3 586	3 965	4 327	4 720	5 071	19	21	2.1
Power generation	3 568	6 984	8 613	9 187	9 640	10 057	10 440	100	100	1.5
Coal	1 901	5 022	6 505	6 911	7 240	7 518	7 800	72	75	1.6
Oil	718	600	485	455	414	362	326	9	3	-2.2
Gas	948	1 362	1 623	1 821	1 986	2 176	2 313	20	22	2.0
TFC	5 314	7 882	9 901	10 557	11 153	11 814	12 434	100	100	1.7
Coal	2 219	3 039	3 927	4 031	4 011	3 984	3 946	39	32	1.0
Oil	2 295	3 586	4 402	4 802	5 256	5 764	6 233	46	50	2.1
Transport	*1 119*	*2 043*	*2 595*	*2 944*	*3 360*	*3 848*	*4 314*	*26*	*35*	*2.8*
Gas	800	1 257	1 572	1 724	1 886	2 066	2 255	16	18	2.2

Non-OECD: Current Policies and 450 Scenarios

	Electricity generation (TWh)						Shares (%)		CAAGR (%)	
	2020	2030	2035	2020	2030	2035	2035		2008-2035	
	Current Policies Scenario			450 Scenario			CPS	450	CPS	450
Total generation	16 061	21 597	24 773	14 992	18 061	19 657	100	100	3.6	2.7
Coal	7 795	10 737	12 443	6 387	4 975	4 089	50	21	3.9	-0.3
Oil	576	499	476	512	333	307	2	2	-1.5	-3.1
Gas	3 239	4 466	5 226	2 993	3 337	3 426	21	17	3.7	2.1
Nuclear	1 090	1 423	1 526	1 186	2 235	2 729	6	14	4.6	6.8
Hydro	2 796	3 326	3 576	2 978	4 024	4 389	14	22	2.4	3.2
Biomass and waste	132	311	460	201	764	1 150	2	6	8.9	12.6
Wind	330	571	699	543	1 558	2 116	3	11	12.3	17.0
Geothermal	50	80	100	66	158	222	0	1	5.5	8.7
Solar PV	39	131	173	52	395	737	1	4	27.5	34.5
CSP	16	52	92	74	279	484	0	2	n.a.	n.a.
Marine	-	0	2	0	2	6	0	0	n.a.	n.a.

	Electrical capacity (GW)						Shares (%)		CAAGR (%)	
	2020	2030	2035	2020	2030	2035	2035		2008-2035	
	Current Policies Scenario			450 Scenario			CPS	450	CPS	450
Total capacity	3 675	4 830	5 525	3 610	4 468	5 109	100	100	3.5	3.2
Coal	1 506	2 080	2 424	1 324	987	931	44	18	3.9	0.3
Oil	227	186	182	222	169	155	3	3	-0.7	-1.3
Gas	811	1 060	1 230	741	871	933	22	18	3.5	2.5
Nuclear	146	189	201	158	295	359	4	7	4.2	6.4
Hydro	754	902	972	808	1 112	1 217	18	24	2.5	3.3
Biomass and waste	28	58	83	39	128	186	1	4	7.2	10.5
Wind	157	235	276	241	550	720	5	14	9.5	13.5
Geothermal	8	13	16	11	23	32	0	1	5.4	8.2
Solar PV	33	93	117	44	255	446	2	9	24.4	30.7
CSP	5	14	24	22	77	129	0	3	21.7	29.5
Marine	-	0	0	0	1	2	0	0	n.a.	n.a.

	CO_2 emissions (Mt)						Shares (%)		CAAGR (%)	
	2020	2030	2035	2020	2030	2035	2035		2008-2035	
	Current Policies Scenario			450 Scenario			CPS	450	CPS	450
Total CO_2	21 937	26 469	29 049	19 703	16 671	14 794	100	100	2.3	-0.2
Coal	12 119	14 450	15 842	10 590	7 222	5 284	55	36	2.4	-1.6
Oil	5 801	7 036	7 653	5 357	5 588	5 635	26	38	2.0	0.8
Gas	4 018	4 984	5 555	3 756	3 861	3 875	19	26	2.4	1.0
Power generation	10 031	12 600	14 164	8 610	6 046	4 421	100	100	2.7	-1.7
Coal	7 701	9 823	11 085	6 458	4 055	2 490	78	56	3.0	-2.6
Oil	478	411	391	430	292	271	3	6	-1.6	-2.9
Gas	1 852	2 366	2 688	1 722	1 700	1 659	19	38	2.6	0.7
TFC	10 851	12 687	13 637	10 118	9 655	9 419	100	100	2.1	0.7
Coal	4 161	4 381	4 508	3 902	2 986	2 631	33	28	1.5	-0.5
Oil	4 947	6 183	6 786	4 571	4 913	4 969	50	53	2.4	1.2
Transport	*3 025*	*4 119*	*4 673*	*2 783*	*3 227*	*3 346*	*34*	*36*	*3.1*	*1.8*
Gas	1 743	2 123	2 344	1 644	1 755	1 820	17	19	2.3	1.4

Eastern Europe/Eurasia: New Policies Scenario

	Energy demand (Mtoe)							Shares (%)		CAAGR (%)
	1990	2008	2015	2020	2025	2030	2035	2008	2035	2008-2035
TPED	1 560	1 151	1 207	1 254	1 302	1 344	1 386	100	100	0.7
Coal	374	228	227	213	213	207	203	20	15	-0.4
Oil	478	231	238	245	251	251	254	20	18	0.4
Gas	606	570	605	627	652	672	682	50	49	0.7
Nuclear	61	78	84	106	111	121	128	7	9	1.8
Hydro	23	24	27	28	30	33	35	2	3	1.4
Biomass and waste	18	19	23	27	31	39	50	2	4	3.7
Other renewables	0	1	4	7	13	21	34	0	2	16.1
Power generation	751	554	572	596	618	638	656	100	100	0.6
Coal	203	146	145	133	135	131	126	26	19	-0.5
Oil	127	18	14	12	10	8	7	3	1	-3.5
Gas	333	282	292	301	307	308	300	51	46	0.2
Nuclear	61	78	84	106	111	121	128	14	19	1.8
Hydro	23	24	27	28	30	33	35	4	5	1.4
Biomass and waste	4	5	7	9	12	18	26	1	4	6.4
Other renewables	0	0	3	7	12	21	33	0	5	17.3
Other energy sector	205	192	193	193	195	197	199	100	100	0.1
Electricity	*36*	*41*	*41*	*42*	*44*	*46*	*47*	*21*	*24*	*0.5*
TFC	1 081	731	785	820	857	887	918	100	100	0.8
Coal	115	42	46	46	46	46	46	6	5	0.3
Oil	285	182	193	202	211	216	222	25	24	0.7
Gas	261	242	265	276	291	305	319	33	35	1.0
Electricity	128	105	116	127	137	146	155	14	17	1.4
Heat	279	146	149	151	152	152	152	20	17	0.1
Biomass and waste	13	13	16	17	19	21	23	2	3	2.0
Other renewables	-	0	0	0	0	1	1	0	0	7.2
Industry	402	222	246	256	266	276	286	100	100	0.9
Coal	57	31	34	35	35	35	35	14	12	0.5
Oil	53	22	24	25	25	26	27	10	9	0.8
Gas	91	65	76	80	85	90	95	29	33	1.4
Electricity	76	49	56	61	65	69	73	22	25	1.4
Heat	126	54	54	54	53	53	53	24	19	-0.0
Biomass and waste	0	1	2	2	2	3	3	1	1	3.4
Other renewables	-	0	0	0	0	0	0	0	0	-4.0
Transport	169	147	158	169	179	186	194	100	100	1.0
Oil	123	97	104	112	120	123	127	66	66	1.0
Electricity	12	9	10	12	13	15	16	6	9	2.3
Biofuels	0	0	1	1	2	3	3	0	2	8.9
Other fuels	34	41	42	43	45	46	47	28	24	0.5
Buildings	389	272	287	299	310	319	326	100	100	0.7
Coal	56	10	10	9	9	8	8	4	2	-0.9
Oil	41	20	19	19	18	17	16	7	5	-0.8
Gas	108	100	109	115	122	127	132	37	40	1.0
Electricity	27	42	45	49	52	56	58	16	18	1.2
Heat	144	89	91	94	95	96	96	33	29	0.3
Biomass and waste	12	11	13	13	14	15	16	4	5	1.2
Other renewables	-	0	0	0	0	0	1	0	0	7.2
Other	122	90	95	97	101	106	112	100	100	0.8

Eastern Europe/Eurasia: Current Policies and 450 Scenarios

| | Energy demand (Mtoe) | | | | | | Shares (%) | | CAAGR (%) | |
| | 2020 | 2030 | 2035 | 2020 | 2030 | 2035 | 2035 | | 2008-2035 | |
	Current Policies Scenario			450 Scenario			CPS	450	CPS	450
TPED	**1 275**	**1 402**	**1 470**	**1 207**	**1 219**	**1 227**	**100**	**100**	**0.9**	**0.2**
Coal	228	249	259	192	153	135	18	11	0.5	-1.9
Oil	251	263	270	236	232	227	18	18	0.6	-0.1
Gas	630	698	740	601	581	554	50	45	1.0	-0.1
Nuclear	106	117	117	110	129	141	8	12	1.5	2.2
Hydro	27	30	31	29	40	44	2	4	0.9	2.2
Biomass and waste	26	33	37	30	51	68	2	6	2.5	4.9
Other renewables	6	12	16	9	33	58	1	5	12.8	18.4
Power generation	**605**	**666**	**699**	**574**	**583**	**596**	**100**	**100**	**0.9**	**0.3**
Coal	145	165	173	116	84	70	25	12	0.6	-2.7
Oil	12	8	7	12	8	7	1	1	-3.3	-3.3
Gas	300	321	340	289	264	240	49	40	0.7	-0.6
Nuclear	106	117	117	110	129	141	17	24	1.5	2.2
Hydro	27	30	31	29	40	44	4	7	0.9	2.2
Biomass and waste	9	13	15	10	25	38	2	6	4.2	7.9
Other renewables	6	12	15	8	31	55	2	9	13.9	19.5
Other energy sector	**196**	**205**	**210**	**187**	**176**	**169**	**100**	**100**	**0.3**	**-0.5**
Electricity	*43*	*48*	*50*	*40*	*39*	*38*	*24*	*22*	*0.8*	*-0.3*
TFC	**837**	**931**	**981**	**790**	**799**	**796**	**100**	**100**	**1.1**	**0.3**
Coal	49	52	53	43	39	37	5	5	0.9	-0.5
Oil	207	228	239	193	196	193	24	24	1.0	0.2
Gas	280	314	332	264	268	267	34	34	1.2	0.4
Electricity	130	155	169	123	135	140	17	18	1.8	1.1
Heat	154	163	167	146	133	127	17	16	0.5	-0.5
Biomass and waste	17	20	21	19	26	29	2	4	1.7	2.9
Other renewables	0	0	1	1	1	2	0	0	6.0	10.4
Industry	**260**	**288**	**306**	**241**	**240**	**237**	**100**	**100**	**1.2**	**0.2**
Coal	36	38	40	32	29	28	13	12	1.0	-0.4
Oil	26	28	30	23	23	23	10	10	1.1	0.1
Gas	80	91	96	75	77	76	32	32	1.5	0.6
Electricity	62	73	80	58	61	62	26	26	1.8	0.8
Heat	55	56	57	51	47	45	19	19	0.2	-0.7
Biomass and waste	2	2	3	2	3	4	1	2	2.4	3.9
Other renewables	0	0	0	0	0	0	0	0	-4.0	-4.0
Transport	**171**	**193**	**203**	**162**	**169**	**168**	**100**	**100**	**1.2**	**0.5**
Oil	115	132	139	107	110	106	68	63	1.4	0.3
Electricity	12	14	15	12	17	21	8	12	2.0	3.1
Biofuels	1	2	2	2	4	5	1	3	7.7	11.0
Other fuels	43	46	47	41	39	36	23	22	0.5	-0.4
Buildings	**308**	**341**	**357**	**291**	**285**	**282**	**100**	**100**	**1.0**	**0.1**
Coal	10	10	10	9	7	6	3	2	0.3	-1.8
Oil	19	19	18	18	15	14	5	5	-0.2	-1.2
Gas	118	134	142	111	112	111	40	39	1.3	0.4
Electricity	50	59	63	47	49	50	18	18	1.5	0.6
Heat	96	103	106	91	83	80	30	28	0.6	-0.4
Biomass and waste	13	15	16	15	18	19	4	7	1.2	2.0
Other renewables	0	0	1	1	1	2	0	1	6.2	10.5
Other	**98**	**108**	**115**	**95**	**104**	**110**	**100**	**100**	**0.9**	**0.7**

Eastern Europe/Eurasia: New Policies Scenario

	1990	2008	2015	2020	2025	2030	2035	Shares (%) 2008	Shares (%) 2035	CAAGR (%) 2008-2035
Electricity generation (TWh)										
Total generation	1 924	1 713	1 848	1 985	2 125	2 253	2 366	100	100	1.2
Coal	448	438	438	404	417	408	394	26	17	-0.4
Oil	271	33	18	14	8	5	4	2	0	-7.6
Gas	706	655	740	790	844	861	830	38	35	0.9
Nuclear	231	299	320	405	426	462	489	17	21	1.8
Hydro	269	284	309	328	351	378	409	17	17	1.4
Biomass and waste	0	3	11	19	30	53	84	0	4	12.9
Wind	-	1	8	18	33	63	122	0	5	22.2
Geothermal	0	0	3	6	11	17	25	0	1	16.0
Solar PV	-	0	1	2	3	5	9	0	0	33.0
CSP	-	-	-	-	-	-	-	-	-	n.a.
Marine	-	-	-	0	0	0	1	-	0	n.a.

	2008	2015	2020	2025	2030	2035	Shares (%) 2008	Shares (%) 2035	CAAGR (%) 2008-2035
Electrical capacity (GW)									
Total capacity	416	452	463	479	502	538	100	100	1.0
Coal	109	109	98	92	84	77	26	14	-1.3
Oil	28	24	17	10	7	7	7	1	-5.0
Gas	145	170	175	185	192	196	35	37	1.1
Nuclear	43	45	56	59	63	66	10	12	1.6
Hydro	89	97	102	108	116	124	21	23	1.2
Biomass and waste	1	2	4	6	9	14	0	3	10.1
Wind	0	4	8	13	23	43	0	8	18.1
Geothermal	0	1	1	2	2	3	0	1	14.3
Solar PV	0	1	2	4	5	7	0	1	30.0
CSP	-	-	-	-	-	-	-	-	n.a.
Marine	-	-	0	0	0	0	-	0	n.a.

	1990	2008	2015	2020	2025	2030	2035	Shares (%) 2008	Shares (%) 2035	CAAGR (%) 2008-2035
CO_2 emissions (Mt)										
Total CO_2	4 039	2 679	2 781	2 799	2 869	2 888	2 894	100	100	0.3
Coal	1 364	831	842	791	794	773	755	31	26	-0.4
Oil	1 266	582	597	616	631	634	644	22	22	0.4
Gas	1 409	1 266	1 342	1 391	1 444	1 481	1 495	47	52	0.6
Power generation	2 007	1 331	1 335	1 296	1 305	1 277	1 232	100	100	-0.3
Coal	823	606	604	552	555	533	511	46	42	-0.6
Oil	405	61	45	39	32	26	23	5	2	-3.6
Gas	779	664	686	705	718	718	697	50	57	0.2
TFC	1 918	1 228	1 315	1 368	1 424	1 465	1 509	100	100	0.8
Coal	530	215	228	230	231	232	235	18	16	0.3
Oil	795	464	494	519	541	549	561	38	37	0.7
Transport	*363*	*285*	*307*	*332*	*355*	*364*	*376*	*23*	*25*	*1.0*
Gas	594	548	592	619	652	684	713	45	47	1.0

Eastern Europe/Eurasia: Current Policies and 450 Scenarios

	Electricity generation (TWh)						Shares (%)		CAAGR (%)	
	2020	2030	2035	2020	2030	2035	2035		2008-2035	
	Current Policies Scenario			450 Scenario			CPS	450	CPS	450
Total generation	2 034	2 379	2 565	1 926	2 049	2 096	100	100	1.5	0.8
Coal	469	562	594	327	233	149	23	7	1.1	-3.9
Oil	13	8	6	17	8	6	0	0	-6.0	-6.3
Gas	788	926	1 034	769	630	491	40	23	1.7	-1.1
Nuclear	405	448	450	420	496	541	18	26	1.5	2.2
Hydro	317	345	358	342	466	512	14	24	0.9	2.2
Biomass and waste	17	34	43	21	77	125	2	6	10.1	14.6
Wind	18	43	64	20	105	215	2	10	19.3	24.8
Geothermal	5	9	10	7	25	41	0	2	12.2	18.1
Solar PV	2	4	6	3	9	14	0	1	30.9	35.4
CSP	-	-	-	-	-	-	-	-	n.a.	n.a.
Marine	-	0	0	0	0	1	0	0	n.a.	n.a.

	Electrical capacity (GW)						Shares (%)		CAAGR (%)	
	2020	2030	2035	2020	2030	2035	2035		2008-2035	
	Current Policies Scenario			450 Scenario			CPS	450	CPS	450
Total capacity	474	503	529	450	502	549	100	100	0.9	1.0
Coal	110	107	106	93	71	59	20	11	-0.1	-2.3
Oil	17	9	8	17	6	5	2	1	-4.3	-5.9
Gas	178	192	206	158	154	146	39	27	1.3	0.0
Nuclear	56	61	61	58	68	74	11	13	1.3	2.0
Hydro	98	106	109	107	140	153	21	28	0.8	2.0
Biomass and waste	4	6	8	4	13	21	1	4	7.6	11.7
Wind	7	17	24	8	38	74	4	14	15.6	20.6
Geothermal	1	1	2	1	3	5	0	1	11.2	16.2
Solar PV	2	4	5	3	7	12	1	2	28.4	32.3
CSP	-	-	-	-	-	-	-	-	n.a.	n.a.
Marine	-	0	0	0	0	0	0	0	n.a.	n.a.

	CO_2 emissions (Mt)						Shares (%)		CAAGR (%)	
	2020	2030	2035	2020	2030	2035	2035		2008-2035	
	Current Policies Scenario			450 Scenario			CPS	450	CPS	450
Total CO_2	2 882	3 159	3 309	2 634	2 256	2 071	100	100	0.8	-0.9
Coal	853	949	990	707	482	395	30	19	0.7	-2.7
Oil	630	676	701	591	566	542	21	26	0.7	-0.3
Gas	1 399	1 534	1 619	1 336	1 208	1 133	49	55	0.9	-0.4
Power generation	1 344	1 459	1 526	1 205	903	777	100	100	0.5	-2.0
Coal	602	683	712	485	310	244	47	31	0.6	-3.3
Oil	39	27	24	41	27	24	2	3	-3.4	-3.4
Gas	703	749	789	679	566	508	52	65	0.6	-1.0
TFC	1 402	1 546	1 622	1 299	1 224	1 167	100	100	1.0	-0.2
Coal	242	257	268	214	165	144	17	12	0.8	-1.5
Oil	531	585	610	493	481	460	38	39	1.0	-0.0
Transport	_340_	_389_	_410_	_316_	_324_	_312_	_25_	_27_	_1.4_	_0.3_
Gas	629	705	744	592	578	563	46	48	1.1	0.1

Caspian: New Policies Scenario

	\multicolumn{7}{c}{Energy demand (Mtoe)}	Shares (%)		CAAGR (%)
	1990	2008	2015	2020	2025	2030	2035	2008	2035	2008-2035
TPED	**198**	**169**	**205**	**220**	**234**	**241**	**247**	**100**	**100**	**1.4**
Coal	48	33	40	41	42	40	39	19	16	0.7
Oil	63	28	35	37	39	40	42	17	17	1.5
Gas	83	102	123	133	144	149	152	60	62	1.5
Nuclear	-	1	1	2	2	2	2	0	1	4.9
Hydro	4	5	5	6	6	6	7	3	3	1.3
Biomass and waste	1	1	1	1	1	2	2	0	1	5.4
Other renewables	0	0	0	0	0	1	1	0	1	18.3
Power generation	**72**	**54**	**66**	**72**	**77**	**78**	**78**	**100**	**100**	**1.4**
Coal	27	21	25	26	27	25	25	40	32	0.7
Oil	14	1	1	1	0	0	0	2	0	-3.9
Gas	26	26	34	38	42	43	42	48	54	1.8
Nuclear	-	1	1	2	2	2	2	1	3	4.9
Hydro	4	5	5	6	6	6	7	9	9	1.3
Biomass and waste	-	-	0	0	0	0	1	-	1	n.a.
Other renewables	0	0	0	0	0	1	1	0	1	35.8
Other energy sector	**18**	**34**	**38**	**40**	**42**	**43**	**45**	**100**	**100**	**1.0**
Electricity	*3*	*5*	*6*	*6*	*6*	*7*	*7*	*14*	*15*	*1.3*
TFC	**149**	**112**	**138**	**147**	**157**	**162**	**166**	**100**	**100**	**1.5**
Coal	21	10	13	13	13	13	12	9	7	0.8
Oil	42	23	29	31	35	36	38	20	23	2.0
Gas	49	55	65	70	74	77	79	49	47	1.4
Electricity	19	13	17	19	21	22	23	12	14	2.1
Heat	18	12	12	12	13	12	12	10	7	0.2
Biomass and waste	1	1	1	1	1	1	2	0	1	4.1
Other renewables	-	0	0	0	0	0	1	0	0	14.6
Industry	**50**	**38**	**50**	**54**	**58**	**59**	**61**	**100**	**100**	**1.8**
Coal	18	9	12	12	12	11	11	23	18	0.8
Oil	7	3	4	5	5	5	5	9	9	1.8
Gas	13	16	21	24	27	29	30	42	49	2.4
Electricity	11	6	7	8	9	9	10	15	17	2.1
Heat	1	4	5	5	5	5	5	12	8	0.1
Biomass and waste	-	-	-	-	-	-	-	-	-	n.a.
Other renewables	-	-	-	-	-	-	-	-	-	n.a.
Transport	**14**	**13**	**17**	**18**	**20**	**22**	**24**	**100**	**100**	**2.3**
Oil	12	11	15	16	18	19	22	84	90	2.6
Electricity	1	0	0	1	1	1	1	3	3	2.2
Biofuels	-	-	0	0	0	0	0	-	0	n.a.
Other fuels	0	2	2	2	2	2	2	12	7	0.0
Buildings	**74**	**53**	**60**	**62**	**65**	**67**	**68**	**100**	**100**	**0.9**
Coal	3	0	0	0	0	0	0	1	0	-3.6
Oil	17	5	6	6	6	6	6	9	8	0.7
Gas	34	35	39	40	42	43	44	67	64	0.8
Electricity	3	4	6	7	8	8	9	8	13	2.5
Heat	17	7	8	8	8	8	8	13	11	0.3
Biomass and waste	1	1	1	1	1	1	2	1	2	4.0
Other renewables	-	0	0	0	0	0	1	0	1	14.6
Other	**11**	**9**	**12**	**12**	**13**	**14**	**14**	**100**	**100**	**1.6**

Caspian: Current Policies and 450 Scenarios

	Energy demand (Mtoe)						Shares (%)		CAAGR (%)	
	2020	2030	2035	2020	2030	2035	2035		2008-2035	
	Current Policies Scenario			450 Scenario			CPS	450	CPS	450
TPED	224	256	267	205	211	208	100	100	1.7	0.8
Coal	42	44	45	29	22	21	17	10	1.1	-1.7
Oil	40	46	49	35	37	38	18	18	2.1	1.1
Gas	134	156	162	131	137	131	61	63	1.7	0.9
Nuclear	2	2	2	2	2	2	1	1	4.9	4.9
Hydro	5	6	6	6	7	8	2	4	0.9	2.0
Biomass and waste	1	2	2	1	2	3	1	1	4.6	6.3
Other renewables	0	1	1	1	3	4	0	2	16.9	23.7
Power generation	71	84	87	66	70	70	100	100	1.8	1.0
Coal	26	29	30	16	11	10	34	15	1.3	-2.7
Oil	0	0	0	0	0	0	0	0	-7.9	-8.0
Gas	37	46	48	41	47	44	55	63	2.3	2.0
Nuclear	2	2	2	2	2	2	3	3	4.9	4.9
Hydro	5	6	6	6	7	8	7	12	0.9	2.0
Biomass and waste	0	0	1	0	1	1	1	1	n.a.	n.a.
Other renewables	0	0	1	1	2	4	1	5	33.9	43.8
Other energy sector	40	45	47	38	39	39	100	100	1.2	0.5
Electricity	7	7	8	6	6	6	16	15	1.8	0.6
TFC	153	175	183	137	140	137	100	100	1.8	0.7
Coal	14	13	13	12	10	8	7	6	1.1	-0.5
Oil	34	41	45	29	32	33	25	24	2.6	1.3
Gas	71	80	82	65	65	62	45	45	1.5	0.4
Electricity	20	25	27	18	21	21	15	16	2.7	1.9
Heat	13	14	15	12	11	10	8	7	0.9	-0.5
Biomass and waste	1	1	1	1	2	2	1	1	3.4	4.8
Other renewables	0	0	0	0	0	1	0	1	13.6	16.3
Industry	54	61	62	47	43	39	100	100	1.8	0.1
Coal	13	12	12	11	8	7	19	19	1.1	-0.7
Oil	5	6	6	4	4	4	9	9	1.9	0.3
Gas	24	27	28	20	19	16	46	43	2.2	0.2
Electricity	9	11	12	8	8	8	19	22	2.7	1.5
Heat	5	5	5	4	3	3	8	8	0.2	-1.6
Biomass and waste	-	-	-	-	-	-	-	-	n.a.	n.a.
Other renewables	-	-	-	-	-	-	-	-	n.a.	n.a.
Transport	20	26	30	17	20	21	100	100	3.1	1.9
Oil	18	24	27	15	17	18	91	85	3.4	1.9
Electricity	1	1	1	1	1	1	3	6	2.3	4.2
Biofuels	0	0	0	0	0	0	0	1	n.a.	n.a.
Other fuels	2	2	2	2	2	2	6	8	0.2	0.2
Buildings	66	74	77	61	64	64	100	100	1.4	0.7
Coal	0	0	0	0	0	0	0	0	-1.2	-1.9
Oil	6	7	8	6	6	6	10	10	1.7	0.9
Gas	43	47	48	40	41	40	63	63	1.1	0.4
Electricity	7	9	9	7	8	8	12	12	2.8	2.0
Heat	9	9	10	8	8	7	13	11	1.2	0.1
Biomass and waste	1	1	1	1	1	2	2	3	3.1	4.4
Other renewables	0	0	0	0	0	1	1	1	13.6	16.2
Other	13	14	15	12	13	14	100	100	1.8	1.5

Caspian: New Policies Scenario

Electricity generation (TWh)	1990	2008	2015	2020	2025	2030	2035	Shares (%) 2008	Shares (%) 2035	CAAGR (%) 2008-2035
Total generation	239	209	262	289	314	332	344	100	100	1.9
Coal	68	66	78	79	81	77	80	31	23	0.8
Oil	45	4	2	1	1	1	1	2	0	-4.1
Gas	75	80	116	136	155	167	166	38	48	2.7
Nuclear	-	2	3	6	6	9	9	1	3	4.9
Hydro	51	57	63	65	67	73	79	27	23	1.3
Biomass and waste	-	-	0	1	1	1	2	-	1	n.a.
Wind	-	0	0	1	2	3	5	0	1	33.3
Geothermal	-	0	0	0	0	0	0	0	0	21.0
Solar PV	-	-	0	0	0	1	1	-	0	n.a.
CSP	-	-	-	-	-	-	-	-	-	n.a.
Marine	-	-	-	-	-	-	-	-	-	n.a.

Electrical capacity (GW)	2008	2015	2020	2025	2030	2035	Shares (%) 2008	Shares (%) 2035	CAAGR (%) 2008-2035
Total capacity	55	67	73	78	82	86	100	100	1.7
Coal	15	18	20	20	19	17	28	20	0.4
Oil	4	3	2	2	1	1	6	1	-4.9
Gas	23	31	35	38	41	44	41	52	2.6
Nuclear	0	0	1	1	1	1	1	2	4.6
Hydro	13	14	14	15	16	18	23	21	1.3
Biomass and waste	0	0	0	0	0	0	0	0	15.0
Wind	0	0	1	1	1	2	0	2	18.0
Geothermal	0	0	0	0	0	0	0	0	15.9
Solar PV	-	0	0	1	1	1	-	1	n.a.
CSP	-	-	-	-	-	-	-	-	n.a.
Marine	-	-	-	-	-	-	-	-	n.a.

CO_2 emissions (Mt)	1990	2008	2015	2020	2025	2030	2035	Shares (%) 2008	Shares (%) 2035	CAAGR (%) 2008-2035
Total CO_2	548	412	502	534	569	576	586	100	100	1.3
Coal	184	118	149	154	159	150	147	29	25	0.8
Oil	177	76	93	98	107	111	117	19	20	1.6
Gas	186	217	260	282	304	315	322	53	55	1.5
Power generation	212	147	179	192	204	201	199	100	100	1.1
Coal	105	80	98	101	105	99	99	54	50	0.8
Oil	46	6	2	2	1	1	1	4	1	-5.4
Gas	61	62	80	90	98	100	99	42	50	1.8
TFC	320	239	287	305	324	332	341	100	100	1.3
Coal	79	38	51	53	54	51	48	16	14	0.8
Oil	124	64	84	90	98	103	109	27	32	2.0
Transport	35	32	43	47	53	57	63	13	19	2.6
Gas	117	137	152	162	172	179	184	57	54	1.1

Caspian: Current Policies and 450 Scenarios

	Electricity generation (TWh)						Shares (%) 2035		CAAGR (%) 2008-2035	
	2020	2030	2035	2020	2030	2035	CPS	450	CPS	450
	Current Policies Scenario			450 Scenario						
Total generation	311	376	402	281	310	323	100	100	2.5	1.6
Coal	104	124	130	52	33	34	32	11	2.6	-2.4
Oil	1	0	0	1	0	0	0	0	-8.8	-8.8
Gas	134	169	184	147	170	164	46	51	3.1	2.7
Nuclear	6	9	9	6	9	9	2	3	4.9	4.9
Hydro	63	68	72	70	86	97	18	30	0.9	2.0
Biomass and waste	1	1	2	1	2	3	0	1	n.a.	n.a.
Wind	1	3	4	1	5	10	1	3	32.7	37.2
Geothermal	0	0	0	1	2	3	0	1	17.1	31.0
Solar PV	0	1	1	1	2	2	0	1	n.a.	n.a.
CSP	-	-	-	-	-	-	-	-	n.a.	n.a.
Marine	-	-	-	-	-	-	-	-	n.a.	n.a.

	Electrical capacity (GW)						Shares (%) 2035		CAAGR (%) 2008-2035	
	2020	2030	2035	2020	2030	2035	CPS	450	CPS	450
	Current Policies Scenario			450 Scenario						
Total capacity	75	82	87	74	82	86	100	100	1.7	1.7
Coal	23	24	22	18	18	16	25	18	1.3	0.1
Oil	2	1	1	2	1	1	1	1	-5.5	-5.5
Gas	35	38	44	35	37	37	50	43	2.5	1.8
Nuclear	1	1	1	1	1	1	2	2	4.6	4.6
Hydro	14	15	16	16	20	23	19	27	0.9	2.3
Biomass and waste	0	0	0	0	0	1	0	1	14.2	16.7
Wind	1	2	2	1	2	4	2	5	17.8	21.2
Geothermal	0	0	0	0	0	0	0	1	12.8	25.5
Solar PV	0	1	1	1	2	3	1	3	n.a.	n.a.
CSP	-	-	-	-	-	-	-	-	n.a.	n.a.
Marine	-	-	-	-	-	-	-	-	n.a.	n.a.

	CO_2 emissions (Mt)						Shares (%) 2035		CAAGR (%) 2008-2035	
	2020	2030	2035	2020	2030	2035	CPS	450	CPS	450
	Current Policies Scenario			450 Scenario						
Total CO_2	547	622	649	477	466	448	100	100	1.7	0.3
Coal	156	166	168	108	78	70	26	16	1.3	-1.9
Oil	106	128	139	91	98	100	21	22	2.3	1.0
Gas	285	328	342	278	291	278	53	62	1.7	0.9
Power generation	191	220	229	159	151	144	100	100	1.6	-0.1
Coal	101	113	116	60	41	40	51	28	1.4	-2.6
Oil	1	0	0	1	0	0	0	0	-9.4	-9.5
Gas	88	107	113	97	110	104	49	72	2.3	2.0
TFC	317	357	371	282	276	264	100	100	1.6	0.4
Coal	55	54	52	48	37	31	14	12	1.1	-0.8
Oil	96	116	127	82	88	89	34	34	2.6	1.2
Transport	53	69	79	44	51	53	21	20	3.4	1.9
Gas	167	186	192	152	151	144	52	55	1.3	0.2

Russia: New Policies Scenario

	Energy demand (Mtoe)							Shares (%)		CAAGR (%)
	1990	2008	2015	2020	2025	2030	2035	2008	2035	2008-2035
TPED	880	688	710	735	757	781	805	100	100	0.6
Coal	191	117	119	114	114	111	110	17	14	-0.2
Oil	264	141	140	143	145	143	144	21	18	0.1
Gas	367	366	378	387	397	406	407	53	51	0.4
Nuclear	31	43	49	61	64	70	74	6	9	2.1
Hydro	14	14	15	16	18	20	22	2	3	1.6
Biomass and waste	12	6	7	8	10	15	21	1	3	4.6
Other renewables	0	0	3	6	10	17	27	0	3	16.8
Power generation	444	361	374	391	403	415	427	100	100	0.6
Coal	105	75	79	77	79	77	78	21	18	0.1
Oil	62	13	10	9	8	6	5	3	1	-3.0
Gas	228	212	215	217	218	215	206	59	48	-0.1
Nuclear	31	43	49	61	64	70	74	12	17	2.1
Hydro	14	14	15	16	18	20	22	4	5	1.6
Biomass and waste	4	4	4	5	6	10	16	1	4	5.4
Other renewables	0	0	3	6	10	17	26	0	6	16.8
Other energy sector	127	121	120	119	119	120	120	100	100	-0.0
Electricity	*21*	*25*	*26*	*26*	*28*	*29*	*30*	*21*	*25*	*0.6*
TFC	625	436	453	469	486	502	518	100	100	0.6
Coal	55	17	16	16	15	14	14	4	3	-0.6
Oil	145	107	107	111	115	117	120	24	23	0.4
Gas	143	134	144	149	157	165	173	31	33	1.0
Electricity	71	62	69	75	80	86	91	14	17	1.4
Heat	203	114	114	115	115	115	114	26	22	0.0
Biomass and waste	8	2	3	3	4	5	5	1	1	3.0
Other renewables	-	-	0	0	0	0	0	-	0	n.a.
Industry	209	125	135	137	141	147	154	100	100	0.8
Coal	15	10	10	10	10	10	11	8	7	0.3
Oil	25	13	14	14	14	14	15	10	10	0.5
Gas	30	30	34	35	37	40	43	24	28	1.3
Electricity	41	31	35	38	40	42	45	25	29	1.4
Heat	98	41	41	40	40	39	39	32	26	-0.1
Biomass and waste	-	0	0	0	1	1	1	0	1	4.2
Other renewables	-	-	-	-	-	-	-	-	-	n.a.
Transport	116	97	100	106	112	116	120	100	100	0.8
Oil	73	55	55	59	62	63	65	56	54	0.6
Electricity	9	7	8	10	11	12	14	7	12	2.6
Biofuels	-	-	0	0	0	1	1	-	1	n.a.
Other fuels	34	35	37	38	39	40	40	36	34	0.5
Buildings	228	155	158	165	170	172	174	100	100	0.4
Coal	40	6	5	5	4	4	3	4	2	-2.6
Oil	12	10	9	8	7	6	6	6	3	-1.9
Gas	57	44	47	52	55	58	61	28	35	1.2
Electricity	15	23	24	26	27	28	29	15	17	0.8
Heat	98	70	70	72	73	73	72	45	42	0.1
Biomass and waste	7	2	2	2	3	3	3	1	2	1.9
Other renewables	-	-	0	0	0	0	0	-	0	n.a.
Other	72	59	60	61	63	66	70	100	100	0.7

Russia: Current Policies and 450 Scenarios

	Energy demand (Mtoe)						Shares (%) 2035		CAAGR (%) 2008-2035	
	2020	2030	2035	2020	2030	2035	CPS	450	CPS	450
	Current Policies Scenario			450 Scenario						
TPED	740	805	847	715	709	715	100	100	0.8	0.1
Coal	122	140	149	114	90	80	18	11	0.9	-1.4
Oil	143	143	146	139	134	132	17	18	0.1	-0.3
Gas	386	417	441	368	342	324	52	45	0.7	-0.4
Nuclear	61	68	70	63	71	77	8	11	1.9	2.2
Hydro	16	18	18	16	25	28	2	4	1.0	2.6
Biomass and waste	8	10	11	9	22	32	1	4	2.2	6.2
Other renewables	5	9	11	6	24	42	1	6	13.1	18.8
Power generation	395	431	454	382	380	390	100	100	0.9	0.3
Coal	83	102	112	78	59	51	25	13	1.5	-1.4
Oil	9	6	6	10	7	6	1	2	-3.0	-2.7
Gas	216	222	231	204	178	161	51	41	0.3	-1.0
Nuclear	61	68	70	63	71	77	16	20	1.9	2.2
Hydro	16	18	18	16	25	28	4	7	1.0	2.6
Biomass and waste	5	6	6	5	16	25	1	6	2.0	7.2
Other renewables	5	9	11	6	24	42	2	11	13.1	18.8
Other energy sector	120	124	127	116	105	99	100	100	0.2	-0.7
Electricity	27	30	31	25	24	23	25	23	0.8	-0.4
TFC	473	519	548	456	451	448	100	100	0.9	0.1
Coal	17	17	18	15	12	12	3	3	0.3	-1.3
Oil	112	119	124	107	107	106	23	24	0.6	-0.0
Gas	149	167	178	144	145	145	33	32	1.1	0.3
Electricity	75	89	98	73	79	82	18	18	1.7	1.0
Heat	117	122	125	112	101	96	23	21	0.4	-0.6
Biomass and waste	3	4	5	4	6	7	1	1	2.7	4.0
Other renewables	0	0	0	0	0	1	0	0	n.a.	n.a.
Industry	138	152	163	134	136	137	100	100	1.0	0.3
Coal	10	11	12	10	9	9	7	7	0.6	-0.3
Oil	14	15	16	13	13	14	10	10	0.8	0.2
Gas	34	40	43	35	38	40	26	29	1.3	1.0
Electricity	38	45	49	36	39	39	30	28	1.7	0.8
Heat	41	41	42	39	36	35	26	25	0.1	-0.6
Biomass and waste	0	1	1	0	1	1	0	1	3.0	4.8
Other renewables	-	-	-	-	-	-	-	-	n.a.	n.a.
Transport	106	116	121	102	105	103	100	100	0.8	0.2
Oil	59	64	67	57	57	54	55	53	0.7	-0.0
Electricity	10	12	13	10	14	17	11	16	2.2	3.3
Biofuels	0	1	1	1	1	2	1	2	n.a.	n.a.
Other fuels	38	40	40	35	32	30	33	29	0.5	-0.6
Buildings	168	185	192	161	146	140	100	100	0.8	-0.4
Coal	6	6	6	5	3	2	3	1	-0.3	-4.1
Oil	8	7	6	8	5	4	3	3	-1.6	-2.8
Gas	52	61	65	50	48	48	34	34	1.5	0.3
Electricity	26	30	32	25	24	23	17	17	1.3	0.1
Heat	74	78	80	71	62	59	42	42	0.5	-0.7
Biomass and waste	2	3	3	3	3	4	1	3	1.8	2.6
Other renewables	0	0	0	0	0	1	0	0	n.a.	n.a.
Other	61	67	72	59	64	68	100	100	0.7	0.6

Russia: New Policies Scenario

	Electricity generation (TWh)							Shares (%)		CAAGR (%)
	1990	2008	2015	2020	2025	2030	2035	2008	2035	2008-2035
Total generation	1 082	1 038	1 118	1 194	1 272	1 347	1 416	100	100	1.2
Coal	157	197	214	210	220	218	223	19	16	0.5
Oil	129	16	9	8	5	2	1	2	0	-11.2
Gas	512	495	526	535	558	554	510	48	36	0.1
Nuclear	118	163	186	234	245	268	284	16	20	2.1
Hydro	166	165	172	186	206	227	251	16	18	1.6
Biomass and waste	0	3	5	7	12	28	50	0	4	11.6
Wind	-	0	4	8	16	33	72	0	5	42.5
Geothermal	0	0	3	6	10	16	23	0	2	15.6
Solar PV	-	-	0	1	1	1	2	-	0	n.a.
CSP	-	-	-	-	-	-	-	-	-	n.a.
Marine	-	-	-	-	-	-	0	-	0	n.a.

	Electrical capacity (GW)						Shares (%)		CAAGR (%)
	2008	2015	2020	2025	2030	2035	2008	2035	2008-2035
Total capacity	229	252	262	269	282	306	100	100	1.1
Coal	52	56	53	50	44	43	23	14	-0.7
Oil	6	6	5	4	4	3	3	1	-2.1
Gas	99	111	112	112	113	112	43	37	0.4
Nuclear	23	26	33	34	36	38	10	13	1.9
Hydro	47	49	53	58	64	70	21	23	1.4
Biomass and waste	1	1	2	3	5	9	0	3	8.6
Wind	0	1	3	6	12	25	0	8	31.1
Geothermal	0	0	1	1	2	3	0	1	13.9
Solar PV	-	0	1	1	2	2	-	1	n.a.
CSP	-	-	-	-	-	-	-	-	n.a.
Marine	-	-	-	-	-	0	-	0	n.a.

	CO$_2$ emissions (Mt)							Shares (%)		CAAGR (%)
	1990	2008	2015	2020	2025	2030	2035	2008	2035	2008-2035
Total CO$_2$	2 179	1 580	1 614	1 624	1 650	1 650	1 648	100	100	0.2
Coal	687	422	436	419	422	412	411	27	25	-0.1
Oil	625	337	328	335	337	334	336	21	20	-0.0
Gas	866	821	850	870	891	905	900	52	55	0.3
Power generation	1 162	861	873	863	865	841	812	100	100	-0.2
Coal	432	323	338	324	329	319	318	37	39	-0.1
Oil	198	42	34	31	26	21	18	5	2	-3.0
Gas	532	497	502	508	509	501	476	58	59	-0.2
TFC	960	645	666	684	706	726	748	100	100	0.6
Coal	253	93	93	89	87	86	88	14	12	-0.2
Oil	389	257	256	265	272	273	277	40	37	0.3
Transport	*217*	*161*	*162*	*174*	*183*	*186*	*190*	*25*	*25*	*0.6*
Gas	318	295	317	330	347	366	383	46	51	1.0

Russia: Current Policies and 450 Scenarios

Electricity generation (TWh)							Shares (%)		CAAGR (%)	
	2020	2030	2035	2020	2030	2035	2035		2008-2035	
	Current Policies Scenario			450 Scenario			CPS	450	CPS	450
Total generation	1 206	1 404	1 523	1 166	1 216	1 232	100	100	1.4	0.6
Coal	225	295	327	194	145	91	21	7	1.9	-2.8
Oil	8	2	1	11	4	3	0	0	-10.4	-5.9
Gas	537	599	656	508	365	268	43	22	1.0	-2.2
Nuclear	234	261	269	240	272	293	18	24	1.9	2.2
Hydro	182	205	214	191	296	330	14	27	1.0	2.6
Biomass and waste	6	12	15	8	47	81	1	7	6.7	13.7
Wind	8	20	31	9	64	126	2	10	38.2	45.5
Geothermal	5	8	10	6	21	36	1	3	11.9	17.4
Solar PV	1	1	1	1	2	3	0	0	n.a.	n.a.
CSP	-	-	-	-	-	-	-	-	n.a.	n.a.
Marine	-	-	-	-	0	0	-	0	n.a.	n.a.

Electrical capacity (GW)							Shares (%)		CAAGR (%)	
	2020	2030	2035	2020	2030	2035	2035		2008-2035	
	Current Policies Scenario			450 Scenario			CPS	450	CPS	450
Total capacity	267	284	297	245	285	314	100	100	1.0	1.2
Coal	56	58	61	47	36	31	20	10	0.6	-1.8
Oil	5	4	4	5	2	2	1	1	-2.0	-4.3
Gas	115	116	119	98	92	84	40	27	0.7	-0.6
Nuclear	33	36	36	34	37	40	12	13	1.7	2.0
Hydro	52	58	60	54	82	91	20	29	0.9	2.4
Biomass and waste	2	3	3	2	8	14	1	4	4.5	10.4
Wind	3	7	11	3	23	45	4	14	27.3	33.9
Geothermal	1	1	1	1	3	4	0	1	10.8	15.5
Solar PV	1	1	1	1	2	4	0	1	n.a.	n.a.
CSP	-	-	-	-	-	-	-	-	n.a.	n.a.
Marine	-	-	-	-	0	0	-	0	n.a.	n.a.

CO$_2$ emissions (Mt)							Shares (%)		CAAGR (%)	
	2020	2030	2035	2020	2030	2035	2035		2008-2035	
	Current Policies Scenario			450 Scenario			CPS	450	CPS	450
Total CO$_2$	1 655	1 803	1 902	1 576	1 271	1 157	100	100	0.7	-1.1
Coal	451	534	577	423	272	226	30	20	1.2	-2.3
Oil	337	342	349	325	300	283	18	24	0.1	-0.6
Gas	867	927	976	828	699	647	51	56	0.6	-0.9
Power generation	887	967	1 021	843	595	514	100	100	0.6	-1.9
Coal	351	429	466	332	208	171	46	33	1.4	-2.3
Oil	31	21	18	33	22	20	2	4	-3.0	-2.7
Gas	505	516	537	479	364	322	53	63	0.3	-1.6
TFC	690	750	790	659	605	574	100	100	0.7	-0.4
Coal	95	99	106	86	59	50	13	9	0.5	-2.3
Oil	267	281	289	255	240	227	37	40	0.4	-0.5
Transport	174	190	197	166	167	160	25	28	0.8	-0.0
Gas	329	370	395	319	305	297	50	52	1.1	0.0

A

Non-OECD Asia: New Policies Scenario

	Energy demand (Mtoe)							Shares (%)		CAAGR (%)
	1990	2008	2015	2020	2025	2030	2035	2008	2035	2008-2035
TPED	**1 593**	**3 545**	**4 609**	**5 104**	**5 552**	**6 038**	**6 540**	**100**	**100**	**2.3**
Coal	697	1 821	2 420	2 581	2 681	2 770	2 856	51	44	1.7
Oil	312	761	952	1 048	1 163	1 298	1 423	21	22	2.3
Gas	72	280	409	483	559	662	773	8	12	3.8
Nuclear	10	33	89	171	230	274	315	1	5	8.7
Hydro	24	72	104	133	159	176	186	2	3	3.6
Biomass and waste	472	546	575	593	621	668	731	15	11	1.1
Other renewables	7	32	59	94	139	190	255	1	4	8.0
Power generation	**331**	**1 334**	**1 857**	**2 171**	**2 450**	**2 742**	**3 057**	**100**	**100**	**3.1**
Coal	229	1 042	1 399	1 518	1 608	1 688	1 776	78	58	2.0
Oil	45	49	39	38	32	29	23	4	1	-2.8
Gas	16	108	164	204	244	304	367	8	12	4.6
Nuclear	10	33	89	171	230	274	315	2	10	8.7
Hydro	24	72	104	133	159	176	186	5	6	3.6
Biomass and waste	0	5	13	25	55	104	169	0	6	13.9
Other renewables	7	26	50	81	122	166	222	2	7	8.3
Other energy sector	**166**	**378**	**479**	**515**	**546**	**577**	**606**	**100**	**100**	**1.8**
Electricity	*24*	*84*	*119*	*140*	*158*	*176*	*196*	*22*	*32*	*3.2*
TFC	**1 221**	**2 354**	**3 075**	**3 383**	**3 669**	**3 982**	**4 296**	**100**	**100**	**2.3**
Coal	390	618	807	833	832	830	824	26	19	1.1
Oil	240	649	846	942	1 061	1 200	1 328	28	31	2.7
Gas	33	124	194	229	266	309	356	5	8	4.0
Electricity	85	375	597	735	859	988	1 124	16	26	4.2
Heat	14	53	73	76	79	80	80	2	2	1.5
Biomass and waste	459	529	550	555	554	551	550	22	13	0.1
Other renewables	0	6	9	13	17	24	33	0	1	6.7
Industry	**393**	**943**	**1 331**	**1 467**	**1 564**	**1 663**	**1 758**	**100**	**100**	**2.3**
Coal	234	485	655	683	689	695	700	51	40	1.4
Oil	51	98	121	122	123	124	123	10	7	0.8
Gas	10	53	83	100	119	140	161	6	9	4.2
Electricity	51	224	365	443	505	563	620	24	35	3.8
Heat	11	35	50	52	54	56	57	4	3	1.8
Biomass and waste	36	48	57	66	74	85	97	5	6	2.7
Other renewables	-	-	-	-	-	-	-	-	-	n.a.
Transport	**113**	**309**	**429**	**517**	**632**	**774**	**915**	**100**	**100**	**4.1**
Oil	100	294	405	485	591	718	841	95	92	4.0
Electricity	1	4	7	9	11	14	18	1	2	6.0
Biofuels	-	2	10	15	21	28	37	1	4	11.1
Other fuels	12	9	7	8	10	13	18	3	2	2.6
Buildings	**599**	**842**	**974**	**1 043**	**1 100**	**1 154**	**1 217**	**100**	**100**	**1.4**
Coal	110	88	100	98	92	83	73	11	6	-0.7
Oil	34	90	107	113	116	117	116	11	10	1.0
Gas	5	35	55	69	83	98	115	4	9	4.5
Electricity	24	126	199	252	309	372	443	15	36	4.8
Heat	3	18	23	24	24	24	23	2	2	0.9
Biomass and waste	423	479	482	474	459	438	415	57	34	-0.5
Other renewables	0	6	9	12	17	23	32	1	3	6.6
Other	**115**	**259**	**341**	**357**	**373**	**391**	**407**	**100**	**100**	**1.7**

Non-OECD Asia: Current Policies and 450 Scenarios

	Energy demand (Mtoe)						Shares (%) 2035		CAAGR (%) 2008-2035	
	2020	2030	2035	2020	2030	2035	CPS	450	CPS	450
	Current Policies Scenario			450 Scenario						
TPED	5 285	6 527	7 240	4 974	5 368	5 628	100	100	2.7	1.7
Coal	2 788	3 392	3 742	2 463	1 984	1 762	52	31	2.7	-0.1
Oil	1 076	1 363	1 516	1 011	1 168	1 231	21	22	2.6	1.8
Gas	480	668	790	463	600	694	11	12	3.9	3.4
Nuclear	159	229	254	180	393	498	4	9	7.9	10.6
Hydro	126	151	163	135	192	208	2	4	3.1	4.0
Biomass and waste	581	604	631	602	741	832	9	15	0.5	1.6
Other renewables	75	120	144	119	292	404	2	7	5.8	9.9
Power generation	2 296	3 058	3 516	2 108	2 314	2 479	100	100	3.7	2.3
Coal	1 688	2 182	2 482	1 433	1 025	832	71	34	3.3	-0.8
Oil	39	31	26	34	27	25	1	1	-2.4	-2.5
Gas	202	309	381	194	270	329	11	13	4.8	4.2
Nuclear	159	229	254	180	393	498	7	20	7.9	10.6
Hydro	126	151	163	135	192	208	5	8	3.1	4.0
Biomass and waste	19	56	92	33	156	237	3	10	11.3	15.3
Other renewables	64	100	120	98	252	350	3	14	5.8	10.1
Other energy sector	531	629	678	501	513	518	100	100	2.2	1.2
Electricity	144	192	218	133	153	163	32	31	3.6	2.5
TFC	3 453	4 202	4 617	3 293	3 664	3 853	100	100	2.5	1.8
Coal	861	919	952	808	738	712	21	18	1.6	0.5
Oil	968	1 267	1 424	909	1 075	1 139	31	30	3.0	2.1
Gas	228	309	358	218	283	322	8	8	4.0	3.6
Electricity	757	1 065	1 245	706	886	980	27	25	4.5	3.6
Heat	78	85	87	74	69	65	2	2	1.8	0.7
Biomass and waste	550	535	526	557	572	583	11	15	-0.0	0.4
Other renewables	12	20	25	21	40	54	1	1	5.6	8.7
Industry	1 507	1 793	1 952	1 425	1 526	1 575	100	100	2.7	1.9
Coal	708	773	811	661	615	603	42	38	1.9	0.8
Oil	126	131	132	117	111	105	7	7	1.1	0.3
Gas	102	141	163	99	135	152	8	10	4.2	4.0
Electricity	458	619	710	428	524	563	36	36	4.4	3.5
Heat	54	59	61	52	50	48	3	3	2.1	1.2
Biomass and waste	60	69	74	68	91	104	4	7	1.6	2.9
Other renewables	-	-	-	-	-	-	-	-	n.a.	n.a.
Transport	530	816	978	503	717	829	100	100	4.4	3.7
Oil	500	764	909	464	620	684	93	82	4.3	3.2
Electricity	9	14	17	10	24	40	2	5	5.8	9.1
Biofuels	15	27	36	20	52	73	4	9	11.0	13.9
Other fuels	6	11	16	10	22	33	2	4	2.1	5.0
Buildings	1 054	1 192	1 265	1 008	1 034	1 047	100	100	1.5	0.8
Coal	100	92	86	96	74	61	7	6	-0.1	-1.3
Oil	118	128	130	106	102	101	10	10	1.4	0.4
Gas	67	96	113	57	68	74	9	7	4.4	2.8
Electricity	260	392	471	238	302	336	37	32	5.0	3.7
Heat	24	26	26	22	19	17	2	2	1.3	-0.3
Biomass and waste	474	439	415	469	429	405	33	39	-0.5	-0.6
Other renewables	11	20	24	21	40	53	2	5	5.5	8.6
Other	361	401	422	357	386	401	100	100	1.8	1.6

Non-OECD Asia: New Policies Scenario

Electricity generation (TWh)	1990	2008	2015	2020	2025	2030	2035	Shares (%) 2008	Shares (%) 2035	CAAGR (%) 2008-2035
Total generation	1 273	5 337	8 323	10 168	11 824	13 536	15 348	100	100	4.0
Coal	730	3 621	5 554	6 219	6 726	7 254	7 856	68	51	2.9
Oil	162	183	141	141	121	111	84	3	1	-2.8
Gas	59	515	814	1 048	1 280	1 613	1 959	10	13	5.1
Nuclear	39	126	342	658	884	1 052	1 208	2	8	8.7
Hydro	275	834	1 208	1 552	1 845	2 045	2 168	16	14	3.6
Biomass and waste	2	13	39	81	182	350	574	0	4	15.2
Wind	0	28	181	388	630	843	1 047	1	7	14.4
Geothermal	7	19	30	42	58	76	96	0	1	6.2
Solar PV	0	0	9	28	79	154	267	0	2	30.2
CSP	-	-	5	11	18	37	87	-	1	n.a.
Marine	-	-	-	-	0	0	2	-	0	n.a.

Electrical capacity (GW)	2008	2015	2020	2025	2030	2035	Shares (%) 2008	Shares (%) 2035	CAAGR (%) 2008-2035
Total capacity	1 186	1 870	2 303	2 690	3 106	3 553	100	100	4.1
Coal	697	1 072	1 210	1 293	1 412	1 552	59	44	3.0
Oil	70	72	67	60	57	51	6	1	-1.2
Gas	139	210	258	316	393	478	12	13	4.7
Nuclear	19	45	85	115	137	157	2	4	8.2
Hydro	230	344	449	542	606	646	19	18	3.9
Biomass and waste	6	11	18	35	61	95	0	3	11.0
Wind	22	101	181	258	318	375	2	11	11.1
Geothermal	3	5	7	9	12	15	0	0	6.0
Solar PV	0	9	23	56	100	161	0	5	26.2
CSP	0	2	4	6	11	23	0	1	21.9
Marine	-	-	-	0	0	1	-	0	n.a.

CO_2 emissions (Mt)	1990	2008	2015	2020	2025	2030	2035	Shares (%) 2008	Shares (%) 2035	CAAGR (%) 2008-2035
Total CO_2	3 526	9 579	12 675	13 703	14 521	15 430	16 318	100	100	2.0
Coal	2 532	6 999	9 315	9 891	10 191	10 453	10 713	73	66	1.6
Oil	853	1 962	2 451	2 733	3 076	3 484	3 855	20	24	2.5
Gas	141	618	909	1 079	1 254	1 493	1 750	6	11	3.9
Power generation	1 078	4 523	6 045	6 599	7 013	7 435	7 865	100	100	2.1
Coal	897	4 112	5 534	5 996	6 332	6 622	6 926	91	88	2.0
Oil	143	158	124	121	103	94	73	3	1	-2.8
Gas	38	253	387	483	577	719	866	6	11	4.7
TFC	2 285	4 624	6 139	6 587	6 992	7 460	7 898	100	100	2.0
Coal	1 575	2 706	3 568	3 669	3 650	3 624	3 586	59	45	1.0
Oil	648	1 652	2 156	2 424	2 763	3 159	3 528	36	45	2.8
Transport	*299*	*878*	*1 207*	*1 444*	*1 759*	*2 140*	*2 506*	*19*	*32*	*4.0*
Gas	63	266	415	494	579	676	784	6	10	4.1

Non-OECD Asia: Current Policies and 450 Scenarios

	Electricity generation (TWh)						Shares (%) 2035		CAAGR (%) 2008-2035	
	2020	2030	2035	2020	2030	2035	CPS	450	CPS	450
	Current Policies Scenario			450 Scenario						
Total generation	10 473	14 617	17 015	9 749	12 081	13 283	100	100	4.4	3.4
Coal	6 844	9 537	11 134	5 685	4 467	3 751	65	28	4.2	0.1
Oil	148	118	95	124	98	91	1	1	-2.4	-2.5
Gas	1 010	1 546	1 902	979	1 485	1 848	11	14	5.0	4.8
Nuclear	610	877	974	691	1 508	1 911	6	14	7.9	10.6
Hydro	1 464	1 759	1 891	1 572	2 227	2 416	11	18	3.1	4.0
Biomass and waste	59	187	305	111	540	820	2	6	12.5	16.7
Wind	274	449	525	476	1 236	1 546	3	12	11.5	16.1
Geothermal	35	52	63	48	100	133	0	1	4.5	7.5
Solar PV	25	81	106	34	291	534	1	4	25.8	33.5
CSP	4	10	19	30	128	231	0	2	n.a.	n.a.
Marine	-	0	2	0	1	3	0	0	n.a.	n.a.

	Electrical capacity (GW)						Shares (%) 2035		CAAGR (%) 2008-2035	
	2020	2030	2035	2020	2030	2035	CPS	450	CPS	450
	Current Policies Scenario			450 Scenario						
Total capacity	2 330	3 265	3 799	2 300	2 890	3 310	100	100	4.4	3.9
Coal	1 311	1 865	2 200	1 159	859	824	58	25	4.4	0.6
Oil	67	58	53	66	59	54	1	2	-1.0	-0.9
Gas	273	419	509	246	349	403	13	12	4.9	4.0
Nuclear	79	114	127	90	196	248	3	7	7.3	10.0
Hydro	422	514	556	456	665	728	15	22	3.3	4.4
Biomass and waste	14	36	56	23	91	133	1	4	8.8	12.3
Wind	136	191	214	215	437	523	6	16	8.8	12.4
Geothermal	6	8	10	8	15	20	0	1	4.5	7.2
Solar PV	21	56	70	28	184	316	2	10	22.3	29.4
CSP	2	3	6	9	34	60	0	2	15.7	26.3
Marine	-	0	0	0	0	1	0	0	n.a.	n.a.

	CO_2 emissions (Mt)						Shares (%) 2035		CAAGR (%) 2008-2035	
	2020	2030	2035	2020	2030	2035	CPS	450	CPS	450
	Current Policies Scenario			450 Scenario						
Total CO_2	14 583	18 021	20 060	13 075	10 805	9 410	100	100	2.8	-0.1
Coal	10 693	12 821	14 118	9 426	6 466	4 739	70	50	2.6	-1.4
Oil	2 817	3 693	4 151	2 617	3 047	3 215	21	34	2.8	1.8
Gas	1 073	1 507	1 790	1 032	1 292	1 456	9	15	4.0	3.2
Power generation	7 278	9 440	10 772	6 213	4 262	2 965	100	100	3.3	-1.6
Coal	6 672	8 607	9 786	5 647	3 572	2 185	91	74	3.3	-2.3
Oil	127	100	82	108	86	80	1	3	-2.4	-2.5
Gas	479	733	904	459	603	700	8	24	4.8	3.8
TFC	6 774	8 016	8 691	6 360	6 069	5 980	100	100	2.4	1.0
Coal	3 783	3 985	4 100	3 564	2 727	2 405	47	40	1.6	-0.4
Oil	2 500	3 355	3 805	2 326	2 747	2 905	44	49	3.1	2.1
Transport	1 491	2 275	2 707	1 382	1 848	2 038	31	34	4.3	3.2
Gas	491	675	786	470	596	670	9	11	4.1	3.5

China: New Policies Scenario

	Energy demand (Mtoe)							Shares (%)		CAAGR (%)
	1990	2008	2015	2020	2025	2030	2035	2008	2035	2008-2035
TPED	872	2 131	2 887	3 159	3 369	3 568	3 737	100	100	2.1
Coal	534	1 413	1 879	1 952	1 981	1 990	1 975	66	53	1.2
Oil	114	369	509	558	616	675	716	17	19	2.5
Gas	13	71	142	181	223	277	330	3	9	5.9
Nuclear	-	18	60	136	178	210	233	1	6	10.0
Hydro	11	50	74	94	106	112	116	2	3	3.1
Biomass and waste	200	203	200	195	196	210	240	10	6	0.6
Other renewables	0	7	23	43	69	95	126	0	3	11.6
Power generation	181	867	1 273	1 473	1 632	1 779	1 919	100	100	3.0
Coal	153	777	1 059	1 123	1 171	1 200	1 220	90	64	1.7
Oil	16	9	11	9	7	7	6	1	0	-1.7
Gas	1	11	47	67	90	122	153	1	8	10.2
Nuclear	-	18	60	136	178	210	233	2	12	10.0
Hydro	11	50	74	94	106	112	116	6	6	3.1
Biomass and waste	-	1	5	11	25	51	89	0	5	17.0
Other renewables	-	1	15	32	55	76	101	0	5	18.1
Other energy sector	94	246	324	345	358	367	371	100	100	1.5
Electricity	*12*	*52*	*74*	*85*	*93*	*100*	*106*	*21*	*29*	*2.7*
TFC	668	1 379	1 868	2 029	2 152	2 274	2 369	100	100	2.0
Coal	315	495	629	628	607	582	549	36	23	0.4
Oil	86	323	462	514	574	637	680	23	29	2.8
Gas	10	52	85	103	122	144	167	4	7	4.4
Electricity	43	248	417	512	586	655	719	18	30	4.0
Heat	13	53	72	75	77	79	79	4	3	1.5
Biomass and waste	200	202	195	184	171	159	151	15	6	-1.1
Other renewables	0	5	8	11	14	19	25	0	1	5.8
Industry	242	657	929	999	1 038	1 072	1 094	100	100	1.9
Coal	177	387	505	506	492	476	454	59	42	0.6
Oil	21	50	63	63	63	64	62	8	6	0.8
Gas	3	19	31	38	46	57	68	3	6	4.9
Electricity	30	166	280	339	379	414	444	25	41	3.7
Heat	11	35	50	52	54	56	57	5	5	1.8
Biomass and waste	-	0	0	1	2	6	10	0	1	29.3
Other renewables	-	-	-	-	-	-	-	-	-	n.a.
Transport	38	156	250	304	367	436	492	100	100	4.3
Oil	28	149	240	289	347	410	457	95	93	4.2
Electricity	1	3	5	7	9	11	15	2	3	6.7
Biofuels	-	1	5	7	10	13	17	1	4	9.7
Other fuels	10	3	1	1	1	2	3	2	1	-0.5
Buildings	316	426	512	548	567	580	595	100	100	1.2
Coal	95	64	73	70	63	54	44	15	7	-1.4
Oil	7	45	59	62	63	61	58	11	10	0.9
Gas	2	25	41	51	60	69	78	6	13	4.3
Electricity	9	68	120	154	185	215	245	16	41	4.9
Heat	2	18	22	23	23	23	22	4	4	0.8
Biomass and waste	200	201	190	176	159	140	123	47	21	-1.8
Other renewables	0	5	8	11	14	19	24	1	4	5.7
Other	71	139	177	178	181	185	188	100	100	1.1

China: Current Policies and 450 Scenarios

	Energy demand (Mtoe)						Shares (%)		CAAGR (%)	
	2020	2030	2035	2020	2030	2035	2035		2008-2035	
	Current Policies Scenario			450 Scenario			CPS	450	CPS	450
TPED	3 288	3 907	4 215	3 097	3 094	3 131	100	100	2.6	1.4
Coal	2 104	2 422	2 574	1 878	1 398	1 189	61	38	2.2	-0.6
Oil	567	698	755	543	605	617	18	20	2.7	1.9
Gas	179	270	326	178	266	321	8	10	5.8	5.7
Nuclear	124	174	189	143	288	358	4	11	9.1	11.8
Hydro	92	106	112	94	117	120	3	4	3.0	3.3
Biomass and waste	191	184	196	200	259	308	5	10	-0.1	1.5
Other renewables	32	54	63	61	162	218	1	7	8.8	13.9
Power generation	1 576	2 024	2 256	1 445	1 467	1 520	100	100	3.6	2.1
Coal	1 252	1 551	1 701	1 068	711	559	75	37	2.9	-1.2
Oil	10	8	7	8	6	5	0	0	-1.2	-2.0
Gas	69	120	152	71	129	166	7	11	10.2	10.6
Nuclear	124	174	189	143	288	358	8	24	9.1	11.8
Hydro	92	106	112	94	117	120	5	8	3.0	3.3
Biomass and waste	7	28	51	17	89	136	2	9	14.6	18.9
Other renewables	22	37	44	42	127	175	2	12	14.5	20.5
Other energy sector	357	404	424	336	319	308	100	100	2.0	0.8
Electricity	*88*	*111*	*123*	*81*	*85*	*86*	*29*	*28*	*3.2*	*1.9*
TFC	2 065	2 406	2 573	1 981	2 063	2 092	100	100	2.3	1.6
Coal	644	636	629	615	509	461	24	22	0.9	-0.3
Oil	522	662	722	499	567	581	28	28	3.0	2.2
Gas	100	140	163	97	127	146	6	7	4.3	3.9
Electricity	529	713	810	495	587	626	31	30	4.5	3.5
Heat	77	83	86	73	68	64	3	3	1.8	0.7
Biomass and waste	184	156	145	183	170	171	6	8	-1.2	-0.6
Other renewables	10	16	19	19	34	43	1	2	4.8	7.9
Industry	1 025	1 163	1 235	974	967	956	100	100	2.4	1.4
Coal	519	520	519	494	410	375	42	39	1.1	-0.1
Oil	65	65	65	61	55	49	5	5	0.9	-0.1
Gas	37	54	64	39	57	66	5	7	4.7	4.8
Electricity	350	461	519	328	388	407	42	43	4.3	3.4
Heat	53	59	61	52	50	48	5	5	2.1	1.2
Biomass and waste	1	4	6	1	7	12	1	1	27.0	29.9
Other renewables	-	-	-	-	-	-	-	-	n.a.	n.a.
Transport	306	452	520	297	411	465	100	100	4.6	4.1
Oil	292	426	486	280	359	380	94	82	4.5	3.5
Electricity	6	11	14	7	18	31	3	7	6.5	9.6
Biofuels	7	13	17	7	26	38	3	8	9.6	13.0
Other fuels	0	2	3	3	9	16	1	4	-0.5	5.9
Buildings	555	601	622	533	506	490	100	100	1.4	0.5
Coal	72	62	56	70	51	40	9	8	-0.5	-1.8
Oil	64	65	63	58	52	49	10	10	1.2	0.3
Gas	49	67	76	41	45	45	12	9	4.3	2.3
Electricity	160	226	261	148	169	176	42	36	5.1	3.6
Heat	24	25	25	22	18	16	4	3	1.2	-0.4
Biomass and waste	176	139	122	175	138	121	20	25	-1.8	-1.8
Other renewables	10	16	19	19	34	42	3	9	4.8	7.9
Other	180	190	197	177	179	181	100	100	1.3	1.0

China: New Policies Scenario

	Electricity generation (TWh)							Shares (%)		CAAGR (%)
	1990	2008	2015	2020	2025	2030	2035	2008	2035	2008-2035
Total generation	650	3 495	5 721	6 949	7 900	8 776	9 594	100	100	3.8
Coal	471	2 759	4 199	4 595	4 850	5 060	5 246	79	55	2.4
Oil	49	24	35	26	22	20	17	1	0	-1.2
Gas	3	43	222	332	462	656	841	1	9	11.6
Nuclear	-	68	232	522	684	808	895	2	9	10.0
Hydro	127	585	865	1 094	1 228	1 304	1 348	17	14	3.1
Biomass and waste	-	2	15	36	87	175	306	0	3	19.8
Wind	0	13	142	319	507	633	716	0	7	16.0
Geothermal	-	-	1	3	6	10	15	-	0	n.a.
Solar PV	0	0	6	15	44	82	140	0	1	28.2
CSP	-	-	3	6	11	27	69	-	1	n.a.
Marine	-	-	-	-	-	0	1	-	0	n.a.

	Electrical capacity (GW)						Shares (%)		CAAGR (%)
	2008	2015	2020	2025	2030	2035	2008	2035	2008-2035
Total capacity	780	1 278	1 586	1 803	2 009	2 205	100	100	3.9
Coal	563	843	945	987	1 040	1 091	72	49	2.5
Oil	20	21	19	17	17	16	3	1	-0.8
Gas	29	67	88	117	157	196	4	9	7.3
Nuclear	9	30	67	87	103	114	1	5	9.9
Hydro	145	229	297	337	360	373	19	17	3.6
Biomass and waste	1	3	7	15	29	50	0	2	15.8
Wind	12	80	148	208	244	267	2	12	12.1
Geothermal	0	0	1	1	2	2	0	0	17.8
Solar PV	0	5	13	30	51	80	0	4	26.3
CSP	-	1	2	3	7	17	-	1	n.a.
Marine	-	-	-	-	0	0	-	0	n.a.

	CO_2 emissions (Mt)							Shares (%)		CAAGR (%)
	1990	2008	2015	2020	2025	2030	2035	2008	2035	2008-2035
Total CO_2	2 244	6 550	8 893	9 381	9 709	9 985	10 118	100	100	1.6
Coal	1 911	5 451	7 247	7 488	7 539	7 503	7 385	83	73	1.1
Oil	305	935	1 315	1 470	1 649	1 834	1 961	14	19	2.8
Gas	28	165	331	423	521	648	772	3	8	5.9
Power generation	652	3 137	4 362	4 648	4 867	5 033	5 144	100	100	1.8
Coal	598	3 078	4 208	4 454	4 625	4 715	4 758	98	92	1.6
Oil	52	32	39	30	25	23	19	1	0	-1.9
Gas	2	27	116	164	217	295	367	1	7	10.2
TFC	1 508	3 145	4 222	4 409	4 523	4 635	4 661	100	100	1.5
Coal	1 262	2 204	2 839	2 826	2 720	2 598	2 445	70	52	0.4
Oil	226	825	1 195	1 353	1 529	1 713	1 841	26	39	3.0
Transport	*83*	*444*	*713*	*860*	*1 032*	*1 218*	*1 356*	*14*	*29*	*4.2*
Gas	20	116	188	230	274	323	376	4	8	4.5

China: Current Policies and 450 Scenarios

	Electricity generation (TWh)						Shares (%)		CAAGR (%)	
	2020	2030	2035	2020	2030	2035	2035		2008-2035	
	Current Policies Scenario			450 Scenario			CPS	450	CPS	450
Total generation	7 186	9 586	10 848	6 709	7 822	8 283	100	100	4.3	3.2
Coal	5 037	6 605	7 440	4 200	3 081	2 456	69	30	3.7	-0.4
Oil	32	24	21	23	16	15	0	0	-0.4	-1.7
Gas	320	587	759	342	730	965	7	12	11.2	12.2
Nuclear	475	667	726	550	1 105	1 375	7	17	9.1	11.8
Hydro	1 068	1 227	1 299	1 097	1 356	1 397	12	17	3.0	3.3
Biomass and waste	23	95	173	60	310	472	2	6	17.3	21.7
Wind	209	328	356	397	940	1 095	3	13	13.0	17.8
Geothermal	2	5	6	3	14	23	0	0	n.a.	n.a.
Solar PV	15	39	47	18	183	333	0	4	23.1	32.4
CSP	3	9	17	19	86	153	0	2	n.a.	n.a.
Marine	-	0	1	-	1	1	0	0	n.a.	n.a.

	Electrical capacity (GW)						Shares (%)		CAAGR (%)	
	2020	2030	2035	2020	2030	2035	2035		2008-2035	
	Current Policies Scenario			450 Scenario			CPS	450	CPS	450
Total capacity	1 601	2 131	2 408	1 593	1 834	2 040	100	100	4.3	3.6
Coal	1 011	1 348	1 540	906	619	585	64	29	3.8	0.1
Oil	19	17	16	19	18	17	1	1	-0.8	-0.6
Gas	98	156	185	89	159	192	8	9	7.1	7.2
Nuclear	61	85	92	70	140	175	4	9	9.0	11.6
Hydro	289	337	358	298	375	387	15	19	3.4	3.7
Biomass and waste	5	17	30	11	51	76	1	4	13.6	17.6
Wind	104	142	150	178	337	378	6	19	9.7	13.6
Geothermal	0	1	1	1	2	3	0	0	14.9	19.4
Solar PV	13	26	30	15	110	187	1	9	21.9	30.4
CSP	1	2	4	5	23	39	0	2	n.a.	n.a.
Marine	-	0	0	-	0	0	0	0	n.a.	n.a.

	CO_2 emissions (Mt)						Shares (%)		CAAGR (%)	
	2020	2030	2035	2020	2030	2035	2035		2008-2035	
	Current Policies Scenario			450 Scenario			CPS	450	CPS	450
Total CO_2	9 993	11 711	12 561	9 030	6 617	5 164	100	100	2.4	-0.9
Coal	8 080	9 171	9 718	7 195	4 441	2 875	77	56	2.2	-2.3
Oil	1 495	1 906	2 078	1 419	1 604	1 641	17	32	3.0	2.1
Gas	419	634	765	416	571	648	6	13	5.9	5.2
Power generation	5 174	6 461	7 130	4 425	2 671	1 493	100	100	3.1	-2.7
Coal	4 972	6 144	6 737	4 226	2 376	1 155	94	77	2.9	-3.6
Oil	35	26	23	28	19	18	0	1	-1.3	-2.2
Gas	167	291	370	172	275	321	5	21	10.2	9.6
TFC	4 487	4 917	5 094	4 292	3 674	3 416	100	100	1.8	0.3
Coal	2 892	2 821	2 774	2 770	1 914	1 587	54	46	0.9	-1.2
Oil	1 372	1 783	1 955	1 306	1 492	1 527	38	45	3.2	2.3
Transport	*867*	*1 266*	*1 443*	*831*	*1 065*	*1 129*	*28*	*33*	*4.5*	*3.5*
Gas	223	313	365	216	268	302	7	9	4.4	3.6

India: New Policies Scenario

	Energy demand (Mtoe)							Shares (%)		CAAGR (%)
	1990	2008	2015	2020	2025	2030	2035	2008	2035	2008-2035
TPED	**319**	**620**	**778**	**904**	**1 039**	**1 204**	**1 405**	**100**	**100**	**3.1**
Coal	106	261	327	386	426	478	547	42	39	2.8
Oil	61	145	178	206	245	298	362	23	26	3.5
Gas	11	36	67	82	98	121	149	6	11	5.4
Nuclear	2	4	11	17	30	39	50	1	4	10.0
Hydro	6	10	13	19	28	33	35	2	2	4.8
Biomass and waste	133	164	177	187	199	214	229	26	16	1.3
Other renewables	0	1	4	8	13	21	33	0	2	12.6
Power generation	**73**	**232**	**287**	**352**	**415**	**490**	**584**	**100**	**100**	**3.5**
Coal	58	190	222	259	278	306	347	82	59	2.3
Oil	4	10	10	10	10	10	9	4	1	-0.5
Gas	3	16	25	35	46	60	77	7	13	6.1
Nuclear	2	4	11	17	30	39	50	2	9	10.0
Hydro	6	10	13	19	28	33	35	4	6	4.8
Biomass and waste	-	1	2	5	12	24	37	0	6	13.8
Other renewables	0	1	4	7	12	19	29	1	5	12.6
Other energy sector	**20**	**53**	**77**	**91**	**107**	**125**	**146**	**100**	**100**	**3.8**
Electricity	*7*	*20*	*30*	*38*	*46*	*55*	*66*	*38*	*45*	*4.4*
TFC	**251**	**408**	**525**	**603**	**694**	**807**	**943**	**100**	**100**	**3.2**
Coal	42	56	86	103	120	139	162	14	17	4.0
Oil	52	122	151	177	214	265	326	30	35	3.7
Gas	6	15	32	36	41	48	57	4	6	5.0
Electricity	18	52	81	105	132	164	202	13	21	5.2
Heat	-	-	-	-	-	-	-	-	-	n.a.
Biomass and waste	133	162	175	182	187	190	192	40	20	0.6
Other renewables	0	0	1	1	1	2	4	0	0	12.7
Industry	**70**	**115**	**173**	**210**	**246**	**290**	**340**	**100**	**100**	**4.1**
Coal	29	37	65	82	99	119	144	32	42	5.2
Oil	10	22	30	30	32	33	35	20	10	1.6
Gas	0	4	9	10	12	15	17	3	5	6.1
Electricity	9	24	39	52	65	81	99	21	29	5.4
Heat	-	-	-	-	-	-	-	-	-	n.a.
Biomass and waste	23	28	32	35	39	42	45	25	13	1.7
Other renewables	-	-	-	-	-	-	-	-	-	n.a.
Transport	**27**	**45**	**56**	**75**	**106**	**153**	**213**	**100**	**100**	**5.9**
Oil	24	42	50	68	96	139	191	93	90	5.8
Electricity	0	1	2	2	2	2	3	2	1	3.6
Biofuels	-	0	2	4	5	8	12	0	5	17.6
Other fuels	2	2	2	2	3	5	7	4	3	4.8
Buildings	**137**	**194**	**218**	**231**	**245**	**258**	**274**	**100**	**100**	**1.3**
Coal	11	19	21	21	21	20	18	10	7	-0.2
Oil	11	23	26	29	31	34	37	12	13	1.8
Gas	0	1	2	2	3	5	7	0	2	8.7
Electricity	4	17	27	36	46	58	74	9	27	5.5
Heat	-	-	-	-	-	-	-	-	-	n.a.
Biomass and waste	111	134	141	143	143	140	135	69	49	0.0
Other renewables	0	0	0	1	1	2	3	0	1	11.9
Other	**17**	**53**	**78**	**88**	**97**	**106**	**116**	**100**	**100**	**2.9**

India: Current Policies and 450 Scenarios

	Energy demand (Mtoe)						Shares (%)		CAAGR (%)	
	2020	2030	2035	2020	2030	2035	2035		2008-2035	
	Current Policies Scenario			450 Scenario			CPS	450	CPS	450
TPED	934	1 285	1 535	869	1 106	1 243	100	100	3.4	2.6
Coal	414	584	716	357	365	366	47	29	3.8	1.3
Oil	216	323	395	200	277	319	26	26	3.8	3.0
Gas	82	119	143	79	114	138	9	11	5.3	5.1
Nuclear	17	33	41	19	62	87	3	7	9.1	12.3
Hydro	15	22	25	20	39	45	2	4	3.5	5.8
Biomass and waste	183	195	202	185	215	230	13	18	0.8	1.3
Other renewables	6	10	13	10	34	58	1	5	8.7	15.0
Power generation	367	530	660	332	426	481	100	100	4.0	2.7
Coal	282	389	485	237	201	176	73	37	3.5	-0.3
Oil	10	10	8	10	10	9	1	2	-0.5	-0.1
Gas	34	56	70	33	56	70	11	15	5.7	5.7
Nuclear	17	33	41	19	62	87	6	18	9.1	12.3
Hydro	15	22	25	20	39	45	4	9	3.5	5.8
Biomass and waste	4	13	21	5	27	41	3	8	11.3	14.2
Other renewables	5	9	11	9	32	52	2	11	8.5	15.0
Other energy sector	94	133	157	87	114	128	100	100	4.1	3.3
Electricity	39	57	69	36	49	57	44	44	4.6	3.9
TFC	618	848	999	585	764	870	100	100	3.4	2.9
Coal	108	155	185	98	132	154	19	18	4.6	3.8
Oil	187	290	360	172	246	286	36	33	4.1	3.2
Gas	37	49	58	35	47	56	6	6	5.1	4.9
Electricity	106	170	212	99	149	180	21	21	5.4	4.7
Heat	-	-	-	-	-	-	-	-	n.a.	n.a.
Biomass and waste	180	183	182	180	188	189	18	22	0.4	0.6
Other renewables	1	1	2	1	3	6	0	1	10.1	14.4
Industry	214	303	359	201	279	325	100	100	4.3	3.9
Coal	87	135	167	79	116	141	46	43	5.8	5.1
Oil	32	36	38	29	32	33	11	10	1.9	1.4
Gas	11	15	17	9	14	17	5	5	6.1	5.9
Electricity	52	83	103	49	74	88	29	27	5.5	4.9
Heat	-	-	-	-	-	-	-	-	n.a.	n.a.
Biomass and waste	33	34	34	35	43	46	10	14	0.7	1.8
Other renewables	-	-	-	-	-	-	-	-	n.a.	n.a.
Transport	80	169	234	74	140	180	100	100	6.3	5.2
Oil	74	156	215	66	122	154	92	85	6.2	4.9
Electricity	2	2	3	2	3	5	1	3	3.5	6.5
Biofuels	4	8	11	4	9	13	5	7	17.4	18.1
Other fuels	1	3	5	2	5	7	2	4	3.7	5.0
Buildings	234	267	285	222	238	249	100	100	1.4	0.9
Coal	21	20	19	19	16	13	7	5	-0.1	-1.4
Oil	31	39	43	27	32	35	15	14	2.4	1.6
Gas	2	5	8	2	4	6	3	2	9.2	8.3
Electricity	36	61	78	32	49	59	27	24	5.7	4.6
Heat	-	-	-	-	-	-	-	-	n.a.	n.a.
Biomass and waste	143	141	136	141	135	129	48	52	0.0	-0.1
Other renewables	0	1	2	1	2	6	1	2	9.0	13.9
Other	89	110	121	88	107	117	100	100	3.1	3.0

India: New Policies Scenario

	Electricity generation (TWh)							Shares (%)		CAAGR (%)
	1990	2008	2015	2020	2025	2030	2035	2008	2035	2008-2035
Total generation	289	830	1 281	1 652	2 062	2 538	3 106	100	100	5.0
Coal	192	569	866	1 039	1 164	1 355	1 615	69	52	3.9
Oil	10	34	36	36	36	35	31	4	1	-0.3
Gas	10	82	140	203	263	345	446	10	14	6.5
Nuclear	6	15	42	66	115	150	191	2	6	10.0
Hydro	72	114	154	226	328	385	408	14	13	4.8
Biomass and waste	-	2	6	14	37	76	119	0	4	16.4
Wind	0	14	34	57	90	134	189	2	6	10.2
Geothermal	-	-	0	0	1	1	2	-	0	n.a.
Solar PV	-	0	2	7	21	47	87	0	3	36.4
CSP	-	-	2	4	6	10	17	-	1	n.a.
Marine	-	-	-	-	0	0	1	-	0	n.a.

	Electrical capacity (GW)						Shares (%)		CAAGR (%)
	2008	2015	2020	2025	2030	2035	2008	2035	2008-2035
Total capacity	163	275	352	452	574	722	100	100	5.7
Coal	84	156	178	200	246	312	52	43	5.0
Oil	7	8	9	9	9	9	5	1	0.7
Gas	17	29	42	56	74	97	10	13	6.7
Nuclear	4	7	10	17	22	28	3	4	7.3
Hydro	39	51	74	107	125	133	24	18	4.7
Biomass and waste	2	3	4	8	14	21	1	3	9.3
Wind	10	20	28	38	50	64	6	9	7.3
Geothermal	-	0	0	0	0	0	-	0	n.a.
Solar PV	0	2	6	16	31	53	0	7	25.7
CSP	0	1	1	2	3	5	0	1	14.7
Marine	-	-	-	0	0	0	-	0	n.a.

	CO_2 emissions (Mt)							Shares (%)		CAAGR (%)
	1990	2008	2015	2020	2025	2030	2035	2008	2035	2008-2035
Total CO_2	591	1 428	1 830	2 162	2 456	2 856	3 371	100	100	3.2
Coal	406	978	1 230	1 453	1 597	1 788	2 050	68	61	2.8
Oil	164	374	458	534	647	803	992	26	29	3.7
Gas	21	76	142	175	213	265	329	5	10	5.6
Power generation	245	804	950	1 118	1 215	1 354	1 553	100	100	2.5
Coal	226	737	861	1 004	1 077	1 185	1 346	92	87	2.3
Oil	11	30	32	32	31	30	27	4	2	-0.5
Gas	8	36	58	82	107	140	181	5	12	6.1
TFC	328	573	801	953	1 138	1 379	1 675	100	100	4.1
Coal	175	238	367	443	516	598	699	41	42	4.1
Oil	144	307	375	443	545	688	863	53	52	3.9
Transport	*72*	*127*	*151*	*204*	*290*	*418*	*576*	*22*	*34*	*5.8*
Gas	9	29	60	67	78	93	113	5	7	5.2

India: Current Policies and 450 Scenarios

	Electricity generation (TWh)						Shares (%) 2035		CAAGR (%) 2008-2035	
	2020	2030	2035	2020	2030	2035	CPS	450	CPS	450
	Current Policies Scenario			450 Scenario						
Total generation	1 673	2 635	3 256	1 556	2 291	2 740	100	100	5.2	4.5
Coal	1 137	1 785	2 229	934	875	847	68	31	5.2	1.5
Oil	36	34	31	36	36	35	1	1	-0.4	0.1
Gas	188	300	375	183	322	415	12	15	5.8	6.2
Nuclear	65	125	156	71	240	335	5	12	9.1	12.3
Hydro	175	252	286	235	450	520	9	19	3.5	5.8
Biomass and waste	11	39	64	14	86	134	2	5	13.7	16.9
Wind	55	76	84	62	171	242	3	9	6.9	11.2
Geothermal	0	1	1	1	3	4	0	0	n.a.	n.a.
Solar PV	5	21	30	9	68	131	1	5	31.2	38.5
CSP	0	1	1	11	40	76	0	3	n.a.	n.a.
Marine	-	0	1	0	0	1	0	0	n.a.	n.a.

	Electrical capacity (GW)						Shares (%) 2035		CAAGR (%) 2008-2035	
	2020	2030	2035	2020	2030	2035	CPS	450	CPS	450
	Current Policies Scenario			450 Scenario						
Total capacity	354	583	723	347	535	658	100	100	5.7	5.3
Coal	201	346	441	166	147	147	61	22	6.3	2.1
Oil	9	9	9	8	9	10	1	1	0.7	1.0
Gas	41	69	88	40	67	84	12	13	6.3	6.1
Nuclear	10	18	23	11	34	48	3	7	6.5	9.5
Hydro	58	83	94	77	147	170	13	26	3.4	5.6
Biomass and waste	3	8	13	4	15	22	2	3	7.3	9.6
Wind	27	34	36	30	60	79	5	12	5.0	8.1
Geothermal	0	0	0	0	0	1	0	0	n.a.	n.a.
Solar PV	4	15	20	8	44	79	3	12	21.2	27.6
CSP	0	0	0	3	11	20	0	3	3.8	21.3
Marine	-	0	0	0	0	0	0	0	n.a.	n.a.

	CO_2 emissions (Mt)						Shares (%) 2035		CAAGR (%) 2008-2035	
	2020	2030	2035	2020	2030	2035	CPS	450	CPS	450
	Current Policies Scenario			450 Scenario						
Total CO_2	2 300	3 318	4 089	2 027	2 219	2 316	100	100	4.0	1.8
Coal	1 559	2 175	2 674	1 341	1 242	1 171	65	51	3.8	0.7
Oil	566	885	1 102	518	728	843	27	36	4.1	3.1
Gas	175	257	313	168	249	303	8	13	5.4	5.2
Power generation	1 203	1 668	2 068	1 024	919	849	100	100	3.6	0.2
Coal	1 091	1 508	1 878	915	758	655	91	77	3.5	-0.4
Oil	32	30	26	31	31	29	1	3	-0.5	-0.1
Gas	80	130	164	77	130	165	8	19	5.7	5.7
TFC	1 003	1 520	1 868	914	1 188	1 344	100	100	4.5	3.2
Coal	461	661	789	420	479	510	42	38	4.5	2.9
Oil	473	764	964	429	618	723	52	54	4.3	3.2
Transport	*223*	*469*	*648*	*198*	*369*	*464*	*35*	*35*	*6.2*	*4.9*
Gas	69	96	114	65	92	111	6	8	5.2	5.1

Middle East: New Policies Scenario

	Energy demand (Mtoe)							Shares (%)		CAAGR (%)
	1990	2008	2015	2020	2025	2030	2035	2008	2035	2008-2035
TPED	219	596	735	798	871	940	1 006	100	100	2.0
Coal	3	10	12	12	12	16	20	2	2	2.9
Oil	141	304	360	382	401	415	429	51	43	1.3
Gas	73	280	354	389	436	478	508	47	51	2.2
Nuclear	-	-	2	5	5	5	5	-	1	n.a.
Hydro	1	1	3	3	4	4	5	0	0	6.8
Biomass and waste	1	1	2	3	4	7	10	0	1	8.2
Other renewables	0	1	2	4	7	14	28	0	3	12.4
Power generation	63	198	206	232	257	283	307	100	100	1.6
Coal	2	8	10	10	11	14	19	4	6	3.4
Oil	29	70	61	62	63	56	54	35	17	-1.0
Gas	32	120	129	148	167	187	193	60	63	1.8
Nuclear	-	-	2	5	5	5	5	-	2	n.a.
Hydro	1	1	3	3	4	4	5	0	1	6.8
Biomass and waste	-	0	1	1	2	4	7	0	2	34.3
Other renewables	0	0	1	3	5	11	24	0	8	30.5
Other energy sector	18	76	103	110	126	136	145	100	100	2.4
Electricity	*4*	*13*	*16*	*18*	*19*	*21*	*22*	*17*	*15*	*1.9*
TFC	159	388	509	552	597	645	692	100	100	2.2
Coal	0	1	1	1	1	1	1	0	0	1.6
Oil	109	215	283	303	323	345	362	55	52	2.0
Gas	32	117	155	166	178	191	206	30	30	2.1
Electricity	17	53	68	78	90	103	116	14	17	3.0
Heat	-	-	-	-	-	-	-	-	-	n.a.
Biomass and waste	1	1	1	2	2	2	3	0	0	3.6
Other renewables	0	1	1	2	2	3	4	0	1	4.6
Industry	45	100	131	141	148	154	159	100	100	1.7
Coal	0	0	1	1	1	1	1	0	0	1.6
Oil	22	40	49	51	52	51	49	40	31	0.7
Gas	20	48	67	72	77	81	86	48	54	2.2
Electricity	3	10	14	16	18	21	23	10	14	3.0
Heat	-	-	-	-	-	-	-	-	-	n.a.
Biomass and waste	0	0	0	0	0	0	0	0	0	2.3
Other renewables	-	-	-	-	-	-	-	-	-	n.a.
Transport	50	111	156	173	189	208	226	100	100	2.7
Oil	50	109	154	170	186	205	221	99	98	2.7
Electricity	-	0	0	0	0	0	0	0	0	0.2
Biofuels	-	-	-	-	-	-	-	-	-	n.a.
Other fuels	-	2	2	2	3	3	5	1	2	4.3
Buildings	36	105	123	134	148	164	181	100	100	2.0
Coal	-	0	-	-	-	-	-	0	-	n.a.
Oil	19	25	27	27	27	27	27	24	15	0.3
Gas	3	38	42	46	49	54	60	36	33	1.7
Electricity	13	40	52	59	68	78	88	38	49	3.0
Heat	-	-	-	-	-	-	-	-	-	n.a.
Biomass and waste	1	1	1	1	2	2	2	1	1	3.9
Other renewables	0	1	1	2	2	3	4	1	2	4.6
Other	27	73	99	105	112	119	125	100	100	2.0

Middle East: Current Policies and 450 Scenarios

	Energy demand (Mtoe)						Shares (%)		CAAGR (%)	
	2020	2030	2035	2020	2030	2035	2035		2008-2035	
	Current Policies Scenario			450 Scenario			CPS	450	CPS	450
TPED	**830**	**1 034**	**1 124**	**764**	**822**	**838**	**100**	**100**	**2.4**	**1.3**
Coal	14	23	25	11	6	4	2	0	3.7	-3.1
Oil	398	458	477	370	351	336	42	40	1.7	0.4
Gas	404	531	594	362	407	409	53	49	2.8	1.4
Nuclear	5	5	5	5	12	15	0	2	n.a.	n.a.
Hydro	3	4	4	3	4	5	0	1	6.7	6.9
Biomass and waste	2	4	4	6	16	24	0	3	4.5	11.5
Other renewables	3	9	14	7	25	46	1	5	9.5	14.4
Power generation	**242**	**318**	**359**	**219**	**258**	**278**	**100**	**100**	**2.2**	**1.3**
Coal	13	21	23	9	5	3	7	1	4.2	-3.6
Oil	62	62	64	61	45	42	18	15	-0.3	-1.8
Gas	157	217	248	134	164	162	69	58	2.7	1.1
Nuclear	5	5	5	5	12	15	1	5	n.a.	n.a.
Hydro	3	4	4	3	4	5	1	2	6.7	6.9
Biomass and waste	1	2	2	1	5	10	1	3	28.8	35.5
Other renewables	2	6	11	5	22	42	3	15	26.8	33.2
Other energy sector	**114**	**147**	**159**	**106**	**115**	**113**	**100**	**100**	**2.8**	**1.5**
Electricity	*18*	*23*	*25*	*17*	*18*	*19*	*16*	*17*	*2.5*	*1.4*
TFC	**574**	**706**	**764**	**530**	**562**	**574**	**100**	**100**	**2.5**	**1.5**
Coal	1	1	1	1	1	1	0	0	2.5	0.7
Oil	318	384	403	290	290	279	53	49	2.4	1.0
Gas	171	203	222	158	164	169	29	29	2.4	1.4
Electricity	82	114	133	75	95	107	17	19	3.5	2.7
Heat	-	-	-	-	-	-	-	-	n.a.	n.a.
Biomass and waste	1	1	1	5	10	14	0	2	1.1	9.9
Other renewables	2	2	3	2	3	4	0	1	3.2	4.8
Industry	**146**	**172**	**185**	**135**	**130**	**128**	**100**	**100**	**2.3**	**0.9**
Coal	1	1	1	1	1	1	1	0	2.5	0.7
Oil	55	60	61	50	44	41	33	32	1.5	0.1
Gas	74	88	95	68	65	65	52	50	2.5	1.1
Electricity	16	23	27	15	19	21	15	17	3.6	2.7
Heat	-	-	-	-	-	-	-	-	n.a.	n.a.
Biomass and waste	0	0	0	0	0	0	0	0	2.0	2.9
Other renewables	-	-	-	-	-	-	-	-	n.a.	n.a.
Transport	**179**	**231**	**245**	**166**	**177**	**176**	**100**	**100**	**3.0**	**1.7**
Oil	177	228	241	160	163	153	98	87	3.0	1.3
Electricity	0	0	0	0	3	6	0	4	0.2	23.6
Biofuels	-	-	-	3	8	11	-	6	n.a.	n.a.
Other fuels	2	3	4	2	4	6	2	3	3.9	5.1
Buildings	**140**	**179**	**203**	**128**	**148**	**160**	**100**	**100**	**2.5**	**1.6**
Coal	-	-	-	-	-	-	-	-	n.a.	n.a.
Oil	28	31	31	26	25	24	15	15	0.8	-0.1
Gas	48	60	68	43	50	54	33	34	2.2	1.4
Electricity	61	86	101	56	68	75	49	47	3.5	2.3
Heat	-	-	-	-	-	-	-	-	n.a.	n.a.
Biomass and waste	1	1	1	1	2	2	1	1	0.8	3.8
Other renewables	2	2	3	2	3	4	1	3	3.2	4.7
Other	**108**	**123**	**131**	**101**	**107**	**110**	**100**	**100**	**2.2**	**1.5**

Middle East: New Policies Scenario

	Electricity generation (TWh)							Shares (%)		CAAGR (%)
	1990	2008	2015	2020	2025	2030	2035	2008	2035	2008-2035
Total generation	240	771	980	1 120	1 276	1 446	1 613	100	100	2.8
Coal	10	36	47	46	55	78	105	5	7	4.1
Oil	114	279	243	243	247	220	208	36	13	-1.1
Gas	104	447	642	751	861	976	1 023	58	63	3.1
Nuclear	-	-	7	18	20	20	20	-	1	n.a.
Hydro	12	9	31	39	46	51	53	1	3	6.8
Biomass and waste	-	0	2	4	7	14	25	0	2	34.1
Wind	0	0	4	8	17	37	77	0	5	24.5
Geothermal	-	-	-	-	-	-	-	-	-	n.a.
Solar PV	-	-	1	3	9	20	33	-	2	n.a.
CSP	-	-	3	8	14	30	69	-	4	n.a.
Marine	-	-	-	-	-	-	0	-	0	n.a.

	Electrical capacity (GW)						Shares (%)		CAAGR (%)
	2008	2015	2020	2025	2030	2035	2008	2035	2008-2035
Total capacity	201	306	328	338	358	402	100	100	2.6
Coal	5	8	9	10	12	15	3	4	4.2
Oil	68	83	81	73	58	51	34	13	-1.0
Gas	117	195	208	213	225	238	58	59	2.7
Nuclear	-	1	2	3	3	3	-	1	n.a.
Hydro	12	16	20	23	25	26	6	6	2.9
Biomass and waste	0	0	1	1	2	4	0	1	20.8
Wind	0	1	3	6	13	27	0	7	27.9
Geothermal	-	-	-	-	-	-	-	-	n.a.
Solar PV	-	1	2	6	13	20	-	5	n.a.
CSP	0	1	2	4	8	17	0	4	31.3
Marine	-	-	-	-	-	0	-	0	n.a.

	CO$_2$ emissions (Mt)							Shares (%)		CAAGR (%)
	1990	2008	2015	2020	2025	2030	2035	2008	2035	2008-2035
Total CO$_2$	592	1 476	1 793	1 934	2 094	2 237	2 354	100	100	1.7
Coal	11	34	44	43	47	62	80	2	3	3.2
Oil	418	834	989	1 051	1 110	1 149	1 189	57	51	1.3
Gas	163	608	760	840	937	1 026	1 085	41	46	2.2
Power generation	172	530	533	579	628	667	688	100	100	1.0
Coal	9	31	40	39	43	58	76	6	11	3.4
Oil	89	219	192	193	196	176	168	41	24	-1.0
Gas	74	280	301	347	389	432	443	53	64	1.7
TFC	367	820	1 073	1 156	1 234	1 318	1 395	100	100	2.0
Coal	2	2	3	3	3	3	3	0	0	1.5
Oil	297	567	742	800	852	909	954	69	68	1.9
Transport	*150*	*323*	*455*	*504*	*550*	*607*	*654*	*39*	*47*	*2.7*
Gas	68	251	329	353	378	406	438	31	31	2.1

Middle East: Current Policies and 450 Scenarios

	Electricity generation (TWh)						Shares (%)		CAAGR (%)	
	2020	2030	2035	2020	2030	2035	2035		2008-2035	
	Current Policies Scenario			450 Scenario			CPS	450	CPS	450
Total generation	1 165	1 600	1 851	1 068	1 320	1 477	100	100	3.3	2.4
Coal	58	107	126	42	24	14	7	1	4.8	-3.3
Oil	245	246	253	242	174	159	14	11	-0.4	-2.1
Gas	790	1 123	1 316	691	845	863	71	58	4.1	2.5
Nuclear	18	20	20	18	47	57	1	4	n.a.	n.a.
Hydro	37	49	51	40	52	53	3	4	6.7	6.9
Biomass and waste	4	7	8	4	18	32	0	2	28.8	35.4
Wind	7	17	27	11	69	122	1	8	19.8	26.6
Geothermal	-	-	-	-	-	-	-	-	n.a.	n.a.
Solar PV	2	12	16	3	26	50	1	3	n.a.	n.a.
CSP	4	19	35	17	65	126	2	9	n.a.	n.a.
Marine	-	-	-	-	0	0	-	0	n.a.	n.a.

	Electrical capacity (GW)						Shares (%)		CAAGR (%)	
	2020	2030	2035	2020	2030	2035	2035		2008-2035	
	Current Policies Scenario			450 Scenario			CPS	450	CPS	450
Total capacity	325	366	420	333	393	455	100	100	2.8	3.1
Coal	9	16	19	8	4	4	4	1	4.9	-0.7
Oil	81	62	63	81	60	54	15	12	-0.3	-0.8
Gas	208	242	281	210	236	252	67	55	3.3	2.9
Nuclear	2	3	3	2	6	8	1	2	n.a.	n.a.
Hydro	19	24	25	20	25	26	6	6	2.8	3.0
Biomass and waste	1	1	2	1	3	5	0	1	16.5	21.8
Wind	2	6	9	4	25	44	2	10	23.0	30.3
Geothermal	-	-	-	-	-	-	-	-	n.a.	n.a.
Solar PV	2	8	10	3	17	30	2	7	n.a.	n.a.
CSP	1	5	9	5	17	32	2	7	28.1	34.3
Marine	-	-	-	-	0	0	-	0	n.a.	n.a.

	CO_2 emissions (Mt)						Shares (%)		CAAGR (%)	
	2020	2030	2035	2020	2030	2035	2035		2008-2035	
	Current Policies Scenario			450 Scenario			CPS	450	CPS	450
Total CO_2	2 018	2 506	2 711	1 833	1 809	1 735	100	100	2.3	0.6
Coal	54	87	98	39	23	14	4	1	4.0	-3.2
Oil	1 091	1 284	1 342	1 012	946	896	50	52	1.8	0.3
Gas	873	1 135	1 270	782	841	825	47	48	2.8	1.1
Power generation	611	781	869	542	531	502	100	100	1.9	-0.2
Coal	50	83	93	35	20	11	11	2	4.2	-3.6
Oil	194	195	201	192	142	133	23	26	-0.3	-1.8
Gas	367	503	575	314	370	358	66	71	2.7	0.9
TFC	1 204	1 456	1 550	1 101	1 072	1 034	100	100	2.4	0.9
Coal	3	4	4	3	2	2	0	0	2.4	-0.1
Oil	838	1 019	1 069	764	749	711	69	69	2.4	0.8
Transport	524	674	714	475	481	452	46	44	3.0	1.3
Gas	363	434	477	335	320	321	31	31	2.4	0.9

Africa: New Policies Scenario

	Energy demand (Mtoe)							Shares (%)		CAAGR (%)
	1990	2008	2015	2020	2025	2030	2035	2008	2035	2008-2035
TPED	388	655	735	781	824	868	904	100	100	1.2
Coal	74	104	106	111	113	115	112	16	12	0.3
Oil	86	140	147	151	155	160	169	21	19	0.7
Gas	30	84	114	125	130	135	138	13	15	1.9
Nuclear	2	3	3	3	9	14	16	1	2	6.0
Hydro	5	8	11	13	16	20	24	1	3	4.0
Biomass and waste	190	314	349	368	386	400	410	48	45	1.0
Other renewables	0	1	5	8	15	23	34	0	4	13.3
Power generation	69	132	160	175	191	209	225	100	100	2.0
Coal	39	60	63	65	65	64	62	46	27	0.1
Oil	11	19	13	11	9	8	7	14	3	-3.4
Gas	11	40	63	69	70	70	69	30	31	2.1
Nuclear	2	3	3	3	9	14	16	3	7	6.0
Hydro	5	8	11	13	16	20	24	6	10	4.0
Biomass and waste	0	1	3	5	7	10	14	0	6	12.4
Other renewables	0	1	5	8	14	22	33	1	15	13.2
Other energy sector	57	93	96	102	107	113	115	100	100	0.8
Electricity	*6*	*10*	*12*	*13*	*14*	*14*	*15*	*11*	*13*	*1.6*
TFC	289	484	547	580	612	642	668	100	100	1.2
Coal	19	16	18	18	17	17	17	3	2	0.1
Oil	70	117	131	139	147	156	166	24	25	1.3
Gas	9	28	33	36	38	39	41	6	6	1.5
Electricity	21	44	57	64	72	81	89	9	13	2.6
Heat	-	-	-	-	-	-	-	-	-	n.a.
Biomass and waste	169	279	308	324	338	348	353	58	53	0.9
Other renewables	0	0	0	0	1	1	2	0	0	14.0
Industry	61	86	104	108	111	115	116	100	100	1.1
Coal	13	9	10	10	10	10	10	11	9	0.4
Oil	14	14	16	16	15	15	14	16	12	-0.1
Gas	5	15	19	20	20	21	21	17	18	1.3
Electricity	12	20	24	26	27	29	30	23	26	1.5
Heat	-	-	-	-	-	-	-	-	-	n.a.
Biomass and waste	16	28	34	36	38	41	42	32	36	1.5
Other renewables	-	-	-	-	-	-	-	-	-	n.a.
Transport	37	72	81	89	96	105	114	100	100	1.7
Oil	36	70	79	85	92	100	109	98	95	1.6
Electricity	0	0	1	1	1	1	1	1	1	2.7
Biofuels	-	-	1	1	1	2	2	-	2	n.a.
Other fuels	0	1	1	2	2	2	3	2	2	2.7
Buildings	177	304	334	354	373	390	403	100	100	1.1
Coal	3	6	6	5	5	5	5	2	1	-0.7
Oil	13	20	22	22	23	25	26	7	6	1.0
Gas	1	5	6	7	8	8	9	2	2	2.2
Electricity	8	23	30	35	41	48	55	7	14	3.4
Heat	-	-	-	-	-	-	-	-	-	n.a.
Biomass and waste	152	250	271	285	296	303	307	82	76	0.8
Other renewables	0	0	0	0	0	0	1	0	0	10.9
Other	15	23	27	29	31	33	35	100	100	1.5

Africa: Current Policies and 450 Scenarios

	Energy demand (Mtoe)						Shares (%)		CAAGR (%)	
	2020	2030	2035	2020	2030	2035	2035		2008-2035	
	Current Policies Scenario			450 Scenario			CPS	450	CPS	450
TPED	792	898	948	759	810	829	100	100	1.4	0.9
Coal	124	151	161	105	94	80	17	10	1.6	-1.0
Oil	153	166	180	138	128	126	19	15	0.9	-0.4
Gas	127	144	150	121	103	94	16	11	2.2	0.4
Nuclear	3	8	8	3	28	36	1	4	3.1	9.1
Hydro	13	18	21	13	20	24	2	3	3.6	4.1
Biomass and waste	365	395	405	367	398	408	43	49	0.9	1.0
Other renewables	7	15	23	11	39	61	2	7	11.6	15.7
Power generation	183	225	249	169	190	203	100	100	2.4	1.6
Coal	76	93	104	60	45	33	42	16	2.1	-2.2
Oil	11	9	9	9	4	4	3	2	-2.9	-5.9
Gas	70	74	74	67	43	32	30	16	2.3	-0.8
Nuclear	3	8	8	3	28	36	3	18	3.1	9.1
Hydro	13	18	21	13	20	24	9	12	3.6	4.1
Biomass and waste	3	8	12	6	12	17	5	8	11.5	13.1
Other renewables	7	15	22	11	38	58	9	28	11.5	15.6
Other energy sector	104	120	123	101	107	107	100	100	1.0	0.5
Electricity	*13*	*16*	*17*	*12*	*13*	*13*	*14*	*12*	*2.0*	*1.0*
TFC	584	656	690	564	602	615	100	100	1.3	0.9
Coal	18	18	18	17	15	14	3	2	0.4	-0.5
Oil	141	164	178	127	127	126	26	20	1.6	0.3
Gas	36	41	44	35	38	40	6	6	1.8	1.4
Electricity	66	87	98	63	77	84	14	14	3.0	2.4
Heat	-	-	-	-	-	-	-	-	n.a.	n.a.
Biomass and waste	322	345	351	322	344	349	51	57	0.8	0.8
Other renewables	0	1	1	0	1	3	0	1	12.3	16.5
Industry	109	120	126	105	109	110	100	100	1.4	0.9
Coal	10	11	11	10	9	8	8	7	0.6	-0.4
Oil	17	18	18	14	12	11	14	10	0.8	-0.9
Gas	20	22	23	19	19	20	18	18	1.6	1.0
Electricity	27	31	33	25	27	28	27	26	2.0	1.3
Heat	-	-	-	-	-	-	-	-	n.a.	n.a.
Biomass and waste	35	39	41	37	41	43	33	39	1.4	1.6
Other renewables	-	-	-	-	-	-	-	-	n.a.	n.a.
Transport	89	107	119	80	83	84	100	100	1.9	0.6
Oil	85	103	114	75	75	74	95	88	1.8	0.2
Electricity	1	1	1	1	1	2	1	2	2.3	4.6
Biofuels	1	2	2	2	4	5	2	6	n.a.	n.a.
Other fuels	2	2	3	2	3	3	2	4	2.8	3.6
Buildings	357	394	410	350	378	387	100	100	1.1	0.9
Coal	6	6	5	5	5	5	1	1	-0.1	-0.8
Oil	23	26	29	22	23	24	7	6	1.4	0.6
Gas	7	9	10	7	8	8	2	2	2.5	1.7
Electricity	36	51	59	34	45	51	14	13	3.7	3.0
Heat	-	-	-	-	-	-	-	-	n.a.	n.a.
Biomass and waste	284	302	306	282	296	298	75	77	0.7	0.6
Other renewables	0	0	1	0	1	2	0	1	9.3	15.0
Other	30	34	36	29	33	34	100	100	1.6	1.4

A

Africa: New Policies Scenario

Electricity generation (TWh)	1990	2008	2015	2020	2025	2030	2035	Shares (%) 2008	Shares (%) 2035	CAAGR (%) 2008-2035
Total generation	316	621	789	885	985	1 098	1 204	100	100	2.5
Coal	165	260	279	295	294	295	290	42	24	0.4
Oil	43	74	51	44	37	32	31	12	3	-3.2
Gas	43	176	294	328	341	348	347	28	29	2.6
Nuclear	8	13	13	13	33	55	63	2	5	6.0
Hydro	56	95	125	155	191	231	274	15	23	4.0
Biomass and waste	0	1	9	16	24	34	46	0	4	16.5
Wind	-	1	7	13	21	34	52	0	4	14.6
Geothermal	0	1	3	4	7	11	17	0	1	10.4
Solar PV	-	0	2	6	14	24	36	0	3	30.7
CSP	-	-	6	12	23	34	47	-	4	n.a.
Marine	-	-	-	-	-	0	0	-	0	n.a.

Electrical capacity (GW)	2008	2015	2020	2025	2030	2035	Shares (%) 2008	Shares (%) 2035	CAAGR (%) 2008-2035
Total capacity	132	182	210	240	270	304	100	100	3.1
Coal	41	50	55	54	53	53	31	17	0.9
Oil	23	25	25	22	19	18	17	6	-0.9
Gas	41	64	73	82	87	91	31	30	3.0
Nuclear	2	2	2	5	8	9	1	3	5.8
Hydro	24	32	40	49	59	70	18	23	4.0
Biomass and waste	1	2	4	5	7	8	1	3	8.7
Wind	1	3	5	8	12	18	0	6	13.9
Geothermal	0	0	1	1	2	3	0	1	10.7
Solar PV	0	1	4	9	15	21	0	7	31.4
CSP	-	2	3	6	10	13	-	4	n.a.
Marine	-	-	-	-	0	0	-	0	n.a.

CO_2 emissions (Mt)	1990	2008	2015	2020	2025	2030	2035	Shares (%) 2008	Shares (%) 2035	CAAGR (%) 2008-2035
Total CO_2	546	890	999	1 049	1 069	1 094	1 111	100	100	0.8
Coal	234	300	318	329	321	316	300	34	27	0.0
Oil	248	408	427	443	461	483	511	46	46	0.8
Gas	64	182	253	277	287	295	299	20	27	1.9
Power generation	212	384	432	450	442	432	414	100	100	0.3
Coal	152	232	244	254	248	243	229	60	55	-0.1
Oil	35	59	41	34	29	25	23	15	6	-3.4
Gas	25	93	147	162	165	164	162	24	39	2.1
TFC	303	460	518	544	569	598	628	100	100	1.2
Coal	82	67	74	74	72	71	70	15	11	0.1
Oil	202	333	371	393	416	441	469	72	75	1.3
Transport	*105*	*209*	*234*	*253*	*273*	*297*	*324*	*45*	*52*	*1.6*
Gas	19	60	73	78	82	86	90	13	14	1.5

Africa: Current Policies and 450 Scenarios

	Electricity generation (TWh)						Shares (%)		CAAGR (%)	
	2020	2030	2035	2020	2030	2035	2035		2008-2035	
	Current Policies Scenario			450 Scenario			CPS	450	CPS	450
Total generation	**916**	**1 181**	**1 327**	**862**	**1 031**	**1 116**	**100**	**100**	**2.9**	**2.2**
Coal	339	433	490	272	210	154	37	14	2.4	-1.9
Oil	46	37	36	38	15	15	3	1	-2.7	-5.7
Gas	333	373	382	315	228	166	29	15	2.9	-0.2
Nuclear	13	29	30	13	106	138	2	12	3.1	9.1
Hydro	147	211	250	157	237	282	19	25	3.6	4.1
Biomass and waste	10	27	38	19	41	56	3	5	15.7	17.3
Wind	12	25	34	15	69	118	3	11	12.8	18.2
Geothermal	4	8	12	4	15	23	1	2	8.9	11.5
Solar PV	5	19	27	7	34	55	2	5	29.3	32.8
CSP	7	17	29	23	74	108	2	10	n.a.	n.a.
Marine	-	-	-	-	0	0	-	0	n.a.	n.a.

	Electrical capacity (GW)						Shares (%)		CAAGR (%)	
	2020	2030	2035	2020	2030	2035	2035		2008-2035	
	Current Policies Scenario			450 Scenario			CPS	450	CPS	450
Total capacity	**214**	**278**	**312**	**210**	**291**	**340**	**100**	**100**	**3.2**	**3.6**
Coal	61	74	82	52	42	35	26	10	2.6	-0.6
Oil	25	20	19	25	20	19	6	6	-0.8	-0.7
Gas	75	94	99	69	77	76	32	22	3.4	2.4
Nuclear	2	4	4	2	14	19	1	6	3.0	8.9
Hydro	37	53	63	40	60	72	20	21	3.7	4.2
Biomass and waste	3	5	7	4	8	10	2	3	8.1	9.3
Wind	4	9	12	5	24	41	4	12	12.2	17.4
Geothermal	1	1	2	1	2	3	1	1	9.7	11.6
Solar PV	3	12	16	5	21	33	5	10	30.1	33.5
CSP	2	5	7	7	23	32	2	10	n.a.	n.a.
Marine	-	-	-	-	0	0	-	0	n.a.	n.a.

	CO$_2$ emissions (Mt)						Shares (%)		CAAGR (%)	
	2020	2030	2035	2020	2030	2035	2035		2008-2035	
	Current Policies Scenario			450 Scenario			CPS	450	CPS	450
Total CO$_2$	**1 106**	**1 261**	**1 354**	**974**	**771**	**648**	**100**	**100**	**1.6**	**-1.2**
Coal	373	440	482	304	176	85	36	13	1.8	-4.6
Oil	453	512	552	402	380	375	41	58	1.1	-0.3
Gas	280	310	320	268	214	188	24	29	2.1	0.1
Power generation	**494**	**562**	**603**	**417**	**225**	**110**	**100**	**100**	**1.7**	**-4.5**
Coal	294	361	403	231	118	34	67	31	2.1	-6.9
Oil	36	28	27	29	12	12	4	10	-2.9	-5.9
Gas	164	173	173	157	95	65	29	59	2.3	-1.3
TFC	**555**	**631**	**678**	**504**	**489**	**480**	**100**	**100**	**1.4**	**0.2**
Coal	76	76	76	72	56	50	11	10	0.4	-1.1
Oil	400	465	506	358	353	348	75	73	1.6	0.2
Transport	*253*	*305*	*337*	*224*	*223*	*220*	*50*	*46*	*1.8*	*0.2*
Gas	79	90	97	75	80	82	14	17	1.8	1.2

A

Latin America: New Policies Scenario

	Energy demand (Mtoe)							Shares (%)		CAAGR (%)
	1990	2008	2015	2020	2025	2030	2035	2008	2035	2008-2035
TPED	344	569	667	723	767	812	855	100	100	1.5
Coal	17	24	34	32	30	30	28	4	3	0.6
Oil	157	253	274	281	284	288	293	44	34	0.6
Gas	53	112	143	167	180	189	197	20	23	2.1
Nuclear	2	6	7	11	14	17	18	1	2	4.4
Hydro	31	58	66	72	78	85	91	10	11	1.7
Biomass and waste	82	115	137	151	167	184	200	20	23	2.1
Other renewables	1	3	5	8	13	19	28	0	3	9.0
Power generation	69	146	175	193	209	226	243	100	100	1.9
Coal	5	9	18	16	14	14	13	6	5	1.3
Oil	14	33	26	22	17	13	12	22	5	-3.5
Gas	14	31	44	53	59	61	62	21	25	2.6
Nuclear	2	6	7	11	14	17	18	4	7	4.4
Hydro	31	58	66	72	78	85	91	40	37	1.7
Biomass and waste	2	7	9	11	14	18	22	5	9	4.4
Other renewables	1	3	5	8	12	17	25	2	10	8.8
Other energy sector	59	80	91	102	105	106	107	100	100	1.1
Electricity	*8*	*18*	*21*	*22*	*23*	*24*	*25*	*22*	*23*	*1.2*
TFC	260	436	511	550	586	624	659	100	100	1.5
Coal	7	11	12	12	12	12	12	2	2	0.3
Oil	126	203	229	239	246	254	259	47	39	0.9
Gas	25	60	74	82	89	97	104	14	16	2.1
Electricity	36	74	90	100	110	120	129	17	20	2.1
Heat	-	-	-	-	-	-	-	-	-	n.a.
Biomass and waste	66	87	106	116	128	140	151	20	23	2.1
Other renewables	-	0	0	1	1	2	3	0	0	11.8
Industry	87	152	184	200	214	226	237	100	100	1.7
Coal	7	11	12	12	12	12	11	7	5	0.3
Oil	20	33	38	38	39	39	39	22	17	0.6
Gas	15	32	41	45	49	53	57	21	24	2.2
Electricity	17	33	41	47	51	56	60	22	25	2.2
Heat	-	-	-	-	-	-	-	-	-	n.a.
Biomass and waste	28	43	52	58	62	66	70	28	30	1.8
Other renewables	-	-	-	-	-	-	-	-	-	n.a.
Transport	75	135	158	169	180	195	209	100	100	1.6
Oil	68	117	131	138	142	148	153	86	73	1.0
Electricity	0	0	0	0	1	1	1	0	0	3.8
Biofuels	6	12	20	23	31	38	45	9	21	4.9
Other fuels	0	6	6	7	7	9	11	4	5	2.3
Buildings	71	98	109	115	122	129	136	100	100	1.2
Coal	0	0	0	0	0	0	0	0	0	1.2
Oil	18	19	20	21	22	22	22	20	16	0.5
Gas	6	12	13	15	16	18	20	12	14	2.0
Electricity	18	39	45	49	53	58	63	39	46	1.8
Heat	-	-	-	-	-	-	-	-	-	n.a.
Biomass and waste	28	28	29	29	29	29	28	29	21	-0.1
Other renewables	-	0	0	1	1	2	3	0	2	11.8
Other	28	51	61	66	70	74	77	100	100	1.6

Latin America: Current Policies and 450 Scenarios

	Energy demand (Mtoe)						Shares (%)		CAAGR (%)	
	2020	2030	2035	2020	2030	2035	2035		2008-2035	
	Current Policies Scenario			450 Scenario			CPS	450	CPS	450
TPED	**737**	**852**	**914**	**691**	**750**	**773**	**100**	**100**	**1.8**	**1.1**
Coal	35	38	38	29	23	18	4	2	1.6	-1.2
Oil	290	313	326	266	244	233	36	30	0.9	-0.3
Gas	174	221	246	150	145	136	27	18	3.0	0.7
Nuclear	11	13	14	11	20	21	1	3	3.4	5.1
Hydro	71	83	88	75	90	97	10	13	1.6	1.9
Biomass and waste	147	169	181	150	197	222	20	29	1.7	2.5
Other renewables	8	16	22	10	30	47	2	6	8.1	11.2
Power generation	**205**	**255**	**285**	**188**	**208**	**219**	**100**	**100**	**2.5**	**1.5**
Coal	19	21	21	13	9	4	7	2	3.1	-2.8
Oil	26	19	18	19	8	7	6	3	-2.1	-5.4
Gas	59	89	106	49	28	12	37	5	4.6	-3.5
Nuclear	11	13	14	11	20	21	5	10	3.4	5.1
Hydro	71	83	88	75	90	97	31	44	1.6	1.9
Biomass and waste	11	15	18	12	25	34	6	16	3.6	6.1
Other renewables	7	15	20	9	28	43	7	20	7.9	11.1
Other energy sector	**104**	**109**	**111**	**92**	**98**	**100**	**100**	**100**	**1.2**	**0.8**
Electricity	*23*	*26*	*28*	*22*	*23*	*23*	*25*	*23*	*1.7*	*1.0*
TFC	**556**	**645**	**691**	**531**	**580**	**598**	**100**	**100**	**1.7**	**1.2**
Coal	13	13	13	11	11	10	2	2	0.8	-0.2
Oil	243	271	284	227	217	207	41	35	1.3	0.1
Gas	82	99	108	78	91	96	16	16	2.2	1.8
Electricity	104	131	146	98	114	122	21	20	2.5	1.9
Heat	-	-	-	-	-	-	-	-	n.a.	n.a.
Biomass and waste	113	129	137	115	146	160	20	27	1.7	2.3
Other renewables	1	1	2	1	2	4	0	1	9.8	12.1
Industry	**202**	**234**	**249**	**190**	**215**	**222**	**100**	**100**	**1.9**	**1.4**
Coal	13	13	13	11	10	10	5	4	0.8	-0.3
Oil	40	43	44	36	35	34	18	15	1.1	0.1
Gas	45	55	59	42	49	52	24	23	2.3	1.8
Electricity	48	60	66	45	53	57	26	26	2.5	2.0
Heat	-	-	-	-	-	-	-	-	n.a.	n.a.
Biomass and waste	57	64	67	56	67	70	27	32	1.7	1.8
Other renewables	-	-	-	-	-	-	-	-	n.a.	n.a.
Transport	**169**	**200**	**216**	**161**	**172**	**176**	**100**	**100**	**1.8**	**1.0**
Oil	140	160	169	129	118	108	78	62	1.4	-0.3
Electricity	0	1	1	1	1	3	0	2	3.7	8.6
Biofuels	22	30	35	24	44	55	16	31	4.0	5.7
Other fuels	7	9	11	7	9	10	5	6	2.4	2.1
Buildings	**118**	**137**	**148**	**114**	**122**	**126**	**100**	**100**	**1.5**	**0.9**
Coal	0	0	0	0	0	0	0	0	1.4	1.0
Oil	22	24	25	21	20	20	17	16	0.9	0.1
Gas	15	19	21	15	17	18	14	14	2.2	1.7
Electricity	51	64	72	49	54	57	49	45	2.3	1.4
Heat	-	-	-	-	-	-	-	-	n.a.	n.a.
Biomass and waste	29	29	28	29	29	28	19	22	-0.1	-0.1
Other renewables	1	1	2	1	2	4	1	3	9.8	12.1
Other	**66**	**74**	**77**	**66**	**71**	**74**	**100**	**100**	**1.6**	**1.4**

Latin America: New Policies Scenario

Electricity generation (TWh)								Shares (%)		CAAGR (%)
	1990	2008	2015	2020	2025	2030	2035	2008	2035	2008-2035
Total generation	507	1 069	1 283	1 421	1 545	1 668	1 788	100	100	1.9
Coal	16	37	79	71	64	66	61	3	3	1.9
Oil	66	157	128	105	83	62	59	15	3	-3.6
Gas	46	146	230	289	326	344	346	14	19	3.2
Nuclear	10	21	27	43	55	65	69	2	4	4.4
Hydro	363	673	766	839	912	985	1 054	63	59	1.7
Biomass and waste	7	30	37	44	54	65	79	3	4	3.7
Wind	-	1	11	18	28	42	60	0	3	16.3
Geothermal	1	3	5	6	9	12	16	0	1	6.5
Solar PV	-	-	1	4	10	19	28	-	2	n.a.
CSP	-	-	-	1	5	9	13	-	1	n.a.
Marine	-	-	-	-	-	-	1	-	0	n.a.

Electrical capacity (GW)							Shares (%)		CAAGR (%)
	2008	2015	2020	2025	2030	2035	2008	2035	2008-2035
Total capacity	237	287	320	353	385	420	100	100	2.1
Coal	6	13	13	13	13	12	3	3	2.6
Oil	34	35	35	33	29	28	14	7	-0.7
Gas	43	58	68	74	81	87	18	21	2.7
Nuclear	3	4	6	7	9	9	1	2	4.4
Hydro	146	165	179	195	209	224	62	53	1.6
Biomass and waste	5	6	7	9	10	12	2	3	3.5
Wind	1	5	7	11	15	20	0	5	14.1
Geothermal	1	1	1	1	2	2	0	1	5.6
Solar PV	-	1	4	9	15	21	-	5	n.a.
CSP	-	-	0	1	2	3	-	1	n.a.
Marine	-	-	-	-	-	0	-	0	n.a.

CO_2 emissions (Mt)								Shares (%)		CAAGR (%)
	1990	2008	2015	2020	2025	2030	2035	2008	2035	2008-2035
Total CO_2	605	1 057	1 229	1 290	1 311	1 341	1 363	100	100	0.9
Coal	56	93	140	130	120	120	113	9	8	0.8
Oil	430	714	768	782	786	795	809	68	59	0.5
Gas	119	250	321	377	406	425	441	24	32	2.1
Power generation	99	216	268	263	252	246	241	100	100	0.4
Coal	21	41	83	71	62	62	57	19	24	1.3
Oil	46	102	83	68	54	41	39	47	16	-3.5
Gas	32	73	102	124	137	143	145	34	60	2.6
TFC	441	750	856	901	934	973	1 003	100	100	1.1
Coal	31	48	54	55	55	54	52	6	5	0.3
Oil	354	570	639	667	684	706	720	76	72	0.9
Transport	_203_	_348_	_392_	_411_	_423_	_441_	_455_	_46_	_45_	_1.0_
Gas	56	132	163	180	195	213	231	18	23	2.1

Latin America: Current Policies and 450 Scenarios

| Electricity generation (TWh) | | | | | | Shares (%) 2035 | | CAAGR (%) 2008-2035 | |
| Current Policies Scenario | | | 450 Scenario | | | CPS | 450 | CPS | 450 |
2020	2030	2035	2020	2030	2035				
Total generation 1 473	1 820	2 016	1 387	1 581	1 685	100	100	2.4	1.7
Coal 84	97	99	60	41	21	5	1	3.7	-2.1
Oil 125	91	87	93	39	35	4	2	-2.2	-5.4
Gas 318	497	593	239	148	58	29	3	5.3	-3.4
Nuclear 43	49	53	43	78	82	3	5	3.4	5.1
Hydro 831	963	1 026	868	1 042	1 126	51	67	1.6	1.9
Biomass and waste 42	56	66	46	88	118	3	7	2.9	5.2
Wind 19	37	49	22	79	116	2	7	15.4	19.1
Geothermal 6	11	14	7	18	26	1	2	5.9	8.3
Solar PV 4	15	19	5	34	84	1	5	n.a.	n.a.
CSP -	5	10	4	13	19	0	1	n.a.	n.a.
Marine -	-	-	-	0	2	-	0	n.a.	n.a.

| Electrical capacity (GW) | | | | | | Shares (%) 2035 | | CAAGR (%) 2008-2035 | |
| Current Policies Scenario | | | 450 Scenario | | | CPS | 450 | CPS | 450 |
2020	2030	2035	2020	2030	2035				
Total capacity 332	418	465	316	391	455	100	100	2.5	2.4
Coal 15	18	18	11	10	9	4	2	4.0	1.3
Oil 37	37	39	32	23	22	8	5	0.6	-1.6
Gas 77	113	135	58	55	56	29	12	4.4	1.0
Nuclear 6	7	7	6	11	11	2	2	3.4	5.1
Hydro 178	205	218	186	221	238	47	52	1.5	1.8
Biomass and waste 7	9	11	7	13	17	2	4	2.9	4.8
Wind 7	13	17	9	26	37	4	8	13.5	16.8
Geothermal 1	2	2	1	3	3	0	1	5.2	7.2
Solar PV 4	13	16	5	25	56	3	12	n.a.	n.a.
CSP -	1	2	1	3	5	1	1	n.a.	n.a.
Marine -	-	-	-	0	0	-	0	n.a.	n.a.

| CO_2 emissions (Mt) | | | | | | Shares (%) 2035 | | CAAGR (%) 2008-2035 | |
| Current Policies Scenario | | | 450 Scenario | | | CPS | 450 | CPS | 450 |
2020	2030	2035	2020	2030	2035				
Total CO_2 1 349	1 521	1 616	1 187	1 031	931	100	100	1.6	-0.5
Coal 146	153	154	114	75	50	10	5	1.9	-2.3
Oil 809	870	907	735	649	608	56	65	0.9	-0.6
Gas 394	498	555	338	306	273	34	29	3.0	0.3
Power generation 304	358	395	233	126	67	100	100	2.3	-4.2
Coal 84	89	89	60	35	16	23	24	2.9	-3.4
Oil 82	60	57	59	25	23	15	34	-2.1	-5.4
Gas 139	209	248	114	66	28	63	42	4.6	-3.5
TFC 917	1 037	1 096	853	800	759	100	100	1.4	0.0
Coal 57	59	59	50	37	30	5	4	0.8	-1.7
Oil 679	758	796	631	582	544	73	72	1.2	-0.2
Transport 417	476	505	386	350	323	46	43	1.4	-0.3
Gas 181	220	241	172	182	185	22	24	2.3	1.3

Brazil: New Policies Scenario

	Energy demand (Mtoe)							Shares (%)		CAAGR (%)
	1990	2008	2015	2020	2025	2030	2035	2008	2035	2008-2035
TPED	**138**	**245**	**301**	**336**	**362**	**386**	**411**	**100**	**100**	**1.9**
Coal	10	14	19	17	15	14	14	6	3	0.2
Oil	59	96	110	115	117	118	118	39	29	0.8
Gas	3	21	36	50	56	59	65	9	16	4.2
Nuclear	1	4	4	6	8	11	12	1	3	4.6
Hydro	18	32	35	38	41	43	45	13	11	1.3
Biomass and waste	48	79	96	108	123	136	149	32	36	2.4
Other renewables	-	0	1	1	3	4	6	0	2	13.5
Power generation	**22**	**52**	**67**	**75**	**83**	**92**	**102**	**100**	**100**	**2.5**
Coal	1	3	8	6	4	4	4	6	4	0.6
Oil	1	4	2	3	3	3	3	7	3	-1.2
Gas	0	5	12	15	17	18	21	11	21	5.1
Nuclear	1	4	4	6	8	11	12	7	12	4.6
Hydro	18	32	35	38	41	43	45	62	45	1.3
Biomass and waste	1	4	5	6	8	10	11	7	11	4.3
Other renewables	-	0	1	1	2	4	5	0	5	18.4
Other energy sector	**26**	**41**	**50**	**58**	**59**	**57**	**55**	**100**	**100**	**1.1**
Electricity	*3*	*8*	*10*	*10*	*11*	*12*	*12*	*20*	*22*	*1.5*
TFC	**112**	**195**	**237**	**261**	**283**	**306**	**328**	**100**	**100**	**1.9**
Coal	4	7	8	8	8	8	8	3	2	0.5
Oil	53	86	100	106	107	108	109	44	33	0.9
Gas	2	11	16	20	25	30	35	6	11	4.4
Electricity	18	35	43	48	52	57	62	18	19	2.1
Heat	-	-	-	-	-	-	-	-	-	n.a.
Biomass and waste	34	56	70	80	91	102	113	29	34	2.6
Other renewables	-	0	0	0	1	1	1	0	0	7.9
Industry	**40**	**77**	**95**	**107**	**117**	**128**	**138**	**100**	**100**	**2.2**
Coal	4	7	7	8	8	8	7	9	5	0.5
Oil	8	12	14	14	15	15	15	16	11	0.8
Gas	1	8	12	16	19	23	27	10	20	4.6
Electricity	10	17	21	23	26	28	31	22	22	2.2
Heat	-	-	-	-	-	-	-	-	-	n.a.
Biomass and waste	17	33	41	46	50	54	58	43	42	2.1
Other renewables	-	-	-	-	-	-	-	-	-	n.a.
Transport	**33**	**62**	**76**	**84**	**91**	**99**	**107**	**100**	**100**	**2.0**
Oil	27	48	57	61	61	61	61	77	57	0.9
Electricity	0	0	0	0	0	0	0	0	0	3.8
Biofuels	6	12	17	20	27	34	40	19	38	4.6
Other fuels	0	2	2	3	3	4	5	3	5	3.3
Buildings	**23**	**33**	**37**	**39**	**42**	**44**	**47**	**100**	**100**	**1.3**
Coal	-	-	-	-	-	-	-	-	-	n.a.
Oil	6	7	7	8	8	8	8	22	17	0.5
Gas	0	0	1	1	1	1	1	1	3	4.6
Electricity	8	17	20	22	24	26	28	51	61	2.0
Heat	-	-	-	-	-	-	-	-	-	n.a.
Biomass and waste	9	8	9	9	9	8	8	26	16	-0.3
Other renewables	-	0	0	0	1	1	1	0	3	7.9
Other	**16**	**24**	**29**	**31**	**33**	**35**	**37**	**100**	**100**	**1.6**

Brazil: Current Policies and 450 Scenarios

| | Energy demand (Mtoe) | | | | | | Shares (%) 2035 | | CAAGR (%) 2008-2035 | |
| | 2020 | 2030 | 2035 | 2020 | 2030 | 2035 | | | | |
	Current Policies Scenario			450 Scenario			CPS	450	CPS	450
TPED	340	397	428	316	357	376	100	100	2.1	1.6
Coal	18	16	16	15	11	10	4	3	0.6	-1.0
Oil	118	129	134	110	100	95	31	25	1.3	-0.0
Gas	54	74	86	36	42	46	20	12	5.3	2.9
Nuclear	6	8	9	6	12	13	2	3	3.6	4.8
Hydro	37	42	44	40	45	48	10	13	1.2	1.5
Biomass and waste	105	124	133	107	141	156	31	42	2.0	2.6
Other renewables	1	4	5	2	6	9	1	2	12.7	14.7
Power generation	79	102	116	71	82	89	100	100	3.0	2.0
Coal	7	5	5	5	2	1	4	1	1.3	-3.8
Oil	3	4	4	2	2	2	4	2	0.5	-2.7
Gas	19	32	41	9	5	5	35	5	7.8	-0.5
Nuclear	6	8	9	6	12	13	8	14	3.6	4.8
Hydro	37	42	44	40	45	48	38	54	1.2	1.5
Biomass and waste	6	7	8	7	12	14	7	16	3.1	5.1
Other renewables	1	3	4	1	5	7	4	8	17.6	19.7
Other energy sector	58	57	56	50	53	54	100	100	1.1	1.0
Electricity	11	13	14	10	11	11	24	20	1.9	1.1
TFC	263	312	338	253	286	302	100	100	2.1	1.6
Coal	8	8	9	7	7	6	3	2	0.9	-0.1
Oil	107	118	123	101	92	87	36	29	1.3	0.0
Gas	20	30	36	20	28	32	11	11	4.5	4.0
Electricity	49	62	69	47	54	58	20	19	2.5	1.9
Heat	-	-	-	-	-	-	-	-	n.a.	n.a.
Biomass and waste	77	93	101	78	105	116	30	39	2.2	2.8
Other renewables	0	1	1	0	1	2	0	1	7.2	9.0
Industry	107	129	140	102	123	131	100	100	2.3	2.0
Coal	8	8	8	7	6	6	6	5	0.9	-0.2
Oil	15	16	17	13	14	14	12	10	1.2	0.4
Gas	16	23	28	15	21	24	20	18	4.7	4.1
Electricity	23	29	31	22	27	29	22	22	2.3	2.1
Heat	-	-	-	-	-	-	-	-	n.a.	n.a.
Biomass and waste	45	53	56	45	54	58	40	44	2.0	2.1
Other renewables	-	-	-	-	-	-	-	-	n.a.	n.a.
Transport	84	99	107	80	89	94	100	100	2.0	1.5
Oil	62	68	71	57	47	43	66	46	1.5	-0.4
Electricity	0	0	0	0	1	1	0	1	3.7	9.0
Biofuels	19	26	31	20	37	45	29	48	3.6	5.0
Other fuels	3	4	5	3	4	5	5	5	3.3	3.2
Buildings	41	49	54	39	40	41	100	100	1.9	0.9
Coal	-	-	-	-	-	-	-	-	n.a.	n.a.
Oil	8	9	10	8	7	7	18	16	1.1	-0.2
Gas	1	1	1	1	1	1	3	3	5.1	4.0
Electricity	23	30	34	22	23	24	63	59	2.7	1.5
Heat	-	-	-	-	-	-	-	-	n.a.	n.a.
Biomass and waste	9	8	8	9	8	8	14	18	-0.3	-0.3
Other renewables	0	1	1	0	1	2	2	4	7.2	9.0
Other	31	35	37	31	34	35	100	100	1.6	1.4

Brazil: New Policies Scenario

Electricity generation (TWh)	1990	2008	2015	2020	2025	2030	2035	Shares (%) 2008	Shares (%) 2035	CAAGR (%) 2008-2035
Total generation	223	463	569	632	693	758	827	100	100	2.2
Coal	5	13	37	26	17	18	18	3	2	1.4
Oil	5	18	11	11	12	12	12	4	1	-1.4
Gas	1	29	68	90	105	111	126	6	15	5.6
Nuclear	2	14	14	24	32	43	47	3	6	4.6
Hydro	207	370	409	440	471	500	528	80	64	1.3
Biomass and waste	4	20	24	29	34	40	44	4	5	3.0
Wind	-	1	6	9	14	20	28	0	3	15.0
Geothermal	-	-	-	-	-	-	-	-	-	n.a.
Solar PV	-	-	1	2	6	11	18	-	2	n.a.
CSP	-	-	-	-	2	4	6	-	1	n.a.
Marine	-	-	-	-	-	-	0	-	0	n.a.

Electrical capacity (GW)	2008	2015	2020	2025	2030	2035	Shares (%) 2008	Shares (%) 2035	CAAGR (%) 2008-2035
Total capacity	106	131	147	163	180	199	100	100	2.4
Coal	2	6	5	5	5	5	2	3	3.5
Oil	5	7	8	8	9	9	5	4	2.0
Gas	9	16	20	22	25	29	8	14	4.4
Nuclear	2	2	3	4	6	6	2	3	4.6
Hydro	85	93	100	107	113	119	80	60	1.3
Biomass and waste	3	3	4	5	6	7	3	3	3.1
Wind	0	2	4	5	7	9	0	5	13.1
Geothermal	-	-	-	-	-	-	-	-	n.a.
Solar PV	-	1	2	5	9	13	-	7	n.a.
CSP	-	-	-	0	1	2	-	1	n.a.
Marine	-	-	-	-	-	0	-	0	n.a.

CO_2 emissions (Mt)	1990	2008	2015	2020	2025	2030	2035	Shares (%) 2008	Shares (%) 2035	CAAGR (%) 2008-2035
Total CO_2	194	365	468	502	508	516	530	100	100	1.4
Coal	29	50	80	67	57	57	57	14	11	0.5
Oil	159	266	304	318	321	324	326	73	62	0.8
Gas	7	49	84	116	130	135	147	13	28	4.1
Power generation	12	41	79	73	68	71	78	100	100	2.4
Coal	8	17	44	30	20	20	21	41	27	0.8
Oil	4	12	8	8	8	8	8	28	11	-1.2
Gas	0	13	28	35	41	43	49	31	63	5.1
TFC	167	295	349	373	385	399	413	100	100	1.2
Coal	18	30	34	35	35	35	34	10	8	0.4
Oil	144	240	278	291	294	296	298	81	72	0.8
Transport	81	145	172	182	182	183	184	49	45	0.9
Gas	5	25	37	46	56	68	81	9	20	4.4

Brazil: Current Policies and 450 Scenarios

| | Electricity generation (TWh) | | | | | | Shares (%) | | CAAGR (%) | |
| | 2020 | 2030 | 2035 | 2020 | 2030 | 2035 | 2035 | | 2008-2035 | |
	Current Policies Scenario			450 Scenario			CPS	450	CPS	450
Total generation	656	821	920	617	706	764	100	100	2.6	1.9
Coal	31	21	21	21	7	5	2	1	2.0	-3.2
Oil	13	17	19	10	8	8	2	1	0.3	-2.8
Gas	114	199	251	55	29	25	27	3	8.3	-0.5
Nuclear	24	32	36	24	45	49	4	6	3.6	4.8
Hydro	435	490	515	461	526	554	56	73	1.2	1.5
Biomass and waste	26	32	35	31	46	53	4	7	2.1	3.7
Wind	9	18	23	11	28	39	3	5	14.2	16.4
Geothermal	-	-	-	-	-	-	-	-	n.a.	n.a.
Solar PV	2	10	14	3	12	21	1	3	n.a.	n.a.
CSP	-	3	5	2	6	8	1	1	n.a.	n.a.
Marine	-	-	-	-	0	1	-	0	n.a.	n.a.

| | Electrical capacity (GW) | | | | | | Shares (%) | | CAAGR (%) | |
| | 2020 | 2030 | 2035 | 2020 | 2030 | 2035 | 2035 | | 2008-2035 | |
	Current Policies Scenario			450 Scenario			CPS	450	CPS	450
Total capacity	152	194	219	146	177	196	100	100	2.7	2.3
Coal	7	6	6	4	4	4	3	2	4.1	2.5
Oil	9	12	14	6	6	6	6	3	3.8	0.4
Gas	24	41	53	15	15	16	24	8	6.8	2.2
Nuclear	3	4	5	3	6	7	2	3	3.6	4.8
Hydro	99	111	116	105	119	125	53	64	1.2	1.4
Biomass and waste	4	5	5	5	7	8	2	4	2.2	3.8
Wind	4	6	8	4	10	13	4	7	12.5	14.4
Geothermal	-	-	-	-	-	-	-	-	n.a.	n.a.
Solar PV	2	8	11	3	9	16	5	8	n.a.	n.a.
CSP	-	1	1	0	2	2	1	1	n.a.	n.a.
Marine	-	-	-	-	0	0	-	0	n.a.	n.a.

| | CO$_2$ emissions (Mt) | | | | | | Shares (%) | | CAAGR (%) | |
| | 2020 | 2030 | 2035 | 2020 | 2030 | 2035 | 2035 | | 2008-2035 | |
	Current Policies Scenario			450 Scenario			CPS	450	CPS	450
Total CO$_2$	526	594	635	443	381	359	100	100	2.1	-0.1
Coal	74	65	65	58	33	27	10	8	1.0	-2.2
Oil	326	360	375	301	260	239	59	67	1.3	-0.4
Gas	126	169	195	84	88	92	31	26	5.3	2.4
Power generation	89	112	133	52	26	23	100	100	4.4	-2.2
Coal	36	25	25	24	8	6	19	27	1.4	-3.7
Oil	9	12	13	6	6	6	10	25	0.5	-2.7
Gas	45	76	95	22	12	11	72	48	7.8	-0.5
TFC	380	434	460	353	317	299	100	100	1.7	0.0
Coal	37	38	38	32	23	19	8	6	0.9	-1.6
Oil	297	327	339	276	237	218	74	73	1.3	-0.4
Transport	185	205	213	171	143	130	46	43	1.4	-0.4
Gas	47	70	84	45	57	62	18	21	4.5	3.4

POLICIES AND MEASURES IN THE NEW POLICIES AND 450 SCENARIOS

The *World Energy Outlook 2010* presents projections for three scenarios. The Current Policies Scenario includes all policies in place and supported through enacted measures as of mid-2010. The New Policies and 450 Scenarios are based on the greenhouse-gas emissions-reductions and other commitments associated with the Copenhagen Accord; on other policies currently under discussion or announced but not yet implemented; and the extension or strengthening of some policies already in force and included under the Current Policies Scenario. Access to international offset credits for countries participating in emissions-trading schemes is assumed in both the New Policies and 450 Scenarios, though the timing, prices of CO_2 and scale of trading differ.

The **New Policies Scenario** takes into account all policies and measures included in the Current Policies Scenario, as well as the following:

- Cautious implementation of the Copenhagen Accord commitments by 2020.

- Continuation of the European Union Emissions Trading Scheme (EU ETS), and introduction of a cap-and-trade system in the rest of the OECD+ after 2020.

- Phase out of fossil-fuel consumption subsidies in all net-importing regions by 2020 (and, as in the Current Policies Scenario, in net-exporting regions where specific policies have already been introduced).

- Extension of nuclear plant lifetimes by 5 to 10 years with respect to the Current Policies Scenario, on a plant-by-plant basis.

- For 2020-2035, additional measures that maintain the pace of the global decline in carbon intensity — measured as emissions per dollar of gross domestic product, in purchasing power parity terms — established in the period 2008-2020.

The **450 Scenario** takes into account all policies and measures included in the New Policies Scenario, some of which are assumed to be substantially strengthened and extended, plus the following:

- Implementation by 2020 of the high-end of the range of the Copenhagen Accord commitments where they are expressed as ranges.

- Cap-and-trade systems in the power and industry sectors, from 2013 in OECD+ countries and after 2020 in Other Major Economies (OME).

- International sectoral agreements for the iron and steel, and the cement industries.

- International agreements on fuel-economy standards for passenger light-duty vehicles (PLDVs), aviation and shipping.

- National policies and measures, such as efficiency standards for buildings and labelling of appliances.

- The complete phase-out of fossil-fuel consumption subsidies in all net-importing regions by 2020 (at the latest) and in all net-exporting regions by 2035 (at the latest), except for the Middle East where it is assumed that the average subsidisation rate declines to 20% by 2035.

- Extension of nuclear plant lifetimes by 5 to 10 years with respect to the New Policies Scenario, on a plant-by-plant basis.

Specific policies by selected countries and regions for the New Policies and 450 Scenarios are outlined below.

Table B.1 ● *Overall targets, policies and measures as modelled in the New Policies Scenario and the 450 Scenario in key regions*

	New Policies Scenario	450 Scenario
OECD		
United States	- Clean Energy Jobs and American Power Act (2010). Policy to increase reliance on domestic energy sources, including gas and biofuels. - American Recovery and Reinvestment Act (2009). Federal funding, loan guarantees and tax credits for renewables, nuclear and energy efficiency.	- 17% reduction in greenhouse-gas emissions by 2020 compared with 2005 (with access to international offset credits).
Japan	- Basic Energy Plan (2010). Implementation of renewable deployment in total primary energy supply and other measures.	- 25% reduction in greenhouse-gas emissions by 2020 compared with 1990 (with access to international offset credits).
European Union	- Climate and Energy Package (2009). 25% reduction in greenhouse-gas emissions by 2020 compared with 1990 (with access to international offset credits). - EU directive on renewables (2009). 20% share of renewables in gross final energy consumption by 2020.	- 30% reduction in greenhouse-gas emissions by 2020 compared with 1990 (with access to international offset credits).
Non-OECD		
Russia	- Energy Strategy of Russia until 2030 (2009). 15% reduction in greenhouse-gas emissions by 2020 compared with 1990.	- 25% reduction in greenhouse-gas emissions by 2020 compared with 1990.
China	- 40% reduction in CO_2 intensity by 2020 compared with 2005 (2009). - Rebalancing of the economy from industry towards services (2009). - Further implementation of the directives of the Renewable Energy Law (2005).	- 45% reduction in CO_2 intensity by 2020 compared with 2005. - 15% share of non-fossil energy in primary energy consumption by 2020.
India	- National Action Plan on Climate Change (2008). 20% reduction in CO_2 intensity by 2020. compared with 2005.	- 25% reduction in CO_2 intensity by 2020 compared with 2005.
Brazil	- National Climate Change Plan (2008) and 2019 Energy Expansion Decennial Plan (2010). 36% reduction in greenhouse-gas emissions by 2020 compared with business-as-usual.	- 39% reduction in greenhouse-gas emissions by 2020 compared with business-as-usual.

Note: Existing policies or measures quoted here are assumed to be extended beyond their current duration, for which they have been considered in the *Current Policies Scenario*. Targets in certain countries are exceeded as already met in the *Current Policies Scenario*.

Table B.2 ● Power sector policies and measures as modelled in the New Policies Scenario and the 450 Scenario in key regions

	New Policies Scenario	450 Scenario
OECD		
United States	- 15% share of renewables in electricity generation by 2020. - Extension of nuclear loan guarantee. - OECD+ Emission trading scheme introduced in the power and industry sectors after 2020. - Large-scale demonstration plants fitted with carbon-capture-and-storage (CCS) technology. - Extension of nuclear plants lifetime beyond 60 years.	- OECD+ Emission trading scheme introduced in the power and industry sectors as of 2013. - Extended support to renewables and nuclear.
Japan	Emissions-trading scheme introduced in power sector as of 2013. - Basic Energy Plan. - 9 nuclear additions by 2020; a minimum of 14 additional reactors built by 2030. - Introduction of CCS to coal-fired power generation by 2030.	- OECD+ emissions-trading scheme introduced in the power and industry sectors as of 2013. - Basic Energy Plan. - Share of low-carbon electricity generation raised to 50% by 2020 and to 70% by 2030. - Reinforcement of governmental support in favour of renewables.
European Union	- Extension of EU ETS in accordance with the 25% GHG reduction target. - Expansion of renewable energy sources. - Cancellation of nuclear phase-out plans in Germany (extending average lifetime to 45 years). - EU Directive on the geological storage of carbon dioxide (2009).	- Aligning with the OECD+ emissions-trading scheme as of 2020. - Reinforcement of governmental support in favour of renewables.
Non-OECD		
Russia	- Optimised heat production systems, reduction of losses in heat distribution. - Switch away from coal and gas and increase in nuclear and renewables capacity.	- Other Major Economies emissions-trading scheme introduced after 2020. - Strengthening of the switch away from coal and gas and increase in nuclear and renewables capacity.
China	- Early closure of inefficient coal plants. - Local pollution reduction goals. - Government capacity targets in 2020 including wind 125 GW, nuclear 65 GW and hydro 300 GW. - 20% renewable share in power generation by 2020. - Fossil-fuel subsidies removal by 2020.	- Wind capacity target extended to 150 GW by 2020. - Nuclear capacity target extended to 70 GW by 2020 and continued support to maintain the rate of growth of nuclear additions post 2020. - Solar capacity target of 20 GW by 2020. - Other Major Economies emissions-trading scheme introduced after 2020.
India	- Various renewable energy support policies and targets, including small hydro and solar targets. - Fossil-fuel subsidies removal by 2020.	- Support to renewables, nuclear and efficient coal. - 30 GW of additional renewable (non-large hydro) capacity by 2020.
Brazil	- Increase of biomass and hydro (small and large) capacity.	- Other Major Economies emissions-trading scheme introduced after 2020.

Table B.3 ● Transport sector policies and measures as modelled in the New Policies Scenario and the 450 Scenario in key regions

New Policies Scenario		450 Scenario		
OECD			OECD+	OME	OC

New Policies Scenario	450 Scenario

OECD

United States	- Renewable Fuel Standard. - Support to natural gas in road freight traffic. - Increase of ethanol blending mandates.

Emission targets for passenger light-duty vehicles in 2035 (in gCO_2/km)

	OECD+	OME	OC
	75	85	105

Japan	- Target shares of new car sales according to Next Generation Vehicle Strategy 2010:

Light commercial vehicles — Full technology spillover from passenger light-duty vehicles.

	2020	2030
Conventional ICE vehicles	50 - 80 %	30 - 50 %
Hybrid vehicles	20 - 30 %	30 - 40 %
Electric vehicles and plug-in hybrids	15 - 20 %	20 - 30 %
Fuel cell vehicles	0 - 1 %	0 - 3 %
Clean diesel vehicles	0 - 5 %	5 - 10 %

Medium- and heavy-freight traffic — 5% more efficient by 2035 than in CPS.

Aviation — Sectoral target of 45% efficiency improvements by 2035 and support to the use of biofuels.

Other sectors such as maritime and rail — National policies and measures.

Fuels — Retail fuel prices kept at a level similar to Current Policies Scenario.

European Union	- Extended emission target for passenger light-duty vehicles (95 gCO_2/km by 2020). - Emission target for light commercial vehicles (135 gCO_2/km by 2020). - Enhanced support to alternative fuels. - Several national EV targets, subsidy extension. - Aviation and international maritime shipping in EU ETS as of 2013.

Alternative clean fuels — Enhanced support to alternative fuels.

Non-OECD

China	- Vehicle fuel economy standard 7 l/100 km by 2015. - Extended subsidies on the purchase of alternative vehicles.
India	- Increased utilisation of natural gas in road transport.
Brazil	- Increase of ethanol blending mandates.

Table B.4 ● Industry sector policies and measures as modelled in the New Policies Scenario and the 450 Scenario in key regions

	New Policies Scenario	450 Scenario
OECD		**OECD+, Other Major Economies, Other Countries**
United States	- Reduced industrial emissions through allowances rebates (Program within Title VII of Clean Air Act). - Tax reduction and funding for efficiency improvement by revolutionary technologies and R&D in low carbon technology.	- OECD+ Emission trading scheme introduced in the power and industry sectors as of 2013. - Other Major Economies emissions-trading scheme introduced after 2020. - Wider hosting international offset projects in Other Countries. - International sectoral agreements with targets for iron, steel and cement industries. - Enhanced efficiency standards or improvements. - Policies to support the introduction of CCS in industry.
Japan	- Maintenance and strengthening of top end/low carbon efficiency standards by: - R&D in revolutionary process and its practical realisation - Higher efficiency CHP technology - Promotion of state-of-the-art technology and faster replacement of aging equipments - Fuel switching to gas with higher efficiency equipments.	
European Union	- EU Directive on energy end-use efficiency and energy efficiency (2009) including the development of: - Inverters for electric motors - High-efficiency co-generation - Mechanical vapour compression - Emergence of significant innovations in industrial processes - Extension of EU ETS.	
Non-OECD		
Russia	- Improvement of the energy and environmental efficiency, including through structural changes and more efficient technologies. - Establishment of a new system for domestic energy prices. - Elaboration of comprehensive federal and regional legislation on energy saving.	
China	- Scrapping of small, energy inefficient plans (less than 10 MW), obsolete iron ore refining plants with a 25 million tonnes capacity, of steel refining plants with a 6 million tonnes capacity, of cement plants with a 50 million tonnes capacity and of electrolytic aluminium plants with a 330 000 tonnes capacity. - Contain the expansion of energy intensive industries.	
India	- Implementation of National Mission for Enhanced Energy Efficiency recommendations including: - Enhancement of cost-effective improvements in energy efficiency in energy-intensive large industries and facilities, through certification of energy savings that could be traded. - Creation of mechanisms that would help financing demand side management programmes in all sectors by capturing future energy savings. - Development of fiscal instruments to promote energy efficiency.	
Brazil	- Copenhagen Accord commitment: More utilisation of charcoal in iron production substituting for coal.	

Table B.5 ● Buildings sector policies and measures as modelled in the New Policies Scenario and the 450 Scenario in key regions

	New Policies Scenario	450 Scenario
OECD		
United States	- Clean Energy Jobs and American Power Act (2010). Mandatory standards for lighting systems and appliances, and for manufactured housing. - Extensions to 2025 of tax credit for energy-efficient equipment (including energy-efficient gas, propane, or oil furnaces or boilers, energy-efficient central air conditioners, air and ground source heat pumps, hot water heaters, and windows), extension of access to tax credits for solar PV and solar thermal water heaters.	- More stringent mandatory building codes by 2020. - Extension of energy-efficiency grants to end of projection period. - Zero-energy buildings initiative.
Japan	- Basic Energy Plan: Environmental Efficiency (CASBEE) for all buildings by 2030 - high efficiency lighting 100% of newly sold by 2020; 100% in use by 2030 - deployment of high-efficiency heating, cooling and water heating systems.	- Net zero-energy buildings and net zero-energy houses by 2025 for new construction. - Mandatory standards on high-efficiency heating, cooling and water heating systems.
European Union	- Energy Performance of Buildings Directive (2006) extension to 2020. - Nearly zero-energy buildings standards mandatory for new construction as of 2020.	- Zero-carbon footprint for all new buildings as of 2018.
Non-OECD		
Russia	- Energy Strategy of Russia through 2030 (extension of existing Energy Strategy through 2020): urban development code, customs code, support to renewable energy sources.	- Extension and reinforcement of all measures for energy efficiency, mandatory building codes by 2030 and phase out inefficient lighting equipment and appliances by 2030.
China	- Renewables for rural and 3-selfs scheme ("self construction", "self-management", and "self consumption") aimed at promoting self-reliance in order to be in line with economy-wide 15% renewables target. - Phase out of incandescent light bulbs by 2025.	- 65% energy conservation standard of the "Civil Construction Energy Conservation Design Standard (Heating Housing Construction Part)": Improvement of buildings insulation intended to save up to 65% of heating energy consumption compared with standard buildings designed in the 1980s.
India	- Part of national solar mission: solar water heating systems: (15 million sq. metre solar thermal collector area by 2017 and 20 million by 2022). - Mandatory minimum efficiency requirements and labelling requirements for equipment and appliances by 2035.	- Mandatory energy conservation standards and labelling requirements for equipment and appliances by 2030. - Phase out of incandescent light bulbs by 2025.

ABBREVIATIONS, ACRONYMS, DEFINITIONS AND CONVERSION FACTORS

This annex provides general information on terminology used throughout *WEO-2010* including: units and general conversion factors; definitions on fuels, processes and sectors; regional and country groupings; and abbreviations and acronyms.

Units

Area	Ha	hectare
	GHa	giga-hectare (1 hectare $\times 10^9$)
	km²	square kilometre
Coal	Mtce	million tonnes of coal equivalent
Emissions	ppm	parts per million (by volume)
	kg CO₂-eq	kilogrammes of carbon-dioxide equivalent
	Gt CO₂-eq	gigatonnes of carbon-dioxide equivalent (using 100 year global warming potentials for different greenhouse gases)
	gCO₂/km	grammes of carbon dioxide per kilometre
	gCO₂/kWh	grammes of carbon dioxide per kilowatt-hour
Energy	toe	tonne of oil equivalent
	Mtoe	million tonnes of oil equivalent
	MBtu	million British thermal units
	MJ	megajoule (1 joule $\times 10^6$)
	GJ	gigajoule (1 joule $\times 10^9$)
	TJ	terajoule (1 joule $\times 10^{12}$)
	kWh	kilowatt-hour
	MWh	megawatt-hour
	GWh	gigawatt-hour
	TWh	terawatt-hour
Gas	mcm	million cubic metres
	bcm	billion cubic metres
	tcm	trillion cubic metres
Mass	kg	kilogramme (1 000 kg = 1 tonne)
	kt	kilotonnes (1 tonne $\times 10^3$)
	Mt	million tonnes (1 tonne $\times 10^6$)
	Gt	gigatonnes (1 tonne $\times 10^9$)

Monetary	$ million	1 US dollar × 10⁶
	$ billion	1 US dollar × 10⁹
	$ trillion	1 US dollar × 10¹²
Oil	b/d	barrels per day
	kb/d	thousand barrels per day
	mb/d	million barrels per day
	Ml/year	million litres per year
	mpg	miles per gallon
Oil and gas	boe	barrels of oil equivalent
Power	W	Watt (1 joule per second)
	kW	kilowatt (1 Watt × 10³)
	MW	megawatt (1 Watt × 10⁶)
	GW	gigawatt (1 Watt × 10⁹)
	GW$_{th}$	gigawatt thermal (1 Watt × 10⁹)
	TW	terawatt (1 Watt × 10¹²)

General conversion factors for energy

To:	TJ	Gcal	Mtoe	MBtu	GWh
From:	*multiply by:*				
TJ	1	238.8	2.388×10^{-5}	947.8	0.2778
Gcal	4.1868×10^{-3}	1	10^{-7}	3.968	1.163×10^{-3}
Mtoe	4.1868×10^{4}	10^{7}	1	3.968×10^{7}	11 630
MBtu	1.0551×10^{-3}	0.252	2.52×10^{-8}	1	2.931×10^{-4}
GWh	3.6	860	8.6×10^{-5}	3 412	1

Definitions

Agriculture

Includes all energy used on farms, in forestry and for fishing.

Biodiesel

Biodiesel is a diesel-equivalent, processed fuel made from the esterification (a chemical process which removes the glycerine from the oil) of both vegetable oils and animal fats.

Biofuels

Biofuels are fuels derived from biomass or waste feedstocks and include ethanol and biodiesel.

Biogas

A mixture of methane and CO_2 produced by bacterial degradation of organic matter and used as a fuel.

Biomass and waste

Solid biomass, gas and liquids derived from biomass, industrial waste and the renewable part of municipal waste. Includes both traditional and modern biomass.

Biomass-to-liquids

Biomass-to-liquids (BTL) refers to a process featuring biomass gasification into syngas followed by synthesis of liquid products (such as diesel, naphtha or gasoline) from the syngas using Fischer-Tropsch catalytic synthesis or a methanol-to-gasoline reaction path. The process is similar to those used in coal-to-liquids or gas-to-liquids.

Brown coal

Includes lignite and sub-bituminous coal where lignite is defined as non-agglomerating coal with a gross calorific value less than 4 165 kilocalories per kilogramme (kcal/kg) and sub-bituminous coal is defined as non-agglomerating coal with a gross calorific value between 4 165 kcal/kg and 5 700 kcal/kg.

Buildings

A sector that includes energy used in residential, commercial and institutional buildings. Building energy use includes space heating and cooling, water heating, lighting, appliances and cooking equipment.

Bunkers

Includes both international marine bunkers and international aviation bunkers (see respective category definitions).

Clean coal technologies

Clean coal technologies (CCTs) are designed to enhance the efficiency and the environmental acceptability of coal extraction, preparation and use.

Coal

Coal includes both primary coal (including hard coal and brown coal) and derived fuels (including patent fuel, brown-coal briquettes, coke-oven coke, gas coke, coke-oven gas, blast-furnace gas and oxygen steel furnace gas). Peat is also included.

Coalbed methane

Methane found in coal seams. Coalbed methane (CBM) is a source of unconventional natural gas.

Coal-to-liquids

Coal-to-liquids (CTL) refers to the transformation of coal into liquid hydrocarbons. It can be achieved through either coal gasification into syngas (a mixture of hydrogen and carbon monoxide), combined with Fischer-Tropsch or methanol-to-gasoline synthesis to produce liquid fuels, or through the less developed direct-coal liquefaction technologies in which coal is directly reacted with hydrogen.

Condensates

Condensates are liquid hydrocarbon mixtures recovered from associated or non-associated gas reservoirs. They are composed of C_5 and higher carbon number hydrocarbons and normally have an API between 50° and 85°.

Electricity generation

Defined as the total amount of electricity generated by power only or combined heat and power plants including generation required for own use. This is also referred to as gross generation.

Ethanol

Although ethanol can be produced from a variety of fuels, in this book, ethanol refers to bio-ethanol only. Ethanol is produced from fermenting any biomass high in carbohydrates. Today, ethanol is made from starches and sugars, but advanced generation technologies will allow it to be made from cellulose and hemicellulose, the fibrous material that makes up the bulk of most plant matter.

Gas

Gas includes natural gas (both associated and non-associated with petroleum deposits, but excluding natural gas liquids) and gas-works gas.

Gas-to-liquids

Gas-to-liquids refers to a process featuring reaction of methane with oxygen or steam to produce syngas followed by synthesis of liquid products (such as diesel and naphtha) from the syngas using Fischer-Tropsch catalytic synthesis. The process is similar to those used in coal-to-liquids or biomass-to-liquids.

Hard coal

Coal of gross calorific value greater than 5 700 kilocalories per kilogramme on an ash-free but moist basis. Hard coal can be further disaggregated into anthracite, coking coal and other bituminous coal.

Heat energy

Heat is obtained from fuel combustion, nuclear reactors, geothermal reservoirs, capture of sunlight, exothermic chemical processes and heat pumps which can extract it from ambient air and liquids. It may be used for heating or cooling or converted into mechanical energy for transport vehicles or electricity generation. Commercial heat sold is reported under total final consumption with the fuel inputs allocated under power generation.

Heavy petroleum products

Heavy petroleum products include heavy fuel oil.

Hydropower

Kinetic energy of water converted into electricity in hydroelectric plants. It excludes output from pumped storage and marine (tide and wave) plants.

Industry

A sector that includes fuel used within the manufacturing and construction industries. Key industry sectors include iron and steel, chemical and petrochemical, non-metallic minerals, and pulp and paper. Use by industries for the transformation of energy into another form or for the production of fuels is excluded and reported separately under other energy sector. Consumption of fuels for the transport of goods is reported as part of the transport sector.

International aviation bunkers

This category includes the deliveries of aviation fuels to aircraft for international aviation. Fuels used by airlines for their road vehicles are excluded. The domestic/ international split is determined on the basis of departure and landing locations and not by the nationality of the airline. For many countries this incorrectly excludes fuels used by domestically owned carriers for their international departures.

International marine bunkers

This category covers those quantities delivered to ships of all flags that are engaged in international navigation. The international navigation may take place at sea, on inland lakes and waterways, and in coastal waters. Consumption by ships engaged in domestic navigation is excluded. The domestic/international split is determined on the basis of port of departure and port of arrival, and not by the flag or nationality of the ship. Consumption by fishing vessels and by military forces is also excluded and included in residential, services and agriculture.

Light petroleum products

Light petroleum products include liquefied petroleum gas (LPG), naphtha and gasoline.

Low-carbon technologies

Refers to technologies that produce low- or zero- greenhouse-gas emissions while operating. In the power sector this includes fossil-fuel plants fitted with carbon capture and storage, nuclear plants and renewable-based generation technologies.

Lower heating value

Lower heating value is the heat liberated by the complete combustion of a unit of fuel when the water produced is assumed to remain as a vapour and the heat is not recovered.

Middle distillates

Middle distillates include jet fuel, diesel and heating oil.

Modern biomass

Includes all biomass with the exception of traditional biomass.

Modern renewables

Includes all types of renewables with the exception of traditional biomass.

Natural decline rate

The base production decline rate of an oil or gas field without intervention to enhance production.

Natural gas liquids

Natural gas liquids (NGLs) are the liquid or liquefied hydrocarbons produced in the manufacture, purification and stabilisation of natural gas. These are those portions of natural gas which are recovered as liquids in separators, field facilities, or gas processing plants. NGLs include but are not limited to ethane, propane, butane, pentane, natural gasoline and condensates.

Non-energy use

Fuels used for chemical feedstocks and non-energy products. Examples of non-energy products include lubricants, paraffin waxes, coal tars, and oils used as timber preservatives.

Nuclear

Nuclear refers to the primary heat equivalent of the electricity produced by a nuclear plant with an average thermal efficiency of 33%.

Observed decline rate

The production decline rate of an oil or gas field after all measures have been taken to maximise production. It is the aggregation of all the production increases and declines of new and mature oil or gas fields in a particular region.

Oil

Oil includes crude oil, condensates, natural gas liquids, refinery feedstocks and additives, other hydrocarbons (including emulsified oils, synthetic crude oil, mineral oils extracted from bituminous minerals such as oil shale, bituminous sand and oils from CTL and GTL) and petroleum products (refinery gas, ethane, LPG, aviation gasoline, motor gasoline, jet fuels, kerosene, gas/diesel oil, heavy fuel oil, naphtha, white spirit, lubricants, bitumen, paraffin waxes and petroleum coke).

Other energy sector

Other energy sector covers the use of energy by transformation industries and the energy losses in converting primary energy into a form that can be used in the final consuming sectors. It includes losses by gas works, petroleum refineries, coal and gas transformation and liquefaction. It also includes energy used in coal mines, in oil and gas extraction and in electricity and heat production. Transfers and statistical differences are also included in this category

Power generation

Power generation refers to fuel use in electricity plants, heat plants and combined heat and power (CHP) plants. Both main activity producer plants and small plants that produce fuel for their own use (autoproducers) are included.

Renewables

Includes biomass and waste, geothermal, hydropower, solar photovoltaics (PV), concentrating solar power (CSP), wind and marine (tide and wave) energy for electricity and heat generation.

Syngas

A synthetic gas primarily composed of hydrogen and carbon monoxide produced by a chemical process.

Total final consumption

Total final consumption (TFC) is the sum of consumption by the various end-use sectors. TFC is broken down into energy demand in the following sectors: industry, transport, buildings (including residential and services) and other (including agriculture and non-energy use). It excludes international marine and aviation bunkers, except at world level where it is included in the transport sector.

Total primary energy demand

Total primary energy demand (TPED) represents domestic demand only and is broken down into power generation, other energy sector and total final consumption.

Traditional biomass

Traditional biomass refers to the use of fuelwood, charcoal, animal dung and agricultural residues in stoves with very low efficiencies.

Transport

Fuels and electricity used in the transport of goods or persons within the national territory irrespective of the economic sector within which the activity occurs. This includes fuel and electricity delivered to vehicles using public roads or for use in rail vehicles; fuel delivered to vessels for domestic navigation; fuel delivered to aircraft for domestic aviation; and energy consumed in the delivery of fuels through pipelines. Fuel delivered to international marine and aviation bunkers is presented only at the world level and is excluded from the transport sector at the domestic level.

Regional and country groupings

Africa

Algeria, Angola, Benin, Botswana, Cameroon, Congo, Democratic Republic of Congo, Côte d'Ivoire, Egypt, Eritrea, Ethiopia, Gabon, Ghana, Kenya, Libya, Morocco, Mozambique, Namibia, Nigeria, Senegal, South Africa, Sudan, United Republic of Tanzania, Togo, Tunisia, Zambia, Zimbabwe and other African countries (Burkina Faso, Burundi, Cape Verde, Central African Republic, Chad, Comoros, Djibouti, Equatorial Guinea, Gambia, Guinea, Guinea-Bissau, Lesotho, Liberia, Madagascar, Malawi, Mali, Mauritania, Mauritius, Niger, Reunion, Rwanda, Sao Tome and Principe, Seychelles, Sierra Leone, Somalia, Swaziland and Uganda).

Annex I Parties to the United Nations Framework Convention on Climate Change

Australia, Austria, Belarus, Belgium, Bulgaria, Canada, Croatia, Czech Republic, Denmark, Estonia, Finland, France, Germany, Greece, Hungary, Iceland, Ireland, Italy, Japan, Latvia, Liechtenstein, Lithuania, Luxembourg, Monaco, Netherlands, New Zealand, Norway, Poland, Portugal, Romania, Russian Federation, Slovak Republic, Slovenia, Spain, Sweden, Switzerland, Turkey, Ukraine, United Kingdom and United States.

ASEAN

Brunei Darussalam, Cambodia, Indonesia, Laos, Malaysia, Myanmar, Philippines, Singapore, Thailand and Vietnam.

Caspian

Armenia, Azerbaijan, Georgia, Kazakhstan, Kyrgyz Republic, Tajikistan, Turkmenistan and Uzbekistan.

China

Refers to the People's Republic of China, including Hong Kong.

Developing countries

Non-OECD Asia, Middle East, Africa and Latin America regional groupings.

Eastern Europe/Eurasia

Albania, Armenia, Azerbaijan, Belarus, Bosnia and Herzegovina, Bulgaria, Croatia, Estonia, Georgia, Kazakhstan, Kyrgyz Republic, Latvia, Lithuania, the former Yugoslav Republic of Macedonia, the Republic of Moldova, Romania, Russian Federation, Serbia[1], Slovenia, Tajikistan, Turkmenistan, Ukraine and Uzbekistan. For statistical reasons, this region also includes Cyprus, Gibraltar and Malta.

European Union

Austria, Belgium, Bulgaria, Cyprus, Czech Republic, Denmark, Estonia, Finland, France, Germany, Greece, Hungary, Ireland, Italy, Latvia, Lithuania, Luxembourg, Malta, Netherlands, Poland, Portugal, Romania, Slovak Republic, Slovenia, Spain, Sweden and United Kingdom.

G8

Canada, France, Germany, Italy, Japan, Russian Federation, United Kingdom and United States.

G20

G8 countries and Argentina, Australia, Brazil, China, India, Indonesia, Mexico, Saudi Arabia, South Africa, Korea, Turkey and the European Union.

Latin America

Argentina, Bolivia, Brazil, Chile, Colombia, Costa Rica, Cuba, the Dominican Republic, Ecuador, El Salvador, Guatemala, Haiti, Honduras, Jamaica, Netherlands Antilles, Nicaragua, Panama, Paraguay, Peru, Trinidad and Tobago, Uruguay, Venezuela and other Latin American countries (Antigua and Barbuda, Aruba, Bahamas, Barbados, Belize, Bermuda, British Virgin Islands, Cayman Islands, Dominica, Falkland Islands, French Guyana, Grenada, Guadeloupe, Guyana, Martinique, Montserrat, St. Kitts and Nevis, Saint Lucia, Saint Pierre et Miquelon, St. Vincent and the Grenadines, Suriname and Turks and Caicos Islands).

Middle East

Bahrain, the Islamic Republic of Iran, Iraq, Israel, Jordan, Kuwait, Lebanon, Oman, Qatar, Saudi Arabia, Syrian Arab Republic, United Arab Emirates and Yemen. It includes the neutral zone between Saudi Arabia and Iraq.

---

1. Serbia includes Montenegro until 2004 and Kosovo until 1999.

Non-OECD Asia

Bangladesh, Brunei Darussalam, Cambodia, China, Chinese Taipei, India, Indonesia, the Democratic People's Republic of Korea, Malaysia, Mongolia, Myanmar, Nepal, Pakistan, the Philippines, Singapore, Sri Lanka, Thailand, Vietnam and other non-OECD Asian countries (Afghanistan, Bhutan, Cook Islands, East Timor, Fiji, French Polynesia, Kiribati, Laos, Macau, Maldives, New Caledonia, Papua New Guinea, Samoa, Solomon Islands, Tonga and Vanuatu).

North Africa

Algeria, Egypt, Libyan Arab Jamahiriya, Morocco and Tunisia.

OECD[2]

Includes OECD Europe, OECD North America and OECD Pacific regional groupings.

OECD Europe

Austria, Belgium, Czech Republic, Denmark, Finland, France, Germany, Greece, Hungary, Iceland, Ireland, Italy, Luxembourg, Netherlands, Norway, Poland, Portugal, Slovak Republic, Spain, Sweden, Switzerland, Turkey and United Kingdom.

OECD North America

Canada, Mexico and United States.

OECD Oceania

Australia and New Zealand.

OECD Pacific

Australia, Japan, Korea and New Zealand.

OECD+

OECD regional grouping and those countries that are members of the European Union but not of the OECD.

Other Major Economies

Brazil, China, Russia, South Africa and the countries of the Middle East.

Other Countries

Comprises all countries not included in OECD+ and Other Major Economies regional groupings, including the African countries (excluding South Africa), the countries of

---

2. Chile, Israel and Slovenia joined the OECD in mid-2010, but, as accession negotiations were continuing when our modelling work commenced, these countries are not included in the OECD in this *Outlook*.

Latin America (excluding Brazil), the countries of non-OECD Asia (excluding China) and the countries of Eastern Europe/Eurasia (excluding Russia).

OPEC

Algeria, Angola, Ecuador, Islamic Republic of Iran, Iraq, Kuwait, Libya, Nigeria, Qatar, Saudi Arabia, United Arab Emirates and Venezuela.

Other Asia

Non-OECD Asia regional grouping excluding China and India.

Sub-Saharan Africa

Africa regional grouping excluding the North African regional grouping and South Africa. In Chapter 8, however, South Africa is included in the sub-Saharan African grouping.

Abbreviations and Acronyms

API	American Petroleum Institute
ASEAN	Association of Southeast Asian Nations
BTL	biomass-to-liquids
CAAGR	compound average annual growth rate
CAFE	corporate average fuel economy (standards in the US)
CBM	coalbed methane
CBTL	Coal-and-biomass-to-liquids
CER	Certified Emission Reduction
CCGT	combined-cycle gas turbine
CCHP	combined cooling, heat and power
CCS	carbon capture and storage
CDM	Clean Development Mechanism (under the Kyoto Protocol)
CDU	crude distillation unit
CFL	compact fluorescent light bulb
CH_4	methane
CHP	combined heat and power; when referring to industrial CHP, the term co-generation is sometimes used
CMM	coal mine methane
CNG	compressed natural gas
CO	carbon monoxide
CO_2	carbon dioxide
CO_2-eq	carbon-dioxide equivalent

COP	Conference of Parties
CPC	Caspian Pipeline Consortium
CSP	concentrating solar power
CSS	cyclic steam stimulation
CTL	coal-to-liquids
DME	dimethyl ether
E&P	exploration and production
EDI	Energy Development Index
EOR	enhanced oil recovery
EPA	Environmental Protection Agency (United States)
EPC	engineering, procurement and construction
ESCO	energy service company
ETBE	ethyl tertiary butyl ether
EU	European Union
EU ETS	European Union Emissions Trading System
FAO	Food and Agriculture Organization of the United Nations
FDI	foreign direct investment
FFV	flex-fuel vehicle
FOB	free-on-board
GDP	gross domestic product
GHG	greenhouse gas
GTL	gas-to-liquids
HDI	Human Development Index
HDV	heavy duty vehicles
HIV/AIDS	human immunodeficiency virus/acquired immune deficiency syndrome
IAEA	International Atomic Energy Agency
ICE	internal combustion engine
IGCC	integrated gasification combined-cycle
IMF	International Monetary Fund
IOC	international oil company
IPCC	Intergovernmental Panel on Climate Change
IPP	independent power producer
LCV	light commercial vehicle
LDV	light-duty vehicle
LHV	lower heating value
LNG	liquefied natural gas

LPG	liquefied petroleum gas
LRMC	long run marginal cost
LULUCF	land use, land use change and forestry
MER	market exchange rate
MDG	Millennium Development Goal
MEPS	minimum energy performance standard
MSC	multiple service contract
MTBE	methyl tertiary butyl ether
N_2O	nitrous oxide
NAMA	nationally appropriate mitigation action
NEA	Nuclear Energy Agency (an agency within the OECD)
NGL	natural gas liquid
NGV	natural gas vehicle
NOC	national oil company
NO_x	nitrogen oxides
OC	Other Countries (see regional and country groupings)
OCGT	open-cycle gas turbine
ODI	outward foreign direct investment
OECD	Organisation for Economic Co-operation and Development
OECD+	OECD countries, plus EU countries not in the OECD (see regional and country groupings)
OME	Other Major Economies (see regional and country groupings)
OPEC	Organization of the Petroleum Exporting Countries
PDS	public distribution system
PLDV	passenger light-duty vehicle
PM	particulate matter
PPP	purchasing power parity
PSA	production-sharing agreement
PV	solar photovoltaics
RD&D	research, development and demonstration
RDD&D	research, development, demonstration and deployment
RPK	revenue passenger kilometers
SAGD	steam-assisted gravity drainage
SCO	synthetic crude oil
SO_2	sulphur dioxide
SRMC	short-run marginal cost

TAME	tertiary amyl methyl ether
TFC	total final consumption
TPED	total primary energy demand
UAE	United Arab Emirates
UCG	underground coal gasification
UNDP	United Nations Development Programme
UNEP	United Nations Environment Programme
UNFCCC	United Nations Framework Convention on Climate Change
UNIDO	United Nations Industrial Development Organization
USC	ultra-supercritical
USGS	United States Geological Survey
WEO	World Energy Outlook
WEM	World Energy Model
WHO	World Health Organization
WTI	West Texas Intermediate (crude oil category)
WTO	World Trade Organization

REFERENCES

PART A: Global energy trends

Chapter 1: Context and analytical framework

IEA (International Energy Agency) (2009), *World Energy Outlook 2009*, OECD/IEA, Paris.

– (2010a), *Medium-Term Oil and Gas Market Report*, OECD/IEA, Paris.

– (2010b), *Energy Technology Perspectives 2010*, OECD/IEA, Paris.

IMF (International Monetary Fund) (2010a), *World Economic Outlook Update: Restoring Confidence without Harming Recovery*, July, IMF, Washington, DC.

– (2010b), *Global Financial Stability Report (GFSR)*, July, IMF, Washington, DC.

OECD (Organisation for Economic Co-operation and Development) (2010), *OECD Economic Outlook No. 87*, May, OECD, Paris.

UNPD (United Nations Population Division) (2009), *World Population Prospects: The 2008 Revision*, UNPD, New York.

– (2010), *World Urbanization Prospects: the 2009 Revision*, UNPD, New York.

Chapter 2: Energy projections to 2035

BGR (*Bundesanstalt für Geowissenschaften und Rohstoffe* – German Federal Institute for Geosciences and Natural Resources) (2009), *Energierohstoffe 2009, Reserven, Ressourcen, Verfügbarkeit, Tabellen [Energy Resources 2009, Reserves, Resources, Availability, Tables]*, BGR, Hannover, Germany.

IEA (International Energy Agency) (2009), *World Energy Outlook 2009*, OECD/IEA, Paris.

NBS (National Bureau of Statistics) (2010), "Communiqué on Energy Consumption per Unit of GDP by Region in 2009", 15 July, National Bureau of Statistics, Beijing.

NEA (Nuclear Energy Agency, OECD) and IAEA (International Atomic Energy Agency) (2009), *Uranium 2009: Resources, Production and Demand*, OECD, Paris.

O&GJ (*Oil and Gas Journal*) (2009), "Worldwide Look at Reserves and Production", *Oil and Gas Journal*, 21 December, Pennwell Corporation, Oklahoma City, Oklahoma.

Chapter 3: Oil market outlook

Barbus, B. (2010), *Current Developments in the Mid-stream*, Menecon Publishing, Paris.

BGR (*Bundesanstalt für Geowissenschaften und Rohstoffe* – German Federal Institute for Geosciences and Natural Resources) (2009), *Energierohstoffe 2009, Reserven, Ressourcen, Verfügbarkeit, Tabellen [Energy Resources 2009, Reserves, Resources, Availability, Tables]*, BGR, Hannover, Germany.

Deutsche Bank (2009), *The Peak Oil Market: Price Dynamics at the End of the Oil Age*, Global Markets Research, 4 October, Deutsche Bank Securities, New York.

IEA (International Energy Agency) (2008), *World Energy Outlook 2008*, OECD/IEA, Paris.

– (2009), *World Energy Outlook 2009*, OECD/IEA, Paris.

– (2010a), "The Contribution of Natural Gas Vehicles to Sustainable Transport", *IEA Working Paper*, OECD/IEA, Paris.

– (2010b), *Medium Term Oil and Gas Markets 2010*, OECD/IEA, Paris.

– (2010c), *Natural Gas Liquids Supply Outlook 2008-2015*, April, OECD/IEA, Paris.

O&GJ (*Oil and Gas Journal*) (2009), "Worldwide Look at Reserves and Production", *Oil and Gas Journal*, 21 December, PennWell Corporation, Oklahoma City, Oklahoma.

SPE (Society of Petroleum Engineers) (2007), "The Petroleum Resources Management System", SPE, *www.spe.org/spe-site/spe/spe/industry/reserves/Petroleum_Resources_Management_System_2007.pdf*, accessed September, 2010.

USGS (United States Geological Survey) (2000), *World Petroleum Assessment*, USGS, Boulder, Colorado.

– (2008), "Circum-Arctic Resource Appraisal: Estimates of Undiscovered Oil and Gas North of the Arctic Circle", *Fact Sheet 2008-3049*, USGS, Boulder, Colorado.

Chapter 4: The outlook for unconventional oil

BGR (*Bundesanstalt für Geowissenschaften und Rohstoffe* – German Federal Institute for Geosciences and Natural Resources) (2009), *Energierohstoffe 2009, Reserven, Ressourcen, Verfügbarkeit, Tabellen [Energy Resources 2009, Reserves, Resources, Availability, Tables]*, BGR, Hannover, Germany.

Biglarbigi, K., J. Killen and M. Carolus (2009), "The Impact of Oil Prices on Oil Shale Development in the United States", SPE/124885, October, SPE (Society of Petroleum Engineers), Richardson, Texas.

Brandt, A. (2008), "Converting Oil Shale to Liquid Fuels: Energy Inputs and Greenhouse Gas Emissions of the Shell in Situ Conversion Process", *Environmental Science & Technology*, Vol. 42, No. 19, American Chemical Society, Washington, DC.

CERA (Cambridge Energy Research Associates) (2009), *Growth in the Canadian Oil Sands: Finding the Right Balance*, IHS CERA special report, IHS Cambridge Energy Research Associates, Massachusetts.

Charpentier, A., J. Bergerson and H. MacLean (2009), "Understanding the Canadian Oil Sands Industry's Greenhouse Gas Emissions", *Environmental Research Letters*, Vol. 4, IOP Publishing Ltd, United Kingdom.

Dyni, J. (2005), *Geology and Resources of Some World Oil-Shale Deposits*, USGS Scientific Investigations Report 2005-5294, U.S. Department of the Interior and U.S Geological Survey, Boulder, Colorado.

Jacobs Consultancy (2009), "Life Cycle Assessment Comparison of North American and Imported Crudes", paper prepared for Alberta Energy Research Institute, Report 1747, July, Jacobs Consultancy, Chicago, Illinois.

IEA (International Energy Agency) (2010), *Natural Gas Liquids Supply Outlook 2008-2015*, OECD/IEA, Paris.

Kelly E., J. Short, D. Schindler, P. Hodson, M. Ma, A. Kwan and B. Fortin (2009), *Oil Sands Development Contributes Polycyclic Aromatic Compounds to the Athabasca River and Its Tributaries*, proceedings of National Academy of Sciences 106; *www.pnas.org/content/106/52/22346*.

O&GJ (*Oil and Gas Journal*) (2009), "Worldwide Look at Reserves and Production", *Oil and Gas Journal*, 21 December, PennWell Corporation, Oklahoma City, OK.

Qian, J. (2008), "Recent Oil Shale Activities in China", *www.kirj.ee/public/oilshale_pdf/2008/issue_4/oil-2008-4-news-2.pdf*.

TIAX (2009), *Comparison of North American and Imported Crude Oil Lifecycle GHG Emissions*, prepared for Alberta Energy Research Institute, TIAX LLC, Lexington, MA.

US DOE (United States Department of Energy) (2009a), *Consideration of Crude Oil Source in Evaluating Transportation Fuel GHG Emissions*, DOE/NETL-2009/1360, March 20, National Energy Technology Laboratory, Pittsburgh, PA.

– (2009b), *Affordable, Low-Carbon Diesel Fuel from Domestic Coal and Biomass*, DOE/NETL-2009/1349, January 14, National Energy Technology Laboratory, Pittsburgh, PA.

USGS (United States Geological Survey) (2009a), "An Estimate of Recoverable Heavy Oil Resources of the Orinoco Oil Belt, Venezuela", *Fact Sheet 2009-3028*, October, USGS, Boulder, Colorado.

– (2009b), "Assessment of In-place Oil Shale Resources of the Green River Formation, Piceance Basin, Western Colorado", *Fact Sheet 2009-3012*, March, USGS, Boulder, Colorado.

– (2010), "Assessment of In-place Oil Shale Resources of the Green River Formation, Uinta Basin, Utah and Colorado", *Fact Sheet 2010-3010*, May, USGS, Boulder, Colorado.

Chapter 5: Natural gas market outlook

BGR (*Bundesanstalt für Geowissenschaften und Rohstoffe* – German Federal Institute for Geosciences and Natural Resources) (2009), *Reserves, Resources and Availability of Energy Resources 2009*, BGR, Hannover, Germany.

Cedigaz (2009), *Natural Gas in the World*, Institut français du pétrole, Rueil-Malmaison.

IEA (International Energy Agency) (2009), *World Energy Outlook 2009*, OECD/IEA, Paris.

– (2010), *Medium-Term Oil and Gas Markets 2010*, OECD/IEA, Paris.

USGS (United States Geological Survey) (2000), *World Petroleum Assessment*, USGS, Boulder, Colorado.

– (2008), "Circum-Arctic Resource Appraisal: Estimates of Undiscovered Oil and Gas North of the Arctic Circle", *Fact Sheet 2008-3049*, USGS, Boulder, Colorado.

Chapter 6: Coal market outlook

BGR (*Bundesanstalt für Geowissenschaften und Rohstoffe* – German Federal Institute for Geosciences and Natural Resources) (2009), *Reserves, Resources and Availability of Energy Resources 2009*, BGR, Hannover, Germany.

IEA (International Energy Agency) (2007), *World Energy Outlook 2007: China and India Insights*, OECD/IEA, Paris.

– (2009), *World Energy Outlook 2009*, OECD/IEA, Paris.

IEA CCC (International Energy Agency Clean Coal Centre) (2009), *Underground Coal Gasification*, IEA CCC, London.

Chapter 7: Power sector outlook

EIA (Energy Information Administration) (2010), *Annual Energy Outlook 2010 with Projections to 2035: U.S. Nuclear Power Plants – Continued Life or Replacement After 60?, www.eia.doe.gov/oiaf/aeo/nuclear_power.html*, accessed 31 August 2010.

IAEA (International Atomic Energy Agency) (2010), *PRIS Database, www.iaea.org/programmes/a2/*, accessed 31 August 2010.

IEA (International Energy Agency) (forthcoming), *IEA Technology Roadmap: Smart Grids*, OECD/IEA, Paris.

Chapter 8: Energy poverty

AGECC (Advisory Group on Energy and Climate Change) (2010), "Energy for a Sustainable Future, the Secretary-General Advisory Group on Energy and Climate Change: Summary Report and Recommendations", 28 April, United Nations, New York.

Asaduzzaman, M., D.F. Barnes and S.R. Khandker (2009), "Restoring Balance: Bangladesh's Rural Energy Realities", *Energy and Poverty Special Report 006/09*, March, The World Bank Group, Washington, DC.

Bazilian, M., A. Cabraal, R. Centurelli, R. Detchon, D. Gielen, H. Rogner, M. Howells, H. McMahon, V. Modi, N. Nakicenovic, P. Nussbaumer, B. O'Gallachoir, M. Radka, K. Rijal, M. Takada and F. Ziegler (2010), "Measuring Energy Access: Supporting a Global Target", *Earth Institute Working Paper*, Columbia University, New York.

Birol, F. (2007), "Energy Economics: A Place for Energy Poverty in the Agenda?", *The Energy Journal*, Vol. 28, No. 3, Elsevier, Amsterdam.

ECLAC (Economic Commission for Latin America and the Caribbean), Club de Madrid, Vundesministerium fur wirtschaftliche Zusammenarbeit und Entwicklung [Ministry for Economic Cooperation and Development], GTZ and UNDP (2010), *Contribution of Energy Services to the Millennium Development Goals and to Poverty Alleviation in Latin America and the Caribbean*, United Nations, New York.

Hutton, G., E. Rehfuess and F. Tediosi (2007), "Evaluation of the Costs and Benefits of Interventions to Reduce Indoor Air Pollution", *Energy for Sustainable Development*, Vol. 11, No. 4, Elsevier, Amsterdam.

IEA (International Energy Agency) (2002), *World Energy Outlook 2002*, OECD/IEA, Paris.

– (2004), *World Energy Outlook 2004*, OECD/IEA, Paris.

– (2006) *World Energy Outlook 2006*, OECD/IEA, Paris.

– (2008) *World Energy Outlook 2008*, OECD/IEA, Paris.

Luoma, J. (2010), "World's Pall of Black Carbon Can Be Eased with New Stoves", *Yale Environment 360*, 8 March, Yale School of Forestry and Environmental Studies, New Haven.

Mathers, C.D. and D. Loncar (2006), "Projections of Global Mortality and Burden of Disease from 2002 to 2030", *PLoS Medicine*, Vol. 3, No. 11, November 28, doi: 10.1371/journal.pmed.0030442, Public Library of Science, Cambridge.

Ministry of Infrastructure, Republic of Rwanda (2010), *Biomass Energy Strategy*, Ministry of Infrastructure, Kigali.

Modi, V., S. McDade, D. Lallement and J. Saghir (2005), *Energy Services for the Millennium Development Goals*, Energy Sector Management Assistance Programme, United Nations Development Programme, UN Millennium Project, and World Bank, New York.

Niez, A. (2010), "Comparative Study on Rural Electrification Policies in Emerging Economies: Keys to Successful Policies", *IEA Information Paper*, March, OECD/IEA, Paris.

Smith K.R., S. Mehta and M. Feuz (2004), "Indoor Air Pollution from Household Use of Solid Fuels", in M. Ezzati, A.D. Lopez, A. Rodgers and C.J.L. Murray (eds.), *Comparative Quantification of Health Risks: Global and Regional Burden of Disease Attributable to Selected Major Risk Factors*, WHO, Geneva.

Tian, Y. and D. Song (forthcoming), "Development and Promotion of Biogas Utilization in Rural China", May, UNDP Regional Centre, Bangkok.

UNDP (United Nations Development Programme) (2006), *Expanding Access to Modern Energy Services: Replicating, Scaling Up and Mainstreaming at the Local Level – Lessons from Community-based Energy Initiatives*, UNDP, New York.

UNDP and AEPC (Alternative Energy Promotion Centre) (2010), *Capacity Development in Scaling up Decentralized Energy Access Programmes. Lessons from Nepal on Its Role, Costs and Financing*, UNDP, New York.

UNDP and WHO (World Health Organization) (2009), *The Energy Access Situation in Developing Countries - A Review on the Least Developed Countries and Sub-Saharan Africa*, UNDP, New York.

UN-Energy (United Nations-Energy) (2005), *The Energy Challenge for Achieving the Millennium Development Goals*, http://esa.un.org/un-energy/pdf/UN-ENRG%20paper.pdf.

UNEP (United Nations Environment Programme) (2003), *Energy subsidies: lessons learned in assessing their impact and designing policy reforms*, UNEP/Earthprint, Nairobi.

Venkataraman, C., A.D. Sagar, G. Habib, N. Lam and K.R. Smith (2010), "The Indian National Initiative for Advanced Biomass Cookstoves: The Benefits of Clean Combustion", *Energy for Sustainable Development*, Vol. 14, No. 2, Elsevier, Amsterdam.

World Bank (2007), *Africa Development Indicators*, World Bank, Washington DC.

– (2009), "Africa's Infrastructure: A Time for Transformation", *World Bank Africa Infrastructure Country Diagnostic*, World Bank, Washington, DC.

World Bank Group (2010), *New Private Infrastructure Activity in Developing Countries Recovered Selectively in the Third Quarter of 2009: Assessment of the Impact of the Crisis on New PPI Projects - Update 5*, February, The World Bank Group, Washington, DC.

WHO (World Health Organization) (2004), "Indoor Smoke from Solid Fuel Use: Assessing the Environmental Burden of Disease", *Environmental Burden of Disease Series*, No. 4, WHO, Geneva.

– (2008), *The Global Burden of Disease: 2004 Update*, WHO, Geneva.

PART B: Outlook for renewable energy

Chapter 9: How renewable energy markets are evolving

BNEF (Bloomberg New Energy Finance) (2010), *Green Stimulus Update, May 2010, Analyst Reaction*, BNEF (available by subscription).

BTM Consult (2009), *International Wind Energy Development, World Market Update 2008*, BTM Consult, Ringkøbing, Denmark.

BTM Consult (2010), *International Wind Energy Development, World Market Update 2009*, BTM Consult, Ringkøbing, Denmark.

Doornbosch, R. and R. Steenblik (2007), "Biofuels: Is the Cure Worse Than the Disease?", paper presented at the OECD Round Table on Sustainable Development, OECD, Paris, 11-12 September.

Fischer, G. and L. Schrattenholzer (2001), "Global Bioenergy Potentials through 2050," *Biomass and Bioenergy*, Vol. 20, Elsevier, Amsterdam.

FAO (Food and Agriculture Organization) (2010), *Global Forest Resources Assessment 2010*, FAO, Rome.

Fraunhofer Institute, Ecofys, Energy Economics Group, Austria, Rütter + Partner, Switzerland, SEURECO (Société Européenne d'Économie), France and Lithuanian Energy Institute (2009), *The Impact of Renewable Energy Policy on Economic Growth and Employment in the European Union*, http://ec.europa.eu/energy/renewables/studies/index_en.htm.

Greenpeace and EREC (European Renewable Energy Council) (2010), *Energy [R] evolution: A Sustainable World Energy Outlook*, Greenpeace International and EREC, www.energyblueprint.info.

Heinrich Böll Stiftung Cambodia, WWF (World Wide Fund for Nature) Denmark and International Institute for Sustainable Development (2008), *Rethinking Investments in Natural Resources: China's Emerging Role in the Mekong Region*, Heinrich Böll Stiftung, WWF and International Institute for Sustainable Development, Winnipeg, Canada.

Hirshman, W.P. (2009), "Little Smiles on Long Faces", *Photon International – The Solar Power Magazine*, March, Photon Europe GmbH, Aachen, Germany.

– (2010), "Surprise, Surprise: Solar Cell Production for 2009 Hits 12 GW", *Photon International – The Solar Power Magazine*, March, Photon Europe GmbH, Aachen, Germany.

IEA (International Energy Agency) (2009a), "The Impact of the Financial and Economic Crisis on Global Energy Investment", *IEA Background Paper for the G8 Energy Ministers' Meeting, 24-25 May*, OECD/IEA, Paris, www.iea.org/papers/2009/g8_fincrisis_impact.pdf.

– (2009b), *Transport, Energy and CO$_2$ – Moving toward Sustainability*, OECD/IEA, Paris.

– (2010a), *Sustainable Production of Second-generation Biofuels – Potential and Perspectives in Major Economies and Developing Countries*, OECD/IEA, Paris.

– (2010b), *Energy Technology Perspectives 2010*, OECD/IEA, Paris.

IMF (International Monetary Fund) (2010), *Global Financial Stability Report*, IMF, Washington, DC.

KPMG (2010), *Powering Ahead: 2010, An Outlook for Renewable Energy M&A*, KPMG International Cooperative, Switzerland.

Parthan B., M. Osterkorn, M. Kennedy, St.J. Hoskyns, M. Bazilian and P. Monga (2010), "Lessons for Low-carbon Energy Transition: Experience from the Renewable Energy and Energy Efficiency Partnership (REEEP)", *Energy for Sustainable Development*, Vol. 14, No. 2, Elsevier, Amsterdam.

PWC (Price Waterhouse Coopers) (2010), *Renewables Deals, 2009 Annual Review*, www.pwc.com/energy.

REN21 (2010), *Renewables 2010 Global Status Report*, REN21, Paris.

Smeets, E., A. Faaij, I. Lewandowski and W. Turkenburg (2007), "A Bottom-up Assessment and Review of Global Bioenergy Potentials to 2050", *Energy and Combustion Science*, Vol. 33, No. 1, Elsevier, Amsterdam.

UNEP (United Nations Environment Programme) (2008), *Green Jobs: Towards Decent Work in a Sustainable, Low-carbon World*, UNEP, Nairobi.

– (2009), *Towards Sustainable Production and Use of Resources: Assessing Biofuels*, UNEP, Nairobi.

UNEP, SEFI (Sustainable Energy Finance Initiative), BNEF (Bloomberg New Energy Finance) and Chatham House (2009), *Private Financing of Renewable Energy – a Guide for Policymakers*, Chatham House, London.

Chapter 10: Renewables for electricity

BTM Consult (2010), *International Wind Energy Development, World Market Update 2009*, BTM Consult, Ringkøbing, Denmark.

CAISO (California Independent System Operator) (2007), *Integration of Renewable Energy Resources*, CAISO, Folsom, CA, www.caiso.com.

Canton, J. and Å. Johannesson Lindén (2010), "Support Schemes for Renewable Electricity in the EU", *Economic Paper 408*, European Commission, Brussels.

CEC (Commission of the European Communities) (2009), *The Renewable Energy Progress Report, Commission Report in Accordance with Article 3 of Directive 2001/77/EC, Article 4(2) of Directive 2003/30/EC and on the Implementation of the EU Biomass Action Plan, COM(2005)628*, http://eur-lex.europa.eu.

DCENR (Department of Communications, Energy and Natural Resources) (Ireland) and Department of Enterprise, Trade and Investment (DETI) (UK) (2008), *All Island Grid Study, Work Stream 4: Analysis of Impacts and Benefits*, www.dcenr.gov.ie.

DENA (Deutschen Energie-Agentur) (2005), *Energiewirtschaftliche Planung für die Netzintegration von Windenergie in Deutschland an Land und Offshore bis zum Jahr 2020*, [Planning of the Grid Integration of Wind Energy in Germany Onshore and Offshore by 2020], DENA, www.dena.de.

EnerNex Corporation (2006), *Minnesota Wind Integration Study, Volume I*, Minnesota Public Utilities Commission, www.puc.state.mn.us.

– (2010), *Eastern Wind Integration and Transmission Study*, NREL, *www.nrel.org*.

EWEA (European Wind Energy Association) (2005), *Large Scale Integration of Wind Energy in the European Power Supply: Analysis, Issues and Recommendations*, EWEA, Brussels, *www.ewea.org*.

GE Energy (2008), *Analysis of Wind Generation Impact on ERCOT Ancillary Service Requirements*, Electric Reliability Council of Texas, Austin, *www.ercot.com*.

– (2010), *Western Wind and Solar Integration Study*, NREL, Golden, CO.

GWEC (Global Wind Energy Council) (2010), *Global Wind 2009 Report*, GWEC, Brussels.

Guarrera L., J. Kappauf, E. Menichetti, M. Urbani and R. Vigotti (2010), *Mediterranean Energy Prospects*, Observatoire Méditerranéen de l'Energie (OME), Nanterre, France.

Holttinen, H., P. Meibom, A. Orthis, F. van Hulle, B. Lange, M. O'Malley, J. Pierik, B. Ummels, J.O. Tande, A. Estanqueiro, M. Matos, E. Gomez, L Soder, G. Strbac, A. Shakoor, J. Ricardo, J.C. Smith, M. Milligan, E. Ela (2009), "Design and Operation of Power Systems with Large Amounts of Wind Power", Final Report, IEA Wind Task 25, Phase One 2006-2008, OECD/IEA, Paris.

IEA (International Energy Agency) (2009), *IEA Technology Roadmap: Wind Energy*, OECD/IEA, Paris.

– (forthcoming, a), *Empowering Electricity Customers: Customer Choice and Demand Response in Competitive Markets*, OECD/IEA, forthcoming.

– (forthcoming, b), GIVAR (Grid Integration of Variable Resources), OECD/IEA, Paris.

– (forthcoming, c), *IEA Technology Roadmap: Smart Grids*, OECD/IEA, Paris.

Inage, S. (2009), "Prospects for Large-Scale Energy Storage in Decarbonised Power Grids", *IEA Working Paper*, OECD/IEA, Paris.

Mills, A., R. Wiser and K. Porter (2009), *The Cost of Transmission for Wind Energy: A Review of Transmission Planning Studies*, Ernest Orlando Lawrence Berkeley National Laboratory, Berkeley, California.

North American Electric Reliability Council (NERC) (2009), *Accommodating High Levels of Variable Generation*, NERC, *www.nerc.com*.

REN21 (2010), *Renewables 2010 Global Status Report*, REN21, Paris.

Transpower Stromübertragungs-Gmbh (2010), "European Wind Integration Study: towards a Successful Integration of Large Scale Wind Power into European Electricity Grids", *www.wind-integration.eu/downloads/library/EWIS_Final_Report.pdf*.

UK ERC (United Kingdom Energy Research Centre) (2006), *The Costs and Impacts of Intermittency: An Assessment of the Evidence on the Costs and Impacts of Intermittent generation on the British electricity network*, UK ERC, *www.ukerc.ac.uk*.

US DOE/EIA (United States Department of Energy/Energy Information Administration) (2009), *Annual Energy Outlook 2010*, US DOE, Washington, DC.

VTT Technical Research Centre of Finland (2009), *Design and Operation of Power Systems with Large Amounts of Wind Power, Final Report*, IEA Wind Implementing Agreement Task 25, Phase One 2006 – 2008, Julkaisija– Utgivare, Finland.

WEC (World Energy Council) (2010), *Survey of Energy Resources 2010*, WEC, London.

Chapter 11: Renewables for heat

Bürger, V., S. Klinski, U. Lehr, U. Leprich, M. Nast and M. Ragwitz (2008), "Policies to Support Renewable Energies in the Heat Market", *Energy Policy*, Vol. 36, No. 8, Elsevier, Amsterdam.

Connor, P., V. Bürger, L. Beurskens, K. Ericsson and C. Egger (2009), *Overview of RES-H/RES-C Support Options, D4 of WP2 Report of the EU Project RES-H Policy*, www.res-h-policy.eu.

EcoHeatCool (2006), *The European Heat Market, Workpackage 1*, EuroHeat&Power, Brussels.

EurObserv'ER (2009), *Heat Pump Barometer*, www.eurobserv-er.org/pdf/baro193.pdf.

IEA (International Energy Agency) (2007), *Renewables for Heating and Cooling*, OECD/IEA, Paris.

– (2008), *Deploying Renewables, Principles for Effective Policies*, OECD/IEA, Paris.

– (2010a), *Energy Technology Perspectives 2010*, OECD/IEA, Paris.

– (2010b), "Renewable Energy Essentials: Geothermal", IEA Working Paper *www.iea.org/papers/2010/Geothermal_Essentials.pdf*.

REN21 (2009), *Background Paper: Chinese Renewables Status Report*, REN21, Paris.

Taibi, E., D. Gielen and M. Bazilian (forthcoming), *Renewable Energy in Industrial Applications - An Assessment of the 2050 Potential*, UNIDO, Vienna.

Weiss, W. (2010), *Solar Heat Worldwide Edition 2010*, AEE INTEC, Gleisdorf, Austria.

Chapter 12: Renewables for transport

Biofuels International (2010), *Towards Greener Aviation*, Vol. 4, No. 4, Horseshoe Media Limited, Surrey, UK.

FO Licht (2010a), "The German Sustainability Decree - a Blueprint for the EU Biofuels Sector?", *World Ethanol & Biofuels Report*, February, Agra Informa Ltd., Kent, England.

– (2010b), "Feedstocks for Biofuels - the Outlook for 2010", *World Ethanol & Biofuels Report*, May, Agra Informa Ltd., Kent, England.

IEA (International Energy Agency) (2009a), *Sustainable Production of Second-generation Biofuels – Potential and Perspectives in Major Economies and Developing Countries*, OECD/IEA, Paris.

– (2009b), *Transport, Energy and CO$_2$ – Moving Toward Sustainability*, OECD/IEA, Paris.

– (2010), *Medium-Term Oil and Gas Markets 2010*, OECD/IEA, Paris

IEA Bioenergy (2009), "Bioenergy – The Impact of Indirect Land Use Change", *www.ieabioenergy.com/DocSet.aspx?id=6214*.

IFPRI (International Food Policy Research Institute) (2010), *Global Trade and Environmental Impact Study of the EU Biofuels Mandate*, March, IFPRI, Washington, DC.

OECD (Organisation for Economic Co-operation and Development) (2010), *OECD-FAO Agricultural Outlook 2010*, OECD, Paris.

Schlumberger, C.E. (2010), "Are Alternative Fuels an Alternative: A Review of the Opportunities and Challenges of Alternative Fuels for Aviation", *XXXV Annals of Air and Space Law*, McGill University, Montreal.

UNEP (United Nations Environment Programme) (2009), *Towards Sustainable Production and Use of Resources: Assessing Biofuels*, UNEP, Nairobi.

PART C: Achieving the 450 Scenario After Copenhagen

Chapter 13: Energy and the ultimate climate change target

Ackerman, F. and E.A. Stanton (2008), *"The Cost of Climate Change: What We'll Pay If Global Warming Continues Unchecked"*, May, Natural Resources Defence Council, New York, *www.nrdc.org/globalwarming/cost/contents.asp*.

Garnaut, R. (2008), *The Garnaut Climate Change Final Report*, Cambridge University Press, Melbourne.

Global Burden of Disease Study (forthcoming), www.globalburden.org/index.html.

IEA (International Energy Agency) (2009), *World Energy Outlook 2009*, OECD/IEA, Paris.

IIASA (International Institute for Applied Systems Analysis) (2010), *Emissions of Air Pollutants for the World Energy Outlook 2010 Energy Scenarios,* report prepared for the IEA using the GAINS model, IIASA, Laxenberg, *www.worldenergyoutlook.org*.

IPCC (Intergovernmental Panel on Climate Change) (2007), "Climate Change 2007: Synthesis Report", contribution of Working Groups I II, and III to the *Fourth Assessment Report of the IPCC*, R.K. Pachauri and A. Reisinger (eds.), IPCC, Geneva.

Meinshausen, M., N. Meinshausen, W. Hare, S.C.B. Raper, K. Frieler, R. Knutti, D.J. Frame and M. Allen (2009), "Greenhouse Gas Emission Targets for Limiting Global Warming to 2°C", *Nature*, Vol. 458, doi: 10.1038/nature08017, Nature Publishing Group, London.

Parry, M., N. Arnall, P. Berry, D. Dodman, S. Fankhauser, C. Hope, S. Kovats, R. Nicholls, D. Satterthwaite, R. Tiffin and T. Wheeler (2009), *Assessing the Costs of Adaptation to Climate Change: A Review of the UNFCCC and Other Recent Estimates*, International Institute for Environment and Development, London.

UNFCCC (UN Framework Convention on Climate Change) (2007), *Investment and Financial Flows to Address Climate Change*. Bonn: Climate Change Secretariat, Bonn.

World Bank (2007), "Cost of Pollution in China: Economic Estimates of Physical Damages", Working Paper, World Bank, Report No. 39236, Washington, DC., *http://go.worldbank.org/7LM8L9FAV0*.

Chapter 14: The energy transformation by sector

IEA (International Energy Agency) (2008), *Energy Technology Perspectives 2008*, OECD/IEA, Paris.

– (2009a), *IEA Technology Roadmap: Carbon Capture and Storage*, OECD/IEA, Paris.

– (2009b), *Transport, Energy and CO_2 – Moving Toward Sustainability*, OECD/IEA, Paris.

– (2009c), *Energy Technology Transitions for Industry*, OECD/IEA, Paris.

– (2010), *Energy Technology Perspectives 2010*, OECD/IEA, Paris.

Chapter 15: Implications for oil markets

IEA (International Energy Agency) (2009), *Sustainable Production of Second-generation Biofuels – Potential and Perspectives in Major Economies and Developing Countries*, OECD/IEA, Paris.

– (2010), *Energy Technology Perspectives 2010*, OECD/IEA, Paris.

IIASA (International Institute for Applied Systems Analysis) (2010), "Emissions of Air Pollutants for the World Energy Outlook 2010 Energy Scenarios", report prepared for the IEA using the GAINS model, IIASA, Laxenberg, *www.worldenergyoutlook.org*.

IPCC (Intergovernmental Panel on Climate Change) (2007), "Climate Change 2007: Synthesis Report", contribution of Working Groups I II, and III to the *Fourth Assessment Report of the IPCC*, in R.K. Pachauri and A. Reisinger (eds.), IPCC, Geneva.

PART D: Outlook for Caspian energy

Chapter 16: Caspian domestic energy prospects

ADB (Asian Development Bank) (2009), *Energy Outlook for Asia and the Pacific*, ADB, Manila.

CAREC (The Regional Environment Centre for Central Asia) (2009), "Gap Analysis in the Area of Climate Change and Energy Efficiency in Central Asia: Defining Opportunities for CAREC", CAREC, Almaty.

CASE (Centre for Social and Economic Research) (2008), "The Economic Aspects of the Energy Sector in CIS Countries", *Economic Papers 327*, European Commission, Brussels.

Government of the Republic of Tajikistan (2008), "The Second National Communication of the Republic of Tajikistan under the United Nations Framework Convention on Climate Change", Government of the Republic of Tajikistan, Dushanbe.

IEA (International Energy Agency) (2009), *World Energy Outlook 2009*, OECD/IEA, Paris.

IMF (International Monetary Fund) (2010), W*orld Economic Outlook Update: Restoring Confidence without Harming Recovery*, July, IMF, Washington, DC.

Ministry of Energy of Georgia (2010), official website of the Ministry of Energy of Georgia, *www.minenergy.gov.ge*, accessed September 2010.

President of Kazakhstan (2010), "On the Strategic Development Plan of the Republic of Kazakhstan to 2020", Decree No. 922 of the President of Kazakhstan, 1 February, Astana.

Ramani, K.V. (2009), *Kazakhstan Sustainable Energy Program: Outcome Evaluation*, United Nation Development Programme, Kazakhstan.

REEEP (Renewable Energy & Energy Efficiency Partnership) (2009), "Fresh Wind from Kazakhstan: New Renewable Energy Law", *www.reeep.org*.

SPECA (UN Special Programme for the Economies of Central Asia) (2007), "Sub-regional Program on Energy Efficiency for Central Asian Economies: Forming Possibilities for Control of Air Quality and Use of Clean Technologies of Burning Coal in Central Asia", UNECE, Geneva.

Transparency International (2009), *Corruption Perceptions Index*, Berlin, *www.transparency.org/policy_research/surveys_indices/cpi/2009*.

UNCTAD (UN Conference on Trade and Development) (2009), *World Investment Review*, Geneva, Switzerland.

UNECE (UN Economic Commission for Europe) (2008), "Environmental Performance Review: Kazakhstan (2nd Review)", United Nations, New York and Geneva.

UNESCAP (UN Economic and Social Commission for Asia and the Pacific) (2008), "Study on Central Asia Energy Efficiency Potential", United Nations, New York and Geneva.

Von Moltke, A., C. McKee and T. Morgan (2003), *Energy Subsidies*, Greenleaf/United Nations Environment Programme, Sheffield, United Kingdom.

World Bank (2009), *Worldwide Governance Indicators 1996-2008*, World Bank, Washington, DC., *http://info.worldbank.org/governance/wgi/index.asp*.

– (2010), *Lights Out? The Outlook for Energy in Eastern Europe and Central Asia*, World Bank, Washington, DC.

Chapter 17: Hydrocarbon resource and supply potential

BGR (Bundesanstalt für Geowissenschaften und Rohstoffe – German Federal Institute for Geosciences and Natural Resources) (2009), *Reserves, Resources and Availability of Energy Resources 2009*, BGR, Hannover, Germany.

Cedigaz (2009), *Natural Gas in the World*, Institut français du pétrole, Rueil-Malmaison.

EC (European Commission) (2009), *Second Strategic Energy Review - an EU Energy Security and Solidarity Action Plan*, European Commission, Brussels.

Fredholm, M. (2008), "The World of Central Asian Oil and Gas", *Asian Cultures and Modernity Research Report No. 16*, University of Stockholm, Stockholm.

Government of Russia (2009), *The Energy Strategy of Russia for the Period to 2030*, Government of Russia, Moscow.

Government of Turkmenistan (2006), *Oil and Gas Industry Development Programme to 2030*, Government of Turkmenistan, Ashgabat.

IEA (International Energy Agency) (2008a), *World Energy Outlook 2008*, OECD/IEA, Paris.

– (2008b), *Perspectives on Caspian Oil and Gas Development*, IEA Working Paper Series, OECD/IEA, Paris.

– (2009), *World Energy Outlook 2009*, OECD/IEA, Paris.

Janusz, B. (2005), "The Caspian Sea, Legal Status and Regime Problems", *Chatham House Briefing Paper*, Chatham House, London.

NOAA (National Oceanic and Atmospheric Administration) (2009), "National Geophysical Data Center, Global Gas Flaring Estimates", Washington, DC. *www.ngdc.noaa.gov/ dmsp/interest/gas_flares.html*, accessed September 2010.

O&GJ (*Oil and Gas Journal*) (2009), "Worldwide Look at Reserves and Production", *Oil and Gas Journal*, 21 December, PennWell Corporation, Oklahoma City, OK.

USGS (United States Geological Survey) (2000), *World Petroleum Assessment*, USGS, Boulder, Colorado.

– (2008), "Circum-Arctic Resource Appraisal: Estimates of Undiscovered Oil and Gas North of the Arctic Circle", *Fact Sheet 2008-3049*, USGS, Boulder, Colorado.

Chapter 18: Regional and global implications

ADB (Asian Development Bank) (2010), *Proposal for a Central Asia Regional Economic Cooperation: Power Sector Regional Master Plan*, ADB, Manila.

CASE (Centre for Social and Economic Research) (2008), "The Economic Aspects of the Energy Sector in CIS Countries", *Economic Papers 327*, European Commission, Brussels.

EC (European Commission) (2009), *Second Strategic Energy Review – an EU Energy Security and Solidarity Action Plan*, European Commission, Brussels.

EDB (Eurasian Development Bank) (2009), *The Impact of Climate Change on Water Resources in Central Asia*, Eurasian Development Bank, Almaty.

Energy Charter (2002), *The Energy Charter Treaty, a Reader's Guide*, Energy Charter Secretariat, Brussels.

Kalyuzhnova, Y. (2008), *Economics of the Caspian Oil and Gas Wealth*, Palgrave Macmillan, Basingstoke.

UNDP (United Nations Development Programme) (2009), *Central Asia Regional Risk Assessment: Responding to Water, Energy and Food Insecurity*, UNDP, New York.

PART E: Focus on fossil-fuel subsidies

Chapter 19: Analysing fossil-fuel subsidies

Birol F., A.V. Aleagha and R. Ferroukhi (1995), "The Economic Impact of Subsidy Phase-out in Oil Exporting Developing Countries: a Case Study of Algeria, Iran and Nigeria", *Energy Policy*, Vol. 23, No. 3, Elsevier, Amsterdam.

Coady, D., R. Gillingham, R. Ossowski, J. Piotrowski, S. Tareq and J. Tyson (2010), "Petroleum Product Subsidies: Costly, Inequitable, and Rising", *IMF Staff Position Note*, SPN/10/05, IMF, Washington, DC.

Ellis, J. (2010), *Untold Billions: Fossil-fuel Subsidies, Their Impacts and the Path to Reform- The Effects of Fossil-fuel Subsidy Reform, a Review of Modelling and Empirical Studies*, April, Global Subsidies Initiative of the International Institute for Sustainable Development, Geneva.

GSI (Global Subsidies Initiative) (2010), *Defining Fossil-Fuel Subsidies for the G-20: Which Approach is Best*, Global Subsidies Initiative of the International Institute for Sustainable Development, Geneva.

Hale, T.N. (2008), "Transparency, Accountability and Global Governance", in T. Farer and T.D. Sisk (eds), *Global Governance*, Vol. 14, No. 1, Lynne Rienner Publishers, Inc., Boulder, CO.

IEA (International Energy Agency) (2008), *Energy Policy Review of Indonesia*, OECD/IEA, Paris.

IEA, OECD (Organisation for Economic Co-operation and Development), World Bank and OPEC (Organization of the Petroleum Exporting Countries) (2010), *Analysis of the Scope of Energy Subsidies and Suggestions for the G-20 Initiative*, Joint Report prepared for submission to the G-20 Meeting of Finance Ministers and Central Bank Governors, Busan, Korea, 5 June.

IMF (International Monetary Fund) (2008), *Fuel and Food Price Subsidies: Issues and Reform Options*, prepared by the Fiscal Affairs Department (in consultation with other departments), Washington DC., 8 September.

Khatib, H. (2010), *Electricity Subsidies in Arab Countries*, presentation to Arab Energy Forum, Doha, 13 May.

Koplow, D. (2009), *Measuring Energy Subsidies Using the Price-Gap Approach: What Does it Leave Out?*, International Institute for Sustainable Development, Winnipeg.

Kosmo, M. (1987), *Money to Burn? The High Costs of Energy Subsidies*, World Resources Institute, Washington, DC.

Laan, T. (2010), *Untold Billions: Fossil-Fuel Subsidies, Their Impacts and the Path to Reform – Gaining Traction: the Importance of Transparency in Accelerating the Reform of Fossil-fuel Subsidies*, Global Subsidies Initiative of the International Institute for Sustainable Development, Geneva.

Laan, T., B. Christopher, and P. Bertille (2010), *Strategies for Reforming, Fossil-fuel Subsidies: Practical Lessons from Ghana, France and Senegal*, Global Subsidies Initiative of the International Institute for Sustainable Development, Geneva.

Larsen, B., and A. Shah (1992), "World Fossil Fuel Subsidies and Global Carbon Emissions", *Policy Research Working Papers*, WPS 1002, World Bank, Washington, DC.

Government of India (2010), *Report of the Expert Group on a Viable and Sustainable System of Pricing of Petroleum Products*, Government of India, New Delhi.

Shenoy, B.V. (2010), *Lessons Learned from Attempts to Reform India's Kerosene Subsidy*, International Institute for Sustainable Development, Geneva.

Steenblik, R. and P. Coroyannakis (1995), "Reform of Coal Policies in Western and Central Europe", *Energy Policy*, Vol. 23, No. 6, Elsevier, Amsterdam.

UNEP (United Nations Environment Programme) (2008), *Reforming Energy Subsidies*, UNEP, Nairobi.

UNEP and IEA (International Energy Agency) (2002), *Reforming Energy Subsidies: An Explanatory Summary of the Issues and Challenges in Removing or Modifying Subsidies on Energy That Undermine the Pursuit of Sustainable Development*, UNEP, Nairobi.

Chapter 20: Country subsidy profiles

Amuzegar, J. (2010), "Iran's New Subsidy Reform: a Shot in the Dark", *Middle East Economic Survey*, Vol. 53, No. 22, Middle East Petroleum and Economic Publications Ltd, Cyprus.

BGR (*Bundesanstalt für Geowissenschaften und Rohstoffe* – German Federal Institute for Geosciences and Natural Resources) (2009), *Reserves, Resources and Availability of Energy Resources 2009*, BGR, Hannover, Germany.

China Electricity Council (2010), *China Power Industry Statistics Brief for 2009*, Government of China, Beijing.

Clarke, K. and D. Graczyk (2010), "India's Downstream Petroleum Sector: Refined Product Pricing and Refinery Investment", *IEA Working Paper*, OECD/IEA, Paris.

FACTS (FACTS Global Energy) (2010), *Changing Petroleum Product Prices and Refining Margins in China: An Update*, May, FACTS, Singapore.

Gazprom (2010), "The Board Reviewed Information on the Prospects of Transition to Market Principles of Pricing", press release, *Gazprom*, *http://gazprom.ru/press/news/2010/march/article90450/*, accessed August 2010.

Government of India (2010), *Report of the Expert Group on a Viable and Sustainable System of Pricing of Petroleum Products*, Government of India, New Delhi.

Gonn, A. (2010), "Iran to Invest $46b in Refineries", 28 July, *Iran Daily*, Tehran.

Harris, K. (2010), "The Politics of Subsidy Reform in Iran", *Middle East Report*, No. 254, spring, Middle East Research and Information Project, Washington, DC.

Mercados (2010), *Study on Common Pool Price Mechanism for Natural Gas in the Country*, Mercados Energy Markets India Private Limited, prepared for GAIL (India) Limited, New Delhi.

Ministry of Petroleum and Natural Gas (2009), *Basic Statistics on Indian Petroleum & Natural Gas*, Government of India, New Delhi.

NBS (National Bureau of Statistics) (2010), *Notice on 2009 GDP Data Revision*, Beijing, *www.stats.gov.cn/tjdt/zygg/sjxdtzgg/t20100702_402654527.htm*, accessed July 2010.

Ni, C. (2009), *China Energy Primer*, Ernest Orlando Lawrence Berkeley National Laboratory, Berkeley.

O&GJ (*Oil and Gas Journal*) (2009), "Worldwide Look at Reserves and Production", *Oil and Gas Journal*, 21 December, PennWell Corporation, Oklahoma City, OK.

Russian Federation Decree #333 (2007), *On Improving the State Regulation of Gas Prices*, 28 May, Government of the Russian Federation, Moscow.

Russian Federation Law #36-FZ (2003), *On the Specifics of Electric Power Industry Functioning during the Transition Period, Introduction of Amendments into Certain Legislative Acts of the Russian Federation, and Repeal of Certain Legislative Acts of the Russian Federation due to the Adoption of the Federal Law "On the Electric Power Industry"*, 26 March, Government of the Russian Federation, Moscow.

Russian Federation Resolution #529 (2006), *On Improving Operations of the Wholesale Electricity (Capacity) Market*, 31 August, Government of the Russian Federation, Moscow.

International
Energy Agency

Online
bookshop

The paper used for this document and the forest from which it comes have received
FSC certification for meeting a set of strict environmental and social standards.
The FSC is an international, membership-based, non-profit
organisation that supports environmentally appropriate, socially beneficial,
and economically viable management of the world's forests.

IEA PUBLICATIONS, 9, rue de la Fédération, 75739 PARIS CEDEX 15
PRINTED IN FRANCE BY SOREGRAPH, November 2010
(61 2010 15 1P1) ISBN: 978-92-64-08624-1
Cover design: IEA. Photo credit: © Maciej Frolow, Brand X Pictures